Sunflower

Chemistry, Production, Processing, and Utilization

Sunflower

Chemistry, Production, Processing, and Utilization

Editors

Enrique Martínez Force

Nurhan Turgut Dunford

Joaquín J. Salas

AOCS PRESS

Urbana, Illinois

AOCS Mission Statement
AOCS advances the science and technology of oils, fats, surfactants and related materials, enriching the lives of people everywhere.

ISBN 978-1-893997-94-3 (print)
ISBN 978-1-630670-62-7 (.epub)
ISBN 978-1-630670-63-4 (.mobi)

Library of Congress Cataloging-in-Publication Data
Sunflower : chemistry, production, processing, and utilization / editors, Enrique Martínez Force, Nurhan Turgut Dunford, Joaquín J. Salas.
pages cm
ISBN 978-1-893997-94-3 (hardcover : alk. paper)—ISBN 978-1-63067-062-7 (.epub) — ISBN 978-1-63067-063-4 (.mobi) 1. Sunflowers. 2. Sunflower seed oil. I. Martínez Force, Enrique.
SB299.S9S855 2015
635.9'3399—dc23

2014047012

www.aocs.org

Contents

Preface

Sunflower is one of the most important oilseed crops, and its oil is now the fourth vegetable oil by volume of production in the world after palm, soybean, and rapeseed oils. Sunflower oil production reached 15.29 million metric tons in 2014, accounting for about one-tenth of the total world oil supply. The four largest producers of sunflower seeds (Ukraine, European Union, Russia, and Argentina) account for 76% of global production, which has grown exponentially in the Black Sea region in the last 10 years, with increased acreage and higher yields achieved by replacing old varieties with improved hybrid seeds. Sunflower oil has a high α-tocopherol content (vitamin E) and other health-beneficial components, such as folic acid or selenium, so it is consumed all over the world and recognized as natural, healthy oil.

Although the global market is smaller for sunflower oil than for other oils such as palm or soybean, the demand for sunflower oil is steadily increasing due to several new types of sunflower oils that are now available for consumers. They are mid-oleic, high oleic, or high stearic lines that provide oils with improved properties of stability that could be healthy substitutes for hydrogenated fat in the future.

The AOCS Monograph Series on Oilseeds is an encyclopedic view of the best information on the oilseed field. This monograph, *Sunflower: Chemistry, Production, Processing, and Utilization,* is a new addition to the Monograph Series that includes volumes on soybean, canola, olive, gourmet, and specialty oils. This comprehensive reference book delivers key information on all aspects of sunflower. The book contains 20 chapters and provides an extensive review of the latest developments in sunflower genetics, breeding, processing, quality, and utilization, including food, energy, and industrial applications of bioproducts. Production and processing aspects of sunflower, as well as international and U.S. practices, are also reviewed.

The contributing authors of the book come from the United States, Europe, and Asia and are internationally renowned experts in their fields. All authors are currently involved in some aspects of sunflower research or in the sunflower industry and seed companies. This book is a reference source for scientists, students, governments, and industry personnel interested in the recent developments and future opportunities in sunflower production, procesasing, and utilization.

As in previous monographs of this AOCS Oilseed Series, most of the chapters in this book are presented in a standalone manner without the need to make reference to other chapters in the book, thus allowing the contributors to provide comprehensive reviews of their fields of expertise. The editors have made every effort to select contributors who are internationally recognized as experts in their fields in order to cover all relevant topics. We thank all the authors for their outstanding efforts and the timely submission of their chapters. In addition, the editors would like to thank the editorial staff of AOCS Press for their helpful suggestions toward improving and coordinating the timely publication of this book. We hope you, the readers, will find this book to be a valuable resource on the current developments in sunflower scientific knowledge, production, processing, and utilization.

Enrique Martínez-Force
Nurhan Turgut Dunford
Joaquín J. Salas

About the Editors

Enrique Martínez-Force. Dr. Martínez-Force is a researcher at the Spanish National Research Council (CSIC) and is head of the Department of Biochemistry and Molecular Biology of Plant Products of the Instituto de la Grasa. He has a degree and a Ph.D. in biological sciences from the University of Seville. Since 1995, his research has focused on the field of the metabolism of plant lipids, and more specifically on the genetic, biochemical, and molecular characterization of the fatty acid and lipid biosynthetic pathways in sunflower. His work has led to the publication of more than 80 research articles and book chapters, participation in the development of 9 licensed international patents, and the supervision of 12 doctoral theses related to this field.

Nurhan Turgut Dunford. Dr. Dunford's experience as an engineer and scientist encompasses over 30 years, including positions in Turkey, Canada, and the United States. She is currently a professor at Oklahoma State University in the Biosystems and Agricultural Engineering Department. Dr. Dunford is also on staff at the Robert M. Kerr Food and Agricultural Products Center as the oil and oilseed specialist. She has a BS in chemical engineering, MS in chemistry, Master of Engineering in Food Process Engineering and a Ph.D. in Food Processing/Engineering. She is a registered professional engineer and a certified food scientist, and she is active in several professional organizations including AOCS and IFT in the United States.

Dr. Dunford is internationally known for her research in food, oil, oilseed, microalgae, and bioprocessing. Her research focuses on improving existing and developing new bioprocessing technologies, examining potential of nonfood biomass as biofuel and bioproduct feedstocks, and advancing utilization of byproducts and waste streams for biofuel production and value-added product development. She has established herself as a leading expert in utilization of environmentally benign techniques for edible oil processing focusing on extraction, refining, and value-added product development from vegetable oil and food industry byproducts. Her research has led to collaborations with scholars from South Africa, Sweden, South Korea, Iraq, Mexico, China, Austria, and Turkey.

Joaquín J. Salas. Dr. Salas got his bachelor's degree in chemistry from the University of Seville in 1994 and his Ph.D. from the same university in 1999. Then, he was a postdoc at Michigan State University, where he participated in biotechnology projects in collaboration with Dow Chemical Co. In 2002, he joined the Spanish CSIC, first as a hired researcher and then as a permanent one. There he has participated in several projects on chemistry of oils and biochemistry of sunflower. His research focuses on the production of oils with new applications in the food industry.

Contributors

AOCS Press extends gratitude and appreciation to the *Sunflower: Chemistry, Production, Processing, and Utilization* authors who helped make this title possible.

L.A.N. Aguirrezábal
Comisión Nacional Investigaciones Cientificas y Técnicas, Universidad Nacional de Mar del Plata (FCA-UNMdP), Balcarce, Argentina

C. Alberio
Comisión Nacional Investigaciones Cientificas y Técnicas, Universidad Nacional de Mar del Plata (FCA-UNMdP), Balcarce, Argentina

Igor Balalic
Institute of Field and Vegetable Crops, Novi Sad, Serbia

Charles C. Block
United States Department of Agriculture, Agricultural Research Service, Plant Introduction Station, Ames, IA, USA

Miguel A. Bootello
Instituto de la Grasa (Consejo Superior de Investigaciones Científicas), Sevilla, Spain

Albert J. Dijkstra
Scientific Consultant, St Eutrope-de-Born, France

Carlos E. Feoli
INTA-ASAGIR Agreement, Instituto Nacional de Tecnología Agropecuaria—Asociación Argentina de Girasol, Pergamino, Argentina

Juan Fernández
Limagrain Ibérica, Sevilla, Spain

José M. Fernández-Martínez
Instituto de Agricultura Sostenible (IAS-CSIC), Córdoba, Spain.

Rafael Garcés
Instituto de la Grasa (Consejo Superior de Investigaciones Científicas), Sevilla, Spain

Sergio González-Pérez
Instituto de Recursos Naturales y Agrobiología de Salamanca, Consejo Superior de Investigaciones Cientificas, Salamanca, Spain

Thomas J. Gulya
United States Department of Agriculture, Agricultural Research Service, Northern Crop Science Laboratory, Fargo, ND, USA

James J. Hanzel
Genosys, LLC, Fargo, ND, USA

Robert M. Harveson
University of Nebraska, Panhandle Research and Extension Center, Scottsbluff, NE, USA

Jorge Ingaramo
ASAGIR, Asociación Argentina de Girasol, CABA, Argentina

Lucky Inturrisi
Cargill Australia, Victoria, Australia

N.G. Izquierdo
Comisión Nacional Investigaciones Cientificas y Técnicas, Universidad Nacional de Mar del Plata (FCA-UNMdP), Balcarce, Argentina

Siniša Jocić
Institute of Field and Vegetable Crops, Novi Sad, Serbia

Yalcin Kaya
Trakya Agricultural Research Institute, Edirne, Turkey

Timothy Kemper
Desmet Ballestra, Atlanta, GA, USA

Larry Kleingartner
Retired, National Sunflower Association, USA; Consultant and Adjunct Professor, Bismarck, ND, USA

Etienne Le Clef
Desmet Ballestra, Brussels, Belgium

Alberto León
Center for Biotechnology Research, Advanta Semillas—Nutrisun Business Unit, Balcarce, Argentina

George M. Linz
U.S. Department of Agriculture, Wildlife Services, National Wildlife Research Center, Bismarck, ND, USA

Samuel G. Markell
Department of Plant Pathology, North Dakota State University, Fargo, ND, USA

Dragana Miladinović
Institute of Field and Vegetable Crops, Novi Sad, Serbia

Vladimir Milic
Institute of Field and Vegetable Crops, Novi Sad, Serbia

Justo Pedroche
Instituto de la Grasa, Consejo Superior de Investigaciones Cientifícas, Sevilla, Spain

Begoña Pérez-Vich
Instituto de Agricultura Sostenible (IAS-CSIC), Córdoba, Spain.

Jarrad R. Prasifka
United States Department of Agriculture, Agricultural Research Service, Fargo, ND, USA

M. Victoria Ruiz-Méndez
Instituto de la Grasa (IG-CSIC), Sevilla, Spain

Manuel A. Troncoso-Ponce
Instituto de la Grasa (Consejo Superior de Investigaciones Científicas), Sevilla, Spain

Leonardo Velasco
Instituto de Agricultura Sostenible (IAS-CSIC), Córdoba, Spain.

Monica Venegas-Calerón
Instituto de la Grasa (Consejo Superior de Investigaciones Científicas), Sevilla, Spain

Andrés Zambelli
Trakya Agricultural Research Institute, Edirne, Turkey

List of Abbreviations

3-PGA	3-phosphoglycerate
γ-TMT	γ-tocopherol methyl transferase
ACE	Angiotensin-I converting enzyme
ACP	Acyl carrier protein
ADDC	Antimony dialkyldithiocarbamate
AHAS	Acetohydroxyacid synthase
ALS	Acetolactate synthase
AOCS	American Oil Chemists' Society
AOF	Australian Oilseed Federation
AOM	Active oxygen method
AQ	9,10-anthraquinone
AV	*p*-anisidine value
Az	Sodium azide
BCCP	Biotin carboxyl carrier protein
BHT	Butylatedhydroxyltoluene
BSE	Bovine spongiform encephalopathy
CA	Caffeic acid
CAP	Common agricultural policy
CETIOM	Centre Technique Interprofessionnel des Oléagineux Metropolitains
CGA	Chlorogenic acid
CMS	Cytoplasmic male sterility
cPGI	Cytosolic PGI
CPT	Cholinephosphotransferase
CSIC	Consejo Superior de Investigaciones Científicas
DAG	Diacylglycerol
DHAP	Dihydroxyacetone-phosphate
DAI	Dobrudzha Agricultural Institute
DAP	Delivered at place
DHA	Docosahexanoic acid
DHPLC	Denaturing high-performance liquid chromatography
DMPBQ	2,3-dimethyl-6-phytyl-1,4-benzoquinone
DMS	Dimethyl sulphate

DNAn	Dinitroaniline
DTAB	Dodecyltrimethylammonium bromide
EMS	Ethyl methane sulfonate
ENO	Enolase
EPA	Eicosapentaenoic acid
ER	Endoplasmic reticulum
EST	Expressed sequence tags
EU	European Union
FA	Fatty acid
FAO	Food and Agricultural Organization
FDA	Food and Drug Administration
FFA	Free fatty acids
FOB	Free on board
FTIR	Fourier transform infrared spectroscopy
G3P	sn-glycerol-3-phosphate
GAP	Glyceraldehyde 3-phosphate
GCA	General combining ability
GMO	Genetically modified organisms
HGA	Homogentisic acid
HI	Harvest index
HOHP	High oleic–high palmitic
HOSO	High oleic sunflower oil
HP	High protein
HPHL	High palmitic–high linoleic
HPPD	Hydroxy phenyl-pyruvate dioxygenase
HPT	Homogentisic acid phytyl transferase
HS	High stearic
HSHO	High stearic–high oleic
I	Isoleucine
IAS	Instituto de Agricultura Sostenible
ICAR	Romanian Agronomical Research Institute
IFAPA	Instituto Andaluz de Investigación y Formación Agraria, Pesquera, y Alimentaria
IFVCNS	Institute of Field and Vegetable Crops

IKB	Agricultural Research Institute of Lushnje
IMI	Imidazolinones
INIA	Instituto Nacional de Investigaciones Agrarias
INRA	Institut Nacional pour la Recherche Agronomique
IPM	Integrated pest management
IPO	Inward processing regime
IR	Intercepted solar radiation
IV	Iodine value
L	Linoleic
LAD	Leaf area duration
LAI	Leaf area index
LG	Linkage group
LMWE	Low-molecular weight emulsifier
LP	Low-protein
LPA	Lysophosphatidic acid
LPAAT	Lysophosphatidate acyltransferase
LPCAT	Lysophosphatidylcholine acyltransferase
LTP	Lipid transfer protein
MA	Methyl anthranilate
MAS	Marker assisted selection
MNU	Methyl-nitrosourea
MPBQ	2-methyl-6-phytyl-1,4-benzoquinone
MS	Medium stearic
MUFA	Monounsaturated fatty acid
MW	Molecular weight
NAR	Net assimilation rate
NCBI	National Center for Biotechnology Information
NCRPIS	North Central Regional Plant Introduction Station
NDGA	Nordihydroguaiaretic acid
NDSU	North Dakota State University
NGS	Next-generation sequencing method
NHP	Nonhydratable phosphatides
NOL	Neutral oil loss
NPGS	National plant germplasm system
NSA	National Sunflower Association

O	Oleic
OIT	Oxidation induction time
ONIDOL	Organisation Nationale Interprofessionnelle des Oléagineux
OP	Open pollinated
OPPP	Oxidative pentose phosphate pathway
OSI	Oil stability index
OT	Onset temperatures
P	Palmitic
PA	Phosphatidic acid
PAH	Polycyclic aromatic hydrocarbon
PAO	Polyalphaolefin
PAR	Photosynthetically active radiation
PC	Phosphatidylcholine
PDAT	Phospholipid:diaylglycerol acyltransferase
PDSC	Pressure differential scanning calorimetry
PE	Phosphatidylethanolamine
PEP	Phosphoenolpyruvate
PG	Propyl gallate
PGK	Phosphoglycerate kinase
PGM	Phosphoglycerate mutase
PI	Phosphatidyllinositol
PK	Pyruvate kinase
PL	Phospholipids
PLA	Phospholipase A
PLC	phospholipase C
PLL	1-palmityl-2,3-dilinoleyl-glycerol
PLP	1,3-dipalmityl-2-linoleyl glycerol
PO	Palmitoleic
POB	Pyrimidinyloxybenzoates
POL	1-palmityl-2-oleyl-3-linoleyl glycerol
POO	1-palmityl-2,3-dioleyl-glycerol
POP	1,3-dipalmityl-2-oleyl glycerol
PPR	Prairie pothole region
PS	Phosphatidylserine
PUFAs	Polyunsaturated fatty acids
PV	Peroxide value

PVP	Polyvinyl-pyrrolidone
Pyr	Pyruvate
QA	Quinic acid
Q-RT-PCR	Quantitative reverse transcription polymerase chain reaction
QTL	Quantitative trait loci
RAPD	Random amplification of polymorphic DNA
RBD	Refined, bleached, deodorized
RBOT	Rotary bomb oxidation test
RGCs	Resistance gene candidates
RIL	Recombinant inbred lines
RP-HPLC	Reversed-phase high-performance liquid chromatography
RST	Rotary steam tube
RUE	Radiation use rfficiency
S	Sowing; Stearic
SAD	Stearoyl-ACP desaturase
SCA	Specific combining ability
SC-CO2	Supercritical carbon dioxide
SCD	Soft column deodorization
SCT	Sulfonylaminocarbonyl-triazolinones
SDS	Sodium dodecyl sulfate
SEM	Scanning electron microscope
SFA	Sunflower albumens
SFM	Sunflower meal
SMT2	Sterol methyltransferase II
SNPs	Single-nucleotide polymorphisms
SSR	Simple sequence repeat
StLL	1-stearoyl-2,3-dilinoleoyl-glycerol
StLSt	1,3-distearoyl-2-linoleyl-glycerol
StOL	1-stearoyl-2-linoleyl-3-oleyl-glycerol
StOO	1-stearoyl-2,3-dioleoyl glycerol
SU	Sulfonylureas
SURES	Sulyfonyl urea
SUS	Saturated-unsaturated-saturated
SUU	Saturated-unsaturated-unsaturated
SWOT	Strengths, weaknesses, opportunities, and threat analysis

T	Threonine
TC	Tocopherol cyclase
TAG	Triacylglycerol
TBHQ	Tertiary butylhydroquinone
TBP	Thiamin-binding protein
TEM	Transmission electron microscope
TFA	Trans fatty acid
THBP	2,4,5-trihydroxybutyrophenone
TILLING	Targeting induced local lesions in genomes
TP	Triose phosphate
TPI	Triosephosphate isomerase
TRAP	Target region amplification polymorphism
TZ	Triazolopyrimidines
USDA ARS	U.S. Department of Agriculture Agricultural Research Service
USDA-WS-NWRC	U.S. Department of Agriculture, Wildlife Services, National Wildlife Research Center
WCSP	Wildlife Conservation Sunflower Plots
WS	Wildlife services
WU	Water use
Y	Yield
YPPI	Yuryev Plant Production Institute
ZDDC	Zinc diamyl dithiocarbamate
γ-TMT	γ-tocopherol methyltransferase

1

Breeding and Genetics of Sunflower

Siniša Jocić and Dragana Miladinović ■ *Institute of Field and Vegetable Crops, Novi Sad, Serbia*

Yalcin Kaya ■ *Trakya Agricultural Research Institute, Edirne, Turkey*

Introduction

Sunflower (*Helianthus annuus* L.) is an annual plant. Its botanical name *Helianthus* comes from the Greek words *helios* (sun) and *anthos* (flower). It is a diploid species ($2n = 34$) that belongs to the sub tribe *Helianthineae*, tribe *Helianthea*, subfamily *Asteroideae*, and family *Compositae* (*Asteraceae*) (Panero and Funk, 2002). The genus *Helianthus* is native to temperate North America and contains 14 annual and 37 perennial species (Schilling, 2006).

Archaeological findings show that the American Indians were the first to cultivate sunflower in 4625 B.C. (Crites, 1993). Sunflower is used in the diet (nuts and flour), for oil (skin protection from the sun and to decorate the hair), to obtain colors (yellow and red), for medical purposes (anti-inflammatory and diuretic effects), and as an ornamental plant (religious ceremonies).

Soon after the discovery of America, Spanish explorers brought the sunflower to Europe, to the botanical garden in Madrid. This most probably happened during the Spanish expedition in 1510 (Putt, 1997). The first published record of the sunflower appeared in 1568 by the Belgian Rembert Dodoens, one of the famous herbalists of the era. After introduction into Europe, the sunflower was used solely as an ornamental plant for more than two centuries. The first hint of sunflower use as an oil crop was the registration of patent for extracting oil from sunflower seeds for industrial use in 1716 in England. However, sunflower became an oil crop in Russia in 1697, introduced by the Russian Tsar Peter the Great, who was delighted with its beauty. The invention of D.S. Bokarev from the Belgorod area, who in 1829 discovered a way to extract oil from sunflower seeds, initiated sunflower cultivation as a field crop and its use for oil production (Pustovoit, 1990). Until 1850, a few dozen oil factories were opened in Central and Eastern Europe. In the late 19th century, there was a rapid expansion of sunflower and a number of local varieties were created. During this period, the sunflower was used for human consumption and the production of oil.

In the second half of the 19th century, immigrants from the Russian Empire brought with them seeds of local sunflower varieties and spread them to Canada, the United States, and Argentina. The most famous varieties were Russian Mammoth and Giant of Russia. They were used for human consumption, silage, and poultry feeding. In 1892, sunflower was produced on 315 ha in Argentina (Paniego et al., 2007). The first production of oil from sunflower seeds was recorded in 1926 in Canada.

We can consider 1912 to be the year when sunflower breeding on scientific grounds began. That year Kruglik Plant Breeding and Experimental Station in the province Kruglik Kuban in Russia was opened (Škorić, 1988), as well as stations in the provinces of Saratov and Kharkov. Since 1932, the former Soviet Union established several sunflower breeding centers, the first in Krasnodar (VNIIMK), and then in Rostov-on-Don, Kharkov, and Odessa. High-yield and high-oil sunflower varieties (Peredovik, VNIIMK 8931, Smena, etc.) that were created in these centers made a significant contribution to the spread of sunflower as an oil crop on all continents and influenced the development of sunflower production in the world. Sunflower breeding in North America began in Saskatchewan, Canada, in 1937, and in 1950 in the United States in the Texas Experimental Station (Škorić, 2012). The year 1933 is thought to be the starting year of sunflower breeding in Argentina, when Enrique Klein began his breeding work in his field in Pla, Buenos Aires Province (Romano and Vazquez, 2003). In the middle of the last century, a number of breeding centers were opened around the world, especially in Europe (Romania, Bulgaria, Hungary, Serbia, France, etc.), that created a considerable number of sunflower varieties in a relatively short period of time.

Sunflower (*Helianthus annuus* L.) is one of the four most important oilseed plants in the world (along with palm, soy, and rapeseed) and one of the two most important oil crops in Europe, together with rapeseed. In contrast to other vegetable oils, about 90% of the total production of sunflower oil is used for human consumption, and only 10% is used for biodiesel and industrial applications. It is cultivated on 25.56 million ha, with an annual world production of about 40.64 million tons of seed (Table 1.A) (U.S. Department of Agriculture, 2013), with a slight tendency to increase both in terms of production area and in terms of seed production due to an increase in seed yield per area unit (Kaya et al., 2012). The world's largest producers of sunflower are Russia, with 7.20 million ha, and the Ukraine, with 5.80 million ha. These two countries provide more than 50% of the total world production of sunflower. Other considerable sunflower producers are the European Union, with 4.24 million ha, and Argentina, with 1.82 million ha. Sunflower is also spreading into countries where it has not previously been cultivated, primarily in Asia and Africa. This tendency toward sunflower production is the result of the high quality of sunflower oil compared to other major oil crops. The highest seed yield per hectare was recorded in China (2.46 t/ha) and in Serbia (2.37 t/ha), which is significantly higher than the world average of 1.59 t/ha and the average yield of the largest producers of sunflower. The world's largest producer of sunflower is the Ukraine, with 10.50 million tons of seed in 2012.

Sunflower is mostly used for oil production from seed, but it is also used as a protein crop for human consumption, as well as for feed. Depending on the breeding goals and final use, we could say that there are three basic sunflower types: oilseed, confectionery, and ornamental sunflower.

Table 1.A World Sunflower Area, Production, and Yield for 2012

Country	Area Harvested	Production (million t)	Yield (t/ha)
Ukraine	5.80	10.50	1.81
Russia	7.20	9.63	1.34
European Union-27	4.24	8.37	1.97
Argentina	1.82	3.34	1.83
China	0.94	2.32	2.46
USA	0.59	0.93	1.57
Turkey	0.49	0.93	1.89
Burma	0.86	0.79	0.92
Pakistan	0.50	0.75	1.50
India	0.74	0.62	0.84
South Africa	0.45	0.52	1.15
Moldova	0.28	0.43	1.54
Serbia	0.18	0.42	2.37
Kazakhstan	0.95	0.41	0.43
Bolivia	0.16	0.22	1.38
Australia	0.04	0.05	1.18
Uruguay	0.01	0.02	1.50
Canada	0.01	0.02	1.43
Others	0.29	0.40	1.38
Total	25.56	40.64	1.59

Source: U.S. Department of Agriculture (2013).

Methods of Selection

In its historical development, sunflower breeding has gone through three phases, depending on the breeding method that was predominantly used. These are mass selection, method of individual selection for development of varieties, and method for development of the hybrids.

Mass Selection

Mass selection as a method for sunflower improvement undoubtedly has its origins in the initial domestication of sunflower plants. Since each head was harvested separately, it was possible to observe variations in seed size, and the heads with the biggest seeds were used for sowing the next year. Burke et al. (2002) found by quantitative trait loci (QTL) analysis that the direct selection for increased seed size played a major role in sunflower domestication. Mass selection most likely created the cultivated sunflower as we know it today from the wild *H. annuus*, which is characterized by tiny seeds and branched stalks. Several authors confirmed this hypothesis using different molecular techniques (Arias and Rieseberg, 1995; Cronn et al., 1997; Harter et al., 2004).

Mass selection contributed to the further spreading of sunflower and creation of a number of local varieties that were mostly grown in gardens at the end of 19th century. According to Pustovoit (1967), in 1880, sunflower cultivars Zelenka, Cherenyankia, Fuksinka, and Puzanok were cultivated in Russia. These first sunflower varieties were characterized by great variability, especially in vegetation cycle and seed traits. There were two types of varieties according to the seed type: varieties with well-filled, rounded seeds with thin hulls and oil content of 20–30% that were used for oil production, and varieties with big and long seeds with thicker hulls and oil content of 15–20% for human consumption (Gundaev, 1971). These local varieties were created by mass selection, that is, selecting plants from the population based on their phenotype and sowing a mixture of seeds from the chosen plants with the aim to obtain new varieties or to preserve the purity of the existing varieties. The main achievements of mass selection were the creation of cultivars resistant to sunflower moth (*Homoeosoma nebulella* Denis and Schiffermüller) and broomrape (*Orobanche cumana* Wallr.) at the end of the 19th century and the beginning of 20th century, as this insect and parasite seriously endangered sunflower production (Marinković et al., 2003).

At the beginning of 20th century, sunflower breeding stations were founded in Kuban, Saratov, and Harkov provinces in the former USSR. A number of varieties were created in these centers, the most common of which were Kruglik A-41, Zhdanovsky 8281, and Saratovski 169, which were cultivated at the time on over 1 million ha. Improved methods of mass selection were used in these centers, which were based on selection of phenotypically desirable plants, their self-pollination, and estimation of their value based on the progeny. Selected plants were self-pollinated and their S_0 progeny were sown separately. Chosen progenies were classified into the groups according to tested traits, and those groups were sown as a mixture in spatial isolation. This type of mass selection in sunflower is basically similar to the maize selection "ear to row," so it could be termed "head to row." This method of selection was used in other breeding centers as well, and new varieties were created in Argentina (Luciano and Davreux, 1967), Serbia (Nikolić-Vig, 1976), and Mexico (Robles, 1982).

Mass selection lost its significance in contemporary sunflower breeding, although it is still used in smaller breeding centers. The main advantage of this method is that it is simple and economical. Its efficiency depends on gene effects on selected trait, trait heritability, genotype × environment interaction, and sample size. It is more efficient for the traits that are highly heritable and controlled by additive genes. Mass selection did not enable an increase in sunflower yield, but it enabled a change in sunflower earliness, oil content, and resistance to diseases and insects (Morozov, 1947; Vranceanu, 1974).

Individual Selection

Individual selection with seed reserve preserving was introduced into sunflower breeding around 1920 by V. S. Pustovoit (Pustovoit, 1967), which is why this selection method is also called Pustovoit "method of reserves." This method has been the most common and the most successful method of sunflower variety creation.

It is based on individual selection of the best plants from the initial population. These are separately harvested and the seeds of each plant are divided into two parts: part for sowing and part for reserve. Super elites of the best varieties, intervarietal hybrids, and the best progenies from the previous selection cycles are used as initial populations. At least 15,000–20,000 plants with at least 1500–2000 seeds per plant are selected from these populations. Plant architecture is observed during vegetation and yield and seed characteristics, especially hull and oil content, are estimated after the harvest. Between 1200 and 1500 best progenies are selected, but their seed is not used for the sowing, as these plants were open pollinated and also crossed with the progenies that were discarded. That is why the seeds from reserve are used for the next year sowing. One row of each selected progeny is sown in two repetitions, and in each third row in the plot, an elite variety that is the best in given agro-ecological conditions as control is sown. The same observations as in previous generations are performed, with the emphasis on disease and pest resistance, seed yield, and hull and oil content. It is desirable to select about 200 best progenies. The same procedure is repeated in the third year, but selected progenies with control are also sown in the field infested with either some sunflower disease or broomrape, according to the breeding goals. In the next cycle, several best progenies are multiplied together or divided into the groups, depending on the variability. Reserve seeds from the beginning of the cycle are used for sowing. Sowing is randomized in five to six repetitions, in order to enable crossing of each selected progeny with other selected progenies in open pollination. It is also necessary to ensure spatial isolation of 2–3 km. Crossing between best progenies is enabled in this way and the number of heterozygots increases in the newly created population. During vegetation, phenotypic observations are performed on single plants. Each plant is harvested separately and its seed yield, hull, and oil content are analyzed. Seeds of selected plants within one progeny are mixed and used for preliminary trials in the following year. This seed could also be used as initial material for the new selection cycle. Preliminary trials are set on site as a comparative trial between the best selected progenies and the best varieties for the given region. Based on the productivity results, the best progenies are selected for prevarietal and varietal trials that are set for three years at several sites and that provide information on adaptability and stability of the best newly created varieties.

Galina Pustovoit made significant progress in variety creation through introduction of interspecies hybrids as initial material (Pustovoit, 1963; Pustovoit and Gubin, 1974). Wild relative of cultivated sunflower *Helianthus tuberosus* L. was included into the breeding process. Cultivated sunflower is diploid ($2n$ = 34) and *H. tuberosus* is hexaploid ($2n$ = 102), which is why their interspecies hybrids ($2n$ = 68) were sterile. This problem was overcome by temperature shocks during meiosis. Temperature shock is performed by exposing interspecies hybrids to day temperature of 25–30 °C and night temperature of 3–5 °C for 7–10 days. Plants with $2n$ = 34 are obtained as the result of temperature shock, which enables their backcrossing with cultivated sunflower. One more backcross is performed in the next generation and the obtained progenies are tested in the field conditions for resistance to dominant diseases. The best progenies are crossed among themselves and the same procedure is repeated in the next generation. After that, the best progenies are multiplied manually under the isolation cages. In the next six generations, these progenies are tested in the field and in the greenhouse for disease resistance and their seed yield, while hull and oil content, 1000-seed mass, and other traits are analyzed. At the end of this cycle, the best progenies are manually multiplied under isolation cages and are included as initial material in the Pustovoit "method of reserves."

The main contribution of this selection method to sunflower production is in creation of varieties with increased oil content. Leading sunflower varieties that were cultivated before use of this method had 30–33% oil content. The Pustovoit method based on seed reserve enabled an increase of oil content to 43% in 1935, 46% in 1953, and 51% in 1958, when variety Peredovik was created (Panachenco, 1966). Owing to this, Peredovik spread to North America and Western Europe during the 1960s and cultivation of sunflower as world oil culture started (Fick and Miller, 1997). Significant results were also achieved regarding earliness, disease, and sunflower moth resistance (Gundaev, 1971). Varieties obtained by Pustavoit method were cultivated on about 5 million ha in the former USSR in 1973 (Pustovoit and Gubin, 1974), and the method itself was successfully used for creation of the varieties in other countries, such as Romania (Vranceanu, 1974) and Serbia (Škorić, 1988). Besides that, sunflower genetic variability was significantly increased by introduction of interspecies hybrids into the breeding that was carried out by G. Pustovoit (1963). Varieties created in this way served as a source of resistance to downy mildew, broomrape, rust, *Verticillium,* and other sunflower diseases, as well as initial populations for selection of inbred lines for hybrid creation. This method is still used in breeding centers in the former USSR countries and in some developing countries.

Hybrid Development

The main goal of this sunflower breeding method is utilization of the phenomenon of heterosis, or F_1 vigor, in order to obtain higher yields. From the genetic aspect, heterosis is mostly the result of intra-allelic interaction (domination and super domination), and, to a lesser extent, of inter-allelic interaction (epistasis). This is in fact the state of maximum heterozygosity that is most successfully achieved by crossing genetically divergent self-pollinated homozygous lines (inbred lines). First studies of heterosis use in sunflower were performed in the 1940s, when the yield increase of 60%, compared to varieties, was observed (Morozov, 1947; Unrau and White, 1944). At the initial stages of heterosis use in sunflower, intervarietal hybrids were used, but it was found that the heterosis effect is higher in interline hybrids (Kloczowskii, 1967). However, practical application of hybridization in sunflower breeding started much later due to the absence of an appropriate male sterility system. The first attempt of commercial use of sunflower hybrids was in Canada during the 1950s, where inbred lines with high self-incompatibility were used as female parents of hybrids (Putt, 1962). The first sunflower hybrid, Advance, was created in this way, and later hybrids Advent and Admiral found their place in commercial production. However, only 50% of hybrid seed was generally obtained during seed production, as female lines were not 100% self-incompatible. That is why the main advantage of hybrids compared to the varieties, which is heterosis for seed yield, could not fully be expressed. The next important step in commercial use of sunflower hybrids is the discovery of nuclear male sterility (Kuptok, 1935; Leclercq, 1966; Putt and Heiser, 1966). In the most cases, this trait is controlled by one recessive gene. The discovery by Leclercq (1966), who found the gene for male sterility linked with the gene for anthocyanin-pigmented hypocotyle, lead to practical application of sunflower hybrids in France and Romania in the 1970s. The application of this system enabled production of almost 100% hybrid seed and the obtained hybrids had 24% higher yield than varieties (Vranceanu, 1974). The first registered hybrid of this type was INRA6501 in 1969. The main disadvantage of this system of production was uneconomical seed production that required significant labor for removal of fertile plants with anthocyanin from rows with female plants and green male sterile plants from the rows of male plants. Real commercial use of heterosis in sunflower was possible only after the discovery of cytoplasmatic male sterility (PET1 gene; Leclercq, 1969) and the corresponding gene for fertility restoration (rf gene; Kinman, 1970). The first hybrids based on this type of male sterility (Fransol and Relax) were registered in France in 1974. The process of hybrid creation based on cytoplasmatic male sterility is a complex process that consists of the following stages: inbred line creation and testing of combining abilities of newly created inbred lines.

Development of Inbred Line

The right selection of initial material for inbred line creation is critical for success in sunflower breeding. Local populations; new or commercial varieties; intervarietal, interline, and interspecific hybrids; populations obtained by planned crossing; and populations improved by recurrent selection could be used as initial material for inbred line creation. It is important that initial material has great genetic variability because only then can creation of a number of different inbred lines be expected. The size of the initial population is also important, as there should not be less than 100 plants in self pollination (Škorić and Marinković, 1981). Basically, there are three main concepts in initial population selection: variety concept, trait concept, and gene concept. Variety concept is applied in the first years of sunflower hybrid use and is based on selection of a great number of varieties and local populations as initial populations for the creation of a great number of inbred lines with the presumption that a certain number of these lines will have desirable characteristics. Although this concept gave satisfactory results during creation of the first sunflower hybrids, it is not used anymore in the advanced breeding programs because it is mostly based on coincidence. Trait concept is substantially more successful than variety concept. It is based in preliminary testing of varieties and inbred lines that are used for creation of initial populations. By crossing the lines and varieties of the known properties, higher genetic diversity of the initial population, as well as a reduced rate of undesired traits, could be achieved, which enables higher success in perspective lines creation. Genetic studies that explained the genetic basis of a great number of sunflower traits enabled introduction of the gene concept that is based on the knowledge of genetic constitution of the trait selected. In sunflower, this concept was especially useful in breeding for resistance to diseases such as downy mildew (Jocić et al., 2010, 2012) and broomrape, breeding for oil quality (Škorić et al., 2007), as well as breeding for herbicide tolerance (Jocić et al., 2008). The better we know the genetic constitution of the trait, the more adequately we can choose inbred lines and varieties for planned creation of initial populations, and the greater the success in creating new inbred lines with desirable traits.

Sunflower is an open-pollinated plant which allows self-pollination; that is, the inbred lines are created by the selfing process during six or more generations. The first studies of sunflower inbreeding were carried out in 1920 by Corden (1922) who created the first inbred lines by selfing of the variety Mammoth Russian. Hamilton (1926) found that selfing is possible in sunflower, but with a yield reduction of 15–50% compared to open pollination. Putt (1941) determined that the selfing percentage during sunflower breeding is very different depending on the origin of the initial material. Inbreeding was used for the creation of lines with the increased oil content and resistance to diseases and insects (Jagodkin, 1937; Voskoboinik and Soldatov, 1974). For creation of inbred lines, pedigree method, bulk selection, and single-seed descent method could be used (Fernandez-Martinez et al., 2009). Due to its high efficacy, the most commonly used method is pedigree selection. It is the method of individual plant selection in segregation generations and monitoring of origin or pedigree of selected plants until the creation of inbred lines. During the growing period, it is necessary to test initial populations by artificial inoculation methods for resistance to dominant diseases and to perform phenotypic observations. The best plants from the initial population are selfed by isolation with linen or paper bags immediately before flowering. It is important to exclude extremely self-incompatible genotypes in the first year of selfing, as it is the trait that hampers creation, development, and maintenance of sunflower inbred lines.

Seeds of the first generation of selfing (S_0 or F_2, depending on initial population) are sown "head to row" according to the pedigree method. Plants of the S_1 generation are very different, due to trait segregation that is the result of selfing, as the initial populations were heterozygous for most of the traits. Special attention is paid to the length of the growing period, plant height, seed yield per plant, head angle, 1000-seed weight, hull content, oil content, disease resistance, lodging resistance, and other specific traits that were determined as the breeding goals. The best plants from the best progenies are chosen for further sowing, while the extremely inadequate progenies are discarded. In the S_2 generation, each progeny becomes more uniform, and the difference between different progenies increases. As the result of selfing, dwarfism, leaf yellowing, albinism, and partial sterility could appear in some progenies or plants. In the S_3 generation, progenies (i.e., inbred lines) are mostly uniform and the difference between different lines is further increased. Inbreeding depression is even more pronounced on traits such as plant height, seed yield, and so forth. After six to eight generations of selfing and selection, the negative effect of inbreeding stops and the lines that are very uniform for the most traits (i.e., over 96% homozygotes) are obtained.

The obtained inbred lines could be directly used in sunflower production via synthetic varieties. High-yielding synthetic varieties could be created by mixing three to five inbred lines (Putt, 1966; Voskoboinik and Soldatov, 1974). However, hybrids have higher genetic potential, so the inbred lines are mostly used for hybrid production, except in some countries with less developed sunflower breeding programs (Ado et al., 1991; Shabana, 1990).

Combining Ability Testing of Newly Created Inbred Lines

Newly created inbred lines should be tested in order to determine which ones will give heterosis in the F_1 generation. As heterosis is the state of maximal heterozygosity, heterozygosity of the F_1 generation for the greatest number of alleles is obtained by crossing genetically different inbred lines. This results in increased vigor of the whole organism. Heterosis is not obtained by crossing any pair of inbred lines, since the lines could be genetically similar, which is why it is necessary to test the combining ability of the new lines, as the breeding value of the line is estimated based on the heterosis it gives in combination with other lines. Final estimation of the value of even more carefully selected inbred lines is based on their results in hybrid combinations. Good combining ability is the ability of the inbred line to give superior progeny in combination with another line. There are general and specific combining abilities. General combining ability (GCA) is an average value of the inbred line based on its behavior in crosses with other lines. Specific combining ability (SCA) is the value of the line in crossing in a specific cross.

The combining ability of sunflower inbred lines is mostly tested in the S_4 generation, although early testing of combining ability after the first generation of self-pollination was found to be good for identification of lines with good combining ability (Shein, 1978). General combining ability is most commonly tested by polycross and topcross methods, and specific combining ability is tested by diallel crosses.

Polycross Method

Tested lines are sown in spatial isolation of at least 3 km, in 4 repetitions, with 10 plants per repetition in a complete randomized block system to enable cross pollination between the tested lines. Progeny of each line is tested in comparative trial and GCA is determined based on productivity. The method is based on the presumption that there is a probability that each line will be pollinated by each line. This is very hard to achieve in practical conditions due to different growing periods of the tested lines, different flowering periods, different attractiveness for pollinators, different pollen productions, and so on. Thus, the method is rarely used for testing combining ability in sunflower.

Topcross Method

Estimation of the GCA of new lines is done based on testers that could be either variety or line with good combining ability. The lines that are parental components of the best hybrid are mostly used as testers. Miller et al. (1980) and Dominguez and Fernandez-Martinez (1987) found that the lines of the best combining ability could be identified in this way. There are two variations of this method:

1. Tested lines are sown in spatial isolation together with the tester, one row of the line, and one row of the tester. Artificial male sterility is induced in tested lines with the gibberelic acid solution. Pollen transfer from the tester is enabled by insects, so it is necessary to place beehives on the plot, as bees are the main sunflower pollinators. The obtained hybrids are tested in comparative trials with commercial hybrids and GCA is estimated based on the productivity results (i.e., seed and oil yield). The main disadvantage of this method is that gibberelic acid treatment does not ensure 100% sterility of treated plants. Application of 50–100 ppm solution of gibberelic acid when buttons are 1–1.5 cm generally gives good results (Miller and Fick, 1978). However, different genotypes react differently and optimal concentrations of gibberelic acid and application time vary depending on the genotype. Furthermore, it is necessary to ensure that the plants of the treated genotype are in the same stage of development at the time of gibberelic acid application (Piquemal, 1970). Gibberelic acid treatment

could cause negative effects such as decrease in pollination rate, decrease in head diameter, descrease in number of flowers, flower deformation, and so on (Miller, 1987). Due to these practical problems, this method is not widely used for combining ability testing of sunflower inbred lines.

2. Since mostly inbred lines that are parental components of the best hybrids are used as testers, and because sterile forms of these lines were previously created, these sterile lines could be used as testers. Sterile form is manually crossed with pollen mixture of at least five plants of a newly created line. Obtained hybrids are tested in comparative trial with commercial hybrids and on a plot where there are enough fertile sunflower plants, since most of the obtained hybrids are sterile. The advantage of this method is that in the same trial, presence of fertility restoration genes in new lines is tested. GCA is determined based on productivity, and the inbred lines that gave the highest yield with the tester are selected for the further work and others are discarded. A great number of lines are eliminated in this way, and others are tested for SCA in diallel crosses.

Diallel Crosses

Diallel crosses could be used for estimation of both GCA and SCA, as well as for determination of the effect of reciprocal crosses. Although the most reliable information on combination ability of the tested lines is obtained by this method, this method is based on crossing each line with each line, including reciprocal crosses, and could not be used for testing a great number of lines, due to practical limitations. This is why diallel crosses are used on selected lines with good GCA and other agronomic traits. Diallel crosses are also frequently used in genetic studies for determining mode of inheritance of the examined trait, as well as the number of genes that control the trait and gene effects.

Heterosis in sunflower is used mostly through single-cross hybrids that are created by crossing female lines that are cytoplasmic male sterile and male inbred lines that have a fertility restoration gene. That is why the inbred lines with the best combining ability are converted into sterile form or fertility restoration genes are introduced into them by backcrossing. Three-way and double-cross hybrids are less used, although they are more adaptable and stable than single-cross hybrids due to higher heterogeneity (Fick and Zimmer, 1976; Schuster and Friedt, 1988; Vulpe, 1974). The main advantage of single-cross hybrids is that they are more uniform and have higher yield (Miller, 1987; Škorić, 1988).

Hybrids have 25–30% higher seed yield than varieties. Besides that, hybrids have other advantages over varieties. They are genetically homogenous and uniform regarding plant height and vegetation length, which contributes to lower losses during harvest, and seed with uniform moisture content suitable for storage is obtained. The important advantage of hybrids compared to varieties is easier introduction of genes for resistance to dominant sunflower diseases and consequently higher disease resistance of hybrids compared to varieties.

Directions of Sunflower Breeding

There are several basic directions of sunflower breeding:

1. Creation of oil hybrids (with high seed and oil yield) resistant to dominant diseases and tolerant to drought
2. Creation of hybrids with different oil quality (high oleic acid content and changed tocopherol content)
3. Creation of confectionery hybrids (decreased oil content and increased protein content) for human consumption and poultry and bird feeding
4. Creation of hybrids tolerant to certain groups of herbicides (imidazolynes and tribenuron-methyl)
5. Creation of ornamental hybrids (for cultivation in gardens and parks, cut flowers, and growing in pots)

Highly Productive Oil Sunflower Hybrids

The main direction of sunflower breeding is creation of hybrids with high genetic potential for seed yield (over 5 t/ha) and oil content in seed (>50%) that give high oil yield per hectare (>2.5 t/ha), with changed plant architecture capable of adjusting to regions of cultivation, and resistant to dominant diseases, parasitic weed broomrape, insects, and stress conditions (drought). In creation of new hybrids, significant attention should be paid to increase of adaptability, stability, and attractiveness to pollinators.

The main goal of sunflower production is to obtain high yield of oil per hectare and this trait is the main indicator of productivity of every sunflower hybrid. Oil yield is a product of seed yield per hectare and oil content in seed. Up-to-date, significant results have been achieved and most of contemporary sunflower hybrids have 45–50% oil in seed. The highest recorded oil content in seed is 60–65% (Vear, 2010), which is probably the biological maximum in sunflower. That is why significant increase in oil content in sunflower seeds could not be expected in the future; rather, only a 1–2% increase in future generations of hybrids should be expected. Besides that, it was determined that there is a negative correlation between extremely high oil content and high seed yield.

Seed yield is a quantitative trait that is significantly affected by the environment. It is the so-called super trait where selection to any other trait is either in positive or negative correlation with seed yield. Sunflower breeders have achieved significant results in increasing sunflower yield, as modern single cross CMS hybrids have a mean of 140% of the yield of the varieties grown 30 years ago (Vear et al., 2003). This represents a 1.3% gain in seed yield per year, similar to many results for maize, the other main hybrid crop.

Breeding for disease resistance is a significant part of sunflower breeding for increased yield. Unfortunately, sunflower is the host of more than 50 pathogens and sunflower breeding is frequently considered to be mostly breeding for disease resistance. It is not possible to breed sunflower for resistance to all pathogens and it is necessary that breeders focus their work on the most economically significant diseases in the region where the hybrid is to be grown.

It is necessary to clearly and precisely define breeding goals before starting to work on hybrid creation, and to determine which type of hybrid is going to be created and for which growing region. The breeders should know their initial material well, especially regarding the most important traits that affect yield. Breeders should also be familiar with the genetic basis of sunflower traits, the number of genes that control certain traits, the mode of inheritance of the traits, and heterosis expression. It is also very important to know positive and negative correlations between yield components. Creation of highly productive sunflower hybrids demands building a model hybrid for certain agro-ecological conditions with determination of the priorities in breeding for main traits. Ideotype of the sunflower hybrid should contain the following desirable genes (Škorić, 2012):

- Sunflower genotype
  - Genes for length of growing season: ultra-early (less than 80 days), early (80–90 days), medium early (90–100 days), medium late (100–115 days), and late (115–130 days)
  - Genes for plant height: dwarf (80–90 cm), semi-dwarf (90–100 cm), medium short (100–120 cm), medium (120–140 cm), medium tall (140–160 cm), and tall (160–190 cm)
  - Genes for leaf area: for high-yielding hybrids, 6000–7000 cm^2/plant
  - Genes for increased number of disk florets (1500–2000 per plant)
  - Genes for disease resistance: *Plasmopara halstedii, Diaporthe/Phomopsis helianthi, Puccinia helianthi, Sclerotinia sclerotiorum, Verticillium dahliae, Phoma macdonaldi, Macrophomina phaseoli, Botrytis cinerea, Albugo tragopogis, Rhizopus* spp., *Erysiphe cichoracearum, Septoria helianthi, Fusarium* spp.
  - Genes for resistance to broomrape (*Orobanche cumana*)
  - Genes for resistance to sunflower moth and other insects
  - Genes for resistance to viruses and bacteria
  - Genes for short head incline
  - Genes for tolerance to drought and high temperatures

  - Genes for long leaf area duration and stay green
  - Genes for efficient net assimilation rate (NAR)
  - Genes for efficient translocation of assimilates into the grain (high contents of oil and proteins)
  - Genes for oil and protein quality
  - Genes for high harvest index
  - Genes for tolerance to certain herbicides: imidazolines and tribenuron-methyl
  - Genes for wide adaptability
  - Genes for plump and heavy grains
- Direct yield components: In the case of oil sunflower genotypes, the ultimate goal is high seed yield and oil yield per unit area. The following characteristics are among the direct yield components:
  - Number of plants per unit area (ha): The optimal number of plants ranges from 55,000/ha to 75,000/ha, depending on maturity group and leaf position on the stem.
  - Number of seeds per plant (1500–2000 seeds)
  - Weight of 1000 seeds (up to 80 g)
  - Hectoliter weight (50–55 kg/hl)
  - Low husk percentage (<25%)
  - Oil content in seeds (50–55%)

Main goals in creation of highly productive sunflower hybrids are increased seed yield and oil content per surface unit, assimilative acceptors, harvest index, attractiveness to pollinators (melliferous characteristics), auto fertility, resistance to main disease and insects, early maturation (shortening of vegetative phase), and optimization of plant architecture. All this should enable an increased number of plants per hectare in the conditions of intensive cultivation, and lead to yield increase (Miklič et al., 2008). It is possible to achieve these goals by teamwork, use of proper equipment, and profound knowledge of breeding methods.

Hybrids with Changed Oil Quality

As an industrial plant, sunflower is mainly cultivated for oil, and for a long time scientific studies were mainly focused on increase and expression of genetic potential for seed yield and oil content in seed in new hybrids. Only recently, oil quality became one of the main challenges that market set before science. In contemporary agriculture, there are ever more demands to satisfy the needs of population not only in food quantity, but also in the increased quality of the food. Optimal quality of sunflower oil depends on the way the oil will be used and whether it will be used in food or nonfood industry. Sunflower oil is used as salad oil, cooking oil, or for production of margarine and other products. In nonfood industry, sunflower oil is used for production of biodiesel, different lubricants, varnish, and in the cosmetics industry. There is no optimal sunflower oil quality, as it depends on its final use. The main parameter in defining oil quality is fatty acid composition, arrangement of fatty acid models within the triglyceride molecule, and total content and profile of several polyisoprenoid lipids that are present in the oil, mostly tocopherols and sterols (Fernandez-Martinez et al., 2009).

Standard sunflower oil contains on average about 70% polyunsaturated linoleic acid (18:2). This acid is an essential fatty acid that cannot be synthesized by the human organism but has to be introduced by food. The next is monosaturated oleic acid (18:1) at about 20%. Although the content of these two fatty acids could vary due to the effect of the environment, it is typical for sunflower oil that jointly they make about 90% of total fatty acid content in sunflower oil. Sunflower oil also contains palmitic acid (16:0) at about 4–9% and stearic acid (18:0) at about 1–7%. There are traces of other fatty acids, such as myristic (4:0), myristoleic (14:1), palmitoleic (16:1), arachidic (20:0), and behenic (22:0). All these fatty acids make up 10% of total fatty acid content in sunflower oil. Total tocopherol content in standard sunflower oil is around 700–1000 mg/kg. Natural tocopherols are present in four isomers: α (5,7,8-trimethyltocol), β (5,8-dimethyltocol), γ (7,8-dimethyltocol), and δ (8-methyltocol). Standard sunflower oil contains mostly α-tocopherol (95%), and lower quantities of β-tocopherol (3%) and γ-tocopherol (2%) (Škorić et al., 2008). Tocopherols are considered to be natural antioxidants, but have different in vivo and in vitro antioxidative activities. α-tocopherol expresses maximum in vivo activity, also known as vitamin E activity, but low in vitro protection of extracted oil. In contrast to that, γ- and δ-tocopherols, to a higher extent, and β-tocopherol, to a lower extent, are very important antioxidants, but with low vitamin E values (Packer and Obermuller-Jevic, 2002).

Although sunflower oil is one of the best quality vegetable oils, sunflower breeders have responded to market demands and managed to significantly change sunflower oil quality regarding composition of fatty acids and tocopherols. The first significant change in sunflower oil quality and the first source of increased oleic acid content was created by Soldatov (1976), who treated seeds of variety VNIMK 8931 with 0.5% solution of dimethyl-sulfate (DMS). Soldatov separated single plants with over 50% of oleic acid in the M_3 generation from which he selected variety Pervenets with 80–90% of oleic acid in oil. Pervenets was further used for creation of inbred lines and hybrids with the increased oleic acid content, although the mode of inheritance and the number of the genes controlling this trait has not yet been determined. High oleic acid content in this variety was found to be stable during cultivation in different environments, and this trait could easily be transferred by conventional breeding methods into other genotypes. High oleic inbred lines and hybrids of sunflower were obtained in many breeding centers by classical methods of selection (Jocić et al., 2004b). With the use of physical and chemical mutagens, mutant lines with increased concentration of palmitic (Ivanov et al., 1988; Osorio et al., 1995) and stearic acid (Osorio et al., 1995) were created, as well as mutants with lower palmitic and/or stearic acid content (Cvejić et al., 2011; Miller and Vick, 1999; Vick and Miller, 1996). Sunflower is the first plant species in which the genes that control tocopherol content in seed were identified. These are two independent genes, Tph1 and Tph2 (Popov et al., 1988). Gene Tph1 controls the ratio between α- and β-tocopherol, and the Tph2 gene between α- and γ-tocopherol. Recessive mutated gene tph2 with recessive mutated gene tph1 has an epystatic effect that results in appearance of δ-tocopherol when both of them are homozygous (Demurin, 2012). Generally, changed fatty acid composition and tocopherol composition are controlled with the small number of genes that facilitates classical breeding for these traits.

Sunflower oil quality was significantly improved compared to other oil crops, mostly by classical breeding. Genotypes with high, intermediate, and low saturated acid content, medium and high oleic acid content, as well as genotypes with different tocopherols, enable creation of more different oil profiles than in any other oil crop. This allows creation of a "new oil crop" for specific uses. There are two basic types of sunflower oil in the world market at the moment: standard linoleic type and high oleic or mid oleic type. High oleic sunflower oil has the highest oleic acid content (over 90%) compared to all vegetable oils present in the world market. Different sunflower oil qualities enable production of new oil types for different uses, such as oil for salads and cooking, with low saturated acid content, standard or increased oleic acid content, and with increased α-tocopherol content. The oil for production of healthier margarines has increased stearic acid content, standard or increased oleic acid content, and increased α-tocopherol content. There are also oils for biodiesel production with high oleic acid content and γ- and δ-tocopherols and oils for use in high temperatures with high palmitic acid content and γ- and δ-tocopherols. A new oil type could be used in medicine for prevention against cardiovascular diseases or in diets of people with heart conditions.

Hybrids Tolerant to Herbicides

Increased crop yield could not only be achieved by creation of more productive genotypes, but it is also necessary to improve production technology. Through plant breeding for herbicide resistance, both of these aspects are improved. It is theoretically possible to create plants tolerant to all herbicides, but only economically important plant species and herbicides with favorable characteristics are commercially used (Malidža et al., 1999).

Sunflower was not included in initial programs of cultivated plant breeding for herbicide resistance. However, weeds significantly reduce sunflower yield due to lack of efficient herbicides for broadleaf weeds and postemergence application, since development of herbicides for application in sunflower is slower than in other crops (Malidža et al., 2004). This motivated breeders to initiate the research on development of sunflower genotypes tolerant to herbicides.

The main precondition for development of sunflower tolerant to herbicides was fulfilled when Al-Khatib et al. (1998) discovered wild *Helianthus annuus* L. populations (ANN-PUR) originating from Rossville, Kansas (United States), that were resistant to imidazoline herbicides. After studies of genetics of resistance (Jocić et al., 2001; Miller and Al-Khatib, 2002), this led to creation of the first sunflower hybrids tolerant to imidazoline herbicides in the United States in 2003 and in Serbia and Turkey in 2004 (Jocić et al., 2004a). The discovery of the wild *Helianthus annuus* L. population (ANN-KAN) in Kansas (Al-Khatib et al., 1999) that was resistant to sulphonylurea herbicides (i.e., tribenuron-methyl) led to expansion of the breeding programs for creation of herbicide-tolerant sunflower. The programs were additionally extended when partially dominant nuclear allele *Ahasl1-3* was developed by seed mutagenesis and selection with imazapyr (Sala et al., 2008a).

Acetohydroxyacid synthase (AHAS) is the first enzyme in the pathway for the synthesis of the branched chain amino acids valine, leucine, and isoleucine (Singh, 1999). This enzyme has been shown to be the site of action for different classes of herbicides collectively known as AHAS inhibitors: sulfonylureas, imidazolines, triazolopyrimidines, pyrimidinyloxybenzoates, and sulfonylaminocarbonyl-triazolinones. Tolerance in most of these cases is due to a form of the AHAS large subunit enzyme that is less sensitive to herbicide inhibition and is conferred by a single, partially dominant nuclear gene (Tan et al., 2005; Tranel and Wright, 2002). This reduction in herbicide binding is caused by mutations at key sites in the genes coding for the catalytic subunit of AHAS. Based on molecular studies, Kolkman et al. (2004) identified and characterized three genes coding for the AHAS catalytic subunits in sunflower (*Ahasl1*, *Ahasl2,* and *Ahasl3*). *Ahasl1* is a multi-allelic locus and the only member of this small family where all the induced and natural mutations for herbicide tolerance were described thus far in sunflower (Sala et al., 2008b). *Ahasl1-1* (also known as *Imr1* or *Arpur*) (Bruniard and Miller, 2001; Kolkman et al., 2004, respectively) harbors a C-to-T mutation in codon 205 (*A. thaliana* nomenclature) that confers a moderate tolerance to imidazolinones. *Ahasl1-2* (also known as *Arkan*) shows a C-to-T mutation in codon 197 conferring high levels of sulfonylurea tolerance (Kolkman et al., 2004). *Ahasl1-3* presents a G-to-A mutation in codon 122 that confers high levels of imidazolinone tolerance (Sala et al., 2008b).

These studies enabled creation of sunflower hybrids tolerant to imidazolinone and hybrids tolerant to tribenuron-methyl herbicides. In order to enhance and accelerate breeding processes, different field and laboratory methods for herbicide resistance testing of newly created hybrids were developed and used (Dimitrijević et al., 2012; Jocić et al., 2008).

Confectionery Sunflower

Besides main sunflower cultivation for oil production, non-oil sunflower is widely cultivated in the world. Non-oil type of sunflower *Helianthus annuus* L. var. macrocarpus (DC.) Ckll. is also called confectionery, protein, nibbling, or large-seed sunflower. Seeds of non-oil sunflower are usually black with white stripes or colorful and significantly larger than oil-type seeds, with a thicker hull that is loosely attached to the kernel. The hull is easily separated from the kernel and enables complete seed hulling (Gonzalez-Perez and Vereijken, 2007). Confectionery sunflower seeds also have a great proportion of hull, mostly 40–50% (Jovanović, 2001) and high 1000-seed weight that is frequently higher than 100 g (Dozet and Jovanović, 1997). Work on confectionery sunflower breeding is aimed at increasing genetic potential for yield, yield stability, health safety, and nutritive value, along with increase of production efficiency (Hladni et al., 2011). High-protein sunflower hybrids for human consumption have found their place in production in the United States, Hungary, Slovakia, France, Spain, Turkey, Ukraine, Russia, Serbia, and many other countries. Seed color is one of the main characteristics for confectionery sunflower. While white seeds with grey stripes are mostly preferred in Turkey, grey seeds with stripes are popular in the United States, Spain, and China, and black seed is more preferred in Balkan countries and Russia (Kaya et al., 2008).

Confectionery sunflower breeding is mostly similar to oil sunflower breeding, especially in increase of seed yield, adaptability and stability, and resistance to main disease, but there are also certain specificities. Specific breeding goals for confectionery sunflower are increase of total protein content in seed to over 25%, increase of essential amino acid content, 1000-seed mass, kernel content, and decrease of oil content in the seed to 40% with simultaneous increase of oil stability, uniformity in seed size and color, and improved seed hulling. Additional breeding goals are aimed at creation of different hybrids and seeds that could be used for consumption without heat processing in food and confectionery industries, and production of protein flour.

Confectionery sunflower seed should ideally have over 80 g 1000-seed weight, oil content less than 30%, higher seed size, lower cadmium rate, and higher protein, oleic acid, and vitamin E (tocopherol) content (Jovanović et al., 1998; Lofgren, 1997). The seed size is the main criterion for the quality of confectionery sunflower. While larger size (>15 mm) type goes into the in-shell market to be used as snacks, medium-size seeds are hulled for the kernel market both for consumption as snacks or for bakery use, and smaller seeds go for bird and pet feeding markets (Hofland and Kadrmas, 1989).

In order to produce larger seeds, plants should first of all have good genetic potential for this trait. Larger seeds could also be obtained by irrigation (or enough rain during the growing period) in normal row planting (70 × 40 cm), or decreasing plant population per hectare, especially in normal rain-fed areas. For instance, confectionery sunflower is cultivated at 1 m × 50 cm with only 20,000 plants per hectare in order to obtain larger seed size in fallow areas of the Middle Anatolia region in Turkey (Kaya, 2004). Therefore, increased seed length is one of the main goals in confectionery sunflower breeding and it can be achieved by selection. Sun (2009) found that the polygenic system controls seed length in sunflower, but QTL analysis showed that only one or two major genes play an important role.

Finally, newly developed confectionery hybrids should have higher yield potential, higher self-fertility rate, and larger seeds with high oleic acid and vitamin E (tocopherol) content to increase their nutritional value and prolong seed shelf life.

Ornamental Sunflower

Although today mostly grown as an oil crop, sunflower has never lost its place as a horticultural plant. First ornamental sunflower varieties were extremely high (over 2 m) with yellow flowers. Some of these varieties can still be found in the United States market under the commercial names Mammoth Russian, Russian Giant, Tall Russian, and Mammoth. These varieties are a real oddity, since they have been on the market for over 130 years and the demand for them is still high.

Main goals in ornamental sunflower breeding are desirable plant architecture and number of disc and ray flowers, as well as length of flowering. Modern breeding of ornamental sunflower goes into several directions, depending on the use: ornamental hybrids for cut flowers, ornamental hybrids for growing in gardens and parks, and ornamental hybrids with short stems for growing in pots (Miklič et al., 2008).

The most important breeding direction is the creation of sunflower ornamental hybrids for cut flowers. The main aim of this breeding program is to obtain the highest number possible of useful flowers per hectare. The breeding goes into two directions: creation of unbranched genotypes that could stand dense sowing (15 cm in row and 50 cm between rows) and creation of branched genotypes that are sown at normal density, but due to branching produce four to five first-class flowers and four to five second-class flowers. Unbranched ornamental sunflower hybrids have to have: big flowers and lodging resistance, since they are sown at high density; strong but not thick stems; a length of cut flower of at least 80 cm; short vegetation; resistance to low temperatures; resistance to the diseases that attack leaves, resistance to transport; and a long duration of the flower in the vase. Branched genotypes are branched along the whole stem and the length of lateral branches that are cut should be 70 cm.

Another direction of ornamental sunflower breeding is for cultivation in gardens and parks. These genotypes should be resistant to low temperatures and pathogens that attack leaves, and be branched and have strong habitus. They are generally 50–170 cm tall depending on their use either as hedge or in combination with other plant species.

The third direction of ornamental sunflower breeding is for growing in pots. The genotypes should have branched stems that are 30–40 cm long, small leaves, and short period until flowering. This type of ornamental sunflower is produced in the greenhouses and is transported to flower shops immediately before flowering. That is why it should also support growing conditions in the greenhouse and have leaves that do not wither during transportation.

The genotypes for all three directions of breeding must meet certain common criteria in terms of flower appearance (Cvejić and Jocić, 2010). The disc flowers should be sterile in order to extend the lifespan of blooms in a vase. The color of disc flowers may be yellow or dark red (anthocyanin). The ray flowers should be rounded and they should completely encircle the head, with no space between them. They should be short—not longer than the radius of the head. The outline of the ray flower should not be spiky or jagged, but rather straight. The color of ray flowers should be lemon yellow, yellow, orange, gold, red, or variegated (Kaya et al., 2012). The yellow ray flower coloration is most common. This trait is controlled by complementarily interacting dominant alleles of different genes (Tolmachev, 2006).

References

Ado, S. G.; Zaria, A. A.; Tanimu, B.; Bello, A. Relative Performance of syn 1 and syn 2 Populations of Sunflower Germplasm Materials. *Helia* **1991,** *14,* 37–42.

Al-Khatib, K.; Baumgartner, J. R.; Peterson, D. E.; Currie, R. S. Imazethapyr Resistance in Common Sunflower (*Helianthus annuus*). *Weed Sci.* **1998,** *46,* 403–407.

Al-Khatib, K.; Baumgartner, J. R.; Currie, R. S. In *Survey of Common Sunflower (Helianthus annuus) Resistance to ALS Inhibiting Herbicides in Northeast Kansas,* Proceedings of 21st Sunflower Research Workshop, Bismark, ND, January 14–15, 1999; National Sunflower Association: USA, 1999; pp 210–215.

Arias, D. M.; Rieseberg, L. H. Genetic Relationship among Domesticated and Wild Sunflowers (*Helianthus annuus,* Asteraceae). *Econ. Bot.* **1995**, *49,* 239–248.

Bruniard, J. M.; Miller, J. F. Inheritance of Imidazolinone Herbicide Resistance in Sunflower. *Helia* **2001,** *35,* 11–16.

Burke, J. M.; Tang, S.; Knapp, S.; Rieseberg, L. H. Genetic Analysis of Sunflower Domestication. *Genetics* **2002,** *162,* 1257–1267.

Corden, P. V. Sunflower Studies. *Agronomy J.* **1922,** *14,* 69–72.

Crites, G. D. Domesticated Sunflower in Fifth Millennium B.P. Temporal Context: New Evidence from Middle Tennessee. *Am. Antiq.* **1993,** *58,* 146–148.

Cronn, R.; Brothers, M.; Klier, K.; Bretting, P. J.; Wendel, J. F. Allozyme Variation in Domesticated Annual Sunflower and Its Wild Relatives. *Theor. Appl. Genet.* **1997,** *95,* 532–545.

Cvejić, S.; Jocić S. Development of Ornamental Sunflower Hybrids. *Ratsr. Povrt.* **2010,** *47,* 147–152.

Cvejić, S.; Jocić, S.; Prodanović, S.; Terzić, S.; Miladinović, D.; Balalić, I. Creating New Genetic Variability in Sunflower by Using Induced Mutations. *Helia* **2011,** *34,* 47–54.

Demurin, Y. In *Tocopherol Mutations in Sunflower,* Proceedings of 18th International Sunflower Conference, Mar del Plata, Argentina, Feb 27–March 1, 2012; International Sunflower Association: Paris, France, 2012; pp 820–823.

Dimitrijević, A.; Imerovski, I.; Miladinović, D.; Jocić, S.; Malidža, G.; Miklič, V. In *Laboratory Method for Detection of Tribenuron-Methyl Resistant Sunflower* (*Helianthus annuus* L.), Proceedings of 18th International Sunflower Conference, Mar del Plata, Argentina, Feb 27–March 1, 2012; International Sunflower Association: Paris, France, 2012; pp 519–523.

Dodoens, R. Florum, et Coronariarum Odoratarumque Nonnullarum Herbarum Historia. *Antuerpiae* 1568, p 311.

Dominguez, J.; Fernandez-Martinez, J. M. Evaluation of Inbred Testers in Sunflower Hybrid Breeding. *Helia* **1987,** *10,* 15–19.

Dozet, B.; Jovanović, D. Combination Ability and Mode of Inheritance of Seed Size in F1 Generation of Confectionary Sunflower [*Helianthus annuus* var. macrocarpus Dc (Ckll)]. *Selekcija i Semenarstvo* **1997,** *3–4,* 20–25.

Fernandez-Martinez, J. M.; Perez-Vich, B.; Velasco, L. Mutation Breeding for Oil Quality Improvement in Sunflower. In *Induced Plant Mutations in the Genomics Era;* Shu, Y.Q., Ed.; FAO: Rome, Italy, 2009; pp 177–181.

Fick, G. N.; Zimmer, D. E. In *Yield Stability of Sunflower Hybrids and Open Pollinated Varieties,* Proceedings of 7th International Sunflower Conference, Krasnodar, USSR, July 23–27, 1976; International Sunflower Association: Paris, France, 1976; pp 253–258.

Fick, G. N.; Miller, J. F. Sunflower Breeding. In *Sunflower Technology and Production*; Schneiter, A. A., Ed.; Agronomy Monograph 35; American Society of Agronomy: Madison, WI, 1997; pp 395–439.

Gonzalez-Perez, S.; Vereijken, J. M. Sunflower Proteins: Overview of Their Physicochemical, Structural and Functional Properties. *J. Sci. Food Agric.* **2007,** *87,* 2173–2191.

Gundaev, A. I. Basic Principles of Sunflower Selection. In *Genetic Principles of Plant Selection;* Nauka: Moscow, Russia, 1971; pp 417–465.

Hamilton, R. I. Improving Sunflower by Inbreeding. *Sci. Agric.* **1926,** *6,* 190–192.

Harter, A. V.; Gardner, K. A.; Falush, D.; Lentz, D. L.; Bye, R. A.; Rieseberg, L. H. Origin of Existent Domesticated Sunflowers in Eastern North America. *Nature* **2004,** *430,* 201–205.

Hladni, N., Miklič, V.; Jocić, S.; Sakač, Z.; Radić, V.; Radeka, I. New NS Confectionery Sunflower Hybrids. *Ratsr. Povrt.* **2011,** *48,* 49–56.

Hofland, C.; Kadrmas, N. *Confectionary Sunflower Handbook,* 2nd ed.; National Sunflower Association: Bismarck, ND, 1989.

Ivanov, P.; Petakov, V.; Nikolova, V.; Petchev, E. In *Sunflower Breeding for High Palmitic Acid Content in the Oil,* Proceedings of 12th International Sunflower Conference, Novi Sad, Yugoslavia, July 25–29, 1988; International Sunflower Association: Paris, France, 1988; pp 463–465.

Jagodkin, I. G. Application of Method of Inbreeding and Diallel Crossing in Sunflower Growing. *Selektsiya i Semenovodstvo* **1937,** *1,* 21–27.

Jocić, S.; Škorić, D.; Malidža, G. Sunflower Breeding for Herbicide Tolerance. *Ratsr. Povrt.* **2001,** *35,* 223–233.

Jocić, S.; Malidža, G.; Škorić, D. Sunflower Tolerant to Imidazoline Herbicides. *J. Sci. Agric. Res.* **2004a,** *229,* 81–89.

Jocić, S.; Škorić, D.; Lečić, N.; Molnar, I. In *Development of Inbred Lines of Sunflower with Various Oil Quality,* Proceedings of the 15th International Sunflower Conference, Toulouse, France, June 12–15, 2004b; International Sunflower Association: Paris, France, 2004; pp A43–A48.

Jocić, S.; Miklič, V.; Malidža, G.; Hladni, N.; Gvozdenović, S. In *New Sunflower Hybrids Tolerant to Tribenuron-Methyl,* Proceedings of 17th International Sunflower Conference, Cordoba, Spain, June 8–12, 2008; International Sunflower Association: Paris, France, 2008; Vol. II, pp 505–508.

Jocić, S.; Cvejić, S.; Hladni, N.; Miladinović, D.; Miklič, V. Development of Sunflower Genotypes Resistant to Downy Mildew. *Helia* **2010,** *33,* 173–180.

Jocić, S.; Miladinović, D.; Imerovski, I.; Dimitrijević, A.; Cvejić, S.; Nagl, N.; Kondić-Špika, A. Towards Sustainable Downy Mildew Resistance in Sunflower—A Review. *Helia* **2012,** *35,* 61–72.

Jovanović, D. Possibilities of Using Sunflower and Breeding for Specific Purposes. *Ratsr. Povrt.* **2001,** *35,* 209–221.

Jovanović, D.; Škorić, D.; Dozet, B. In *Confectionery Sunflower Breeding,* Proceedings of 2nd Balkan Symposium on Field Crops, Novi Sad, Yugoslavia, June 16–20, 1998; Institute of Field and Vegetable Crops: Novi Sad, Serbia, 1998; pp 349–352.

Kaya, Y. In *Confectionery Sunflower Production in Turkey,* Proceedings of 16th International Sunflower Conference, Fargo, ND, Aug. 29–Sep. 2, 2004; International Sunflower Association: Paris, France, 2004; pp 817–822.

Kaya, Y.; Evci, G.; Pekcan, V.; Gucer, T.; Yilmaz, I. M. In *Yield Relationships in Confectionery Sunflower (Helianthus annuus L.),* Proceedings of Annual Conference University of Rousse, Bulgaria, October 31–1 November, 2008; University of Rousse: Rousse, Bulgaria, 2008; pp 7–11.

Kaya, Y.; Jocić, S.; Miladinović, D. Sunflower. In *Technological Innovations in Major Oil Crops, Breeding;* Gupta, S. K., Ed.; Springer Science and Business Media: New York, 2012; Vol. 1, pp 85–129.

Kinman, M. L. In *New Developments in the USDA and State Experiment Station Sunflower Breeding Programs,* Proceedings of 4th International Sunflower Conference, Memphis, TN, July 25–29, 1970; International Sunflower Association: Paris, France, 1970; pp 181–183.

Kloczowskii, Z. Investigation on Methods of Obtaining Heterosis in the Estimation of Combining Ability of Certain Varieties and Lines of Oil Sunflower in Reciprocal Crosses. *Aklimatyz Nasiennityo Hodowla Roslin* **1967,** *15,* 235–257.

Kolkman, J. M.; Slabaugh, M. B.; Bruniard, J. M.; Berry, S.; Shaun Bushman, B.; Olungu, C.; Maes, N.; Abratti, G.; Zambelli, A.; Miller, J. F.; Leon, A.; Knapp, S.J. Acetohydroxyacid Synthase Mutations Conferring Resistance to Imidazolinone or Sulfonylurea Herbicides in Sunflower. *Theor. Appl. Genet.* **2004,** *109,* 1147–1159.

Kuptok, A. L. A Unisexual Female Sunflower. *Bot. Leningrad Ser.* A 14, 149–150. *Plant Breed. Abstr.* **1935,** *6,* 981.

Leclercq, P. Une Sterilite Male Utilisable pour la Production des Hybrides Simples de Tournesol. Ann. *Amelior. Plant* **1966,** *16,* 135–144.

Leclercq, P. Une Sterilite Cytoplasmique chez le Tournesol. Ann. Amelior. Plant **1969,** *19,* 99–106.

Lofgren, J. R. 1997. Sunflower for Confectionary Food, Bird Food, and Pet Food. In *Sunflower Technology and Production*; Schneiter, A. A., Ed.; Agronomy Monograph 35; American Society of Agronomy: Madison, WI, 1997; pp 747–764.

Luciano, A.; Davreux, M. *Produccion de Girasol en Argentina.* Publ. Tech. 37 INTA, Estancion Exp. Regional Agropecuaria: Pergamino, Argentina, 1967.

Malidža, G.; Ivanović, D.; Bekavac, G.; Jasnić, S. Significance of GM Plants in Pest Control. *Pesticidi* **1999,** *14,* 125–152.

Malidža, G.; Jocić, S.; Škorić, D.; Orbović, B. Clearfield System of Sunflower Production. *Ratsr. Povrt.* **2004,** *40,* 279–290.

Marinković, R.; Dozet, B.; Vasić, D. *Sunflower Breeding;* Školska Knjiga: Novi Sad, Serbia, 2003; p 368.

Miklič, V.; Hladni, N.; Jocić, S.; Marinković, R.; Atlagić, J.; Miladinović, D.; Dušanić, N.; Gvozdenović, S. Sunflower Breeding at Institute of Field and Vegetable Crops. *Ratsr. Povrt.* **2008,** *45,* 31–63.

Miller, J. F. Sunflower. In *Principles of Cultivar Development;* Ferh, W. R., Ed.; Macmillan: New York, NY, 1987; Vol. 2, pp 626–668.

Miller, J. F.; Fick, G. N. Adaptation of Reciprocal Full-Sib Selection in Sunflower Breeding using Gibberellic Acid Induced Male Sterility. *Crop Sci.* **1978,** *18,* 161–162.

Miller, J. F.; Hammond, J. J.; Roath, W. W. Comparison of Inbred vs. Single-Cross Testers and Estimation of Genetic Effect in Sunflower. *Crop Sci.* **1980,** *20,* 703–706.

Miller, J. F., Vick. B. A. Registration of Three Low Palmitic Acid and Five Low Stearic Acid Sunflower Genetic Stocks. *Crop Sci.* **1999,** *39,* 305.

Miller, J. F.; Al-Khatib, K. Registration of Imidazolinone Herbicide-Resistant Sunflower Maintainer (HA 425) and Fertility Restorer (RHA 426 and RHA 427) Germplasms. *Crop Sci.* **2002,** *42,* 988–989.

Morozov, V. K. *Sunflower Selection in USSR.* Pishchepromizdat: Moscow, 1947; pp 1–272.

Nikolić-Vig, V. *Official Gazette of SFRJ* **1976,** *64,* 1401.

Osorio, J.; Fernandez-Martinez, J.; Mancha, M.; Garces, R. Mutant Sunflowers with High Concentration of Saturated Fatty Acids in the Oil. *Crop Sci.* **1995,** *35,* 739–742.

Packer, L.; Obermuller-Jevic, U. Vitamin E in Disease Prevention and Therapy: Future Perspectives. In *The Antioxidant Vitamins C and E;* Packer, L., Traber, M., Kraemer, K., Frei, B., Eds.; AOCS Press: Urbana, IL, 2002; pp 255–288.

Panachenco, A. Y. In Sunflower Production and Breeding in the USSR, Proceedings of 2nd International Sunflower Conference, Morden, Manitoba, Canada, Aug 17–18, 1966; International Sunflower Association: Paris, France, 1966; pp 15–29.

Panero, J. L.; Funk, V. A. Toward a Phylogenetic Subfamilial Classification for the *Compositae* (*Asteraceae*). *P. Biol. Soc. Wash.* **2002,** *115,* 909–922.

Paniego, N.; Heinz, R.; Fernandez, P.; Talia, P.; Nishinakamasu, V.; Hopp, H. E. Sunflower. In *Genome Mapping and Molecular Breeding in Plants—Oilseeds;* Kole, C., Ed.; Springer-Verlag: Berlin Heidelberg, Germany, 2007; Vol. 2, pp 153–178.

Piquemal, G. In *How to Produce Hybrid Sunflower Seeds by Inducing Male Sterility with Gibberelic Acid,* Proceedings of 8th International Sunflower Conference, Minneapolis, MN, July 23–27, 1970; International Sunflower Association: Paris, France, 1970; pp 127–135.

Popov, P. S.; Dyakov A. B.; Borodulina A. A.; Demurin, Y. N. Genetic Analysis of Tocopherol and Fatty Acid Composition in Sunflower Seeds. *Genetics* **1988,** *24,* 518–527.

Pustovoit, G. V. Sunflower Selection for Group Resistance by Interspecies Hybridization. *Masličnie i Efiromasličnie Kulturi* **1963,** 75–93.

Pustovoit, G. V.; Gubin I. A. In *Results and Prospects in Sunflower Breeding for Group Immunity by Using the Interspecific Hybridization Method,* Proceedings of 6th International Sunflower Conference, Bucharest, Romania, July 22–24, 1974; International Sunflower Association: Paris, France, 1974; pp 373–381.

Pustovoit, V. S. *Handbook of Selection and Seed Growing of Oil Plants.* U.S. Department of Commerce: Springfield, IL, 1967.

Pustovoit, V. S. *Selected Works.* Agropromizdat: Moscow, Russia, 1990; p 367.

Putt, E. D. Investigations of Breeding Technique for the Sunflower (*Helianthus annuus* L.). *Sci. Agric.* **1941,** *21,* 689–702.

Putt, E. D. The Value of Hybrids and Synthetics in Sunflower Seed Production. *Can. J. Plant Sci.* **1962,** *42,* 488–500.

Putt, E. D. Heterosis, Combining Ability and Predicted Synthetics from a Diallel Cross in Sunflower (*Helianthus annuus* L.). *Can. J. Plant Sci.* **1966,** *46,* 59–67.

Putt, E. D. Sunflower Early History. In *Sunflower Technology and Production*; Schneiter, A. A., Ed.; Agronomy Monograph 35; American Society of Agronomy: Madison, WI, 1997; pp 1–21.

Putt, E. D.; Heiser, C. B. Male Sterility and Partial Male Sterility in Sunflowers. *Crop Sci.* **1966,** *6,* 165–168.

Robles, S. R. In *Stratified Masal Selection on Sunflower as a Breeding Method for Synthetic Varieties for Forage or Grain,* Proceedings of 10th International Sunflower Conference, Surfers Paradise, Australia, March 14–18, 1982; International Sunflower Association: Paris, France, 1982; pp 257–258.

Romano, A. B.; Vazquez, A. N. Origin of the Argentinian Sunflower Varieties. *Helia* **2003,** *26,* 127–136.

Sala, C. A.; Bulos, M.; Echarte, A. M. Genetic Analysis of an Induced Mutation Conferring Imidazolinone Resistance in Sunflower. *Crop Sci.* **2008a,** *48,* 1817–1822.

Sala, C. A.; Bulos, M.; Echarte, A. M.; Whitt, S. R.; Ascenzi, R. Molecular and Biochemical Characterization of an Induced Mutation Conferring Imidazolinone Resistance in Sunflower. *Theor. App. Genet.* **2008b,** *108,* 105–112.

Schilling, E. E. Helianthus. In *Flora of North America North of Mexico;* Flora of North America Editorial Committee, Eds.; Oxford University Press: New York, 2006; Vol. 21, pp 141–169.

Schuster, W.; Friedt, W. Results and Trends in Breeding and Cultivation of Sunflower (*Helianthus annuus*) in West Germany. *Helia* **1988,** *11,* 85–91.

Shabana, R. Performance of a New Synthetic Sunflower Stock Developed from Local and Introduced Germplasm and Further Improvement via Population Improvement Method. *Helia* **1990,** *13,* 11–16.

Shein, S. E. In *An Evaluation of Early Generation Testing for General Combining Ability in Sunflower,* Proceedings of 8th International Sunflower Conference, Minneapolis, MN, July 23–27, 1978; International Sunflower Association: Paris, France, 1978; pp 432–436.

Singh, B. K. 1999. Biosynthesis of Valine, Leucine and Isoleucine. In *Plant Amino Acids;* Singh, B. K, Ed.; Marcel Dekker: New York, 1999; pp 227–247.
Škorić, D. Sunflower Breeding. *J. Edible Oil Ind.* **1988,** *25,* 1–90.
Škorić, D. Sunflower Breeding. In *Sunflower Genetics and Breeding;* Škorić, D., Ed; Serbian Academy of Science and Arts: Branch in Novi Sad, Serbia, 2012; pp 165–354.
Škorić, D.; Marinković, R. Significance of Citoplasmic Male Sterility in Sunflower Breeding. *Zbornik Matice Srpske* **1981,** *61,* 559–567.
Škorić, D.; Jocić, S.; Lečić, N.; Sakač, Z. Development of Sunflower Hybrids with Different Oil Quality. *Helia* **2007,** *30,* 205–212.
Škorić, D.; Jocić, S.; Sakač, Z.; Lečić, N. Genetic Possibilities for Altering Sunflower Oil Quality to Obtain Novel Oils. *Can. J. Physiol. Pharmacol.* **2008,** *86,* 215–221.
Soldatov, K. I. In *Chemical Mutagenesis in Sunflower Breeding,* Proceedings of 7th International Sunflower Conference, Krasnodar, USSR, July 23–27, 1976; International Sunflower Association: Paris, France, 1976; pp 352–357.
Sun, Y. Genetic Analysis of Important Traits for Confectionery Sunflower in Australia. Ph.D. Thesis, School of Land, Crop and Food Sciences, The University of Queensland, 2009.
Tan, S.; Evans, R. R.; Dahmer, M. L.; Singh, B. K.; Shaner, D. L. Imidazolinone Tolerant Crops: History, Current Status and Future. *Pestic. Manag. Sci.* **2005,** *61,* 246–257.
Tolmachev, V. V. Genetic Control of Coloration of Pseudo-Ligulate Flower in Sunflower. *Nauch.-Tekhn. Byul. VNIIMK* **2006,** *134,* 10–15.
Tranel, P. J.; Wright, T. R. Resistance of Weeds to AHAS Inhibiting Herbicides: What Have We Learned? *Weed Sci.* **2002,** *50,* 700–712.
Unrau, J.; White, W. J. The Yield and Other Characters of Inbred Lines and Single Crosses of Sunflower. *Sci. Agric.* **1944,** *24,* 516–528.
U.S. Department of Agriculture. http://www.fas.usda.gov/wap/current (accessed July 10, 2013).
Vear, F. Classic Genetics and Breeding. In *Genetics, Genomics and Breeding of Sunflower;* Hu, J., Seliler, G., Kole, C., Eds.; Science Publisher: Enfield, NH, 2010; pp 51–78.
Vear, F.; Bony, H.; Joubert, G.; Tourvieille de Labrouhe, D.; Pauchet, I.; Pinochet, X. 30 Years of Sunflower Breeding in France. *Oléagineux, Corps Gras, Lipides* **2003**, *10,* 66–73.
Vick, B. A., Miller, J. In *Utilization of Mutagens for Fatty Acid Alteration in Sunflower,* Proceedings 18th NSA Sunflower Research Workshop, Fargo, ND, January 11–12, 1996; National Sunflower Association: USA, 1996; pp 11–17.
Voskoboinik, L. K.; Soldatov, K. I. In *The Research Trends in the Field of Sunflower Breeding for Heterosis At All Union Research Institute for Oil Crops (VNIIMK),* Proceedings of 6th International Sunflower Conference, Bucharest, Romania, July 22–24, 1974; International Sunflower Association: Paris, France, 1974; pp 383–389.
Vranceanu, A. V. *Sunflower;* Academy of Romanian Socialist Republic: Bucharest, 1974.
Vulpe, V. V. In *Single, Three-Way and Double-Crosses in Sunflower,* Proceedings of 6th International Sunflower Conference, Bucharest, Romania, July 22–24, 1974; International Sunflower Association: Paris, France, 1974; pp 443–449.

2

Mutagenesis in Sunflower

Andrés Zambelli and Alberto León ▪ *Center for Biotechnology Research, Advanta Semillas—Nutrisun Business Unit, Balcarce, Argentina*

Rafael Garcés ▪ *Instituto de la Grasa (CSIC), Sevilla, Spain*

Introduction

Feeding the world's rapidly increasing population is a global problem for which breeding varieties of food crops and developing farming systems to produce crops is mandatory. To face this challenge, the objectives to be reached can be summarized as higher yields, improved reproduction, tolerance to droughts and to extreme environmental conditions, and greater resistance to pests and diseases. Among the obstacles against crop production are the changing climate (increasing temperatures and more erratic rainfall), which most often compromises crop productivity (Parry et al., 2005), and the need to produce additional food and crops for bioenergy while minimizing the carbon costs of production (Powlson et al., 2005). There is, therefore, an urgent requirement for new higher-yielding varieties with improved agronomic performance (Lea and Azevedo, 2006; Parry et al., 2007; Reynolds et al., 2009; Richards, 2000).

Sunflower (*Helianthus annuus* L.), 2n = 34, belongs to family Asteraceae, has an estimated genome size of 3000 Mbp, and is the fourth most important oilseed crop (Kolkman et al., 2007). It is a native of North America (Harter et al., 2004) and is widely cultivated in the world with an annual grain production (2012–2013) of 36.4 million tons (United States Department of Agriculture, 2013). Its genome size is large compared to species like *Arabidopsis* (125 Mbp), rice (430 Mbp), sorghum (750 Mbp), soybean (1100 Mbp), or tomato (950 Mbp) (Rashid et al., 2011). The major drivers of this large genome size are the recent polyploidization event and the important amplification of transposable elements (Harter et al., 2004; Staton et al., 2012). Many wild sunflower species were used as genetic sources for improvement of cultivated varieties by conventional breeding (Seiler, 1992). For many decades, plant breeders have concentrated their efforts on improvement of sunflower through traditional breeding and, recently, molecular mapping has been successfully undertaken for marker-assisted breeding (Micic et al., 2005). The understanding of the sunflower genome and its manipulation has contributed to continuous improvement in yields, quality, and agronomic performance of this crop in the past decades. As a result of several rounds of selection for targeted traits, many useful alleles that existed in natural sunflower populations have been eliminated. Furthermore, cultivated sunflower in general has limited genetic variability, especially for major agronomic traits. Therefore, creation of novel genetic resources using biotechnological tools is critical in sunflower improvement. Among others, mutagenesis has gained popularity in plant genetics research in the last century as a powerful tool for inducing genetic variation. Induced mutations have been applied to produce mutant cultivars in sunflower, significantly increasing crop yield and quality, and many characteristics of modern genotypes have been obtained through this approach (Cvejić et al., 2011; Jambhulkar and Shitre, 2009). Induced mutations were widely used in sunflower breeding programs due to its low cost and technology requirement, and its adaptation to a wide spectrum of breeding objectives. Even though very few among thousands of induced mutations could be valuable for a breeder's use, this small input contributes to generate a wider genetic variability. Thus, mutagenesis does not substitute for conventional breeding methods, but rather complements them.

Mutations can be classified as: (1) point mutations, (2) structural mutations, or (3) mutations changing the chromosome number (for a review see Pathirana, 2012). The most usual kind of point mutation is the base pair substitution, which can result in a changed codon or a changed amino acid. Among substitutions, transitions are the more usual type where a purine base is changed by another purine base, or a pyrimidine base is changed by another pyrimidine base (G-A or C-T transitions). Structural mutations occur as a consequence of chromosomes breaks and rearrangements and include inversions, translocations, duplications, and deletions.

Mutagenesis has allowed for the development of several features of modern sunflower genotypes, but it has been particularly important for the progress of the sunflower oil market and in the creation of herbicide-resistant genotypes, as will be assessed throughout this chapter.

Mutagenic Agents

Mutagens can be classified as physical (i.e., forms of energetic radiations) or chemical substances. Thus, a suitable mutagenic agent should ideally lead to maximal genomic variation by increasing the frequency of the mutations in the genome of the exposed species, minimizing detrimental effects such as decreased plant viability or fertility and seed viability. Induction of mutations with ionizing radiation is a worldwide method for developing mutant varieties. Among the radiation-based methods, X-ray has been the most extensively used. Other radiation, such as γ-ray, is also very popular and is less destructive and its use causes point mutations and small deletions. Both types of radiations cause chromosome breakages, usually damaging the genome on a large scale and thus severely reducing viability. Chemical mutagens have gained popularity as an easy-to-use technique providing a very high mutation frequency. Chemical mutagens usually cause single base pair substitutions or single-nucleotide polymorphisms (SNPs), which in some cases can result in amino acid changes with no complete loss of function of the protein as deletions do. Among the chemical mutagens may be found alkylating compounds such as ethyl methane sulfonate (EMS) and dimethyl sulfate (DMS), EMS being the most extensively used in sunflower. This mutagen selectively alkylates guanine bases causing the DNA polymerase to favor placing a thymine residue over a cytosine residue opposite to the O-6-ethyl guanine during DNA replication, which results in a random point mutation. Most of the changes in EMS-mutated populations are G-C to A-T base pair transitions (Till et al., 2007). Other mutagens such as sodium azide (Az) and methyl-nitrosourea (MNU) are also used and often combined into an Az-MNU solution. Sodium azide produces an organic metabolite of azide compound that enters into the nucleus, interacts with DNA, and creates a point mutation in the genome (Al Qurainy and Khan, 2009). Genetically, Az-MNU predominantly causes G-C to A-T shifts, or A-T to G-C shifts. In contrast to EMS, nucleotide transitions may occur in either direction.

To increase the efficiency of mutation detection, it is advisable to optimize the mutagen dose to achieve a high mutation rate while avoiding serious effects on germination and plant development. In diploid species this can be difficult, as even relatively low levels of mutation can be associated with the almost complete loss of male or female fertility, resulting in a need for very large populations if saturated coverage is required. By contrast, polyploidy species, such as wheat, have a very high tolerance to mutations due to the complementation of essential genes by homeologous copies. Thus, populations saturated with mutations can be much smaller, resulting in more cost-effective screening for mutations. However, a consequence of this genetic buffering is that recessive mutations in single homeologues of genes in polyploid species are less likely to show a phenotype, and it becomes necessary to identify mutations in each homeologous copy of the target gene and bring these together by crossing. The choice of chemical mutagen will also influence the maximum permissible mutation rate achievable: EMS creates a larger proportion of nonsense mutations, involving the introduction of novel stop codons, than a mutagen such as MNU, due to the specificity of EMS in creating mainly G-A and C-T transitions, and any individual mutations is therefore more likely to have a phenotypic effect (Parry et al., 2009).

Induced Mutations Modifying Sunflower Oil Quality

Vegetable oils are one of the most valuable commodities in world trade. They are linked to specific quality requirements, both for food and nonfood uses, and new oil types are in continuous demand. Vegetable oils are mainly triacylglycerols (92–98%), polar lipids (phospholipids and galactolipids), monoacylglycerols, diacylglycerols, and minor amounts of free fatty acids and polyisoprenoids. The latter group comprises sterols and sterol derivates (phytosterols), tocopherols, carotenoids, and chlorophylls, with some of these compounds having important vitamin and antioxidant properties. Focusing on seed oil quality improvement, the three main features that are usually considered as targets in breeding programs for oilseed crops are fatty acid (FA) composition, triacylglycerol (TAG) composition, and the quantity and composition of antioxidants present in the oil.

Seed oils principally used for human consumption are comprised of five main fatty acids, the saturated palmitic (16:0) and stearic (18:0) acids, the monounsaturated oleic acid (18:1), and the polyunsaturated linoleic (18:2) and α-linolenic (18:3) acids. Fatty acids differ in their number of carbon atoms and/or their number and position of double bonds (unsaturations) and functional groups. Clinical and animal studies have reported that both the monounsaturated FA (MUFA) oleic acid and polyunsaturated FA (PUFA) of the n-6 family linoleic acid decrease serum cholesterol. A prospective study on elderly Italian subjects showed that a typical Mediterranean diet rich in MUFA increased survival. Multiple beneficial effects on decreasing cardiovascular disease risk were reported for the PUFAs of the n-3 family linoleic acid, eicosapentaenoic acid (EPA), and docosahexanoic acid (DHA). Together with a daily intake of PUFAs in the range of 6–10% of energy, it is also health-beneficial for the diet to include an n-6/n-3 ratio of about 5 (Crupkin and Zambelli, 2008, and references therein).

As said, vegetable oils are mainly triacylglycerols, which consist of a glycerol molecule having three fatty acids esterified at the hydroxyl residues: one in the central position of the glycerol molecule, *sn*-2, and the other two at the terminal positions, *sn*-1 and *sn*-3. The most common fatty acids forming these triacylglycerols in standard sunflower oil are: diunsaturated (linoleic), monounsaturated (oleic), and minor amounts of saturated (palmitic and stearic). The final use of each type of oil is defined by both physical and chemical characteristics depending on its fatty acids and triacylglycerol composition. For instance, the physical difference between oils (liquid) and fats (solid) is due to the amount of saturated fatty acids. The thermo-oxidative stability of vegetable oils depends mainly on the amount of polyunsaturated fatty acids they contain (oils with a high content of these unsaturated fatty acids are more unstable), as well as on the content and type of tocopherols.

In recent years, considerable focus has been placed on genetically engineering oilseed plants to incorporate additional fatty acids of nutritional importance that are usually obtained from other sources, or to provide a low-cost production of industrial fatty acids that are currently sourced from petrochemicals or from low-yielding plants. The ability to introduce such new fatty acids into plants is built on a growing understanding of the metabolic pathways for the synthesis of individual fatty acids, an increasing availability of genes that encode the biosynthetic enzymes involved, and the now well-established capacity to metabolically engineer plants to express genes that encode single or multistep biosynthetic pathways (Singh et al., 2005). Sunflower is the second-largest source of vegetable oil in Europe, after rapeseed, and the fourth-largest global source after soybean, palm, and rapeseed, with a total production (2012–2013) of 13.7 million tons (United States Department of Agriculture, 2013). Sunflower oil has been appreciated as a high quality commodity in the world oil market. However, new emerging markets, together with some concerns about health risks of foods and an increasing request for vegetable oils for specific nonfood uses, are demanding changes in oil quality. Traditionally, the concept of oil quality has been almost exclusively associated with the fatty acid composition of the oil. However, this concept is constantly evolving and oil chemists and nutritionists are emphasizing other components of vegetable oils that influence their nutritional and technological properties. As in other vegetable oils, the objective of modifying sunflower seed oil quality has been to develop oils with enhanced nutritional and functional properties that require little, if any, processing for specific end-use markets. In many cases, oil processing results in detrimental implications for human health. For example, common uses of sunflower oil, such as deep frying or margarine production, require previous hydrogenation or hardening of the oil. This process produces artificial *trans* fatty acid (TFA) and positional isomers that have been proved to be negatively related to heart disease. Thus, fatty acid configuration is of great relevance from a nutritional point of view, ever since TFA has been demonstrated to have a detrimental effect on human health. This is considered one of the greatest discoveries in human nutrition research during the last three decades (Katan et al., 2007; Mozaffarian et al., 2006). From a nutritional point of view, increased oleic acid (70%) and decreased linoleic acids (20%) are desirable. Therefore, breeding for oil quality in sunflower has mainly focused on the modification of the relative amount of fatty acids, increasing oleic acid to have stable and healthy oil and increasing stearic acid to have a stable and healthy fat.

The biosynthesis of fatty acids in plants begins with the formation of acetyl-CoA from pyruvate. The acetyl-CoA produced in plastids is activated to malonyl-CoA; the malonyl group is subsequently transferred to acyl carrier protein (ACP) giving rise to malonyl-ACP, the primary substrate of the fatty acid synthase complex. The formation of malonyl-CoA is the committed step in fatty acid synthesis and is catalyzed by the highly regulated plastidic acetyl-CoA carboxylase complex (Nikolau et al., 2003). De novo fatty acid synthesis in the plastids occurs through a repeated series of condensation, reduction, and dehydration reactions that add two carbon units derived from malonyl-ACP to the elongating fatty acid chain. A series of condensation reactions proceed with acetyl-CoA and malonyl-ACP, then acyl-ACP acceptors. Three separate condensing enzymes or 3-ketoacyl-ACP synthases (KAS I–III) are necessary for the production of an 18-carbon fatty acid. In sunflower, the main products of intraplastidial fatty acid synthesis are, first, palmitoyl-ACP (16:0-ACP), which is further elongated by the FAS II complex to produce stearoyl-ACP (18:0-ACP). Three desaturases are involved in fatty acid biosynthesis pathways in plants: the intraplastidial stearoyl-ACP desaturase (SAD), or Δ9-desaturase, which catalyzes the first desaturation of stearic acid (18:0) to oleic acid (18:1); oleoyl-CoA desaturase, or Δ12-desaturase, which catalyzes the second desaturation of 18:1 to 18:2; and linoleyl-CoA desaturase, or Δ15-desaturase, which catalyzes the third desaturation or 18:2 to 18:3. These two last desaturases are membrane-bound at endoplasmic reticulum (ER) (Shanklin and Cahoon, 1998). The palmitic, stearic, and oleic fatty acids are synthesized in the plastid and can be exported to the ER after the hydrolysis between the acyl molecule and ACP by the acyl-ACP thioesterase. Two types of acyl-ACPs thioesterases have been identified in higher plants such as sunflowers: FatA and FatB. FatA thioesterases preferentially act on long chain fatty acids, and have particularly high specificity for 18:1-ACP and a lower affinity for 16:0-ACP and 18:0-ACP. In contrast, FatB thioesterases exhibit a higher affinity for saturated acyl-ACPs (Pleite et al., 2006).

The oleoyl-CoA desaturase (oleate desaturase) is a membrane-bound enzyme, that catalyzes the first extraplastidial desaturation in plants and converts oleic acid esterified to phosphatidylcholine, mainly in the *sn*-2 position, to linoleic acid. This reaction involves the concomitant reduction of molecular oxygen to water and requires the presence of NADH, NADH:cytochrome b5 reductase, and cytochrome b5 (Smith et al., 1990). Not one but several different microsomal oleate desaturase genes (also known as *FAD2*) may be present, depending on the particular species. In sunflower, three different *FAD2* genes were reported (*FAD2-1*, *FAD2-2*, and *FAD2-3*) encoding sunflower microsomal oleate desaturases. Genomic analysis revealed that at least one copy of each of these genes is present in the sunflower genome, except for the *FAD2-1* gene from the high oleic mutant, which might be duplicated. The *FAD2-2* and *FAD2-3* genes were weakly expressed in all tissues studied from both varieties, normal and high oleic sunflowers. In contrast, the *FAD2-1* gene was expressed strongly and exclusively in developing embryos of normal-type sunflower, whereas its expression in high oleic developing embryos was drastically reduced. Functional expression of the corresponding cDNAs in yeast confirmed that they encode microsomal oleate desaturases (Martínez-Rivas et al., 2001).

Mutagenic treatments are usually applied to the seeds, which after treatment are named M_1 seeds. Mutants can be detected already in the M_2 generation. In sunflower, the fatty acids and the tocopherol profiles are mainly under gametophytic control, that is, they are governed by the genotype of the developing embryo. Therefore, mutants are detected by analyzing M_2 half seeds (Garcés and Mancha, 1993; Ruiz-López et al., 2003). It requires the removal of a small portion of the seed in the distal extreme of the embryo in such a way that the germination capacity of the seed is not affected. The excised half seed is used for chemical analysis, whereas the other half seed containing the embryo can be sown to generate a viable plant. The half seed technique has been also used in sunflower for the nondestructive analysis of tocopherols composition and total tocopherols content (Velasco et al., 2004a). Thus, research efforts in the sunflower oil field are being put into the following aspects:

1. More stable sunflower oils have been obtained by increasing the content of monounsaturated (oleic acid) and decreasing the content in polyunsaturated fatty acids (linoleic acid); this type of oil is also suitable for obtaining biolubricants. Oil stability can also be improved by increasing the tocopherols content.

2. Healthy substitutes for higher melting animal, tropical, or hydrogenated fats required by food industry are being obtained by increasing their content in saturated fatty acids, mostly stearic acid, whose intake does not modify the serum cholesterol levels in humans (Garcés et al., 2009). Table 2.A summarizes the different sunflower induced mutants so far available in which the seed oil fatty acid composition was modified.

Development of High Oleic Sunflower Oils

Although polyunsaturated fatty acids are beneficial for human consumption, their susceptibility to oxidation makes them undesirable at high levels in vegetable oils. Oleic acid is nowadays considered the preferred fatty acid for edible purposes because it combines a hypocholesterolemic effect and a high oxidative stability. One of the most valuable and successful crop products obtained by mutagenesis was the sunflower variety Pervenets, having an oleic acid content above 75%. This high oleic material was produced at the All-Union Research Institute of Oil Crops in the former USSR, after treatment with the chemical mutagen DMS (Soldatov, 1976; Table 2.A). Several inbred lines have been derived from this mutant, with oleic acid contents of up to 90% (Fernández-Martínez et al., 1993), which are extensively used in breeding programs and commercialized around the world. Even with the increased oleic content in the seed oil, the *FAD2-1* gene from the high oleic varieties does not show mutations in the coding region that is expected to codify a fully active enzyme. These results suggest that the high oleic mutation in sunflower interferes with the regulation of the transcription of the seed-specific *FAD2* gene (Martínez-Rivas et al., 2001). Later, it was demonstrated that this allele is organized in two parts: the first one, conserved in both high oleic and low oleic genotypes carries a normal *FAD2-1* gene; the second part is specific to high oleic genotypes and carries *FAD2-1* gene duplications. The study of mRNA accumulation in low oleic and high oleic seeds revealed that the mutation is dominant and induces an *FAD2-1* mRNA down regulation. Furthermore, *FAD2-1* small interfering RNA, characteristic of gene silencing, was shown to be accumulated specifically in high oleic seeds. Considered together, these observations show that the mutation is associated with oleate desaturase gene duplications, leading to gene silencing of the *FAD2-1* gene and, consequently, to oleic acid accumulation. This finding allowed the development of molecular markers characterizing the mutation that can be used in breeding programs to facilitate the selection of high oleic genotypes (Lacombe et al., 2009).

Recently, additional high oleic sunflower varieties having induced mutations affecting oleate desaturase gene *FAD2-1* were reported. Two different high oleic sunflower mutants were generated by treatment of seeds with X-rays (Leon et al., 2013a; Table 2.A). The isolated mutants (identified as 29065 and 29066) underwent independent polynucleotide insertions on the *FAD2-1* gene induced by the

Table 2.A Sunflower Induced Mutant Showing Modified Fatty Acid Composition in Seed Oil

Mutant Name	Type	Mutagen Agent	Fatty Acid Composition (%)					Reference
			P (16:0)	Po (16:1)	S (18:0)	O (18:1)	L (18:2)	
CAS-3	HS	EMS	7.8	0	28.8	21.5	41.9	Osorio (1995)
CAS-4	HS	Az	8.8	0	16.1	40.8	34.3	Osorio (1995)
CAS-8	MS	Az	7.8	0	14.8	28.8	48.6	Osorio (1995)
CAS-5	HPHL	X-rays	31.0	5.0	3.0	13.0	48.0	Osorio (1995)
CAS-12	HPHO	X-rays	32.0	6.0	4.0	54.0	4.0	Fernández-Martínez (1997)
Pervenets	HO	DMS	5.0	0	5.0	85.0	5.0	Soldatov (1976)
29065	HO	X-rays	2.9	0	2.7	93.3	1.1	Leon (2013a)
29066	HO	X-rays	4.4	0	3.7	90.6	1.3	Leon (2013a)
29074	HO	EMS	3.6	0	2.9	91.0	2.5	Leon (2013b)
29075	HO	EMS	3.6	0	3.5	91.9	1.0	Leon (2013b)
29076	HO	EMS	3.7	0	3.9	87.3	5.0	Leon (2013b)
29077	HO	EMS	3.9	0	3.1	86.6	6.3	Leon (2013b)
29078	HO	EMS	3.8	0	2.9	89.3	3.9	Leon (2013b)
29079	HO	EMS	3.4	0	2.8	90.7	3.1	Leon (2013b)
29082	HO	EMS	3.7	0	3.3	90.6	2.4	Leon (2013b)
29081	HO	EMS	3.4	0	3.4	91.5	1.7	Leon (2013b)
39096	HO	EMS	5.0	0	2.9	86.3	5.7	Leon (2013b)

EMS: ethyl methane sulphonate; Az: sodium azide; DMS: dimethyl sulphate; P: palmitic; Po: palmitoleic; S: stearic; O: oleic; L: linoleic; HS: high stearic; MS: medium stearic; HPHL: high palmitic-high linoleic; HPHO: high palmitic-high oleic; HO: high oleic.

mutagenic treatment. The mutant 29065 showed a gene insertion 785 bp long at nucleotide position 310, while the mutant 29066 exhibited a longer insertion of 4872 bp at nucleotide position 201. In both cases, the consequence of the insertions was the generation of premature stop codons producing truncated oleate desaturase protein, explaining the accumulation of oleic acid in seed oil at the expense of linoleic acid (Table 2.A). Unlike Perventes high oleic mutation, which produces a down regulation of the *FAD2-1* transcription, 29065 and 29066 high oleic mutations are structural, affecting enzyme activity by producing a truncated protein, and both of them showed an oleic acid percentage higher than 90% (Leon et al., 2013a).

An extra group of nine independent high oleic mutants were obtained by treatment with EMS (Leon et al., 2013b; Table 2.A). Sequencing of the *H. annuus FAD2-1* (*HaFAD2-1*) gene showed that these materials have nucleotide point mutations producing amino acid substitutions in eight of the high oleic mutants and a premature stop codon in the remaining one (Leon et al., 2013b; Tables 2.A and 2.B). Based on secondary structure predictions using the TMHMM server for prediction of transmembrane helices in proteins (http://www.cbs.dtu.dk/services/TMHHMW), and PredictProtein (http://www.expasy.org/proteomics), the sequence stretches of *HaFAD2-1* have been identified corresponding to the membrane-spanning and cytosolic domains previously postulated by Hoffman et al. (2007) for membrane-bound fatty acid desaturases. In analogy, the sequence of *HaFAD2-1* was divided into the consecutive domains I through VIII, as follows: domain I is equivalent to the cytosolic N-terminal region; II includes two transmembrane helices (TM1 and TM2) connected by a short ER luminal loop; III is comprised of the subsequent short cytosolic loop containing the first histidine box contributing to the catalytic site; IV enfolds the first peripheral membrane-associated segment MPS1; V forms the second short cytosolic loop and contains the second histidine box; VI comprises the second peripheral membrane-associated region MPS2; VII enfolds the third short cytosolic loop; VIII forms the second set of transmembrane helices (TM3 and TM4) connected by a short ER-luminal loop; and domain IX is comprised of the cytosolic C terminus containing the third histidine motif. On the base of Hoffman's protein model, the location of the induced point mutations on *HaFAD2-1* was predicted (Figure 2.1; Table 2.B).

Development of Sunflower Oils with Modified Saturated Fatty Acid Content

Artificial *trans* fats are formed during the partial hydrogenation of vegetable oils, a process that converts vegetable oils into semisolid fats for use in margarines, commercial and industrial cooking, and manufacturing processes. From the perspective of the food industry, partially hydrogenated vegetable oils are attractive because of their long shelf life, their stability during deep frying, and their semisolidity, which can be

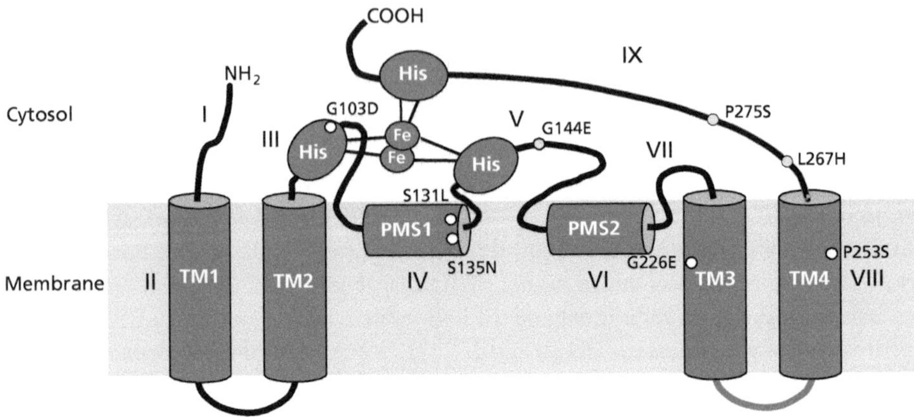

Figure 2.1 Proposed secondary structure of the native sunflower oleate desaturase protein *HaFAD2-1*. Circles indicate amino acid substitutions generated by the nucleotide point mutations induced by EMS treatment.

Source: Adapted from Hoffman et al. (2007).

Table 2.B High Oleic Sunflower Mutants Exhibiting Different Point Mutations on *HaFAD2-1* Gene Induced EMS Treatment

Mutant Name	Nucleotide Position (np) of Mutation	Amino Acid Substitution	Protein Domain Location of Amino Acid Substitution
29074	C by T at np 392	S131L	PMS1
29075	G by A at np 243	W81stop	
29076	C by T at np 757	P253S	TM4
29077	C by T at np 823	P275S	Domain IX
29078	G by A at np 404	S135N	Near histidine box II (PMS1)
29079	G by A at np 431	G144E	Near histidine box II (domain V)
29082	G by A at np 308	G103D	Histidine box I (domain III)
29081	C by T at np 799	L267F	Domain IX
39096	G by A at np 677	G226E	TM3

customized to enhance the palatability of baked goods and sweets. Therefore, there is a need to find suitable solid fat substitutes that do not increase cardiovascular disease risk. For solid fat applications, stearic acid–rich fat arises as an excellent substitute for *trans* fats and cholesterol-raising saturated fatty acids like myristic and palmitic acids (Crupkin and Zambelli, 2008; Hunter et al., 2010). To solve the problem regarding the use of hydrogenated vegetable fat, animal fat, or tropical fats, a research project was carried out with the aim of obtaining natural, semisolid sunflower oils that could be used directly in the food industry for the production of margarine and similar products without the need for any chemical manipulation. Sunflower lines having high stearic acid content together with oleic or linoleic acids are a healthy alternative to these unhealthy fats and, for this purpose, new lines have been selected by mutagenesis. From a practical point of view, high stearic, high oleic sunflower oils are suitable for manufacturing margarine. The triacylglycerols composition of these new oils is different from those of the standard sunflower oil, making them appropriate for industry demands. High stearic lines contain a considerable percentage of the triacylglycerol species Stearic-Oleic-Stearic (esterification of those three fatty acids with hydroxyl groups in position *sn-1, sn-2* and *sn-3* of glycerol molecule, respectively), which makes them appropriate for the production of healthy margarines, spreads, bakery items, and other products where plastic fats are demanded (Garcés et al., 2009). Sunflower oil with high levels of stearic acid (>25%) have been obtained through induced mutagenesis with both EMS and sodium azide (Fernández-Moya et al., 2002; Osorio et al., 1995). Also, sunflower materials with high levels of palmitic acid were obtained by treatment of sunflower seeds with physical mutagens (X- and γ-rays) (Fernández-Martínez et al., 1997; Ivanov et al., 1988; Osorio et al., 1995; Salas et al., 2004) and EMS (Velasco et al., 2008).

Modifying Minor Components of Sunflower Oil

Breeders have also invested efforts in oilseed crop improvement for minor oil compounds with nutritional and antioxidant value, especially tocopherols and phytosterols. As first assessed by Demurin (1993), sunflower tocopherol profiles can be altered using mutagenic techniques. Increased levels of γ-tocopherol (>95%) have been obtained following chemical mutagenesis with EMS (Velasco et al., 2004a, 2004b). Tocopherols are the most important antioxidants in sunflower seeds, exerting an antioxidant protective action both in biological systems (in vivo or by vitamin E activity) and in oils, fats, and food (in vitro activity). The tocopherols occur as a family of four derivatives named α-, β-, γ-, and δ-tocopherol, which differ in their in vitro and in vivo antioxidant efficiency. α-tocopherol has a maximum vitamin E activity, but in vitro activity is relatively low. Conversely, β-, γ-, and δ-tocopherol possess lower vitamin E values, but they are better in vitro antioxidants than α-tocopherol (Eitenmiller et al., 2004; Hunter and Cahoon, 2007; Warner, 2007). Sunflower seeds mainly contain α-tocopherol, which accounts for more than 90% of the total tocopherols. β- and γ-tocopherol can be present in sunflower seeds, usually in amounts below 5% of the total tocopherols (Demurin, 1993; Velasco et al., 2002). The modification of the tocopherol profile through a partial substitution of α-tocopherol by other tocopherol derivatives is an important goal for developing sunflower oil with improved in vitro oxidative stability (Škorić, 1992). Breeding efforts in this direction during the past decade have led to the development and characterization of several sources of modified tocopherol profiles in sunflower. Through a strategy based on evaluation of germplasm of wild and cultivated sunflower, Demurin (1993) and Demurin et al. (1996) isolated two lines with contrasting tocopherol profiles. However, novel variation for tocopherol profiles in sunflower was created through the utilization of mutagenesis and genetic recombination. The open pollinated cultivar Peredovik had revealed variability for genes involved in tocopherol biosynthesis (Velasco et al., 2004a) for which four accessions of this cultivar were selected for EMS mutagenesis. Single-seed screening in the M_2 generation resulted in two M_2 family seeds, derived from different M_1 plants, with increased γ-tocopherol contents of 19.2% and 96.7%. M_3 progeny from the M_2 seed with the 96.7% content bred true for high γ-tocopherol content, having more than 90% γ-tocopherol. M_3 progeny from the M_2 seed with 19.2% γ-tocopherol segregated in a range from 0 to 84.6%. Selection for high γ-tocopherol content produced an $M_{4:5}$ line, designated IAST-1, with a stable high concentration of γ-tocopherol. Crosses between IAST-1 and T589, with increased β-tocopherol content, produced F_2 segregants with transgressive levels of up to 77% β-tocopherol or up to 68% δ-tocopherol. Thus, two lines with unique tocopherol profiles: IAST-4 (high δ-tocopherol) and IAST-5 (high β-tocopherol) were isolated (Velasco et al., 2004b).

Induced Mutations Conferring Herbicide Resistance in Sunflower

Sulfonylurea (SU) and imidazolinone (IMI) herbicides are specific and potent inhibitors of acetohydroxyacid synthase (AHAS, EC 2.2.1.6), also known as *acetolactate synthase* (ALS). This is an octameric enzyme with four catalytic and four regulatory subunits (Duggleby et al., 2008). AHAS catalyzes two parallel reactions in the synthesis of branched chain amino acids. The first reaction is condensation of two pyruvate molecules to yield acetolactate, leading to the production of valine and leucine. The second reaction is the condensation of pyruvate and α-ketobutyrate to yield acetohydroxybutyrate, leading to the production of isoleucine (Singh, 1999). The AHAS-inhibiting herbicides are known to bind at the substrate access channel, blocking the path of substrate to the active site (McCourt et al., 2006; Pang et al., 2002, 2003). When AHAS is inhibited, deficiency of the amino acids causes a decrease in protein synthesis, which in turn slows down cell division rate. This process eventually kills the plant after showing symptoms in meristematic tissues, where biosynthesis of amino acids primarily takes place (Zhou et al., 2007). Resistant plants, in most cases, depend on reduced sensitivity to these herbicides by an isoform of AHAS, which does not severely affect its catalytic activity. Most AHAS isoenzymes resistant to the herbicides carry substitutions for the amino acid residues Ala122, Pro197, Ala205, Asp376,

Trp574, or Ser653 (amino acid numbering refers to the sequence in *Arabidopsis thaliana*), which are conserved among wild types (herbicide-susceptible genotypes) and even across genera. The amino acid residues Ala122, Pro197, and Ala205 are located at the N-terminal end of the enzyme, whereas Asp376, Trp574, and Ser653 are located at the C-terminal end (Pozniak et al., 2004; Tranel and Wright, 2002).

Spontaneous mutations conferring resistance to AHAS-inhibiting herbicides can rapidly increase in frequency in wild populations under strong herbicide selection (Tranel and Wright, 2002). The first weed biotype resistant to AHAS-inhibiting herbicides was discovered in prickly lettuce (*Lactuca serriola* L.) after five generations of chlorsulfuron treatment (Mallory-Smith et al., 1990). Weed biotypes resistant to AHAS-inhibiting herbicides have been identified in many species, more than any other herbicide group (Heap, 2013). Typically, resistant biotypes are selected in populations chronically exposed to specific AHAS-inhibiting herbicides. Common sunflower populations with cross-resistance to both SU and IMI herbicides were first discovered in 1996 in Kansas (Al-Khatib et al., 1998) and South Dakota (White et al., 2002) in fields that had been repeatedly treated with herbicides for seven to eight years. Presently, AHAS resistance has been confirmed in common sunflower from Kansas, South Dakota, Missouri, and Iowa (Heap, 2013). White et al. (2003) described an Ala205Val mutation in an AHAS gene from common sunflower herbicide-resistant biotypes from South Dakota. Genes for resistance to AHAS-inhibiting herbicides in sunflower have been introgressed from resistant wild populations (ANN-PUR and ANN-KAN) into elite inbred lines for the purpose of developing and deploying herbicide resistant cultivars and hybrids (Al-Khatib and Miller, 2000; Miller and Al-Khatib, 2002, 2004). Traditionally, sunflower producers have had few herbicides for controlling broadleaf weeds. Resistance to AHAS-inhibiting herbicides has greatly increased the spectrum of herbicides for controlling broadleaf weeds in sunflower. On the basis of molecular studies, Kolkman et al. (2004) identified and characterized three AHAS large subunit genes (*AHASL* genes) in sunflower (*AHASL1*, *AHASL2*, and *AHASL3*) and demonstrated that the IMI-resistant and SU-resistant genes were allelic variants of the same locus (*AHASL1*). Only six of the 48 SNPs discovered in *AHAS1* would cause amino acid substitutions. Two of these, Pro197 and Ala205, have previously been shown to confer resistance to AHAS-inhibiting herbicides in other plant genera (Tranel and Wright, 2002). The ANN-PUR *AHAS1* allele, as identified from IMISUN-1, IMISUN-2, and HA425 *AHAS1* alleles, harbored an alanine (GCG) to valine (GTG) mutation in codon 205 (IMI resistant), whereas the ANN-KAN *AHAS1* allele, as identified from SURES-1 and SURES-2 *AHAS1* alleles, harbored a proline (CCC) to leucine (CTC) mutation in codon 197 (SU resistant).

With the objectives of identifying and characterizing induced mutations conferring resistance to AHAS-inhibiting herbicides, independent EMS mutagenesis on sunflower was performed, which resulted in three different mutants. Two patents were

```
           1                                                                                               100
       Ath MAAATTTTTTSSSISFSTKPSPSSSKSPLPISRFSLPFSLNPNKSSSSSRRRGIKSSSPSSISAVLNTTTNVTTTPSPTKPTKPETFISRFAPDQPRKGA
      HA89 -----MAAPPNPSISFK----PPSPAAALPPRSAFLPRFALPITSTTQKRHR----LHISN--VLSDSKSTTTTTTTTQRPLPVQPFVSRYAPDQPRKGA
     MUT28 -----MAAPPNPSISFK----PPSPAAALPPRSAFLPRFALPITSTTQKRHR----LHISN--VLSDSKSTTTTTTTTQRPLPVQPFVSRYAPDQPRKGA
    RHA266 -----MAAPPNPSISFK----PPSPAAALPPRSAFLPRFALPITSTTQKRHR----LHISN--VLSDSKS-TTTTTTTQPPLQAQPFVSRYAPDQPRKGA
      MUT9 -----MAAPPNPSISFK----PPSPAAALPPRSAFLPRFALPITSTTQKRHR----LHISN--VLSDSKS-TTTTTTTQPPLQAQPFVSRYAPDQPRKGA
     BTK47 ----------------------------------------------------------------------------------------------------
 CLHA-Plus ----------------------------------------------------------------------------------------------------
           101                 *A122T                                                                    *P197L
       Ath DILVEALERQGVETVFAYPGGASMEIHQALTRSSSIRNVLPRHEQGGVFAAEGYARSSGKPGICIATSGPGATNLVSGLADALLDSVPLVAITGQVPRRM
      HA89 DVLVEALEREGVTDVFAYPGGASMEIHQALTRSSTIRNVLPRHEQGGVFAAEGYARASGLPGVCIATSGPGATNLVSGLADALLDSVPMVAITGQVPRRM
     MUT28 DVLVEALEREGVTDVFAYPGGASMEIHQALTRSSTIRNVLPRHEQGGVFAAEGYARASGLPGVCIATSGPGATNLVSGLADALLDSVPMVAITGQVLRRM
    RHA266 DVLVEALEREGVTDVFAYPGGASMEIHQALTRSNTIRNVLPRHEQGGVFAAEGYARASGLPGVCIATSGPGATNLVSGLADALLDSVPMVAITGQVPRRM
      MUT9 DVLVEALEREGVTDVFAYPGGASMEIHQALTRSNTIRNVLPRHEQGGVFAAEGYARASGLPGVCIATSGPGATNLVSGLADALLDSVPMVAITGQVPRRM
     BTK47 ---------------FAYPGGASMEIHQALTRSSTIRNVLPRHEQGGVFAAEGYARASGLPGVCIATSGPGATNLVSGLADALLDSVPMVAITGQVPRRM
 CLHA-Plus ---------------FAYPGGTSMEIHQALTRSSTIRNVLPRHEQGGVFAAEGYARASGLPGVCIATSGPGATNLVSGLADALLDSVPMVAITGQVPRRM
                                                                                                       300
             *T203I
       Ath IGTDAFQETPIVEVTRSITKHNYLVMDVEDIPRIIEEAFFLATSGRPGPVLVDVPKDIQQQLAIPNWEQAMRLPGYMSRMPKPPEDSHLEQIVRLISESK
      HA89 IGTDAFQETPIVEVTRSITKHNYLVLDVEDIPRIVREAFYLASSGRPGPVLIDVPKDIQQQLVVPKWDEPMRLPGYLSRMPKPQYDGHLEQIVRLVGEAK
     MUT28 IGTDAFQETPIVEVTRSITKHNYLVLDVEDIPRIVREAFYLASSGRPGPVLIDVPKDIQQQLVVPKWDEPMRLPGYLSRMPKPQYDGHLEQIVRLVGEAK
    RHA266 IGTDAFQETPIVEVTRSITKHNYLVLDVEDIPRIVREAFYLASSGRPGPVLIDVPKDIQQQLVVPKWDEPMRLPGYLSRMPKPQYDGHLEQIVRLVGEAK
      MUT9 IGIDAFQETPIVEVTRSITKHNYLVLDVEDIPRIVREAFYLASSGRPGPVLIDVPKDIQQQLVVPKWDEPMRLPGYLSRMPKPQYDGHLEQIVRLVGEAK
     BTK47 IGTDAFQETPIVEVTRSITKHNYLVLDVEDIPRIVREAFYLASSGRPGPVLIDVPKDIQQQLVVPKWDEPMRLPGYLSRMPKPQYDGHLEQIVRLVGEAK
 CLHA-Plus IGTDAFQETPIVEVTRSITKHNYLVLDVEDIPRIVREAFYLASSGRPGPVLIDVPKDIQQQLVVPKWDEPMRLPGYLSRMPKPQYDGHLEQIVRLVGEAK
           301                                                                                             400
       Ath KPVLYVGGGCLNSSDELGRFVELTGIPVASTLMGLGSYPCDDELSLHMLGMHGTVYANYAVEHSDLLLAFGVRFDDRVTGKLEAFASRAKIVHIDIDSAE
      HA89 RPVLYVGGGCLNSDDELRRFVELTGIPVASTLMGLGAYPASSDLSLHMLGMHGTVYANYAVDKSDLLLAFGVRFDDRVTGKLEAFASRAKIVHIDIDPAE
     MUT28 RPVLYVGGGCLNSDDELRRFVELTGIPVASTLMGLGAYPASSDLSLHMLGMHGTVYANYAVDKSDLLLAFGVRFDDRVTGKLEAFASRAKIVHIDIDPAE
    RHA266 RPVLYVGGGCLNSDDELRRFVELTGIPVASTLMGLGAYPASSDLSLHMLGMHGTVYANYAVDKSDLLLAFGVRFDDRVTGKLEAFASRAKIVHIDIDPAE
      MUT9 RPVLYVGGGCLNSDDELRRFVELTGIPVASTLMGLGAYPASSDLSLHMLGMHGTVYANYAVDKSDLLLAFGVRFDDRVTGKLEAFASRAKIVHIDIDPAE
     BTK47 RPVLYVGGGCLNSDDELRRFVELTGIPVASTLMGLGAYPASSDLSLHMLGMHGTVYANYAVDKSDLLLAFGVRFDDRVTGKLEAFASRAKIVHIDIDPAE
 CLHA-Plus RPVLYVGGGCLNSDDELRRFVELTGIPVASTLMGLGAYPASSDLSLHMLGMHGTVYANYAVDKSDLLLAFGVRFDDRVTGKLEAFASRAKIVHIDIDPAE
           401                                                                                             500
```

```
      Ath IGKNKTPHVSVCGDVKLALQGMNKVLENRAEELKLDFGVWRNELNVQKQKFPLSFKTFGEAIPPQYAIKVLDELTDGKAIISTGVGQHQMWAAQFYNYKK
     HA89 IGKNKQPHVSICGDIKVALQGLNKILEEKNSVTNLDFSTWRKELDEQKMKFPLSFKTFGEAIPPQYAIQVLDELTGGNAIISTGVGQHQMWAAQFYKYNK
    MUT28 IGKNKQPHVSICGDIKVALQGLNKILEEKNSVTNLDFSTWRKELDEQKMKFPLSFKTFGEAIPPQYAIQVLDELTGGNAIISTGVGQHQMWAAQFYKYNK
   RHA266 IGKNKQPHVSICGDIKVALQGLNKILEEKNSVTNLDFSNWRKELDEQKVKFPLSFKTFGEAIPPQYAIQVLDELTGGNAIISTGVGQHQMWAAQFYKYNK
     MUT9 IGKNKQPHVSICGDIKVALQGLNKILEEKNSVTNLDFSNWRKELDEQKVKFPLSFKTFGEAIPPQYAIQVLDELTGGNAIISTGVGQHQMWAAQFYKYNK
    BTK47 IGKNKQPHVSICGDIKVALQGLNKILEEKNSVTNLDFSTWRKELDEQKMKFPLSFKTFGEAIPPQYAIQVLDELTGGNAIISTGVGQHQMWAAQFYKYNK
CLHA-Plus IGKNKQPHVSICGDIKVALQGLNKILEEKNSVTNLDFSTWRKELDEQKMKFPLSFKTFGEAIPPQYAIQVLDELTGGNAIISTGVGQHQMWAAQFYKYNK
          501                                                                                              600
      Ath PRQWLSSGGLGAMGFGLPAAIGASVANPDAIVVDIDGDGSFIMNVQELATIRVENLPVKVLLLNNQHLGMVMQWEDRFYKANRAHTFLGDPAQEDEIFPN
     HA89 PRQWLTSGGLGAMGFGLPAAIGAAVARPDAVVVDIDGDGSFMMNVQELATIRVENLPVKILLLNNQHLGMVVQWEDRFYKANRAHTYLGNPSKESEIFPN
    MUT28 PRQWLTSGGLGAMGFGLPAAIGAAVARPDAVVVDIDGDGSFMMNVQELATIRVENLPVKILLLNNQHLGMVVQWEDRFYKANRAHTYLGNPSKESEIFPN
   RHA266 PRQWLTSGGLGAMGFGLPAAIGAAVARPDAVVVDIDGDGSFMMNVQELATIRVENLPVKILLLNNQHLGMVVQWEDRFYKANRAHTYLGNPSKESEIFPN
     MUT9 PRQWLTSGGLGAMGFGLPAAIGAAVARPDAVVVDIDGDGSFMMNVQELATIRVENLPVKILLLNNQHLGMVVQWEDRFYKANRAHTYLGNPSKESEIFPN
    BTK47 PRQWLTS---------------------------------------------------------------------------------------------
CLHA-Plus PRQWLTS---------------------------------------------------------------------------------------------
          601                                                                 670
      Ath MLLFAAACGIPAARVTKKADLREAIQTMLDTPGPYLLDVICPHQEHVLPMIPSGGTFNDVITEGDGRIKY
     HA89 MVKFAEACDIPAARVTQKADLRAAIQKMLDTPGPYLLDVIVPHQEHVLPMIPAGGGFSDVITEGDGRTKY
    MUT28 MVKFAEACDIPAARVTQKADLRAAIQKMLDTPGPYLLDVIVPHQEHVLPMIPAGGGFSDVITEGDGRTKY
   RHA266 MLKFAEACDIPAARVTQKADLRAAIQKMLDTPGPYLLDVIVPHQEHVLPMIPAGGGFSDVITEGDGRTKY
     MUT9 MLKFAEACDIPAARVTQKADLRAAIQKMLDTPGPYLLDVIVPHQEHVLPMIPAGGGFSDVITEGDGRTKY
    BTK47 ----------------------------------------------------------------------
CLHA-Plus ----------------------------------------------------------------------
```

Figure 2.2 Alignment of AHASL-1 amino acidic sequence of the three EMS-induced sunflower mutants—MUT28, MUT9, and CLHA-plus—with tolerance to AHAS-inhibiting herbicides. The amino acidic substitutions associated with herbicide resistance for each mutant are indicated taken as reference sequence AHASL from *A. thaliana* (Ath): P197L, T203I, and A122T, respectively. In the alignment are included AHASL-1 sequences from their respective mutagenized original inbred line: HA89, RHA266, and BTK47, respectively.

published describing two different induced sunflower mutants having resistance to AHASL-inhibiting herbicides, which were named MUT9 and MUT28. MUT9 was screened from a mutant population generated by EMS treatment of sunflower seeds from inbred line RHA266, and the induced mutation on *AHASL* gene produced a threonine (T) to isoleucine (I) amino acid substitution on codon 203 (referring to *A. thaliana* sequence). T203I mutation conferred resistance to imazamox (imidazolinone group) (Leon et al., 2007). MUT28 was identified from a mutant population generated by EMS treatment of seeds from HA89 sunflower inbred line; in this case, the induced mutation on *AHASL* gene produced a proline (P) to leucine (L) amino acid substitution on codon 197 (Figure 2.2), conferring resistance to thifensulfuron (sulfonylurea group) and a moderate resistance to imazamox (Leon et al., 2006).

A partially dominant nuclear gene conferring resistance to the imidazolinone herbicides was identified in the mutant cultivated sunflower line named CLHA-Plus, developed by EMS mutagenesis of seeds from experimental maintainer BTK47 (Sala et al., 2008). CLHA-Plus showed a complete susceptibility to sulfonylureas (metsulfuron, tribenuron, and chlorsulfuron), but it showed a complete resistance to imidazolinones (imazamox, imazapyr, and imazapic) at two rates of herbicide application. This pattern was in close association with the AHAS-inhibition kinetics of protein extracts of CLHA-Plus challenged with different doses of imazamox and chlorsulfuron. Nucleotide and deduced amino acid sequence comparisons between resistant and susceptible lines indicated that the imidazolinone-resistant AHAS large subunit of CLHA-Plus has a threonine codon (ACG) at position 122, whereas the herbicide-susceptible enzyme from BTK47 has an alanine residue (GCG) at this position (A122T mutation) (Figure 2.2). The higher tolerance level to imidazolinones and the lack of cross-resistance to other AHAS-inhibiting herbicides indicate that this induced mutation can be used to develop commercial hybrids with superior levels of tolerance and, at the same time, to assist weed management where control of weedy common sunflower is necessary (Sala et al., 2008). Imidazolinone tolerance observed in IMISUN (with the A205V mutation) and CLHA-Plus is determined by the expression of different alleles at the same locus, *AHASL1-1* and *AHASL1-3*, respectively. It was reported that the level of tolerance expressed by plants containing both alleles in a homozygous, heterozygous, and heterozygous stacked state. Homozygous CLHA-Plus (*Ahasl1-3/Ahasl1-3*) genotypes and materials containing a combination of both tolerant alleles (IMISUN/CLHA-Plus heterozygous stack, *Ahasl1-1/Ahasl1-3*) showed the highest values of the estimated dose of imazapyr required to reduce biomass accumulation by 50% (GR_{50}), 300 times higher than the susceptible genotypes and more than 2.5 times higher than homozygous IMISUN materials (*Ahasl1-1/Ahasl1-1*). The degree of dominance for the accumulation of biomass after herbicide application calculated for the *Ahasl1-1* allele indicated can vary from co-dominant to recessive, depending on the imazapyr dose used. On the contrary, the *Ahasl1-3* allele showed dominance to semidominance according to the applied dose. This last allele is dominant over *Ahasl1-1* over the entire range of herbicide rates tested. At the level of enzymatic activity, however, both alleles showed recessivity to semirecessivity with respect to the wild-type allele, even though the *Ahasl1-3* allele is dominant over *Ahasl1-1* at all the herbicides rates used (Sala et al., 2012).

Mutagenesis, Genomics, and Reverse Genetics

One of the most direct ways of establishing gene function is to identify a mutation in the specific gene and to link this mutation to the phenotypic change in the mutated organism. In the forward genetics approach (from mutation through phenotype to the gene), large mutated populations have been created and screened for alterations in the trait or biological process of interest. Over the decades, large mutant collections have been developed for many model organisms and the generated mutants have then served for the identification of the genes underlying the change in phenotype. The sequence of the gene responsible for the altered phenotype can be isolated using the process of map-based cloning. Although this approach is both time-consuming and labor intensive, it has been successfully applied for cloning several genes, even in species with large-genomes, such as barley and wheat (Krattinger et al., 2009; Zhang et al., 2009). In this century, there has been a dramatic increase in the amount of genome sequence data available for major world food crops. For instance, complete genome sequences have been reported for rice (Matsumoto et al., 2005) and sorghum (Paterson et al., 2009).

The exploitation of sequence data for crop improvement is limited by the complexity of many of the traits that determine agronomic performance (Parry and Reynolds, 2007; Parry et al., 2005). However, reverse genetics approaches allow progress to be made on the major challenge of linking sequence information to the biological function of genes and on determining their contribution to important characteristics and traits. Typically, these approaches rely on the disruption of candidate genes by mutagenesis, transposons, and T-DNA tagging or RNA interference. Many efforts in sunflower plant breeding have been directed toward the improvement of agronomic traits. Genetic engineering (Lucas et al., 2000), traditional breeding approaches (Jan and Seiler, 2007), in vitro breeding techniques (somaclonal variation), and conventional mutation technologies (Encheva et al., 2008; Škorić et al., 2008) have been used to improve yield, oil quality, and disease, salt, and pest resistance.

Targeting Induced Local Lesions in Genomes (TILLING) is a reverse genetic technique suitable for most plants for mutation screening in a given mutant population. In a TILLING project, mutagenic treatment with EMS is coupled with PCR-based mutation detection on the targeted locus. Thus, with only a small population, multiple alleles of a specific gene may be obtained regardless of the gene size, combining the high frequency of mutations induced by traditional mutagenesis (inducing missense DNA changes, truncation, and mutations in splice junction sequences) with sensitive techniques for discovering single nucleotide mutations. The user must know the genomic sequence of the gene chosen to determine whether it is a good candidate for TILLING (Greene et al., 2003; McCallum et al., 2000; Till et al., 2007). In contrast to insertional mutagenesis that generates mostly gene knockouts, using TILLING it is possible to induce and select a series of alleles in a targeted locus. In addition to loss-of-function alleles, chemical mutagens generate gain-of-function and hypomorphic alleles that can provide a range of phenotypes (Alonso and Ecker, 2006). The great mutagenic potential of chemical agents has been proven by the high density of mutations reported for established TILLING populations, including sunflower (Table 2.C). The success of the TILLING approach relies on the construction of high-quality DNA mutant libraries, in which DNA sampling and preparation are the most critical steps. An accurate evaluation of the genetic uniformity both of seed stock used for the mutagenic treatment and of the mutagenized material is extremely important (Sabetta et al., 2011).

The first strategy that was described for the TILLING procedure (McCallum et al., 2000) included the EMS treatment of Arabidopsis seeds, DNA isolation and pooling, PCR reaction of the fragment of interest, heteroduplex formation, and the identification of heteroduplexes using denaturing high-performance liquid chromatography (DHPLC). Since then, TILLING has been used with many different organisms and many modifications to the original procedure have been introduced that help to automate the screening of mutations and reduce its cost. Over time, the DHPLC method for detecting mutations in the TILLING approach was, in most cases, replaced by the digestion of heteroduplexes using specific endonucleases followed by polyacrylamide electrophoresis and visualization in the very sensitive LI-COR gel analyzer system (LI-COR Biosciences). The most popular mutation detection

Table 2.C Comparison of Frequency of EMS Induced Mutations Reported in Sunflower (*Helianthus annuus*) and Different Species

Species	Ploidy	Mutation Frequency	Reference
Arabidopsis thaliana	2x	1/300 kb	Greene (2003)
Brassica napus	4x	1/41.5 kb	Wang et al. (2008)
Glycine max	4x	1/140 to 1/550 kb	Cooper et al. (2008)
Helianthus annuus	2x	1/480 kb 1/475 kb	Kumar (2013) Sabetta (2011)
Hordeum vulgare	2x	1/1000 kb	Caldwell et al. (2004)
Pisum sativum	2x	1/200 kb	Dalmais et al. (2008)
Sorghum bicolor	2x	1/526 kb	Xin et al. (2008)
Zea mays	2x	1/485 kb	Till et al. (2004)

method for TILLING is a mismatch cleavage of heteroduplex DNA by using the single-stranded specific endonuclease CEL I from celery. ENDO I from *A. thaliana* is another mismatch-specific endonuclease that belongs to the S1 type. The application of ENDO I has been reported in mutation diagnostics in humans, the fingerprinting of complex populations of viruses, TILLING in *P. sativum* and *S. lycopersicum,* and EcoTILLING in *A. thaliana*. The advances in DNA sequencing technologies have been used in two ways in the TILLING strategy. First, the growing number of whole-genome sequencing projects in plants for both crop and model species has led to an increase in the possible choices of target genes in TILLING. Second, new strategies for sequencing, so-called next-generation sequencing methods (NGS), which have been applied to a limited number of sequenced individuals using Sanger's method, can be used for direct mutation determination without any prescreening. The advent of NGS platforms has dramatically increased the speed at which a DNA sequence can be acquired and has also reduced the cost of sequencing by more than two orders of magnitude (for a review, see Kurowska et al., 2011) (Figure 2.3).

Chemical mutagenesis has been used by breeders to create variability but has been mostly restricted to dominant traits. Therefore, many desirable mutations that are recessive have been missed during selection (Barkley and Wang, 2008) in traits such as oil quality that cannot be selected by visual inspection of plants. Novel mutants having useful mutation in genes can be found by screening the population through high throughput allele discovery platform technique. To investigate the efficiency of the mutagenic treatment on sunflower, a pilot screening was carried out by Sabetta et al. (2011) on 1152 M_2 lines focusing on four genes, three involved in the fatty acid biosynthetic pathway and one for downy mildew resistance. A total of nine mutant lines were identified and confirmed by sequencing; thereby, the estimated overall mutation frequency for the pilot assay resulted as 1/475 kb. One of these mutations was found on the *KASII* gene (encoding the 3-ketoacyl-ACP synthetase II) producing a premature stop codon. Another mutation screening carried by Kumar et al. (2013) on 5000 M_2 sunflower populations (with 4763 M_3 plants with fertile seed sets) resulted in a mutation frequency of 1/480 kb. In this case, screening focused on two proof-of-concept genes participating in fatty acid synthesis: stearoyl-ACP desaturase and acyl-ACP thioesterase A, allowing different mutants for these genes to be found.

Conclusion

During the past 70 years, many sunflower varieties, derived either as direct mutants or from their progenies, have been released worldwide. The prime strategy in mutation-based sunflower breeding has been to upgrade the well-adapted plant varieties by altering one or two major traits that limit their productivity or enhance their quality value. The economic contribution of the selected mutant varieties of sunflower was highly

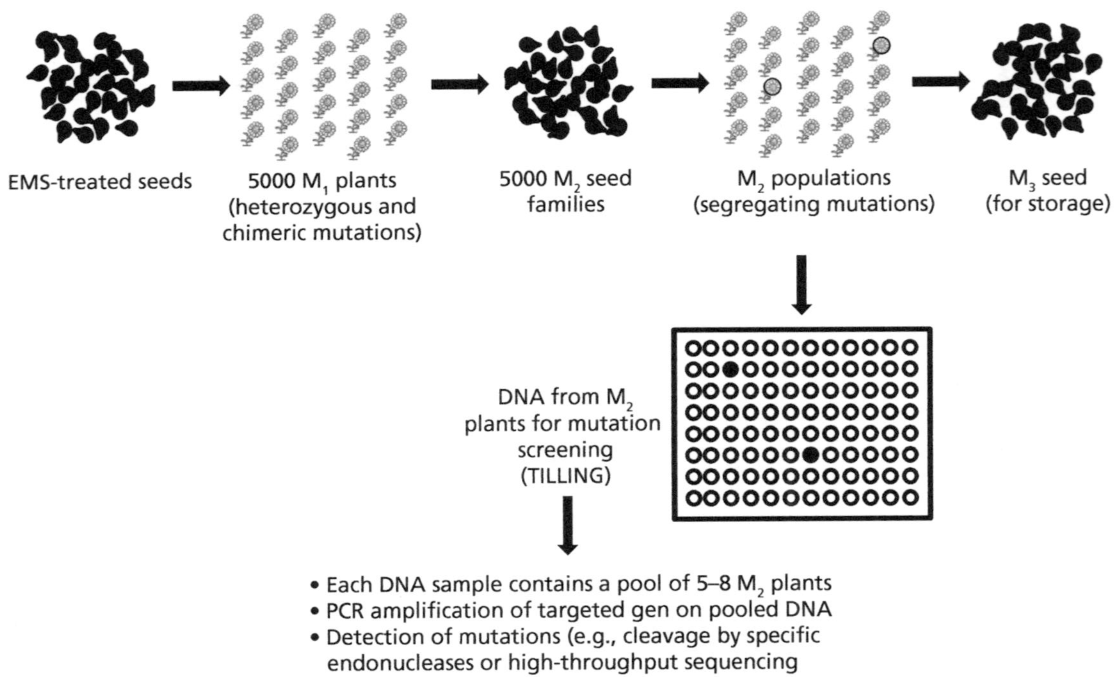

Figure 2.3 A diagram showing a typical TILLING strategy using sunflower seed mutagenesis. The first generation (M_1) is chimeric because of the multicellular composition of the seed at mutagenesis, making it unsuitable for mutation discovery. M_2 plants are nonchimeric and suitable for mutation screening. DNA from each M_2 is isolated and M_3 seeds are collected and stored. DNA concentrations are normalized, arrayed in 96 well plates, and pooled (5 to 8 plants) to increase screening efficiency. DNAs are PCR-amplified using gene-specific primers followed by mutation discovery via endonuclease mismatch cleavage (followed by polyacrylamide electrophoresis) or by NGS of PCR products.

valuable and has resulted in synergistic effects on increasing the yield and quality of the crop, improving agronomic inputs, crop rotation, and consumer acceptance. Other agronomical uses of induced sunflower mutants beyond oil quality or herbicide resistance may arise in the future. For instance, chemical mutagenesis allowed researchers to obtain sunflower with improved metal phytoextraction properties (Kolbas et al., 2011). Additional applications of induced mutations will grow in importance as the sunflower genome knowledge increases. Thus, it is expected that reverse genetics approaches will contribute to improving the agronomic value of sunflower crop.

References

Al-Khatib, K.; Baumgartner, J. R.; Peterson, D. E.; Currie, R. S. Imazethapyr Resistance in Common Sunflower (*Helianthus annuus*). *Weed Sci.* **1998,** *46,* 403–407.

Al-Khatib, K.; Miller, J. F. Registration of Four Genetic Stocks of Sunflower Resistant to Imidazolinone Herbicides. *Crop Sci.* **2000,** *40,* 869–870.

Alonso, J. M.; Ecker, J. R. Moving Forward in Reverse: Genetic Technologies to Enable Genome-Wide Phenomic Screens in Arabidopsis. *Nat. Rev. Genet.* **2006,** *7,* 524–536.

Al-Qurainy, F.; Khan, S. Mutagenic Effects of Sodium Azide and Its Application in Crop Improvement. *World Appl. Sci. J.* **2009,** *6,* 1589–1601.

Barkley, N. A.; Wang, M. L. Application of TILLING and EcoTILLING as Reverse Genetic Approaches to Elucidate the Function of Genes in Plants and Animals. *Curr. Gen.* **2008,** *9,* 212–226.

Caldwell, D. G.; McCallum, N.; Shaw, P.; Muehlbauer, G. J.; Marshall, D. F.; Waugh, R. A Structured Mutant Population for Forward and Reverse Genetics in Barley (*Hordeum vulgare* L.). *Plant J.* **2004,** *40,* 143–150.

Cooper, J. L.; Till, B. J.; Laport, R. G.; Darlow, M. C.; Kleffner, J. M.; Jamai, A.; El-Mellouki, T.; Liu, S.; Ritchie, R.; Nielsen, N.; et al. TILLING to Detect Induced Mutations in Soybean. *BMC Plant Biol.* **2008,** *8,* 9–18.

Crupkin, M.; Zambelli, A. Detrimental Impact of *Trans* Fats on Human Health: Stearic Acid–Rich Fats as Possible Substitutes. *Compr. Rev. Food Sci. Food Saf.* **2008,** *7,* 271–279.

Cvejić, S.; Jocić, S.; Prodanović, S.; Terzić, S.; Miladinović, D.; Balalić, I. Creating New Genetic Variability in Sunflower Using Induced Mutations. *Helia* **2011,** *34,* 47–54.

Dalmais, M.; Schmidt, J.; Signor, C. L.; Moussy, F.; Burstin, J.; Savois, V.; Aubert, G.; Brunaud, V.; Oliveira, Y.; Guichard, C.; et al. UTILLdb, a *Pisum sativum* in Silico Forward and Reverse Genetics Tool. *Genome Biol.* **2008,** *9,* R43. DOI: 10.1186/gb-2008-9-2-r43.

Demurin, Y. Genetic Variability of Tocopherol Composition in Sunflower Seeds. *Helia* **1993,** *16,* 59–62.

Demurin, Y.; Škorić, D.; Karlovic, D. Genetic Variability of Tocopherol Composition in Sunflower Seeds as a Basis of Breeding for Improved Oil Quality. *Plant Breed.* **1996,** *115,* 33–36.

Duggleby, R. G.; McCourt, J. A.; Guddat, L. W. Structure and Mechanism of Inhibition of Plant Acetohydroxyacid Synthase. *Plant Physiol. Biochem.* **2008,** *46,* 309–324.

Eitenmiller, R.; Lee, J. *Vitamin E: Food Chemistry, Composition, and Analysis.* Marcel Dekker: New York, 2004.

Encheva, J.; Shindrova, P.; Penchev, E. Developing Mutant Sunflower Lines (*Helianthus annuus* L.) through Induced Mutagenesis. *Helia* **2008,** *31* (Suppl. 48), 61–72.

Fernández-Martínez, J.; Muñoz, J.; Gómez-Arnau, J. Performance of Near-Isogenic High and Low Oleic Acid Hybrids of Sunflower. *Crop Sci.* **1993,** *33,* 1158–1163.

Fernández-Martínez, J. M.; Osorio, J.; Mancha, M.; Garcés, R. Isolation of High Palmitic Mutants on High Oleic Background. *Euphytica* **1997,** *97,* 113–116.

Fernández-Moya, V.; Martínez-Force, E.; Garcés, R. Temperature Effect on a High Stearic Acid Sunflower Mutant. *Phytochem.* **2002,** *59,* 33–38.

Garcés, R.; Mancha, M. One-Step Lipid Extraction and Fatty Acid Methyl Esters Preparation from Fresh Plant Tissues. *Anal. Biochem.* **1993,** *211,* 139–143.

Garcés, R.; Martínez-Force, E.; Salas, J. J.; Venegas-Calerón, M. Current Advances in Sunflower Oil and Its Applications. *Lipid Technol.* **2009,** *21,* 79–82.

Greene, E. A.; Codomo, C. A.; Taylor, N. E.; Henikoff, J. G.; Till, B. J.; Reynolds, S. H.; Enns, L. C.; Burtner, C.; Johnson, J. E.; Odden, A. R.; et al. Spectrum of Chemically Induced Mutations from a Large-Scale Reverse-Genetic Screen in Arabidopsis. *Genetics* **2003,** *164,* 731–740.

Harter, A. V.; Gardner, K. A.; Falush, D.; Lentz, D. L.; Bye, R. A.; Rieseberg, L. H. Origin of Extant Domesticated Sunflowers in Eastern North America. *Nature* **2004,** *430,* 201–205.

Heap, I. The International Survey of Herbicide Resistant Weeds. http://www.weedscience.org (accessed Jun 10, 2013).

Hoffmann, M.; Hornung, E.; Busch, S.; Kassner, N.; Ternes, P.; Braus, G. H.; Feussner, I. A Small Membrane-Peripheral Region Close to the Active Center Determines Regioselectivity of Membrane-Bound Fatty Acid Desaturases from *Aspergillus nidulans. J. Biol. Chem.* **2007,** *282,* 26666–26674.

Hunter, J. E.; Zhang, J.; Kris-Etherton, P. M. Cardiovascular Disease Risk of Dietary Stearic Acid Compared with *Trans*, Other Saturated, and Unsaturated Fatty Acids: A Systematic Review. *Am. J. Clin. Nutr.* **2010,** *91,* 1–18.

Hunter, S. C.; Cahoon, E. B. Enhancing Vitamin E in Oilseeds: Unravelling Tocopherol and Tocotrienol Biosynthesis. *Lipids* **2007,** *42,* 97–108.

Ivanov, P.; Petakov, V.; Nikolova, V.; Petchev, E. Sunflower Breeding for High Palmitic Acid Content in the Oil. In *Proceedings of the 12th International Sunflower Conference;* International Sunflower Association: Novi Sad, Serbia, 1998.

Jambhulkar, S. J.; Shitre, A. S. Development and Utilization of Genetic Variability through Induced Mutagenesis in Sunflower (*Helianthus annuus* L.). Induced Plant Mutations in the Genomics Era. In *Proceedings of an International Joint FAO/IAEA Symposium;* Shu, Q. Y., Ed.; Vienna, Austria, 2009; pp 104–105.

Jan, C. C.; Seiler, G. J. Oilseed Crops. *Genetic Resources, Chromosome Engineering and Crop Improvement;* Singh, R. J., Ed.; Taylor: New York, 2007; pp 103–168.

Katan, M. B.; Boekschoten, M. V.; Connor, W. E.; Mensink, R. P.; Seidell, J.; Vessby, B.; Willett, W. Which Are the Greatest Recent Discoveries and the Greatest Future Challenges in Nutrition? *Eur. J. Clin. Nutr.* **2007,** *63,* 2–10.

Kolbas, A.; Mench, M., Herzig, R.; Nehnevajova, E; Bes, C. M. Copper Phytoextraction in Tandem with Oilseed Production Using Commercial Cultivars and Mutant Lines of Sunflower. *Int. J. Phytoremediation* **2011,** *13* (Suppl. 1), 149–165.

Kolkman, J. M.; Slabaugh, M. B.; Bruniard, J. M.; Berry, S.; Bushman, B. S.; Olungu, C.; Maes, N.; Abratti, G.; Zambelli, A.; Miller, J. F.; et al. Acetohydroxyacid Synthase Mutations Conferring Resistance to Imidazolinone or Sulfonylurea Herbicides in Sunflower. *Theor. Appl. Genet.* **2004,** *109,* 1147–1159.

Kolkman, J. M.; Berry, S. T.; Leon, A. J.; Slabaugh, M. B.; Tang, S.; Gao, W.; Shintani, D. K.; Burke, J. M.; Knapp, S. J. Single Nucleotide Polymorphisms and Linkage Disequilibrium in Sunflower. *Genetics* **2007,** *177,* 457–468.

Krattinger, S. G.; Wicker, T.; Keller, B. Map-Based Cloning of Genes in Triticeae (Wheat and Barley). In *Genetics and Genomics of the Triticeae, Plant Genetics and Genomics: Crops and Models;* Muehlbauer, G., Feuillet, C., Eds.; Springer US: New York, 2009; pp 337–357.

Kumar, A. P. K.; Boualem, A.; Bhattacharya, A.; Parikh, S.; Desai, N.; Zambelli, A.; Leon, A.; Chatterjee, M.; Bendahmane, A. SMART—Sunflower Mutant Population and Reverse Genetic Tool for Crop Improvement. *BMC Plant Biol.* **2013,** *13,* 38.

Kurowska, M.; Daszkowska-Golec, A.; Gruszka, D.; Marzec, M.; Szurman, M.; Szarejko, I.; Maluszynski, M. TILLING: A Shortcut in Functional Genomics. *J. Appl. Genet.* **2011,** *52,* 371–390.

Lacombe, S.; Souyris, I.; Bervillé, A. J. An Insertion of Oleate Desaturase Homologous Sequence Silences via siRNA the Functional Gene Leading to High Oleic Acid Content in Sunflower Seed Oil. *Mol. Genet. Genomics* **2009,** *281,* 43–54.

Lea, P. J.; Azevedo, R. A. Nitrogen Use Efficiency. 1. Uptake of Nitrogen from the Soil. *Ann. Appl. Biol.* **2006,** *149,* 243–247.

Leon, A. J.; Morata, M. M.; Zambelli, A. D. Herbicide-Resistant Sunflower Plants, Polynucleotides Encoding Herbicide-Resistant Acetohydroxy Acid Synthase Large Subunit Proteins, and Methods of Use. WIPO Patent WO/2006/024351A1, March 9, 2006.

Leon, A. J.; Morata, M. M.; Olungu, C. Herbicide-Resistant Sunflower Plants with a Novel Mutation in the Gene Encoding the Large Subunit of Acetohydroxyacid Synthase, Isolated Polynucleotides, and Methods of Use. U.S. Patent US 2007/0118920 A1, May 24, 2007.

Leon, A. J.; Zambelli, A. D.; Reid, R. J.; Morata, M. M.; Kaspar, M. Nucleotide Sequences Mutated by Insertion that Encode a Truncated Oleate Desaturase Protein, Proteins, Methods and Uses. WIPO Patent WO/2013/004281, Jan 10, 2013a.

Leon, A. J.; Zambelli, A. D.; Reid, R. J.; Morata, M. M.; Kaspar, M.; Martinez-Force, E.; Garcés, R.; Salas, J. J.; Venegas-Caleron, M. Isolated Mutated Nucleotide Sequences That Encode a Modified Oleate Destaurase Sunflower Protein, Modified Protein, Methods and Uses. WIPO Patent WO/2013/004280, Jan 10, 2013b.

Lucas, O.; Kallerhoff, J.; Alibert, G. Production of Stable Transgenic Sunflowers (*Helianthus annuus* L.) from Wounded Immature Embryos by Particle Bombardment and Co-Cultivation with *Agrobacterium tumefaciens*. *Mol. Breeding* **2000,** *6,* 479–487.

Mallory-Smith, C. A.; Thill, D. C; Dial M. J. Identification of Sulfonylurea Herbicide–Resistant Prickly Lettuce (*Lactuca serriola*). *Weed Technol.* **1990,** *4,* 163–168.

Martínez-Rivas, J. M; Sperling, P.; Lühs, W.; Heinz, E. Spatial and Temporal Regulation of Three Different Microsomal Oleate Desaturase Genes (FAD2) from Normal-Type and High-Oleic Varieties of Sunflower (*Helianthus annuus* L.). *Mol. Breeding* **2001,** *8,* 159–168.

Matsumoto, T.; Wu, Z. J.; Kanamori, H.; Katayose, Y.; Fujisawa, M.; Namiki, N.; Mizuno, H.; Yamamoto, K.; Antonio, B. A; Baba, T.; et al. The Map-Based Sequence of the Rice Genome. *Nature* **2005,** *436,* 793–800.

McCallum, C. M.; Comai, L.; Greene, E. A.; Henikoff, S. Targeting Induced Local Lesions in Genomes (TILLING) for Plant Functional Genomics. *Plant Physiol.* **2000,** *123,* 439–442.

McCourt, J. A.; Pang, S. S.; King-Scott, J.; Guddat, L. W.; Duggleby, R. G. Herbicide-Binding Sites Revealed in the Structure of Plant Acetohydroxyacid Synthase. *Proc. Natl. Acad. Sci. USA* **2006,** *103,* 569–573.

Micic, Z.; Hahn, V.; Bauer, E.; Schon, C.; Melchinger, A. QTL Mapping of Resistance to Sclerotinia Midstalk Rot in RIL of Sunflower Population NDBLOSsel × CM625. *Theor. Appl. Genet.* **2005,** *110* (Suppl. 8), 1490–1498.

Miller, J. F.; Al-Khatib, K. Registration of Imidazolinone Herbicide-Resistant Sunflower Maintainer (HA425) and Fertility Restorer (RHA426 and RHA427) Germplasms. *Crop Sci.* **2002,** *42,* 988–989.

Miller, J. F.; Al-Khatib, K. Registration of Two Oilseed Sunflower Genetic Stocks, SURES-1 and SURES-2, Resistant to Tribenuron Herbicide. *Crop Sci.* **2004,** *44,* 1037–1038.

Mozaffarian, D.; Katan, M. B.; Ascherio, A.; Stampfer, M. J.; Willett, W. *Trans* Fatty Acids and Cardiovascular Disease. *N. Eng. J. Med.* **2006,** *354,* 1601–1613.

Nikolau, B. J.; Ohlrogge, J. B.; Wurtele E. S. Plant Biotin-Containing Carboxylases. *Arch. Biochem. Biophys.* **2003,** *414,* 211–222.

Osorio, J.; Fernández-Martínez, J.; Mancha, M.; Garcés, R. Mutant Sunflowers with High Concentration of Saturated Fatty Acids in the Oil. *Crop Sci.* **1995,** *35,* 739–742.

Pang, S. S.; Duggleby, R. G.; Guddat, L. W. Crystal Structure of Yeast Acetohydroxyacid Synthase: A Target for Herbicidal Inhibitors. *J. Mol. Biol.* **2002,** *317,* 249–262.

Pang, S. S.; Guddat, L. W.; Duggleby, R. G. Molecular Basis of Sulfonylurea Herbicide Inhibition of Acetohydroxyacid Synthase. *J. Biol. Chem.* **2003,** *278,* 7639–7644.

Parry, M. A. J.; Flexas, J.; Medrano H. Prospects for Crop Production under Drought: Research Priorities and Future Directions. *Ann. Appl. Biol.* **2005,** *147,* 211–226.

Parry, M. A. J.; Madgwick, P. J.; Carvalho, J. F. C.; Andralojc, P. J. Prospects for Increasing Photosynthesis by Overcoming the Limitations of Rubisco. *J. Agric. Sci.* **2007**, *145,* 31–43.

Parry, M. A. J.; Reynolds, M. P. Improving Resource Use Efficiency. *Ann. Appl. Biol.* **2007,** *151,* 133–135.

Parry, M. A. J.; Madgwick, P. J.; Bayon, C.; Tearall, K.; Hernandez-Lopez, A.; Baudo, M.; Rakszegi, M.; Hamada, W.; Al-Yassin, A.; Ouabbou, H.; et al. Mutation Discovery for Crop Improvement. *J. Exp. Bot.* **2009,** *60,* 2817–2825.

Paterson, A. H.; Bowers, J. E.; Bruggmann, R.; Dubchak, I.; Grimwood, J.; Gundlach, H.; Haberer, G.; Hellsten, U.; Mitros, T.; Poliakov, A.; et al. The Sorghum Bicolor Genome and the Diversification of Grasses. *Nature* **2009,** *457,* 551–556.

Pathirana, R. *Plant Mutation Breeding in Agriculture. Plant Science Reviews;* Hemming, D., Ed; CAB International: Oxfordshire, UK, 2012; 107–126.

Pleite, R.; Martínez-Force, E.; Garcés, R. Inhibitors of Fatty Acid Biosynthesis in Sunflower Seeds. *J. Plant Physiol.* **2006,** *163,* 885–894.

Powlson, D. S.; Riche, A. B.; Shield, I. Biofuels and Other Approaches for Decreasing Fossil Fuel Emissions from Agriculture. *Ann. Appl. Biol.* **2005,** *146,* 193–201.

Pozniak, C. J.; Birk, I. T.; O'Donoughue, L. S.; Menard, C.; Hucl, P. J.; Singh, B. K. Physiological and Molecular Characterization of Mutation-Derived Imidazolinone Resistance in Spring Wheat. *Crop Sci.* **2004,** *44,* 1434–1443.

Rashid, M.; He, G.; Guanxiao, Y.; Khurram, Z. Relevance of Tilling in Plant Genomics. *Aust. J. Crop Sci.* **2011,** *5,* 411–420.

Reynolds, M.; Foulkes, M. J.; Slafer, G. A.; Berry, P.; Parry, M. A. J.; Snape, J. W.; Angus, W. J. Raising Yield Potential in Wheat. *J. Exp. Bot.* **2009,** *60,* 1899–1918.

Richards, R. A. Selectable Traits to Increase Crop Photosynthesis and Yield of Grain Crops. *J. Exp. Bot.* **2000,** *51,* 447–458.

Ruiz-López, N.; Martínez-Force, E.; Garcés, R. Sequential One-Step Extraction and Analysis of Triacylglycerols and Fatty Acids in Plant Tissues. *Anal. Biochem.* **2003,** *317,* 247–254.

Sabetta, W.; Alba, V.; Blanco, A.; Montemurro, C. SunTILL: A TILLING Resource for Gene Function Analysis in Sunflower. *Plant Methods* **2011,** *7,* 20.

Sala, C. A.; Bulos, M.; Echarte, A. M. Genetic Analysis of an Induced Mutation Conferring Imidazolinone Resistance in Sunflower. *Crop Sci.* **2008,** *48,* 1817–1822.

Sala, C. A.; Bulos, M.; Altieri, E.; Weston, B. Response to Imazapyr and Dominance Relationships of Two Imidazolinone-Tolerant Alleles at the Ahasl1 Locus of Sunflower. *Theor. Appl. Genet.* **2012,** *124,* 385–396.

Salas, J. J.; Martínez-Force, E.; Garcés, R. Biochemical Characterization of a High Palmitoleic Acid *Helianthus annuus* Mutant. *Plant Physiol. Biochem.* **2004,** *42,* 373–381.

Seiler, G. J. Utilization of Wild Sunflower Species for the Improvement of Cultivated Sunflower. *Field Crop Res.* **1992,** *30* (Suppl. 3–4), 195–230.

Shanklin, J.; Cahoon, E. B. Desaturation and Related Modifications of Fatty Acids. *Annu. Rev. Plant Physiol. Plant Mol. Biol.* **1998,** *49,* 611–641.

Singh, B. K. *Plant Amino Acids;* Singh, B. K., Ed.; Dekker: New York, 1999; pp 227–247.

Singh, S. P.; Zhou, X. R.; Liu, Q.; Stymne, S.; Green, A. G. Metabolic Engineering of New Fatty Acids in Plants. *Curr. Opin. Plant Biol.* **2005,** *8,* 197–203.

Škorić, D. Achievements and Future Directions of Sunflower Breeding. *Field Crop Res.* **1992,** *30,* 231–270.

Škorić, D.; Jocić, S.; Sakac, Z.; Lecić, N. Genetic Possibilities for Altering Sunflower Oil Quality to Obtain Novel Oils. *Can. J. Physiol. Pharmacol.* **2008,** *86,* 215–221.

Smith, M. A.; Cross, A. R.; Jones, O. T. G.; Griffiths, W. T.; Stymne, S.; Stobart, K. Electron-Transport Components of the 1-acyl-2-oleoyl-sn-glycero-3-phosphocholine Δ12-desaturase (Δ12-desaturase) in Microsomal Preparations from Developing Safflower (*Carthamus tinctorius* L.) Cotyledons. *Biochem. J.* **1990,** *272,* 23–29.

Soldatov, K. I. Chemical Mutagenesis in Sunflower Breeding. In *Proceedings of the 7th International Sunflower Conference;* International Sunflower Association: Vlaardingen, The Netherlands, 1976; pp 352–357.

Staton, S. E.; Bakken, B. H.; Blackman, B. K.; Chapman, M. A.; Kane, N. C.; Tang, S.; Ungerer, M. C.; Knapp, S. J.; Rieseberg, L. H.; et al. The Sunflower (*Helianthus annuus* L.) Genome Reflects a Recent History of Biased Accumulation of Transposable Elements. *Plant J.* **2012.** DOI:10.1111/j.1365-313X.2012.05072.

Till, B. J.; Reynolds, S. H.; Weil, C.; Springer, N.; Burtner, C.; Young, K.; Bowers, E.; Codomo, C. A.; Enns, L. C.; Odden, A. R.; et al. Discovery of Induced Point Mutations in Maize Genes by TILLING. *BMC Plant Biol.* **2004,** *4,* 12.

Till, B. J.; Comai, L.; Henikoff, S. TILLING and EcoTILLING for Crop Improvement. In *Genomics-Assisted Crop Improvement;* Varshney, R. K., Tuberosa, R., Eds.; Springer: Dordrect, The Netherlands, 2007; Vol.: Genomics Approaches and Platforms, pp 333–349.

Tranel, P. J.; Wright, T. R. Resistance of Weeds to AHAS-Inhibiting Herbicides: What Have We Learned? *Weed Sci.* **2002,** *50,* 700–712.

United States Department of Agriculture, Foreign Agricultural Service, 2013. http://www.fas.usda.gov/psdonline/psdQuery.aspx (accessed Jul 21, 2013).

Velasco, L.; Fernández-Martínez, J. M.; García-Ruíz, R.; Domínguez, J. Genetic and Environmental Variation for Tocopherol Content and Composition in Sunflower Commercial Hybrids. *J. Agric. Sci.* **2002,** *139,* 425–429.

Velasco, L.; Domínguez, J.; Fernández-Martínez, J. M. Registration of T589 and T2100 Sunflower Germplasms with Modified Tocopherol Profiles. *Crop Sci.* **2004a,** *44,* 361–362.

Velasco, L.; Pérez-Vich, B.; Fernández-Martínez, J. M. Novel Variation for the Tocopherol Profile in a Sunflower Created by Mutagenesis and Recombination. *Plant Breed.* **2004b,** *123,* 490–492.

Velasco, L.; Pérez-Vich, B.; Fernández-Martínez, J. M. A New Sunflower Mutant with Increased Levels of Palmitic Acid in Seed Oil. *Helia* **2008,** *31,* 55–60.

Wang, N.; Wang, Y. J.; Tian, F.; King, G. J.; Zhang, C. Y.; Long, Y.; Shi, L.; Meng, J. L. A Functional Genomics Resource for *Brassica napus*: Development of an EMS Mutagenized Population and Discovery of FAE1 Point Mutations by TILLING. *New Phytol.* **2008,** *180,* 751–765.

Warner, K. Increasing Gamma- and Delta-Tocopherols in Oils Improves Oxidative Stability. *Lipid Technol.* **2007,** *19,* 229–231.

White, A. D.; Owen, M. D.; Hartzler, R. G.; Cardina, J. Common Sunflower Resistance to Acetolactate-Inhibiting Herbicides. *Weed Sci.* **2002,** *50,* 432–437.

White, A. D.; Graham, M. A.; Owen, M. D. K. Isolation of Acetolactate Synthase Homologs in Common Sunflower. *Weed Sci.* **2003,** *51,* 845–853.

Xin, Z. G.; Wang, M. L.; Barkley, N. A.; Burow, G.; Franks, C.; Pederson, G.; Burke, J. Applying Genotyping (TILLING) and Phenotyping Analyses to Elucidate Gene Function in a Chemically Induced Sorghum Mutant Population. *BMC Plant Biol.* **2008,** *8,* 103. DOI: 10.1186/1471-2229-8-103.

Zhang, J. Y.; Lu, Y.; Yuan, Y.; Zhang, X.; Geng, J.; Chen, Y.; Cloutier, S.; McVetty, P. B. E.; Li, G. Map-Based Cloning and Characterization of a Gene Controlling Hairiness and Seed Coat Color Traits in *Brassica rapa. Plant Mol. Biol.* **2009,** *69,* 553–563.

Zhou, Q.; Liu, W.; Zhang, Y.; Liu, K. K. Action Mechanisms of Acetolactate Synthase-Inhibiting Herbicides. *Pestic. Biochem. Physiol.* **2007,** *89,* 89–96.

3

Sunflower Crop Physiology and Agronomy

C. Alberio, N.G. Izquierdo, and L.A.N. Aguirrezábal ■ *Comisión Nacional Investigaciones Cientificas y Técnicas (CONICET), Facultad de Ciancias Agrarias, Universidad Nacional de Mar del Plata (FCA-UNMdP), Balcarce, Argentina*

Introduction

Agronomy is the science and technology of producing and using plants for food, fuel, and fiber. It presents agriculture from an integrated, holistic perspective (American Society of Agronomy, 2014). Crop physiology studies the structure and function of crops in relation to productivity and quality product for different uses. Also, in recent years, it is considered that efficiency of plant breeding could be increased by applying knowledge and tools from crop physiology. Plant breeding aims to adapt the genetics of plants to develop crops better adapted to human needs.

Compared to previous reviews about sunflower crop physiology, this chapter examines crop physiology and agronomy together, giving the basis for the crop management and genetic improvement of sunflower from a physiological point of view. Also, some differences among the physiology of sunflower and other crops will be specially highlighted in this chapter. First, current knowledge on sunflower physiology (crop phenology and growth) is presented. Emphasis is given to determination of yield and product quality as well as ecophysiological components underlying its determination. Second, the effect of some main abiotic and biotic stresses affecting sunflower crop growth, yield, and ecophysiological components around the world are given. Nutrients and water crop requirement are briefly reviewed. The effects of some of the main weeds, pests, and diseases affecting sunflower yield and ecophysiological components around the world are presented. Third, the basis for adjusting some main crop management practices that could help to obtain good yields and quality product with an efficient use of environmental resources in a framework of sustainable production are conferred. Finally, the main avenues for a physiological-based management and genetic improvement of the sunflower crop are given.

Physiology

Crop Phenology and Crop Growth

Several phenological scales have been established to describe the phenology of the sunflower crop (e.g., Merrien and Milan, 1992; Schneiter and Miller, 1981). These scales describe the phenological stages considering macro or micro indicators of the plant. The phenological scale from Marc and Palmer (1981) describes microscopic aspects of the apex development during the plant cycle. They are very useful for several research purposes. The phenological scale from Schneiter and Miller (1981) considers external macro indicators that make it easier to use in field conditions (e.g., for adjusting crop management practices).

Phenological stages during the sunflower cycle can be classified into vegetative and reproductive stages. As an example of some phenological stages, the scale of Schneiter and Miller (1981) is presented in Figure 3.1. The most conspicuous phenological stages during the vegetative period are emergence (occurring when cotyledons are fully expanded) and the number of developed leaves (length ≥ 4 cm). These first stages during which canopy is established are critical for crop yield (Box 3.1, p. 56).

Reproductive stages begin when a miniature floral head appears, with progressive growth until the capitulum opens and anthesis begins. Under no limiting conditions, the duration of the period of flower differentiation is between 7 and 12 days. The differentiated flowers finally develop and grow until anthesis (Figure 3.1).

The anthesis stage is completed in 10–15 days from its beginning and occurs from the outer to the inner rings of the capitulum. This stage can be divided into 10 subphases according to the percentage of flowers in the capitulum in that stage (e.g., R5.1 and R5.5 indicate that 10% and 50% of the capitulum is in anthesis, respectively). When the inner part of the capitulum presents flowers in anthesis stage, flowers from the outer rings are being fecundated. In this period, pollen formation and fecundation are crucial for the subsequent development of the grain. Potential grain weight is determined in preanthesis (Cantagallo et al., 2004; Lindstrom et al., 2006), and potential grain number is determined in the early postanthesis (Lindstrom et al., 2006). When anthesis finishes, fecundated grains start to grow and the grain filling period begins. At the beginning of the anthesis, the demand of resources by the grains is low. Then, grains start to fill and their weight quickly increases. Oil accumulation is determined during this phase (Rondanini et al., 2003) and it ends when the maximum weight per grain is obtained, at physiological maturity stage. Also, water concentration in the grains starts to decline during grain filling and continues during physiological maturity; the latter is reached when water concentration attains 38% (Rondanini et al., 2007). Physiological maturity is visually identified by the brown color of bracts in the sunflower capitula (Farizo et al., 1982).

Temperature affects the duration of all phenological stages. Their duration is often estimated as cumulated °C day. In sunflower, the base temperature for calculating daily cumulated °C day has been determined for various processes ranging from 4.0 to 8.5 °C (Aguirrezábal and Tardieu, 1996; Granier and Tardieu, 1998; Sadras and Hall, 1988; Villalobos and Ritchie 1992). Typical differences in duration of phenological stages of sunflower hybrids with different cycle durations are presented in Figure 3.2 (p. 57). Within a hybrid, differences between sowing dates affect mostly the duration of the vegetative stage. Late sowing dates shorten the vegetative stage, whereas the

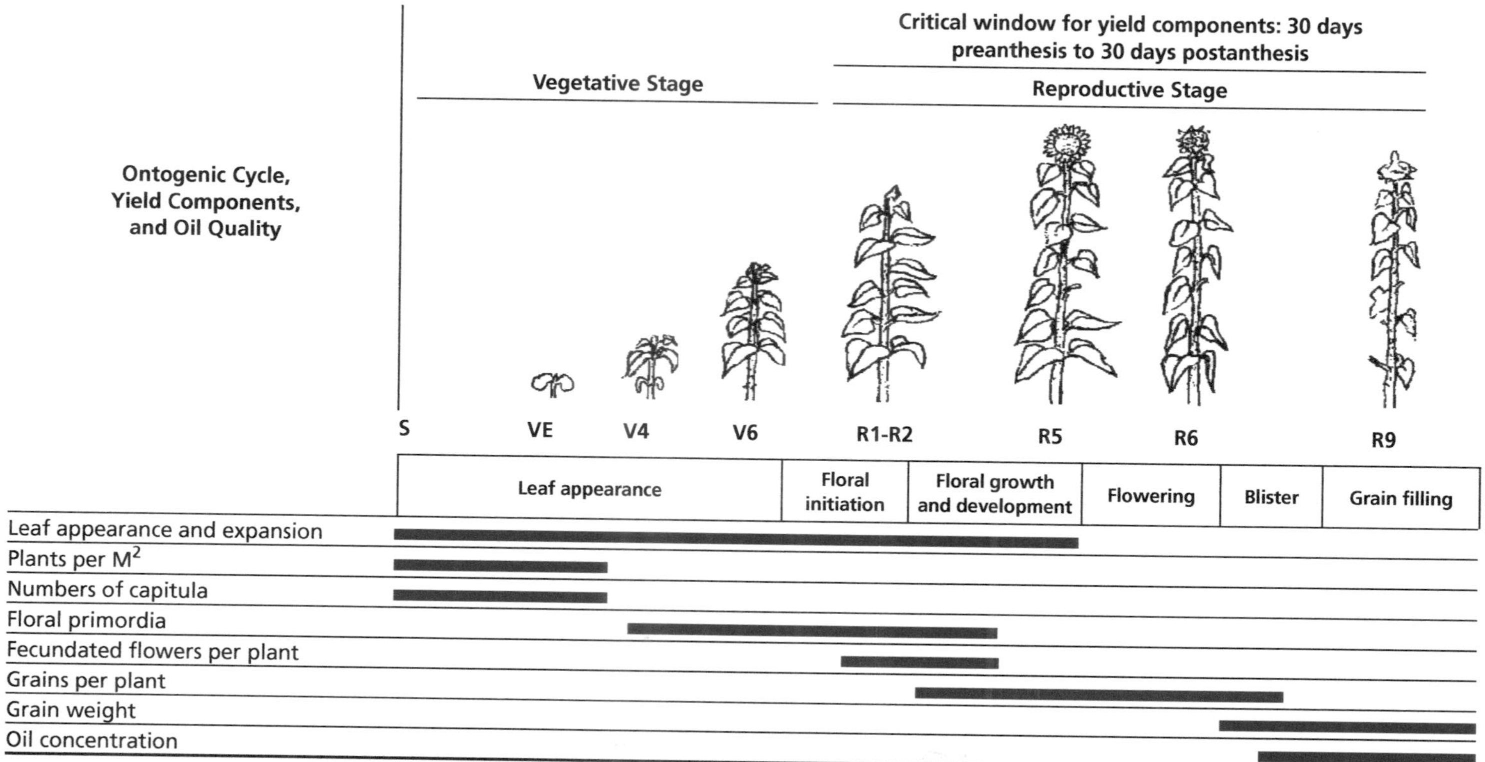

Figure 3.1 Phenological periods according to Schneiter and Miller (1981). Periods of sunflower crop cycle where yield components and oil quality are defined (30 days preanthesis and 30 days postanthesis for the yield components and 100–300 °C day period (base temperature: 6 °C) for fatty acid composition). Black lines describe the critical period for each yield and oil quality component. S: sowing; VE: vegetative emergence; V4–V6: vegetative stage determined by counting 4 or 6 true leaves (≥4 cm length); R1–R9: reproductive stages (R1: terminal bud forms a miniature floral head; R2: the bud elongates 0.5 to 2.0 cm above the nearest leaf attached to the stem; R5: the beginning of anthesis; R6: end of anthesis; R9: physiological maturity).

BOX 3.1 A Good Establishment of the Plant Stand Is the Key for Crop Performance

Germination and seedling emergence should occur quickly in order to obtain a homogenous canopy with plants of similar sizes to minimize intraspecific competence and to maximize the use of environmental resources. The longer the length of the crop establishment period, the higher the risks of seedling losses due to different environmental factors (e.g., damages by pests or diseases). Variations in the number of plants per unit area account for 25% of the variations in yield in Argentina (Mercau et al., 2001).

The environment (e.g., soil moisture and temperature) and crop management (e.g., preparation of seedbed) could affect the seed germination and seedling emergence. Some factors related to the seed (e.g., seed viability, its germination power and vigor) also affect the germination. It is known that seed composition partially affects its ability to germinate under different environmental conditions (González Belo et al., 2014; see section "Sunflower Crop Management"). Therefore, to achieve a good canopy set, it is necessary to select the sowing date according to soil temperature, to ensure water supply, and to sow seeds in an adequate seedbed.

Most leaves emerge on the shoot apex during the first stages of crop development (Pereyra-Irujo et al., 2008), until the reproductive stage is attained. Leaf appearance occurs with some delay. The main function of leaves is to capture the solar radiation and carbon dioxide of the air required for the photosynthesis process. In nonlimiting water and mineral conditions, the rates of leaf emergence, appearance, and expansion largely depend on the temperature. Base growth temperatures during the growing season are about 5 °C for leaf expansion (Goyne et al., 1982; Granier and Tardieu, 1998).

Meanwhile, roots are responsible for water and nutrient absorption. The architecture of the root system is defined in the early stages of the crop. Soil temperature affects the rate of appearance and elongation of roots in early stages (Aguirrezábal et al., 1996). The sunflower root system growth has a high sensitivity to soil strength, so poor soil preparation can affect the absorption of water and mineral nutrients during later stages of the crop cycle.

inverse response is found in the early sowings. There are also differences in the length of the reproductive stage between short and long sunflower cycles.

Sunflower response to photoperiod is less known, largely because it behaves as a short-day plant in one stage and as a long-day in another (Hall, 2001). After a juvenile phase during which plants are insensitive to photoperiod, the rate of development (calculated on a temperature basis) increases with increasing photoperiod until the stage of floral initiation, which is attained earlier during days with longer daylight hours. In contrast, between floral initiation and flowering the response is reversed (the stage would be shorter with short days). Also there are some photoperiod-insensitive genotypes (Sadras et al., 2000c).

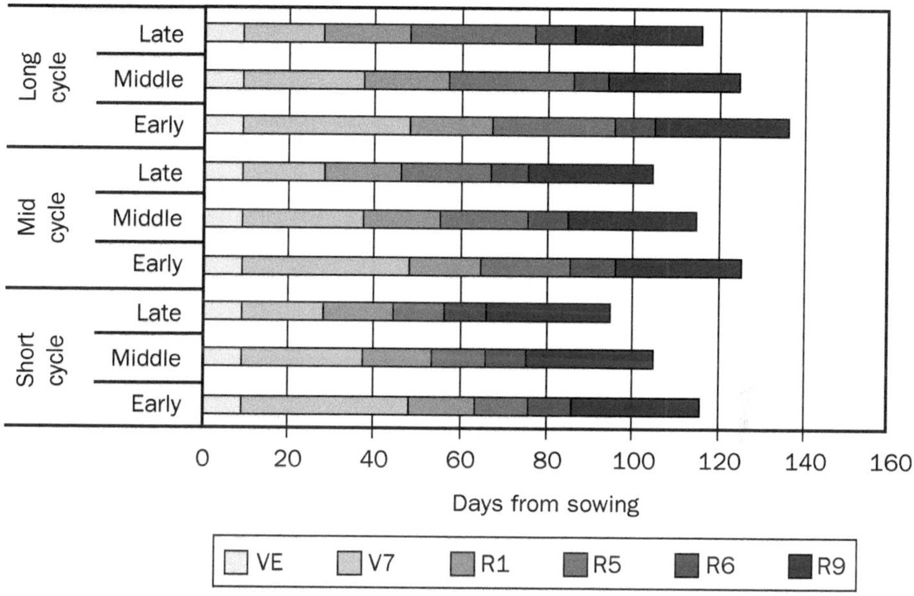

Figure 3.2 Time (days) from sowing to each phenological stage for short, mid, and long sunflower crop cycles in early, middle, and late sowing dates in Argentina (Pedraza et al., 2000). Legends correspond to phenological stages described in Figure 3.1 on p. 55.

Ecophysiological Components

Ecophysiological components are useful to understand variations in crop yield in order to achieve a better crop management and a high efficiency of genetic improvement. There are two sources of energy that can be used for biomass production: one is the incident solar radiation and the other one is water.

Regarding solar radiation, yield, as total grain weight (Y), has been defined as a function of intercepted solar radiation (IR), radiation use efficiency (RUE), and harvest index (HI) (Monteith, 1977) as follows:

$$Y = IR \times RUE \times HI \qquad \text{(Eq. 3.1)}$$

RUE is defined as crop biomass produced per unit of total solar radiation or photosynthetically active radiation (PAR) intercepted by the canopy (Hammer and Wright, 1994; Montieth, 1977; Sinclair and Horie, 1989; Sinclair and Muchow, 1999), and harvest index is the ratio of grain yield to above-ground dry matter (Passioura, 1996).

Intercepted solar radiation is modeled via the Lambert-Beer Law. Light extinction in a canopy is described as a function of two parameters: (1) the leaf area index (LAI, leaf m^2 per soil m^2) and (2) the extinction coefficient (k) (Monsi et al., 2005). The extinction coefficient k could vary depending on the species and plant architecture (e.g., maize vs. sunflower); lower k is associated with more erect leaves (Monsi et al., 2005) that can distribute light more efficiently over the layers of the leaf area within the canopy (Duncan et al., 1967). The relation between these variables could be expressed as:

$$IR = 1 - e^{-kLAI} \qquad \text{(Eq. 3.2)}$$

Water is the other source to obtain energy, as was previously mentioned. Passioura (1977) proposed that Y under water-limited conditions is related to crop seasonal water use (WU), seasonal water use efficiency (WUE), and the harvest index (HI):

$$Y = WU \times WUE \times HI \qquad \text{(Eq. 3.3)}$$

Water use efficiency is used to describe the biomass–transpiration ratio and it relates yield with evapotranspiration (Fischer and Turner, 1978; Hall and Richards, 2013; Sinclair et al., 1984; Steduto et al., 2007). The components are independent of each other, so an increase in any of them is likely to increase yield.

Variation of Ecophysiological Components and its Underlying Variables during the Crop Cycle

Evolution of IR (Eq. 3.2) largely depends on LAI evolution. LAI is the result of the interaction between the dynamics of leaf emergence, expansion, and senescence. Crop LAI changes with the development stage and the environmental conditions. It increases during the vegetative stages as leaves emerge and expand, and decreases in the reproductive stages (mainly after R6) due to leaves' senescence. Temperature directly affects crop LAI and expansion rate through its effects on the crop phenology and leaf growth rate.

The number of leaves is a genetically fixed trait and it is slightly affected by environmental factors. The rate of leaf initiation and appearance depends on the temperature. The duration of the vegetative stage, during which leaves are initiated, also depends on the temperature, decreasing with its increase. The duration of the vegetative stage also depends on the photoperiod, but its response is poorly known and the literature is contradictory about this point (see section "Crop Phenology and Crop Growth") (Hall, 2001).

The growth rate of phytomers (the unit of vegetative development consisting of a leaf and its subtended node, internode, nodal roots, and tiller bud) (Hoshikawa, 1989) is high during the first 100 °C day after leaf initiation (base temperature 4.8 °C) (Figure 3.3; Dosio et al., 2003). This phase is typically followed by a second exponential phase that includes leaf expansion. In the course of time, the duration between the appearances of each phytomer, expressed per unit thermal time, increases.

Leaf expansion is affected by water (Pereyra-Irujo et al., 2008) and nutrient (Davies et al., 2000) deficits. Water deficit alters leaf growth by reducing expansion rate (between 44% and 67%) and by increasing the duration of leaf growth (between 2% and 26%). This factor has an influence on the LAI; therefore, it has an effect in the IR (Eq. 3.2). It also has an influence in the final yields obtained by affecting WU (Eq. 3.3). Nutrient deficit, as nitrogen deficit, significantly reduces leaf area but has a small effect on leaf expansion rate compared to water deficit (Trápani and Hall, 1996).

Crop management also affects LAI evolution (Figure 3.4). For example, high plant population increase LAI and often decrease the duration of the active photosynthetic leaf area during the postanthesis stage (Aguirrezábal et al., 1996).

In the planophile sunflower crop, the critical LAI (when the canopy intercepts the 95% of the incident solar radiation) is lower (2.5–3) than in erectophile crops such as maize (4.5–5), because of its higher k. Although k (Eq. 3.2) is not strictly considered an ecophysiological component, variations thereof affect final yields mainly via its effects on IR and RUE (Eq. 3.1). It also depends on the stage of the cycle (the canopies during vegetative stages have usually higher k than during reproductive stages) and the canopy structure (associated with the position of leaves in the canopy area). In addition, crop management (e.g., plant density, Figure 3.4) could determine the time during which critical LAI is attained.

Radiation use efficiency (Eq. 3.2) varies along crop development stages in sunflower. RUE is high during the vegetative stage, due to its high efficiency of carbon dioxide fixation (particularly for C3 species). RUE decreases sharply during the reproductive stage as (1) the chemical composition of the dry matter synthesized changes (the synthesis of grain lipids requires a higher energy cost than carbohydrates or protein synthesis), (2) the leaf senescence increases (and therefore the leaf photosynthetic efficiency decreases), and (3) the radiation intercepted by lower leaf strata is low. This is accounted for the low k (Eq. 3.2) of the planophile species, resulting in low-lighted leaves, especially in the lower strata of the canopy surface (Trápani et al., 1992).

Harvest index is quite stable in sunflower as compared with other crops such as maize (Andrade, 1995). It is remarkably stable against large changes in plant population (Hernández and Orioli, 1985). Harvest index slightly varies as a result of environmental stresses applied from early stages until physiological maturity. Sunflower crops subjected to a stress period shortly before anthesis often generate a higher partitioning to the capitulum at the expense of vegetative organs (Merrien, personal communication). As it could be expected, crops subjected to environmental stresses during anthesis or grain filling showed a decrease in harvest index (Eq. 3.3) mainly accounted by a decrease in grain number or weight per grain (Alahdadi et al., 2011).

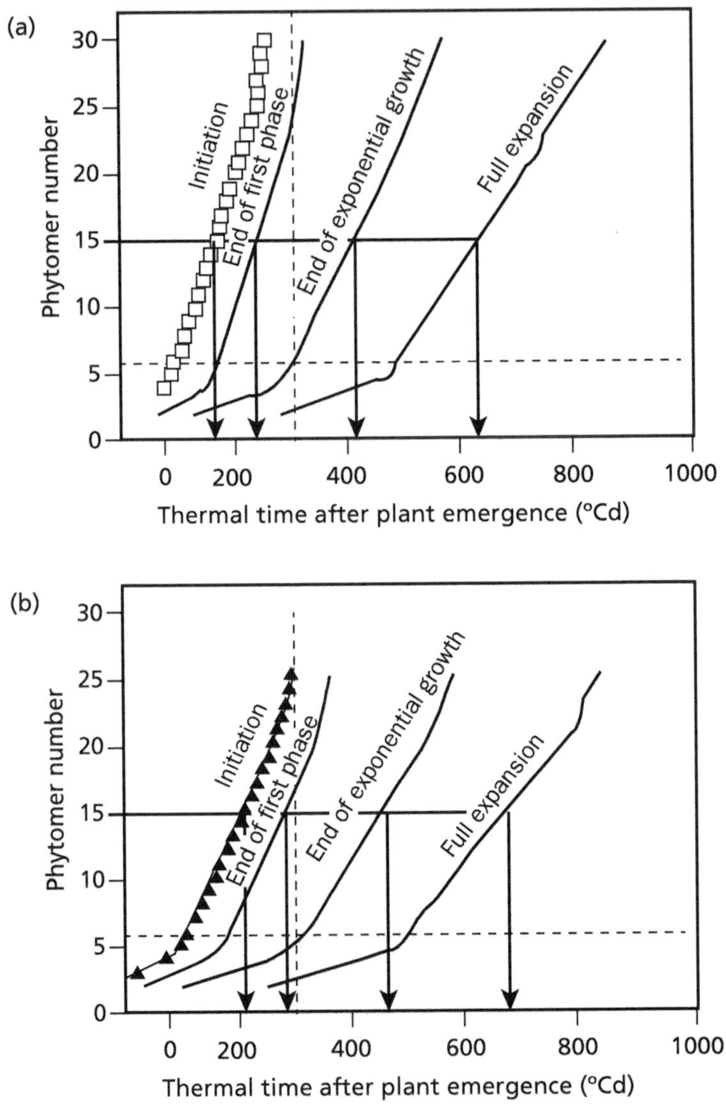

Figure 3.3 Number of initiated phytomers, number of phytomers beyond first exponential phase, number of phytomers beyond exponential growth, and number of phytomers with full expanded leaves as a function of thermal time after plant emergence in sunflower hybrids Albena (a) and G100 (b). Base temperature = 4.8 °C.

Reprinted from Dosio et al. (2003).

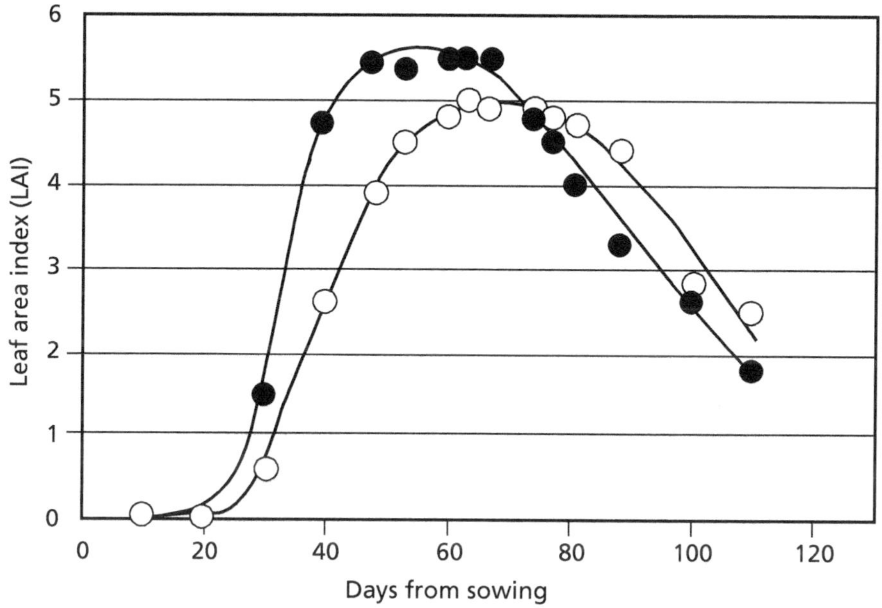

Figure 3.4 LAI evolution during crop cycle in two densities: 5.6 pl m^{-2} (empty circles) and 16.6 pl m^{-2} (full circles).

Reprinted from Hernández and Orioli (1982).

Sunflower can develop a deep root system (Figure 3.5; Dardanelli et al., 1997). The growth of sunflower root system is faster compared with leaf development. When soil resistance to root growth is low (e.g., in disturbed soil) the taproot elongation rates can attain 12 cm per day (Figure 3.5a). In undisturbed soil, the root system can attain a growth rate of 3 cm per day (Figure 3.5b).

Besides, under no limiting conditions roots can reach a depth deeper than of 3 m to extract water (Figure 3.5c). The water and nutrient absorption is more closely associated with root system architecture (the number, length, and position of each root axis) than to total root length. As root architecture is defined early during the crop cycle, poor soil tillage, associated to soil layers with high mechanical impedance, can affect crop water and nutrient absorption during later phases of the cycle (Aguirrezábal et al., 1996).

Under good soil water availability conditions, sunflower has low WUE as a consequence of the high water consumption during the vegetative stage (Eq. 3.3) and low growth rates during the reproductive stage. The high water consumption during the vegetative stage is the consequence of a fast increase in IR (because of fast LAI increase [Eq. 3.2]) and the low resistance imposed by the sunflower plant to water flow from

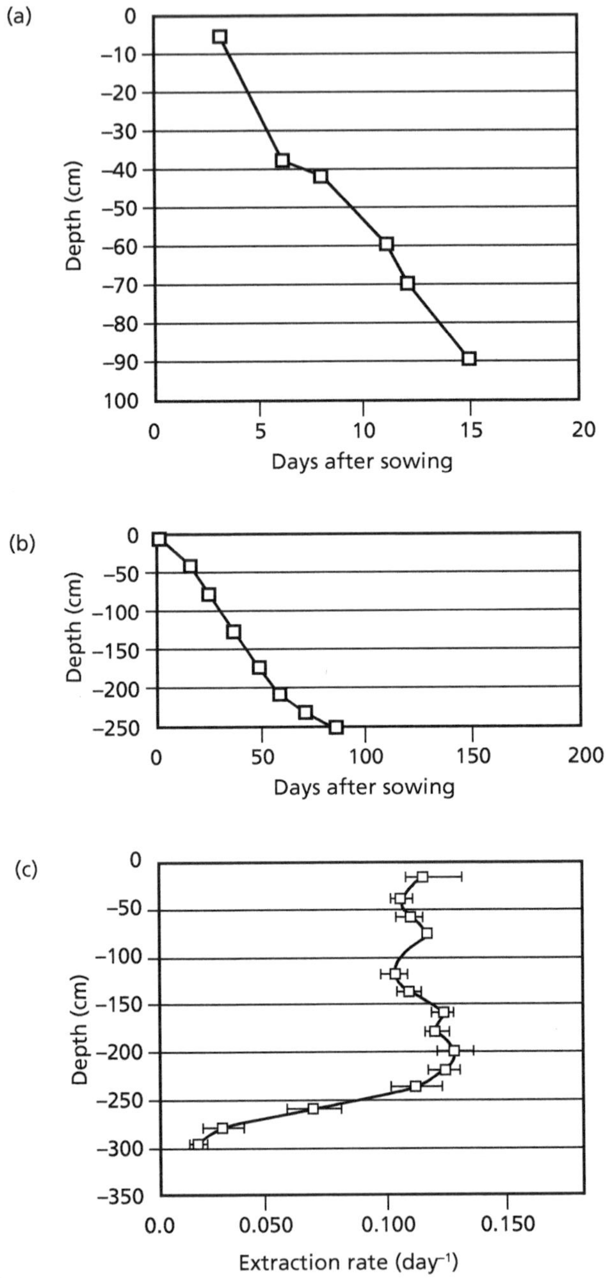

Figure 3.5 (a) Length of primary root as a function of time after sowing during the first 15 days (Aguirrezábal et al., 1996), (b) length of sunflower crop root system as a function of time from sowing during the whole cycle (Andriani, 2000), (c) and water extraction capacity root system (Dardanelli et al., 1997). Plants grew under no limiting conditions.

the soil to the atmosphere. Water use efficiency decreases during reproductive stages because of the low growth rates of the canopy during this period.

Components of Yield and Grain and Oil Quality

As in other crops, grain yield per area in sunflower is determined by grain number per area and grain weight. As an oilseed crop, oil concentration is an important oil yield component in sunflower (Box 3.2). Grain number, weight, and oil concentration are determined sequentially during the crop cycle. The phenological periods during which these yield components are determined partially overlaps (see Figure 3.1). Grain yield per area is mainly driven by the grain number per area. Grain number per area depends on the capitula number per unit area and grain number per capitulum. Because commercial sunflower hybrids produce only one capitulum per plant, the number of capitula per unit area depends on plant population, except when plants or capitula are loosed by the effect of stresses (see section "Sowing Date and Crop Establishment: Management of Sunflower When Sown as Winter or a Spring Crop"). The number of florets is determined early in the crop cycle (between 30 and 50 days after sowing). The size of the capitulum and its development rate are the most important factors in determining the number of flowers. Resources availability per plant affects the potential number of florets (Valentinuz et al., 1996). However, the grain number is infrequently limited by the florets number. As a common ecological strategy of the Asteraceae family, the number of effective grain set is lower than the floret number. The environmental conditions and the growth rate per plant during the period between 30 days before anthesis (Vega et al., 2001) and 24–30 days after anthesis are the key in determining the number of grains (Cantagallo et al., 1997).

In different crops and cropping systems, yields obtained on farm are usually lower than yields obtained by applying the best management practices (attainable yields, see Hall et al., 2013 and references therein). These authors recently quantified these yield gaps for different regions in Argentina. These gaps were in most of cases higher than 25%, suggested by Fischer et al. (2009) as the likely lower limit yield gaps in commercial farming. Variations of farm yield in Argentina highly depend on variations in grain number per unit surface (Mercau et al., 2001). Hall et al. (2013) argued that is necessary to identify constraints in the view to manage crops to decrease the gaps. Also, based on results from this work it could be suggested that in Argentina the yield gap is higher in sunflower than in other crops. Similar conclusions could be attained in France by analyzing local results (Merrien, personal communication). Identifying biotic and abiotic stresses as well as failing in management practices responsible for these gaps could be useful to improve sunflower yield.

Vegetable oils are composed mainly of triglycerides (or triacylglycerol, TAG), which consist of three fatty acids bound to a glycerol backbone. The fatty acid composition defines the nutritional, industrial, and organoleptic quality of the oil. Sunflower exhibits a high degree of plasticity in its fatty acid composition, which can be modified by breeding and mutagenesis (Haddadi et al., 2010; Fernández-Martinez et al., 1989; Lacombe and Bervillé, 2001; Lagravère et al., 2000) as well as by the interaction among genotype and environment. Oleic acid percentage in the oil is 10–50% in traditional hybrids (Aguirrezábal et al., 2009). Oleic and linoleic acids represent nearly 90% of the fatty acid content, with stearic and palmitic acids making up most of the remainder. High oleic cultivars display an increased oleic acid content of above 80%, resulting from a mutation associated with silencing of oleate desaturase gene (Alberio et al., 2013; Garcés and Mancha, 1991; Lacombe et al., 2009). Mid-oleic sunflower oil cultivars (60–79% oleic acid) are usually produced by crossing a high oleic and a traditional line. Sunflower lines with increased concentrations of saturated fatty acids (high stearic or high palmitic) were also developed (Fernández-Moya et al., 2002; Osorio et al., 1995). Inbred lines and hybrids combining the high stearic with the high oleic trait (high stearic/high oleic) have also been obtained (Martínez et al., 2012; Fernández-Moya et al., 2005; Serrano-Vega et al., 2005).

Yield and Quality Components Determination

Components of oil yield (grain number, weight per grain, grain oil concentration) and quality (fatty acid composition) are genetically determined, but they may be affected to a different extent by environmental factors that could highly vary during the crop cycle. In well water and nutrient feed crops, temperature and intercepted solar radiation play a main role in determining yield components and grain quality (Eq. 3.1 and 3.3) (Box 3.3).

As previously discussed, the grain number depends on the number of capitula and on the number of plant set per unit area. Often it is considered that the number of plants (and therefore the potential number of capitula per unit area) is set when seedlings attain the fourth leaf stage. The risk of plant (and capitulum) losses is lower after this stage (Merrien, 1983).

Main Abiotic and Biotic Factors Frequently Limiting Potential Yield

The potential yield of a crop depends mainly on incident solar radiation and temperature. Very few crops grow to their full potential, and thus the understanding of abiotic and biotic factors that limit yield is useful to adjust crop management and to assist genetic improvement with the objective to obtain attainable yields (see section "Components of Yield and Grain and Oil Quality").

A classical approach in crop physiology (and in ecophysiological-based prediction models of crop growth and yield) is to understand and estimate "actual yields"

BOX 3.2 Grain Oil Concentration: An Important Component of Sunflower Crop Yield

Most of sunflower is sown around the world to obtain oil from its grains. Grain oil concentration (usually expressed in percent of grain dry mass) mainly determines the industrial yield of the grains. As a consequence, in some countries, grains with oil concentration above a threshold are paid a premium over the regular price. Grain oil concentration is genetically determined, and plant breeding has improved it. Low-oil varieties and hybrids (38–47% oil) have been replaced by modern high-oil hybrids (47–53% oil). The improvement in oil concentration was mainly achieved by increasing the proportion of tissue in which oil is stored (Tang et al., 2006) rather than by increasing the oil concentration in this tissue. The difference in oil concentration between low-oil and high-oil hybrids is caused by differences in the duration rather than the rate of oil accumulation (Izquierdo et al., 2008; Mantese et al., 2006). Grain oil is synthesized from carbohydrates either from current photosynthesis or from the remobilization of storage carbohydrates (Hall et al., 1990). Therefore, grain oil accumulation greatly depends on the carbon economy of the crop during grain filling (Andrade and Ferreiro, 1996; Dosio et al., 2000). In sunflower, grain oil concentration is mainly determined by the amount of the photosynthetically active radiation intercepted per plant during the grain filling period (Andrade and Ferreiro, 1996; Dosio et al., 2000; Izquierdo et al., 2008).

Proteins represent the main subproduct of the sunflower oil industry and its concentration in the grain varies between 15% and 19%, depending on the oil content (oil and protein concentration are inversely related); so, there is also some interest in protein content. The composition of oil and proteins determine sunflower grain quality. Sunflower is also sown in a minor scale as a confectionery crop. Varieties sown for that purpose produce grains with lower oil concentration than oil crop varieties (Campbell, 1983; Mantese et al., 2006). For confectionery sunflower, yield and quality product are mainly determined by the grain size and uniformity and not by its oil concentration.

considering the effects of abiotic and biotic factors that affect ecophysiological components and limit the obtaining of attainable yields. Crop yields are influenced by a large variety of abiotic and biotic factors. Many of these factors can affect the growth, development, yield, and/or the quality of the crop. However, sunflower crops are limited only by a few environmental factors at a global scale. In this section, we will briefly review some of these factors, trying to identify the ecophysiological components more frequently affected by each factor.

Environmental factors are classified as abiotic and biotic factors. The interactions between biotic and abiotic factors on sunflower yield are not presented in this review.

Box 3.3 Critical Periods for Yield and Quality Components

The relative effects of environmental factors on a given component vary according to the moment of the crop cycle because oil yield components and grain quality are defined in different phenological stages.

The number of seeds per unit area, which is the main yield component in cereals and oilseed crops (Cantagallo et al., 1997), is closely related to the crop growth rate during the critical period for grain set (Villalobos et al., 1992). The critical period for the seed number determination is considered between 24 and 30 days after anthesis (Cantagallo et al., 1997). During this stage, grain number per capitulum is determined.

During grain filling, grain weight and grain oil concentration are determined. Aguirrezábal et al. (2003) defined the critical period for these components as a function of cumulative intercepted radiation during the 250–450 °C daa (base temperature 6 °C) (see Figure 3.6 on the following page for oil concentration). The grain filling period is the only stage during which the determinations of the three oil yield components are partially overlapped. So, environmental conditions during this period highly affect yield.

Water Requirements: Tolerance to Water Deficit and Related Physiological Traits

Sunflower crop presents a high transpiration rate under no water limiting conditions. In the Western Pampas, maximum cumulative seasonal crop evapotranspiration for sunflower was close to the order of 630 mm, corresponding to an estimated yield of near 5000 kg ha^{-1}. Support for this assumption derives from different approaches used by Grassini et al. (2009) and supported by other bibliographical sources. Under this maximum value, a maximum water productivity observed was 8 kg grain ha^{-1} mm^{-1}. Many crops did not reach the expected yield for the amount of seasonal water that was available because of (1) the timing of the water availability for the crop (the crop was under water deficit during the critical periods for yield, see Box 3.3) or (2) other environmental factor were limiting yield.

Water stress is the major limitation to sunflower yields worldwide. However, the crop presents several features to mitigate water stress and yield losses under water deficit. An example is the sequential development of flowers that gives the crop some plasticity in supporting short stress periods (Aguirrezábal et al., 1996). Also, there are reports of osmotic adjustment under drought stress and genetic variability among inbred lines of sunflower. Higher osmotic stress expressed by F3 families, obtained from the crosses of contrasting genotypes selected after the screening of 25 inbred lines, accounted for a higher water extraction from the soil profile during the

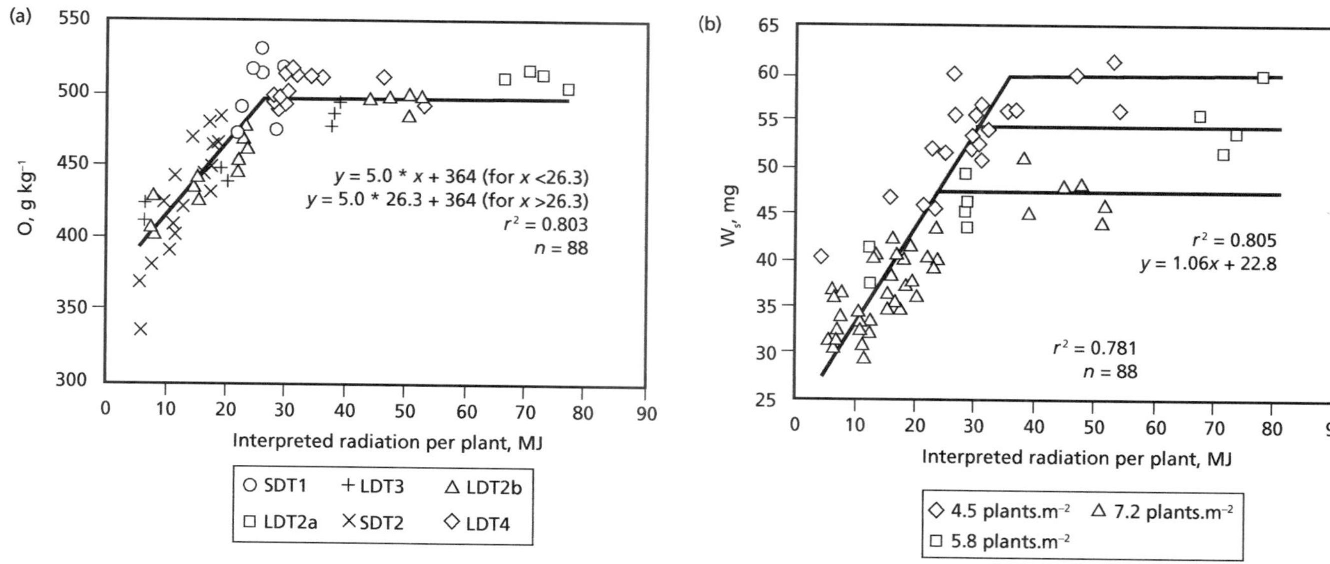

Figure 3.6 Oil concentration (O) and weight per seed (Ws) as a function of cumulative intercepted radiation (PAR) per plant from 250 to 450 °C daa (base temperature = 6 °C) for six short and long duration experiments (SDT1, SDT2, LDT2a, LDT2b, LDT3, and LDT4). Line represents the adjusted function. The regression in (b) represents the relationship established with intercepted radiation <37MJ plant^{-1} and the horizontal lines represent the maximum W, calculated for each plant density ($W_s = 81.64 - 4.75x$). Data points represent values obtained in each experimental plot.

Reprinted from Aguirrezábal et al. (2003).

stress period, a greater shoot biomass and higher HI and grain yield at physiological maturity (Eq. 3.3) (Chimenti et al., 2002). The deep root system also helps to avoid water deficit when water is available in deeper soil layer (Figure 3.5b and c). Sunflower leaf wilting contributes in a genotype-dependent manner to the response of whole-plant transpiration rate to soil drying by changing the shape of the response function. A significant part of the genotypic difference in this response was explained by wilting. This mechanism allows sunflower plants to conserve more soil water and protects them from thermal stress and high radiation loads (Velázquez et al., 2012). Sunflower also reduces HI in a lower extent compared to other crops (Andrade et al., 2000). Also, under water stress conditions, WUE (Eq. 3.3) increases as a consequence of a decrease in water consumption. Recently, nine QTLs controlling WUE under drought scenarios were reported (Adiredjo et al., 2014). This could represent the basis for the improvement of WUE and Y (Eq. 3.1 and 3.3).

Salt Stress

Salinity is a serious constraint to agricultural productivity. Most of crops are sensitive to this condition (Maas and Hoffman, 1977), and sunflower is not an exception to this rule. This species has been rated as moderately salt-tolerant (Katerji et al., 2000). Although the best option is not cultivating under saline conditions, the area of soils affected by salinity is growing with the expansion of irrigation (Wang et al., 2003). Genetic variability for saline tolerance has been reported in sunflower (Ashraf and Tufail, 1995). Osmotic effects of salinity are observed immediately after stress imposition, compared to nonexposed sunflower. This effect is evidenced by a sharp decrease of leaf expansion which results in a decrease in LAI and IR. Other effects of salinity on sunflower physiology (excess of ions affecting nutrient uptake and balance and high buildup of potentially toxic ions) progress later and they can also affect other physiological mechanisms, like floral development and leaf senescence (Céccoli et al., 2012), underlying the RUE and dry mass partitioning (see section "Ecophysiological Components").

Three concurrent and nonexcluding general mechanisms contribute to saline tolerance: osmotic tolerance, ion-uptake control and compartmentalization, and tissue ion tolerance (Munns and Tester, 2008). Recently, Céccoli et al. (2012) found that variability for saline resistance mechanisms exists in sunflower. Osmotic tolerance and tissue Na^+ tolerance were detected in different inbred lines, highlighting that such variability may be exploited for increasing saline resistance in sunflower crop. Interestingly, the inbred lines better ranked for tolerance to leaf expansion to osmotic stress of salinity were the same than those better ranked for water deficit tolerance (Pereyra-Irujo et al., 2008), suggesting that common genetic improvement for tolerance to both abiotic stresses could be performed in some cases.

Nutrient Requirements: Criteria for Optimizing Soil and Fertilizer Nutrient Management

Sunflower requires six macronutrients (N, K, P, Ca, Mg, and S) and seven micronutrients (Fe, Mn, Zn, Cu, B, Cl, and Mo). Uptake depends on root exploration, soil water content, and nutrient availability, rather than total nutrient content in the soil.

As for other crops, N, K, and P are the major macronutrients required by sunflower. Considerable quantities of these macronutrients are required to support crop growth. The greater proportion is absorbed in the first crop vegetative stage; the crop reproductive stage is characterized mostly by redistribution of nutrients among different organs of the plant. Each nutrient has an individual pattern of uptake and redistribution and different proportions of each nutrient are removed by harvest of seeds.

For a sunflower crop cultivated under well-watered conditions, 41.3 kg N ton^{-1} biomass and 26.8 kg N ton^{-1} grain were required (Andrade et al., 2002). Nitrogen is involved in development and growth of leaves and florets. Nitrogen deficiency in early vegetative stages may reduce the leaf expansion (Trápani and Hall, 1996), significantly affecting LAI (Eq. 3.2), and therefore yield, by reducing IR (Eq. 3.1). Higher N deficit could also reduce RUE because most of foliar N is a constituent of the photosynthetic enzymes (mainly of Rubisco) and chloroplasts. Nitrogen concentration in plants decreases with the increase of plant biomass (by dilution), so the N nutrition index considers the relationship between the actual tissue N concentration and the N concentration required for maximum crop growth rate (Figure 3.7) (Debaeke et al., 2012). Ensuring an adequate supply of N before floral initiation affects mostly the number of grains (by reducing HI; Eq. 3.1 and 3.3) and root biomass. Late applications only partially modify the weight per grain, affecting mainly the protein concentration and decreasing its oil concentration. When fertilization takes place at early stages of crop development, it can stimulate lush biomass development affecting LAI (Eq. 3.2) after flowering by proliferation of diseases or excessive water consumption that limits their normal provision during grain filling in conditions of limited water input. An excess of N availability can cause a reduction in grain oil concentration and oil quality without increasing seed yield (Merrien and Milán, 1992). Fertilizing rates must then be adequately adjusted to avoid N availability for the crop that is higher than the optimum amount. Recently a specific sunflower N critical dilution curve (Figure 3.7) has been established. It is usefull for accurate diagnosis of crop N status and the fertilization management. It would also facilitate applications of crop growth simulation models in crop management (Debaeke et al., 2012).

Sunflower is very sensitive to variations in K. Potassium has the major role of maintaining the turgor and cell extension in sunflower organs, particularly stems, leaves, and roots (Connor and Hall, 1997), which affects final LAI. Consequently, yield increases by increasing K (Abbadi et al., 2008). Estimated K requirements for a sunflower crop were 28.8 kg K ton^{-1} biomass and 6.8 kg K ton^{-1} grain (Andrade

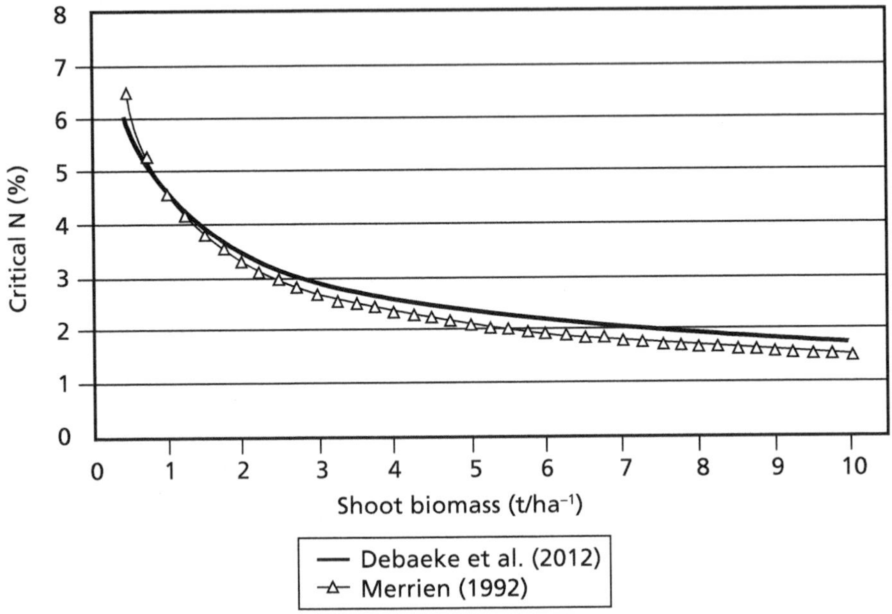

Figure 3.7 Sunflower N critical dilution curve. Data of Merrien and Milan (1992) are based on a limited dataset, and data of Debaeke et al. (2012) are based on a larger dataset.

et al., 2002). Potassium deficiency is associated with 11 to 24 g kg^{-1} K in leaf tissue, while less than 19 g kg^{-1} K in the upper leaves is associated with yield decline (Blamey et al., 1997).

Phosphorus is required in smaller amounts than K and N. For well-watered sunflower, 6.1 kg P ton^{-1} biomass and 3.8 kg P ton^{-1} grain are required (Andrade et al., 2002). A P concentration of 2 g kg^{-1} in the plant appears critical for sunflower growth (Blamey et al., 1997). This nutrient is an essential component in metabolically active cells. However, sunflower plants grown in P-enriched media enhanced biomass (Padmanabhan and Sahi, 2011), increasing LAI (Eq. 3.2) and yield (Eq. 3.1) (Amanullah and Khan, 2010). If P is available in the soil, it could be uptaken by sunflower in later phenological phases (Connor and Hall, 1997). There is extensive mobilization of P within the plant (as it occurs with N); the stem, roots, or leaves are important sources of this nutrient for seed development under restricted P uptake conditions.

Calcium deficiency is evident as a reduction in stem elongation and wilting leaves. Magnesium deficiency is observed as interveinal chlorosis and eventually necrotic patches. Sulfur deficiencies appear mainly with chlorosis in young leaves, delay anthesis, and reduce grain number per plant because soils where sunflower is cultivated are not generally deficient in Ca, Mg, and S. For sunflower crop, plant deficiencies of these three macronutrients are not frequent and also they should not limit sunflower production.

Micronutrients are required in small amounts (lower than 0.028 kg K ton^{-1} biomass and 0.03 kg K ton^{-1} grain) (Andrade et al., 2002). They are also infrequently deficient in sunflower crop, but when recognized, the deficiencies can be rapidly corrected with fertilizer. Of all micronutrients necessary for sunflower crop, B is the most frequently deficient. Deficiencies of B produce malformed leaves (consequently LAI is reduced; Eq. 3.2) and nonviable seeds, but the most frequent symptom appears in anthesis with the "knife cut" when the whole capitula is loosed (consequently HI and yield is reduced; Eq. 3.1 and 3.3) (Connor and Hall, 1997).

Low and High Temperatures

The effect of nonstressing temperature (typical range 8–27 °C) on ecophysiological components underlying the determination of sunflower yield and quality has been presented in previous sections. In this section, the effect of temperatures outside this range will be presented. The tolerance of freezing and low temperatures during the seedling stage is especially significant when early sowing dates are selected for maximizing the IR during the critical periods when seeds are exposed to low soil temperatures. Also, for typical sowing dates in a given region, it could help with good establishment of the crop (Box 3.1) despite climate variability (e.g., when cold periods occur during the early growth season). Cold and freezing temperatures cause mechanical damages and affect many physiological processes (e.g., photosynthesis, cell membrane stability) (Xiong et al., 2002). As a consequence, the survival, growth, and reproduction of plants are affected by cold temperatures. In some cases, a heterogeneous stand of plants is obtained because of seedling death and losses of apical dominance of survival seedlings. Resistance to low and freezing temperatures is associated with the accumulation of osmotic active substances that reduce the osmotic potential by decreasing the freezing temperature of the tissues (Levitt, 1980). Also, the transcriptome from different sunflower hybrids of early and late anthesis under low temperature were similar (Hewezi et al., 2006). A study about physiological changes on leaves from plants of four young sunflower genotypes (evaluated in growth chambers) revealed some interesting differences (Hejnák et al, 2014). Despite the importance of this topic, studies about the variability of the resistance of a wide number of sunflower genotypes to low temperatures are still lacking. As for other plant species, this could be explained by the fact that reproducibility of these kinds of studies are low due to the confounding effects of the conditions prior to the exposure of seedlings to low temperatures, the timing, the intensity, and the duration of the stress period, as well as their interactions.

Low temperature during reproductive stages of sunflower can affect ecophysiological components underlying yield determination. Harvest index follows a similar linear increase with time regardless of genotypes, sowing date, N level, and solar irradiation level. However, this relationship is not stable at temperatures lower than 9 °C during grain filling (Bange et al., 1998). These authors suggested that this could be due to possible limitations to biomass accumulation and translocation.

Climate models predict an increase in the mean global surface temperature of about 1.0–3.5 °C in the next century and an increase in the frequency of episodes of high temperature (Christensen et al., 2007). Sunflower crops could also be exposed to periods of high temperatures, especially during the reproductive stage, both in temperate and subtropical regions. On sunny days, grain temperature is 5 °C warmer than air temperature (Ploschuk and Hall, 1995). This can be mainly explained because sunflower heads are exposed to direct sunlight. Exposure of sunflower capitula to constant temperatures over 34 °C for seven days reduced grain weight and quality (Rondanini et al., 2003). Interestingly, Rondanini et al. (2006) applied alternate day/night temperature treatments that simulated the temperature regime of sunflower grains and capitula in the field. During the grain filling period, four days of alternating day/night temperatures, resulting in mean daily grain temperatures of over 30 °C, can reduce sunflower grain yield and quality depending on the time of exposure. The effects of high temperature on grain weight and oil percentage trend to decrease with the increase of time after anthesis. In contrast, fatty acid composition tends to recover when high temperatures are applied in the early and middle, but not late, grain filling periods. Some experiments similar to those of Rondanini et al. (2006) carried out in a temperate region must be performed in subtropical regions to confirm the temperature threshold found in this work. In some cases, sunflower crops exposed to higher temperatures than those applied by Rondanini et al. (2006) during the whole grain filling registered high yields and oil percentage (Angeloni, personal communication); this suggests that a genetic variability could exist in the tolerance of grain exposure to high temperatures, or that sunflower plants grown under high temperature during most of the cycle could acclimate, increasing their tolerance.

Also, yield can be decreased via the effects of high temperatures on ecophysiological components (see section "Ecophysiological Components"). For instance, RUE could be expected to decrease under high temperatures because it frequently affects photosynthesis to a higher extent than respiration. Working in a very warm environment, Kalyar et al. (2013) recently described an interesting new trait that allows the increase of leaf tolerance to high temperature. Genotypes with upward inclined leaves were more tolerant than those with downward inclined leaves as they maintained lower post-noon temperature and heat injury. More recently they also proposed that the leaf and capitulum orientation at anthesis is a mechanism of resistance to heat stress in sunflower (Kalyar et al., 2014a, 2014b).

Crop Biotic Stresses

There are many biotic stresses that can affect sunflower yield. However, only a few works have considered how the ecophysiological components are affected by these stresses (e.g., Quiroz et al., 2014; Sadras et al., 2000a). In Table 3.A, some of the most important biotic stresses affecting sunflower crop around the world are presented. They are grouped according to the effect they produce on yield determination. These groups are: (1) biotic stresses that affect the whole plant, so the number of plants per area and the number of grains per area are reduced; (2) biotic stresses that affect green leaf area (by reducing LAI or RUE; Eq. 3.1 and 3.2) resulting in grain production with low weight and also yield losses; and (3) biotic stresses that directly affect the capitulum, in which grain number and/or grain weight are reduced, reducing the HI (Eq. 3.1 and 3.3).

When the stress is a disease, identifying the causal agent and knowing its lifecycle are essential for deciding the most adapted management to reduce the damage. The strategies orientated to avoid the impact of diseases will depend on the disease characteristics, but they can be summarized as: (1) controlling the host through the use

Table 3.A Most Important Biotic Stresses of Sunflower Crop and Their Effects on Yield Determination

Effect on Yield Determination	Biotic Stress	Causal Agent
Effect on the whole plant	Pathogen	*Sclerotinia sclerotiorum* *Verticilium dahlia* *Alternaria helianthi* *Phomopsis helianthi* *Plasmopara halstedii*
	Insects	Lepidopteras (*Agriotes* spp., *Agrotis* spp., *Euxoa* spp., *Feltia ducens, Suleima helianthana*) Coleopteras (*Apion occidentale, Baris strenua, Cylindrocopturus adspersus, Dextes texanus, Lethrus apterus, Ligyrus gibbosus, Opatum sabulossum, Psalidium maxillosum, Pseudoheteronyx* spp., *Pterohelaeus* spp., *Tanymecus dilatiollis*) Dipteras (*Strauzia longipennis*) Orthopteras (*Acheta deserta, Lacolampra* spp.)

(Continued)

Table 3.A *Continued*

Effect on Yield Determination	Biotic Stress	Causal Agent
Effect on green leaf area	Pathogen	*Verticilium dahlia* *Phoma oleracea* *Albugo tragopogonis* *Puccinia helianthi*
	Insects	Orthopteras (*Cammula pellucida, Melanoplus* spp.) Homopteras (*Aphis helianthi, Brachycaudus helichrysy, Masonaphis masoni, Vanessa cardui*) Coleopteras (Diabrotica speciosa, Zigogramma exclamationis) Lepidoteras (*Chlosyne lacinia saundersii, Loxostege sticticalis, Mamestra brassicae, Plusia sp., Rachiplusia nu*) Hemipteras (*Galeatus helianthi*) Thysanopteran (*Frankliniella* sp., *Thrips* spp.)
Effect on the capitulum or grains	Pathogen	*Sclerotinia sclerotiorum*
	Insects	Dipteras (*Acanthiophilus helianthi, Contarinia schulzi, Gymnocarena diffusa, Neothephritis finalis*) Coleopteras (*Cyclocephala melanocephala, Haplothynchites aeneus, Smicronyx* spp.) Lepidopteran (*Cochylis hosoes, Helicoverpa* spp., *Homoeosoma* spp.) Hemipteran (*Eurydema ventral, Lygus rugulipennes, Nezara viridula, Nysius* spp.)
	Birds	Sparrows (*Passer domesticus* L., *P. hispaniolensis* L.) Crows (C. splendens) Parrots (P. *eupatria*) Doves (*Zenaida auriculata, Streptopelia decaocto* L., *S. turtur*) Parakeets (*Psittacula krameri Scopoli*)

of cultivars with resistance or high tolerance to the disease, and (2) controlling the source of the pathogen through the use of certified seeds and crop rotation.

For some diseases, breeding has been successful for improving plant resistance, as is the case with Verticilium (De la Vega et al., 2007; Sadras et al., 2000b) or downy mildew (Delmotte et al., 2012). However, the appearance of new races implies continued research in some cases (Delmotte et al., 2012).

When the stress is caused by insects, a periodic pest monitoring is necessary to identify the species and know the intensity of the infection. According to the intensity and the crop stage (Critical Periods, Box 3.3), the adequate chemical control can be chosen. For insects that attack the seed during the emergence, it is best to use certified and chemically treated seeds. When birds attack the crop, control is much more difficult; no effective control techniques have been developed yet.

Weeds are other biotic adversities that can reduce sunflower yield. Their effects mainly occur via competition for nutrients and water, and in minor cases for light. They can reduce yield and produce poor plant growth by affecting LAI (Eq. 3.2) and HI (Eq. 3.1 and 3.3). Chemical control is more effective for annual than for perennial weeds. A particular parasitic weed that is very important for sunflower crop is Broomrape, which is produced by *Orobanche cumana* Wallr. Seeds of this weed germinate and infect sunflower roots absorbing the plants' water, nutrients, and carbon. This infection can easily produce yield losses of 50%, and severe infections can result in total yield loss. This parasite is mainly distributed in Europe and Asia (Miladinović et al., 2012). Chemical control can be used, but genetic resistance is constantly being investigated because new races of this weed have appeared (Delmotte et al., 2012).

Sunflower Crop Management

Selecting the Hybrid to Be Sowed

Although it is often considered trivial, the selection of the hybrid that will be sown is a key decision in crop management. This involves choosing potential aspects of the future crop, such as: (1) the yield, (2) the percentage of oil, (3) the oil quality and other grain quality traits, (4) the cycle, and (5) the tolerance and resistance to different diseases. As discussed here, many of these characteristics will interact with other management practices, which largely determine the performance of the crop.

Sowing Date and Crop Establishment: Management of Sunflower When Sown as Winter or a Spring Crop

The sowing date defines the environmental conditions to which the crop will be exposed in key moments of its developmental cycle (e.g., critical periods for yield and quality components, Box 3.3). Indeed, different sowing dates might subject the crop to different environmental conditions during grain filling and oil synthesis (Connor and Sadras, 1992) determining different phenological stages depending on the temperature, radiation and day length. It is important to match the critical periods for grain number, grain weight, and oil percentage with the moment of the growing season during which the conditions are most favorable for incident radiation, temperature, and other environmental factors. Selecting different sowing dates requires a coherent selection of plant density and genotype to maximize the use of environmental resources during the growing season.

In high latitude regions, early planting dates of hybrids with an appropriate length of the cycle allow critical stages to occur when incident radiation levels are high (near the summer solstice) (Pereyra-Irujo and Aguirrezábal, 2007). Also, in some environments, early sowing dates help avoid terminal drought because grain filling occurs before the water deficit period (Hejnák et al, 2014). However, in high latitudes, good crop establishment is often more difficult to achieve with early sowing dates because low temperatures could decrease the rate of germination and seedling emergence. Crop establishment could be still more difficult in early sowing dates when using a no tillage system with a high quantity of stubble in the soil surface because of even lower soil temperatures in the seedbed (Fabrizzi et al., 2005). Under such conditions, better seedling establishment could be obtained by using row stubble cleaner. Overall, sowing seeds with high vigor are essential to decrease the risk of bad seedling establishment in early sowings in high latitude environments where temperatures during the spring are often cold. Under these conditions, seeds with high unsaturation levels have the advantage of germinate faster than seeds with less unsaturation because of their low base temperature (González Belo et al., 2014).

The advantage of decreasing the risk of a bad crop establishment by delaying the date of sowing depends, however, on the magnitude of the delay in the sowing date and on the potential length of the growing season in the location where the crop is cultivated. Late sowing dates often help to obtain a better crop establishment but hasten the development because the crop is subjected to higher temperatures during the vegetative stage. When the vegetative stage is hastened, the crop cycle is shortened, so the LAI (Eq. 3.2) and cumulative IR during the whole crop cycle decrease and consequently the crop biomass per unit area decrease (Eq. 3.1) (De la Vega and Hall, 2002). Later sowing dates reduce oil yield (Roche et al., 2004; Zheljazkov et al., 2011) of high oleic and conventional cultivars (Barros et al., 2003; D'Amato and Giordano, 1992) due to a decrease in the number of grains per area, grain weight, and oil concentration (Andrade, 1995; Pereyra-Irujo and Aguirrezábal, 2007).

In latitudes where the crop is sown at the end of the winter (e.g., latitudes lower than 27°) delaying sowing date exposes the crop to higher temperatures during initial vegetative reproductive stages. The length of the vegetative period could be highly reduced. Photoperiod-sensitive hybrids could be important to maintain a stable flowering time and cycle duration in these subtropical environments. Unfortunately, the knowledge of the response of sunflower phenology to photoperiod is still fragmentary, and these kinds of genotypes are not still available in the market (De la Vega et al., 2001). Also, early sowing dates and appropriate selection of hybrid cycle duration are essential in these environments to avoid high temperatures during grain filling (see section "Crop Phenology and Crop Growth"). Indeed, high temperatures at the beginning of grain filling cause a decrease in grain weight and oil concentration (Rondanini et al., 2003). In general, long-cycle hybrids are favored for early sowing dates, depending on the length of the season, location, and weather conditions, and show the least yield reduction when the sowing date is delayed (López Pereira, 2010).

In addition to the effects of sowing date on ecophysiologycal components and oil yield, the response of fatty acid composition to sowing dates depends on the latitude. In low latitudes the oleic acid increases when sowing dates are delayed, and in high latitudes it remains almost constant within the season (Izquierdo et al., 2006; Pereyra-Irujo and Aguirrezábal, 2007). In this way, both temperature and IR (Eq. 3.1) modified independently but additively the oleic acid concentration in traditional sunflower crops (Echarte et al., 2010).

Plant Population and Row Spacing: Management of Canopy Structure

Optimum density and row spacing depend on the environmental conditions during the crop cycle, management practices, and hybrid response. A wide range of plant densities for the achievement of optimum yields in sunflower can be found in literature (e.g., Andrade, 1995; Villalobos et al., 1994). For instance, a plateau for yield response to plant density from 3 pl m^{-2} up to 7 pl m^{-2} has been described for a large region of the Argentinean Pampas (Aguirrezábal et al., 1996). This response to plant density is evidence of a high plasticity of sunflower to adapt the plant size to available resources. As for other crops, the optimum plant density in sunflower is influenced by several factors such as temperature, soil fertility, water availability, and genotype (Diepenbrock et al., 2001; Villalobos et al., 1994). However, in a given environment, the high plasticity of sunflower (compared to several crop species) allows it to obtain the maximal yield in a wide range a plant population.

Intraspecific competition to capture the essential resources for growth affects the structure of crops. From the moment when plants start to compete, plant density is higher than the optimum, causing a reduction in the plant size that is mainly reflected in the leaf area per plant. Canopy structure is also affected by the variation of the density. At higher densities, the k coefficient is lower than at lower densities (associated with erectophile and planophile plants; see Eq. 3.2) (Monsi et al., 2005).

The effect of the plant population leaf senescence increases when density increases because radiation is more attenuated and the red/far red ration is lower in lower strata of the canopy (Rousseaux et al., 2000), affecting LAI and IR (Eq. 3.1).

Plant density is a factor that also impacts head growth as well as the yield components and the grain oil percentage. Crops with low populations (and high plant growth rates during the critical periods) in general present high grain numbers per plant, but the response of this yield component to plant population is not linear because there is a morphogenetic restriction to generate grains under favorable conditions (Vega et al., 2001). Even when IR is maximized during the critical periods (Box 3.3), plant population previous to anthesis determines the maximum grain weight (plateau value of Figure 3.6) (Aguirrezábal et al., 2003).

Another important aspect is that at low densities, WUE and yield tend to decrease. In contrast, high densities increase oil concentration, WUE, and Y; however, the protein concentration is reduced.

The distance between rows can strongly affect the crop yield by affecting IR. Reducing the distance between rows at constant density reduces the intraspecific competition for light, water, and nutrients during early stages of the crop cycle. The optimum spacing is that which maximizes light interception. However, the increase or decrease of yield components depends not only on row spacing but also on the hybrid cycle. Long-cycle hybrids achieve the maximum IR when flowering in wide rows, whereas short-cycle hybrids achieve the lowest IR (Andrade et al., 2002). Yield increases in response to narrow rows in short-cycle hybrids, whereas in long-cycle hybrids yield showed no response (Andrade et al., 2002). The canopy with smaller plants from short-cycle hybrids, as a result of reducing rows, do not reach the maximum IR during the critical period for grain number (Box 3.3). When low size per plant is expected (e.g., sowing short cycle hybrids at early dates by applying no tillage), sowing in narrower rows could help to attain maximum IR, to increase early vegetative biomass production, and to improve the ability of the crop to compete with weeds for environmental resources compared to sowing in larger rows.

Management Practices to Improve Water and Nutrient Availability for the Crop

Most of management practices to improve water availability for the rain-fed sunflower crop have the goal to accumulate water in the crop before and after crop sowing. Those practices are: (1) selecting a plot to cultivate the soil with an adequate soil depth and without compacted soil layers to profit a high soil exploration by the roots, (2) preparing the seedbed to obtain a good contact between the soil and the seed to promote root growth, (3) selecting an adequate preceding crop, (4) performing fallow as long as possible and keeping it without weeds, and (5) obtaining a good soil coverage with stubble when cultivating the crop under no tillage. Most of these practices are targeted toward obtaining a higher WU or WUE (Eq. 3.3).

Adjustments of sowing date, plant population, and row spacing, as well as the selection of hybrid cycle, are also important agronomical practices to improve the performance of rain-fed sunflower crops when sown in environments with middle and high probabilities of facing drought periods. Combining an adequate sowing date with a suitable crop cycle could avoid water deficit during critical periods for oil yield and quality (e.g., terminal drought affecting grain filling; Box 3.3). Managing sunflower crops to attain at early stages a LAI of 2.5–3.0 (that allows to maximize IR (Eq. 3.2) and yield) is a goal in scenarios in which no water deficit or short water deficit periods are expected. In order to increase harvest security, plant density is decreased and/or row spacing is increased to delay the moment when high LAI is reached. These are practices to be considered when long periods of water deficit are expected.

The use of low or no tillage systems has been increasingly applied in the last years in several regions of the world. Some of the main goals of these reduced tillage systems is to increase and stabilize crop yields and sustain agricultural production. Also, the accumulation of residues on the surface (frequently associated with no tillage system) favors greater water storage in the soil (e.g., during fallow) and diverts to crop transpiration a larger fraction of evapotranspiration. Only a few scientific works have compared sunflower performance under both systems. In most of cases, the main advantage seems to be the increase of the sustainability of the agricultural production. Also, a high level of soil water storage during the fallow period is often obtained. For instance, in a dry region, combining experiments and simulations showed that when the fallow period was initially wet, but dry later, minimum and no tillage systems gave the best water storage, but the differences in WU between tillage methods were small (Aboudrare et al., 2006). A higher initial soil water content could be especially important when the crop is sown in environments where water storage in the soil at sowing accounts for a significant proportion of total water availability for the crop (i.e., the total rainfall during the growth season is low). The differential effects of stubble mulch in soil water storage tend to decrease or disappear with the increase of crop ground coverage in crops grown under traditional tillage systems. For these reasons, in most of cases, no differences were found in estimated WU, WUE, and yield (Eq. 3.1 and 3.3) over the whole season between no-tillage systems compared to traditional tillage systems (e.g., Aparicio et al., 2002). Also, increased soil mechanical resistance, at least in some soil layers, mainly during the first years of continuous no tillage systems, could be high and can decrease the growth of roots (Ferreras et al., 2000). In spite of its high potential root system growth (Box 3.1 and section "Ecophysiological Components"), sunflower root elongation is severely affected by soil mechanical resistance (e.g., a compacted soil layer is often present in some soil during the first years of continuous no tillage system; Merrien, personal communication). For this reason soil exploration and water extraction from the subsoil by sunflower crops could be affected in those cases.

Management of irrigated crops is basically similar to those that are rain fed. Irrigation potentially allows supplying the crop with enough water to attain an adequate LAI to maximize IR, as well as good plant water conditions during critical periods for oil yield (Box 3.3). However, some cautions must be considered when managing irrigated crops. It is important not to overwater during the leaf area development stage because the crop could reach a very high LAI. Indeed, LAI higher than 3 do not intercept more IR, but crop water consumption could be higher. Moreover, crops with very high LAI during grain filling require an important supply of water for irrigation to avoid plants suffering water deficit (which in turn could cause a faster senescence of leaves and a decrease of IR). It must also be considered that the incidence of some diseases could be higher when irrigating during specific crop phenological stages because of a wetter microenvironment (e.g., head *Sclerotinia sclerotiorum* in R5). Finally, it is important to manage the crop to avoid other factors limiting yield to valorize it as high as possible.

Practices to increase nutrient availability for sunflower crops are, in general, similar to those presented in the first paragraph of this section in order to increase water availability. Compared to some crops (e.g., maize), the response of sunflower to fertilizer is generally low, and in some cases it is erratic because its requirements are relative low (see section “Nutrient Requirements: Criteria for Optimizing Soil and Fertilizer Nutrient Management”) and it could be mostly supported by soil supply. This is the case in the Argentinean Pampas, for example. The response of sunflower crop to N fertilizer depends on water availability, the tillage type (e.g., traditional, no tillage), the plot potential, and the location where the crop is being cultivated (e.g., the level of other factors limiting crop yield). Therefore, a good diagnosis of N deficiencies and potential N supply from soil are important to maximize the response to N fertilizing. For instance, a reduced availability of N as a consequence of reduced organic matter mineralization and higher soil compaction, which would affect the root growth and water and nutrient uptake by roots, have been frequently observed in the first years of the implementation of continuous no tillage with a good rate of stubble coverage of the soil surface (Aparicio et al., 2002). Under these conditions, the soil temperature is lower and the response to N fertilizer is higher compared to traditional tillage systems. Because soil solution concentration of P is low, an adequate diagnosis of P supply by soil is needed to assess the attainable response to P fertilization. For instance, in the southeast of the Argentinean Pampas, no response to P fertilizer has been found when P is higher than 10–12 ppm (Díaz-Zorita, 2006). However, independently of the P soil supply, a low rate of P fertilizer applied at sowing (30 kg ha^{-1}) has been shown to increase the growth rate of seedlings and the development of the root system (Díaz-Zorita, 2010), improving the establishment of the crop (Aguirrezábal et al., 1996).

Fertilizing sunflower crop with K is not frequent. Fertilization with this nutrient is recommended at sowing. Yield responses to fertilization were observed only in tropical and subtropical environments (Díaz-Zorita, 2010).

One way to avoid the loss of water and nutrients is to have proper weed control. Chemical control is the most effective and most used method for weed control in sunflower crops. This control should begin during the seedbed preparation to eliminate the emerged seedlings so that the crop emergence can begin with no weeds in the surface. In addition, broad-spectrum action herbicides are used before sowing or before emergence (Bedmar, 2010). These herbicides are selected considering their low toxicity on sunflower seedlings. In general, post-emergence herbicides are specific and do not present a broad-spectrum action. However, IMI hybrids (IMI resistant genotypes) were developed that present genetic resistance to Imazapir, a broad-spectrum herbicide. This technology has most of the advantages of the glyphosate tolerant soybean (RoundUp Ready® genotypes). But because IMI tolerance in sunflower has been developed naturally (Al-Khatib et al., 1998) and then improved by breeding (Sala et al., 2008), IMI resistant sunflower hybrids are not genetically modified organisms and are consequently better accepted by several markets. It facilitates the weed control by allowing the applications of herbicide in early post-emergence, delaying weed competence with the crop and avoiding reductions in LAI or HI.

Conclusion

As it was previously presented, applying ecophysiological knowledge could be helpful to optimize sunflower crop management to obtain high yields and high oil quality; to save soil water and nutrients; and to reduce the application of chemical products. Ecophysiological knowledge allows for understanding the effects of various environmental factors and agronomic practices and their interactions with the hybrid on the determination of yield and quality product. This information could be also helpful to analyze the most promising perspectives for improving sunflower management in the future.

It could be expected that sunflower production and crop system sustainability could be improved by a greater adoption of precision agriculture. Information systems could help to take into account the spatial heterogeneity of soil and crop properties as a decision support for process optimization (e.g., variable rate technology). The response of sunflower crops to fertilizer is low and erratic, so the adoption of such support, combined with precise application systems, could have a significant impact (the potential for fertilizer saving is huge, as is the potential for improving fertilizer use efficiency).

Most of the more promising avenues for sunflower improvement of yield and quality product through a better management seem to depend on genetic improvement. For instance, a higher tolerance of seedlings to low temperatures allows for the earliest sowing date and, in turn, critical windows for yield occur when incident radiation levels are higher (near the summer solstice). The impact of this practice could be especially important in high latitudes where the growth season is short.

Recently, Hall and Richards (2013) looked at unexploited opportunities for enhancing current rates of genetic improvement in major grain crops. They examined three main avenues: (1) genetic engineering of photosynthesis, (2) above-ground ideotype design, and (3) improvement of root capacity for water (and nutrient) uptake.

Improving sunflower yield by applying the first issue is unlikely because sunflower is currently a crop for which the economic and human resources available for research are scarce compared to other crops. Concerning the second issue, sunflower root system has a high potential for exploration of deep soil layers. Soil exploration can however, be limited by a high impedance of soil layers (Merrien, personal communication). An adequate functioning of root system is the key for the improvement of nutrient (e.g., N, P, K) and water use efficiency. Better understanding of root performance is needed as input for the development of novel varieties. Optimal root performance is strongly dependent on production systems. Thus, novel production systems (e.g., no till, improving plant population) need optimizing root systems.

One of the largest challenges in plant sciences is to understand how the structure and function of an organism (the phenotype) result from processes determined by its genetic material in combination with the prevailing environmental characteristics (Köhl et al., 2010). Filling the current gap between genotype and phenotype is required to increase the efficiency of genetic improvement of cultivated plants, especially when complex traits are targeted (e.g., tolerance to abiotic stresses, improvement of root system architecture and root functioning). Advances in genotyping technologies have lowered the cost per genotype to levels that have enabled an explosion of genetic studies in many fields. Phenotyping has hence become a bottleneck for plant variety improvement. Unfortunately, a realistic phenotyping of sunflower root systems is in its infancy, especially for field-grown plants.

In different crops, increases in plant density (within a certain range) have led to increase in yield through an increase in the amount of incident solar radiation, water, and nutrient interception. The limit for this increase is given by the competition for a limited number of environmental resources such as light, water, nutrients, and so forth. In some species, this limit has tried to be overcome by overexpression of a photoreceptor, which resulted in a 25% increase in yield at high density (Boccalandro et al. 2003, in potato) by modifying common plant responses to the increment of population pressure (the HI increased). Increasing plant population in sunflower could be an interesting issue for increasing yield. However, it would need to get the natural genetic variability of sunflower to tolerate high population pressure (delivery of genetically modified sunflower hybrids does not seem to be a probable option due to restrictions of the sunflower seed market) (Sala, 2010) together with the adequate agronomic practices to cultivate the crop at high densities (e.g., control of plant diseases). Increasing sunflower yield by increasing plant population could be combined with a modification of the above-ground plant architecture (the third issue proposed by Hall and Richards, 2013) in order to attain higher yields. Specifically, lowering K to get more erectophile sunflower plants could increase RUE, one of the main ecophysiological components of Y, due to a higher lightening of green area once the critical LAI is attained. Moreover, combining this modification in above-ground architecture with the stay-green trait could be helpful to attain higher RUE (the photosynthesis of the older basal leaves could be maintained during a higher period) (De la Vega et al., 2011). Genetic variability for canopy architecture and stay-green traits exists in sunflower (e.g., Hernández and Orioli, 1983). The application of these proposals to increase sunflower yield can be obviously limited by the amount of resources needed and the length of time between the moment an "idea" is presented and a hybrid is released into the seed market (see Hall and Richards, 2013).

References

Abbadi, J.; Gerendás, J.; Sattelmacher, B. Effects of Potassium Supply on Growth and Yield of Safflower as Compared to Sunflower. *J. Plant Nutr. Soil Sci.* **2008,** *171,* 272–280.

Aboudrare, A.; Debaeke, P.; Bouaziz, A.; Chekli, H. Effects of Soil Tillage and Fallow Management on Soil Water Storage and Sunflower Production in a Semi-arid Mediterranean Climate. *Agricultural Water Manage.* **2006,** *83,* 183–196.

Adiredjo, A. L.; Navaud, O.; Muños, S.; Langlade, N. B.; Lamaze, T.; Grieu, P. Genetic Control of Water Use Efficiency and Leaf Carbon Isotope Discrimination in Sunflower (*Helianthus annuus* L.) Subjected to Two Drought Scenarios. *PLoS ONE,* **2014,** *9* (7), e101218.

Aguirrezábal, L. A.; Lavaud, Y.; Dosio, G.A.; Izquierdo, N.G.; Andrade, F. H.; González, L. M. Weight per Seed and Oil Concentration in a Sunflower Hybrid Are Accounted for by Intercepted Solar Radiation during a Definite Period of Seed Filling. *Crop Sci.* **2003,** *43,* 153–161.

Aguirrezábal, L. A. N.; Tardieu, F. An Architectural Analysis of the Elongation of Field-grown Sunflower Root Systems. Elements for Modelling the Effects of Temperature and Intercepted Radiation. *J. Exp. Bot.* **1996,** *47,* 411–420.

Aguirrezábal, L. A. N.; Orioli, G. A.; Hernández, L. F.; Pereyra, V. R.; Miravé, J. P. *Girasol: aspectos fisiológicos que determinan el rendimiento,* 1st ed.; Unidad Integrada Balcarce: Buenos Aires, Argentina, 1996.

Aguirrezábal, L. A. N.; Martre, P.; Pereyra-Irujo, G. A.; Izquierdo, N. G.; Allard, V. Management and Breeding Strategies for the Improvement of Grain and Oil Quality. In *Crop Physiology: Applications for Genetic Improvement and Agronomy*; Sadras, V. O., Calderini, D., Eds.; Academic Press: San Diego, CA, 2009; pp 387–421.

Alahdadi, I.; Oraki, H.; Khajani, F. P. Effect of Water Stress on Yield and Yield Components of Sunflower Hybrids. *Af. J. Biotech.* **2011,** *10,* 6504–6509.

Alberio, C.; Izquierdo, N. G.; Galella, T.; Reid, R.; Zambelli, A.; Zuil, S. Aguirrezábal, L. *Books of Abstracts,* XLII Argentine Congress og Genetics and II Regional SAG-NOA Meeting, Salta, Argentina, October 20–23, 2013.

Al-Katib, K.; Baumgartner, J. R.; Peterson, D. E.; Currie, R. S. Imazethapyr Resistance in Common Sunflower (*Helianthus annuus*). *Weed Sci.* **1998**, 46, 403–407.

Amanullah, Khan, M. W. Interactive Effects of Potassium and Phosphorus on Phenology and Grain Yield of Sunflower in Northwest Pakistan. *Pedosphere,* **2010,** *20,* 674–680.

American Society of Agronomy (2014). About Agronomy. Retrieved from https://www.agronomy.org/

Andrade, F. H. Analysis of Growth and Yield of Maize, Sunflower and Soybean Grown at Balcarce, Argentina. *Field Crops Res.* **1995,** *41,* 1–12.

Andrade, F. H.; Ferreiro, M. A. Reproductive Growth of Maize, Sunflower and Soybean at Different Source Levels during Grain Filling. *Field Crops Res.* **1996,** *48,* 155–165.

Andrade, F. H.; Aguirrezábal, L. A. N.; Rizzalli, R. H. Crecimiento y rendimiento comparados. In *Bases para el manejo del maíz, el girasol y la soja;* Andrade, F. H., Sadras, V. O., Eds.; Unidad Integrada Balcarce: Buenos Aires, Argentina. 2000; pp 57–96.

Andrade, F. H.; Calviño, P.; Cirilo, A.; Barbieri, P. Yield Responses to Narrow Rows Depend on Increased Radiation Interception. *Agron. J.* **2002,** *94,* 975–980.

Andriani, J. M. El agua en los sistemas productivos para mejorar la producción. Ph.D Dissertation, INTA EEA Oliveros, Santa Fe, Argentina, 2000.

Aparicio, V.; Costa, J. L.; Echeverría, H.; Caviglia, O. Evaluación de Propiedades Edáficas y Crecimiento del Maíz Bajo Diferentes Sistemas de Labranza en Cuatro Sitios del Sudeste Bonaerense. *RIA*, **2002,** *31* (3) 55–71.

Ashraf, M.; Tufail, M. Variation in Salinity Tolerance in Sunflower (*Helianthus annuus* L.). *J. Agron. Crop Sci.* **1995,** *174,* 351–362.

Bange, M. P.; Hammer, G. L.; Rickert, K. G. Temperature and Sowing Date Affect the Linear Increase of Harvest Index. *Agron. J.* **1998,** *90,* 324–328.

Barros, J. F. C.; Carvalho, M.; Basch, G. Response of Sunflower (*Helianthus annuus*) to Sowing Date and Plant Density under Mediterranean Conditions. *Eur. J. Agron.* **2003,** *21,* 347–356.

Bedmar, F. Malezas del cultivo de girasol y su manejo. In *Producción de Girasol,* Balsategui, J., Ed.; Asociación Argentina de Consorcios Regionales de Experimentación Agrícola-AACREA: Buenos Aires, Argentina, 2010; pp 33–36.

Blamey, F. P. C.; Zollinger, R. K.; Schneiter, A. A. Sunflower Production and Culture. In *Sunflower Technology and Production*; Schneiter, A. A., Ed; ASA, CSSA and SSSA: Madison, WI, 1997; pp 595–670.

Boccalandro, H. E.; Ploschuk, E. L.; Yanovsky, M. J.; Sánchez, R. A.; Gatz, C.; Casal, J. J. Increased Phytochrome B Alleviates Density Effects on Tuber Yield of Field Potato Crops. *Plant Physiol.* **2003,** *133,* 1539–1546.

Campbell, E. J. Sunflower oil. *J. Am. Oil Chem. Soc.* **1983,** *60,* 387–392.

Cantagallo, J. E.; Chimenti, C. A.; Hall, A. J. Number of Seeds per Unit Area in Sunflower Correlates Well with a Photothermal Quotient. *Crop Sci.* **1997,** *37,* 1780–1786.

Cantagallo, J. E.; Medan, D.; Hall, A. J. Grain Number in Sunflower as Affected by Shading during Floret Growth, Anthesis and Grain Setting. *Source of the Document Field Crops Research* **2004,** *85,* 191–202.

Céccoli, G.; Eugenia Senn, M.; Bustos, D.; Ismael Ortega, L.; Córdoba, A.; Vegetti, A.; Taleisnik, E. Genetic Variability for Responses to Short- and Long-term Salt Stress in Vegetative Sunflower Plants. *J. Plant Nutr. Soil Sci.* **2012,** *175,* 882–890.

Chimenti, C. A.; Pearson, J.; Hall, A. J. Osmotic Adjustment and Yield Maintenance under Drought in Sunflower. *Field Crops Res.* **2002,** *75,* 235–246.

Christensen, J. H.; Carter, T. R.; Rummukainen, M.; Amanatidis, G. Evaluating the Performance and Utility of Regional Climate Models: The PRUDENCE Project. *Climatic Change* **2007,** *81,* 1–6.

Connor, D. J.; Sadras, V. O. Physiology of Yield Expression in Sunflower. *Field Crops Res.* **1992,** *30,* 333–389.

Connor, D. J.; Hall, A. J. Sunflower Physiology. In *Sunflower Technology and Production;* Schneiter, A. A., Ed.; ASA, CSSA and SSSA: Madison, WI, 1997; pp 113–182.

D'Amato, A.; Giordano, I. Effects of Climate on the Response of Sunflower (*Helianthus annuus* L.) in Relation to Sowing Time in a Southern Environment of Italy. *Proceedings of the XIII International Sunflower Conference,* Pisa, Italy, **1992,** *1,* 106–112.

Dardanelli, J. L.; Bachmeier, O. A.; Sereno, R.; Gil, R. Rooting Depth and Soil Water Extraction Patterns of Different Crops in a Silty Loam Haplustoll. *Field Crops Res.* **1997,** *54,* 29–38.

Davies, W. J.; Bacon, M. A.; Thompson, D. S.; Sobeih, W.; González Rodríguez, L. Regulation of Leaf and Fruit Growth in Plants Growing in Drying Soil: Exploitation of the Plants' Chemical Signalling System and Hydraulic Architecture to Increase the Efficiency of Water Use in Agriculture. *J. Exp. Bot.* **2000,** *51,* 1617–1626.

Debaeke, P.; van Oosterom, E. J.; Justes, E.; Champolivier, L.; Merrien, A.; Aguirrezábal, L. A. N.; González-Dugo, V.; Massignam, A. M.; Montemurro, F. A Species-specific Critical Nitrogen Dilution Curve for Sunflower. *Field Crops Res.* **2012,** *136,* 76–84.

De la Vega, A. J.; Chapman, S. C.; Hall, A. J. Genotype by Environment Interaction and Indirect Selection for Yield Sunflower I. Two-mode Pattern Analysis of Oil and Biomass Yield across Environments in Argentina. *Field Crops Res.* **2001,** *72,* 17–38.

De la Vega, A. J.; Hall, A. J. Effects of Planting Date, Genotype, and Their Interactions on Sunflower Yield: I. Determinants of Oil-corrected Grain Yield. *Crop Sci.* **2002,** *42,* 1191–1201.

De la Vega, A. J.; De Lacy, I. H.; Chapman, S. C. Progress over 20 Years of Sunflower Breeding in Central Argentina. *Field Crops Res.* **2007,** *100,* 61–72.

De la Vega, A. J.; Cantore, M. A.; Sposaro, M. M.; Trápani, N.; López Pereyra, M.; Hall, A. J. Canopy Stay-green and Yield in Non-stressed Sunflower. *Field Crops Res.* **2011,** *121,* 175–185.

Delmotte, F.; Tourvieille de Labrouhe, D.; Ahmed, S. *Emerging Races Arising from Recombination Facilitated by Multiple Introductions of the Sunflower Downy Mildew* Plasmopara halstedii *in France.* Paper presented at the 18th International Sunflower Conference, Mar del Plata and Balcarce, Argentina, 2012; p 70.

Díaz-Zorita, M. Girasol. In *Fertilidad de suelos y fertilización de cultivos;* Echeverría, H. E., García, F. O., Eds.; Ediciones INTA: Buenos Aires, Argentina, 2006; pp 301–315.

Díaz-Zorita, M. Bases de la nutrición del cultivo de girasol. In *Producción de Girasol;* Balsategui, J., Ed.; Buenos Aires, Argentina: Asociación Argentina de Consorcios Regionales de Experimentación Agrícola-AACREA, 2010; pp 33–36.

Diepenbrock, W.; Long, M.; Feil, B. Yield and Quality of Sunflower as Affected by Row Orientation, Row Spacing and Plant Density. *Bodenkultur* **2001,** *52,* 29–36.

Dosio, G. A. A.; Aguirrezábal, L. A. N.; Andrade, F. H.; Pereyra V. R. Solar Radiation Intercepted during Seed Filling and Oil Production in Two Sunflower Hybrids. *Crop Sci.* **2000,** *40,* 1637–1640.

Dosio, G. A. A.; Rey, H.; Lecoeur, J.; Izquierdo, N.; Aguirrezábal, L. A. N.; Tardieu, F.; Turc, O. A Whole-plant Analysis of the Dynamics of Expansion of Individual Leaves of Two Sunflower Hybrids. *J. Exp. Bot.* **2003,** *54,* 2541–2552.

Duncan, W. G.; Loomis, R. S.; Williams, W. A.; Hanau, R. A Model for Simulating Photosynthesis in Plant Communities. *Hilgardia* **1967,** *38,* 181–205.

Echarte, M.; Puntel, L.; Aguirrezábal, L. Assessment of the Critical Period for the Effect of Intercepted Solar Radiation on Sunflower Oil Fatty Acid Composition. *Field Crops Res.* **2013,** *149,* 213–222.

Echarte, M. M.; Angeloni, P.; Jaimes, F.; Tognetti, J.; Izquierdo N. G.; Valentinuz, O.; Aguirrezábal, L. A. N. Night Temperature and Intercepted Solar Radiation Additively Contribute to Oleic Acid Percentage in Sunflower Oil. *Field Crops Res.* **2010,** *119,* 27–35.

Fabrizzi, K.; García, F.; Costa, J. L.; Picone, L. Soil Water Dynamics, Physical Properties and Corn and Wheat Responses to Minimum and No-tillage Systems in the Southern Pampas of Argentina. *Soil and Tillage Research* **2005,** *81,* 57–69.

Farizo, C. L.; Pereyra, V. R.; Cardinali, F.; Orioli, G. Determination of Physiological and Harvest Maturity in Sunflower. In *Proceedings of the 10th International Sunflower Conference,* Surfers Paradise, QLD, Australia, 1982; pp 42–44.

Fernandez-Martinez, J.; Jimenez, A.; Dominguez, J.; Garcia, J. M.; Garces, R. Genetic Analysis of the High Oleic Acid Content in Cultivated Sunflower (*Helianthus annuus* L.). *Euphytica* **1989,** *41,* 39–51.

Fernández-Moya, V.; Martínez-Force, E.; Garcés, R. Temperature Effect on a High Stearic Acid Sunflower Mutant. *Phytochemistry* **2002,** *59,* 33–37.

Fernández-Moya, V.; Martínez-Force, E.; Garcés, R. Oils from Improved High Stearic Acid Sunflower Seeds. *J. Agric. Food Chem.* **2005,** *53,* 5326–5330.

Ferreras, L. A.; Costa, J. L.; Garcia, F. O.; Pecorari, C. Effect of No-tillage on Some Soil Physical Properties of a Structural Degraded Petrocalcic Paleudoll of the Southern "Pampa" of Argentina. *Soil and Tillage Research* **2000,** *54,* 31–39.

Fischer, R. A.; Turner, N. C. Plant Productivity in Arid and Semiarid Zones. *Annu. Rev. Plant Physiol.* **1978**, *29,* 277–317.

Fischer, R. A.; Byerlee, D.; Edmeades, G. O. Can Technology Deliver on the Yield Challenge to 2050? In *FAO Expert Meeting on How to Feed the World in 2050,* FAO, Rome, June 24–26, 2009.

Garcés, R.; Mancha, M. In Vitro Oleate Desaturase in Developing Sunflower Seeds. *Phytochemistry* **1991,** *30,* 2127–2130.

González Belo, R.; Tognetti, J.; Benech-Arnold, R.; Izquierdo, N. G. Germination Responses to Temperature and Water Potential as Affected by Seed Oil Composition in Sunflower. *Ind Crop Prod.* **2014**, 62, 537–544.

Goyne, P. J.; Hammer, G. L.; Woodruff, D. R. Phenology of Sunflower Cultivars. I. Classification of Responses (*Helianthus annuus*). *Austr. J. Agric. Res. Research* **1982,** *33,* 243–250.

Granier, C.; Tardieu, F. Is Thermal Time Adequate for Expressing the Effects of Temperature on Sunflower Leaf Development? *Plant Cell Environ.* **1998,** *21,* 695–703.

Grassini, P.; Hall, A. J.; Mercau, J. L. Benchmarking Sunflower Water Productivity in Semiarid Environments. *Field Crops Res.* **2009,** *110,* 251–262.

Haddadi, P.; Yazdi-Samadi, B.; Langlade, N. B.; Naghavi, M. R.; Berger, M.; Kalantari, A.; Calmon, A.; Maury, P.; Vincourt, P.; Sarrafi, A. Genetic Control of Protein, Oil and Fatty Acids Content under Partial Drought Stress and Late Sowing Conditions in Sunflower (*Helianthus annuus*). *Afr. J. Biotech.* **2010,** *9,* 6768–6782.

Hall, A. J. Sunflower Ecophysiology: Some Unresolved Issues. *Oleagineux Corps gras Lipides* **2001,** *8,* 15–21.

Hall, A. J.; Richards, R. A. Prognosis for Genetic Improvement of Yield Potential and Water-limited Yield of Major Grain Crops. *Field Crops Res.* **2013,** *143,* 18–33.

Hall, A. J.; Whitfield, D. M.; Connor, D. J. Contribution of Pre-anthesis Assimilates to Grain-filling in Irrigated and Water Stressed Sunflower Crops II. Estimates from a Carbon Budget. *Field Crops Res.* **1990,** *24,* 273–294.

Hall, A. J.; Feoli, C.; Ingaramo, J.; Balzarini, M. Gaps between Farmer and Attainable Yields across Rainfed Sunflower Growing Regions of Argentina. *Field Crops Res.* **2013,** *143,* 119–129.

Hammer, G. L.; Wright, G. C. A Theoretical Analysis of Nitrogen and Radiation Effects on Radiation Use Efficiency in Peanut. *Aust. J. Agric. Res.* **1994,** *45,* 575–589.

Hejnák, V.; Nemcová, O.; Matejovî, M.; Martinková, J.; Hnilîka, F.; Skalický, M.; Grieu, P. Physiological Responses as Influenced by Night Freeze Stress at the Beginning of Vegetative Growth of Sunflower. *Res. Crops* **2014,** *12,* 473–480.

Hernández, L. F.; Orioli, G. A. Growth Analysis of Irrigated Sunflower at Two Plant Populations. In *International Sunflower Conference Proceedings,* Surfers Paradise, Australia, 1982; pp 18–21.

Hernández, L. F.; Orioli, G. A. Estudio comparativo de la estructura foliar, intercepción de luz y rendimiento en girasol. *Anales de Edafología y Agrobiología* (Madrid) **1983,** *42,* 2137–2148.

Hernández, L. F.; Orioli, G. A. Evaluación del esfuerzo reproductivo en el girasol cultivado. In *Proceedings of the 11th International Sunflower Conference,* Mar del Plata, Argentina, March 10–13, 1985; pp 57–60.

Hewezi, T.; Léger, M.; El Kayal, W.; Gentzbittel, L. Transcriptional Profiling of Sunflower Plants Growing under Low Temperatures Reveals an Extensive Down-regulation of Gene Expression Associated with Chilling Sensitivity. *J. Exp. Bot.* **2006,** *57,* 3109–3122.

Hoshikawa, K. The Growing Rice Plant: An Anatomical Monograph. In *Icones, Anatomy and morphology*; Nobunkyo: Tokyo, Japan,1989; Vol. 16, p 310.

Izquierdo, N.; Aguirrezábal, L.; Andrade, F.; Pereyra, V. Night Temperature Affects Fatty Acid Composition in Sunflower Oil Depending on the Hybrid and the Phenological Stage. *Field Crops Res.* **2002,** *77,* 115–126.

Izquierdo, N.; Aguirrezábal, L.; Andrade, F.; Cantarero, M. Modeling the Response of Fatty Acid Composition to Temperature in a Traditional Sunflower Hybrid. *Agron. J.* **2006,** *98,* 451–461.

Izquierdo, N.; Dosio, G. A. A.; Cantarero, M.; Aguirrezábal, L. A. N. Weight per Grain, Oil Concentration and Solar Radiation Intercepted during Grain Filling in Black Hull and Stripped Hull Sunflower Hybrids. *Crop Sci.* **2008,** *48,* 688–699.

Izquierdo, N.; Aguirrezábal, L.; Andrade, F.; Geroudet, C.; Pereyra Iraola, M.; Valentinuz, O. Intercepted Solar Radiation Affects Oil Fatty Acid Composition in Crop Species. *Field Crops Res.* **2009,** *114,* 66–74.

Izquierdo, N. G.; Aguirrezábal, L. A. N.; Martínez-Force, E.; Garcés, R.; Paccapelo, V., Andrade, F.; Reid R.; Zambelli, A. Effect of Growth Temperature on the High Stearic Sunflower Trait. *Crop and Pasture Sci.* **2013,** *64,* 18–25.

Kalyar, T.; Rauf, S.; Teixeira da Silva, J. A.; Iqbal, Z. Variation in Leaf Orientation and its Related Traits in Sunflower (*Helianthus annuus* L.) Breeding Populations under High Temperature. *Field Crops Res.* **2013,** *150,* 91–98.

Kalyar, T.; Rauf, S.; Teixeira Da Silva, J. A.; Shahzad, M. Handling Sunflower (*Helianthus annuus* L.) Populations under Heat Stress. *Arch. Agron. Soil Sci.* **2014a,** *60,* 655–672.

Kalyar, T.; Rauf, S.; Teixeira da Silva, J. A.; Shahzad, M. Erratum to Handling Sunflower (*Helianthus annuus* L.) Populations under Heat Stress. *Arch. Agron. Soil Sci.* **2014b,** *60,* 1029.

Katerji, N.; Van Hoorn, J. W.; Hamdy, A.; Mastrorilli, M. Salt Tolerance Classification of Crops According to Soil Salinity and to Water Stress Day Index. *Agric. Water Mgt.* **2000,** *43,* 99–109.

Köhl, J.; van Tongeren, C. A. M.; Groenenboom-de Haas, B. H.; van Hoof, R. A.: Driessen, R.; van der Heijden, L. Epidemiology of Dark Leaf Spot Caused by Alternaria brassicicola and A. brassicae in Organic Seed Production of Cauliflower. *Plant Path.* **2010,** *59,* 358–367.

Lacombe, S.; Bervillé, A. A Dominant Mutation for High Oleic Acid Content in Sunflower (*Helianthus annuus* L.) Seed Oil Is Genetically Linked to a Single Oleate-desaturase RFLP Locus. *Mol. Breed.* **2001,** *8,* 129–137.

Lacombe, S.; Souyris, I.; Bervillé, A. J. An Insertion of Oleate Desaturase Homologous Sequence Silences via siRNA the Functional Gene Leading to High Oleic Acid Content in Sunflower Seed Oil. *Mol. Gen. Genom.* **2009,** *281,* 43–54.

Lagravere, T.; Champolivier, L.; Lacombe, S.; Kleiber, D.; Berville, A.; Dayde, J. Effects of Temperature Variations on Fatty Acids Composition in Oleic Sunflower Oil (*Helianthus annuus* L.) Hybrids. In *Proceedings of the 15th International Sunflower Conference,* Toulouse, France, 2000; p A-73.

Levitt, J. *Responses of plants to environmental stresses. Water, radiation, salt, and other stresses*; 2nd ed.; Academic Press: New York, **1980**.

Lindstrom, L. I.; Pellegrini, C. L., Aguirrezábal, L. A. N.; Hernández, L. F. Growth and Development of Sunflower Fruits under Shade During Pre and Early Post-anthesis Period. *Field Crops Res.* **2006,** *96,* 151–156.

López Pereira, M. Fecha y densidad de siembra. In *Producción de Girasol;* Aguirrezábal, L., Ed.; Asociación Argentina de Consorcios Regionales de Experimentación Agrícola-AACREA: Buenos Aires, Argentina, 2010; pp 33–36.

Maas, E. V.; Hoffman, G. J. Crop Salt Tolerance—Current Assessment. *J. Irrig. Drain. Div.* **1977,** *103,* 115–134.

Mantese, A. I.; Medan, D.; Hall, A. J. Achene Structure Development and Lipid Accumulation Is Sunflower Cultivars Differing in Oil Content at Maturity. *Annals of Botany* **2006,** *97,* 999–1010.

Marc, J.; Palmer, J. H. Photoperiodic Sensitivity of Inflorescence Initiation and Development in Sunflower. *Field Crops Res.* **1981,** *4,* 155–164.

Martínez, R. D.; Izquierdo, N. G.; González Belo, R.; Aguirrezábal, L.; Andrade, F. H.; Reid, R. Oil Yield Components and Oil Quality of High Stearic-High Oleic Sunflower Genotypes as Affected by Intercepted Solar Radiation during Grain Filling. *Crop and Past. Sci.* **2012,** *63,* 330–337.

Mercau, J. L.; Sadras, V. O.; Satorre, E. H.; Messina, C.; Balbi, C.; Uribelarrea, M.; Hall, A. J. On-farm Assessment of Regional and Seasonal Variation in Sunflower Yield in Argentina. *Agric. Syst.* **2001,** *67,* 83–103.

Merrien, A., *Physiologie de la formation du rendement chez le tournesol,* 1st ed.; Centre Tech. Interprofessionel des Oleagineux Metropolitains: Paris, France, 1983.

Merrien, A.; Milan, M. J. *Physiologie du tournesol,* 1st ed.; CETIOM: Paris, France, 1992.

Miladinović, D.; Cantamutto, M.; Vasin, J.; Dedić, B.; Alvarez, D.; Poverene, M. Exploring Environmental Determinations of the Geographic Distribution of Broomrape (*Orobanche Cumana* Wallr.). *Helia* **2012,** *35,* 79–88.

Monsi, M.; Saeki, T.; Schortemeyer, M. On the Factor Light in Plant Communities and Its Importance for Matter Production. *Ann. Bot.* **2005,** *95,* 549–567.

Monteith, J. L. Climate and Efficiency of Crop Production in Britain. *Philosophical Transactions of the Royal Society of London, Series B-Biological Sciences* **1977,** *281,* 277–294.

Munns, R.; Tester, M. Mechanisms of Salinity Tolerance. *Ann. Rev. Plant Biol.* **2008,** *59,* 651–681.

Osorio, J.; Fernández-Martinez, J.; Mancha, M.; Garcés, R. Mutant Sunflowers with High Concentration of Saturated Fatty Acids in the Oil. *Crop Sci.* **1995,** *35,* 739–742.

Padmanabhan, P.; Sahi, S. V. Suppression Subtractive Hybridization Reveals Differential Gene Expression in Sunflower Grown in High P. *Plant Phys. Biol.* **2011,** *49,* 584–591.

Passioura, J. B. Grain Yield, Harvest Index, and Water Use of Wheat. *J. Aust. Inst. Agric. Sci.* **1977,** *43,* 117–121.

Passioura, J. B. Drought and Drought Tolerance. *Plant Growth Reg.* **1996,** *20,* 79–83.

Pedraza, M. V.; Pereyra, V. R.; Aguirrezábal, L.; Laurlund, A. *Manual de estimación de pérdidas de rendimiento en girasol. Efectos de reducciones en el área foliar, en la densidad de plantas y en el número de capítulos.* INTA Balcarce-Mercantil Andina: Balcarce, Buenos Aires, Argentina, 2000.

Pereyra-Irujo, G.; Aguirrezábal, L. A. N. Sunflower Yield and Oil Quality Interactions and Variability: Analysis through a Simple Simulation Model. *Agr. Forest. Meteor.* **2007,** *143,* 252–265.

Pereyra-Irujo, G. A.; Velázquez L.; Lechner L.; Aguirrezábal, L. A. N. Genetic Variability for Leaf Growth Rate and Duration under Water Deficit in Sunflower. Analysis of Responses at Cell, Organ and Plant Level. *J. Exp. Bot.* **2008,** *59,* 2221–2232.

Ploschuk, E. L.; Hall, A. J. Capitulum Position in Sunflower Affects Grain Temperature and Duration of Grain Filling. *Field Crop Res.* **1995,** *44,* 111–117.

Quiroz, F.; Edwards Molina, J. P.; Dosio, G. Black Stem by Phoma macdonaldii Affected Ecophysiological Components That Determine Grain Yield in Sunflower (*Helianthus annuus* L.). *Field Crops Res.* **2014,** *160,* 31–40.

Roche, J.; Essahat, A.; Bouniols, A.; El Asri, M.,Mouloungui, Z.; Mondies, M.; Alghoum, M. Diversified Composition of Sunflower (*Helianthus annuus* L.) Seeds within Cultural Practices and Genotypes (Hybrids and Populations). *Helia* **2004,** *27,* 73–98.

Rondanini, D.; Savin, R.; Hall, A. Dynamics of Fruit Growth and Oil Quality of Sunflower (*Helianthus annuus* L.) Exposed to Brief Intervals of High Temperature during Grain Filling. *Field Crop Res.* **2003,** *83,* 79–90.

Rondanini, D.; Mantese, A.; Savin, R.; Hall, A. J. Responses of Sunflower Yield and Grain Quality to Alternating Day/Night High Temperature Regimes during Grain Filling: Effects of Timing, Duration and Intensity of Exposure to Stress. *Field Crops Res.* **2006,** *96,* 48–62.

Rondanini, D.; Savin, R.; Hall, A. J. Estimation of Physiological Maturity in Sunflower as a Function of Fruit Water Concentration. *Eur. J. Agron.* **2007,** *26,* 295–309.

Rousseaux, M. C.; Hall, A. J.; Sánchez, R. A. Basal Leaf Senescence in a Sunflower (*Helianthus annuus*) Canopy: Responses to Increased R/FR Ratio. *Physiologia Plantarum* **2000,** *110,* 477–482.

Sadras, V. O.; Hall, A. J. Quantification of Temperature, Photoperiod and Population Effects on Plant Leaf Area in Sunflower Crops. *Field Crops Res.* **1988,** *18,* 185–196.

Sadras, V.; Quiroz, F.; Echarte, L.; Escande, A.; Pereyra, V. Effect of Verticillium dahliae on Photosynthesis, Leaf Expansion and Senescence of Field-grown Sunflower. *Ann. Botany* **2000a,** *86,* 1007–1015.

Sadras, V. O.; Trápani, N.; Pereyra, V. R.; López Pereira, M.; Quiroz, F.; Mortarini, M. Intraspecific Competition and Fungal Diseases as Sources of Variation in Sunflower Yield. *Field Crops Res.* **2000b,** *67,* 51–58.

Sadras, V. O.; Echarte, L.; Andrade, F. H. Profiles of Leaf Senecence during Reproductive Growth of Sunflower and Maize. *Ann. Botany* **2000c,** *85,* 187–195.

Sala, C. A.; Bulos, M.; Echarte, A. M. Genetic Analysis of an Induced Mutation Conferring Imidazolinone Resistance in Sunflower. *Crop Sci.* **2008**, *48,* 1817–1822.

Sala, C. A. *Nuevas Tecnologías en Tolerancia a Herbicidas.* Paper presented at the 17th International Sunflower Conference, Buenos Aires, Argentina, 2010; p 129.

Schneiter, A. A.; Miller, J. F. Description of Sunflower Growth Stages. *Crop Sci.* **1981,** *21,* 901–903.

Serrano-Vega, M., Martínez-Force, E., Garcés, R. Lipid Characterization of Seed Oils from High-palmitic, Low Palmitoleic and Very High-stearic Acid Sunflower Lines. *Lipids* **2005,** *40,* 369–374.

Sinclair, T. R.; Muchow R. C. Radiation Use Efficiency. *Adv. Agron.* **1999,** *65,* 15–263.

Sinclair, T. R.; Tanner, C. B.; Bennet, J. M. Water Use Efficiency in Crop Production. *Biol. Sci.* **1984,** *34,* 36–40.

Sinclair, T. R.; Horie, T. Leaf Nitrogen, Photosynthesis, and Crop Radiation Use Efficiency: A Review. *Crop Sci.* **1989,** *29,* 90–98.

Steduto, P.; Hsiao, T. C.; Fereres, E. On the Conservative Behavior of Biomass Water Productivity. *Irrig. Sci.* **2007,** *25,* 189–207.

Tang, S.; León, A.; Bridges, W. C.; Knapp, S. J. Quantitative Trait Loci for Genetically Correlated Seed Traits Are Tightly Linked to Branching and Pericarp Pigment Loci in Sunflower. *Crop Sci.* **2006,** *46,* 721–734.

Trápani, N; Hall, A. J. Effects of Leaf Position and Nitrogen Supply on the Expansion of Leaves of Field Grown Sunflower (*Helianthus annuus* L.). *Plant Soil.* **1996,** 184, 331–340.

Trápani, N.; Hall, A. J.; Sadras, V. O.; Vilella, F. Ontogenic Changes in Radiation Use Efficiency of Sunflower (*Helianthus annuus* L.) Crops. *Field Crops Res.* **1992,** *29,* 301–316.

Valentinuz, O. R.; Uhart, S. A.; Andrade, F. H.; Vega, C. R. *Books of Abstracts,* VII Agentine Congress of Meteorology and VII Latin American and Iberian Congress of Meteorology, Buenos Aires, Argentina, September, 1996; pp 39–40.

Vega, C.R. C.; Andrade, F. H.; Sadras, V. O.; Uhart, S. A.; Valentinuz, O. R. Seed Number as a Function Of Growth. A Comparative Study in Soybean, Sunflower, and Maize. *Crop Sci.* **2001,** *41,* 748–754.

Velázquez, L.; Alberdi, I.; Aguirrezábal, L.; Pereyra-Irujo, G. *Books of Abstracts,* XVIII International Sunflower Conference, Mar del Plata and Balcarce, February 27–28–March 1, 2012, pp 129.

Villalobos, F.; Sadras, V.; Soriano, A.; Fereres, E. Planting Density Effects on Dry Matter Partitioning and Productivity of Sunflower Hybrids. *Field Crops Res.* **1994,** *36,* 1–11.

Villalobos, F. J.; Ritchie, J. T. The Effect of Temperature on Leaf Emergence Rates of Sunflower Genotypes. *Field Crops Res.* **1992,** *29,* 37–46.

Villalobos, F. J.; Soriano, A.; Fereres, E. Effects of Shading on Dry Matter Partitioning and Yield of Field-grown Sunflower. *Eur. J. Agron.* **1992,** *1,* 109–115.

Wang, L. X.; Zhang, Y. S.; Zhang, H. Q.; Xu, J. T.; Li, X. Y. Changes of Soil Characteristics in Process of Land Desertification. *Proceedings of SPIE. The International Society for Optical Engineering* **2003,** *4890,* 187–192.

Xiong, L.; Schumaker, K. S.; Zhu, J.-K. Cell Signaling during Cold, Drought, and Salt Stress. *Plant Cell* **2002,** *14,* 165–183.

Zheljazkova, V. D.; Vick, B. A.; Baldwinc, B.; Buehringa, N.; Coker, C.; Astatkiee, T.; Johnson, B. Oil Productivity and Composition of Sunflower as a Function of Hybrid and Planting Date. *Ind. Crop Prod.* **2011,** *33,* 537–543.

Zuil, S. G.; Izquierdo, N. G.; Luján J.; Cantarero M.; Aguirrezábal, L. A. N. Oil Quality of Corn and Soybean Genotypes with Increased Oleic Acid Percentage as Affected by Intercepted Solar Radiation and Temperature. *Field Crops Res.* **2012,** *127,* 203–214.

4

Sunflower Diseases

Samuel G. Markell ■ *Department of Plant Pathology, North Dakota State University, Fargo, ND, USA*

Robert M. Harveson ■ *University of Nebraska, Panhandle Research and Extension Center, Scottsbluff, NE, USA*

Charles C. Block ■ *United States Department of Agriculture, Agricultural Research Service, Plant Introduction Station, Ames, IA, USA*

Thomas J. Gulya ■ *United States Department of Agriculture, Agricultural Research Service, Northern Crop Science Laboratory, Fargo, ND, USA*

Introduction

Sunflowers are native to North America, with 51 species of annual and perennial types found, but the primary cultivated sunflower worldwide is the herbaceous annual *Helianthus annuus.* The range of sunflower cultivation spread from its origins in the United States and Canada to become widely grown in temperate zones of Europe, Asia, South America, and Australia. Sunflower is a known host for over 30 pathogens, but the relative importance of specific diseases varies with geographic region. Differences in climate, pathogen distribution, and cropping practices affect the prevalence of individual diseases in each region. For example, charcoal rot is mainly found in warmer regions where high temperatures and low soil moisture predispose plants to infection. In contrast, downy mildew is found in regions where emerging seedlings are exposed to low temperatures and wet soils. Some diseases, such as rust, Sclerotinia head rot, and Sclerotinia stalk rot and wilt are widespread, threatening sunflower production in nearly every sunflower growing region. In this chapter, two tables of selected diseases are presented. Table 4.A lists diseases considered to be of widespread importance. Table 4.B lists selected diseases of regional importance or those that may occur sporadically. Individual diseases and their effects on sunflower production are discussed in more detail. These include downy mildew, Phomopsis stem canker, rust, Rhizopus head rot, Sclerotinia head rot, Sclerotinia stalk rot and wilt, and Verticillium wilt.

Downy Mildew

Introduction and Disease History

Downy mildew of sunflower is caused by the obligate fungal pathogen *Plasmopara halstedii.* The pathogen of sunflower was first found in the late 1890s in the Northeastern

Table 4.A Major Sunflower Diseases of Widespread Economic Importance

Common Name	Pathogen	Order/Family	Key References
Downy mildew	*Plasmopara halstedii*	Peronosporales/ Peronosporaceae	Berglund (2007), Gulya et al. (1997), Kolte (1985)
Phomopsis stem canker	*Phomopsis helianthi, Diaporthe gulyae*	Diaporthales/ Diaporthaceae	Berglund (2007), Gulya et al. (1997), Thompson et al. (2011)
Sunflower rust	*Puccinia helianthi*	Uredinales/ Pucciniaceae	Berglund (2007), Gulya et al. (1997), Kolte (1985)
Sclerotinia head rot, stalk rot, and wilt	*Sclerotinia sclerotiorum*	Helotiales/ Sclerotiniaceae	Berglund (2007), Gulya et al. (1997), Kolte (1985)
Verticillium wilt	*Verticillium dahliae*	Hypocreales/ Plectosphaerellaceae	Berglund (2007), Gulya et al. (1997), Kolte (1985)

Table 4.B Sunflower Diseases of Regional or Sporadic Importance

Common Name	Pathogen	Order/Family	Key References
Alternaria leaf and stem spot	*Alternariaster (Alternaria) helianthi, Alternaria zinniae*	Pleosporales/ Leptosphaeriaceae, Pleosporales/ Pleosporaceae	Berglund (2007), Gulya et al. (1997), Kolte (1985)
Bacterial stalk/ head rot	*Pectobacterium (Erwinia) carotovorum* subsp. *carotovorum, Pect. carotovorum* subsp. *atrosepticum*	Enterobacteriales/ Enterobacteriaceae	Berglund (2007), Gulya et al. (1997), Kolte (1985)
Broomrape	*Orobanche* spp.	Lamiales/ Orobanchaceae	Kolte (1985)
Charcoal rot	*Macrophomina phaseolina*	Botryosphaeriales/ Botryosphaeriaceae	Berglund (2007), Gulya et al. (1997), Kolte (1985)

Table 4.B *Continued*

Common Name	Pathogen	Order/Family	Key References
Phoma black stem	*Phoma macdonaldii*	Pleosporales/ Leptosphaeriaceae	Berglund (2007), Gulya et al. (1997), Kolte (1985)
Powdery mildew	*Golovinomyces cichoracearum* (syn: *Erysiphe cichoracearum)*	Erysiphales/ Erysiphaceae	Berglund (2007), Gulya et al. (1997), Kolte (1985)
Rhizopus head rot	*Rhizopus stolonifer, Rhizopus* spp.	Mucorales/ Rhizopodaceae	Berglund (2007), Gulya et al. (1997), Kolte (1985)
Septoria leaf spot	*Septoria helianthi*	Capnodiales/ Mycosphaerellaceae	Berglund (2007), Gulya et al. (1997), Kolte (1985)
Southern blight	*Sclerotium rolfsii*	Atheliales/Atheliaceae	Berglund (2007), Gulya et al. (1997)

United States on *Eupatorium,* a flowering plant in the family Compositae (Harveson, 2011a). Like the cultivated sunflower (*Helianthus annuus* L.), *P. halstedii* is a native of North America and the Western Hemisphere. With the exception of Australia, the pathogen is now known to be distributed worldwide wherever sunflowers are grown. Severe losses of 50–95% have been documented in cool, wet years. Thus, it is considered to be one of the most important diseases of sunflower due to its great potential for destructiveness (Gulya et al., 1997).

Downy Mildew Pathogen

The majority of downy mildew pathogens have restricted host ranges, but *P. halstedii* causes disease on at least 80 species in 35 genera of the Composite family, including both weed and horticultural species (Harveson, 2011a). As of 2010, 17 physiological races of the pathogen were known worldwide; however, more than 75% of the identified species actually consisted of three primary races (Gulya et al., 2012). The appearance of a new physiological race indicates that new cultivars containing known resistance genes will be susceptible to disease (Gulya et al., 1991, 2013).

Downy Mildew Signs and Symptoms

Downy mildew infections can be subdivided into two major categories based on symptoms. Systemic symptoms are produced when young seedlings are infected through the root system, often resulting in death of plants (Friskop et al., 2009; Harveson, 2011a). If seedlings survive, they are severely stunted (Figure 4.1) with chlorotic, puckered leaves.

The chlorosis connected with downy mildew infection indicates the portion of the leaves being colonized by the pathogen. This is often associated with the leaf veins (Figure 4.2), but can also cover entire areas of the leaf (Figure 4.3). When extended periods of cool (15–18 °C), wet, or humid weather are realized, a dense cottony, white growth will form, generally on the underside of leaves (Figure 4.4, p. 98).

If sufficient leaf moisture is present, secondary infections occur when windblown zoospores from adjacent plants land on leaves of uninfected plants; these are referred to as *local lesions.* These lesions are generally chlorotic and angular in shape (Figure 4.5, p. 98), being limited by leaf veins, but they do not become systemic. Directly beneath the yellow lesions, on the underside of leaves, the white cottony sporulation of the fungus is formed, which is a unique sign of downy mildew infection (Figure 4.6, p. 99). Plants are susceptible to secondary infection for longer periods of time than with the early systemic root infections, but these infections do not result in yield loss (Friskop et al., 2009; Harveson, 2011a).

Figure 4.1 A young sunflower plant infected very early as a seedling showing effects of systemic infection, including stunting, chlorosis, and leaf puckering.
Source: Harveson (2010).

Figure 4.2 Systemic infection showing chlorotic leaf veins, indicating pathogen colonization of vascular system.

Source: Harveson (2009).

Figure 4.3 Systemic infection exhibiting chlorosis of large areas of the surfaces of newly developing leaves.

Source: Harveson (2007).

Figure 4.4 White cottony growth of pathogen (sporulation) on bottom leaf surface as a result of extended periods of cool and wet weather.

Source: Harveson (2010).

Figure 4.5 Local infections from windblown zoospores. Note the angular appearance of chlorotic lesions on upper leaf surface.

Source: Harveson (2008).

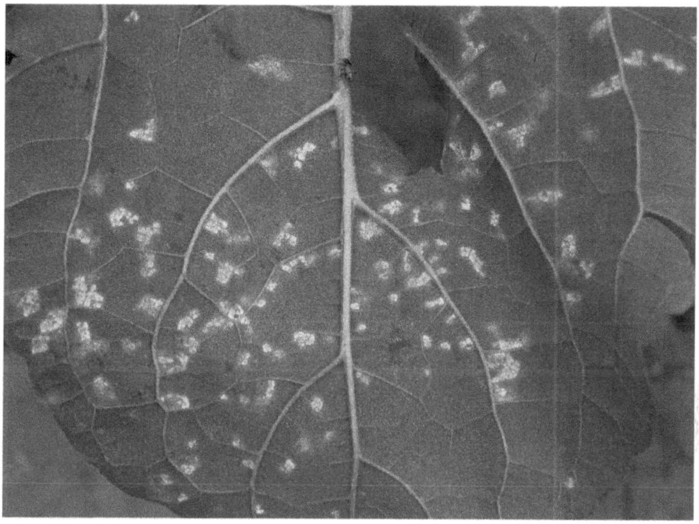

Figure 4.6 Local infection from windblown zoospores. Note sporulation of the pathogen on the lower leaf surface.
Source: Harveson (2013).

Downy Mildew Disease Cycle and Damage

The pathogen resides in soil as sexually produced oospores, which are thick-walled structures that are formed in root tissues shortly after infection. These also serve as survival structures. They are highly resistant to adverse environmental conditions and are capable of remaining dormant in soils for up to 10 years. When soils are saturated and cool (12–14 °C), oospores will germinate, forming zoosporangia that release motile zoospores capable of swimming through soil water (Gulya et al., 1997; Sackston, 1981).

Yield losses from downy mildew depend on numerous factors, including incidence and distribution of the pathogen in soils, percentage of infected plants in a field, and the prevailing weather conditions during the season (Gulya et al., 1997). Surviving plants can often compensate, in terms of yield, for neighboring plants that prematurely die after early infection. However, when highly concentrated areas in a field are affected, such as low spots, the combination of plant mortality, reduced head sizes, lighter and fewer seeds produced by surviving plants, and lowered oil contents negatively impact yield and quality (Figure 4.7).

Downy Mildew Management

Like many soil-borne pathogens, management of downy mildew must be preventative. Two significant challenges to management of downy mildew are the high

Figure 4.7 Low spot in a field with a high incidence of the downy mildew pathogen.
Source: Harveson (2009).

variability of the pathogen, which can result in race or fungicide sensitivity changes, and the long-lived nature of the pathogen, which makes management using cultural practices largely ineffective.

- Genetic resistance is often the most effective means of control. However, the pathogen has numerous races and the longevity of resistance is limited by frequent pathogen race changes (Gulya et al., 2012). Consequently, it is important to monitor fields for downy mildew development even when utilizing a hybrid with a resistance gene.
- Fungicide seed treatments can be an effective management tool for downy mildew, but in many areas the pathogen has developed resistance to certain fungicides. In the United States, isolates insensitive to FRAC4 fungicides, mefenoxam and metalaxyl, were identified and determined to be widespread in the late 1990s and early 2000s, rendering them ineffective (Gulya et al., 2013).
- Wild and volunteer sunflowers can harbor the downy mildew pathogen and provide a potential source of inoculum and new races. Wild and volunteer sunflowers in close proximity with fields (for example, ditches, roadsides, and abandoned farmsteads) should be destroyed (Friskop et al., 2009).

Phomopsis Stem Canker

Introduction and Disease History

Phomopsis stem canker is a relatively newly identified disease of sunflower, and was first documented in Europe and the United States in the 1980s (Herr et al., 1983; Muntanola-Cvetkovic et al., 1981). The disease was then confirmed in other European countries and became quite serious, especially in France and the former Yugoslavia, where 50–80% of incidence has been observed in fields (Gulya et al., 1997). Similarly, the disease is established throughout the Midwestern states, and is one of the most serious yield-limiting factors in the U.S. Northern Great Plains (Kandel, 2013). Worldwide, the disease has been confirmed in Europe (11 countries), Asia (Pakistan), Africa (Morocco), North America (Canada, United States, and Mexico), and South America (Argentina and Venezuela) (Commonwealth Agricultural Bureau, 1999; Thompson et al., 2011).

Phomopsis Stem Canker Pathogen

The causal agent of Phomopsis stem canker was initially ascribed to a new species, *Phomopsis/Diaporthe helianthi,* but several Serbian pathologists suspected that other *Phomopsis* species might be involved based on morphological differences between isolates (Masirevic and Gulya, 1992). As early as 1990, mycologists dealing with the *Phomopsis* genus felt that the species could not be identified by morphological characters alone. In a recent taxonomic review of the genus (Udayanga et al., 2011), mycologists have come to rely on molecular techniques, as well as cultural and morphological characters, to delineate species and biotypes (Rekab et al., 2004; Vukojević et al., 2001). The use of molecular taxonomic tools facilitated the identification of three new *Phomopsis* species (*P. kochmanii, P. kongii,* and *P. gulyae*) from diseased sunflower in Australia in 2011 (Thompson et al., 2011). Parallel work in the United States identified five species causing disease on sunflower (*P. helianthi, P. gulyae, P. stewartii, P. longicolla,* and *P. phaseolorum*), with *P. helianthi* being the dominant species (Mathew et al., 2012). *P. phaseolorum* has also been isolated from sunflower lesions in Croatia (Vrandecic et al., 2010). Sunflower is also not the only host for *P. helianthi;* the pathogen has been isolated from lesions on many weed genera, including *Amaranthus, Arctium, Artemesia, Cichorium, Cirsium, Daucus, Iva, Solidago,* and *Xanthium* (Vrandecic et al., 2009). *Phomopsis helianthi* also exists as a latent epiphyte on non-hosts, such as grapes and fruit trees (Brum et al., 2012).

Phomopsis Stem Canker Signs and Symptoms

Disease symptoms appear first as *Diaporthe* ascospores infecting the leaf margins, and then they progress down the main veins (Gulya et al., 1997; Masirevic and Gulya,

1992). The most diagnostic feature is a band of necrotic tissue on the vein bordered by chlorotic tissue. The leaf lesion will often create a “bronzed” appearance on the leaf and quickly degrade the initial infected leaf quadrant (Figure 4.8). Initial leaf symptoms can be confused with the early stage of Verticillium wilt, where the necrotic area occurs between and not on the veins. The *Phomopsis* infection then proceeds down the petiole and, upon reaching the stem, it produces a light to dark brown lesion girdling the stem, sometimes reaching more than 15 cm in length (Figures 4.9 and 4.10). The fungus will eventually rot the pith, which leads to lodging as the weight of the sunflower head increases (Figure 4.11, p. 104). Pith damage can easily be determined by simple thumb pressure on the lesion. As the plant matures and the stem turns brown, the Phomopsis lesion turns a silvery gray color and small black pycnidia can be seen with the aid of a 10-times magnifying lens.

Phomopsis Stem Canker Disease Cycle and Damage

Phomopsis helianthi, the asexual stage of *Diaporthe helianthi,* may produce two types of spores in pycnidia immersed in stem tissue, but neither of these infect sunflower (Masirevic and Gulya, 1992; Muntanola-Cvetkovic et al., 1981). After infected stems overwinter, the sexual *Diaporthe* stage produces perithecia, from which ascospores ooze out in a gelatinous matrix. These spores are splashed by rain and also carried by the wind to initiate foliar infection. In the United States, based on annual surveys

Figure 4.8 Advanced leaf lesion of Phomopsis stem canker.
Source: Markell (2010).

Figure 4.9 Typical stem lesion of Phomopsis stem canker, originating from the dead petiole.

Source: Gulya (2010).

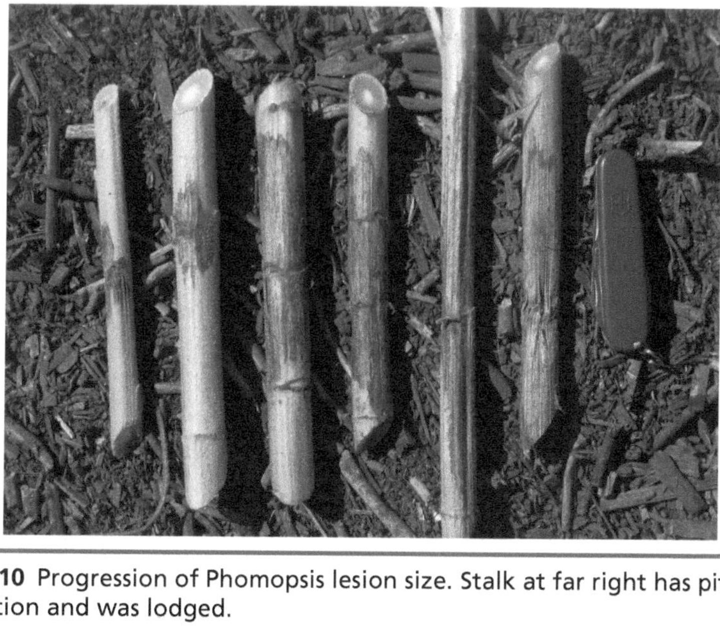

Figure 4.10 Progression of Phomopsis lesion size. Stalk at far right has pith degradation and was lodged.

Source: Gulya (2003).

Figure 4.11 Lodging caused by Phomopsis stem canker.
Source: Markell (2010).

coordinated by the National Sunflower Association across eight Midwestern states, the incidence of Phomopsis stem canker has increased from 30% in 2001 to 70% of inspected fields in 2012, while the percentage of the crop affected has increased from 2% to 25% in the same time period. Yield losses are due to premature plant death and are aggravated by lodging.

Phomopsis Stem Canker Management

Management of Phomopsis stem canker has proven difficult, and even the best management tools can be overcome when the environment is highly conducive to disease development. Consequently, it is best to use as many management tools as possible.

- Genetic resistance to *P. helianthi* is available in commercial hybrids. Since several genes are involved, these hybrids have varying degrees of resistance rather than total immunity. At this time, it is unclear whether resistance to *P. helianthi* confers resistance to the other *Phomopsis* species.
- Fungicides are commonly used to manage Phomopsis stem canker in Europe but fungicide usage in the United States is limited.

- High plant populations and dense canopy, fostered by high nitrogen fertility, will increase disease severity (Debaeke et al., 2003; Desanlis et al., 2013); therefore, plant populations should be reduced.
- Because *P. helianthi* is able to infect many common weeds, the control of weedy hosts will minimize the amount of inoculum (Vrandecic et al., 2010). In addition, because *P. helianthi* and other species can exist latently on other plants, such as soybeans, it may be advisable to not plant sunflowers and soybeans adjacent to each other.

Rhizopus Head Rot

Introduction

Rhizopus head rot has been reported from several areas of sunflower production in the world, including Australia, South Africa, and Mediterranean areas in Europe and India. In the United States, Rhizopus head rot has historically caused only minor and localized damage to sunflowers. However, as the production of sunflowers has increased in the Central High Plains region (Colorado, Nebraska, and Wyoming), Rhizopus head rot has become more important (Harveson, 2013).

Rhizopus Head Pathogen and Disease History

The genus *Rhizopus* head rot is a common fungus that occurs naturally in soils and as airborne spores. Head rot disease is caused by several species of this genus, including *R. stolonifer, R. arrhizus, R. oryzae,* and *R. microsporus.* These species may occur singly or in a complex (Schipper, 1984a, 1984b). This genus is well known for causing soft rots of fruits, vegetables, and root crops, especially in postharvest storage situations. Rhizopus head rot is one of the few diseases caused by this group of fungi that occurs in field crops during the cropping season. It has historically been considered to be of minor importance in the United States and Canada; however, surveys of sunflower diseases in California have found Rhizopus head rot to be the most severe and commonly identified disease (Klisiewicz and Beard, 1976). It has also been documented to cause severe damage annually to sunflower grown in Israel and Texas (Rogers et al., 1978; Yang et al., 1980).

Rhizopus Head Rot Signs and Symptoms

The disease first appears as dark spots of varying sizes on the back of ripening heads as a result of some type of wound, followed by a watery, soft rot that later turns dark brown (Figure 4.12). As the disease progresses, heads dry prematurely and shrivel, and tissues appear to shred. The fungus can be observed inside shredded heads as coarse,

Figure 4.12 Advanced stage of disease showing both watery soft rot and dried, shredded tissues.
Source: Harveson (2013).

threadlike mycelial strands that are later followed by the appearance of fungal reproductive structures (fruiting structures) called sporangia that look like small black dots about the size of pinheads. The disease can also be recognized on the flower side of the head by the appearance of a grayish, fuzzy substance (Figure 4.13) covered with sporangia (Harveson, 2013). These dark fruiting bodies are sacks filled with spores that are easily broken open to release spores spread on wind currents to infect neighboring plants. Symptoms can often seem to appear simultaneously on opposite sides of the head. The fungus also causes a systemic effect in the plant on the side where the head became infected. If the peduncle becomes infected, it can allow the head to fall completely off, increasing the potential for severe yield loss (Gulya et al., 1997).

Rhizopus Head Rot Disease Cycle and Damage

Even though *Rhizopus* spp. spores are found everywhere in the environment, mechanical injury on heads is a prerequisite for infection and disease development. Afterward, the disease becomes more severe in warm, humid environments, especially under irrigation. Thus, as spores are carried to sunflower plants through air currents and under conditions of high humidity, infection is initiated through wounds created by hail (Figure 4.14), birds, or insects (Harveson, 2013; Rogers et al., 1978; Yang et al., 1980).

Damage and economic losses are highly dependent on the creation of wounds, the environmental conditions present, and the time of season that wounding and

Figure 4.13 Grayish, fuzzy fungal growth on flower side of head.
Source: Harveson (2013).

Figure 4.14 Mechanical damage due to hail, followed by infection by Rhizopus spp.
Source: Harveson (2011).

infection occur. Research has demonstrated that head rot rarely occurs before flowering, so it appears that mature tissues are required to support the growth of *Rhizopus* spp. (Gulya et al., 1997). Yields are negatively affected because seeds in infected heads fail to fill properly and have reduced weights. This can affect profits for confectionary seed growers because payment is based on seed size. Yield can also be affected even if infection occurs late in the season due to the loss of seeds that have fallen to the ground from affected heads. Oilseed growers may also be adversely affected by head rot due to bitter or poor quality oils obtained from infected plants (Harveson, 2013).

Rhizopus Head Rot Management

Rhizopus head rot is directly linked to physical injury of heads, and as a consequence, anything that can be done to minimize head damage is important. However, controlling hail, birds, and insects may be at best difficult, if not impossible.

- Avoid or limit mechanical damage at or after flowering.
- Control head moth infestation before or at flowering.
- Control bird feeding. Select varieties with head types that turn down after flowering. Avoid planting sunflowers near water that consistently harbors many birds.
- Remove wild sunflowers that may serve as reservoirs for insects and pathogens before they produce seed.

Sunflower Rust

Introduction

Sunflower rust occurs in every sunflower-growing region in the world (Gulya et al., 1997). Although the disease tends to be sporadic and often localized, it is capable of causing high levels of yield loss when left unmanaged. In one case, yield in nontreated strips in a naturally infected field in North Dakota, USA, was seven-fold lower than yield in areas of the field treated with fungicides (Markell et al., 2009). Although management tools are often available (resistance, fungicides), the explosive nature of rust epidemics and frequent pathogen race changes make management of the disease a challenge (Friskop et al., 2011, 2012; Markell et al., 2013).

Sunflower Rust Pathogen and Disease History

Sunflower rust is caused by the basidiomycete fungal pathogen *Puccinia helianthi*. The pathogen is macrocyclic (five spore stages) and autoecious (all spore stages occur on one host) rust (Gulya et al., 1997). The host range of the pathogen includes all perennial and annual *Helianthus* species, *Heliopsis helianthoides,* and *Iva xanthiifolia.*

Numerous physiological races have been documented in many geographic locations; prevalence and distribution of the races change frequently (Friskop, 2013; Friskop et al., 2012; Gulya et al., 1997; Klisiewicz and Beard, 1976).

Sunflower Rust Signs and Symptoms

The first stage observed is the pycnia, which appears to the naked eye as a small (often 4–6 mm) yellow-orange bump on the top side of a leaf (Figure 4.15) or cotyledon (Figure 4.16) (Friskop et al., 2011). Aecia may form directly opposite the pycnia on the underside of the leaf, and appear as clusters of small orange cups similar in size to the pycnia (Figure 4.17) (Friskop et al., 2011). If pycnia and aecia are observed, they are usually seen early in the season, and they are difficult to find unless one is specifically looking for them. Rust is most commonly observed as uredinia, which are small (2–4 mm) raised pustules filled with dusty cinnamon-brown urediniospores (Figure 4.18, p. 111) (Friskop et al., 2011). Uredinia can be produced on any green plant part (leaves, bracts, stem, petiole). Early in the season, it is not uncommon to see leaves infected with uredinia and aecia (Figure 4.19, p. 111) and sometimes with evidence of pycnia still on the leaf. Hard and black telia will form at the season's end (Figure 4.20, p. 112) (Friskop et al., 2011).

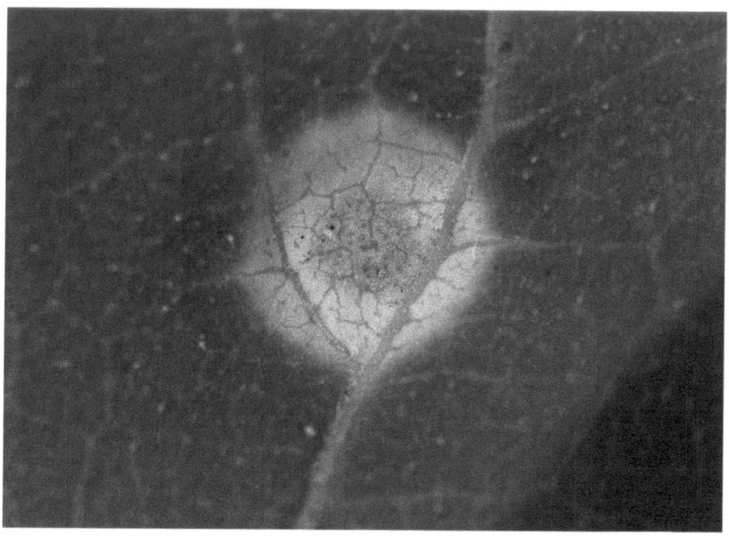

Figure 4.15 Pycnia of *Puccinia helianthi* on a leaf.
Source: Markell (2008).

Figure 4.16 Pycnia of *Puccinia helianthi* on a cotyledon.
Source: Markell (2013).

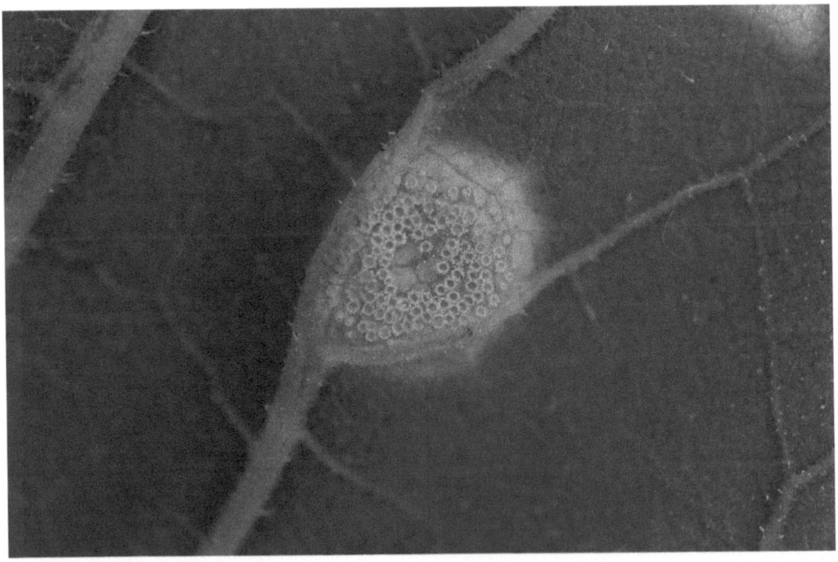

Figure 4.17 Aecia of *Puccinia helianthi*.
Source: Markell (2008).

Figure 4.18 Uredinia of *Puccinia helianthi*.
Source: Markell (2010).

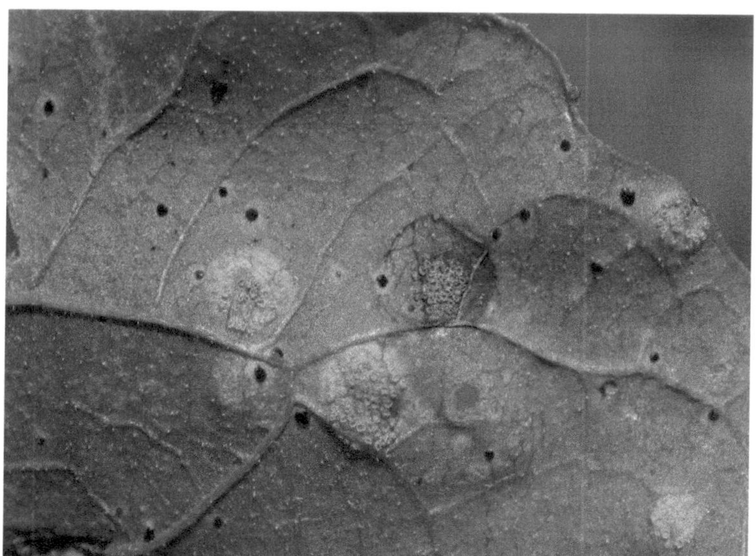

Figure 4.19 Uredinia and aecia of *Puccinia helianthi* on a heavily infected leaf.
Source: Markell (2010).

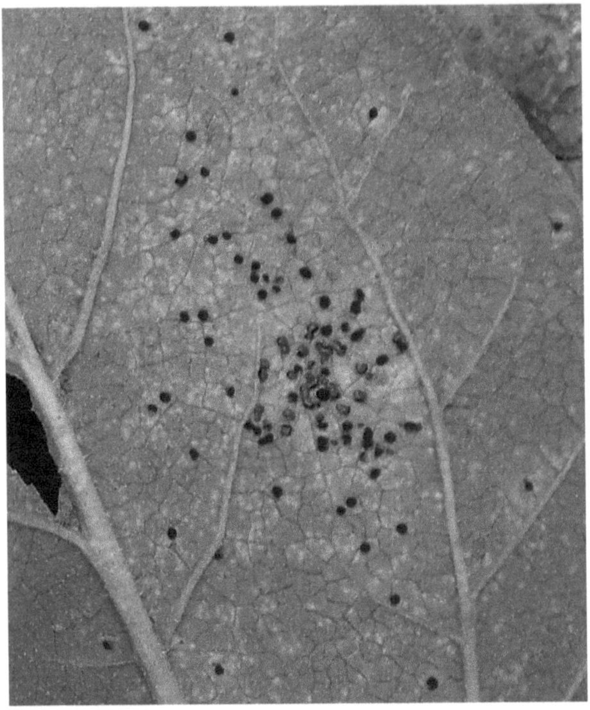

Figure 4.20 Telia of *Puccinia helianthi.*
Source: Friskop (2009).

Sunflower Rust Disease Cycle and Damage

The sexual cycle begins when telia germinate to produce four haploid basidospores that infect sunflower and produce pycnia (Gulya et al., 1997). Crossing of the strains generates aecia, which generate binucleate aeciospores that infect sunflower tissue and give rise to dikaryotic urediniospores in uredinia pustules. The uredia are a repeating stage that will continue to cycle until production of telia (Gulya et al., 1997).

Yield loss depends heavily on time of disease onset (Friskop et al., 2011; Markell, 2013; Markell et al., 2009). When the pathogen undergoes sexual reproduction in a commercial field, epidemics begin early and can have devastating potential to reduce yield (Markell et al., 2009). When an epidemic begins as a result of urediniospore dispersal from other fields, onset is often later and yield loss potential may be lower (Friskop et al., 2011). Wild *Helianthus* species and volunteer *H. annuus* serve as important reservoirs for the pathogen and can serve as foci for epidemics (Friskop et al., 2011; Gulya et al., 1997).

A minimum of six hours of free moisture is needed for germination and infection by uredniospores (Gulya et al., 1997). The optimal temperatures for uredinospore development are 20–35 °C, but the optimal temperatures for uredinisopore germination are 10–25 °C. However, germination and development can occur at a wider temperature range (Gulya et al., 1997).

Sunflower Rust Management

Rust can occur at any time, and in a favorable environment, can increase in severity very quickly. Additionally, the pathogen has numerous races and a propensity for race changes in response to resistance genes, making utilization of resistance as a sole management tool risky. However, fungicides offer an effective in-season management option if applying at optimal timing. Consequently, scouting for rust is the most critical component of management in season.

- The utilization of resistant hybrids reduces the chances of an epidemic by reducing the likely number of races that can infect the crop. However, it is unlikely that a resistant hybrid will be resistant to all pathogen races in the population (Friskop et al., 2012).
- Recent research in the United States suggests a fungicide action threshold occurs when 1% rust severity occurs at or before R5 (early bloom) on the upper four fully emerged leaves (Friskop, 2013; Markell and Khan, 2012), while earlier research in Israel suggested an action threshold of 3% was more appropriate (Shtienberg, 1995).
- Control unwanted hosts (Friskop et al., 2011) such as volunteer sunflowers and wild *Helianthus* species. These hosts can serve as a source of inoculum early in the season and as a breeding ground for new races.
- Rotate to a non-host, and avoid planting sunflowers adjacent to the previous years residue.

Sclerotinia sclerotiorum Diseases of Sunflower: Wilt, Stalk Rot, and Head Rot

Introduction

Sclerotinia sclerotiorum (Lib.) de Bary is a fungal pathogen that occurs in all sunflower growing regions in the world (Gulya et al., 1997). Depending on the location and mechanism of infection, the pathogen causes three different diseases: Sclerotinia head rot occurs on the sunflower head, Sclerotinia stalk rot occurs anywhere on the stalk, and Sclerotinia wilt occurs through infected roots and causes premature plant death (Berglund, 2007; Harveson, 2011b). When conditions are conducive for infection and disease development, the disease can be very destructive. In North America, Sclerotinia head rot is the most common and damaging in the U.S. Northern Great Plains and Prairie provinces of Canada, while the other forms of the disease are common in the U.S. Central and High Plains (Gulya et al., 2012; Harveson, 2011b; Kandel, 2013; Markell et al., 2013). Colloquially, the diseases are collectively referred to as "white mold" (Berglund, 2007).

Sclerotina sclerotiorum Pathogen and Disease History

The pathogen has a very large host range and is capable of causing infection on many dicotyledonous crops that are commonly grown in rotation with sunflower (for example, canola, dry edible beans, field peas, and soybeans) and broadleaf weeds that are present in sunflower production fields (Berglund, 2007; Gulya et al., 2007; Harveson, 2011). Disease epidemics caused by *Sclerotina sclerotiorum* in other crops can facilitate devastating epidemics in commercial sunflower, which is particularly sensitive to the pathogen (Maserivic and Gulya, 1992). Frequent disease epidemics and limited management options have challenged agricultural production enough in some areas to facilitate cropping pattern changes. Variation in aggressiveness among different *Sclerotinia sclerotiorum* strains has been documented, but evidence of physiological speciation does not exist (Rimmer et al., 2007).

Sclerotina sclerotiorum Signs and Symptoms

Symptoms of *S. sclerotiorum* are often first observed as white to tan colored lesions with concentric rings (Harveson, 2011b). Regardless of location on the plant, white mycelia (hence the common name "white mold") and hard black sclerotia will be produced both inside and outside of the lesion (Berglund, 2007; Harveson, 2011b).

Sclerotinia wilt is typically first noticed near the bloom initiation, when plants suddenly wilt and die. Sclerotinia frequently occurs on a single plant, in a small patch, or in a row (Figure 4.21). At the time of wilting, the characteristic lesion will be present at the base of the stem (Figure 4.22). The basal lesion will weaken the stem and may result in lodging. Abundant sclerotia often are produced on the outside of the basal lesion (Figure 4.23) (Berglund, 2007; Harveson, 2011b).

Sclerotinia stalk rot is often initiated through infection on the leaves, which can be observed as tan lesions with concentric rings, often surrounding a floret (Figure 4.24). Lesions on the stem surround a deceased petiole and rapidly expand up and down the plant (Figure 4.25, p. 117). As the lesion matures, the host tissue shreds and may lodge, often revealing abundant sclerotia inside the stem (Figure 4.26, p. 117) (Berglund, 2007; Harveson, 2011b).

Sclerotinia head rot infection can be observed as small water soaked spots on the back of the head that quickly enlarge and become tan colored (Figure 4.27, p. 118). Abundant white mycelium is produced internally and large sclerotial mats often cover the face of the head. Shredding and decapitation frequently occur (Figures 4.28, p. 118; and 4.29, p. 119) (Berglund, 2007; Harveson, 2011b).

Figure 4.21 Typical pattern of Sclerotinia wilt in a commercial sunflower field.
Source: Markell (2009).

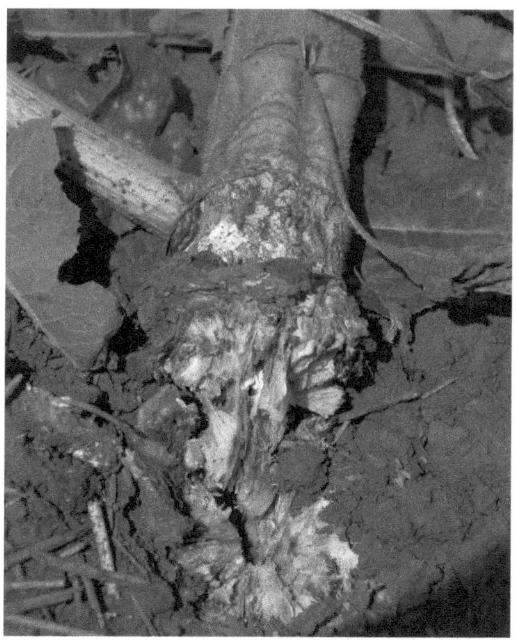

Figure 4.22 Typical basal lesion of Sclerotinia wilt.
Source: Markell (2008).

Figure 4.23 Mycelia and sclerotia of *Sclerotinia sclerotiorum* surrounding basal lesion.

Source: Markell (2007).

Figure 4.24 *Sclerotinia sclerotiorum* infection radiating from a floret on a sunflower leaf.

Source: Markell (2009).

Figure 4.25 Sclerotinia stalk rot lesion on sunflower.
Source: Markell (2009).

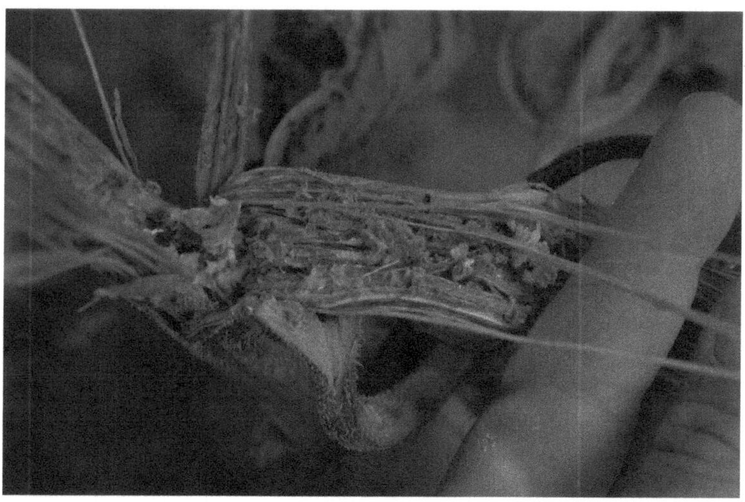

Figure 4.26 Sclerotia produced in a sunflower stem.
Source: Markell (2010).

Figure 4.27 Water soaking on the back of a sunflower head infected with *Sclerotinia sclerotiorum*.
Source: Markell (2010).

Figure 4.28 A skeletonized sunflower head typical of Sclerotinia head rot.
Source: Markell (2009).

Figure 4.29 Decapitated sunflower head typical of Sclerotinia head rot.
Source: Markell (2009).

Sclerotinia sclerotiorum Disease Cycle and Damage

Sclerotinia sclerotiorum produces robust sclerotia that can survive for many years in the soil. Sclerotia can infect sunflowers in two distinct ways. Myceliogenic germination of sclerotia produces mycelium that can directly infect sunflower roots, causing Sclerotinia wilt (Gulya et al., 1997). Carpogenic germination of sclerotia results in apothecia that discharge airborne ascospores. Ascospores most frequently cause infection when they land on the most susceptible host tissue, commonly florets or senescing plant material. Once infection occurs, damage spreads rapidly though the plant and plant-to-plant spread can occur (Figure 4.30). Sclerotinia stalk rot occurs when infection begins on the leaf, spreads through the petiole, and eventually produces a lesion on the stalk (Harveson, 2011b). Sclerotinia head rot occurs when infection begins on the head (often through a floret). Ascospore discharge, mycelium growth, and infection are favored at temperatures from 20–25 °C, although each can occur at a wider temperature range (Gulya et al., 1997). High soil moisture favors germination of sclerotia, and free moisture is necessary for infection by ascospores (Maserivic and Gulya, 1992).

Figure 4.30 Plant to plant spread of *Sclerotinia sclerotiorum*.
Source: Markell (2010).

Sclerotinia sclerotiorum *Management*

Sclerotinia diseases are among the most difficult diseases to manage. Fungicides have limited, if any, efficacy; if resistance is available it is only partial resistance; and rotation and tillage are challenged by the number of susceptible hosts commonly rotated with sunflowers and the long-lived nature of the pathogen.

- Rotate sunflowers with a non-host crop (Berglund, 2007; Harveson, 2011b); because most broadleaf crops are susceptible, a cereal grain is a preferred crop choice.
- Plant a hybrid with partial resistance, if available (Harveson, 2011b).
- Lower plant populations to increase air flow in canopy (Gulya et al., 2007).
- Fungicides have limited effectiveness and are generally not recommended (Harveson, 2011b; Markell, 2013; Markell and Khan, 2012).

Verticillium Wilt

Introduction and Disease History

Verticillium wilt, originally known as leaf mottle, is caused by the fungus *Verticillium dahliae*. Its presence as a new disease was first reported and described in North America in the late 1940s, and erroneously attributed to *V. albo-atrum*. Its true identity was recognized as *V. dahliae* in 1970, after the fungus was found to grow at temperatures exceeding 32 °C and produce true microsclerotia, traits not associated with *V. albo-atrum* (Gulya et al. 1997; Harveson, 2009). Today the disease is known to occur in most sunflower-producing areas of the world and is potentially serious enough to be considered a high-priority issue for the U.S. National Sunflower Association (NSA)

(Gulya et al., 1997). Significant economic losses have been routinely documented, particularly with confection hybrids, which generally lack genetic resistance. U.S. NSA surveyor teams found the disease occurring in 25% of inspected fields, with those from South Dakota and Manitoba suffering the most severe disease levels (Kandel et al., 2013).

Verticillium Wilt Signs and Symptoms

Symptoms of the disease are first noted when the crop nears flowering. Generally the first visible symptoms consist of small yellow areas between veins that merge and increase in size with time, and eventually become necrotic (Figure 4.31) (Harveson, 2009). Symptom development begins on lower leaves as interveinal yellowing of tissues that continues to develop progressively as the pathogen moves up the stems to newer leaves, giving leaves a mottled appearance (Figure 4.32). This provides the origin for the disease name "leaf mottle" for describing the contrast between yellow and persisting green areas of the leaves (Gulya et al., 1997). The yellow areas become brown and necrotic, often with bright yellow zones surrounding necrotic tissues (Figure 4.33). Affected lower leaves will eventually dry and die.

Infected stems in a longitudinal section may show reddish-brown vascular discoloration (Figure 4.34, p. 123). In severely diseased plants, black patches are observed late in the season lining the lower stems, consisting of thousands of tiny microsclerotia (Fig-

Figure 4.31 Early Verticillium wilt symptoms consisting of interveinal yellowing followed by necrosis of leaf tissue.

Source: Harveson (2008).

Figure 4.32 Interveinal "mottling" symptoms characteristic of Verticillium wilt. Note the progression up the stem to new leaves.

Source: Harveson (2009).

Figure 4.33 Interveinal necrotic patches surrounded by wavy, yellow haloes.

Source: Harveson (2009).

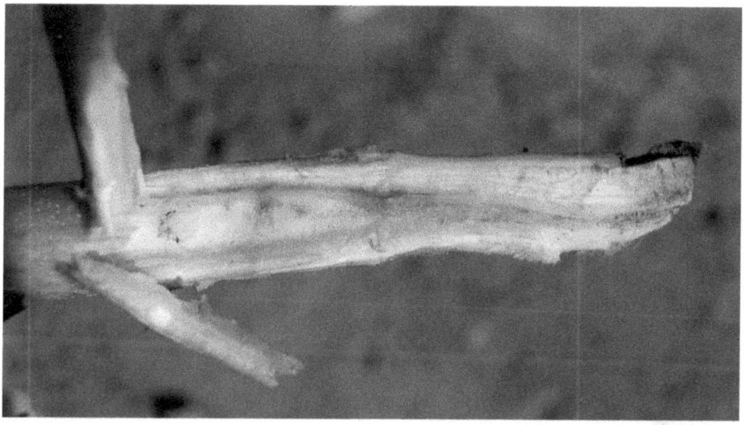

Figure 4.34 Mild vascular discoloration of Verticillium-infected sunflower stalk.
Source: Harveson (2008).

ure 4.35). Due to the systemic nature of the disease, the pathogen may be distributed throughout the plant and may be found in roots, stems, petioles, leaf midribs, or heads.

Verticillium Wilt Pathogen and Disease Cycle

Verticillium dahliae has a large host range consisting of more than 350 plant species; however, all are tap-rooted plants with no grass hosts. In addition, the pathogen is soilborne and survives in soil or plant residue as microsclerotia, which allows survival of the fungus for several years without a host (Harveson, 2009). The microscleotia typically remain dormant until being stimulated by host root exudates, resulting in germination. Germinating microsclerotia produce hyphae that infect roots and then invade the plant's water-conducting system, causing a vascular wilt disease (Gulya et al., 1997; Markell, 2013). As disease progresses, the fungus can further colonize and develop and serve as a primary source of later infection. Movement of microsclerotia associated with seed is likely the mechanism for the pathogen's spread from its origin in the Americas throughout the rest of the world (Gulya et al., 1997).

Verticillium Wilt Management

Limited management tools for Verticillium wilt are available. Where genetic resistance is available, it should be utilized.

- Three- to four-year rotations with small grains or other non-host crops may reduce the pathogen population in the field and reduce the likelihood of an epidemic (Harveson, 2009).

Figure 4.35 Blackened pith of severely affected plant consisting of thousands of microsclerotia.
Source: Harveson (2006).

- Proper seed sanitation will reduce the possibility of microsclerotia occurring with seed, which could lead to introduction of the pathogen into new fields, into new areas in fields, and/or higher severity in the current crop (Harveson, 2009).
- Use cultivars with genetic resistance where available.
- Avoid fields with a history of the disease.

Concluding Remarks and Future Developments

Sunflower producers face several challenges when managing sunflower diseases, most notably pathogens that can cause significant yield loss under a variety of environmental conditions (for example, sunflower rust or Verticillium wilt cause significant damage when it is dry, compared to Sclerotinia head rot and Phomopsis stem canker, which do more damage when the weather is wet), variable pathogens [for example, the pathogens that cause rust and downy mildew undergo frequent race changes or fungicide sensitivity shifts (downy mildew only)], and limited management tools to the most devastating diseases (for example, Sclerotinia head rot and Phomopsis stem canker). However, pathologists and breeders are continually working to develop and deploy new management tools that can mitigate yield losses to diseases. In the future, there may be more effective management tools available.

In our view, there are four critical things sunflower producers can do currently, to maximize their disease management.

1. *Identify the diseases that you are at greatest risk of suffering yield loss to.* Answer such questions as what diseases have occurred in the past, what diseases do the current and future environmental trends favor and what management strategy is lacking in my production system that increases risk? With this information in mind, monitoring and proactive management will be most effective.
2. *Scout for and correctly identify the diseases in your field.* Proper identification of diseases is the most important step for in-season management (for example, fungicides for rust) and determination of future management options (for example, resistant hybrid selection). Conversely, lack of scouting or accurate identification can lead to significant yield losses that may be avoidable (for example, rust) or incursion of an economic loss by incorporation of the wrong management strategy (for example, fungicides applied for Verticillium wilt, which is not effective).
3. *Use an Integrated Pest Management (IPM) strategy.* IPM involves the utilization of the most current management information with multiple appropriate management tools in an environmental sensitive way. While this includes disease risk assessment, scouting, and proper disease identification, it also includes simultaneous use of multiple tools to manage the same disease. This reduces the likelihood of a management failure that could occur if a single management tool fails. This is critical because several sunflower pathogens have races that overcome resistance and only partially effective resistance may be available to other pathogens.
4. *Stay engaged.* Available management tools, the best information to use them (for example, fungicide thresholds), and pathogens change abruptly and frequently. Utilization of the most effective management tools with the most current information will increase your chance of crop success, while ignoring pathogen changes (for example, new races) will increase the likelihood of economic loss due to disease epidemics or unnecessary (and often costly) management inputs.

References

Berglund, D. R. Sunflower Production. *North Dakota State University Ext. Ser. Bull. 25,* 2007.

Brum, M. C.; Araujo, W. L.; Maki, C. S.; Azevedo, J. L. Endophytic Fungi from *Vitis labrusca* and Its Potential for the Biological Control of *Fusarium oxysporum. Genet. Mol. Res.* **2012,** *3,* 4187–4197.

Commonwealth Agricultural Bureau. *Diaporthe helianthi.* Distribution Maps of Plant Diseases, Map. No. 776, 1999.

Debaeke, P.; Estragnat, A.; Reau, R. Influence of Crop Management on Sunflower Stem Canker (*Diaporthe helianthi*). *Agronomie* **2003,** *23,* 581–592.

Desanlis, M., Aubertot, J. N.; Mestries, E.; Debaeke, P. Analysis of the Influence of a Sunflower Canopy on *Phomopsis helianthi* Epidemics as a Function of Cropping Practices. *Field Crops Res.* **2013,** *149,* 63–75.

Friskop, A. *Development of Management Strategies to Control Sunflower Rust Using Genetic Resistance and Fungicides.* Ph.D. Thesis, North Dakota State University, December 2013.

Friskop, A.; Markell, S.; Gulya, T. *Downy Mildew of Sunflower.* North Dakota State University Extension Service Publication PP-1402, 2009.

Friskop, A.; Markell, S.; Gulya, T.; Halley, S.; Schatz, B.; Schaefer, J.; Wunsch, M.; Meyer, S.; Kandel, H.; Acevedo, M.; et al. *Sunflower Rust.* North Dakota Cooperative Extension Service Publication PP-1557, 2011.

Friskop, A.; Gulya, T.; Acevedo, M.; Harveson, R.; Humann, R.; Markell, S. Race Diversity of *Puccinia helianthi* (Sunflower Rust) in the Northern Great Plains in 2011. 2012 Annual Meeting of the American Phytopathological Society, Providence, RI, August 4–8, 2012. *Phytopathology* **2012,** *102,* S4.42.

Gulya, T.; Hulke, B.; Seiler, G.; Markell, S.; Kandel, H. *Changes in Disease Incidence over the Last Decade in the United States and Around the World.* Abstracts of the 18th International Sunflower Conference, Mar del Plata, Argentina, February 27–March 1, 2012.

Gulya, T.; Kandel, H.; McMullen, M.; Knodel, J.; Mathew, F.; Lamey, H.; Nowatzki, J.; Markell, S. Prevalence and Incidence of Sunflower Downy Mildew in North Dakota between 2001 and 2011. *Plant Health Progress,* 2013. Online. DOI: 10.1094/PHP-2013-0522-01-RS.

Gulya, T. J.; Sackston, W. E.; Viranyi, F.; Masirevic, S.; Rashid, K. Y. New Races of the Sunflower Downy Mildew Pathogen (*Plasmopora halstedii*) in Europe and North and South America. *J. Phytopathol.* **1991,** *132,* 303–311.

Gulya, T. J.; Rashid, K. Y.; Masirevic, S. M. Sunflower Diseases. In *Sunflower Technology and Production;* Schneiter, A. A., Ed.; Agronomy 35; American Society of Agronomy: Madison, WI, 1997; 263–379.

Harveson, R. M. Verticillium Wilt of Sunflowers in Nebraska. NebGuide G1963, 2009.

Harveson, R. M. Downy Mildew of Sunflower in Nebraska. NebGuide G2075, 2011a.

Harveson, R. M. Sclerotinia Diseases of Sunflower in Nebraska. NebGuide G2107, 2011b.

Harveson, R. M. Rhizopus Head Rot of Sunflower in Nebraska. NebGuide (revised) G1677, 2013.

Herr, L. J.; Lipps, P. E.; Watters, B. L. Diaporthe Stem Canker of Sunflower. *Plant Disease,* **1983,** *67,* 911–913.

Kandel, H. *2012 National Sunflower Association Survey: Yield, Cultural Practices and Yield Limiting Factors.* National Sunflower Association: Mandan, ND, 2013. http://www.sunflower nsa.com/uploads/research/1201/kandel_2012.nsa.survey_13.pdf. Accessed 03/08/2014.

Klisiewicz, J. M.; Beard, B. H. Diseases of Sunflower in California. *Plant Dis. Rep.* **1976,** *60,* 298–301.

Kolte, S. J. Diseases of Annual Edible Oilseed Crops. *Vol. III: Sunflower, Safflower and Nigerseed Diseases.* CRC Press: Boca Raton, FL, 1985.

Markell, S.; Khan, M. *2013 North Dakota Field Crop Fungicide Guide.* North Dakota Cooperative Extension Service Publication PP-622, 2012.

Markell, S.; Gulya, T.; McKay, K.; Hutter, M.; Hollingsworth, C.; Ulstad, V.; Kock, R.; Knudsvig, A. Widespread Occurrence of the Aecial Stage of Sunflower Rust Caused by *Puccinia helianthi* in North Dakota and Minnesota in 2008. *Plant Disease* **2009,** *93,* 668–669.

Markell, S. G. Sunflowers. In *Fungicides for Field Crops;* Mueller, D., Wise, K. A., Dufault, N. S., Bradley, C. A., Chilvers, M. I., Eds.; APS Press: St. Paul, MN, 2013; pp 95–97.

Masirevic, S. M.; Gulya, T. Sclerotinia and Phomopsis—Two Devastating Sunflower Pathogens. *Field Crops Research* **1992,** *30,* 271–300.

Mathew, F.; Alananbeh, K.; Balbyshev, N.; Heitkamp, E.; Castlebury, L.; Gulya, T.; Markell, S. *Reevaluation of* Phomopsis *Species Affecting Sunflowers in the United States.* Abstracts of the 18th International Sunflower Conference, Mar del Plata, Argentina, February 27–March 1, 2012.

Muntanola-Cvetkovic, M.; Mihaljcevic, M.; Petrov, M. On the Identity of the Causative Agent of a Serious Phomopsis-Diaporthe Disease in Sunflower Plants. *Nova Hedwigia* **1981,** *34,* 417–435.

Rekab, D.; del Sorbo, G.; Reggio, C.; Zoina, A.; Firrao, G. Polymorphisms in Nuclear rDNA and mtDNA Reveal the Polyphyletic Nature of Isolates of Phomopsis Pathogenic to Sunflower and a Tight Monophyletic Clade of Defined Geographic Origin. *Mycol. Res.* **2004,** *108,* 393–402.

Rimmer, S. R.; Shattuck, V. I.; Buchwaldt, L. *Compendium of Brassica Diseases.* The American Phytopathological Society, APS Press: St. Paul, MN, 2007; pp 43–47.

Rogers, C. E.; Thompson, T. E.; Zimmer, D. E. Rhizopus Head Rot of Sunflower: Etiology and Severity in the Southern Plains. *Plant Dis. Rep.* **1978,** *62,* 769–771.

Sackston, W. E. Downy Mildew of Sunflower. In *The Downy Mildews;* Spencer, D. M., Ed.; Academic Press: London, 1981; pp 545–575.

Schipper, M. A. A Revision of the Genus Rhizopus. I. The *Rhizopus stolonifer*-group and *Rhizopus oryzae. Stud. Mycol.* **1984a,** *25,* 1–19.

Schipper, M. A. A Revision of the genus Rhizopus. II. The Rhizopus microsporus Group. *Stud. Mycol.* **1984b,** *25,* 20–34.

Schtienberg, D. Rational Suppressioin of Sunflower Rust: Development and Evaluation of an Action Threshold. *Plant Dis.* **1995,** *79,* 506–510.

Thompson, S. M.; Tan, Y. P.; Young, A. J.; Neate, S. M.; Aitken, E. A. B.; Shivas, R. G. Stem Cankers on Sunflower (*Helianthus annuus*) in Australia Reveal a Complex of Pathogenic Diaporthe (*Phomopsis*) Species. *Persoonia* **2011,** *27,* 80–89.

Udayanga, D.; Xingzhong, L.; McKenzie, E. H. C.; Chukeatirote, E.; Bahkali, A. H. A.; Hyde, K. D. The Genus *Phomopsis:* Biology, Applications, Species Concepts and Names of Common Pathogens. *Fungal Diversity* **2011,** *50,* 189–225.

Vrandecic, K.; Cosic, J.; Jurkovic, D. First Report of *Diaporthe phaseolorum* on Sunflower (*Helianthus annuus*) in Croatia. *Plant Dis.* **2009,** *93,* 1074.

Vrandecic, K.; Jurkovic, D.; Riccioni, L.; Cosic, J.; Duvnjak, T. *Xanthium italicum, Xanthium strumarium* and *Arctium lappa* as New Hosts for *Diaporthe helianthi. Mycopathologia* **2010,** *170,* 51–60.

Vukojević, J.; Mihaljčević, M.; Franić-Mihajlović, D. Variability of *Phomopsis* Populations in Sunflowers (*Helianthus annuus* L.). *Helia* **2001,** *24* (34), 69–76.

Yang, S. M.; Morris, J. B.; Unger, P. W.; Thompson, T. E. Rhizopus Head Rot of Cultivated Sunflower in Texas. *Plant Dis.* **1980,** *63,* 833–835.

5

Sunflower Broomrape (*Orobanche cumana* Wallr.)

José M. Fernández-Martínez, Begoña Pérez-Vich, and Leonardo Velasco ■ *Instituto de Agricultura Sostenible (IAS-CSIC), Córdoba, Spain*

Introduction

Parasitic plants obtain some or all their nutritional requirements from a host plant. They develop a special organ known as the *haustorium,* which invades host plant tissues and serves to withdraw water and nutrients (Joel, 2013). Some of them are *hemiparasitic,* which means they have photosynthetic capacity but rely on the host for water and soil nutrients. Others are *holoparasitic,* which means they do not have photosynthetic capacity and entirely depend on the host for nutrition (Mohamed et al., 2006). Some parasitic plants are noxious weeds. Some of the most damaging parasitic weeds belong to the genera *Striga* (witchweeds) and *Orobanche*/*Phelipanche* (broomrapes), both of them of the Orobanchaceae family (Musselman, 1980).

The closely related genera *Orobanche* and *Phelipanche,* known as broomrapes, encompass around 170 holoparasitic plant species mainly distributed in the northern hemisphere (Gevezova et al., 2012), with greatest diversity in the Mediterranean region and Western Asia (Mohamed et al., 2006). Some of them are noxious weeds on a more or less restricted spectrum of hosts. The most important weedy species are *O. crenata* Forsk., mainly parasitizing legume crops and safflower; *O. cernua* L., mainly found on carrot, tomato, and tobacco; *P. ramosa* (L.) Pomel, common on rapeseed/canola and many Solanaceae crops; *P. aegyptiaca* (Pers.) Pomel, which possesses ability to parasitize a broad range of hosts; and *O. cumana* Wallr., which parasitizes exclusively sunflower crops (Mohamed et al., 2006; Parker, 2013). Due to its host specificity, *O. cumana* is commonly named as sunflower broomrape. Broomrapes are very destructive weeds, with yield losses close to 100% under high infestations (Bernhard et al., 1998; Bülbül et al., 2009).

For many years, there has been some confusion in the literature about the scientific name of sunflower broomrape, with *O. cumana* Wallr. and *O. cernua* Loefl. being used indistinctly. The reason for such confusion can be traced back to several classical taxonomic studies in which *O. cumana* was treated as an infraspecific taxon of *O. cernua* (Beck-Mannagetta, 1930; Rechinger, 1943). Nevertheless, comparative studies between both taxa clearly support their treatment as separate species (Pujadas Salvà and Velasco, 2000; Román et al., 2003). Accordingly, *O. cumana* is the accepted scientific name for sunflower broomrape, though other *Orobanche* species such as *O. aegyptiaca* Pers. also parasitize sunflower at a local scale (Eizenberg et al., 2003).

Figure 5.1 Sunflower broomrape growing in the wild on *Artemisia maritima.*

Sunflower broomrape is found in the wild parasitizing wild *Compositae* spp., mainly *Artemisia* spp. (Figure 5.1), from Central Asia to South-Eastern Europe (Pujadas Salvà and Velasco, 2000). After the introduction of the North American–native sunflower as a crop in the Old World, *O. cumana* was first observed parasitizing sunflower (Figure 5.2) in the Saratov district of Russia in the 1890s (Skoric, 2012). The first sunflower cultivar resistant to *O. cumana* was reported in 1918, after which a new *O. cumana* race that overcame sunflower genetic resistance appeared in 1926 (Skoric, 2012). Unlike the other *Orobanche* and *Phelipanche* weedy species, for which genetic resistance in the host is of quantitative nature (horizontal), genetic resistance to *O. cumana* in sunflower is in most cases qualitative or vertical (Pérez-Vich et al., 2013). For this reason, *O. cumana* populations are commonly classified into physiological races (Vranceanu et al., 1980), which is not the case for the other *Orobanche* and *Phelipanche* spp.

Figure 5.2 Sunflower broomrape parasitizing sunflower.

Sunflower broomrape has great capacity of dispersion and mutation. Individual broomrape plants produce thousands of tiny seeds that are easily dispersed by wind and other agents, including sunflower seeds, to which broomrape seed can be found attached. Broomrape seeds may remain viable for more than 15–20 years and will only germinate in the presence of the host plant (Skoric, 2012). The main way to control broomrape attacks has been the development of resistant sunflower cultivars based on vertical resistance mechanisms. However, such vertical resistance mechanisms have been repeatedly surpassed by the parasite, which has led to focus on broomrape resistance as one of the most important breeding traits in sunflower areas prone to broomrape parasitization. Alternatively, sunflower broomrape can be controlled by postemergence application of imidazolinone (IMI) herbicides on non-GMO (genetically modified organism), IMI-tolerant hybrids, derived from resistance alleles identified in wild populations (Al-Khatib et al., 1998) or developed by mutagenesis

(Sala et al., 2008). In the next sections, we will discuss the current status of sunflower broomrape infestations and race evolution around the world, recent research advances in the biology and genetics of the parasite, and current strategies and future prospects for sunflower broomrape control through exploiting sunflower genetic resistance to either broomrape or IMI herbicides.

Current Status of Broomrape Infestations around the World

Sunflower broomrape represents one of the most serious production constraints in many sunflower-producing countries around the world, particularly in Central and Eastern Europe, Spain, Turkey, Israel, Iran, Kazakhstan, and China (Skoric et al., 2010). Moreover, the parasite has spread to new areas in recent years. Of particular relevance is the recent observation of broomrape attacks in France, one of the main sunflower producing countries in Europe (Jouffret and Lecomte, 2010). It has also been recently detected in Tunisia (Amri et al., 2012). In countries such as Spain (Fernández-Escobar et al., 2009; Pineda-Martos et al., 2013) and Russia (Antonova et al., 2013), virulent races are spreading to areas where broomrape was not present few years ago. In America, sunflower broomrape is not present (Miladinovic et al., 2012). According to estimations of commercial seed companies (unpublished data), between 4 and 5 million ha of the world cultivated area (15–20%) are infested with sunflower broomrape.

The current racial situation of broomrape in the main infested areas remains unclear, as there is little information on the correspondence of races with the same name reported in different countries. Races D and E were predominant till the middle 1990s, and they were satisfactorily controlled by resistance gene *Or5*, widely used in commercial hybrids. Populations overcoming *Or5* resistance were detected in 1995 in Spain (Alonso et al., 1996) and shortly after in Romania (Pacureanu-Joita et al., 2004), Turkey (Kaya et al., 2004), and several other countries (Antonova et al., 2013; Fernández-Martínez et al., 2012; Skoric, 2012). Currently, reports on races named as F, G, and H are common in several countries, but studies to clarify and unify this nomenclature are required. It seems clear that new races more virulent than race E are now predominant in some regions of Spain (Molinero-Ruiz et al., 2008), Turkey (Kaya et al., 2004), Romania (Pacureanu-Joita et al., 2009a), and Russia (Antonova et al., 2013), and these races are rapidly spreading to new areas (Antonova et al., 2013).

Biological Aspects of Broomrape

Broomrape–Sunflower Interaction

Individual broomrape plants produce an impressive number of seeds: from 50,000 to 500,000 (Gevezova et al., 2012). Seeds are of near-microscopic size: from 250 to 380 μm long and from 150 to 240 μm wide (Joel, 1987), with a weight between 1.0 and 2.5 μg (Castejón et al., 1991; Skoric, 2012). The seeds maintain their viability in the soil for up to 20 years (Skoric, 2012). Seed germination is dependent on compounds secreted by the host plant, as broomrape seedlings cannot survive in absence of a host root (Louarn et al., 2012). Germination stimulation requires a previous conditioning period of several days under adequate temperature and humidity, after which the root exudates from the host induce germination (Louarn et al., 2012). Optimal soil temperature for germination is between 20 and 25 °C (Skoric, 2012). In most *Orobanche* spp., the main host compounds that stimulate germination are strigolactones (Yoneyama et al., 2010). However, although sunflower produces strigolactones as well (Yoneyama et al., 2011), the main natural germination stimulant from sunflower root exudates is dehydrocostus lactone, a guaianolide sesquiterpene lactone (Joel et al., 2011).

After germination, the radicle grows toward the host root and forms an attachment structure known as *appresorium.* Afterward, the broomrape tissue known as haustorium penetrates the cortex and reaches the vascular cylinder, where it draws water and nutrients necessary for broomrape survival (Parker and Riches, 1993; Rispail et al., 2007). Haustorium initiation is triggered by specific exogenous molecules exuded by the host roots, called *haustoria-inducing factors* (Thorogood and Hiscock, 2010). The outer part of the seedling develops into a bulb-like structure, known as a *nodule* or *tubercle,* which in turn develops into a shoot that emerges from the soil giving rise to a flowering spike that produces mature seeds (Rispail et al., 2007), starting a new cycle. Sunflower broomrape develops from one to three shoots per tubercle (Antonova et al., 2012). Figure 5.3 shows sunflower broomrape shoots at different stages of development.

Reproductive Biology

Orobanche spp. have been traditionally classified into two main groups according to their mating system: species pollinated by insects, particularly bumblebees and bees, and self-fertilizing species (Musselman et al., 1981). Sunflower broomrape has been habitually classified within the second group, mainly due to its flower morphology, with bent tubular corollas having small lower lips (Figure 5.4) that do not facilitate pollinator landing (Satovic et al. 2009), as well as to the existence of low intrapopulation genetic diversity (Gagne et al., 1998). However, experiments conducted using a mutant lacking anthocyanins (Figure 5.5) as a monogenic morphological marker (Rodríguez-Ojeda et al., 2011) revealed rates of natural cross-fertilization between 14.8% and 40.0%, which suggests that the species is not strictly self-pollinated and some extent of cross-pollination should be expected (Rodríguez-Ojeda et al., 2013a).

Figure 5.3 Sunflower broomrape shoots at different stages of development.

Figure 5.4 Flowering plant of sunflower broomrape.

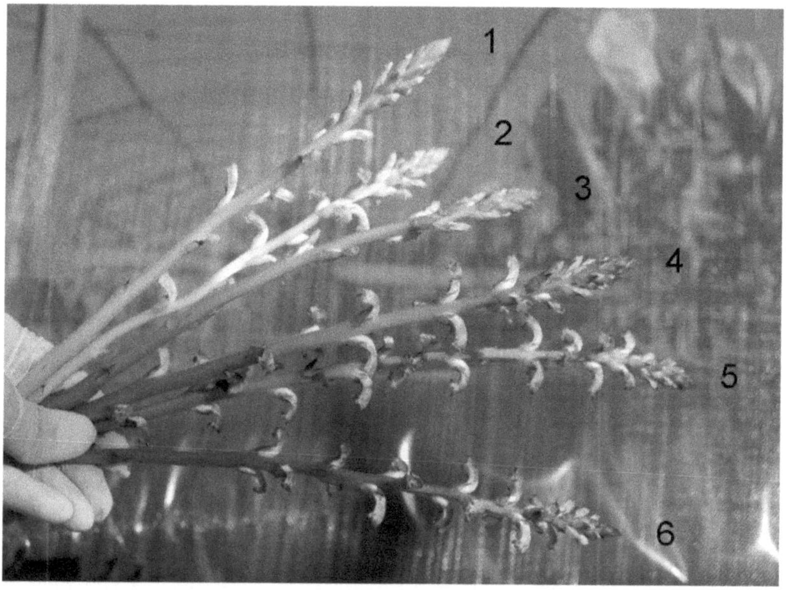

Figure 5.5 Unpigmented mutant plants (1–2), partially pigmented heterozygotes (3–4), and fully pigmented plants (5–6) of sunflower broomrape.

Genetics, Race Evolution, and Population Dynamics

There are few genetic studies on sunflower broomrape. Castejón-Muñoz et al. (1991) found low intrapopulation genetic diversity in four populations from Southern Spain using isoenzyme markers, but greater genetic diversity in a fifth population from the same area. The results were similar to those reported by Pineda-Martos et al. (2013) after studying 50 populations from different geographical areas of Spain using simple sequence repeats (SSRs) markers. They found that the populations fell into two main genetic pools, one concentrated in the Guadalquivir Valley (Southern Spain) and another one in Cuenca Province (Central Spain). Both groups were genetically distant, but both intra- and interpopulation genetic variations were, in general, extremely low within each genetic pool due to a founder effect. Exceptionally, a few populations exhibited larger genetic diversity, which was explained on the basis of simultaneous occurrence of individuals of both genetic pools in some fields followed by genetic recombination between them. Molinero-Ruiz et al. (2013) confirmed the existence of great genetic distance between populations of Southern and Central Spain using random amplification of polymorphic DNA (RAPD) markers, though greater genetic diversity within each genetic pool was detected. Gagne et al. (1998) studied genetic diversity in eight populations from several countries—three from Bulgaria, one from

Turkey, one from Romania, and three from Southern Spain—using RAPD markers. They identified large interpopulation genetic variation, concluding the existence of two main genetic pools, one comprising the populations from Eastern Europe and another one including the populations from Southern Spain. As in the other studies, intrapopulation diversity was low. The occurrence of low genetic diversity in most sunflower broomrape populations suggests the monophyletic origin of the genetic pools identified thus far (Gagne et al., 1998; Pineda-Martos et al., 2013). Molinero-Ruiz et al. (2013) found that one population from Hungary and four populations from Turkey were closer to populations from Southern Spain than to populations from Central Spain.

The genetic broomrape–sunflower interaction is in most cases governed by the gene-for-gene theory that describes plant–pathogen interactions in which resistance reactions are governed by the interaction of host genes for resistance and the corresponding pathogen genes for avirulence (Flor, 1971). This genetic system determines the occurrence of physiological races of broomrape that are controlled by dominant resistant genes of sunflower (Fernández-Martínez et al., 2012). Sunflower broomrape was first detected on sunflower in the Saratov district of Russia in the 1890s, from where it spread across significant areas of Russia at the beginning of the 20th century. Resistance to those initial populations, later named as race A, was identified in 1918, although a new race B overcoming race A resistance appeared in few years (Skoric, 2012). After resistance to race B was developed, a new race C was detected in Moldova and Ukraine in the mid 1960s. Resistance to this new race was developed around 10 years later (Alonso, 1998). The occurrence of a race C, probably different than that reported in Moldova and Ukraine, overcoming both race A and race B resistant cultivars, was observed in Bulgaria in 1970 (Skoric, 2012). Vranceanu et al. (1980) conducted an experiment in Braila County, Romania, which was highly infested with virulent populations. Seeds from several populations were collected at several locations, mixed, and incorporated into the soil in an experimental plot. By evaluating a high number of sunflower cultivars, the authors concluded the existence of five races named A through E and developed a set of differential lines to identify them. Later studies confirmed the prevalence of races D and E in several countries such as Spain (Melero-Vara et al., 1989), Serbia (Masirevic and Medic-Pap, 2009), Turkey (Bulbul et al., 1991), and Bulgaria (Shindrova, 2006). New races appeared from the middle 1990s onward in several countries. Initially, all of them were named as race F, although the relationship between the different races of F has not been studied. They have been reported in Spain (Alonso et al., 1996), Romania (Pacureanu-Joita et al., 2004), Turkey (Kaya et al., 2004), Bulgaria (Shindrova, 2006), Ukraine (Burlov and Burlov, 2010), and Russia (Antonova et al., 2013). Populations overcoming resistance sources to race F, named as races G and H, have been identified in most of these countries (Antonova et al., 2013; Kaya et al., 2009; Pacureanu-Joita et al., 2009a;

Shindrova and Penchev, 2012). As mentioned for race F, no comparative studies have been conducted between races G and H reported in different countries.

In sunflower, genetic resistance to broomrape has been found in most cases to be controlled by single dominant genes (Velasco et al., 2012; Vranceanu et al., 1980). A genetic study consisting of crossing plants of sunflower broomrape races E and F and studying avirulence/virulence reactions of the F_1 and F_3 plant generations on the race F differential line P-1380, resistant to race E and susceptible to race F, concluded that race E avirulence is dominant and controlled by a single gene (Rodríguez-Ojeda et al., 2013b). This confirmed at the parasite level the validity of the gene-for-gene theory in the sunflower–broomrape interaction. The discovery of the monogenic control of avirulence genes has important implications for understanding race evolution in sunflower broomrape, as new races may evolve as the result of a single mutation. Taking into account the large number of seeds produced by each single plant, the probability of functional mutations related to virulence should not be underestimated. In the previously mentioned study of Pineda-Martos et al. (2013) on broomrape populations from Spain, the authors found that clustering of populations into two main genetic pools was not associated with virulence groups, as populations within each cluster included several virulence groups, including race F. This suggested the occurrence of independent mutational events at both genetic pools. The existence of a relatively high rate of cross-fertilization in sunflower broomrape (Rodríguez-Ojeda et al., 2013a) may also represent an important driving force for race evolution if populations with different virulence genes get into contact with subsequent genetic recombination, as demonstrated in Spanish populations by Pineda-Martos et al. (2013).

Sunflower broomrape is becoming an increasingly important problem for sunflower production in the world, not only due to the occurrence of more virulent races, but also due to the spread of the parasite to new areas. New infestations are continually being observed in countries such as Russia (Antonova et al., 2013) and Spain (Pineda-Martos et al., 2013), where the parasite has been present for many years, and also in countries where the parasite has not been observed before, such as France (Jouffret and Lecomte, 2010) and Tunisia (Amri et al., 2012). The seeds are easily dispersed by water, wind, animals, humans, machinery, or through attachment to sunflower seeds (Castejón et al., 1991; Pacureanu-Joita et al., 2009b).

Sunflower Breeding for Broomrape Control

Genetic Resistance in Sunflower

Sources of Resistance

Selection for sunflower resistance to broomrape started in the early 1910s using individual selection methods. The first cultivars resistant to broomrape were announced in 1915 by the researchers of the Saratov Experimental Station (former

USSR), E. M. Plachek and A. I. Steboot (Gorbachenko et al., 2011). One of these cultivars, Saratovsky 169, was grown for several years at a plot that exceeded 1 million ha (Skoric, 2012). After the appearance of race B in the middle 1920s, several experimental stations in Russia became involved in the development of germplasm resistant to the new race. The first resistant cultivars were developed by L. A. Zhdanov, in particular Zhdanovsky 6432 and Zhdanovsky 8281, which were widely cultivated (Gorbachenko et al., 2011). Zhdanov was a pioneer in the use of wild *Helianthus* spp., particularly *H. tuberosus*, for sunflower breeding for broomrape resistance. A number of sources of resistance to race B were developed in the following years in Russia (Skoric et al., 2010).

In more recent times, the spread of broomrape to most of the main sunflower-producing regions in Europe led to the inclusion of broomrape resistance as one of the most important traits in both private and public breeding programs. In the late 1970s, the most valuable sources of resistance in Romania were cultivars derived from resistant germplasm developed in the former USSR, as well as populations derived from interspecific crosses between *H. annuus* and *H. tuberosus* (Vranceanu et al., 1980). The appearance of new races considerably reduced the available sources of resistance in the germplasm of cultivated sunflower. In Turkey, Gulya et al. (1994) found only 22 resistant entries in a field evaluation of 903 accessions, whereas in Spain, Domínguez et al. (1996) found 8 resistant and 33 segregating entries in the evaluation of 429 accessions of different origins for resistance to race E. Sources of resistance to the latest races have been scarce in germplasm of cultivated sunflower, although valuable resistant germplasm has been identified in breeding programs conducted in Spain (Fernández-Martínez et al., 2004; Rodríguez-Ojeda et al., 2001a), Romania (Pacureanu-Joita et al., 2004, 2009b), Turkey (Kaya et al., 2009), and Russia (Gontcharov et al., 2004; Gontcharov, 2009). In most cases, the germplasm exhibited vertical or qualitative genetic resistance. In addition, genetic sources of horizontal or quantitative genetic resistance have been developed (Pérez-Vich et al., 2005).

Wild *Helianthus* spp. are a major reservoir of genes controlling resistance to the new virulent races of *O. cumana* (Christov et al., 2009). Resistance is mainly found in perennial *Helianthus* spp., though resistant annual species have been identified as well. For race E, Ruso et al. (1996) evaluated 24 perennial and 11 annual species, identifying resistance in most of the perennial and two annual species, *H. anomalus* and *H. debilis*. Similar results were reported by Fernández-Martínez et al. (2000) for resistance to race F, although in this case a higher number of perennial species showed some incidence, whereas accessions of the annual species *H. agrestis* and *H. anomalus* were identified as resistant. Extensive evaluations of wild *Helianthus* spp. for resistance to the latest broomrape races have been conducted in recent years by several research groups (Antonova et al., 2011; Christov et al., 2009; Petcu and Pacureanu, 2011; Terzic et al., 2010).

Introgression of broomrape resistance genes from annual *Helianthus* spp. into elite germplasm can be easily achieved in most cases by conventional crossing and backcrossing, although embryo rescue may be required in early generations and accurate testing for broomrape resistance is required at each generation (Velasco et al., 2012). Conversely, introgression of resistance genes from perennial species is generally more difficult due to problems associated with early hybrid embryo abortion and sterility in early generations and may require chromosome doubling of the F_1, in addition to embryo rescue (Jan and Fernández-Martínez, 2002). Successful development of resistant sunflower germplasm incorporating genes from wild *Helianthus* spp. has been reported both for annual (Hladni et al., 2009; Velasco et al., 2012) and perennial species (Christov et al., 2009; Cvejic et al., 2012; Jan et al., 2002; Petcu and Pacureanu, 2011). Resistant sources derived from intergeneric hybridization between cultivated sunflower and species of the genera *Aster*, *Inula*, *Tithonia*, and *Verbesina* have been developed as well (Christov et al., 2010).

Genetics of Resistance

Resistance to *Orobanche* spp. in crop plants is generally under polygenic, non-race-specific genetic control. Sunflower is a notable exception where resistance to broomrape has been found in most cases to be monogenic, dominant, and race specific (Pérez-Vich et al., 2013). The genetic control of broomrape resistance by a single dominant gene was first reported by Pogorletsky and Geshele (1976). Shortly after, Vranceanu et al. (1980) identified five broomrape races named A to E that were controlled by five dominant resistant genes named *Or1* to *Or5*, respectively. The authors found that each dominant gene conferred resistance to the corresponding race and also to the previous ones. Several studies confirmed monogenic dominant resistance to these races (Ish-Shalom-Gordon et al., 1993; Lu et al., 2000; Pérez-Vich et al., 2004b; Sukno et al., 1999). However, other studies reported genetic control by two independent dominant genes (Domínguez, 1996) or one recessive gene (Ramaiah, 1987). Similarly, several different results have been reported for races overcoming *Or5* resistance, including one dominant gene (Pacureanu-Joita et al., 2004; Velasco et al., 2012), two independent dominant genes (Pacureanu-Joita et al., 2008), one dominant and one modifying gene (Pérez-Vich et al., 2004a; Velasco et al., 2007), and two recessive genes (Akhtouch et al., 2002; Rodríguez-Ojeda et al., 2001a). In these cases, gene nomenclature *Or6* and *Or7* has been used in several studies concluding different types of genetic control, which may lead to confusion.

Several studies have been conducted to map broomrape resistance genes in the sunflower genetic map, to develop molecular markers for marker-assisted selection, and to determine their function. The use of marker-assisted selection is of particular relevance for breeding for broomrape resistance in sunflower, as phenotypic evaluation is sometimes inaccurate due to the existence of escapes determined by genetic background effects, environmental factors, and genotype × environment interactions (Pérez-Vich et al., 2013). Most of the molecular studies have been conducted on the *Or5* gene conferring resistance to races A to E, which has been mapped to a terminal, probably telomeric region of linkage group (LG) 3 of the sunflower genetic map (Lu et al., 2000; Pérez-Vich et al., 2004b; Tang et al., 2003). The closest marker was identified at around 6 centimorgan downstream of *Or5* (Lu et al., 2000; Tang et al., 2003), but no flanking markers were found in the upper part of the chromosome. Márquez-Lema et al. (2008) identified a telomere-associated target region amplification polymorphism (TRAP) marker linked to *Or5*, probably flanking the gene in the upper telomeric side. Several resistance gene candidates (RGCs) have been identified in the chromosomal region to which *Or5* was mapped, some of them derived from *H. tuberosus* (Radwan et al., 2008), though *Or5* has not been unequivocally associated with any of those RGCs yet. In addition to the major role of *Or5* in race E resistance, Pérez-Vich et al. (2004b) also identified a quantitative component of the resistance determined by four quantitative trait loci (QTL) with minor effect associated with the number of broomrape shoots per plant in susceptible genotypes. Imerovski et al. (2013) demonstrated that simple SSR markers of LG 3 were strongly associated with resistance genes *Or2*, *Or4*, and *Or6*, the latter also monogenic dominant. Hassan et al. (2011) reported a broomrape resistance gene in LG 4, but no details on its race specificity were provided. For race F resistance, QTL analysis in a population derived from line P-96, for which phenotypic analysis suggested the presence of two recessive loci (Akhtouch et al., 2002), revealed the presence of six QTL with small to moderate effects on reducing the number of broomrape shoots per plant, three of them being non–race specific (Pérez-Vich et al., 2004b).

Mechanisms of Resistance

One of the most important aspects when selecting for broomrape resistance is to get information on the physiological basis of the different resistant sources, with the ultimate goals of conducting physiology-based breeding and pyramiding resistance genes underlying different resistance mechanisms (Pérez-Vich et al., 2013). Broomrape resistance mechanisms may operate at preattachment, prehaustorial, or posthaustorial stages of the parasite–host interaction (Pérez-de-Luque et al., 2008).

At the preattachment level, one of the resistance mechanisms is a low exudation of germination stimulants by sunflower roots (Labrousse et al., 2001). Another preattachment mechanism is the exudation by sunflower roots of seed germination inhibitors and/or inhibitors of radicle exoenzymes (Höniges et al., 2008). Phytoalexins, in particular 7-hydroxilated simple coumarins, have been suggested as playing a defensive role by preventing broomrape germination and subsequent connection with sunflower roots (Serghini et al., 2001).

Prehaustorial resistance mechanisms operate from the contact of broomrape radicle with the host root to the formation of the functional haustorium. They can take place at the cortex, endodermis, or central cylinder level (Pérez-de-Luque et al., 2008). Studies in resistant *H. annuus* (Dörr et al., 1994) and *H. debilis* (Labrousse et al., 2001) have identified the formation of an encapsulation layer that prevents intrusion of broomrape radicle into the cortex. Similarly, Echevarría-Zomeño et al. (2006) found that broomrape growth suppression in the cortex was produced by suberization of cell walls in contact with broomrape tissue, reinforcement of adjacent cell walls by protein cross-linking, and release of phenolic compounds toxic for broomrape tissues. Jorrín et al. (1996) observed an increase of lignin content in sunflower roots in response to the presence of broomrape tissue. Peroxidases have been suggested to play an important role in defensive mechanisms such as suberization and lignification of cell walls (Echevarría-Zomeño et al., 2006). Antonova and Ter Borg (1996) suggested that broomrape races C and D differed for the extracellular excretion of peroxidases from the apex of broomrape radicles, with absence of excretion in genotypes of race D. The extracellular peroxidases excreted by broomrape race C genotypes react with phenolic compounds produced in the roots of sunflower genotypes carrying the *Or3* resistance gene, resulting in host resistance due to the formation of lignin layers. The absence of extracellular peroxidases in broomrape race D genotypes prevents lignin formation and enables parasite penetration into sunflower root tissues. In perennial *Helianthus* spp., the high levels of resistance have been attributed to physiological–biochemical features of the cortex cells of the roots (Antonova et al., 2011).

Posthaustorial mechanisms of resistance produce the death of broomrape tubercles during their initial developmental stages (Pérez-de-Luque et al., 2008). Dörr et al. (1994) and Labrousse et al. (2001) observed necrosis of tubercles in resistant genotypes caused by occlusion of sunflower xylem vessels in close connection with the tubercle, due to deposits of an unknown chemical nature. Tubercle necrosis in resistant genotypes has been associated with overexpression of callose synthase cDNA *HaGSL1* and subsequent intense callose deposition at the host–parasite interface (Letousey et al., 2007).

Some of the previously mentioned studies have revealed the simultaneous occurrence of several resistance mechanisms in genotypes exhibiting complete resistance (Echevarría-Zomeño et al., 2006; Labrousse et al., 2001; Letousey et al., 2007). Letousey et al. (2007) observed strong induction of the pathogenesis-related gene *def.* as a response to the presence of broomrape tissue. The authors hypothesized that the function of *def.* could be to organize an overall defensive response in the resistant genotypes.

Screening Methods

Breeding for resistance to sunflower broomrape requires the use of suitable screening methods to measure the degree of resistance of the breeding material, as well as to determine the physiological mechanisms underlying resistance. The former can be used at a large scale at each breeding generation, whereas the latter are better suited for discriminating between different resistance sources at advanced breeding generations.

Two indexes are widely used for measuring resistance in breeding programs: incidence, which measures the percentage of parasitized plants, and degree of attack, which measures the number of parasite structures per infected plant. Parasite structures may refer to emerged shoots or to root attachments, depending on the method of evaluation used. A common procedure is the evaluation of sunflower plants in fields showing a high degree of natural (Pacureanu-Joita et al., 2008) or artificial infestation (Vranceanu et al., 1980). The use of naturally infested fields is usually less reliable since the parasite inoculum is not equally spread and escapes cannot be precluded. In these cases, it is advisable to surround each row by rows of a susceptible check to be used as reference for each entry evaluated (Skoric, 2012). Artificial infestation in field experiments can be done by incorporating the seed into the soil using basic tillage (Vranceanu et al., 1980) or by inoculating individual plants in small pots to be transplanted into the field after 2–3 weeks in the growth chamber (Velasco et al., 2007).

Screening for resistance can be also conducted in pots of varying sizes under growth chambers, greenhouse, or open air conditions. Evaluation in pots has several advantages over field evaluation. A first advantage is that the evaluations can be conducted during the whole year on very large numbers of plants, particularly when small pots are used (Figure 5.6). Experiments in small pots based on counting the number of root attachments can be evaluated between 28 and 45 days after sowing (Hladni et al., 2012; Rodríguez-Ojeda et al., 2013b). Another advantage is that it is possible to conduct the evaluation under controlled conditions, particularly in the case of growth chambers. And last, the methods of evaluation in pots are highly reliable and allow the complete control of the inoculum, which is not possible under field conditions (Hladni et al., 2012). Methods for artificial inoculation in pots, including those intended to be transplanted into the field, consist basically of sowing sunflower seeds or planting two-day-old seedlings in pots filled with a soil mixture homogeneously mixed with broomrape seeds (Skoric, 2012). These methods are well suited for both large pots and small pots. Rodríguez-Ojeda et al. (2013b) obtained well correlated results using 3 L and 0.3 L pots, counting emerged shoots in the former case and root attachments in the latter.

Several methods have been reported for screening for different physiological mechanisms of resistance. One most commonly used by breeders is the evaluation of germination percentage of broomrape seeds in presence of sunflower roots. Even though there are many variations in the practical implementation of the method, it consists basically broomrape forcing sunflower roots to develop on filter paper or glass fiber paper in which broomrape seeds have been homogeneously distributed. This can be done in two-layer filter paper rolls wrapped with plastic film (Antonova et al., 2011; Rodríguez-Ojeda et al., 2010) or in Petri dishes covered with glass fiber paper (Echevarría-Zomeño et al., 2006). After a minimum of 20 days in a growth chamber, the percentage of germinated seeds is counted using a stereoscopic microscope

Figure 5.6 Evaluation of sunflower resistance to broomrape in multipot tray.

or a magnifying glass. This method can also be used to monitor the initial steps of broomrape parasitization and is useful to elucidate the stage at which resistance operates. More complex biochemical, histological, and molecular techniques are required for identifying the specific resistance mechanisms that are activated in resistant lines. The description of such techniques, which is out of the scope of this review, can be found in Pérez-de-Luque et al. (2008).

Use of Herbicide-Tolerant Cultivars

Sunflower broomrape can be controlled by treatment with certain herbicides. García-Torres et al. (1995) reported the effective control of sunflower broomrape through postemergence application at a low dose of imazapyr, a herbicide of the IMI family. In 1996, a wild population of *H. annuus* survived in a soybean field in Kansas (USA) after treatment with another IMI herbicide, imazethapyr (Alonso et al., 1998; Al-Khatib et al., 1998). Initial experiments using broomrape artificial inoculation on plants of the wild *H. annuus* population demonstrated that postemergence IMI herbicide treatment on 40-day-old plants produced necrosis of most of the broomrape nodules without affecting sunflower plants (Alonso et al., 1998). Introgression of the genes underlying the IMI tolerance trait was successful and genetic stocks and breeding lines with IMI tolerance, known as *Imisun,* were released shortly after (Al-Khatib and Miller, 2000; Miller and Al-Khatib, 2002). Similar to sulfonylurea herbicides, IMI herbicides control weeds by inhibiting acetohydroxyacid synthase (AHAS), a key enzyme in the branched-chain amino acid biosynthetic pathway (Tan et al., 2005). However, broomrape control by sulfonylurea herbicides is less efficient than by IMI herbicides (Semirci and Kaya, 2009). The first genetic analysis of IMI tolerance suggested that the trait was controlled by two additive dominant genes (Alonso et al., 1998). Later studies found that one of the genes was partially dominant and the second gene was a modifier. Completely tolerant hybrids require having tolerance genes in both parents (Bruniard and Miller, 2001). Kolkman et al. (2004) identified three *Ahas* loci. Tolerance to both sulfonylurea and IMI herbicides was explained by single nucleotide polymorphisms at locus *Ahas1*, mapped to LG 9 of the sunflower genetic map. A second genetic source of IMI tolerance in sunflower was developed through ethyl methanesulfonate mutagenesis, showing higher levels of herbicide tolerance than Imisun genotypes (Sala et al., 2008). IMI tolerance in the new genetic source, designated CLHA-PLUS or CLPlus, was controlled by a single, partially dominant mutation at *Ahasl1* (=*Ahas1*) locus (Sala et al., 2008). The allele controlling CLPlus tolerance was designated *Ahasl1-3* and was dominant over allele *Ahasl1-1* controlling Imisun tolerance (Sala et al., 2012). In addition, a new genetic source of tolerance to four types of AHAS herbicides, including IMI, was developed from a wild sunflower population in Argentina. The trait was controlled by a single, partially dominant allele designated *Ahasl1-4* (Sala and Bulos, 2012). Other sources of tolerance to IMI herbicides have been identified in populations of wild *H. annuus* and *H. argophylus* (Christov et al., 2010).

The first IMI tolerant hybrids entered the market as Sunflower Clearfield hybrids in 2003 in the United States, Argentina, and Turkey (Pfenning et al., 2008), and in 2004 in Serbia (Jocic et al., 2011). One of the main advantages of Clearfield hybrids is the simultaneous control of broomrape and a broad spectrum of weeds (Pfening et al., 2008). IMI-tolerant hybrids have become very popular in some areas; in the Thrace region of Turkey, for example, they are cultivated in more than 50% of sunflower acreage (Semerci et al., 2010).

Agronomic Management and Biological Control

Preventing the movement of broomrape seeds from infested into uninfested areas is a key component of control (Rubiales and Fernández-Aparicio, 2012). The efficacy of agronomic practices for the control of sunflower broomrape in infested areas is constrained by the existence of large seed banks in the soils. Broomrape plants produce an immense number of tiny seeds with extreme longevity (Goldwasser and Rodenburg, 2013). Once a field is infested, agronomic management generally has a low short-term impact on broomrape control, although adequate management strategies may contribute to reducing broomrape seedbank if maintained in the long term (Rubiales et al., 2009). For example, minimum tillage has been suggested as an effective strategy to reduce the number of seeds incorporated into the soil (Rubiales et al., 2009). One of the most promising strategies to reduce broomrape seedbank is the incorporation of trap crops into the rotation. Trap crops stimulate germination of broomrape seeds but do not allow parasitism, as they impede further development of parasitic structures, resulting in the death of germinated seeds. Rodríguez-Ojeda et al. (2001b) reported that corn, sorghum, cotton, rice, eggplant, and cauliflower, in decreasing order, stimulated the germination of sunflower broomrape seeds. Fernández-Aparicio et al. (2009) reported cotton, oat, wheat, sorghum, pearl millet, chickpea, fava bean, rapeseed, and triticale, in decreasing order, as the main crops inducing higher germination percentages. Ma et al. (2013) found that corn hybrids and lines differed in their capacity to stimulate sunflower broomrape seed germination, with maximum germination percentages close to 40%. Similar results were obtained by Zhang et al. (2013) using soybean cultivars. Ma et al. (2012) reported some additional plants that stimulated sunflower broomrape seed germination. Several strategies for biological control of sunflower broomrape are also under study. Most remarkable are the development of bioherbicides based on fungal spores of *Fusarium oxysporum* f. sp. *orthoceras* (Shabana et al., 2003) and the use of arbuscular mycorrhizal fungi, the latter having a negative impact on sunflower broomrape germination (Louarn et al., 2012).

Conclusion

Broomrape-resistant hybrids of sunflower are mainly based on single dominant *Or* genes. This is a practical approach for seed companies, as the resistance gene is only introgressed in one of the parents. However, this approach, based on vertical resistance, has contributed to the rapid evolution of virulence in *O. cumana* populations. Alternative breeding strategies are required to increase the durability of genetic resistance to broomrape. In this sense, it is important to develop new sources of genetic resistance that are not only governed by single dominant genes, but that also contain horizontal resistance mechanisms. Wild *Helianthus* spp. remain as the main genetic reservoir for new sources of resistance. Durable resistance will necessarily require pyramiding resistance genes. One approach is pyramiding more than one dominant *Or* gene. Another alternative is to combine genes underlying vertical and horizontal resistance mechanisms. In both cases, gene pyramiding should be based on physiological characterization of resistance and molecular-assisted selection. Accordingly, it will be crucial to develop a detailed characterization of the physiological mechanisms of resistance of all available sources of broomrape resistance. Also, efforts to identify genes underlying such mechanisms of resistance and to develop reliable molecular markers for efficient selection will be required. These breeding strategies should be coupled with a better understanding of sunflower broomrape biology, genetics, and population dynamics. A comprehensive evaluation of genetic diversity of sunflower broomrape across the main infested areas is urgently required. Also, the mechanisms of avirulence/virulence and their underlying genetic bases should be studied in parallel to the studies on resistance/susceptibility in sunflower. Last but not least, molecular diagnostic tests for identifying the main broomrape genetic pools and the different physiological races need to be developed. Breeding strategies for genetic resistance to broomrape need to be complemented by adequate cultural practices, particularly prevention of broomrape seed movement from infested to uninfested areas, and the incorporation of herbicide tolerance genes into genetically resistant hybrids. The combination of both strategies of broomrape control, genetic resistance to broomrape and herbicide tolerance, will contribute to a more durable control of broomrape while simultaneously controlling a wide spectrum of weeds.

References

Akhtouch, B.; Muñoz-Ruz, J.; Melero-Vara, J. M.; Fernández-Martínez, J. M.; Domínguez, J. Inheritance of Resistance to Race F of Broomrape (*Orobanche cumana* Wallr.) in Sunflower Lines of Different Origin. *Plant Breeding* **2002,** *121*, 266–269.

Al-Khatib, K.; Miller, J. F. Registration of Four Genetic Stocks of Sunflower Resistant to Imidazolinone Herbicides. *Crop Sci.* **2000,** *40*, 869–870.

Al-Khatib, K.; Baumgartner, J. R.; Peterson, D. E.; Currie, R. S. Imazethapyr Resistance in Common Sunflower (*Helianthus annuus*). *Weed Sci.* **1998,** *46*, 403–407.

Alonso, L. C. Resistance to *Orobanche* and Resistance Breeding: A Review. In *Current Problems of Orobanche Research, Proceedings of the 4th International Workshop on Orobanche,* Albena, Bulgaria, Sept 23–26, 1998; Wegmann, H., Musselman, L. J., Joel, D. M., Eds.; Institute of Wheat and Sunflower Research: General Toshevo, Bulgaria, 1998; pp 233–237.

Alonso, L. C.; Fernández-Escobar, J.; López, G.; Rodríguez-Ojeda, M. I.; Sallago, F. New Highly Virulent Sunflower Broomrape (*Orobanche cernua* Loefl.) Pathotype in Spain. In *Advances in Parasitic Plant Research, Sixth International Parasitic Weed Symposium,* Córdoba, Spain, April 16–18, 1996; Moreno, M. T., Cubero, J. I., Berner, D., Joel, D., Musselman, L. J., Parker, C., Eds; Junta de Andalucia: Sevilla, Spain, 1996; pp 639–644.

Alonso, L. C.; Rodríguez-Ojeda, M. I.; Fernández-Escobar, J.; López-Calero, G. Chemical Control of Broomrape (*Orobanche cernua* Loefl.) in Sunflower (*Helianthus annuus* L.) Resistant to Imazethapyr Herbicide. *Helia* **1998,** *29*, 45–54.

Amri, M.; Abbes, Z.; Ben Youssef, S.; Bouhadida, M.; Ben Salah, H.; Kharrat, M. Detection of the Parasitic Plant, *Orobanche cumana* on Sunflower (*Helianthus annuus* L.) in Tunisia. *Afr. J. Biotechnol.* **2012,** *11*, 4163–4167.

Antonova, T. S.; Ter Borg, S. The Role of Peroxidase in the Resistance of Sunflower against *Orobanche cumana* in Russia. *Weed Res.* **1996,** *36*, 113–121.

Antonova, T. S.; Araslanova, N. M.; Strelnikov, E. A.; Ramazanova, S. A.; Tchelustnikova, T. A.; Guchetl, S. Z. Screening of Wild *Helianthus* Species for Resistance to High Virulent *Orobanche cumana* Wallr., Affecting Sunflower in the Rostov Region of the Russian Federation. *Helia* **2011,** *55*, 115–124.

Antonova, T. S.; Araslanova, N. M.; Strelnikov, E. A.; Ramazanova, S. A.; Guchetl, S. Z.; Tchelustnikova, T. A. Some Peculiarities of Ontogenesis of *Orobanche cumana* Wallr., Parasitizing on Sunflower in Rostov Region of Russian Federation. *Helia* **2012,** *56*, 99–110.

Antonova, T. S.; Araslanova, N. M.; Strelnikov, E. A.; Ramazanova, S. A.; Guchetl, S. Z.; Chelyustnikova, T. A. Distribution of Highly Virulent Races of Sunflower Broomrape (*Orobanche cumana* Wallr.) in the Southern Regions of the Russian Federation. *Russ. Agr. Sci.* **2013,** *39*, 46–50.

Beck-Mannagetta, G. Orobanchaceae. In *Das Pflanzenreich;* Engler, A., Ed.; Verlag von Wilhelm Engelmann: Leipzig, Germany, 1930; pp 1–348.

Bernhard, R. H.; Jensen, J. E.; Andreasen, C. Prediction of Yield Losses Caused by *Orobanche* spp. in Carrot and Pea Crops Based on the Soil Seedbank. *Weed Res.* **1998,** *38*, 191–197.

Bruniard, J. M.; Miller, J. F. Inheritance of Imidazolinone-Herbicide Resistance in Sunflower. *Helia* **2001,** *35*, 11–16.

Bulbul, A.; Salihoglu, M.; Sari, C.; Aydin, A. Determination of Broomrape (*Orobanche cumana* Wallr.) Races of Sunflower in the Thrace Region of Turkey. *Helia* **1991,** *15*, 21–26.

Bülbül, F.; Aksoy, E.; Uygur, S.; Uygur, N. Broomrape (*Orobanche* spp.) Problem in the Eastern Mediterranean Region of Turkey. *Helia* **2009,** *51*, 141–152.

Burlov, V.; Burlov, V. Breeding of Sunflower Resistant to New Races of Broomrape (*Orobanche cumana* Wallr.). *Helia* **2010,** *53*, 165–172.

Castejón, M.; Romero-Munoz, F.; Garcia-Torres, L. *Orobanche cernua* Seed Dispersal through Sunflower Achenes. *Helia* **1991,** *14*, 51–54.

Castejón-Muñoz, M.; Suso, M. J.; Romero-Muñoz, F.; García-Torres, L. Isoenzymatic Study of Broomrape (*Orobanche cernua*) Populations Infesting Sunflower (*Helianthus annuus*). In *Proceedings of the 5th International Symposium on Parasitic Weeds*, Nairobi, Kenya, June 24–30, 1991; Ransom, J. K., Musselman, L. J., Worsham, A. D., Parker, C., Eds.; International Maize and Wheat Improvement Center (CIMMYT): Nairobi, Kenya, 1991; pp 313–319.

Christov, M.; Batchvarova, R.; Hristova-Cherbadzhi, M. Wild Species of *Helianthus* L. Sources of Resistance to the Parasite *Orobanche cumana* Wallr. *Helia* **2009,** *51*, 65–74.

Christov, M.; Hristova-Cherbadzi, M.; Entcheva, V.; Shindrova, P.; Kiryakov, I.; Valkova, D.; Sabev, G.; Batchvarova, R. Creating New Sunflower Forms and Lines, Resistant to Diseases, the Broomrape Parasite and Certain Types of Herbicides Applying the Interspecific and Intergeneric Hybridization. In *Proceedings of the International Symposium on Sunflower Breeding on Resistance to Diseases,* Krasnodar, Russia, June 23–24, 2010; All-Russia Research Institute of Oil Crops: Krasnodar, Russia, 2010; pp 73–85.

Cvejic, S.; Dedic, B.; Jocic, S.; Miladinovic, D.; Miklic, V. Broomrape Resistance in Newly Developed Sunflower Inbred Lines. In *Proceedings of the 18th International Sunflower Conference,* Mar del Plata, Argentina, Feb 27–March 1, 2012. http://www.asagir.org.ar/asagir2008/buscar_congreso.asp (accessed July 2, 2013).

Domínguez, J. R-41, a Sunflower Restorer Inbred Line, Carrying Two Genes for Resistance against a Highly Virulent Spanish Population of *Orobanche cernua. Plant Breeding* **1996,** *115,* 203–204.

Domínguez, J.; Melero-Vara, J. M.; Ruso, J.; Miller, J.; Fernández-Martínez, J. M. Screening for Resistance to Broomrape (*Orobanche cernua*) in Cultivated Sunflower. *Plant Breeding* **1996,** *115*, 201–202.

Dörr, I.; Staack, A.; Kollmann, R. Resistance of *Helianthus* to *Orobanche.* Histological and Cytological Studies. In *Biology and Management of Orobanche, Proceedings of the 3rd International Workshop on Orobanche and related* Striga *Research,* Amsterdam, The Netherlands, Nov 8–12, 1994; Pieterse, A. H., Verkleij, J. A. C., Ter Borg, S. T., Eds; KIT Press: Amsterdam, The Netherlands, 1994; pp 276–289.

Echevarría-Zomeño, S.; Pérez-de-Luque, A.; Jorrín, J.; Maldonado, A. Pre-Haustorial Resistance to Broomrape (*Orobanche cumana*) in Sunflower (*Helianthus annuus*): Cytochemical Studies. *J. Exp. Bot.* **2006,** *57*, 4189–4200.

Eizenberg, H.; Plakhine, D.; Hershenhorn, J.; Kleifeld, Y.; Rubin, B. Resistance to Broomrape (*Orobanche* spp.) in Sunflower (*Helianthus annuus* L.) is Temperature Dependent. *J. Exp. Bot.* **2003,** *54*, 1305–1311.

Fernández-Aparicio, M.; Flores, F.; Rubiales, D. Recognition of Root Exudates by Seeds of Broomrape (*Orobanche* and *Phelipanche*) Species. *Ann. Bot.* **2009,** *103*, 423–431.

Fernández-Escobar, J.; Rodríguez-Ojeda, M. I.; Fernández-Martínez, J. M.; Alonso, L. C. Sunflower Broomrape (*Orobanche cumana* Wallr.) in Castilla-León, a Traditionally Non-Broomrape Infested Area in Northern Spain. *Helia* **2009,** *51*, 57–64.

Fernández-Martínez, J. M.; Melero-Vara, J. M.; Muñoz-Ruz, J.; Ruso, J.; Domínguez, J. Selection of Wild and Cultivated Sunflower for Resistance to a New Broomrape Race that Overcomes Resistance to *Or5* Gene. *Crop Sci.* **2000,** *40*, 550–555.

Fernández-Martínez, J. M.; Pérez-Vich, B.; Akhtouch, B.; Velasco, L.; Muñoz-Ruz, J.; Melero-Vara, J. M.; Domínguez, J. Registration of Four Sunflower Germplasms Resistant to Race F of Broomrape. *Crop Sci.* **2004,** *44,* 1033–1034.

Fernández-Martínez, J. M.; Velasco, L.; Pérez-Vich, B. Progress in Research on Breeding for Resistance to Broomrape. In *Proceedings of the 18th International Sunflower Conference,* Mar del Plata, Argentina, Feb 27–March 1, 2012. http://www.asagir.org.ar/asagir2008/buscar_congreso.asp (accessed July 2, 2012).

Flor, H. H. Current Status of the Gene-for-Gene Concept. *Ann. Rev. Phytopathol.* **1971,** *9,* 275–296.

Gagne, G.; Roeckel-Drevet, P.; Grezes-Besset, B.; Shindrova, P.; Ivanov, P.; Grand-Ravel, C.; Vear, F.; Tourvieille de Labrouhe, D.; Charmet, G.; Nicolas, P. Study of the Variability and Evolution of *Orobanche cumana* Populations Infesting Sunflower in Different European Countries. *Theor. Appl. Genet.* **1998,** *96,* 1216–1222.

García-Torres, L.; Castejón-Muñoz, M.; López-Granados, F.; Jurado-Expósito, M. Imazapyr Applied Postemergence in Sunflower (*Helianthus annuus*) for Broomrape (*Orobanche cernua*) Control. *Weed Technol.* **1995,** *9,* 819–824.

Gevezova, M.; Dekalska, T.; Stoyanov, K.; Hristeva, T.; Kostov, K.; Batchvarova, R.; Denev, I. Recent Advances in Broomrapes Research. *J. Biosci. Biotech.* **2012,** *1,* 91–105.

Goldwasser, Y.; Rodenburg, J. Integrated Agronomic Management of Parasitic Weed Banks. In *Parasitic Orobanchaceae*; Joel, D. M., Gressel, J., Musselman, L. J., Eds.; Springer: New York, 2013; pp 393–413.

Gontcharov, S. V. Sunflower Breeding for Resistance to the New Broomrape Race in the Krasnodar Region of Russia. *Helia* **2009,** *51,* 75–80.

Gontcharov, S. V.; Antonova, T. S.; Araslanova, N. M. Sunflower Breeding for Resistance to the New Broomrape Race. *Helia* **2004,** *40,* 193–198.

Gorbachenko, F. I.; Usatenko, T. V.; Gorbachenko, O. F. Results of Sunflower Breeding in Resistance to Broomrape on Don. *Helia* **2011,** *54,* 9–18.

Gulya, T. J.; Aydin, A.; Brothers, M. Evaluation of Broomrape (*Orobanche cumana*) Resistance in Sunflower Germplasm of the USDA Plant Introduction collection. In *Proceedings of the 16th Sunflower Research Workshop,* Fargo, ND, Jan 13–14, 1994; Seiler, E. J., Ed.; National Sunflower Association: Bismarck, ND, 1994; pp 53–55.

Hassan, E. A. R. S.; Hoeft, E.; Tulsieram, L. Genetic Markers for *Orobanche* Resistance in Sunflower. U.S. Patent 7,872,170, Jan 18, 2011.

Hladni, N.; Jocic, S.; Miklic, V.; Saftic-Pankovic, D.; Skoric, D. Using New *Rf* Inbred Lines Originating from an Interspecific Population with *H. deserticola* for Development of Sunflower Hybrids Resistant to Broomrape. *Helia* **2009,** *51,* 81–90.

Hladni, N.; Dedic, B.; Jocic, S.; Miklic, V.; Dusanic, N. Evaluation of Resistance of New Sunflower Hybrids to Broomrape in the Breeding Programs in Novi Sad. *Helia* **2012,** *56,* 89–98.

Höniges, A.; Wegmann, K.; Ardelean, A. *Orobanche* Resistance in Sunflower. *Helia* **2008,** *49,* 1–12.

Imerovski, I.; Dimitrijevic, A.; Miladinovic, D.; Dedic, B.; Jocic, S.; Kovacevic, B.; Obreht, D. Identification of PCR Markers Linked to Different *Or* Genes in Sunflower. *Plant Breeding* **2013,** *132,* 115–120.

Ish-Shalom-Gordon, N.; Jacobsohn, R.; Cohen, Y. Inheritance of Resistance to *Orobanche cumana* in Sunflower. *Phytopathology* **1993,** *83,* 1250–1252.

Jan, C. C.; Fernández-Martínez, J. M. Interspecific Hybridization, Gene Transfer, and the Development of Resistance to Broomrape Race F in Spain. *Helia* **2002,** *36,* 123–136.

Jan, C. C.; Fernández-Martínez, J. M.; Ruso, J.; Muñoz-Ruz, J. Registration of Four Sunflower Germplasm Populations Resistant to Broomrape Race F. *Crop Sci.* **2002,** *42,* 2217–2218.

Jocic, S.; Malidza, G.; Cvejic, S.; Hladnil, N.; Miklic, V.; Skoric, D. Development of Sunflower Hybrids Tolerant to Tribenuron Methyl. *Genetika* **2011,** *43,* 175–182.

Joel, D. M. Identification of *Orobanche* Seeds. In *Proceedings of the 4th International Symposium on Parasitic Flowering Plants,* Marburg, Germany, August 8–12, 1987; Weber, H. C., Forstreuter, W. L., Eds.; Philipps-Universität: Marburg, Germany, 1987; pp 437–444.

Joel, D. M. The Haustorium and the Life Cycles of Parasitic *Orobanchaceae.* In *Parasitic Orobanchaceae*; Joel, D. M., Gressel, J., Musselman, L. J., Eds.; Springer: New York, 2013; pp 21–23.

Joel, D. M.; Chaudhury, S.; Plakhine, D.; Ziadna, H.; Steffens, J. Dehydrocostus Lactone Is Exuded from Sunflower Roots and Stimulates Germination of the Root Parasite *Orobanche cumana. Phytochemistry* **2011,** *72,* 624–634.

Jorrín, J.; De Ruck, E.; Serghini, K.; Pérez de Luque, A.; Muñoz-García, J.; García-Torres, L.; Castejón-Muñoz, M. Biochemical Aspects of the Parasitism of Sunflower by *Orobanche.* In *Advances in Parasitic Plant Research, Sixth International Parasitic Weed Symposium,* Córdoba, Spain, April 16–18, 1996; Moreno, M. T., Cubero, J. I., Berner, D., Joel, D., Musselman, L. J., Parker, C., Eds; Junta de Andalucia: Sevilla, Spain, 1996; pp 551–558.

Jouffret, P.; Lecomte, V. Orobanche Sur Tournesol. Un Parasite en Extension. *Lettre d'Informations Regionales CETIOM, Zone Sud* [Online] **2010.** http://www.cetiom.fr/fileadmin/cetiom/regions/Sud/PDF/2010/oleomail/OleoMail_TO_Prophylaxie_O_cumana_sud_07092010.pdf (accessed July 2, 2013).

Kaya, Y.; Evci, G.; Pekcan, V.; Gucer, T. Determining New Broomrape-Infested Areas, Resistant Lines and Hybrids in Trakya Region in Turkey. *Helia* **2004,** *40,* 211–218.

Kaya, Y.; Evci, G.; Pekcan, V.; Gucer, T.; Yilmaz, M. I. Evaluation of Broomrape Resistance in Sunflower Hybrids. *Helia* **2009,** *51,* 161–170.

Kolkman, J. M.; Slabaugh, M. B.; Bruniard, J. M.; Berry, S.; Bushman, B. S.; Olungu, C.; Maes, N.; Abratti, G.; Zambelli, A.; Miller, J. F.; et al. Acetohydroxyacid Synthase Mutations Conferring Resistance to Imidazolinone or Sulfonylurea Herbicides in Sunflower. *Theor. Appl. Genet.* **2004,** *109,* 1147–1159.

Labrousse, P.; Arnaud, M. C.; Serieys, H.; Bervillé, A.; Thalouan, P. Several Mechanisms Are Involved in Resistance of *Helianthus* to *Orobanche cumana* Wallr. *Ann. Bot.* **2001,** *88,* 859–868.

Letousey, P.; De Zélicourt, A.; Vieira Dos Santos, C.; Thoiron, S.; Monteau, F.; Simier, P.; Thalouarn, P.; Delavault, P. Molecular Analysis of Resistance Mechanisms to *Orobanche cumana* in Sunflower. *Plant Pathol.* **2007,** *56,* 536–546.

Louarn, J.; Carbonnel, F.; Delavault, P.; Bécard, G.; Rochange, S. Reduced Germination of *Orobanche cumana* Seeds in the Presence of Arbuscular Mycorrhizal Fungi or Their Exudates. *Plos One* **2012,** *7,* e49273.

Lu, Y. H.; Melero-Vara, J. M.; García-Tejada, J. A.; Blanchard, P. Development of SCAR Markers Linked to the Gene *Or5* Conferring Resistance to Broomrape (*Orobanche cumana* Wallr.) in Sunflower. *Theor. Appl. Genet.* **2000,** *100,* 625–632.

Ma, Y.; Zhang, W.; Dong, S.; Ren, X.; An, Y.; Lang, M. Induction of Seed Germination in *Orobanche* spp. by Extracts of Traditional Chinese Medicinal Herbs. *Sci. China Life Sci.* **2012,** *55,* 250–260.

Ma, Y.; Jia, J.; An, Y.; Wang, Z.; Mao, J. Potential of Some Hybrid Maize Lines to Induce Germination of Sunflower Broomrape. *Crop Sci.* **2013,** *53,* 260–270.

Márquez-Lema A., Delavault P., Letousey P., Hu J., Pérez-Vich B. Candidate Gene Analysis and Identification of TRAP and SSR Markers Linked to the *Or5* Gene Conferring Sunflower Resistance to Race E of Broomrape (*Orobanche cumana* Wallr.). In *Proceedings of the 17th International Sunflower Conference,* Córdoba, Spain, June 8–12, 2008; Velasco, L., Ed.; Consejería de Agricultura y Pesca, Junta de Andalucía: Sevilla, Spain, 2008; pp 661–666.

Masirevic, S.; Medic-Pap, S. Broomrape in Serbia from Its Occurence till Today. *Helia* **2009,** *51,* 91–100.

Melero-Vara, J. M.; Domínguez, J.; Fernández-Martínez, J. M. Evaluation of Differential Lines and a Collection of Sunflower Parental Lines for Resistance to Broomrape (*Orobanche cernua*) in Spain. *Plant Breeding* **1989,** *102,* 322–326.

Miladinovic, D.; Dedic, B.; Quiróz, F.; Álvarez, D.; Poverone, M.; Cantamutto, M. *Orobanche cumana* Wallr. Resistance of Commercial Sunflower Cultivars Grown in Argentina. *J. Basic Appl. Genet.* **2012,** *23,* 37–41.

Miller, J. F.; Al-Khatib, K. Registration of Imidazolinone Herbicide-Resistant Sunflower Maintainer (HA 425) and Fertility Restorer (RHA 426 and RHA 427) Germplasms. *Crop Sci.* **2002,** *42,* 988–989.

Mohamed, K. I.; Papes, M.; Williams, R.; Benz, B. W.; Peterson, A. T. Global Invasive Potential of 10 Parasitic Witchweeds and Related Orobanchaceae. *Ambio* **2006,** *35,* 281–288.

Molinero-Ruiz, M. L.; Pérez-Vich, B.; Pineda-Martos, R.; Melero-Vara, J. M. Indigenous Highly Virulent Accessions of the Sunflower Root Parasitic Weed *Orobanche cumana. Weed Res.* **2008,** *48,* 169–178.

Molinero-Ruiz, M. L.; García-Carneros, A. B.; Collado-Romero, M.; Raranciuc, S.; Domínguez, J.; Melero-Vara, J. M. Pathogenic and Molecular Diversity in Highly Virulent Populations of the Parasitic Weed *Orobanche cumana* from Europe. *Weed Res.* **2013,** 54, 87–96.

Musselman, L. J. The Biology of *Striga, Orobanche,* and Other Root-Parasitic Weeds. *Ann. Rev. Phytopathol.* **1980,** *18,* 463–489.

Musselman, L. J.; Parker, C.; Dixon, N. Notes on the Autogamy and Flower Structure in Agronomically Important Species of *Striga* (Scrophulariaceae) and *Orobanche* (Orobanchaceae). *Beitr. Biol. Pflanz.* **1981,** *56,* 329–343.

Pacureanu-Joita, M.; Veronesi, C.; Raranciuc, S.; Stanciu, D. Parasite–Host Plant Interaction of *Orobanche cumana* Wallr. (*Orobanche cernua* Loefl) with *Helianthus annuus.* In *Proceedings of the 16th International Sunflower Conference,* Fargo, ND, Aug 29–Sept 2, 2004; Seiler, G. J., Ed.; International Sunflower Association: Paris, France, 2004; pp 171–177.

Pacureanu-Joita, M.; Raranciuc, S.; Stanciu, D.; Sava, E.; Nastase, D. Virulence and Aggressiveness of Sunflower Broomrape (*Orobanche cumana* Wallr.) Populations, in Romania. *Romanian Agric. Res.* **2008,** *25,* 47–50.

Pacureanu-Joita, M.; Raranciuc, S.; Sava, E.; Stanciu, D.; Nastase, D. Virulence and Aggressiveness of Sunflower Broomrape (*Orobanche cumana* Wallr.) Populations in Romania. *Helia* **2009a,** *51,* 111–118.

Pacureanu-Joita, M.; Fernandez-Martinez, J. M.; Sava, E.; Raranciuc, S. Broomrape (*Orobanche cumana* Wallr.), the Most Important Parasite in Sunflower. *Analele I.N.C.D.A. Fundulea,* **2009b,** *77,* 49–56.

Parker, C. The Parasitic Weeds of the Orobanchaceae. In *Parasitic Orobanchaceae*; Joel, D. M., Gressel, J., Musselman, L. J., Eds.; Springer: New York, 2013; pp 313–344.

Parker C.; Riches, C. R. *Parasitic Weeds of the World: Biology and Control;* CAB International: Wallingford, UK, 1993.

Pérez-de-Luque, A.; Moreno, M. T.; Rubiales, D. Host Plant Resistance against Broomrapes (*Orobanche* spp.): Defence Reactions and Mechanisms of Resistance. *Ann. Appl. Biol.* **2008,** *152,* 131–141.

Pérez-Vich, B.; Akhtouch, B.; Mateos, A.; Velasco, L.; Jan, C. C.; Fernández, J.; Domínguez, J.; Fernández-Martínez, J. M. Dominance Relationships for Genes Conferring Resistance to Sunflower Broomrape (*Orobanche cumana* Wallr.). *Helia* **2004a,** *40,* 183–192.

Pérez-Vich, B.; Akhtouch, B.; Knapp, S. J.; Leon, A. J.; Velasco, L.; Fernández-Martínez, J. M.; Berry, S. T. Quantitative Trait Loci for Broomrape (*Orobanche cumana* Wallr.) Resistance. *Theor. Appl. Genet.* **2004b,** *109,* 92–102.

Pérez-Vich, B.; Velasco, L.; Muñoz-Ruz, J.; Domínguez, J.; Fernández-Martínez, J. M. Registration of Three Sunflower Germplasms with Quantitative Resistance to Race F of Broomrape. *Crop Sci.* **2005,** *46,* 1406–1407.

Pérez-Vich, B.; Velasco, L.; Rich, P. J.; Ejeta, G. Marker-Assisted and Physiology-Based Breeding for Resistance to Orobanchaceae. In *Parasitic Orobanchaceae*; Joel, D. M., Gressel, J., Musselman, L. J., Eds.; Springer: New York, 2013; pp 369–391.

Petcu, E.; Pacureanu, J. M. Developing Drought and Broomrape Resistant Sunflower Germplasm Utilizing Wild *Helianthus* Species. *Helia* **2011,** *54,* 1–8.

Pfenning, M.; Palfay, G.; Guillet, T. The Clearfield® Technology—A New Broad-Spectrum Post-Emergence Weed Control System for European Sunflower Growers. *J. Plant Dis. Protect.* **2008,** *Special Issue XXI,* 647–651.

Pineda-Martos, R.; Velasco, L.; Fernández-Escobar, J.; Fernández-Martínez, J. M.; Pérez-Vich, B. Genetic Diversity of Sunflower Broomrape (*Orobanche cumana*) Populations from Spain. *Weed Res.* **2013,** 53, 279–289.

Pogorletsky, P. K.; Geshele, E. E. Sunflower Immunity to Broomrape, Downey Mildew and Rust. In *Proceedings of the 7th International Sunflower Conference,* Krasnodar, Russia, June 27–July 3, 1976; International Sunflower Association: Paris, France, 1976; pp 238–243.

Pujadas Salvà, A. J.; Velasco, L. Comparative Studies on *Orobanche cernua* L. and *O. cumana* Wallr. (Orobanchaceae) in the Iberian Peninsula. *Bot. J. Linn. Soc.* **2000,** *134,* 513–527.

Radwan O.; Gandhi, S.; Heesacker, A.; Whitaker, B.; Taylor, C.; Plocik, A.; Kesseli, R.; Kozik, A.; Michelmore, R. W.; Knapp, S. J. Genetic Diversity and Genomic Distribution of Homologs Encoding NBS-LRR Disease Resistance Proteins in Sunflower. *Mol. Genet. Genomics* **2008,** *280,* 111–125.

Ramaiah, K. V. Control of *Striga* and *Orobanche* Species. A Review. In *Parasitic Flowering Plants;* Weber, H. C., Forstreuter, W., Eds.; Philipps-Universität: Marburg, Germany, 1987; pp 637–664.

Rechinger, K. H. Flora Aegaea. Flora der Inseln und Halbinseln des ägäischen Meeres. *Kaiserl. Akad. Wiss. Wien, Math. Naturwiss. Kl., Denkschr.* **1943**, 105/1.

Rispail, N.; Dita, M. A.; González-Verdejo, C.; Pérez-de-Luque, A.; Castillejo, M. A.; Prats, E.; Román, B.; Jorrín, J.; Rubiales, D. Plant Resistance to Parasitic Plants: Current Approaches for an Old Foe. *New Phytol.* **2007,** *173*, 703–712.

Rodríguez-Ojeda, M. I.; Fernández-Escobar, J.; Alonso, L. C. Sunflower Inbred Line (KI-374) Carrying Two Recessive Genes for Resistance against a Highly Virulent Spanish Population of *Orobanche cernua* Loefl./*O. cumana* Wallr. Race F. In *Proceedings of the 7th International Parasitic Weed Symposium,* Nantes, France, June 5–8, 2001; Musselman, L. J., Parker, C., Verkleij, J. A. C., Eds.; Faculté des Sciences, Université de Nantes: France, 2001a; pp 208–211.

Rodríguez-Ojeda, M. I.; Alonso, L. C.; Fernández-Escobar, J. Effect of Different Crops on the Germination of *Orobanche cernua* Loefl. (*O. cumana* Wallr.) Seeds. In *Proceedings of the 7th International Parasitic Weed Symposium,* Nantes, France, June 5–8, 2001; Musselman, L. J., Parker, C., Verkleij, J. A. C., Eds.; Faculté des Sciences, Université de Nantes: France, 2001b; p 124.

Rodríguez-Ojeda, M. I.; Pérez-Vich, B.; Alonso, L. C.; Fernández-Escobar, J. The Influence of Flowering Plant Isolation on Seed Production and Seed Quality in *Orobanche cumana. Weed Res.* **2010,** *50,* 515–518.

Rodríguez-Ojeda, M. I.; Velasco, L.; Alonso, L. C.; Fernández-Escobar, J.; Pérez-Vich, B. Inheritance of the Unpigmented Plant Trait in *Orobanche cumana. Weed Res.* **2011,** *51,* 151–156.

Rodríguez-Ojeda, M. I.; Fernández-Martínez, J. M.; Velasco, L.; Pérez-Vich, B. Extent of Cross-Fertilization in *Orobanche cumana* Wallr. *Biol. Plant.* **2013a** *57*, 559–562.

Rodríguez-Ojeda, M. I.; Pineda-Martos, R.; Alonso, L. C.; Fernández-Escobar, J.; Fernández-Martínez, J. M.; Pérez-Vich, B.; Velasco, L. A Dominant Avirulence Gene in *Orobanche cumana* Triggers *Or5* Resistance in Sunflower. *Weed Res.* **2013b,** *53*, 322–327.

Román, B.; Alfaro, C.; Torres, A. M.; Moreno, M. T.; Satovic, Z.; Pujadas, A.; Rubiales, D. Genetic Relationships among *Orobanche* Species as Revealed by RAPD Analysis. *Ann. Bot.* **2003,** *91,* 637–642.

Rubiales, D.; Fernández-Aparicio, M. Innovations in Parasitic Weeds Management in Legume Crops. A Review. *Agron. Sustain. Dev.* **2012,** *32*, 433–449.

Rubiales, D.; Fernández-Aparicio, M.; Wegmann, K.; Joel, D. M. Revisiting Strategies for Reducing the Seedbank of *Orobanche* and *Phelipanche* spp. *Weed Res.* **2009,** *49* (Suppl. 1), 23–33.

Ruso, J.; Sukno, S.; Domínguez-Giménez, J.; Melero-Vara, J. M.; Fernández-Martínez, J. M. Screening of Wild *Helianthus* Species and Derived Lines for Resistance to Several Populations of *Orobanche cernua. Plant Dis.* **1996,** *80,* 1165–1169.

Sala, C. A.; Bulos, M. Inheritance and Molecular Characterization of Broad Range Tolerance to Herbicides Targeting Acetohydroxyacid Synthase in Sunflower. *Theor. Appl. Genet.* **2012,** *124,* 355–364.

Sala, C. A.; Bulos, M.; Echarte, M. Genetic Analysis of an Induced Mutation Conferring Imidazolinone Resistance in Sunflower. *Crop Sci.* **2008,** *48,* 1917–1922.

Sala, C. A.; Bulos, M.; Altieri, E.; Weston, B. Response to Imazapyr and Dominance Relationships of Two Imidazolinone-Tolerant Alleles at the *Ahasl1* Locus of Sunflower. *Theor. Appl. Genet.* **2012,** *124,* 385–396.

Satovic, Z.; Joel, D. M.; Rubiales, D.; Cubero, J. I.; Román, B. Population Genetics in Weedy Species of *Orobanche. Austral. Plant Pathol.* **2009,** *38,* 228–234.

Semerci, A.; Kaya, Y.; Sahin, I.; Citak, N. Determination of the Performances and Adoption Levels of Sunflower Cultivars Based on Resistance to Broomrape in Farm Conditions in Thrace Region. *Helia* **2010,** *53,* 69–76.

Semirci, M.; Kaya, Y. Status of *Orobanche cernua* Loefl. and Weeds in Sunflower Production in Turkey. *Helia* **2009,** *51,* 153–160.

Serghini, K.; Pérez-de-Luque, A.; Castejón-Muñoz, M.; García-Torres, L.; Jorrín, J. V. Sunflower (*Helianthus annuus* L.) Response to Broomrape (*Orobanche cernua* Loefl.) Parasitism: Induced Synthesis and Excretion of 7-Hydroxilated Simple Coumarins. *J. Exp. Bot.* **2001,** *52,* 2227–2234.

Shabana, Y. M.; Müller-Stöver, D.; Sauerborn, J. Granular Pesta Formulation of *Fusarium oxysporum* F. Sp. *Orthoceras* for Biological Control of Sunflower Broomrape: Efficacy and Shelf-Life. *Biol. Control* **2003,** *26,* 189–201.

Shindrova, P. Broomrape (*Orobanche cumana* Wallr.) in Bulgaria. Distribution and Race Composition. *Helia* **2006,** *44,* 111–120.

Shindrova, P.; Penchev, E. Race Composition and Distribution of Broomrape (*Orobanche cumana* Wallr.) in Bulgaria During 2007–2011. *Helia* **2012,** *57,* 87–94.

Skoric, D. The Genetics of Sunflower. In *Sunflower Genetics and Breeding*; Skoric, D., Sakac, Z., Eds.; Serbian Academy of Sciences and Arts: Novi Sad, Serbia, 2012; pp 1–163.

Skoric, D.; Pacureanu-Joita, M.; Sava, E. Sunflower Breeding for Resistance to Broomrape (*Orobanche cumana* Wallr.). *Analele I.N.C.D.A. Fundulea* **2010,** *78,* 63–79.

Sukno, S.; Melero-Vara, J. M.; Fernández-Martínez, J. M. Inheritance of Resistance to *Orobanche cernua* Loefl. in Six Sunflower Lines. *Crop Sci.* **1999,** *39,* 674–678.

Tan, S.; Evans, R. R.; Dahmer, M. L.; Singh, B. K.; Shaner, D. L. Imidazolinone-Tolerant Crops: History, Current Status and Future. *Pest Manag. Sci.* **2005,** *61,* 246–257.

Tang, S.; Heesacher, A.; Kishore, V. K.; Fernández, A.; Sadik, E. S.; Cole, G.; Knapp, S. J. Genetic Mapping of the *Or5* Gene for Resistance to *Orobanche* Race E in Sunflower. *Crop Sci.* **2003,** *43,* 1021–1028.

Terzic, S.; Dedic, B.; Atlagic, J.; Jocic, S.; Tancic, S. Screening Wild Sunflower Species and F_1 Interspecific Hybrids for Resistance to Broomrape. *Helia* **2010,** *53,* 25–30.

Thorogood, C.; Hiscock, S. Specific Developmental Pathways Underlie Host Specificity in the Parasitic Plant *Orobanche. Plant Signal Behav.* **2010,** *5,* 275–277.

Velasco, L.; Pérez-Vich, B.; Jan, C. C.; Fernández-Martínez, J. M. Inheritance of Resistance to Broomrape (*Orobanche cumana* Wallr.) Race F in a Sunflower Line Derived from Wild Sunflower Species. *Plant Breeding* **2007,** *126*, 67–71.

Velasco, L.; Pérez-Vich, B.; Yassein, A. A. M.; Jan, C. C.; Fernández-Martínez, J. M. Inheritance of Resistance to Sunflower Broomrape (*Orobanche cumana* Wallr.) in an Interspecific Cross between *Helianthus annuus* and *Helianthus debilis* subsp. *tardiflorus*. *Plant Breeding* **2012,** *121*, 220–221.

Vranceanu, A. V.; Tudor, V. A.; Stoenescu, F. M.; Pirvu, N. Virulence Groups of *Orobanche cumana* Wallr. Differential Hosts and Resistance Sources and Genes in Sunflower. In *Proceedings of the 9th International Sunflower Conference*, Torremolinos, Spain, July 8–13, 1980; International Sunflower Association, Paris, France, 1980; pp 74–80.

Yoneyama, K.; Awad, A. A.; Xie, X.; Yoneyama, K.; Takeuchi, Y. Strigolactones as Germination Stimulants for Root Parasitic Plants. *Plant Cell Physiol.* **2010,** *51*, 1095–1103.

Yoneyama, K.; Xie, X.; Kisugi, T.; Nomura, T.; Sekimoto, H.; Yokota, T.; Yoneyama, K. Characterization of Strigolactones Exuded by Asteraceae Plants. *Plant Growth Regul.* **2011,** *65*, 495–504.

Zhang, W.; Ma, Y.; Wang, Z.; Ye, X.; Shu, J. Some Soybean Cultivars Have Ability to Induce Germination of Sunflower Broomrape. *Plos One* **2013,** *8*, e59715.

6

Sunflower Insect Pests

Jarrad R. Prasifka ■ *United States Department of Agriculture, Agricultural Research Service, Fargo, ND, USA*

Introduction

Cultivated sunflowers have conspicuous heads, abundant pollen and nectar, and are grown together in large plots, all of which contribute to the attractiveness of sunflowers to insects. In part, the visual and olfactory properties of sunflowers are useful; beneficial insects are drawn to the crop (Adedipe and Park, 2010; Ballal and Singh, 1999), including predators that feed on insect pests, contributing to the health and productivity of the sunflowers. The most identifiable beneficial insects in sunflowers are bees, which improve sunflower yields by increasing pollination (i.e., seed set), even for some hybrids considered to be self-fertile (Parker, 1981a). Attraction of bees also is essential for production of hybrid seed. Although honey bees are used as commercial pollinators, diverse native bee species in North America are more effective pollinators for seed production, likely because honey bees often prefer to visit male-sterile lines and, consequently, acquire little pollen from male-fertile plants (Parker, 1981b).

On the other hand, sunflowers also attract an astounding diversity of herbivorous (plant-feeding) insects. Most of the insect species that feed on sunflower are not consistent economic pests because the damage they cause is usually less than the cost of management (e.g., an insecticide application). However, reports on sunflower production worldwide still document dozens of insects that can cause significant losses in sunflower (Rogers, 1992). Crop losses and management challenges with insect pests are most severe in the native range of sunflowers. Although insects on other continents have recently adapted to feed on sunflowers, several species in North America are believed to have evolved feeding primarily on the *Helianthus* spp. As cultivated sunflower production increased in Canada and the United States into the 1970s, insects that previously fed on small patches of wild sunflower took advantage of large fields of suitable hosts. In addition to the concentration of host plants, morphological and chemical changes produced by domestication of sunflower appear to make the crop more susceptible to insect damage than wild sunflowers (Chen and Welter, 2005; Michaud and Grant, 2009).

Previous publications on insect pests of sunflower have described the biology of dozens of species that feed on cultivated sunflower worldwide (Charlet et al., 1997; Rogers, 1992). Rather than produce a similarly exhaustive description of sunflower insect pests, this chapter attempts to provide updated information on the biology and management of a few key pests of oilseed sunflowers in North America. Additional topics include the costs of insect pests, sunflower insects in other settings (e.g., outside of North America, in non-oilseed sunflowers), and the potential for breeding to reduce costs associated with sunflower insects.

Insect Pests of Sunflower in North America

Sunflowers in North America host insects that feed throughout the development of the crop and on virtually all parts of the plant. Although most species are inconsistent or occasional pests, this diversity of herbivores in sunflower means that the crop may require some monitoring throughout the season to permit timely management of insect pests. For example, just after crop emergence, caterpillars called cutworms (larvae of several noctuid moths) may reduce plant stand by severing sunflower stems just above or below the soil line. Other species of insect pests may attack during later vegetative growth stages, including insects that feed on foliage or within stems. However, as with other crops, the most significant insect pests are those that attack the fruit (i.e., sunflower achenes) and the seeds contained inside during the reproductive stages of plant growth.

Shifts in the significance of sunflower insect pests in North America are noteworthy. For example, insects such as the sunflower stem weevil, *Cylindrocopturus adspersus,* and sunflower midge, *Contarinia schulzi,* were once considered important pests in the northern sunflower growing states of Minnesota, South Dakota, and North Dakota. However, both pests have declined in importance for sunflower growers in the northern United States for different reasons. For sunflower stem weevil, the geographic range over which weevils are numerous enough to be considered consistent economic pests has moved southward, reducing the concern for this pest over most much of sunflower acreage. Conversely, significance of the sunflower midge, whose feeding can cause severe distortion of heads, has declined as sunflower crop production shifted. Because severe damage by sunflower midge was largely confined to the Red River Valley, westward movement of North Dakota sunflower production has reduced concern over this once serious insect pest.

Head-Infesting Insect Pests

All of the species discussed here as head-infesting insect pests of sunflower in North America are considered primary pests because they consistently require management or produce losses over a significant portion of the production area. For each of the three head-infesting insect pests discussed here, synopses of their biology, damage, and management are presented, with an emphasis placed on research published since the last thorough review of sunflower insects (Charlet et al., 1997).

The sunflower moth, *Homoeosoma electellum,* is generally considered the most damaging insect pest of sunflowers, particularly in the southern and central parts of the United States. Moths are small (about 1 cm), grey, and inconspicuous unless they are very numerous, in which case they may be seen aggregated on blooming sunflowers. Depending on location, there may be one to several generations per year. When moths locate sunflowers, the presence of pollen induces females to lay eggs among sunflower florets, with larvae hatching two to four days later. After hatch, growing larvae feed in succession on pollen, florets, and seeds. If larvae receive environmental cues indicating the end of summer, they enter diapause, crawling into the soil to overwinter. Because sunflower moth larvae are relatively cold intolerant, the species is believed to be incapable of year-round survival in the United States north of Kansas. However, significant infestations occur well north of Kansas because (1) the actual host range of sunflower moth includes several genera of Asteraceae, and (2) warm temperatures and long days suppress larval diapause. Consequently, several generations may develop on various hosts in southern parts of North America, leading to a gradual, wind-aided migration of moths into the primary sunflower-producing U.S. states (North Dakota, South Dakota, Minnesota) and Canada.

Feeding by a single sunflower moth larva can cause considerable plant injury. On average, a single larva may damage more than 90 florets (which may prevent fertilization and reduce seed set) and 20 or more achenes. Infestation by many larvae becomes visually apparent because they deposit silk and excrement across the face of the head, producing a distinct appearance often described as "trashy." By the time this symptom is noted, larvae may be feeding within achenes and shielded from insecticides. Infestations of sunflower moth larvae can lead to further losses because their feeding promotes development of *Rhizopus* head rot, which lowers seed yield and oil quality. If populations of adult moths are high when anthesis begins, infestations can be greater than 100 larvae per head. Consequently, where sunflower moth is common, management before and during the growing season is important. In general, late planting is useful to limit the severity of sunflower moth infestations. Before and during bloom, moth populations can be monitored using pheromone traps, which help determine whether insecticides are needed and when any treatment should be applied. If fields are scouted for adult moths, the high level of damage on a per-larva basis results in guidelines for insecticide use that are well below one moth per five plants; in areas where sunflower moth is a consistent pest, informal recommendations are to apply insecticide if any moths are present at the onset of bloom. When high moth populations are present, multiple insecticide applications spanning the bloom period may be used. Because insecticide applications to manage sunflower moth (and other head-infesting pests) generally correspond to the days when pollinators are most active in the crop, selection of an insecticide that is less toxic to bees or applying insecticides at times of the day when bees are inactive may balance the need for pollination and crop protection.

Insecticides are usually effective in limiting losses to sunflower moth. However, weather often constrains timing of insecticide applications, and even timely applications may be ineffective because of drift or other factors. When limits of insecticide efficacy are combined with the direct costs of repeated insecticide applications and adverse effects on pollinators (and other beneficial insects), the need to more broadly manage sunflower moth is apparent. Increasing knowledge of the relationship between sunflower moth and its primary host suggests plant resistance is a viable complementary tool to limit damage, and research to facilitate breeding of resistant sunflowers is ongoing. There appear to be at least three complementary ways to reduce sunflower moth damage via host plant resistance. The first option is to make sunflowers less attractive to adult moths. Although the precise compounds that stimulate oviposition are not known, breeding materials show clear differences in the relative attractiveness for egg-laying by female moths (Mphosi and Foster, 2010). The second possible defense is to utilize terpenoid compounds sequestered in glandular trichomes on florets to kill or delay development of small sunflower moth larvae. Although the presence and utility of these glands in wild sunflowers have long been known (Rogers et al., 1987), only recently has it become apparent that the glandular trichomes and terpenoid compounds are present in some public germplasm at levels comparable to wild sunflowers (Figure 6.1) (Prasifka, 2014). Finally, a third defense against the sunflower moth is a strong pericarp, which makes feeding on achenes by older larvae much more difficult. Material identified as resistant to sunflower moth (PI 170415 sunflower line) (Charlet et al., 2008) has been found with exceptional achene strength recently determined as the mechanism of resistance (Prasifka, et al., 2014). Perhaps predictably, there are also nutritional and morphological differences in cultivated sunflower that make the crop more susceptible to sunflower moth than wild *Helianthus* (Chen and Welter, 2007), but the concerted use of other defensive traits may significantly reduce susceptibility in future sunflower hybrids.

The banded sunflower moth, *Cochylis hospes,* is another moth whose larvae feed on sunflower florets and seeds. Banded sunflower moth adults are small (<1 cm) and tan with a dark brown patch visible on the forewings. This species is often compared to (and occasionally confused with) the sunflower moth. However, there are distinct differences in distribution and life history relative to sunflower moths. Banded moths can be found throughout the sunflower-producing areas of the United States and Canada, but they are most common in the northern part of their range. Multiple generations of banded moths are possible in the southern United States, but where this species is common (e.g., North Dakota), only one complete generation occurs. Adult moths are attracted to lay eggs before bloom begins, with eggs deposited on the outer whorl of bracts and the back of the capitulum. Immediately after hatch, larvae

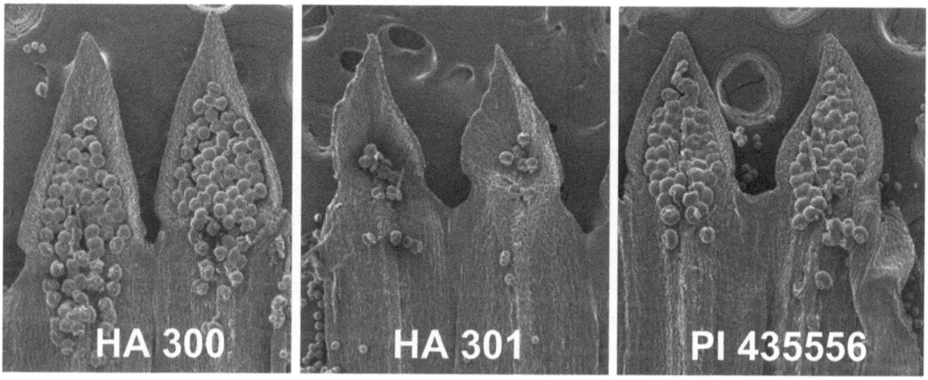

Figure 6.1 Micrographs of glandular trichomes containing insect-repellent and insect-toxic compounds at the tip of sunflower anthers. USDA inbred lines HA 300 and HA 301 illustrate the variation in gland number in public germplasm. For comparison, PI 435556 is a wild sunflower from an area where sunflower moths are common.

may be found between bracts, but later they move inward to feed on pollen, florets, and achenes. Occasionally infestations attributed to banded sunflower moth are actually caused by the similar but less common Arthur's sunflower moth, *Cochylis arthuri.*

Individual larvae of banded sunflower moth are considered less damaging than the sunflower moth. Larvae reduce seed set by feeding on florets and each larva will consume all or part of several seeds. One attempt to estimate the effect of damage to florets by banded sunflower moth (simulated using manual removal of florets) suggests each 10% of floret loss produces a 10% or greater loss of yield and quality (i.e., oil content) (Charlet and Miller, 1993), but other research suggests sunflowers are capable of compensating for some floret damage. Banded sunflower moth is typically managed with a combination of late planting and, if needed, insecticides. Pheromone traps may be used for monitoring, and there are thresholds for insecticide use based on the number of adult moths present within sunflower fields. However, because moths often spend daylight hours in other vegetation outside of field borders, accurate population estimates may be difficult. Alternatively, decisions on insecticide use may be based on the number of banded sunflower moth eggs on plants, as determined by scouting in each field. Relative to sunflower moth, application of insecticides for banded sunflower moths is distinct because it primarily targets larvae rather than adults, and a single application is generally considered sufficient to limit losses from larval feeding.

Recent research on banded sunflower moth has provided insights into alternatives to conventional, broad-spectrum insecticides. Some formulations of the entomopathogenic bacterium *Bacillus thuringiensis* (hereafter, *Bt*) may have efficacy equivalent to conventional insecticides, but with reduced risks to pollinators and other beneficial insects (Jyoti and Brewer, 1999a). However, as larvae develop, they become less susceptible to *Bt* toxins, making timing of applications important (Barker, 1998). One inventive application strategy is the use of honey bee colonies to disseminate a powdered formulation of *Bt* to heads, which appeared to increase yields both by suppression of banded moths and improved seed set (from bee pollination) (Jyoti and Brewer, 1999b). More recent field trials show some evidence for host plant resistance in wild and cultivated sunflower germplasm (Charlet et al., 2009), but information on the mechanisms of resistance (which would facilitate its use in breeding) is generally absent. However, the identification of compounds that stimulate egg-laying by female banded sunflower moths (Morris et al., 2005, 2009) suggests that sunflowers with decreased attractiveness for oviposition could be identified or developed.

The red sunflower seed weevil, *Smicronyx fulvus,* was once a severe pest of sunflower in the northern production area, but recently has been sporadic as an economic pest in North and South Dakota. As its name suggests, the small (0.4 cm) beetle is red with a long snout. There is only one generation of this weevil each year. Weevil adults emerge from the soil in early summer and move to pre-bloom sunflower, feeding on bracts. No damage is caused by this early adult feeding, but it sustains adult weevils until bloom, when pollen becomes a primary food. Pollen is necessary for females to reach reproductive maturity, and weevils will move off of a plant that has completed anthesis to find other sunflowers still in bloom. Female weevils typically lay eggs singly inside the pericarp of developing achenes, and all larval feeding occurs within the single achene. Once a weevil larva is mature, it chews an exit hole in the pericarp and drops to the soil to overwinter. The closely related gray sunflower seed weevil, *Smicronyx sordidus,* also feeds within sunflower achenes in sunflower-producing areas of the United States. However, the gray sunflower seed weevil is far less abundant in the northern Great Plains and is not generally considered a significant pest.

Damage caused by red sunflower seed weevils is limited by the fact that each larva consumes just one developing seed. This produces estimated economic thresholds for oilseed sunflower that are high (about five adults per head) in comparison to thresholds used for head-infesting moths. Unlike most other sunflower insect pests, early planting can be used to manage sunflower seed weevils because heads in early-planted fields may be too mature (i.e., without pollen and with hardened achenes) for weevils. Insecticides are also effective for management, but following order-of-magnitude decreases in red sunflower seed weevil abundance in the mid-1990s, insecticides have been needed less frequently.

Research on red sunflower seed weevils has declined concurrently with their populations. Repeated field surveys in North Dakota document a significant weevil decline between 1992 and 1994 (Charlet, 2002). Although the precise cause of the drop in populations is not known, increased precipitation (and soil moisture) since the early 1990s is the most common theory. Field research using artificial infestation of weevils or known areas where greater weevil populations have persisted has allowed some work documenting host plant resistance in sunflower germplasm (Charlet et al., 2010; Gao and Brewer, 1998). The most promising source of resistance, PI 431542, is being used to develop seed weevil–resistant germplasm because a return of the population levels formerly found in North Dakota (>100 adults per head, with about 1000 seeds per head destroyed) would require rapid changes in management to minimize losses.

Typical seed damage from the primary head-infesting pests differs in amount of seed consumed or location of entry and exit holes in the pericarp (Figure 6.2). However, the location of seed feeding can vary, and the amount of feeding is influenced by the size of developing larvae (for sunflower moth and banded sunflower moth). Consequently, damage to any achene may not always be reliably assigned to an insect based solely on appearance.

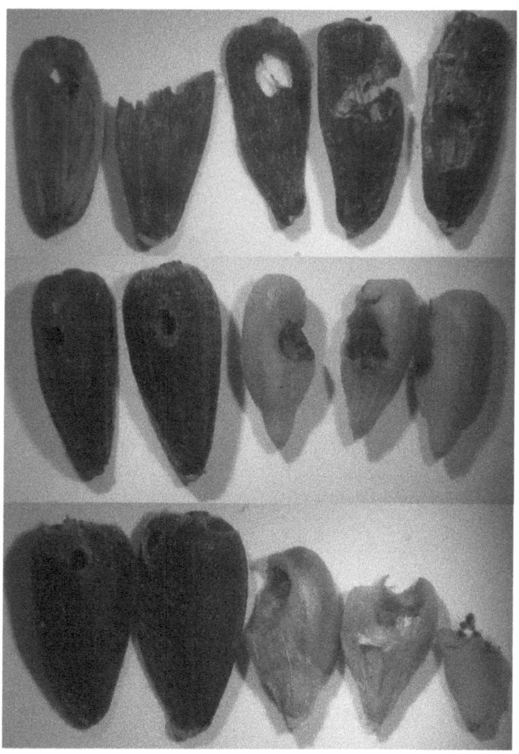

Figure 6.2 Damage to achenes and seeds by sunflower moth (top), red sunflower seed weevil (middle), and banded sunflower moth (bottom). No seeds are shown for sunflower moth because feeding resulted in complete consumption of seed.

Stem-Infesting Insect Pests

In any part of North America, several insect species may occur within stems of cultivated sunflower, but few cause injury sufficient to reduce yields. However, some species (which cause apparently trivial injury within stems) may have hidden importance in either transmitting or promoting the occurrence of pathogens. For the two beetle species discussed here as primary pests, the main concern is physical weakening of stems, which leads to lodging and a reduction of harvestable yields.

Sunflower stem weevils, *Cylindrocopturus adspersus,* occur throughout the Great Plains of the United States and in Saskatchewan. As with some other insect pests, the area where sunflower stem weevil is a concern has shifted, moving from primary production areas of North Dakota and Minnesota southward into Colorado, Kansas, and Nebraska. Adult weevils are small (0.5 cm) and brown with white spots. Only one generation of stem weevils occurs each year, with most adult weevils appearing between April (southern) and June (northern). Weevils typically enter sunflower fields when plants are in early vegetative growth stages. Though adults feed on leaf and stem tissue, no significant damage is produced by adult feeding. Female weevils later lay eggs just beneath the epidermis near the base of sunflower stems. Weevil larvae move vertically within stems, feeding on vascular tissue and pith. When larvae are mature, they enter diapause and overwinter inside sunflower stems, with most larvae found less than 10 cm above soil level.

Sunflower plants are affected by sunflower stem weevils in multiple ways. Damage to vascular tissues by larval feeding can reduce seed yield and oil content, but this is believed to occur only with very high densities (>70) of weevil larvae per stem. The most significant risk of sunflower stem weevil infestation is that larval feeding may weaken stems and promote lodging, reducing the number of plants that can be mechanically harvested. Lodging can be caused by relatively modest numbers (<40) of weevil larvae per stem. In addition to the potential for direct and harvest losses, sunflower stem weevils also introduce and promote diseases such as charcoal stem rot and Phoma black stem.

Delayed planting limits the risk of losses from sunflower stem weevil. Reducing planting density also reduces lodging, as greater plant spacing produces thicker, more robust stalks. Insecticides have been effective in the past, but the favored active ingredient, carbofuran, is no longer available for use in sunflowers. The shift of weevil populations southward and loss of one effective insecticidal management option has led to additional work on management of sunflower stem weevil. The use of a popular insecticidal seed treatment seemed to reduce weevil numbers, but the effect was not consistent and notably less effective than carbofuran (Charlet et al., 2007a). The effects of water and weed management have also been examined; weeds did not appear to affect the number of sunflower stem weevils per stem (Michaud, 2009), but the use of irrigation had a small additional benefit in reduced stem weevil infestation (Charlet, 2007b). Recent efforts have also included examining the effect of parasitoids, which appears to slightly suppress sunflower stem weevil populations (Charlet, 2002), and identification of germplasm that is potentially resistant to stem-infesting insects (Charlet et al., 2009). Due in part to the labor-intensive process of quantifying resistance (by dissection of plants to estimate per-stem weevil densities) and the confounding effect of stem density on weevil numbers (Ode et al., 2011), efforts to produce sunflowers with resistance to sunflower stem weevils have made little progress. The use of time-saving tools to more efficiently assess infestations (e.g., digital X-rays; Figure 6.3) may improve the likelihood of success in future work with sunflower stem weevils.

The Dectes stem borer, *Dectes texanus,* is another beetle whose larvae feed within sunflower stems and is an increasing concern to growers in the United States. Weevil populations appear to be gradually expanding from central states such as Nebraska and Kansas into northern areas with greater sunflower production. Adult beetles are large (1–2 cm), grey, and conspicuous by their antennae, which are as long or longer than the rest of the body. Only one generation occurs each year, with adults appearing

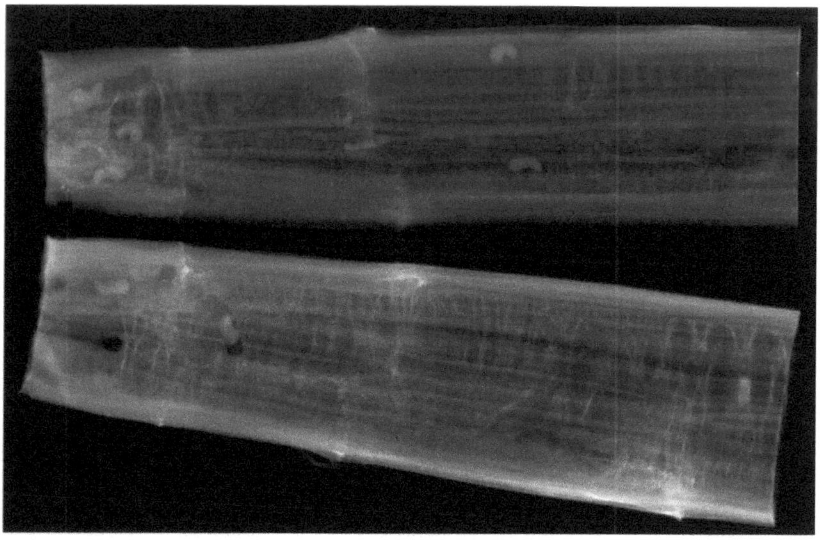

Figure 6.3 Digital X-ray of sunflower stem weevil larvae in the lowest 8 cm of a stem. Stem section has been split longitudinally to facilitate accurate counting of weevils, which are concealed in woody tissue along the circumference of the stem.

in mid to late summer. Dectes stem borer adults will feed on sunflower stems and petioles, but like adult feeding by sunflower stem weevils, this has no consequence for crop yields. Females lay eggs in petioles of recently emerged leaves, and larvae initially feed within petioles. As larvae grow, they enter stems and tunnel through the pith. When plants near physiological maturity, large larvae descend within the stalk and consume a ring of internal stem tissue (i.e., girdle) that weakens the base of the stem. Larvae then move further down and pack the portion of tunnel above with excrement and bits of stem tissue. If the stem later breaks, the process of stem-girdling and sealing off their tunnel results in a well-protected chamber in which larvae spend the winter. Because the larvae are strongly cannibalistic, most infested stems eventually contain only one surviving larva.

Feeding by the Dectes stem borer appears to have no direct effect on sunflowers (Michaud et al., 2007), although resultant lodging can greatly impact harvested yields. Economic thresholds are not used for several reasons. First, adult beetles emerge over a long period and are long-lived, confounding efforts to recommend specific timing of insecticide application. Second, the effect of a given larval infestation (e.g., 70% of stalks infested) may not be predictive of eventual lodging. Relative to densely planted sunflowers, greater within-row plant spacing (which promotes thicker stems) may greatly reduce the ability of larvae to girdle stems completely enough to cause lodging (Michaud et al., 2009). Also, insecticide applications (including carbofuran and another compound shown to be effective in laboratory trials) are unable to consistently reduce the percentage of plants infested by more than 50% (Michaud et al., 2007). Consequently, reduced plant density and use of hybrids that have previously shown good resistance to lodging are probably the best management practices for areas where Dectes stem borers have been a problem.

There are some limited possibilities for using host plant resistance to improve management. Unlike most sunflower pests, the Dectes stem borer may be better known as a pest of another major annual crop: soybean. Although adult beetles strongly prefer cultivated sunflower over soybean (Michaud and Grant, 2005), wild sunflowers are significantly less acceptable (Michaud and Grant, 2009). Increased resin production is believed to be partially responsible for resistance in wild sunflower, but this quality is not acceptable in commercial sunflowers. However, the use of hybrids with the stay green trait (in which stems remain green into physiological maturity) may delay the downward movement and girdling behavior of larvae, and has other benefits to sunflower production (Cukadar-Olmedo and Miller, 1997).

Costs of Sunflower Insect Pests

Estimating the costs of insect pests to sunflower growers is difficult at both small and large scales. For example, at a field level, considerable losses of stand (from cutworms or wireworms) or foliage (from sunflower beetle or defoliating caterpillars) may occur with little or no impact on yields. While compensation by sunflowers for this type and degree of damage is a benefit, the appearance of insect feeding at levels that may not impact yields still may induce application of insecticides, and those applications represent a cost to growers. On a national level, assembly of data on direct losses from insects and management practices requires considerable resources and organization; this is especially true in the United States because of regional differences in occurrence and abundance of sunflower insect pests.

However, as one element in understanding the costs of sunflower pests (insects, weeds, birds, and diseases), the National Sunflower Association (NSA) conducts an annual survey that typically includes 150–200 fields, with each field intended to represent 10,000 planted acres of sunflower. For insects, some data on plant lodging is taken and aggregate seed samples from several heads per field are sent to the United States Department of Agricultural Research Service (USDA-ARS) sunflower entomology lab for rating. Seed ratings, which estimate the percentage of achenes damaged by sunflower moth, banded sunflower moth, and red sunflower seed weevils, generally show low (1–2%) damage in the survey. While the seed damage data provide a way to assess changes in the severity or geographic distribution of insect pests over time, as quantitative estimates of cost, they have a severe bias. The survey underestimates costs of insect pests for several reasons, including: (1) management with cultural practices and insecticides are effective in reducing insect damage; (2) scoring seed samples does not show losses in seed set due to floret-feeding caterpillars; (3) actual costs from one or more insecticide applications are not factored in; and (4) opportunity costs from cultural management practices, including late planting, are not included. For comparison, estimates from research trials without cultural or insecticidal management may be considered as more realistic estimates of costs. Although the research trials are limited in geographic scope, data for sunflower moth and banded sunflower moth damage in recent years illustrate the differences in estimates of direct (i.e., seed feeding) costs with corresponding states in the NSA survey (Table 6.A).

Overall, results from NSA surveys show that cultural practices and insecticides combined are very effective at limiting damage from seed-feeding sunflower insects, but costs of management are not known. If more than one insecticide application is used (expected in areas with high populations of sunflower moth), costs of insecticides may equal or exceed $14 per acre. This expense can represent a significant fraction of the returns over costs (Dumler et al., 2007), meaning that limiting management costs could substantially increase profits. The use of insect monitoring (e.g., using traps or in-field scouting) and economic thresholds are both integrated pest management tactics that help limit the costs of sunflower insect pests. If effective host plant resistance for any major sunflower pest is incorporated into hybrids, it could reduce costs (by eliminating the need for some insecticide applications) or improve

Table 6.A Seed Damage to Commercial Sunflower Hybrids in Grower Fields and Unmanaged Research Plots, 2010–2011

Insect	Year	Location	Plot Type (n)[a]	Insect Management	Damage (% ± SE)
Sunflower moth	2010	Kansas	Grower fields (7)	Late planting, insecticides	0.3 ± 0.5
			Research plots (3)	None	13.7 ± 5.1
Banded sunflower moth	2011	North Dakota	Grower fields (62)	Late planting, insecticides	1.1 ± 2.3
			Research plots (3)	None	7.3 ± 0.9

[a]For grower fields, n = number of oilseed sunflower fields in state included in fall survey by National Sunflower Association. In research plots, n = number of commercial hybrids tested in replicated plots.

the efficacy of insecticides when they are needed. Insect management using plant-incorporated *Bt* toxins (i.e., transgenic insect-resistant sunflowers) is possible, but the high potential of transgenes escaping into one or more species of wild *Helianthus* has, so far, discouraged commercial use.

Sunflower Insect Pests outside of North America

Many insects are capable of causing injury to sunflowers outside of North America (Charlet et al., 1997), although few are serious pests. The insect pests that cause consistent or severe economic losses often are species from genera that feed on a small number of related plant species but show a strong affinity for cultivated sunflowers (like the primary insect pests in North America). Other species are generalists, which feed on a wide range of host plants and utilize sunflower opportunistically. A more complete listing of these species can be found elsewhere (Charlet et al., 1997).

Among the insects that are closely associated with cultivated sunflower outside North America, perhaps the best known is the European sunflower moth, *Homoeosoma nebulella.* Damage to sunflowers by larvae of the moth has been noted in Europe since the late 1800s. Because of its relatedness to the sunflower moth in North America, there are many similarities in both species' biology and their effects on sunflower plants, including multiple generations, feeding on florets and seeds, and additional damage caused by promoting plant diseases. However, the European sunflower moth is thought to be somewhat less severe in terms of damage produced by each larva. For oilseed sunflowers, resistance to larval feeding has been associated with phytomelanin, a dark layer present in achenes of most hybrids. Although oilseed hybrids with phytomelanin sometimes have reduced damage compared to confectionary varieties (which lack phytomelanin), losses where the moth occurs remain likely without the use of insecticides. Early planting provides some additional protection for small farms where aerial insecticide applications are not feasible (Szabó et al., 2010). Research on the noncultivated hosts of the European sunflower moth suggests that reducing the weedy Asteraceae used by the first generation of insects could limit damage by the moths that later infest cultivated sunflower (Szabó et al., 2009). Damaging outbreaks of the European sunflower moth have also been noted recently in Mongolia and China (Cao et al., 2010, 2012). Other species of *Homoeosoma* are also present in South America, but their damage is less severe than the two species already noted.

Several small flies in the genus *Melanagromyza* are also associated with sunflowers. In particular, the sunflower head fly, *M. minimoides* (or *mosquita del capítulo*) is a significant pest of sunflower in South America. The tiny black flies can be seen during the day resting on open heads of sunflower, where females lay eggs among the florets. Damage is caused by larval feeding on florets and seeds, and losses of 20–30% have been reported in Argentina and Uruguay (Ves Losada and Figueruelo, 2006; Zerbino, 1991). Detailed observation over several planting dates shows increased damage is strongly associated with late planting (Ves Losada and Figueruelo, 2006). Insecticides may be used to control adults, but as with other insects that attack reproductive-stage sunflowers, possible adverse effects on pollinators are a concern. Among inbred lines with equivalent flowering dates, differences in infestation have been detected, suggesting that host plant resistance to sunflower head fly may be a future management tool. Interestingly, at least three species of *Melanagromyza* associated with sunflower are present in the United States, including a broad geographic distribution for sunflower head fly, but it has not been reported as a significant pest in North America.

The bollworm, *Helicoverpa armigera,* is a generalist caterpillar that feeds on over 100 species of crops, including cotton and sunflower. It has been reported as a pest of sunflower in Europe, Asia, Africa, and Australia. Adult moths lay eggs primarily on florets. Bollworm larvae will feed on sunflower florets and seeds, occasionally causing severe damage. In India, sampling across several locations found that numbers of eggs and larvae in sunflower were comparable to cotton, which has bollworm as a primary pest (Ravi et al., 2005). Populations of bollworm in India from cotton and sunflower were equally susceptible to *Bt* toxins (Gujar et al., 2004). In a comparison of several sunflower hybrids, there were significant differences in the number of eggs deposited on sunflowers and the developmental times for larvae (Sharma and Singh, 2001), suggesting some potential to select hybrid sunflowers with less potential damage by bollworms.

Another generalist species herbivore feeds on sunflower in South Africa, the false chinch bug, *Nysius natalensis.* Like several other *Nysius* species known to attack sunflowers in Africa and Australia, it is a sporadic pest. When present on vegetative or early reproductive stage plants, false chinch bugs feed on vascular tissues, but once seed development begins, nymphs and adults primarily feed on seeds. Reductions in seed yield and oil content have been observed, and young plants may occasionally be killed by large numbers of bugs feeding. Interestingly, sunflowers appear to be an incidental host of the bug, as recent documentation shows they feed on dozens of weedy hosts in eight or more plant families (Du Plessis et al., 2007). Controlled feeding and egg-laying tests indicate that false chinch bugs prefer weedy hosts to sunflower (Du Plessis et al., 2012). As a result, early-season management of weeds may be useful to limit population growth, but weeds should not be destroyed during sunflower seed-fill, as the bugs would likely move into and damage nearby sunflowers. When weed management does not maintain sufficiently low levels of false chinch bug, insecticides can be effective at killing the pest, but there is a risk of fields being recolonized by the pest only a few days later (Du Plessis et al., 2005).

Insect Damage in Non-Oilseed Sunflowers

Relative to oilseed sunflowers, the primary difference in management of insects for other uses (particularly confection) is a lower tolerance for insect damage. When this reduced tolerance is combined with higher market value, growers of confection sunflowers often apply insecticides as a prophylactic (i.e., without having a specific target pest or reaching an economic threshold). One recent example of the low tolerance for damage comes from the symptom referred to as "kernel brown spot" in North America. Starting in the late 1990s, shelled confection sunflower seeds began to appear in higher numbers with a small dark spot on the distal end of the seed; the blemish was apparent on raw seeds, but became even more apparent after roasting. Eventually the tarnished plant bug, *Lygus lineolaris,* was found to be the insect causing the damage. Because of the extremely low (0.5%) tolerance for this cosmetic damage, it was determined that an economic injury level for the bug was one insect per 10–15 plants (Charlet, 2003). The tarnished plant bug and other *Lygus* are considered to be generalists, feeding on many plant species. Consequently, some *Lygus* are also known as pests of sunflower in several other countries. Outside of the United States and Canada, however, their damage is not restricted to confection sunflower, and feeding may reduce oil content and viability of seed (Bujáki et al., 1998; Charlet et al., 1997).

Conclusion

Without management, insect pests of sunflower in North America cause significant damage and limit crop production. Field survey data show that use of cultural practices such as modified planting dates and crop rotation, in combination with insecticides, provide excellent protection against most of the significant insect pests. The use of other management tools, including pest monitoring, can help limit the costs associated with potentially unnecessary insecticide applications. In addition to direct (crop loss) and indirect (management) costs, insect pests may discourage sunflower production if sunflowers are perceived to require more time and attention than other crops with transgenic (*Bt*) pest management provided to growers. To limit the susceptibility of sunflowers to insect pests and enhance the efficacy of needed insecticide applications, work on breeding for host plant resistance is ongoing. Success in breeding for plant resistance seems more likely in the future as tools of modern genetics and crop breeding are applied in sunflower. Outside of North America, challenges with insect pests are fewer, but still present. In many cases, sunflower insects found in South America, Europe, or Asia are closely related to the insects pests of sunflower in North America. Consequently, any advances in insect management, including host plant resistance, should be beneficial to sunflower production worldwide.

References

Adedipe, F.; Park, Y.-L. Visual and Olfactory Preference of *Harmonia axyridis* (Coleoptera: Coccinellidae) Adults to Various Companion Plants. *J. Asia-Pac. Entomol.* **2010,** *13,* 319–323.

Ballal, C. R.; Singh, S. P. Host Plant-mediated Orientational and Ovipositional Behavior of Three Species of Chrysopids (Neuroptera: Chrysopidae). *Biol. Control* **1999,** *16,* 47–53.

Barker, J. F. Effect of *Bacillus thuringiensis* subsp. kurstaki Toxin on the Mortality and Development of the Larval Stages of the Banded Sunflower Moth (Lepidoptera: Cochylidae). *J. Econ. Entomol.* **1998,** *91,* 1084–1088.

Bujáki, G.; Bukovinszky, T.; Csaki, M.; Piukovics, L. The Dominant Heteroptera (Miridae) Pests of the Sunflower. *Novenyvedelem* **1998,** *34,* 45–51.

Cao, L. Z.; Liu, X. X.; Zhang, Z. W. The Sunflower Moth, *Homoeosoma nebulella* (Denis et Schiffermuller) (Lepidoptera: Pyralidae): Outbreaks and Pest Management in Linhe, Inner Mongolia 2007–2008. *Entomol. Fenn.* **2010,** *21,* 65–69.

Cao, L. Z.; Yu, J. H.; Qin, R.; Zhang, Z. W. Genetic Variability and Structure of Populations of *Homoeosoma nebulella* (Denis et Schiffermuller) (Lepidoptera: Pyralidae) in Northern China. *Entomol. Fenn.* **2012,** *23,* 121–126.

Charlet, L. D. Parasitization of the Red Sunflower Seed Weevil (Coleoptera: Curculionidae) by Its Larval Parasitoid *Triaspis aequoris* (Hymenoptera: Braconidae) in Cultivated Sunflower. *Environ. Entomol.* **2002,** *31,* 844–851.

Charlet, L. D. Plant Stage Susceptibility and Economic Injury Level for Tarnished Plant Bug, *Lygus lineolaris* (Heteroptera: Miridae), on Confection Sunflower. *Helia* **2003,** *26,* 83–92.

Charlet, L. D.; Miller, J. F. Seed Production after Floret Removal from Sunflower Heads. *Agron. J.* **1993,** *85,* 56–58.

Charlet, L. D.; Brewer, G. J.; Franzmann, B. A. Sunflower Insects. In *Sunflower Production and Technology*; Schneiter, A. A., Ed.; Agronomy Monograph 35; American Society of Agronomy: Madison, WI, 1997; pp 183–261.

Charlet, L. D.; Armstrong, J. S.; Hein, G. L. Sunflower Stem Weevil and Its Larval Parasitoids in the Central and Northern Plains of the USA. *BioControl* **2002,** *47,* 513–523.

Charlet, L. D.; Aiken, R. M.; Meyer, R. F.; Gebre-Amlak, A. Impact of Combining Planting Date and Chemical Control to Reduce Larval Densities of Stem-infesting Pests of Sunflower in the Central Plains. *J. Econ. Entomol.* **2007a,** *100,* 1248–1257.

Charlet, L. D.; Aiken, R. M.; Meyer, R. F.; Gebre-Amlak, A. Impact of Irrigation on Larval Density of Stem-infesting Pests of Cultivated Sunflower in Kansas. *J. Econ. Entomol.* **2007b,** *100,* 1555–1559.

Charlet, L. D.; Aiken, R. M.; Seiler, G. J.; Chirumamilla, A.; Hulke, B. S.; Knodel, J. J. Resistance in Cultivated Sunflower to the Sunflower Moth (Lepidoptera: Pyralidae). *J. Agric. Urban Entomol.* **2008,** *25,* 245–257.

Charlet, L. D.; Aiken, R. M.; Miller, J. F.; Seiler, G. J. Resistance among Cultivated Sunflower Germplasm to Stem-infesting Pests in the Central Great Plains. *J. Econ. Entomol.* **2009,** *102,* 1281–1290.

Charlet, L. D.; Seiler, G. J.; Grady, K. A.; Hulke, B. S.; Chirumamilla, A. Resistance in Cultivated Sunflower Germplasm to the Red Sunflower Seed Weevil (Coleoptera: Curculionidae) in the Northern Great Plains. *J. Kans. Entomol. Soc.* **2010,** *83,* 51–57.

Chen, Y. H.; Welter, S. C. Crop Domestication Disrupts a Native Tritrophic Interaction Associated with the Sunflower, *Helianthus annuus* (Asterales: Asteraceae). *Ecol. Entomol.* **2005,** *30,* 673–683.

Chen, Y. H.; Welter, S. C. Crop Domestication Creates a Refuge from Parasitism for a Native Moth. *J. Appl. Ecol.* **2007,** *44,* 238–245.

Cukadar-Olmedo, B.; Miller, J. F. Inheritance of the Stay Green Trait in Sunflower. *Crop Sci.* **1997,** *37,* 150–153.

Dumler, T. J.; O'Brien, D. M.; Thompson, C. R.; Olson, B. L. S. Sunflower Cost-Return Budget in Western Kansas. In *Farm Management Guide,* MF-887; Manhattan, KS: Kansas State University, 2007.

Du Plessis, H.; Byrne, M. J.; van den Berg, J. Chemical Control of *Nysius natalensis* Evans (Hemiptera: Orsillidae), a Pest of Sunflower in South Africa. *S. Afr. J. Plant Soil* **2005,** *22,* 94–99.

Du Plessis, H.; Byrne, M. J.; van den Berg, J. Distribution and Host Plant Range of *Nysius natalensis* Evans (Hemiptera: Orsillidae) in the Sunflower Production Area of South Africa. *Afr. Entomol.* **2007,** *15,* 310–318.

Du Plessis, H.; Byrne, M. J.; van den Berg, J. The Effect of Different Host Plants on the Reproduction and Longevity of *Nysius natalensis. Entomol. Exp. Appl.* **2012,** *145,* 209–214.

Gao, H.; Brewer, G. J. Sunflower Resistance to the Red Sunflower Seed Weevil (Coleoptera: Curculionidae). *J. Econ. Entomol.* **1998,** *91,* 779–783.

Gujar, G. T.; Mittal, A.; Kumari, A.; Kalia, V. Host Crop Influence on the Susceptibility of the American Bollworm, *Helicoverpa armigera* to *Bacillus thuringiensis* ssp *kurstaki* HD-73. *Entomol. Exp. Appl.* **2004,** *113,* 165–172.

Jyoti, J. L.; Brewer, G. J. Median Lethal Concentration and Efficacy of *Bacillus thuringiensis* against Banded Sunflower Moth (Lepidoptera: Tortricidae). *J. Econ. Entomol.* **1999a,** *92,* 1289–1291.

Jyoti, J. L.; Brewer, G. J. Honey Bees (Hymenoptera: Apidae) as Vectors of *Bacillus thuringiensis* for Control of Banded Sunflower Moth (Lepidoptera: Tortricidae). *Environ. Entomol.* **1999b**, *28,* 1172–1176.

Michaud, J. P.; Grant, A. K. The Biology and Behavior of the Longhorned Beetle, *Dectes texanus* on Sunflower and Soybean. *J. Insect Sci.* **2005,** *5,* 25.

Michaud, J. P.; Grant, A. K. The Nature of Resistance to *Dectes texanus* (Col., Cerambycidae) in Wild Sunflower, *Helianthus annuus. J. Appl. Entomol.* **2009,** *133,* 518–523.

Michaud, J. P.; Grant, A. K.; Jyoti, J. L. Impact of the Stem Borer, *Dectes texanus,* on Yield of the Cultivated Sunflower, *Helianthus annuus. J. Insect Sci.* **2007,** *7,* 21.

Michaud, J. P.; Stahlman, P. W.; Jyoti, J. L.; Grant, A. K. Plant Spacing and Weed Control Affect Sunflower Stalk Insects and the Girdling Behavior of *Dectes texanus* (Coleoptera: Cerambycidae). *J. Econ. Entomol.* **2009,** *102,* 1044–1053.

Morris, B. D.; Foster, S. P.; Grugel, S.; Charlet, L. D. Isolation of the Diterpenoids, *ent*-kauran-16α-ol and *ent*-atisan-16α-ol, from Sunflowers, as Oviposition Stimulants for the Banded Sunflower Moth, *Cochylis hospes. J. Chem. Ecol.* **2005,** *31,* 89–102.

Morris, B. D.; Charlet, L. D.; Foster, S. P. Isolation of Three Diterpenoid Acids from Sunflowers, as Oviposition Stimulants for the Banded Sunflower Moth, *Cochylis hospes. J. Chem. Ecol.* **2009,** *35,* 50–57.

Mphosi, M. S.; Foster, S. P. Female Preference and Larval Performance of Sunflower Moth, *Homoeosoma electellum,* on Sunflower Pre-breeding Lines. *Entomol. Exp. Appl.* **2010,** *134,* 182–190.

Ode, P. J.; Charlet, L. D.; Seiler, G. J. Sunflower Stem Weevil and Its Larval Parasitoids in Native Sunflowers: Is Parasitoid Abundance and Diversity Greater in the U.S. Southwest? *Environ. Entomol.* **2011,** *40,* 15–22.

Parker, F. D. Sunflower Pollination: Abundance, Diversity and Seasonality of Bees and Their Effect on Seed Yields. *J. Apic. Res.* **1981a,** *20,* 49–61.

Parker, F. D. How Efficient Are Bees in Pollinating Sunflowers? *J. Kans. Entomol. Soc.* **1981b,** *54,* 61–67.

Prasifka, J. R. Variation in Density of Capitate Grandular Trichomes in Wild and Cultivated Sunflower Germplasm and Potential for Use in Host Plant Resistance. *Plant Gen. Res.* **2014,** DOI: 10.1017/S1479262114000653.

Praskifa, J. R.; Hulke, B. S.; Seiler, G. J. Pericarp Strength of Sunflower and Its Value for Plant Defense Against the Sunflower Moth, *Homoeosoma electullum. Arthr. Plant Interact.* **2014,** *8,* 101–107.

Ravi, K. C.; Mohan, K. S.; Manjunath, T. M.; Head, G.; Patil, B. V.; Greba, D. P.; Premalatha, K.; Peter, J.; Rao, N. G .V. Relative Abundance of *Helicoverpa armigera* (Lepidoptera: Noctuidae) on Different Host Crops in India and the Role of These Crops as Natural Refuge for *Bacillus thuringiensis* Cotton. *Environ. Entomol.* **2005,** *34,* 59–69.

Rogers, C. E. Insect Pests and Strategies for Their Management in Cultivated Sunflower. *Field Crops Res.* **1992,** *30,* 301–332.

Rogers, C. E.; Gershenzon, J.; Ohno, N.; Mabry, T. J.; Stipanovic, R. D.; Kreitner, G. L. Terpenes of Wild Sunflowers (*Helianthus*): An Effective Mechanism against Seed Predation by Larvae of the Sunflower Moth, *Homoeosoma electellum* (Lepidoptera: Pyralidae). *Environ. Entomol.* **1987,** *16,* 586–592.

Sharma, M.; Singh, G. D. Development of *Helicoverpa armigera* on Different Hybrids of Sunflower (*Helianthus annuus*). *Indian J. Agric. Sci.* **2001,** *71,* 538–543.

Szabó, B.; Szabó, M.; Varga, C., Tóth, F.; Vagvölgyi, S. Feral Host Plant Range as a Reservoir of European Sunflower Moth (*Homoeosoma nebulellum* Den. et Schiff.) Populations in Nyírség Region. *Northwest. J. Zool.* **2009,** *5,* 290–300.

Szabó, B.; Szabó, M.; Varga, C., Tóth, F.; Vagvölgyi, S. Relationships between Sunflower Variety, Sowing Date and the Extent of Damage Caused by the European Sunflower Moth (*Homoeosoma nebulellum* Den. et Schiff.). *Helia* **2010,** *33,* 37–45.

Ves Losada, J. C.; Figueruelo, A. M. Evaluación del daño provocado por la mosquita del capítulo del girasol, *Melanagromyza minimoides* según le fecha de siembra. Manejo de Plagas y tecnología de cultivos en sistemas mixtos de producción; Boletines de Divulgación Técnica, No. 91; INTA: Anguil, La Pampa, Argentina, 2006; pp 57–62.

Zerbino, M. S. Mosquita del Capítulo del Girasol *Melanagromyza minimoides*, Nueva Plaga. *Agrociencia* **1991,** 5, 90–91.

7

Sunflower Bird Pests

George M. Linz ■ *U.S. Department of Agriculture, Wildlife Services, National Wildlife Research Center, Bismarck, ND, USA*

James J. Hanzel ■ *Genosys, LLC, Fargo, ND, USA*

Introduction

Sunflower (*Helianthus annuus* L.) is annually planted on approximately 26 million hectares in Australia, China, Europe, India, North America, Pakistan, Russia, South America, and Ukraine (National Sunflower Association, 2014). Flocks of granivorous birds, ranging in size from a few to millions, can be found in every sunflower growing region and have been documented to cause serious economic losses (Linz et al., 2011) (Figure 7.1). Avian species that damage sunflower generally belong to the parakeet (Psittacidae), dove (Columbidae), cockatoo (Cacatuidae), sparrow (Passeridae), crow (Corvidae), and blackbird (Icterinae) families (De Grazio, 1989; Linz and Hanzel, 1997; Linz et al., 2011; Rodriquez et al., 1995; van Niekerk, 2009).

Since the 1970s, U.S. scientists have focused on reducing blackbird damage in the Prairie Pothole Region (PPR) of the United States. Population and nesting ecology of blackbirds, the timing and extent of damage, and numerous potential methods of reducing damage were studied (Blackwell et al., 2003; Linz et al., 2011; Peer et al., 2003). Bird damage begins after the ray petals drop from the ripening sunflower heads and achenes begin to develop (Cummings et al., 1989). Sunflower is vulnerable to bird damage until harvest, which can begin 6 weeks or more after plants reaches physiological maturity. Birds seek sunflower because it easy to perch on the heads to quickly access and crack the achenes to obtain the kernel. The high caloric kernel contains proteins and fats needed for breeding season activities, feather replacement, and fat storage prior to migration (Linz et al., 1995). Blackbirds and sparrows prefer oilseed cultivars because they tend to be smaller and have thinner hulls than confectionery varieties (Linz and Hanzel, 1997).

Damage levels vary widely across sunflower growing areas because bird numbers are dependent on landscape features, such as the presence of wetlands and trees that are favored roosting sites and the availability of alternative foods. In the PPR of the United States, which is renowned for wetlands used by migrating birds, many producers have abandoned sunflower and now plant alternative crops that are not subject to serious bird damage and have better options available for managing weeds (Hulke and Kleingartner, 2014; Linz et al., 2011).

Producer complaints from the PPR have motivated scientists to assess bird damage with both models and field surveys. Peer et al. (2003) constructed bioenergetic

Figure 7.1 Blackbirds in ripening sunflower in North Dakota.
Source: North Dakota State University, Fargo.

and economic models to determine the potential effects blackbirds have on the sunflower crop in the northern Great Plains of North America. They estimated that blackbirds annually damaged \$5.2 million (at \$0.26/kg) of sunflower from 1993 to 1997. During 2009 and 2010, Klosterman et al. (2013) conducted field damage surveys and found that blackbirds damaged about 2.7% of the crop, valued at \$3 million. Damage in 15% of the surveyed fields (n = 27) was over 5.0% and 26% of the fields received greater than 1% damage. Damage above 5% is considered economically important by sunflower growers (Linz and Homan, 2011). Thus, solutions to the blackbird–sunflower problem should be focused on methods that reduce the extreme level of damage suffered by a small proportion of growers.

Outside the United States, recent bird damage data have been reported for South Africa, Argentina, and Iran. In South Africa, van Niekerk (2009) surveyed bird damage and reported that doves damaged 12.7% of the crop, whereas quantitative surveys showed that the damage was 8.4%. In 2007, birds damaged 0.9% of the sunflower crop in Argentina (Linz and Homan, 2011). In Iran, birds are considered important pests of sunflower (Khaleghizadeh and Alizadeh, 2009). In 2003 and 2004, Khaleghizadeh and Alizadeh (2009) surveyed damage in Iran and found that the mean damage rates in Karaj and Khoy regions were 45.3% and 21%, respectively. Older data show that damage can be high (15.0–25.0%) in some localities (Khaleghizadeh, 2011) and low (<3%) in others (Linz and Hanzel, 1997).

Although research continues on this especially difficult problem, some advances have been made to help growers cope with birds. Damage management techniques that are effective, are economically viable, and do not compromise the environment are available. Most of these methods and concepts have been used or can be used in nearly all agricultural ecosystems facing problems with flocking granivorous birds. We will also outline current and planned future research direction by scientists from the U.S. Department of Agriculture, Wildlife Services, National Wildlife Research Center (USDA-WS-NWRC) and their university and private industry collaborators.

Cultural Practices

An obvious bird management strategy is to plant crops (e.g., beans) that are not susceptible to bird damage near traditional roosts (Linz and Hanzel, 1997). Some growers, however, recognize the value of sunflower in their crop rotation and have opted to use time-tested and new cultural practices.

One set of time-tested strategies includes synchronizing planting of sunflower to eliminate availability of early-maturing and late-maturing fields, planting large fields to spread the damage over greater areas, delaying the plowing of harvested grain fields to provide an alternate food source, and controlling weeds and insects that may attract birds to feed in sunflower fields prior to achene development (Linz and Hanzel, 1997; Linz et al., 2011).

A second strategy is to advance the harvest date to avoid late-season bird damage. Sunflower growers can use older products, such as paraquat and sodium chloride, or select newer products, such as glyphosate and saflufenacil, to desiccate the crop after it reaches physiological maturity and advance harvest by 7 to 10 days (Linz et al., 2011). Glyphosate and saflufenacil can be mixed in a spray tank and applied with an airplane or a high-clearance ground sprayer to kill both grasses and desiccate sunflower after the achenes are dried to 30% moisture or less. Desiccation and subsequent early harvest reduces the time for birds to cause additional damage without affecting yield or oil content (Howatt et al., 2008).

A third strategy that is garnering attention from growers is the recent development of short-stature sunflower varieties for short growing seasons in northern sunflower growing areas in North America (Mullally, 2012; Trostle et al., 2013). Specifically, sunflower breeders are developing short-stature varieties that are comparable to standard-height sunflower for such agroeconomic characteristics as days to maturity, yield, oil content, and disease tolerance. Short-stature sunflower provides less cover for birds than standard height hybrids and, although an untested concept, might allow for more effective use of scare devices (e.g., propane cannons, pyrotechnics) because the sound will carry farther. In addition, short-stature sunflowers provide less cover from raptors that forage over sunflower fields, and bird repellents could more easily be applied with a high-clearance ground rig.

Hazing

Sunflower producers sometimes resort to shooting firearms to disperse foraging flocks from sunflower. Although few birds are killed with firearms, shooting seems to help disperse flocks from sunflower and might enhance the effectiveness of other scare devices with a similar sound, such as propane cannons and pyrotechnics (Linz and Homan, 2011). A few growers in the United States hire fixed-wing aircraft and helicopters to chase birds from fields (Linz et al., 2011). Blackbirds appear easier to chase from fields in the PPR after their annual feather replacement is nearly complete in mid-September. Prior to that time, the birds tend to find refuge in dense cattail wetlands and in sunflower and corn fields. The cost of using an aircraft probably outweighs the amount of sunflower seed saved, but a quantitative assessment is not available.

In the United States, propane exploders are the most popular of the mechanical scare devices used by growers and wildlife professionals (Bomford and O'Brien, 1990; Conover, 2002). Cummings et al. (1986) reported that the use of propane cannons can be effective where large numbers of blackbirds are congregated in relatively small areas and the birds have not established feeding sites. We recommend that cannons be fitted with timers and motion sensors and be elevated on metal drums or rotating platforms to enhance the effects of the propane explosion and reduce the rate of habituation. Cannons should be moved as often as practical, but at least weekly. In 2012, U.S. Department of Agriculture (USDA), North Dakota Wildlife Services (WS) field specialists distributed 416 propane cannons and 26,700 rounds of pyrotechnics to 135 sunflower growers that were reporting blackbird damage (Phil Mastrangelo, personal communication, January 14, 2013). Field specialists reinforced the effectiveness of these devices with the use of shotguns. Although experimental efficacy data are not available, sunflower producers have responded favorably to this program since it was instituted in 2007.

Habitat Management

Habitat can be manipulated to increase or decrease the suitability of the landscape for breeding and foraging birds. The USDA's Wildlife Services program has carried out an example of decreasing habitat suitability by conducting a cattail roost management program in North Dakota and South Dakota for 20 years (Linz and Homan, 2011). Wildlife Services used a helicopter to spray an aqueous solution containing 2.2 kg ha^{-1}

of glyphosate to reduce the suitability of dense cattails used for roosting blackbirds (Linz and Homan, 2011). They found that spraying approximately 70% of the cattails in strips dispersed the roosting birds and, furthermore, that the strips remained clear of cattails for at least 4 years (Linz and Homan, 2011). The federal program has ended, but sunflower producers can spray cattails at their own expense. Producers have also burned and cut cattails during late summer and fall, especially during dry years when shallow wetlands can sometimes be accessed with farm equipment.

An example of increasing habitat suitability has been demonstrated twice over the last 30 years when scientists experimented with Wildlife Conservation Sunflower Plots (WCSP) (also known as decoy, lure, and trap crops) in strategic locations to attract blackbirds away from commercial sunflower fields (Cummings et al., 1987; Hagy et al., 2008). An alternative quality food source is critical for enhancing the success of nonlethal blackbird deterrent measures in commercial sunflower (Avery and Cummings, 2003). Ideally, the food source should be available until the commercial fields are harvested. However, a WCSP that is available during early seed development when most bird damage occurs would be valuable. Cummings et al. (1987) suggested that decoy crops would have to be planted on public lands to reduce rent costs. Similarly, Hagy et al. (2008) found that annually planting WCSP on private lands that would need to be taken out of commercial production was not cost effective. A few growers have adapted the basic principle of diverting blackbirds from high-value sunflower to lower-valued crops. For example, growers plant corn, a less valuable crop, near wetlands to buffer their sunflower crop from birds. The birds do eat some corn but the economic loss is tolerable for most growers (Klosterman et al, 2013).

To increase the economic benefits of WCSP, Kantar et al. (2014) are developing a perennial sunflower that could be planted near traditional roost sites to attract blackbirds and other wildlife. In 2013, the first improved perennial sunflower was planted in North Dakota (Figure 7.2). The seed will be increased and will become more generally available over the next few years. State resource agencies have expressed strong interest in planting perennial sunflower to benefit all wildlife. Availability of public lands (e.g., refuges, game management areas, Conservation Reserve Program land) for perennial sunflower plantings will further improve cost benefits for sunflower producers and game managers.

Chemical Repellents

Private and government scientists have aggressively pursued the discovery, development, and registration of an effective, environmentally safe bird repellent (Werner et al., 2010, 2011). Wildlife managers are supportive of these efforts because dispersing birds is a way to reduce the severity of localized damage. Several products featuring methyl anthranilate (MA) as the active ingredient have been registered for use in

Figure 7.2 High-clearance ground sprayer applying an experimental bird repellent in a ripening sunflower field in North Dakota.
Source: USDA, Wildlife Services.

ripening sunflower. These products are primary repellents and are effective when the peripheral chemical senses (nostrils, eyes, or mouth) are irritated (Clark, 1998). In field studies, however, MA products failed to reduce sunflower damage (Werner et al., 2005). Rates used in these studies may have been insufficient to elicit a negative response from the birds because the threshold level that induces repellency was not reached. Moreover, high variability in damage among fields within treatments and small sample sizes can contribute to lack of enough evidence to find a statistically significant difference between treatments.

Flock Buster® is another primary repellent that is registered in the United States for use on sunflower. The product is unique because it contains a mixture of items, including lemongrass oil, garlic oil, clove oil, peppermint oil, rosemary oil, thyme oil, and white pepper. Werner et al. (2010) found Flock Buster® to be an ineffective bird deterrent under laboratory conditions. Experimental field trials, however, have not been conducted to test its efficacy.

Another group of possibly effective compounds are secondary repellents that effect gastrointestinal illness and cause the birds to learn to associate illness with a particular food that has a certain appearance or taste (Avery and Cummings, 2003). A secondary repellent produces a stronger effect than do primary repellents, but the birds must sample the food, thus causing some damage before becoming ill from ingesting the food. Of these compounds, 9,10-anthraquinone (AQ) appears to be most promising (Linz et al., 2011). Anthraquinone is synthesized for industrial uses such as dyes and papermaking, but also can be found in plants. Cage trials and field trials with the Avipel® formulation have demonstrated that anthraquinone-based repellents can effectively protect seeded corn, sunflower, and rice seed from blackbirds (Werner et al., 2011). Anthraquinone might prove useful for repelling free-ranging birds from ripening crops, but more testing is needed (Avery and Cummings, 2003). For example, Carlson et al. (2013) showed that treating corn husks with AQ is sufficient to induce red-winged blackbirds to feed on untreated corn. There is some evidence that merely spraying the back of the heads of sunflowers (i.e., not the face) is sufficient to deter blackbirds from feeding on the achenes (Werner et al. 2014). In 2013, scientists sprayed AQ on ripening sunflower at an early bloom stage and placed blackbirds in cages to evaluate efficacy (Figure 7.3). Under the conditions of this experiment, the AQ application failed to deter feeding on treated sunflower (Niner et al. 2014).

Figure 7.3 Perennial sunflower plot planted in North Dakota. Perennial sunflower can provide an alternate food source for birds.
Source: USDA, Wildlife Services.

Population Management

Local and regional attempts to reduce blackbird populations associated with sunflower damage have failed. The idea of using decoy traps to capture and euthanize blackbirds was tested because nontarget birds could be released without harm. Cage traps stocked with decoy blackbirds have been used to remove blackbirds in rice-growing areas (Meanley, 1971). Nevertheless, defending large-scale agriculture by trapping has been proven to be ineffective. For example, Weatherhead et al. (1980) concluded that decoy traps removed less than 2% of the trappable number of blackbirds foraging in ripening corn fields. Linz et al. (2010) evaluated two large-sized, mobile decoy traps (11 × 2.5 × 2.5 m) for capturing blackbirds actively feeding on ripening sunflower fields during late summer and early fall. Despite an intensive effort, only 154 blackbirds were captured and most of those birds were captured after the crop had reached physiological maturity and the achenes had become less palatable and the risk for substantial damage had subsided. Linz and colleagues concluded this method to be economically inefficient for protecting sunflower because of lack of efficacy and labor and travel costs associated with maintaining decoy birds.

U.S. Fish and Wildlife Service, Denver Wildlife Research Center (now USDA-WS-NWRC) scientists discovered and developed the compound DRC-1339 (3-chloro-p-toluidine hydrochloride, also 3-chloro-4-methylbenzenamine hydrochloride) as an avicide to kill starlings and blackbirds. In late winter and early spring, the product is used to kill blackbirds that might damage sprouting rice in the southeastern United States (Glahn and Wilson, 1992). The efficacy of reducing migrating blackbird populations to manage crop damage has not been experimentally tested. Blackbirds that reproduce in the PPR disperse throughout the southern United States during winter. Thus, selecting and targeting specific roosts that harbored blackbirds that damaged sunflower in the PPR was not possible. Scientists also evaluated the use of DRC-1339 for killing spring migrating blackbirds in eastern South Dakota and found that costs exceeded the potential benefit of reducing the population (Blackwell et al., 2003; Linz et al., 2003). Finally, DRC-1339 and related compounds were tested for reducing flocks of fall migrating blackbirds feeding on ripening sunflower (Cummings et al., 1990). Linz and Bergman (1996) concluded that the majority of blackbirds preferred to feed on the ripening achenes rather than forage on dry grains placed on the ground. Linz et al. (2012) recognized that ground-based DRC-1339 plots would not work in sunflower, so they tried attracting blackbirds to elevated food trays placed on cages containing live decoy blackbirds. Linz and colleagues hypothesized that the live decoy blackbirds would attract conspecifics to the bait trays, while decreasing risks of nontarget poisonings. Field observations demonstrated that risks to nontarget species were minimal, but the decoy blackbirds failed to attract sufficient numbers of blackbirds to the trays to make this management strategy cost effective

(Linz et al., 2012). We conclude that managing the population of blackbirds with lethal methods is not practical or cost effective.

Conclusion

Since sunflower became an economically viable crop in North America in the 1970s, scientists have concentrated on chemical and physical frightening agents, aversive repellents, bird-resistant sunflower, decoy crops, habitat management, population management, and cultural modifications in an effort to reduce bird damage. Bird damage to agricultural crops is a vexing worldwide problem that is especially difficult for sunflower producers (Linz et al., 2011). We suggest producers and their crop consultants develop a comprehensive bird management plan. This plan might include modifying roost habitat (e.g., pruning trees, thinning dense vegetation in wetlands); using a plant desiccant (e.g., glyphosate, saflufenacil) to accelerate fall harvest; using propane cannons; planting decoy crops in strategic locations; synchronizing planting time of sunflower with neighbors; leaving stubble, especially sunflower, unplowed for as long as possible to provide alternative feeding sites for blackbirds; and planting short-stature sunflower to facilitate bird-hazing strategies.

In the next decade, we are hopeful that an effective bird repellent will be registered for use on ripening sunflower (and other grain crops) and that a perennial sunflower will be available for use as an alternative food source for blackbirds and other birds. Alternative sources of food, possibly in combination with repellents, should help scientists make significant advances in management of blackbird damage in sunflower (Avery, 2002). We caution, however, that there are no perfect solutions to bird damage conflicts.

Acknowledgments

M. Tobin and M. Lutman provided helpful comments on an earlier draft. We thank all the past and current scientists who contributed to evaluating blackbird management methodologies. Mention of commercial products does not imply endorsement by the U.S. Department of Agriculture or Genosys, LLC.

References

Avery, M. L. Avian Repellents. In *Encyclopedia of Agrochemicals*; Plimmer, J. R., Ed.; John Wiley & Sons: New York, 2002; Vol. 1, pp 122–128.

Avery, M. L.; Cummings, J. L. Chemical Repellents for Reducing Crop Damage by Blackbirds. In *Management of North American Blackbirds;* Linz, G. M., Ed.; National Wildlife Research Center: Fort Collins, CO, 2003; pp 41–48.

Blackwell, B. F.; Huszar, E.; Linz, G. M.; Dolbeer, R. A. Lethal Control of Red-winged Blackbirds to Manage Damage to Sunflower: An Economic Evaluation. *J. Wildl. Manage.* **2003,** *67,* 818–828.

Bomford, M.; O'Brien, P. H. Sonic Deterrents in Animal Damage Control: A Review of Device Tests and Effectiveness. *Wildl. Soc. Bull.* **1990,** *18,* 411–422.

Carlson, J. C.; Tupper, S. K.; Werner, S. J.; Pettit, S. E.; Santer, M. M.; Linz, G. M. Laboratory Efficacy of an Anthraquinone-based Repellent for Reducing Bird Damage to Ripening Corn. *Applied Animal Behaviour Science* **2013,** *145,* 26– 31.

Clark, L. Review of Bird Repellents. *Proceedings of the Vertebrate Pest Conference* **1998,** *18,* 330–337.

Conover, M. R. *Resolving Human-Wildlife Conflicts: The Science of Wildlife Damage Management*; CRC Press: Boca Raton, FL, 2002.

Cummings, J. L; Knittle, C. E.; Guarino, J. L. Evaluating a Pop-up Scarecrow Coupled with Propane Exploder for Reducing Blackbird Damage to Ripening Sunflower. *Proceedings of the Vertebrate Pest Conference* **1986,** *12,* 286–291.

Cummings, J. L.; Guarino, J. L.; Knittle, C. E.; Royal, W. C., Jr. Decoy Plantings for Reducing Blackbird Damage to Nearby Commercial Sunflower Fields. *Crop Protection* **1987,** *6,* 56–60.

Cummings, J. L.; Guarino, J. L; Knittle C. E. Chronology of Blackbird Damage to Sunflowers. *Wildl. Soc. Bull.* **1989,** *17,* 50–52.

Cummings, J. L.; Schafer, E. W., Jr.; Cunningham, D. J. An Evaluation of DRC-2698 Treated Baits for Reducing Blackbird Populations Associated with Sunflower Damage. *Proceedings of the Fourteenth Vertebrate Pest Conference* **1990,** *14,* 357–360.

DeGrazio, J. W. Pest Birds—An International Perspective. In *Quelea Quelea: Africa's Bird Pest*; Bruggers, R. L., Elliot, C. C. H., Eds.; Oxford University Press: New York; 1989; pp 1–8.

Glahn, J. F.; Wilson, E. A. Effectiveness of DRC-1339 Baiting for Reducing Blackbird Damage to Sprouting Rice. *Proceedings Eastern Wildlife Damage Control Conference* **1992,** *5,* 117–123.

Hagy, H. M.; Linz, G. M.; Bleier, W. J. Optimizing the Use of Decoy Plots for Blackbird Control in Commercial Sunflower. *Crop Protection* **2008,** *27,* 1442–1447.

Howatt, K. A.; Jenks, B. M.; Stahlman, P. W.; Moechnig, M. Potential of Saflufenacil for Preharvest Desiccation of Sunflower. *Proc. North Central Weed Science Soc.* **2008,** *63,* 149.

Hulke, B. S.; Kleingartner, L. W. Sunflower. In *Yield Gains in Major U.S. Field Crops;* Smith, S., Diers, B., Specht, J., Carver, B., Eds.; Soil Science Society of America Spec. Publ. 33. American Society of Agronomy, Crop Science Society of America, and Soil Science Society of America: Madison, WI, 2014.

Kantar, M. B.; Betts, K.; Michno, J.-M. S.; Luby, J. J.; Morrell, P. L.; Hulke, B. S.; Stupar, R. M.; Wyse, D. L. Evaluating an Interspecific *Helianthus annuus* × *Helianthus tuberosus* Population for Use in a Perennial Sunflower Breeding Program. *Field Crops Res.* **2014,** *155,* 254–264.

Khaleghizadeh, A. Effect of Morphological Traits of Plant, Head and Seed of Sunflower Hybrids on House Sparrow Damage Rate. *Crop Protection* **2011,** *30,* 360–367.

Khaleghizadeh, A.; Alizadeh, E. Evaluation of Resistance of Sunflower Varieties to Bird Damage. *J. Appl. Entomol. Phytopathology* **2009,** *76,* 115–134 (in Persian with English summary).

Klosterman, M. E.; Linz, G. M.; Slowik, A. A.; Homan, J. J. Comparisons between Blackbird Damage to Corn and Sunflower in North Dakota. *Crop Protection* **2013,** *53,* 1–5.

Linz, G. M.; Bergman, D. L. DRC-1339 Avicide Fails to Protect Ripening Sunflowers. *Crop Protection* **1996,** *15,* 307–310.

Linz, G. M.; Hanzel, J. J. Birds and Sunflower. In *Sunflower Technology and Production*; Schneiter, A. A., Ed.; Agronomy Monograph No. 35; American Society of Agronomy: Madison, WI, 1997; pp 381–394.

Linz, G. M.; Homan, H. J. Current Use of Glyphosate for Managing Invasive Cattail (*Typha* spp.) to Protect Crops Near Blackbird (Icteridae) Roosts. *Crop Protection* **2011,** *30,* 98–104.

Linz, G. M.; Mendoza, L. A.; Bergman, D. L.; Bleier, W. J. Preferences of Three Blackbird Species for Sunflower Meats, Cracked Corn, and Brown Rice. *Crop Protection* **1995**, *5,* 375–378.

Linz, G. M.; Knutsen, G. A.; Homan, H. J.; Bleier, W. J. Baiting Blackbirds (Icteridae) in Stubble Grain Fields during Spring Migration in South Dakota. *Crop Protection* **2003,** *22,* 261–264.

Linz, G. M.; Slowik, A. A.; Homan, H. J.; Byrd, R. Evaluation of Large, Mobile, Decoy Traps for Managing Blackbird Damage to Ripening Sunflower, 2010a. http://www.sunflowernsa.com/research/research-workshop/documents/Linz_DecoyTraps_10.pdf (accessed Feb-ruary 1, 2014).

Linz, G. M.; Homan, H. J.; Werner, S. W.; Hagy, H. M.; Bleier, W. J. Assessment of Bird Management Strategies to Protect Sunflower. *BioScience* **2011,** *61,* 960–970.

Linz, G. M.; Winter, J. B.; Bleier, W. J. Evaluation of Elevated Bait Trays for Attracting Blackbirds (Icteridae) in North Dakota. *Crop Protection* **2012,** *41,* 30–34.

Meanley, B. *Blackbirds and the Southern Rice Crop.* Resource Publication No. 100; U.S. Fish and Wildlife Service, 1971.

Mullally, S. Options Growing with Short-stature Hybrids. *Sunflower Magazine* **2012,** *38,* 22–24.

National Sunflower Association. Sunflower Statistics: World Supply & Disappearance. http://www.sunflowernsa.com/stats/world-supply (accessed February 1, 2014).

Niner, M.; Linz, G., Homan, J.; Clark M. Effectiveness of AV2022 (a.i., 9, 10 Anthraquinone) as an Avian Repellent for Sunflower. In *36th National Sunflower Association Research Forum*, Fargo, ND, 2014.

Peer, B. D.; Homan, H. J.; Linz, G. M.; Bleier, W. J. Impact of Blackbird Damage to Sunflower: Bioenergetic and Economic Models. *Ecological Applications* **2003,** *13,* 248–256.

Rodriguez, E. N.; Bruggers, R. L.; Bullard, R. W.; Cook, R. An Integrated Strategy to Decrease Eared Dove Damage in Sunflower Crops. In *Repellents in Wildlife Management: Proceedings of a Symposium;* Mason, J. R., Ed.; Denver Wildlife Research Center Special Symposium: Denver, CO, 1995; 409–421.

Trostle, C.; Pietsch, D.; Schlegel, A.; Evans, P. Height, Yield, and Oil Content of Short-stature Sunflower (*Helianthus annuus*) vs. Conventional Height Sunflower in the Southern Great High Plains. In 35th National Sunflower Association Research Forum, Fargo, ND, 2013. http://www.sunflowernsa.com/research/research-forum-presentations/2013 (accessed February 1, 2014).

van Niekerk, J. H. Loss of Sunflower Seeds to Columbids in South Africa: Economic Implications and Control Measures. *Ostrich* **2009,** *80,* 47–52.

Weatherhead, P. J.; Greenwood, H.; Tinker, S. H.; Bider, J. R. Decoy Traps and the Control of Blackbird Populations. *Phytoprotection* **1980,** *61,* 65–71.

Werner, S. J.; Homan, H. J.; Avery, M. L.; Linz, G. M.; Tillman, E. A.; Slowik, A. A., Byrd, R. W.; Primus, T. M.; Goodall, M. J. Evaluation of Bird Shield™ as a Blackbird Repellent in Ripening Rice and Sunflower Fields. *Wildl. Soc. Bull.* **2005,** *33,* 251–257.

Werner, S. J.; Linz, G. M.; Tupper, S. K; Carlson, J. C. Laboratory Efficacy of Chemical Repellents for Reducing Blackbird Damage in Rice and Sunflower Crops. *J. Wildl. Manage.* **2010,** *74,* 1400–1404.

Werner, S. J.; Linz, G. M.; Carlson, J. C.; Pettit, S. E.; Tupper, S. K.; Santer, M. M. Anthraquinone-based Bird Repellent for Sunflower Crops. *Appl. Anim. Behav. Sci.* **2011,** *129,* 162–169.

Werner, S. J.; Deliberto, S. K.; Pettit, S. E.; Ellis, J. W.; Carlson, J. C.; Goldade, D. A.; Hofmann, N.M.; Homan, H. J.; Linz, G. M. Experimental Application Strategies for an Avian Repellent on Sunflower Crops. *Crop Protection* **2014,** *59,* 63–70.

8

Sunflower Seed Preparation and Oil Extraction

Etienne Le Clef ■ *Desmet Ballestra, Brussels, Belgium*

Timothy Kemper ■ *Desmet Ballestra, Atlanta, GA, USA*

Introduction

Sunflower is crushed for oil recovery via one of two process methods, hard pressing or prepress solvent extraction. Before 1940, hard pressing was the primary crushing process method used (O'Brien, 2000). Hard pressing relies upon exerting high pressure on the prepared seed to separate oil from the solids fraction, historically recovering about 25% oil from the seed (Williams, 2005).

Prepress solvent extraction relies on exerting medium pressure on the prepared seed to separate the majority of the oil, followed by solvent extraction to separate additional oil from the solids fraction, recovering over 40% oil from the seed in modern crushing plants. Because sunflower oil has a substantially higher economic value than sunflower meal, process economics improve significantly as the oil recovery yield increases. Therefore, prepress solvent extraction, with the highest oil recovery yield, is the process method utilized today for the majority of the sunflower seed that is crushed globally.

In a typical sunflower crushing plant using the prepress solvent extraction method, the initial steps of seed precleaning and seed drying are located in the seed receiving section of the facility, prior to seed storage.

After seed storage, the seed moves to the processing stages of seed cleaning, front-end dehulling, meats flaking, meats cooking, prepressing, prepress oil clarification, and prepress cake conditioning, which take place in a common process area referred to as the *seed preparation area.* The seed preparation area is constructed in a special manner to safeguard against fire and explosion hazards resulting from the potential presence of dust. The intermediate process steps of solvent extraction, meal desolventizing, miscella distillation, and solvent recovery are located in a common process area referred to as the *solvent extraction area.* The solvent extraction area is constructed in a special manner to safeguard against fire and explosion hazards resulting from the potential presence of solvent vapor. The final process steps of tail-end dehulling and meal sizing/pelletizing, for which there is again a potential presence of dust, are typically located in the seed preparation area.

The seed preparation area and the solvent extraction area are physically separated by a safe distance consisting of a controlled area and restricted area (NFPA, 2013). A typical plan view of the seed preparation process area, solvent extraction process area, and surrounding restricted and controlled areas is illustrated in Figure 8.1.

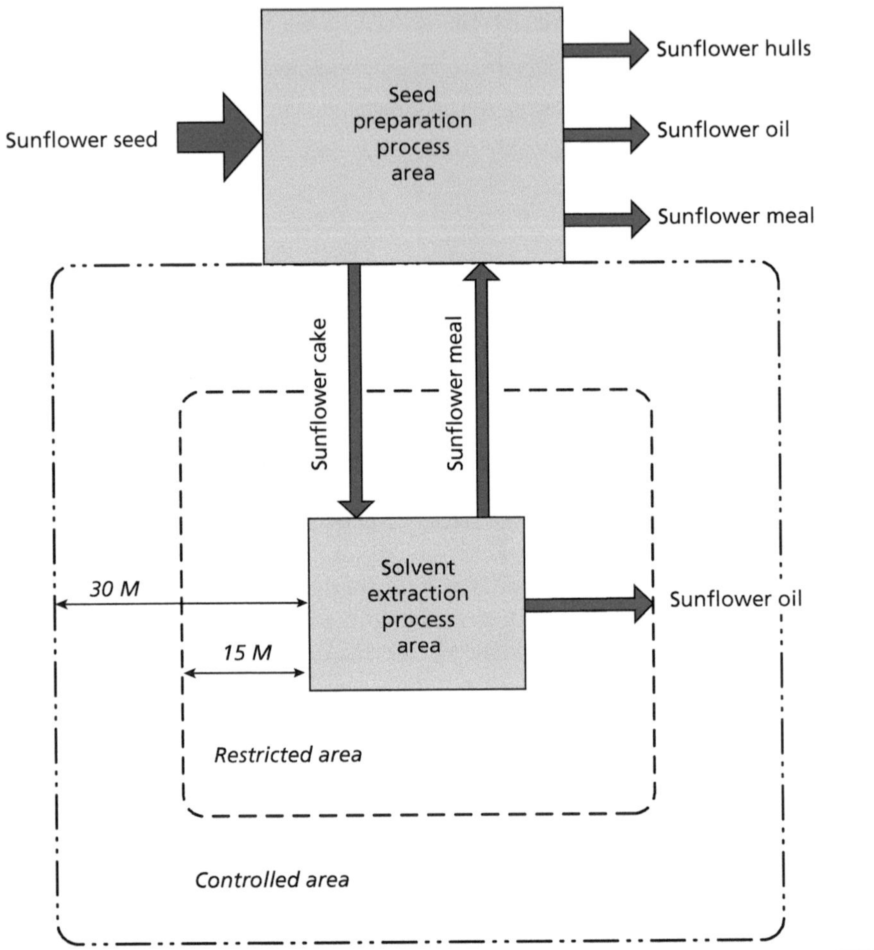

Figure 8.1 Typical sunflower crush plant process layout.

Seed Precleaning and Drying

The marketing or acquisition of sunflower seeds is on "zero foreign matter" basis. Nonetheless, sunflower seed contains foreign matter that needs to be removed in the process. Therefore, the mass of material received is greater than the traded quantity by the added amount of foreign matter. To obtain the equivalent mass on an "as received" basis is a simple arithmetic operation.

Efficient precleaning is highly recommended prior to drying to eliminate blockage or fires in the seed dryer. For precleaning, sunflower processors generally use machines called *scalpers* (USA) or *émotteurs* (France), which are high capacity and made of a perforated drum in which special aspiration channels remove undesirable byproducts from the harvest, such as stems, stalks, chaff, stones, and so forth. The selected equipment should have nonaggressive characteristics on the seed to avoid splitting the seeds open, which consequently causes the seeds to suffer deterioration during storage. A typical seed precleaner is illustrated in Figures 8.2 and 8.3.

The sunflower seed must have a maximum moisture content of 9% for maintaining quality during storage. During storage, there should also be facilities to provide ventilation and control of temperature to maintain quality. Commercially, seed is traded on an 11% moisture content (weight basis); therefore, such seed requires drying prior to storage down to the 9% level.

The acidity of the oil inside the sunflower seed increases in an exponential manner when it exceeds the limits of stability, caused by excess moisture or damage. Such lack of stability is detected by the material temperature rising approximately 10 °C above ambient temperature during storage.

When selecting the heating system for the sunflower seed dryers, one can choose direct combustion gases or indirect heated air. If direct combustion gases are used, one should choose fuels that contain the minimum level of sulfur, because its presence

Figure 8.2 Seed precleaner.

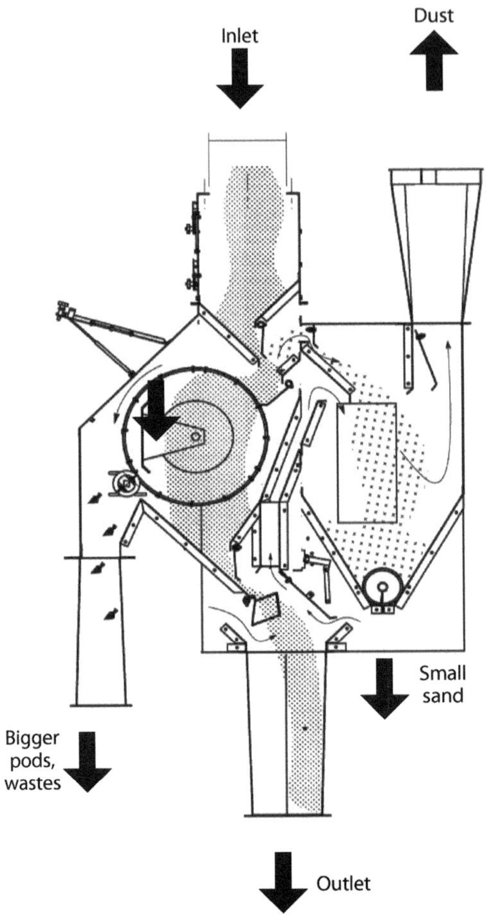

Figure 8.3 Seed precleaner cross section.

in the gases plus the moisture of the seed encourage the formation of undesirable inorganic acids that form in the seed during the downstream meats conditioning stage.

Seed Cleaning and Weighing

After storage, the precleaned and dried seed enters the seed preparation area. The first process steps are cleaning the seed and measuring/regulating the processing rate.

The first cleaning step is composed of a rotating drum sieve, which does not restrict the seed flow, but separates all types of foreign material larger than the seed.

Such large foreign material is commonly referred to as *overs*. Figures 8.4 and 8.5 illustrate a typical rotary drum sieve used for the first cleaning step.

After the first cleaning step, a rotating magnet removes all types of ferrous material. The next cleaning step is performed in a seed screener to remove small foreign material and dust. A specific sunflower rotary distributor is driven by a frequency variator to evenly feed the seed onto the screens. The internal aspiration channel removes most of the dust before the seed crosses the perforated screens. This maintains the whole machine in a slight negative pressure, avoiding dust in the process area and avoiding dust fouling of the screens. The eccentric type movement of the screens is simple and reliable. The screeners are manufactured in one to four superposed stages. This assures large capacities per machine in a very compact space. Large access doors allow an operator to change the screens easily. Figures 8.6 and 8.7 (p. 193) illustrate a typical seed screener used for this cleaning step.

The final cleaning step is to remove any heavy foreign matter with the same approximate size as the seed in a device commonly referred to as a *destoner*. The foreign matter removed is typically comprised of small stones, iron, and stainless steel. An incorporated strong magnet will finalize the cleaning and thus avoid downstream

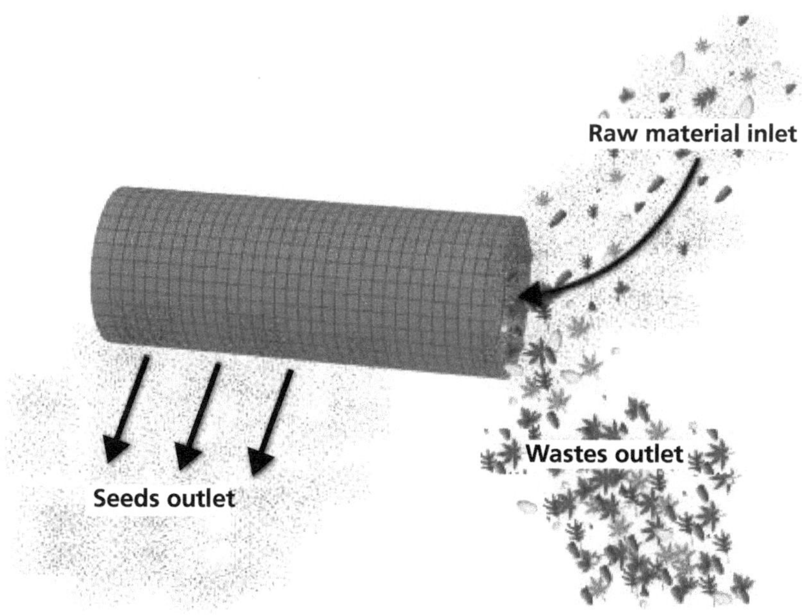

Figure 8.4 Seed cleaner schematic.

Figure 8.5 Rotary sieve type seed cleaner.

Figure 8.6 Screen type seed cleaner.

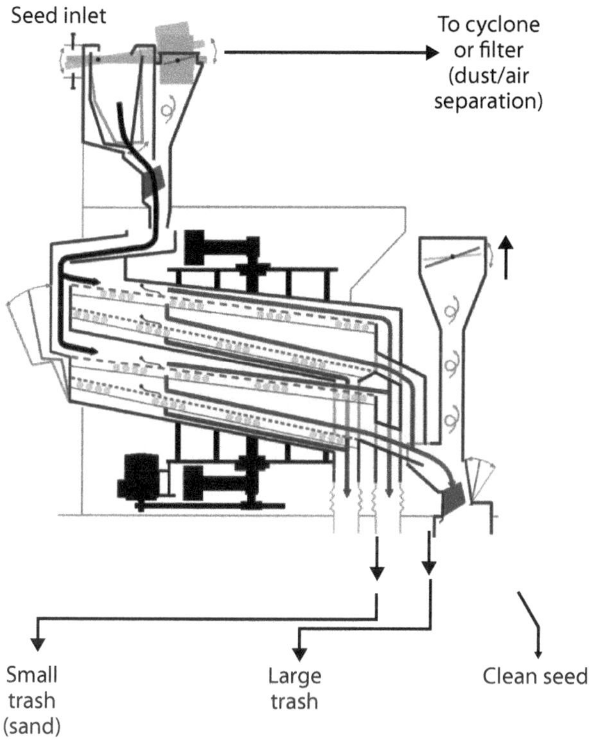

Figure 8.7 Screen type seed cleaner cross section.

damage to machinery. Figure 8.8 illustrates a typical destoner used for this final seed cleaning step.

A semicontinuous weigh scale usually follows the cleaning steps in order to quantify the incoming seed flow to the process. The upstream and downstream hoppers are carefully calculated for the densities of the seed and the processing rate. Such a scale regulates the seed flow and will avoid downstream flow variations that would harm efficiency and throughput of downstream equipment.

After the seed is cleaned and weighed, it will have a composition similar to the following analysis:

Clean Seed (100%)	*Meats Fraction (74%)*	*Hull Fraction (26%)*
Oil: 43.8%	Oil: 58.3%	Oil: 2.5%
Water: 6.8%	Water: 6.0%	Water: 9.1%
Dry matter: 49.4%	Dry matter: 35.7%	Dry matter: 88.4%

In the hulls, 9.1% water represents 35% of the total moisture in the seed, whereas in the meats, 6.0% water represents 65% of the total moisture in the seed. As seen

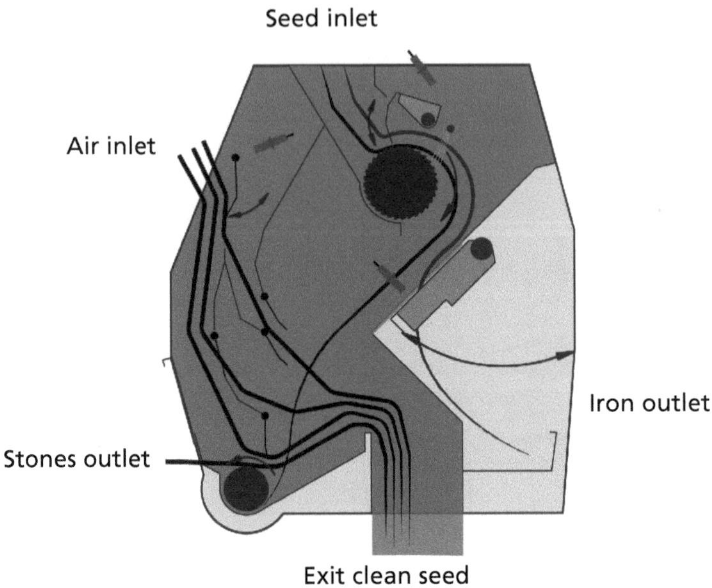

Figure 8.8 Destoner.

from this analysis, the moisture is distributed unequally between the meats and hulls fraction, as the hulls are only 26% of the seed but contain 35% of the water. The fibrous hulls are hygroscopic, which is a natural barrier for moisture transfer to the central oily nucleus (meats).

The distribution of moisture in seed prior to drying is typically 40% in hulls and 60% in meats. The distribution of moisture in the seed is a clear signal of whether the seed is from the field, or whether it has been dried.

Front-End Dehulling

Sunflower seed is characterized by a high oil content of 40–45% with 14–18% proteins and 25–30% hulls that predominantly consist of nondigestible crude fiber. A typical analysis is illustrated in Figure 8.9.

Sunflower seed processing focuses on sunflower oil as the most valuable product. It is used as high-value oil in various food applications. The deoiled meal is considered a byproduct and is mostly used as animal feed. In 2011, annual world production of deoiled sunflower meal was at 16 million tons (FAO, 2013). The deoiled meal composition depends on the efficiency of the crushing process and the seed composition.

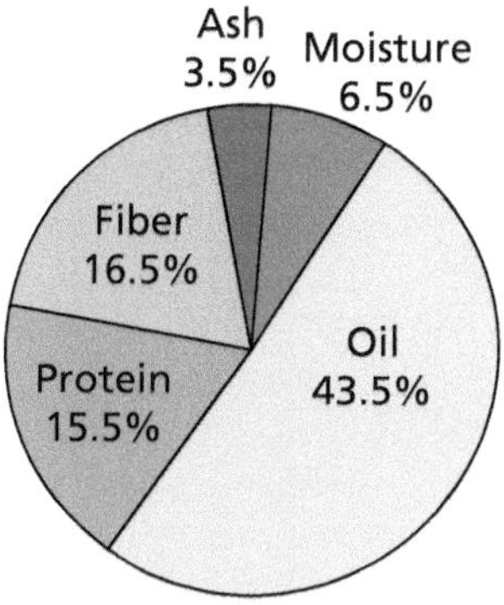

Figure 8.9 Seed composition.

It is characterized by a relative low protein content (25–28%) and high crude fiber content (24–27%), as illustrated in Table 8.A.

The nutritional value of the protein fraction of deoiled sunflower meal is quite good, with only the lysine being low. The deoiled meal also has an absence of known antinutritional factors. However, deoiled sunflower meal has limited applications as a feedstuff because of its high fiber content. Standard deoiled sunflower meal is therefore only used in significant quantities in ruminant feed as an alternative protein source to soybean meal.

The energy value and application range of sunflower meal can be increased by lowering the fiber content. This can be achieved by removing a portion of the hulls from the seed (called *front-end dehulling*) or from the deoiled meal (called *tail-end dehulling*).

"Partial" front-end dehulled (one-stage) or "full" front-end dehulled (two-stage) sunflower meal can also be used in poultry and swine feed, but fiber content is still too high and protein content is too low for use in fish feed. With the growing aquaculture market and the limited availability of fish meal, fish feed producers show an increased interest in sunflower meal as alternative protein. Absence of antinutritional factors is a key reason. Fish feed producers require higher protein content (min. 45%) and lower fiber content. Sunflower meal with such high protein can be obtained by full dehulling of selected seeds with high protein content (18%) or by applying an additional tail-end dehulling on already front-end dehulled seed.

A process flow diagram of a typical front-end dehulling of sunflower seeds is illustrated in Figure 8.10.

Table 8.A Composition of Deoiled Sunflower Meal

	Deoiled Sunflower Meal[a]		
	Undehulled	**Partially Dehulled**	**Fully Dehulled**
Moisture (%)	11–12	11–12	11–12
Oil (%)	1.5	0.85	0.5
Protein (%)	25–28	34–36	38–40
Crude Fiber (%)	24–27	17–21	14–17
Ash (%)	6	6	6

[a]From sunflower seeds with 16% protein and 43% oil.

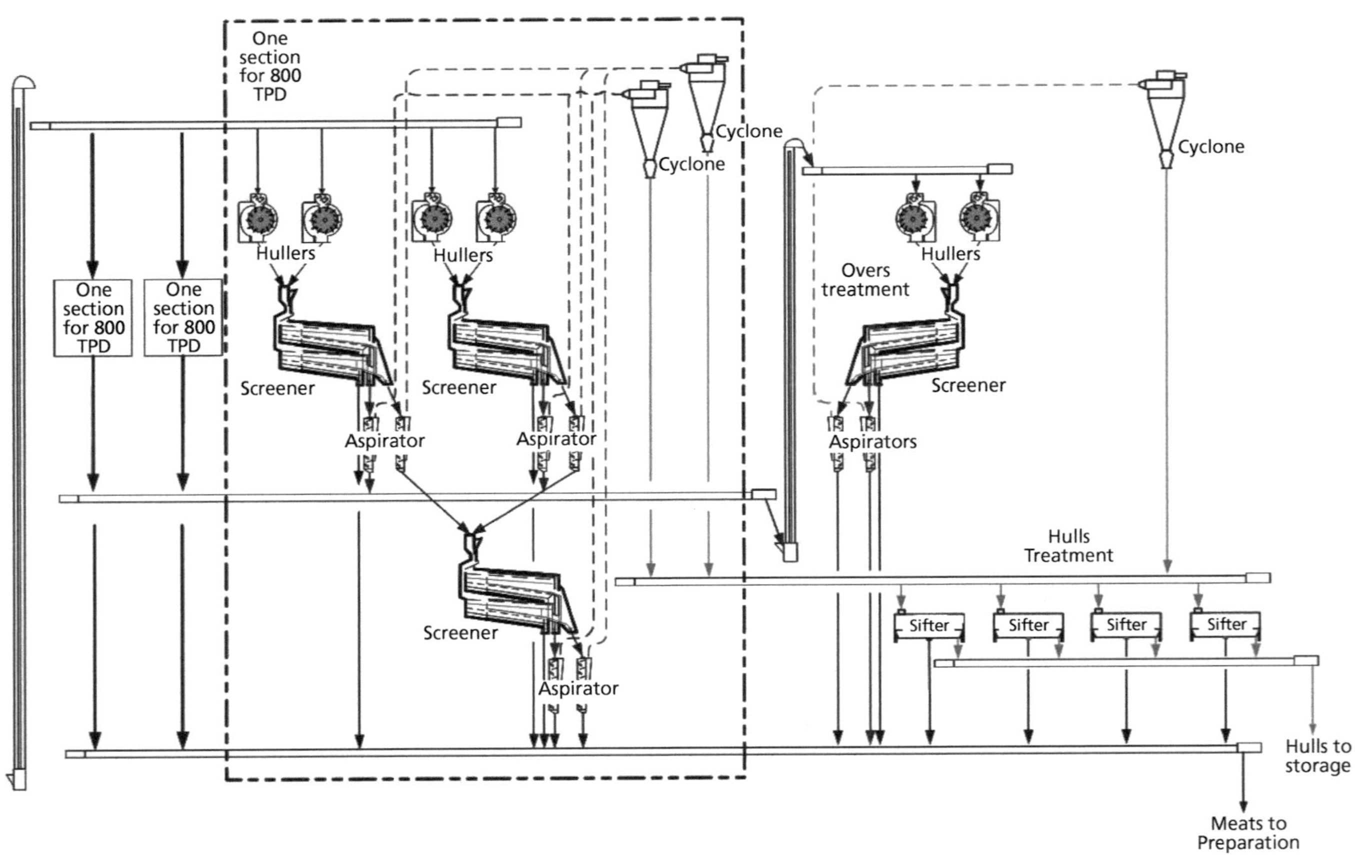

Figure 8.10 Front end dehulling schematic.

The sunflower seeds are first processed through an impactor or decorticator, as illustrated in Figure 8.11. This equipment consists of a rotating blade that propels the seeds against a static corrugated impact plate. The purpose is to detach the hulls from the meats with minimum formation of fines and minimum oil losses in the hull fraction. This can be controlled by optimizing process parameters and also by adjusting the position (inclination) of the impact plate toward the rotor.

The impacted seeds are then transported to a hulls separator. In this equipment, hulls are separated from the meats based on a difference in size (screening). This separation typically results in three fractions: (1) a "fine" dehulled meats fraction that is sent directly to flaking, (2) a "mid" fraction that is either further processed in a hulls separator (in the case of two-stage dehulling) or sent to flaking (in the case of one-stage dehulling), and (3) a "coarse" fraction that is recycled back to the untreated seeds or treated in a separate dehulling section. Figures 8.12 and 8.13 illustrate a typical hulls separator.

Prior to further processing, the mid and coarse fractions are both first sent to an air aspirator for the removal of the light hulls fraction.

Botanical hull content of sunflower seeds varies between 25% and 30% and depends on the type of cultivar, seed size, and also climate conditions during growing.

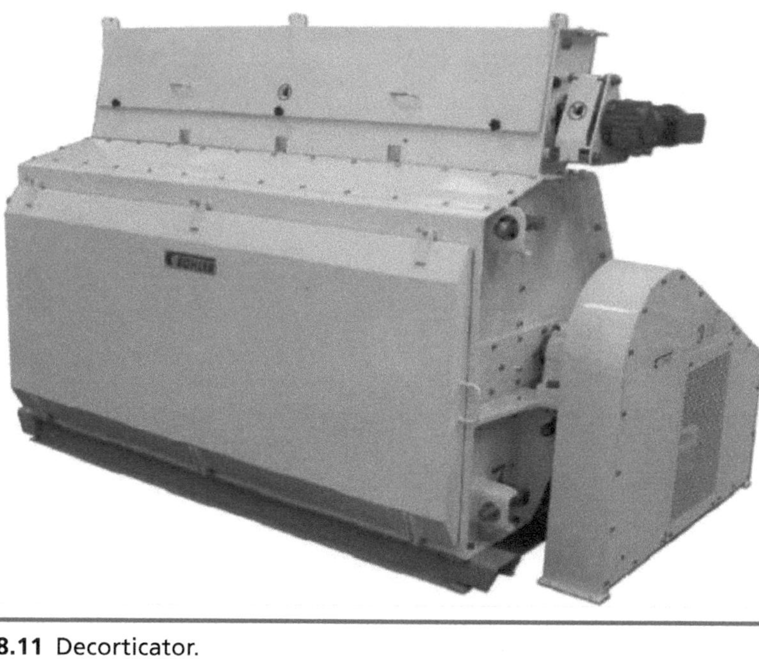

Figure 8.11 Decorticator.

Figure 8.12 Hulls separator.

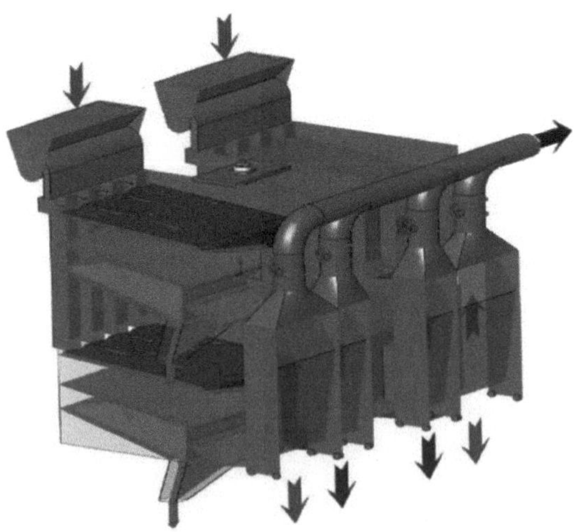

Figure 8.13 Hull separator cross section.

Table 8.B Composition of Sunflower Seed, Meats, and Hulls

	Sunflower Seed		Sunflower Meats	Sunflower Hulls
	Average	Range		
Seed (%)	—	70–75	25–30	—
Moisture (%)	6.7	5.5–8.0	6–10	7–11
Oil (%)	43.5	40–45	55–65	1.5–3.0
Protein (%)	15.5	14–18	16–20	4–6
Crude Fiber (%)	16.5	13–20	2–2.5	50–60
Ash (%)	3.2	3–4	—	5–7

Hulls are very rich in crude fiber (50–60%) and contain only small amounts of proteins (4–6%) and oil (1–2%) as illustrated in Table 8.B.

The challenge of the front-end dehulling system is to minimize additional oil and protein losses (above the botanical losses) in the hulls fraction. In practice, this is done by treating the hulls fraction in a hulls beater, as illustrated in Figure 8.14. The hulls beater mechanically separates protein-rich meats particles sticking to the separated hull stream, so these meats particles can be recovered.

Protein content of the final deoiled sunflower meal depends on the degree of dehulling as illustrated in Figure 8.15. One-stage (partial) dehulling leaves 13–15% hulls in the meats and results in a deoiled meal with 35–38% protein. Higher protein content (38–41%) can be achieved by a two-stage (full) dehulling, which reduces hull content to 10–12%.

Deeper dehulling (hulls < 10%) is not industrially applied for both technical and economic reasons. First of all, meats with very low hull content are difficult to prepress (not enough friction) and the resulting prepress cake has poor percolation

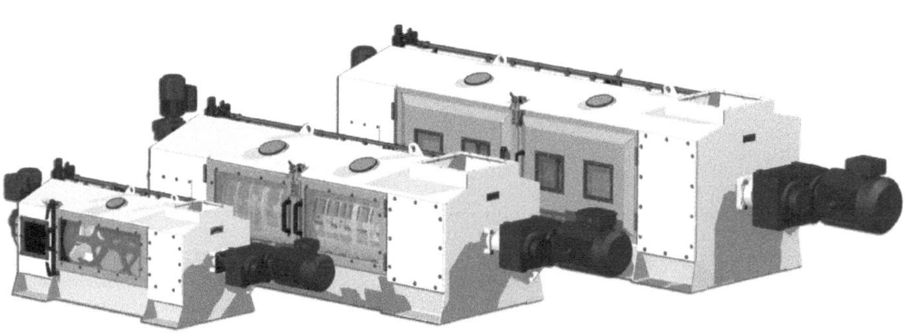

Figure 8.14 Hulls beater.

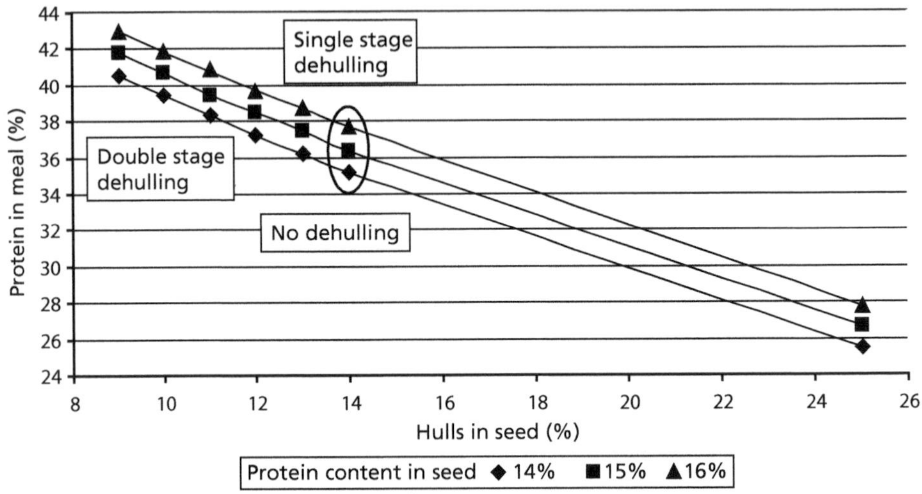

Figure 8.15 Dehulling versus meal protein.

during subsequent solvent extraction. In addition to this technical issue, there is also the economic issue that deeper dehulling gives high oil losses in the hulls that cannot be adequately compensated by the improved value of the higher protein meal.

In addition to the dehulling type, the protein content of the sunflower seed also is influential in the protein content of the dehulled, deoiled meal. Figure 8.16 illustrates this impact.

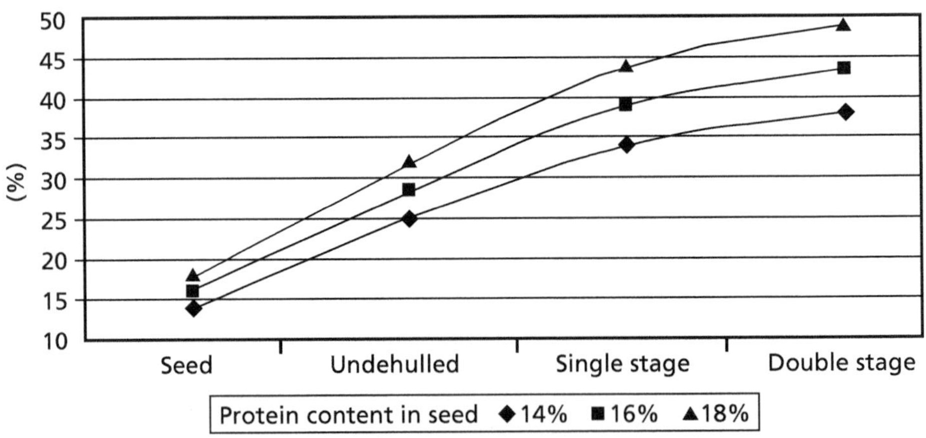

Figure 8.16 Seed quality impact on meal protein.

Full dehulling and deoiling of sunflower seeds with 14% protein content results in a meal with 36–38% protein content, which is not high enough to be marketed as "high protein" meal. Similar processing of sunflower seeds (full dehulling and deoiling) with 18% protein content can give a more valuable meal with protein content similar to that of "low protein" soybean meal (44–45% protein).

Front end dehulling also has a positive effect on the crude sunflower oil quality, specifically on the wax content. Waxes are mainly present in the hulls fraction. Hence, dehulling can significantly reduce wax content, from 1000–1250 ppm in crude oil from undehulled seeds to 600–750 ppm in crude oil from dehulled seeds. This lower wax content will have a positive impact on the refining process because it will enable lower oil losses during winterization.

Meat Flaking, Cooking, and Prepressing

Flaking

In some sunflower processing plants the meats from front-end dehulling go directly to the cooking step, and in some plants, the meats from front-end dehulling are flaked first. Flaking thins the particle profile and distorts the cellular structure to assist the prepress and downstream solvent extraction to be more efficient. The impact of flaking can be measured by a lower milling defect. The milling defect is the percentage remaining oil difference between the "as is" sample laboratory extraction, and the finely ground sample laboratory extraction.

The equipment required to flake the sunflower meats is a flaking mill, similar to the equipment used for flaking soybeans or rapeseed. Figure 8.17 illustrates a typical flaking mill.

For large sunflower plants, flaking mills with 800 mm diameter by 2.1 m long rolls are used for a capacity of up to 450 tons per day (TPD) of seed. A difference of 3–8% in speed between the two rolls assures the shearing effect without consuming excess power. In addition to normal motor starters, frequency variators are now often used to allow adjustment of the roll speeds. The frequency variators allow increasing the speed and throughput, for example, when one flaking mill is stopped for maintenance. Typical installed power on one large sunflower flaking mill is 110 kW.

Cooking

The meats, whether flaked or not, are then subjected to cooking. Cooking comprises a heating and drying step that ruptures the cellular structure of the meats to facilitate efficient downstream prepress oil removal. The meats leaving the cooking step are approximately 4–5% moisture and at 90–100 °C.

Figure 8.17 Flaking mill.

For many years, the typical device for this application was a stacked tray cooker with multiple steam-heated trays. Historically, one cooker was installed above each prepress, typically about 9–11 trays tall with a 4.5 m diameter, for feeding one prepress capable of processing 700 tons per day of sunflower. One disadvantage of such a stacked tray cooker over each prepress is the heavy weight to support the cookers in the building. Another disadvantage is the high power installed (110–132 kW for 700 tons per day), which required high ongoing electrical costs and long-term maintenance costs due to wear.

Nowadays, the row of vertical-stacked cookers has largely been replaced by one large rotary steam tube (RST) cooker placed on the ground floor. In addition to the savings in building costs and reduced power, RST cookers also achieve more uniform drying in less residence time with cocurrent vapor flow. Because the drying air flow is cocurrent, less air is necessary to reduce the moisture and emissions treatment is easier. Typically only one RST cooker is necessary for up to 3000 TPD of sunflower processing capacity. Figure 8.18 illustrates a modern RST cooker.

Prepressing

A prepress is composed of a series of special worms that create friction and pressure to disrupt the cellular structure of the meats and expel the oil. The oil is recovered

Figure 8.18 Rotary steam tube cooker.

through cage bars that surround the special worms along the length of the prepress. Figure 8.19 illustrates a modern prepress.

The expelled meats are formed into a cake at the discharge end of the prepress, which fundamentally utilizes an extrusion effect. The sunflower cake exits the prepress at approximately 100 °C and 7–8% moisture, with 17–20% residual oil. The resultant prepress cake is compressed, yet porous. A cake breaking mechanism at the discharge of the prepress crumbles the cake into granules varying from 40 × 40 × 10 mm down to 2 × 2 × 2 mm in size, making them more ideally suitable for the downstream solvent extraction process.

Prepress Oil Clarification

Prepressing extracts approximately 65% of the oil in the seed. The oil departing the prepress has approximately 10–15% by weight of solids, commonly referred to as *foots*. The prepress oil is first placed in a foots tank for approximately 30 minutes to allow the foots to gravity settle, as the foots specific gravity (1.1) is greater than that of the oil (0.9). The foots tank is equipped with a slow moving chain conveyor that continuously dredges the foots from the bottom of the tank. The foots are recycled back to the cooking step, while the remaining oil, now with less than 5% solids content, is pumped to a centrifugal decanter. The centrifugal decanter uses a higher g-force to further separate the foots from the oil. The foots from the centrifugal decanter are either returned to the cooking step, or put with the prepress cake going to solvent extraction. After the centrifugal decanter, the prepress oil is typically in the range of 0.1% solids content and 0.2% moisture content.

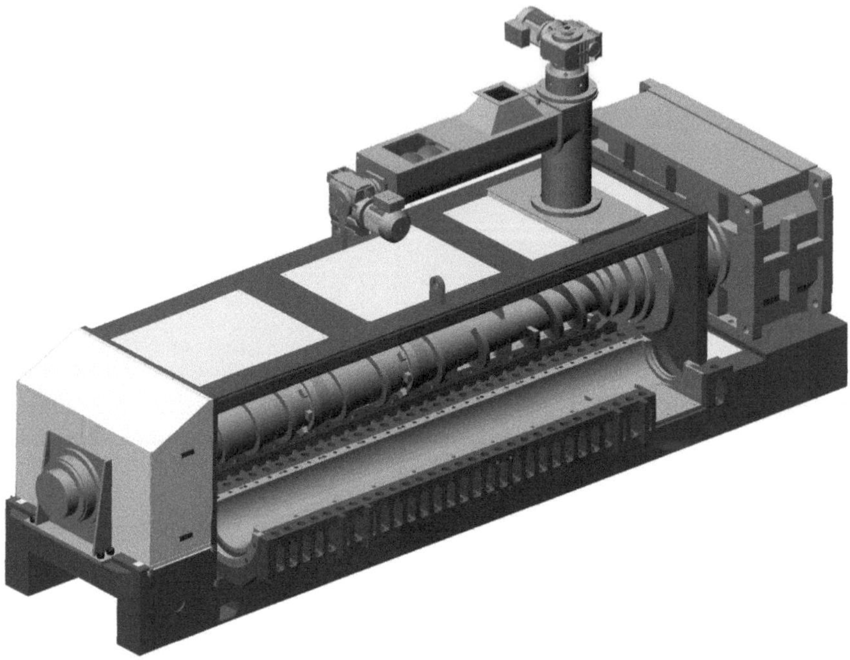

Figure 8.19 Prepress.

If the prepress oil is to be degummed, it is generally mixed with the solvent-extracted oil and then degummed together. If the prepress oil is to be stored prior to further processing, then it is typically passed through a vacuum oil dryer to reduce its moisture below 0.1% and passed through a cooler to reduce its temperature below 50 °C prior to being pumped to storage.

Prepress Cake Conditioning

The prepress cake is very friable at the exit of the screw press and must be conveyed to the cake cooler gently, with a minimum distance and number of drops.

The first purpose of the cake cooler is to allow the cake granules to become more durable. With approximately 3–5 minutes of residence time under aeration, the cake granules become firm and less subject to breaking down into fine powder. This is important for maintaining acceptable miscella drainage rates and solids content in miscella in the downstream extractor.

The second purpose of the cake cooler is to remove water vapor. As the cake undergoes evaporative cooling from 90–100 °C down to 65 °C, the resultant water vapor saturates the air around the prepress cake. As this saturated air cools, it will allow condensation of water onto the surface of the cake in a dew-like manner. Surface moisture impedes downstream miscella drainage rates and extraction efficiency. Therefore, the airflow through the cake cooler must be sufficient to evacuate the water vapor without the air reaching its dew point. A sample of a cooled prepress cake is illustrated in Figure 8.20.

Figure 8.20 Prepress cake.

Solvent Extraction

The prepressed sunflower cake entering the solvent extraction process is in granular form, with the granules composed of a mixture of proteins, oil, carbohydrates, fiber, and water, bound together in a semipermeable matrix. The original cellular structure of the sunflower meats has been largely disrupted by the flaking, cooking, and pre-pressing steps, leaving most of the remaining oil intermixed inside the prepress cake granules, with a minority of the oil remaining inside its initial cellular structure. The degree of cellular disruption is commonly measured via a "milling defect" test. A low milling defect is an indication of substantial cellular disruption, and a low milling defect enables greater solvent extraction efficiency. A typically milling defect test method is described below:

Milling Defect Test Method

1. Determine the percentage moisture (V) of the prepressed cake following AOCS Test Method Ba 2a-38. Discard dried cake after analysis.
2. Precisely weigh 5 g sample of fresh prepressed cake (W_T) and enclose it in a filter paper thimble.
3. Extract the sample for 3 hours using commercial grade hexane in a Butt-type extraction apparatus.
4. Remove the Soxhlet flask and evaporate the solvent from the extracted oil. Deduct the tare flask weight from the gross flask weight to determine the weight of oil extracted (W_A).
5. Empty the full contents of the filter paper thimble into a laboratory mill to first air desolventize, and then grind the sample to less than 0.75 mm particle size (fully passing 24 mesh). Then carefully return the full contents of the milled sample back into the original filter paper thimble.

6. Using a fresh Soxhlet flask, extract the milled sample for 3 hours as explained in step 3.
7. Remove the Soxhlet flask and evaporate the solvent from the extracted oil. Deduct the tare flask weight from the gross flask weight to determine the weight of additional oil extracted (W_B) after milling.
8. Calculate the milling defect as follows:
   a. Volatile matter (W_v) = mass of moisture in the extraction sample

$$W_v = W_T \times V / 100$$

   b. Apparent fatty matter (MGA) = the percentage oil content from the first extraction on a dry weight basis

$$MGA = 100 \times W_A/(W_T - W_V)$$

   c. Total fatty matter (MGT) = the percentage oil content from both extractions on a dry weight basis

$$MGT = 100 \times (W_A + W_B) / (W_T - W_V)$$

   d. Milling defect (ΔB) = the percentage oil content from the second extraction on a dry weight basis

$$\Delta B = 100 \times (W_B) / (W_T - W_V - W_A)$$

The solvent extraction process utilizes a solvent that diffuses into the prepress cake granules. The extraction process is a complex combination of several mechanisms including leaching, washing, diffusion, and dialysis (Johnson, 1997). Leaching and washing remove oil that is free and easily accessible to the solvent. Diffusion and dialysis are slower processes than washing and leaching. They remove oil that is not easily accessable. The solvent diffuses into the cake to solubilize oil and then micella (oil-solvent mixture) diffuses out. Through these various mechanisms, the solvent goes into solution with the oil. This oil–solvent solution is commonly referred to as miscella. Lower oil concentration *miscella* soaks into the prepress cake granule and higher oil concentration miscella exits the prepress cake granule. A continuous downward flow of miscella between the prepress cake granules is used to deliver lower oil concentration miscella to the granules and take away the higher oil concentration miscella from the granules.

The global oilseed crushing industry universally uses a hydrocarbon solvent composed of a blend of normal hexane, hexane isomers, and methylcyclopentane, commonly referred to as *commercial hexane* (Johnson, 2008). Commercial hexane composition varies by region and supplier. A typical commercial hexane at 15.6 °C and 1 bar (absolute) has a boiling point range of 65–70 °C, 79.6 kcal/kg latent heat of vaporization, 0.531 kcal/kg specific heat (as liquid), 0.68 specific gravity, and negligible solubility in water (Kemper, 1997). These thermal and physical properties make commercial hexane a very cost effective solvent for oil extraction. Commercial hexane also has a low flammability range, 1.2–7.7% volume in air and a vapor density three times heavier than air. These properties demand very careful process design and operation to minimize the risk of fire or explosion.

All extractors used today to extract prepressed sunflower cake rely on a similar process design approach (Johnson, 2000). The extractors are operated at 60–62 °C, with an approximately 1:1 ratio of solvent to prepress cake. As illustrated in Figure 8.21, a layer of prepress sunflower cake is physically supported above a screen floor. A miscella circulating pump delivers the miscella to a distributer above the cake layer, where the miscella is dispensed and then flows down through the granules of prepress cake by gravity, drains through the screen floor supporting the prepress cake, and is then collected in a miscella hopper under the screen, which is connected to the miscella circulating pump.

All extractors are designed to operate in a countercurrent mode, wherein the fresh solvent is distributed over the layer of prepress cake with the least oil remaining near the end of the extractor, and the miscella with the greatest oil concentration is distributed onto the prepress cake with the greatest oil content at the beginning of the extractor. All extractors use multiple pumps, distributors, and miscella collection hoppers to keep the oil concentrations separated. Figure 8.22 illustrates a typical oil concentration curve for a sunflower cake extractor.

Despite all extractors adopting a similar process design approach, they are quite different in terms of their overall geometry, method to convey the layer of prepress cake through the extractor, and method to keep adjacent miscella stages from mixing. The following are the principle types of commercial extractors used today.

Rotary Wall Extractor

These extractors use radial vertical walls to convey the layer of prepress cake in a circular path over a fixed screen floor. The layer of prepress cake is generally in the range of 2–3 m deep. The cake layer is fully divided by the radial walls to enable full immersion in miscella without mixing with the neighboring stages. The most common rotary wall extractor supplied today is the Reflex® Extractor (Desmet Ballestra) as illustrated in Figure 8.23.

Linear Chain Extractor

These extractors use parallel chain paddles to convey the layer of prepress cake in a linear path over a fixed screen floor. The layer of prepress cake is generally in the range of 0.5–1.0 m deep. The shallow material layer has a long length that enables sufficient distance between neighboring miscella stages to minimize mixing. The most common linear chain extractor supplied today is the Crown Type III Extractor (Crown Iron Works) as illustrated in Figure 8.24, on page 210.

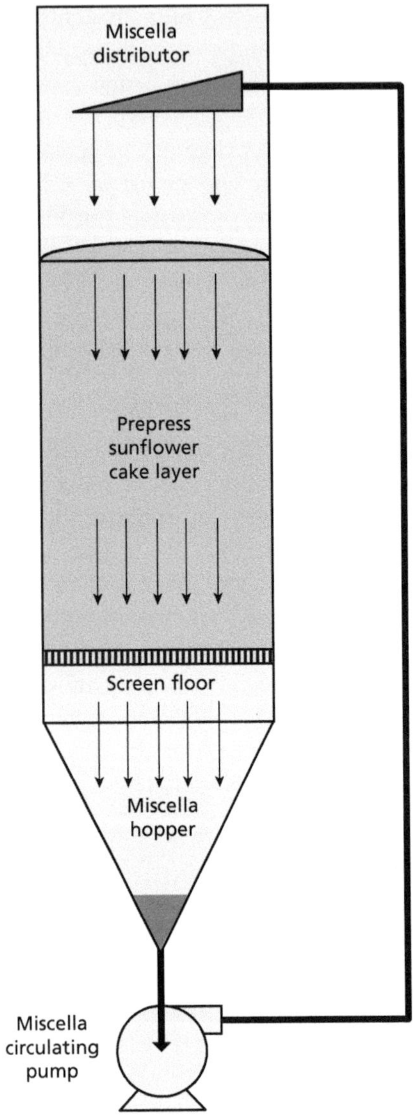

Figure 8.21 Solvent extraction process.

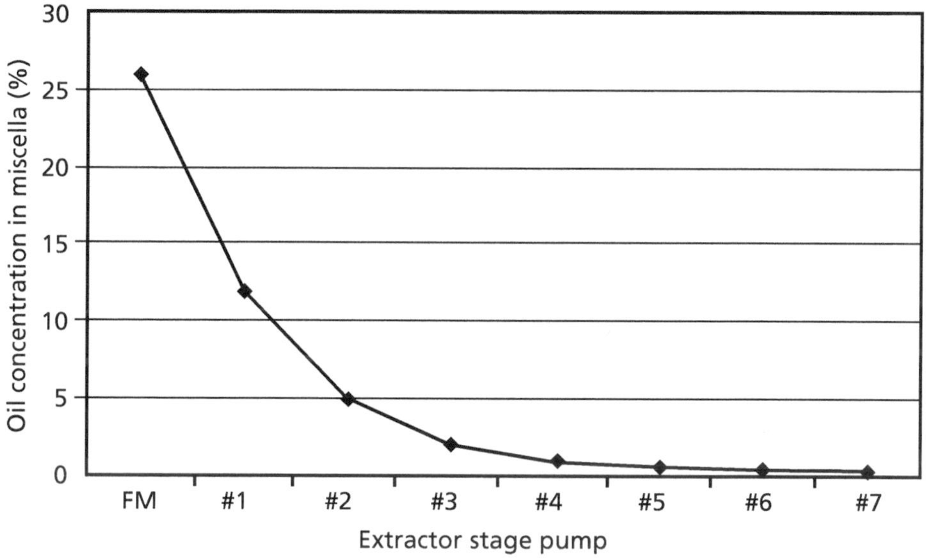

Figure 8.22 Extractor miscella concentration curve.

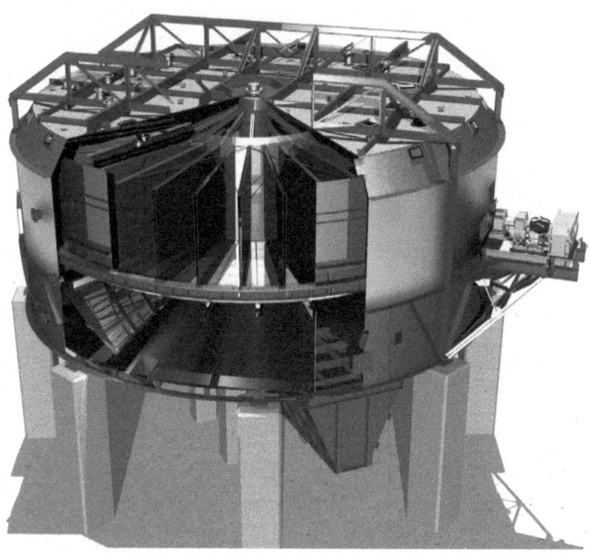

Figure 8.23 Rotary wall extractor.

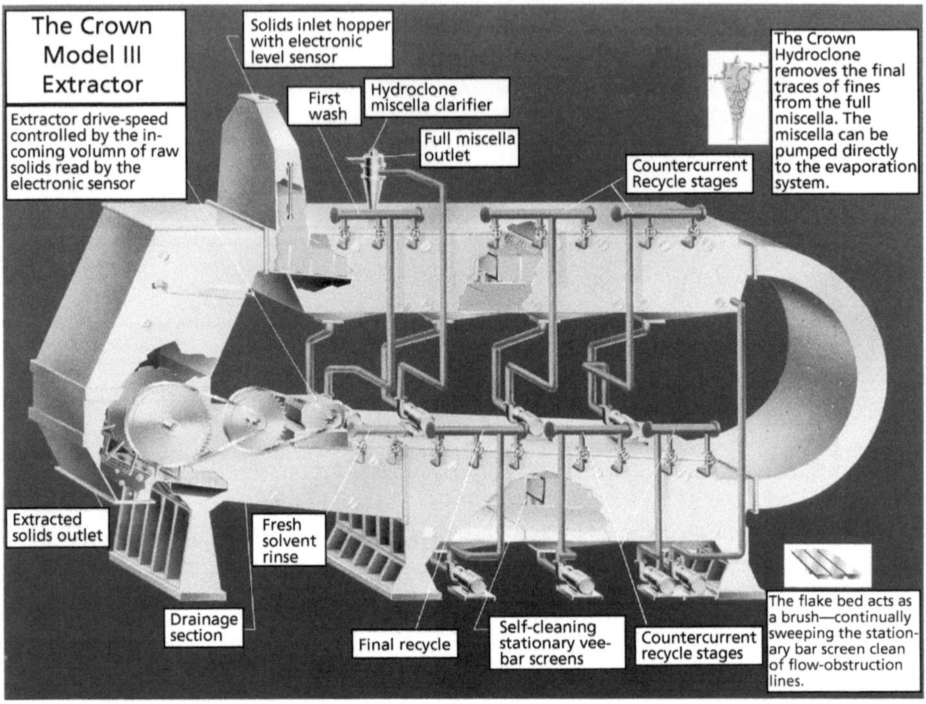

Figure 8.24 Linear chain extractor.

Linear Belt Extractor

These extractors move the screen floor under the layer of prepress cake to carry the material in a linear path. The layer of prepress cake is generally in the range of 1.5–2.0 m deep. The material layer has a long length to enable sufficient distance between neighboring miscella concentration stages to minimize mixing. The most common linear belt extractor supplied today is the LM™ Extractor (Desmet Ballestra), as illustrated in Figure 8.25.

Linear Wall Extractor

These extractors use parallel vertical walls to convey the layer of prepress cake in a linear path over a fixed screen floor. The layer of prepress cake is generally in the range of 1.5–2.5 m deep. The material layer is fully divided by the walls to enable full immersion without mixing of neighboring miscella concentration stages. The most common linear wall extractor supplied today is the Lurgi Extractor (Air Liquide), as illustrated in Figure 8.26.

Figure 8.25 Linear belt extractor.

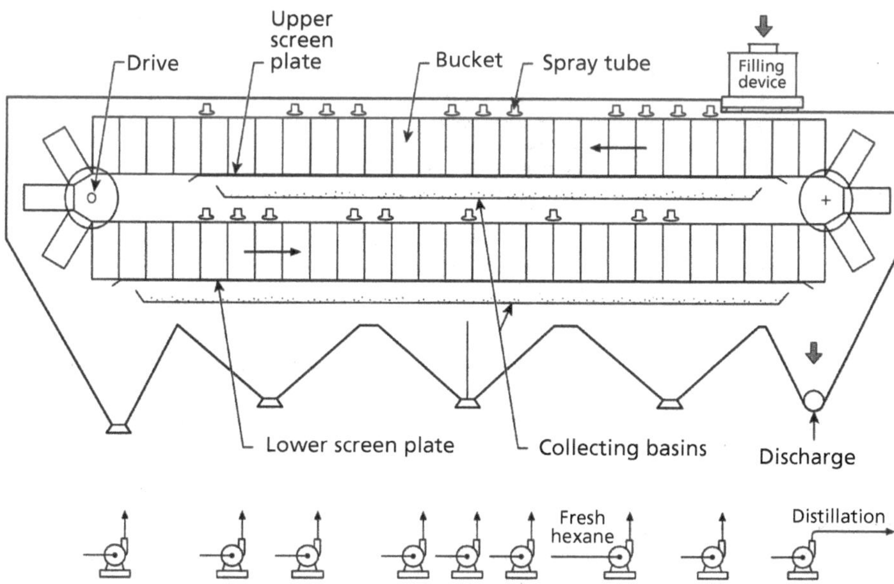

Figure 8.26 Linear wall extractor.

The extracted prepress sunflower cake exiting the extractor is typically in the range of 70–75% meal and 25–30% solvent. The extracted cake is conveyed to the meal desolventizer. The miscella exiting the extractor is typically in the range of 23–27% oil and 73–77% solvent. This miscella is clarified to remove remaining solids via internal extractor screening or external cyclonic separation. The clarified miscella then passes on to miscella distillation.

Meal Desolventizing

The extracted sunflower cake is conveyed to the desolventizer toaster for solvent removal. The desolventizer toaster is a vertical cylindrical vessel with multiple horizontal trays in series. The meal is conveyed in a circular pattern above the trays by rotating stirrers connected to a vertical rotating shaft passing down through the center of the trays to a speed reducer and motor located at the base of the desolventizer toaster.

The desolventizer toaster is the largest energy consumer in the solvent extraction process area. In addition to the considerable electrical energy necessary to convey and stir the meal inside the desolventizer toaster, there is significant thermal energy input. The desolventizer toaster uses the thermal energy of steam to warm the extracted cake from 60 °C up to 109 °C, and to evaporate approximately 25–30% (by weight) of solvent from the extracted cake. The vapors that vent from the desolventizer toaster are a mixture of solvent vapor and excess direct steam used in the desolventizer toaster. The energy efficiency of the desolventizer toaster can be measured by the temperature of the exiting vapors. Lower vapor temperature is an indicator of less direct steam usage and higher efficiency. Typical sunflower desolventizer toasters operate with vent vapor temperature in the range of 70–75 °C. Figure 8.27 illustrates the composition of the desolventizer toaster vent vapors relative to vapor temperature.

The first tray in a sunflower desolventizer toaster is referred to as a *predesolventizing tray.* The purpose of this tray is to increase the meal temperature to approximately 67 °C and flash solvent from the surface of meal. The predesolventizing tray uses pressurized steam inside to conduct heat into a shallow layer of meal stirred above.

The second tray in a sunflower desolventizer toaster is referred to as a *desolventizing tray.* The purpose of this tray is to evaporate most of the remaining solvent from the meal. The desolventizing tray uses pressurized steam inside to conduct heat into a deep layer of meal stirred above. The majority of heat, however, is from direct steam rising up from below and condensing into the meal layer stirred on the desolventizing tray. As the direct steam condenses, its latent heat is used to evaporate solvent from the meal. When the meal exits this tray, it typically has less than 1% solvent and its temperature is near 100 °C.

The next two to four trays in a desolventizer toaster are referred to as *stripping trays.* The purpose of these trays is to provide residence time for ascending, superheated direct steam to strip the remaining solvent from the meal. The stripping trays use pressurized

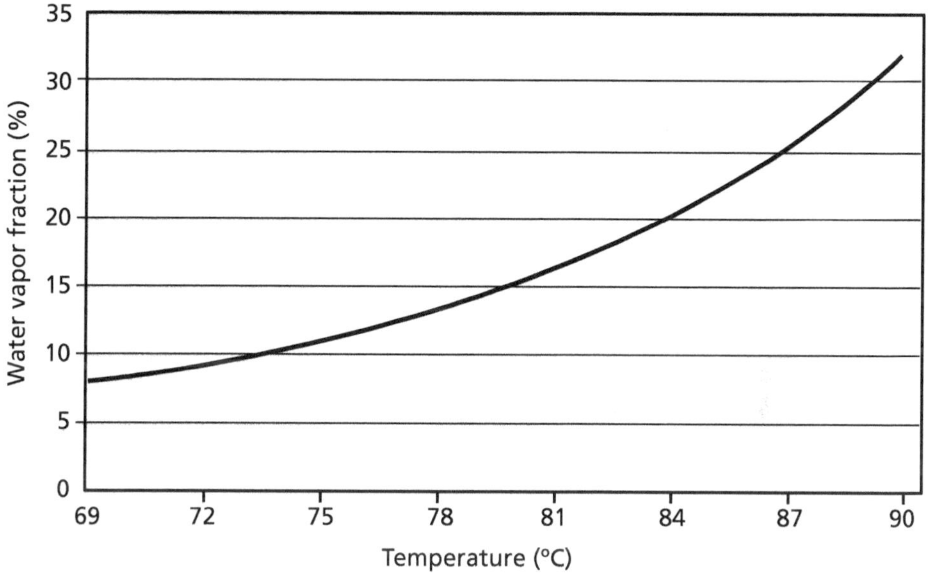

Figure 8.27 Desolventizer toaster vent vapor composition curve.

steam inside to conduct heat into a deep layer of meal stirred above, and to re-superheat the direct steam ascending from tray to tray. When the meal exits these trays, it typically contains 0.01–0.05% solvent and its temperature is near 109 °C.

The final tray in the desolventizer toaster is referred to as the *steam dryer tray.* The purpose of this tray is to dry the meal and capture the water vapor removed for heat recovery. The meal temperature drops from 109 °C to 102 °C, providing heat, and the steam drying tray uses pressurized steam inside to conduct additional heat into a shallow layer of meal stirred above. Approximately 0.5–1.0% of the incoming moisture is removed on this tray and sent to heat recovery as water vapor. The meal typically exits this final tray of the desolventizer toaster at a temperature of 102 °C and moisture of 15% by weight. Figure 8.28 is an illustration of a typical desolventizer toaster.

Sunflower meal for export markets is generally pelletized to maximize bulk density and minimize dusting. The meal temperature and moisture exiting the desolventizer toaster is ideal for pelletizing. If the meal pelletizing and extraction are done in the same plant, the hot meal from the desolventizer toaster is conveyed directly to the meal pelletizing section.

Sunflower meal for local markets is generally sold as ground meal. In this case, the meal must be reduced in temperature and moisture immediately after the desolventizer toaster. This can be accomplished by adding two additional meal cooling trays to the desolventizer toaster. The purpose of meal cooling trays is to reduce the

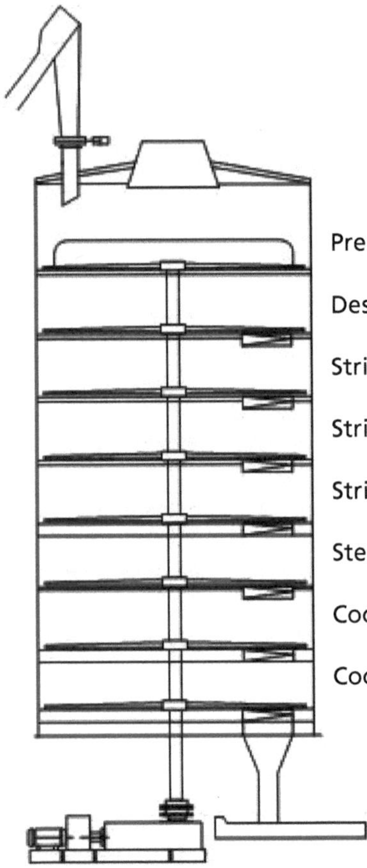

Figure 8.28 Desolventizer Toaster.

meal temperature and moisture. The cooling trays use pressurized air inside and a perforated upper surface to blow air up through a moderate layer of meal stirred above. The air passing through the meal simultaneously absorbs heat and moisture. When the meal exits these cooling trays, it typically has less than 11% moisture and its temperature is less than 38 °C.

Miscella Distillation

The miscella exiting the solvent extractor contains approximately 25–30% oil and 70–75% solvent. The purpose of the miscella distillation process is to remove the residual solvent in the oil to a negligible quantity, with the least amount of energy input. A typical distillation flow is illustrated in Figure 8.29.

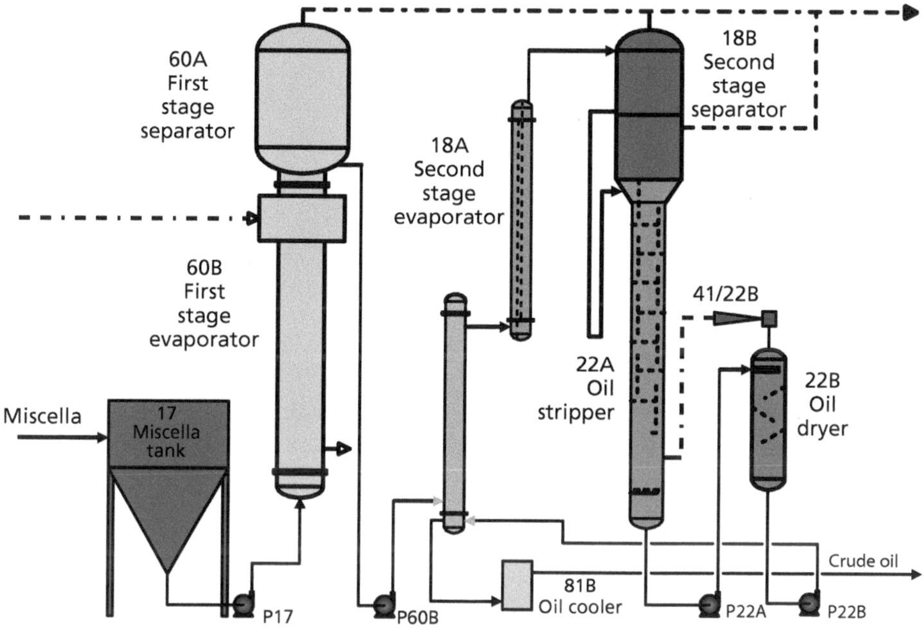

Figure 8.29 Miscella distillation flowsheet.

The first unit process of a typical miscella distillation is a rising film evaporator, which is operated under vacuum on the tube side and uses the vent vapors from the desolventizer toaster as its heat source on the shell side. This evaporator is the single largest opportunity for heat recovery and energy savings in the solvent extraction process area. The miscella leaving the evaporator contains approximately 65–75% oil and 25–35% solvent.

The miscella exiting the first evaporator is typically preheated via heat recovery from the hot oil exiting the miscella distillation process. It then passes to a second rising film evaporator that has miscella on the tube side, under vacuum, and low pressure steam on the shell side. Much of the low pressure steam can be flash steam recovered generated from the hot condensate exiting the trays of the desolventizer toaster. The miscella leaving the evaporator contains approximately 92–98% oil and 2–8% solvent. The miscella temperature will be elevated to 110–120 °C with a downstream disc/donut oil stripper or to 95–100 °C with a downstream sieve tray oil stripper.

After the miscella exits the second evaporator, the solvent content of the oil is as low as can practically be achieved with normal evaporation. The remaining solvent must then be removed via vacuum stripping with superheated steam as the stripping medium. Historically, the apparatus used to perform this stripping step was a disc/donut oil stripper. Such a device relies on a thin film of oil cascading down and across discs and donuts to make contact with rising, superheated stripping steam. Since the steam-to-oil contact is not very effective, it is important that miscella enters the oil stripper at 110–120 °C temperature to aid the separation. A more recent apparatus for this application is the sieve tray oil stripper, as illustrated in Figure 8.30. Such a device relies on several deep trays of frothing oil making contact with rising, superheated steam. With much improved steam-to-oil contact, this type of oil stripper allows a much lower incoming miscella temperature of 95–100 °C. The oil leaving the oil stripper contains approximately 0.001–0.005% solvent and 0.1–0.3% moisture.

After the final solvent stripping step, the oil is often sent to a vacuum oil drying step to reduce the moisture for storage or transit. The vacuum oil dryer is typically operated at 50–100 mbar (absolute) to reduce the moisture in the oil to 0.05–0.10%. The vacuum in oil dryer is created by a single steam ejector and exhausted into the oil stripper for full heat recovery as stripping steam.

After the oil drying step, the crude non-degummed sunflower oil is first cooled via heat recovery and then cooled further in a cooling water heat exchanger to approximately 50 °C prior to storage.

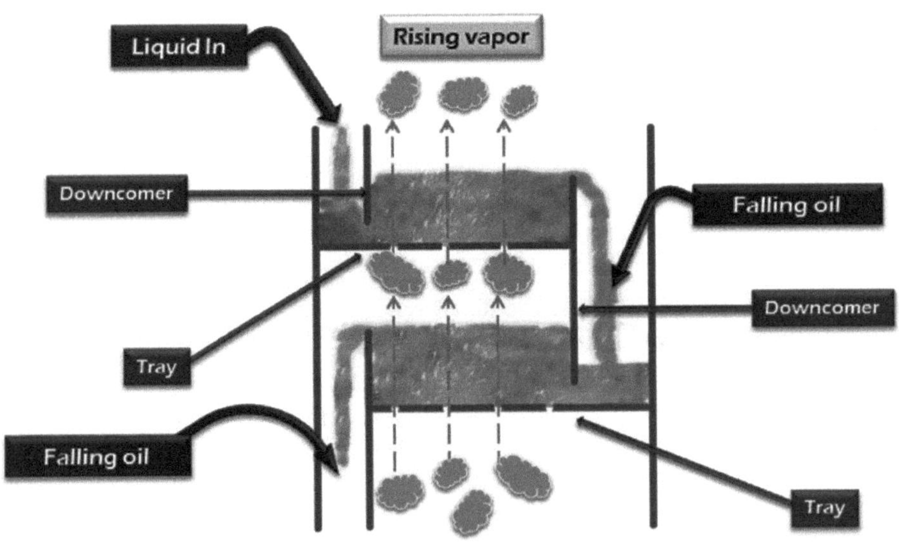

Figure 8.30 Sieve tray oil stripper.

Solvent Recovery

The solvent evaporated from the sunflower meal and the miscella are both recovered in the solvent recovery system within the solvent extraction process area. A typical sunflower plant will recover 99.8–99.9% of the solvent pumped to the extractor.

Solvent vapors from the desolventizer toaster are primarily condensed in the shell side of the first evaporator of miscella distillation for heat recovery purposes. The remaining vapors pass to an atmospheric condenser, to be condensed. Solvent vapors from the first evaporator, second evaporator, and oil stripper of miscella distillation pass to a vacuum condenser to be condensed. The vacuum condenser typically operates at 350–500 mbar (absolute). All the condensed vapor streams from the first evaporator shell, atmospheric condenser and vacuum condenser pass to a solvent/water separating vessel. Due to the negligible solubility of hexane in water, the solvent/water separator vessel simply allows the solvent with specific gravity of approximately 0.68 to float above the water with specific gravity of approximately 1.00.

The waste process water from the solvent/water separator passes to a waste water stripper where direct steam is used to heat the water from 35–45 °C to over 90 °C and strip the residual solvent in the water to less than 0.001% by weight. This water is discharged from the solvent extraction process to a water treatment plant, or in some cases, this waste water is sent to a waste water steam generation system that converts 95–97% of the waste water stream to low pressure steam for reuse in the desolventizer toaster. In this case, a 3–5% purge of the waste water can be sprayed on the desolventized sunflower meal, or sent to water treatment.

The solvent from the solvent/water separator at 35–45 °C passes to a preheater for initial heating. Such a preheater can utilize heat recovered from oil stripper vapors to reduce energy consumption. The preheated solvent then passes to a final solvent heater using low pressure steam to increase the solvent temperature to 60 °C prior to it entering the solvent extractor.

Air enters the solvent extraction process with the sunflower cake. This air passes through the solvent recovery condensers and carries with it an equilibrium quantity of solvent vapor. At 35 °C vapor temperature, 1 part of air will carry 1.5 parts of solvent vapor. As a result, solvent must be further recovered from this vapor stream prior to discharging it to the atmosphere.

The vent vapor recovery system typically consists of a vent gas cooler, followed by a solvent vapor absorber. The vent gas cooler uses the coolest available water source to first minimize the solvent carried with the air stream. The solvent vapor absorber then uses cool mineral oil cascading down across loose packing to absorb solvent vapor from the rising air stream in the vessel. The exiting air stream can typically be reduced to 0.3–1.0% (by weight) solvent in the air.

The mineral oil exiting the solvent vapor absorber will contain 2–5% solvent (by weight). This mineral oil is first preheated via mineral oil to mineral oil heat recovery, and then heated via steam. The heated mineral oil is vacuum steam stripped in an oil stripper vessel. Once stripped, the mineral oil is first precooled via mineral oil to mineral oil heat recovery, and then cooled via cooling water before being reintroduced into the solvent vapor absorber. The solvent vapor stripped from the mineral oil in the mineral oil stripping step is sent to the vacuum condenser for solvent recovery.

Tail-End Dehulling

After the solvent extraction process, an additional increase of the protein content of sunflower meal can be obtained by tail-end dehulling. In this process, hulls are removed from the deoiled meal as a last stage of the crushing process. Tail-end dehulling can be designed in different ways, but it is typically a multistage process consisting of a series of screeners, turbo sifters, and flakers.

After the meal leaves the extraction section, the meal passes through successive screens with aspiration and turbo sifters with aspiration in order to separate the fiber from the protein by density. The number of successive operations depends on the protein level that is required.

In a typical tail-end dehulling system, deoiled, desolventized sunflower meal is first processed over a first turbo sifter that separates the bigger meal particles from the primary, finer meal stream.

The big meal particles are ground in a low speed hammer mill and then mixed back with the finer meal fraction. The combined fine meal stream then flows to a second turbo sifter, equipped with finer screens, where a first high protein meal fraction is collected as "fine fraction" (passing through the screens). The meal fraction that does not pass through the screen ("overs fraction") is collected and processed through a flaker. In this way, the hull is loosened from the meats without creating too many fines (damaged hulls). The flaked meal is then sent to a third set of turbo sifters from which a second high-protein meal is collected. The overs fraction of this turbo sifter step is processed again through a second set of flakers followed by a fourth set of turbo sifters. This gives a third high-protein meal and a remaining overs fraction that is continuously depleted in proteins and enriched in fibrous material. All high-protein fractions are collected as a single stream. The overs fraction of the last turbo sifter is the low protein meal. Figure 8.31 illustrates a schematic flow diagram of the tail-end dehulling process.

Intrinsic advantages of the tail-end dehulling process are its flexibility and also the fact that there's no issue of oil losses involved. Depending on the market situation, the process can be steered toward either maximum protein content in the high-protein (HP) fraction or minimum protein content in the low-protein (LP) fraction. Target for the HP fraction is usually set at 45% protein content. Yield of the HP fraction will then depend on the required protein content in the LP fraction, which will

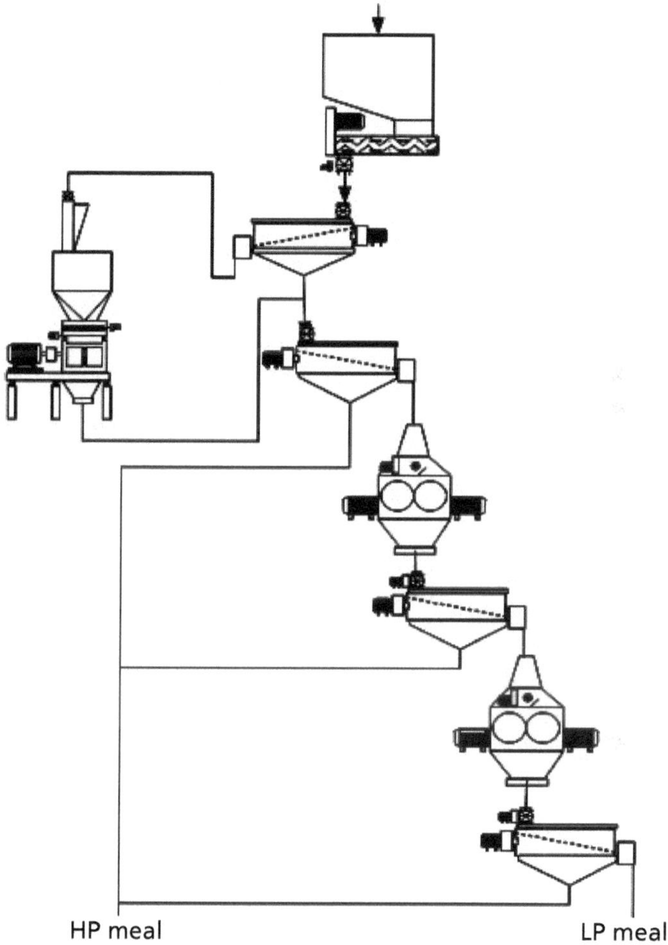

Figure 8.31 Tail end dehulling schematic.

determine its value potential. Table 8.C indicates the protein content versus yield for various HP/LP options.

Sunflower meal with 30% protein content can still be marketed as mid-protein meal, while meal with 27–29% protein content is considered low-protein meal. The price difference between these two meal qualities is usually small. Sunflower meal with less than 27% protein content has little or no value for use in animal feed and is generally used as fuel with a corresponding lower value. If the LP fraction is intended for burning, protein content has no importance and the tail-end dehulling is then usually steered to get the highest yield of a HP fraction meal with 45% protein content.

Table 8.C Sunflower Meal Protein Content and Yield

Sunflower Meal Protein Content (%)	Protein Content (%)		Yield (%)	
	HP Fraction	LP Fraction	HP Fraction	LP Fraction
35.3	45	18	64	36
	45	27	46.1	53.9
	45	28	42.9	57.1
	45	29	39.3	60.7
	45	30	35.3	64.7

Production of such HP sunflower meal is not feasible with only front-end dehulling when starting from sunflower seeds with a normal protein content (14–16%). For such seeds, a combination of one-stage front-end dehulling and tail-end dehulling is a possible technical solution, as indicated in Table 8.D.

The protein enrichment that can be achieved with tail-end dehulling is greatest when starting with one-stage front-end-dehulled meal obtained from seeds with low protein content. However, in such cases, yield of the HP meal fraction will also be lower, which makes the process less economically attractive.

Despite a clear economic potential to improve meal quality and value, tail-end dehulling of sunflower meal is seldom industrially applied today.

Meal Sizing/Pelleting

Sunflower meal for local markets is ground and sold as "loose" meal, while sunflower meal for export markets is generally pelletized. The main purpose of pelletizing the

Table 8.D Sunflower Seed and Meal Protein Content

Protein Content of Sunflower Seed (%)	Protein Content in HP Sunflower Meal (%)[a]		
	One-stage Front-end Dehulling	One-stage Front-end and Tail-end Dehulling	Two-stage Front-end Dehulling
14	32.9	42.8	36.1
15	35.3	45.0	38.7
16	37.6	45.0	41.3
17	40.0	45.0	43.8
18	42.3	45.0	46.4

[a]Protein content LP fraction set at 18% and protein content of HP fraction at 45%.

meal is the increase in bulk density, which thereby reduces the storage and transport cost of the sunflower meal. Pelletizing also increases the physical stability of the meal and increases the possible storage time before consumption.

Meal Sizing

Meal from the solvent extraction process contains a mix of meal lumps, large granular meal, and fine granular meal. Feed mills generally require a uniform, fine granule size profile. Therefore, the sunflower meal from the solvent extraction process is generally screened. The small particles passing through the screen are separated from the large granules and lumps, which pass over the screen and go to a hammer mill. The hammer mill reduces the lumps and large granules to a uniform, fine granule size. The meal stream passing through the screener and the meal stream from the hammer mill are recombined as a single, fine granule meal stream that is ready for storage and commercial sale to a feed mill.

Meal Pelletizing

In the case of export markets, the sunflower meal is generally pelletized. Figure 8.32 illustrates a typical meal pelletizing line.

There are two approaches for producing pelletized meal. The traditional approach is to start with ground meal from the meal sizing operation, and the more recent approach is to start with meal directly from the meal desolventizing step in the solvent extraction process without drying, cooling, or sizing the meal prior to pelletizing.

Sunflower granules coming out of the grinder are hard and generally at around 35 °C with a moisture content of around 11%. The conditions required for the pellet mill are a temperature around 85 °C and the moisture around 15%, with pliable granules. To obtain these conditions, the meal must be processed through a horizontal paddle conditioner using direct and indirect steam. Figure 8.33 (p. 233) illustrates a typical horizontal paddle conditioner. Significant energy is required to increase the temperature of the meal by approximately 50 °C and increase the moisture of the meal by approximately 4% inside the horizontal paddle conditioner.

Sunflower meal direct from the desolventizer toaster is generally around 100 °C with moisture content of around 15% and the meal granules are rather pliable. By the time the meal is conveyed from the desolventizer toaster in the solvent extraction process building over to the pellet mills in the seed preparation process building, the temperature has typically fallen to 85 °C, providing all the correct conditions for pelletizing. This approach saves the thermal energy needed for heating the meal and increasing its moisture. The limitation of the approach of going directly from the desolventizer toaster to meal pelletizing is that tail-end dehulling to further increase meal protein is not possible.

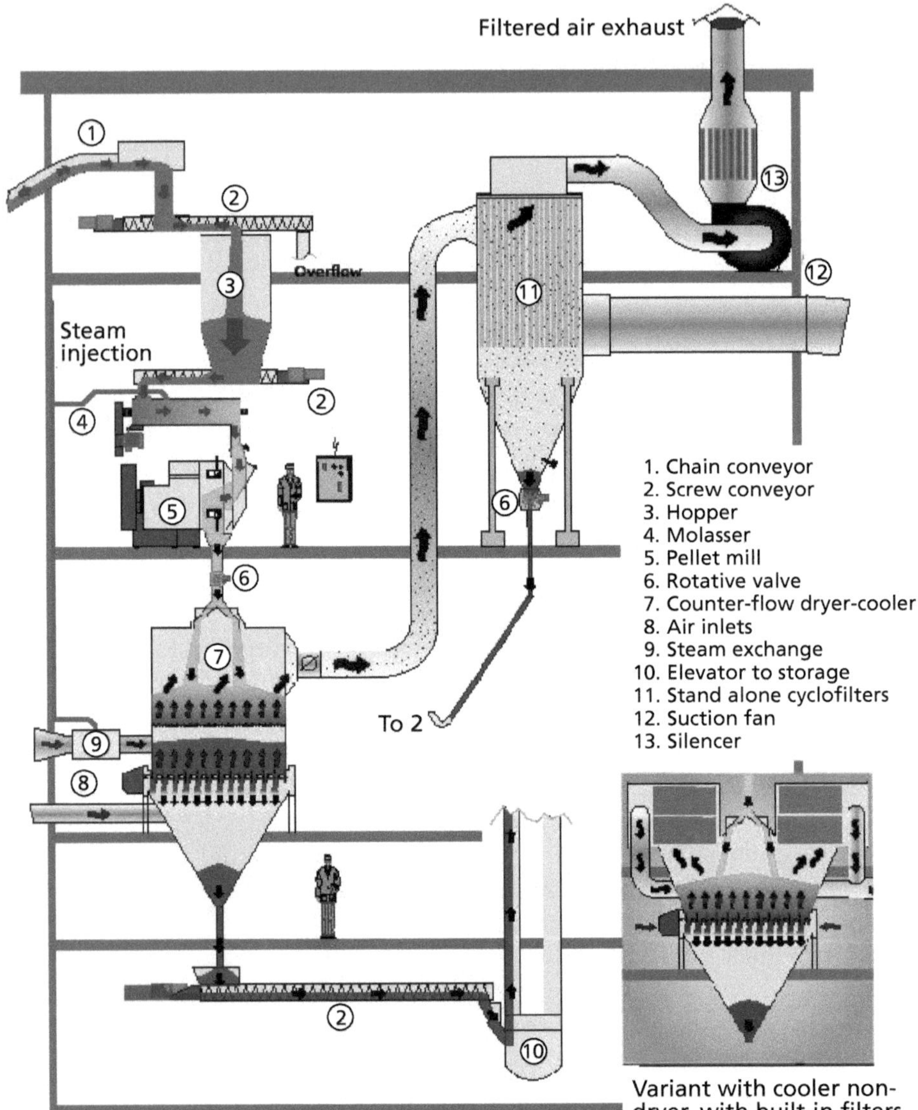

Figure 8.32 Meal pelletizing schematic.

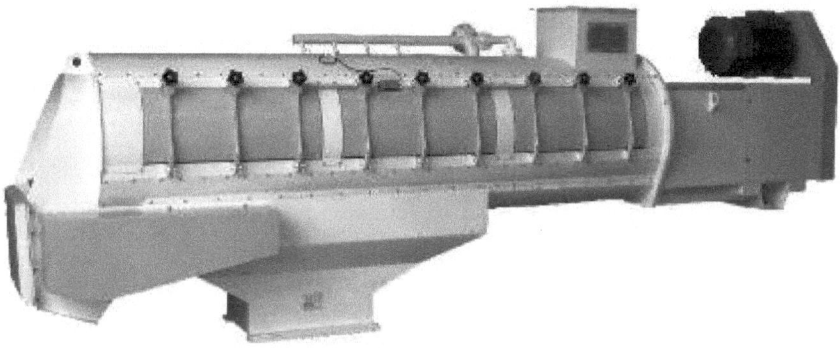

Figure 8.33 Horizontal conditioner.

The pellet mill is a device that presses the meal through holes in a cylindrical die plate using rotating rollers inside to form dense pellets. The dense pellets are generally 6–10 mm in diameter and 1–3 cm long. Different peripheral velocities, from 4 to 8 m/s, and different diameters of the dies allow process flows ranging from 1 to 50 tons per hour of sunflower meal to be pelletized, using from 45 to 400 kW of power. Figure 8.34 illustrates a typical pellet mill used for sunflower meal.

Figure 8.34 Pellet mill.

After the pellet mill, the sunflower meal pellets contain approximately 14% moisture and are at 85 °C and somewhat fragile. The pellets must be conveyed to a pellet cooler with a minimum handling and drops to maintain the pellet integrity. Most pellet mills are located directly over a vertical pellet cooler. The vertical pellet cooler uses forced ambient air to simultaneously cool the pellets, remove moisture via evaporative cooling effect, and give the pellets time to become firm and less friable. Figure 8.35 illustrates a typical vertical pellet cooler used for sunflower meal pellets.

Utilization of Hull Energy

As a fraction of the incoming sunflower seed, approximately 14% sunflower hulls can be removed in the partial hull removal step in the seed preparation process area. These hulls are high in fiber and low in protein, and therefore typically have a very low commercial feed value.

The best economic opportunity is to utilize the sunflower hulls as a fuel source, since the energy value of sunflower hulls is approximately 17,000 kJ/kg (Carbon Friendly Solutions, 2013). The sunflower hulls can be converted to high pressure steam in a biomass boiler, as illustrated in Figure 8.36. The high pressure steam cre-

Figure 8.35 Pellet cooler.

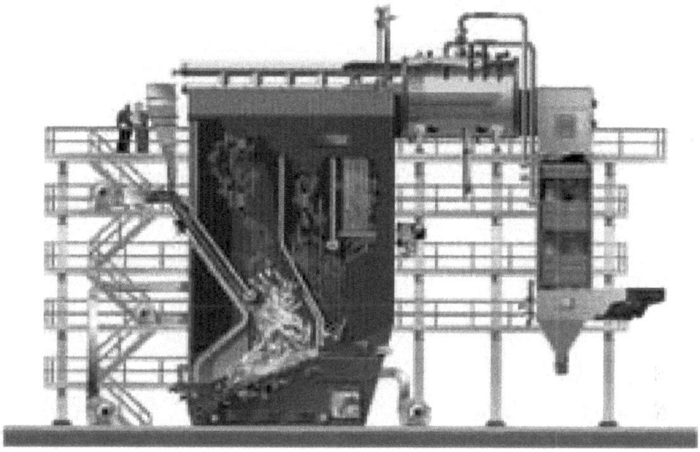

Figure 8.36 Hulls biomass boiler.

ated can then be used for both powering a steam turbine to generate electricity, and for providing the high pressure steam needs in the sunflower crushing plant.

As an example, in a 2000 TPD sunflower crushing plant with 14% hull removal, the 280 TPD of hulls removed have approximately 200 million kJ/h of energy value. This is sufficient fuel energy to create over 50 tons per hour of 20 bar (gauge pressure) steam at 300 °C in a biomass boiler. Approximately 30% of this 20 bar (g) steam can be lowered in pressure and temperature and used directly in the sunflower crushing plant as process steam. The remaining 70% of the 20 bar (g) steam can pass to a turbine to generate electricity. A condensing steam turbine in this example could be expected to produce approximately 2,000 kW of electricity (U.S. Department of Energy, 2012). A typical 2000 TPD sunflower crushing plant will consume approximately 15 tons per hour of steam, so it is possible to supply 100% of the sunflower crushing plant steam demand. The 2000 kW of electricity generated will offset nearly half of the sunflower plant's electrical demand.

As this example illustrates, utilizing the byproduct sunflower hulls as a fuel source covers all of the steam needs and a significant portion of the electrical needs of the facility, making conversion of hulls to energy a critically important element in minimizing the total operational cost of a sunflower crushing plant.

References

Carbon Friendly Solutions (CFS). http://www.carbonfriendly.com/s/CarbioPel_Projects.asp?ReportID=521343 (accessed May 18, 2013).

FAO. FAOSTAT Oilseed Production Volumes, http://faostat.fao.org (accessed May 6, 2013).

Johnson, L. Analysis of Alternative Extraction Technologies. In *Technology and Solvents for Extracting Oilseeds and Non Petroleum Oils;* Wan, P., Wakelyn, P., Eds.; AOCS Press: Urbana, IL, 1997; p 8.

Johnson, L. Recovery of Fats and Oils. In *Introduction to Fats and Oils Technology,* 2nd ed.; O'Brien, R., Farr, W., Wan, P., Eds.; AOCS Press: Urbana, IL, 2000; p 122.

Johnson, L. Oil Recovery from Soybeans. In *Soybeans: Chemistry, Production, Processing and Utilization;* Johnson, L., White, P., Galloway, R., Eds.; AOCS Press: Urbana, IL, 2008; p 349.

Kemper, T. Extraction Principles and Extractor Design. In *Technology and Solvents for Extracting Oilseeds and Non Petroleum Oils;* Wan, P., Wakelyn, P., Eds.; AOCS Press: Urbana, IL, 1997; p 137.

National Fire Protection Association (NFPA). *NFPA-36 Standard for Solvent Extraction Plants;* National Fire Protection Association: Quincy, MA, 2013; p 11.

O'Brien, R. Fats and Oils Processing. In *Introduction to Fats and Oils Technology,* 2nd ed.; O'Brien, R., Farr, W., Wan, P., Eds.; AOCS Press: Urbana, IL, 2000; p 99.

U.S. Department of Energy. http://www4.eere.energy.gov/manufacturing/tech_deployment / amo_steam_tool/equipTurbine (accessed May 18, 2013), Version: v0.9.1-RC, updated Jan 7, 2012.

Williams, M. Recovery of Oils and Fats from Oilseeds and Fatty Materials. In *Bailey's Industrial Oil and Fat Products,* 6th ed.; Shahidi, F., Ed.; John Wiley & Sons: Hoboken, NJ, 2005; Vol. 5, p 101.

9

Oil Refining

Albert J. Dijkstra ■ *Scientific Consultant, St. Eutrope-de-Born, France*

Introduction

Crude sunflower seed oil, as obtained by screw pressing and solvent extraction, contains various impurities that have to be removed to obtain a bland oil with a good shelf life. Since these impurities have quite different chemical properties, different processes have been developed for their removal. These processes can be grouped as follows:

- *Degumming.* Degumming aims at removing the phosphatides present in the crude oil. It also removes water-soluble compounds like sugars and glycolipids, and most particulate impurities like meal residue. The "gums" also contain some oil, the amount of which depends on which degumming process is used. Several different processes are used industrially, but not all of them are suitable for sunflower seed oil.
- *Chemical neutralization.* In this process, the residual phosphatides and the free fatty acids (FFA) present in the crude oil are removed by treating the oil with an aqueous solution of sodium hydroxide that reacts with the FFA to form soaps that are separated from the oil by a centrifuge as soapstock. The soapstock also entrains some neutral oil.
- *Bleaching.* The purpose of the bleaching process is to remove colored compounds like chlorophyll and beta-carotene from the oil by mixing the oil with an adsorbent like bleaching earth. Since the spent earth is removed from the oil by filtration, the bleaching process also removes residual particulate contaminants. The spent bleaching earth contains at least 35% by weight of organics. The literature often quotes lower figures, but these refer to material that is isolated by Soxhlet extraction with hexane. The 35% refers to the oil loss and includes polymerized/oxidized oil that is not extracted by hexane but has to be determined by ashing the sample (Gunstone et al., 2007).
- *Dewaxing.* Sunflower seed oil contains vegetable waxes, esters of fatty acids, and fatty alcohols. The higher melting point waxes will crystallize when the oil is stored at low temperatures and form a deposit. This does not matter if the oil is used in a margarine fat blend, but when the oil is to be sold in a bottle as a salad oil, this deposit is unsightly. The higher melting point waxes are crystallized by cooling the oil and then the crystals are removed from the oil by filtration or centrifugation.

- *Deodorization.* In this process, malodorous compounds are removed from the oil by stripping under vacuum with steam at an elevated temperature. Under these conditions, FFA will also be removed from the oil. It is therefore possible to bypass the chemical neutralization step, provided that the preceding degumming step has been so effective that the residual phosphorus level is less than 10 ppm.

The above processes are performed in an oil refinery. In Europe, the term refining refers to the combination of the above processes. In the United States, the term refining only applies to the removal of FFA, hence the use of the terms chemical refining and physical refining. The latter process is also referred to as steam refining because it is carried out during the vacuum stripping process that uses steam as stripping agent. The advantage of the latter process is that it avoids the neutral oil loss (NOL) by emulsion formation, which is inherent in the chemical refining process. Since NOL is more or less proportional to the FFA being neutralized, the yield advantage of physical refining is more significant for high FFA crude oils, such as palm oil, than for low FFA crude oils, such as soybean oil (Dijkstra, 2013b). Crude sunflower seed oil regularly has an FFA content of over 1% and is therefore a prime candidate for physical refining.

Some properties of crude sunflower seed oil are listed in Table 9.A, together with those of other major oils.

In the table, the iodine value (IV) refers to normal, high linoleic sunflower seed oil. The IV of high oleic sunflower seed oil is of course much lower: 78–90 (Warner et al., 2002). According to Table 9.A, the wax content can vary over a wide range; this reflects the effect of dehulling the seed because the wax coats the seed hull. Removing

Table 9.A Comparison of Various Properties of Several Crude Vegetable Oils

Property	Unit	Sunflower Seed Oil[a]	Soybean Oil[b]	Rapeseed Oil (Canola)[c]	Palm Oil[d]
Iodine Value	% I_2	134–136	131–133	120–122	51–53
Free Fatty Acids	% as oleic acid	0.6–2.0	0.3–0.8	0.4–1.2	1–5
Phospholipids	ppm P	200–500	600–1100	300–1000	0.2–5.0
Waxes	ppm	600–3500	—	100–200	—
Chlorophyll	ppm	0–1	1.0–1.5	5–50	0.3–1.8
Phytosterols	%	0.25–0.50	0.33	0.7–1.8	0.04–0.08
Tocopherols	ppm	440–1520	600–3370	700–1200	150–870
Tocotrienols	ppm	—	—	—	530–740

[a]Warner et al. (2002).

[b]Wang (2008).

[c]Przybylski and Eskin (2011).

[d]Zou et al. (2012).

hulls prior to extraction decreases the wax content of the crude oil. The wax content of the seeds also varies among cultivars.

Degumming

Crude sunflower seed oil contains several phosphatides (phospholipids consisting of phosphatidyl esters and salts) that have to be removed from the oil during the refining operation. The chemical structures of phosphatides are shown in Figure 9.1. The arrows in this figure also indicate which bond is hydrolyzed by the catalysis of different phospholipases. So the A1 arrow pointing at the bond of the ester link of the R1 acyl moiety indicates that phospholipase A1 (PLA) will catalyze the hydrolysis of the acyl ester bond at the *sn*-1 position of the phospholipid. This hydrolysis results in the formation of an FFA and a lysophosphatide.

Formerly, "lecithin" was the trivial name of phosphatidylcholine (PC), but now it tends to refer to the product that is obtained by drying the gums that result from the water degumming process. Lecithin is used as an emulsifier in margarine production; it lowers the viscosity of molten chocolate and thus saves on cocoa butter, and it has a whole range of other, often totally different applications. As shown in Table 9.A, crude soybean oil has a fairly high phosphatide content. It is water degummed on a large scale to prevent deposits from forming during transport and storage of the crude oil, so soybean lecithin was the standard product. This has changed, at least in Europe, where a scare of genetically modified organisms (GMOs) is being orchestrated by a whole range of interested parties and soybean lecithin is gradually being replaced by non-GMO sunflower lecithin.

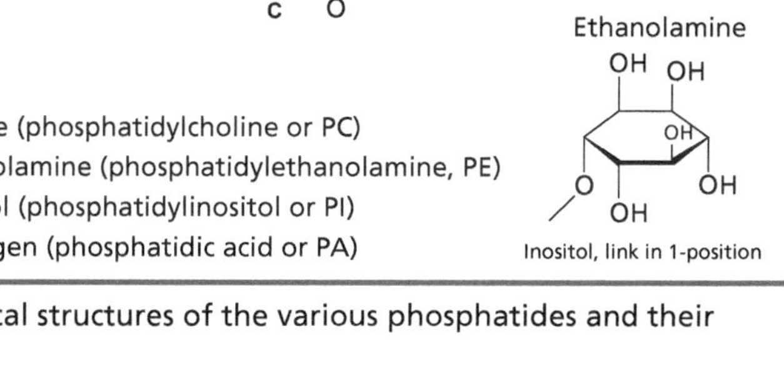

Figure 9.1 Chemical structures of the various phosphatides and their substituents.

Table 9.B Phosphatide Compositions (wt%) of Several Vegetable Lecithins

Phosphatide	Soybean	Sunflower	Rapeseed
Phosphatidylcholine	28.3	39.9	34.4
Phosphatidylethanolamine	30.4	17.8	19.2
Phosphatidylinositol	21.4	30.6	22.2
Phosphatidic acid	7.1	7.2	7.5
Other/lyso-compounds	12.8	4.5	16.7

Source: Data from Doert et al. (2012).

As would be expected, sunflower lecithin contains the same phosphatides as soybean lecithin. It has a higher PC content and a lower phosphatidylethanolamine (PE) content than the other vegetable oil lecithins listed in Table 9.B; this can be considered a performance advantage. Moreover, the "other and lyso-compounds" are quite low in sunflower seed lecithin, which may well be reassuring. Their compositions are shown in Table 9.B, and the fatty acids present in the lecithins are shown in Table 9.C.

Doert et al. (2012) made use of an internal standard when determining the phosphatide compositions of the oils, and we can assume that their data are close to the actual phosphatide contents. However, the best way of measuring the phosphatide content and phosphatide composition of an oil is by means of ^{31}P NMR while using

Table 9.C Fatty Acid Compositions (wt%) of the Individual Phosphatides of Several Vegetable Oils

		Fatty Acid				
Oil Type	Phosphatide	16:0	18:0	18:1	18:2	18:3
Canola	PC	7.12	1.12	56.74	31.54	4.18
	PE	11.43	2.12	42.10	42.23	4.17
	PI	19.44	2.16	31.21	28.98	7.21
Soybean	PC	14.80	3.75	12.29	62.31	6.91
	PE	19.12	3.16	8.64	63.82	5.34
	PI	27.73	6.33	10.62	49.15	6.17
Sunflower seed	PC	10.15	3.43	15.14	71.50	—
	PE	15.74	4.78	11.37	68.64	—
	PI	29.31	7.82	6.29	56.65	—

Source: Smiles et al. (1988).

PC: Phosphatidylcholine; PE: Phosphatidylethanolamine ; PI: Phosphatidylinositol.

triphenylphosphate as the internal standard. Using this NMR method shows that the phosphatides only partially account for the total amount of phosphorus in the sample. Apparently, there are other compounds in addition to phosphatides that contain phosphorus. Salts of phytic acid (inositol hexaphosphate) have been mentioned in this context (De Greyt, 2013).

Hydratability of Phosphatides

There is some controversy with respect to the hydration of phosphatides. Some people quote Sen Gupta (1986), who reported that all phosphatides are hydratable but only differ in their rate of hydration, whereas others prefer a thermodynamic rather than a kinetic approach and look for factors that cause hydratability. I belong to the latter school of thought since the thermodynamic approach provides explanations for all observations of which I am aware. This approach also obviates the need to speculate about encapsulation of slowly hydrating phosphatides by more rapidly hydrating phosphatides (Segers and Van de Sande, 1990).

In short, the hypothesis underlying the thermodynamic approach states that phosphatides are hydratable, or in other words, move into the water phase when the oil is contacted with water, provided that:

- They have a net electrical charge formed by the dissociation of the phosphate group;
- their electrical charges are some distance apart, as in PC, where the three methyl groups prevent the ammonium atom from moving close to the dissociated phosphate group; or
- they have hydrophilic groups like the hydroxyl groups in Phosphatidylinositol (PI).

In Table 9.D, the net charges of the various phosphatides have been listed as a function of the pH of the water phase. In this table, the + sign indicates that almost all molecules of the phosphatide concerned will have a positive charge. In the case of PC, this happens only at very low pH. If the + sign is in parentheses, this means that not all molecules of that phosphatide have a positive charge because some molecules have lost their positive charge. This shift is reinforced at a slightly higher pH when the ± sign is in parentheses, indicating that not all molecules of the phosphatide have both a positive and a negative charge.

As shown in Table 9.D, PE has both a positive charge and a negative charge at around neutral pH. The positive charge is located on the amino group and the negative charge results from the dissociation of the phosphate group. Absence of methyl groups attached to the amino group means that the doubly charged phosphatide can form an internal salt (zwitterion) in which the positive and negative charges are close

Table 9.D Net Charges of Phosphatides as a Function of pH

pH	PC	PE	PI	PA	Calcium Salt of PA
2	+	+	0	0	0
3	(+)	(+)	(0)	(0)	0
4	(±)	(±)	(–)	(–)	0
5–7	±	±	–	–	0
8–9	±	±	–	(2–)	0
>10	±	–	–	2–	0, –, or 2–

PC: Phosphatidylcholine; PE: Phosphatidylethanolamine ; PI: Phosphatidylinositol; PA: Phosphatidic acid

together; this "salt" is therefore poorly hydratable. Consequently, PE is a component of the lecithin obtained by water degumming and is also present in the phosphatides left in the water degummed oil (Dijkstra and Van Opstal, 1989). The poor hydratability of PE also explains why the phosphorus content of water degummed oil may decrease somewhat by a second water degumming treatment.

Therefore, if it turns out to be difficult to reach the specification for residual phosphorus by degumming with deionized water, the water used for degumming is acidified. This reverts the dissociation of the phosphate group and thus leaves a net positive charge on the PE ion, causing hydration. Hydratability is just a matter of charge or hydrophilicity.

PC is hydratable because its quaternary ammonium always has a pH independent net positive charge that is sufficiently far away from the phosphate groups to prevent formation of an internal salt. Phosphatidic acid (PA) is hydratable and is therefore a lecithin component but when combined with divalent cations such as calcium or magnesium, the phosphate group is neutral and the PA salts of these divalent ions are not hydratable.

Water Degumming or Enzymatic PLC Degumming

Figure 9.2 shows the various routes that can be followed from crude oil toward fully refined oil. Oil refining includes the water degumming process, which was formerly only used to prepare the crude sunflower seed oil for sale to refiners. Crude oil delivered to a refiner has to meet a residual phosphorus specification of 200 ppm maximum. This specification ensures that little deposit will be formed during transport and storage and sets an upper limit for the refining losses. The gums resulting from the water degumming process were invariably mixed with the meal and their dry matter was thereby sold at meal value.

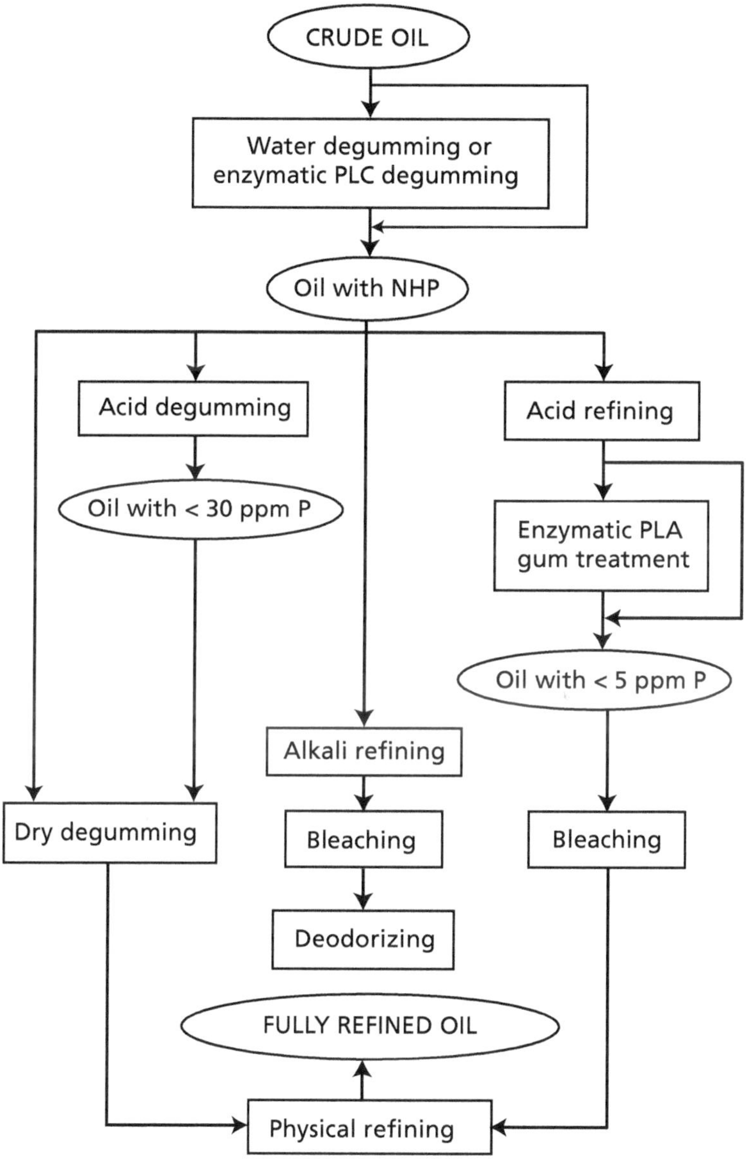

Figure 9.2 Process flow diagram showing various routes from crude oil to fully refined oil.

Nowadays, there is a market for sunflower lecithin as a non-GMO food additive. This means that similar measures to those used in the past for crude soybean oil used for lecithin production must now be taken for crude sunflower oil. These measures may entail miscella or oil filtration to produce a transparent lecithin. They may also entail a preference for crude oil from dehulled sunflower seeds, since this will be less colored and thus require less bleaching.

The water degumming process for crude sunflower seed oil is quite standard. If necessary, the oil is heated to 60–80 °C and an amount of water that is about equal to the amount of lecithin (P × 31) is added and mixed into the oil. (In practice, refiners usually add 3–5% by weight.) This mixing need not be as intensive as acid degumming, which will be discussed later. The mixture is held for 20–30 minutes, but neither the temperature nor the holding period is very critical. The gums are separated from the water degummed oil by using a centrifugal separator.

Sunflower lecithin can be produced by drying the gums in an agitated film evaporator (Erickson, 1995a). To improve its color, the lecithin can be bleached with hydrogen peroxide. Its triacylglyceride oil content will be around 35% (the same as for soybean lecithin) and when the gums from which this lecithin resulted are mixed with the meal, this oil is lost. According to Morrison (1981), about half the 0.5–10% of phospholipids present in crude sunflower seed oil are removed by the water degumming process; however, according to Table 9.B, almost 90% of the phosphatides present in crude sunflower seed oil are hydratable. In practice, the refiner will aim to meet the specification of less than 200 ppm P.

It is important to know this percentage when estimating the benefits that might result from the use of phospholipase C (PLC) (Hitchman, 2009). The benefits of degumming with PLC are twofold. As indicated in Figure 9.1, PLC catalyzes the hydrolysis of the bond between the phosphate group and the glycerol moiety in phosphatides. The PLC that is commercially available (Purifine® PLC) is specific for the hydrolysis of PC and PE, but according to Sein et al. (2013), a PI-specific PLC will be introduced in 2014. The products resulting from the reaction catalyzed by these PLC enzymes are diacylglycerols (DAGs) and a water-soluble phosphate ester. The DAGs move into the oil phase and contribute to oil yield and the phosphate esters hardly retain any triacylglyceride oil; besides, the enzymes decrease the amount of gums, which also lowers oil retention.

Now there are two ways of at least partially recouping this oil loss, the amount of which is not yet accurately known. The enzyme or an enzyme blend can be added to the gums that result from water degumming (Kellens and De Greyt, 2010) or the enzyme solution can be mixed into the degumming water. In both cases, care must be taken to maintain the pH at near neutral and the temperature at around 60 °C. For soybean oil, the use of PLC has been shown to be profitable (De Greyt, 2013) but whether the yield increase of sunflower seed oil justifies the enzyme and other operational costs remains to be seen.

Acid Degumming

The acid degumming process, of which the Unilever SuperDegumming process (Segers, 1982) is the best known, is an early effort to prepare oils for physical refining. In this process, a strong aqueous solution of a degumming acid is mixed with the crude oil and allowed to react with the nonhydratable phosphatides (NHP). This reaction liberates PA, but the subsequent addition of water to hydrate the liberated acid is insufficient (Gunstone et al., 2007) for the PA to be sufficiently dissociated to move into the water phase. Table 9.D shows that the pH of the water phase must be raised more than can be attained by just diluting a strong acid with water. The result of the water addition is only a partial removal of PA; a residual phosphorus content of less than 30 ppm is all that can be guaranteed. Accordingly, the acid degummed oil has to be dry degummed before it can be physically refined (Van Duijn and Den Dekker, 2013). This is also the procedure described by Mag (1983) for canola oil.

On the other hand, the SuperDegumming process has an advantage for sunflower seed oil because the process comprises a cooling stage (below 40 °C) that can be extended to lower temperatures to promote wax crystallization. Consequently, some of the waxes present in the crude oil will crystallize because of the presence of phospholipids with emulsifying properties. The wax crystals will move into the water phase and be removed with the gums by a centrifuge, the capacity of which is much reduced when the oil viscosity increases. The acid degummed oil still has to be treated in a separate dewaxing process. In summary, the acid degumming process can be regarded as rather obsolete now that more effective alternatives have become available.

Dry Degumming

The invention of the dry degumming process precedes the development of the acid refining process, since Sullivan obtained his patent in 1978. The patent is remarkable in that the inventor discloses a process that is eminently suitable for palm oil but only applied in the United States, where palm oil is hardly refined. In this patent, Sullivan discloses the dry degumming process, in which crude oil is treated with a degumming acid to decompose the NHP and, subsequently, bleaching earth is mixed into the oil/acid mixture to adsorb the phosphatides, including the liberated PA and the excess acid in addition to coloring compounds. The spent earth is removed by filtration and the filtrate is physically refined to yield a fully refined oil, which is referred to as an RBD oil (refined, bleached, deodorized).

The refining route employing the dry degumming process has the advantage of consisting of only two steps: dry degumming and physical refining. That is far less than the former route consisting of water degumming, neutralization, bleaching, and deodorization. However, the amount of bleaching earth that the dry degumming process requires to arrive at a sufficiently low residual phosphorus level in the degummed oil is more or less proportional to the phosphatide content of the oil before degumming. Consequently, oils with a somewhat elevated phosphorus content require so much bleaching earth as to make the process uneconomical. Accordingly, the dry degumming process, with its subsequent physical refining process, is limited to oils that have a low phosphatide content. Examples are acid degummed oil, palm oil (Gibon, 2012), lauric oils like palm kernel oil and coconut oil, and animal fats like edible tallow and lard.

Acid Refining

Figure 9.2 represents the latest evolution from previous figures (Gunstone et al., 2007). It now highlights the phosphorus contents of several intermediate products. The acid refining process is not limited to water degummed oils (Segers and Van de Sande, 1990), and the enzymatic treatment with phospholipase A has to be preceded by an acid refining process (Dijkstra, 2010). The acid refining process starts with heating the oil to 90–100 °C. A very fine dispersion of the acid used for degumming (phosphoric, citric, or oxalic acid) into the NHP-containing oil is critical for good oil–acid contact. The degumming acid has to be stronger than phosphatidic acid so that it will liberate this PA from its calcium, magnesium, and iron salts. In addition, the acid has to bind these metal ions so strongly that when the PA becomes an anion by its subsequent partial neutralization, these metal cations are no longer available for recombination with the PA anions. The original patent disclosing the degumming acid treatment (Dijkstra and Van Opstal, 1987) allowed a contact time of not more than 5 minutes (claim 11) but, subsequently, authors like Cowan (2013), who describe the acid refining process as part of an enzymatic degumming process, state that a contact time of about 20 minutes is required.

The gums that result from the acid refining process are rich in PA (Dijkstra and Van Opstal, 1987). They are quite viscous, and this viscosity depends strongly on the pH. Consequently, careful control of both the degumming acid and the caustic streams is crucial for a steady operation of the centrifugal separator used to remove the gums from the oil. In fact, a later patent (Van Opstal et al., 1990) that describes the TOP process discloses how to ensure a minimal oil loss in the gums by operating two separators in series and recycling the water/oil mixture from the second one to the feed of the first one.

Enzymatic Gum Treatment

Oil loss by entrainment with the gums has become less of a concern now that the enzymatic hydrolysis of the gums using phospholipase A1 or A2 can be profitably used to reduce their oil retention. Lysophosphatides and especially glycerol phosphates retain far less oil than the phosphatides originally present in the oil (Cowan, 2013).

Optimal enzyme activity requires a temperature around 55 °C and a pH of 4.8–5.5. This means that the gums resulting from the acid refined oil, or the acid refined oil itself, have to be cooled before the enzyme is mixed into the gums or oil.

Because the enzymatic gum treatment reduces oil loss and alleviates the problem of what to do with the gums that result from the acid refining process, several inventors have zeroed in on this field. Kellens and De Greyt (2010) disclosed the enzymatic treatment of the gums, and Dayton et al. (2008) disclose the simultaneous use of enzymes with PLA and PLC, whereby they also claim that the PLA activity increases because of the presence of the PLC.

However, when experimental data by Dijkstra in 2010 are studied in detail, they clearly show that all experiments leading to a low residual phosphorus content started with an acid refining treatment, and also that experiments that did not start with an acid refining treatment invariably lead to a high residual phosphorus content. However, there is a third category of experiments that start with an acid refining treatment but nevertheless show a high residual phosphorus content. The existence of this last category can only be explained by assuming that the acid refining treatment has not been properly executed. Accordingly, experiments belonging to this last category may not be used to demonstrate that the presence of PLC increases the rate of the hydrolysis that is catalyzed by PLA.

Gum Disposal

What to do with the gums that result as byproducts of a degumming treatment has occasionally been a problem. It was no problem for the oil miller producing and selling lecithin, it was not much of a problem for the oil miller who could incorporate it in the meal—this is allowed provided the agricultural origin of the gums corresponds to that of the meal—and it was no problem for a refinery that also had to treat a soapstock resulting from alkali refining, but it was a problem for the standalone refinery without an outlet for soapstock. More recently, it also was a problem for a biodiesel plant that purchased a crude vegetable oil other than sunflower seed oil that it had to degum one way or another before transesterifying the degummed oil.

Although established utilization or disposal routes may remain attractive, the use of enzymes to recuperate glyceride oil and/or free fatty acids deserves serious consideration. As previously mentioned, the enzymatic treatment can be carried out while the gums to be hydrolyzed are still in the degummed oil or after they have been isolated from the degummed oil (Kellens and De Greyt, 2010). The first possibility has the advantage that it saves a separation step, but the disadvantage that treating the oil requires large tanks for a long period of time. Treating the gums has the additional advantage that existing degumming lines do not require modification (Cowan, 2013). Moreover, the gum separation becomes less critical because oil lost in the gums is recovered anyway. By separating degumming (acid refining) from enzymatic oil recovery, these processes can be optimized individually.

Alkali Refining

Alkali refining is also referred to as caustic refining or chemical refining to distinguish it from the physical refining process. Figure 9.2 shows that the "oil with NHP," which may be water degummed or not, can follow four different routes. The three degumming routes have already been discussed in the previous sections, so now it is time to look at the remaining, fourth route: the alkali refining process. However, before discussing the technical aspects of the alkali refining process, some attention must be given to the choice between physical refining or chemical/alkali refining. Which process is preferred?

There is no simple answer to this question because the answer has to take a large number of factors into account, such as the disposal of byproducts and the possible revenue resulting from their sale; the capital expenditure of a new plant or the modification of an existing one; the degree of flexibility in operation that is deemed to be desirable; quality aspects of the fully refined oil; potential synergies with other processes, such as the partial wax removal during low temperature acid refining (Segers and Van de Sande, 1990); utilities; and many other, often local, factors. However, the main aspects to be taken into account are the oil yield, which is higher for the physical refining process, especially for high FFA oils, and the disposal of waste, which can be the gums resulting from acid degumming and a standalone refinery or the aqueous effluent from soapstock acidulation (Dijkstra, 2013b).

Chemical Neutralization

The oil to be chemically neutralized still contains NHP, so the chemical neutralization process should remove these phosphatides, especially since some of them can contain iron, a known oxidation catalyst (Cleenewerck and Dijkstra, 1992). Although no specific analytical evidence has been reported that this iron is present in the form of iron phosphatidate, there is ample circumstantial evidence that some of the iron present in the oil is indeed in the form of a phosphatidate (Dijkstra and Van Opstal, 1989). Accordingly, as far as removal is concerned, iron can be treated just like calcium and magnesium.

The removal of iron requires treatment of the oil with a strong acid to dislodge the PA from its salts, whereby this acid has to be well-dispersed in the oil. However, in the United States, a different opinion prevails:

> The foregoing discussion suggests that a phosphoric acid treatment is mandatory for removal of NHPs. But a traditional long mix system used on crude oil will remove sufficient phosphatides without any acid pretreatment. However,

many U.S. refiners routinely use an acid pretreat on crude soybean oil to ensure phosphatide removal and to take advantage of a popular theory that the caustic/oil emulsions formed with acid-pretreated oils are less stable and give a better "break." The most convenient method for acid pretreatment of crude oil is addition to the day tank (Erickson, 1995b).

If not finely dispersed in the oil, the acid will not displace any divalent metal ion from NHPs. This raises a most intriguing question: What mechanism ensures removal of NHP in the Long-Mix neutralization process? In this context, some observations by Hvolby (1971) shown in Table 9.E are highly relevant.

Apparently, a pH of greater than or equal to 14 ensures that all NHPs move into the aqueous phase, but at a pH less than or equal to 13, they do this only partially. Accordingly, there must be some kind of equilibrium involving hydroxyl ions and NHPs as a result of which the latter become hydratable; no mechanism explaining this hydratability has been published.

Long-Mix Alkali Refining Process

The Long-Mix alkali refining process is predominantly used in the United States. According to O'Brien (2008), "refining practices in Europe differ from those in the United States mainly because of the need to process all types of oil and typically poor-quality oils." In its most simple form, the oil to be refined according to the Long-Mix process is mixed with caustic soda (10–13% weight/volume; 14–18 °Bé) at ambient temperature and allowed to react for a period of up to 10 minutes. Static mixers can be used to distribute the lye in the oil. Then the emulsion is heated to an oil-dependent temperature to obtain a "break" and separated by a centrifuge. The temperature can

Table 9.E Residual Trace Elements as a Function of pH

	Phosphorus		Magnesium		Calcium		
pH	(mmole/kg)	ppm	(mmole/kg)	ppm	(mmole/kg)	ppm	(Mg + Ca) /P
8	4.18	130	1.25	30	2.08	83	0.80
9	3.83	119	1.33	32	2.16	86	0.91
10	2.55	79	1.02	25	1.54	62	1.00
11	2.00	62	0.79	19	1.37	55	1.08
12	1.83	57	0.71	17	1.07	43	0.97
13	1.11	34	0.42	10	0.66	26	0.97
14	0.00	0	0.00	0	0.00	0	—

Source: Adapted from Hvolby (1971).

be as low as 55 °C or as high as 90 °C. As previously mentioned, phosphoric acid can also be added or dispersed in the oil as a pretreatment.

The oil is washed with 5–10% soft water to remove residual soaps. O'Brien (2008) mentions a range of 10–20% for soybean oil. Washing can be done in two stages, whereby the wash water leaving the second separator is used to wash the oil entering the first separator. Another soap removal method employs silica hydrogel in the so-called modified caustic refining process (Welsh and Bogdanor, 1993); for soybean oil, this is more common than water washing.

The oil may then be dried if it is to be stored as neutral oil, but when the neutral oil is going to be bleached almost immediately, there is no need to dry the oil.

Short-Mix Alkali Refining Process

In the Short-Mix process, the use of a degumming acid is a standard practice. Acid is added and finely dispersed in the oil by using a knife mixer after the oil has been heated to 95 °C. The standard degumming acid is phosphoric acid, and it is used at a level of about 0.1% by volume or less. It is not usual to dilute the acid. The lye used in the Short-Mix process is stronger at 15–22% (weight/volume), but the excess is smaller than that in the Long-Mix process. There is no holding period and the mixture is immediately passed to a centrifugal separator. The oil is washed with water or treated with silica hydrogel to remove residual soap traces.

Next Generation Alkali Refining Process

As pointed out by De Greyt (2013), cavitation flow-through reactors can be used to achieve a fine dispersion of reagents (Gordon et al., 2010). Although the examples in the latter patent only refer to degumming steps, the same Nano Reactor® equipment can also be used in the alkali refining process. Its improved mixing performance allows of a reduction in phosphoric acid usage (up to 90%) and caustic usage (up to 30%) and, according to De Greyt, the resulting lower salt formation also causes a better centrifugal separation of the soapstock and the oil, resulting in an overall increase in neutral oil yield of at least 0.2%. Compare this statement with the "popular theory that the caustic/oil emulsions formed with acid-pretreated oils are less stable and give a better 'break'" from Erickson (1995b). The process has already been adopted in an Argentinian refinery processing sunflower seed oil.

Bleaching

The bleaching process for edible oils aims to remove not only colored compounds from the oil, but also the last traces of particulate matter, residual phosphatides, and soaps, because the latter would cause the oil to foam during the subsequent vacuum stripping process. Accordingly, the bleaching process is carried out after the degumming and neutralization processes and prior to the deodorization or physical refining processes. In general, it is carried out before the dewaxing process, but when dewaxing is part of the neutralization process, as discussed later, and no further dewaxing is necessary, bleaching is followed immediately by vacuum stripping. Otherwise, a separate dewaxing step may be inserted between the bleaching and vacuum stripping steps, whereby this dewaxing step should preferably include heat recovery stages in which outgoing oil is used to precool incoming oil.

The industrial bleaching process has been described as nothing but a scaled-up laboratory experiment that uses a three-necked flask fitted with an agitator, a thermometer, and a connection to vacuum; its contents can be heated and, after the reaction, they are poured into a Buchner filter to remove the spent earth from the oil. The only manner in which the standard industrial bleaching process differs from the above laboratory experiments is that the filter cake may be blown with steam to recover some of the oil entrained in the cake.

Crude sunflower seed oil contains far less chlorophyll than canola and far less beta-carotene than palm oil, so it does not require much bleaching. Depending on the type of bleaching earth used, which will often depend on local availability, an amount of 0.1–0.4 wt% will usually suffice (De Greyt and Kellens, 2000). This means that the savings on bleaching earth that might result from using a counter-current process, such as Öhmi-bleach (Transfeld et al., 1998), is hardly worth the effort. It is not even clear whether the staggered filtration system (Jalalpoor, 2008) is a good idea for sunflower oil. In the latter filtration system, nonbleached oil passes through a filter containing spent earth, and this filter is prevented from clogging by pretreating the nonbleached oil with silica hydrogel. The idea is elegant, but it remains to be seen if it is worthwhile for sunflower seed oil.

When sunflower seeds have been dried by flue gases, there is the possibility that some polyaromatic hydrocarbons (PAH) are adsorbed by the seeds. These PAH will dissolve in the oil on expelling/extraction and thus be present in the crude oil. The way to remove the high molecular weight PAH is by treating the oil with activated carbon. Low molecular weight PAH will be stripped out of the oil during deodorization or physical refining. Monitoring crude oil for PAH is therefore recommended, especially if the origin of the oil is not known or in doubt. The PAH content and past performance data determine how much activated carbon will be needed.

In fact, the main comment to be made on bleaching sunflower seed oil is that the spent earth containing sunflower seed oil may ignite spontaneously. Sunflower seed oil is highly unsaturated and if the bleaching earth has been acid activated, it will catalyze exothermic reactions that cause the temperature of the spent earth to rise so that the earth, where it is exposed to the air, will start to burn. The way to prevent this spontaneous ignition is by spraying the spent earth with water.

In vertically integrated oil mills, spent earth disposal is straightforward. The spent earth can be mixed with the meal and sold at meal value or with expeller cake going into the extractor, so that the oil can be extracted from the spent earth. The latter has an oil yield advantage, but care has to be taken to minimize oil deterioration resulting from the catalytic activity of the bleaching earth.

To summarize, the contact time between the oil and the earth should be kept as short as possible, the temperature should be kept low, and the earth should be diluted by mixing the spent earth with crude oil. If the spent earth is to be extracted, there is no need to blow the filter cake with steam to minimize its oil content. The cake can be dropped, mixed with crude oil, and pumped to the extractor.

Dewaxing

The usual way to remove wax crystals from oil is by filtration, but some of the waxes can also be removed with the gums or the soaps. This makes sense when the wax content of the oil is high, since removing all waxes by filtration would require a large amount of filter aid. The use of filter aid entails oil loss in the filter cake and may also lead to filter cake disposal costs.

Sometimes the filter cake can be sold to the feed industry for inclusion in chicken feed. Chickens need energy in the form of oil, and they do not mind filter aid like diatomaceous earth. However, that does not imply that a local feed manufacturer is going to pay a realistic price to an oil refiner whose only alternative is to pay for filter cake disposal through a landfill or face transport costs to a feed manufacturer who might be willing to pay a bit more but is located much farther away.

In a vertically integrated oil mill with downstream refinery, the filter cake can be mixed into the meal because they both have the same agricultural origin, just as sunflower oil gums can be mixed with sunflower seed meal. Whereas spent earth can be sent to the extractor, filter cake with waxes should not be sent to the extractor because the waxes will redissolve there. Instead, it is possible in some instances to wash the filter cake with hot oil, dissolve the waxes, and use the oil with waxes in oil applications where waxes are tolerated, such as margarine. Then the cleaned filter aid can also be reused, at least for some time.

Wax Composition

Waxes occur naturally in sunflower seeds and reside mostly on the outside of the seed hulls. The wax content of crude sunflower seed oil can therefore be lowered considerably by washing the seeds with hexane prior to oil extraction (Morrison, 1982). These waxes consist mainly of esters of fatty acids and alcohols. Most of them have an even number of carbon atoms but small amounts of odd-numbered chains have also been detected as well. Wax compositions reported in the literature vary widely because some authors (Gracian and Arévalo, 1981; Sindhu Kanya et al., 2004) analyze the wax crystals as obtained by the dewaxing process, whereas others (Carelli et al., 2002) separate all the waxes from the oil by chromatography on a hydrated silica column, whereby they collect the first fraction because wax esters are less polar than triacylglycerides.

These composition differences are most pronounced in the composition of the fatty acids. As shown in Table 9.F, only saturated fatty acids are present in the crystalline waxes, but when the total wax content of the oil is analyzed, several monounsaturated fatty acids, linoleic acid, and some linolenic acid are also found. The fatty alcohols contain only traces of unsaturated chains. This means that waxes containing unsaturated fatty acids do not crystallize during dewaxing. The crystalline waxes that result from the dewaxing process have a melting point of 75–76 °C after purification by recrystallization (Guillaumin and Drouhin, 1966).

Several authors (Sindhu Kanya et al., 2004; Soomro and Sherazi, 2013) have analyzed the waxes by first saponifying them and then esterifying the saponification product with acetic acid and esterifying it with methanol. This yields fatty acid methyl esters with an odd number of carbon atoms and fatty alcohol acetates with an even number. If no odd-numbered chain was present, the alcohols and fatty acids would be separated based on their carbon number by GLC. However, from previous work (Carelli et al., 2002) it is known that some odd-numbered chains are present, so these affect the analytical results.

In 1981, Gracian et al. also compared their own analytical data with results reported in the literature, and they reported huge discrepancies between the various data. However, there seems to be some kind of consensus that the main fatty acids in the crystalline waxes are 20:0, 22:0, and 24:0 and that the main alcohols are 24:0, 26:0, and 22:0. This is to some extent confirmed by the carbon number composition of the waxes themselves. They show a range of 38 to 52 for the crystalline waxes and a somewhat lower range with peaks of around 44 for the waxes that also contain unsaturated waxes.

Wax Crystallization

The wax crystallization process can be carried out in batch or continuous mode, whereby the continuous process has the advantage that it can profit from heat recovery. In the continuous process, the oil to be dewaxed is fed continuously to a crystallizer. This tends to be a tall vessel with compartments and is fitted with cooling coils and an agitator. Oil is fed into the top compartment and flows down from one compartment to the other while being cooled. Finally, the oil leaves the crystallizer to be

Table 9.F Fatty Acid and Fatty Alcohol Compositions of Sunflower Seed Wax According to Various Authors

Fatty Acid	Carelli et al. (2002)[a]		Martini et al. (2008)[b]		Sindhu Kanya et al. (2004)		Gracian and Arévalo (1981)	
	Acid[c]	Alc.	*Acid*	Alc.	*Acid*	Alc.	*Acid*	Alc.
14:0	*1.4*		*1.2*				*1.9*	
15:0							*0.1*	
16:0	*9.8*	4.3	*9.4*		*2.6*		*3.5*	0.3
16:1	*1.2*		*0.8*				*0.1*	
17:0			*1.1*				*0.2*	
18:0	*4.9*	23.1	*6.7*	1.8	*2.3*	0.2	*3.5*	0.4
18:1	*18.6*		*34.6*					
18:2	*44.0*		*4.5*					
18:3	*1.3*							
19:0		18.4					*0.2*	
20:0	*4.3*	2.0	*11.6*	1.4	*46.7*	2.3	*50.9*	1.3
20:1	*1.0*		*0.5*					
20:2			*6.4*					
21:0	*0.2*			0.2			*1.6*	0.1
22:0	*9.7*	7.5	*6.0*	9.0	*25.8*	15.6	*24.9*	8.1
22:1	*0.7*		*1.2*					
23:0		0.6	*1.0*	0.9			*0.9*	0.5
24:0	*0.1*	11.9	*1.7*	27.6	*7.5*	34.8	*6.6*	31.6
24:1			*0.4*					
25:0		1.8	*0.8*	2.3			*0.3*	3.9
26:0	*0.9*	9.3	*0.7*	26.9	*6.6*	26.8	*2.6*	33.7
27:0	*0.2*	0.5	*1.1*	0.9				1.1
28:0	*1.0*	8.3	*2.6*	14.5	*6.8*	9.2	*2.2*	14.1
29:0	*0.3*	0.6	*3.8*	0.8				0.5
30:0	*0.4*	7.9	*3.4*	7.7	*3.4*	3.9		4.3
32:0		3.8		5.4		2.2		

[a]Carelli et al. analyzed the total wax present in the oil.

[b]Martini et al. isolated the waxes from tank settlings; could originate from high oleic acid sunflower seed oil.

[c]For ease of comparison, the columns reporting the acid compositions are in italics.

held in so-called maturation vessels for several (>5) hours, the idea being that small crystals will disappear by Ostwald ripening.

In theory this design looks all right, but in practice the situation is far from ideal. To save on investment, the cooling coil surface tends to be on the small side. This is compensated for by using a lower heat exchange medium temperature, but this causes wax crystals to be deposited onto the cooling coils. Consequently, the heat transfer is suppressed, which means that cooling moves to higher and higher compartments so that the temperature in the top compartment tends to drop below the cloud point of the oil to be winterized. Oil entering the crystallizer is quench-cooled and this causes many small crystals to be formed. They are difficult to filter and require an unnecessarily large amount of filter aid because the crystal maturation previously mentioned does not cause any Ostwald ripening, presumably because of the low solubility of the wax at maturation temperature.

Asbeck and Segers (1991) were aware of this phenomenon and developed a process in which the oil is cooled from a temperature well above its cloud point to a final temperature of 2–10 °C at a rate of less than 15 °C per hour and then filtered immediately. The authors found maturation to be no longer necessary. Later, it was reported (Gunstone et al., 2007) that the low cooling rate does not need to be maintained during the entire cooling process; it is only important around the cloud point of the oil to be dewaxed (see also Figure 8.1 in Dijkstra, 2013a). After the initial crystal formation, a higher cooling rate can be applied without causing small crystal developpment. Applying the above cooling profile saves time and thus requires smaller equipment, and leads to large crystals of mostly uniform size that require less filter aid.

Filter aid can be added at the start of the crystallization process. Claims have been made that filter aid acts as nucleating agent (O'Brien, 2000). However, these claims have never been substantiated, or at least no proof has ever been published. The use of a liquid nucleating agent (Jalalpoor, 2009), such as colloidal silica, has been claimed to improve the cold stability of oil, but since other processes also produce oil with good cold stability, these agents are therefore not essential.

As mentioned above, oil viscosity depends strongly on oil temperature. Wax solubility also depends on oil temperature, but raising the oil temperature from 5 °C to 15–18 °C hardly increases the wax solubility in absolute terms. Accordingly, oil that is saturated with wax at 15–18 °C is almost equally unlikely to develop cloudiness as saturated oil at a lower temperature. Therefore, it is a common practice to raise the oil temperature before separating the wax from the oil. Asbeck and Segers (1991) separated oil from wax by microfiltration using, for instance, a sintered metal filter, but using standard filters and filter aid also works.

Pressure leaf filters are commonly used with a preference for the horizontal, Niagara type of filter as illustrated in Figure 9.3.

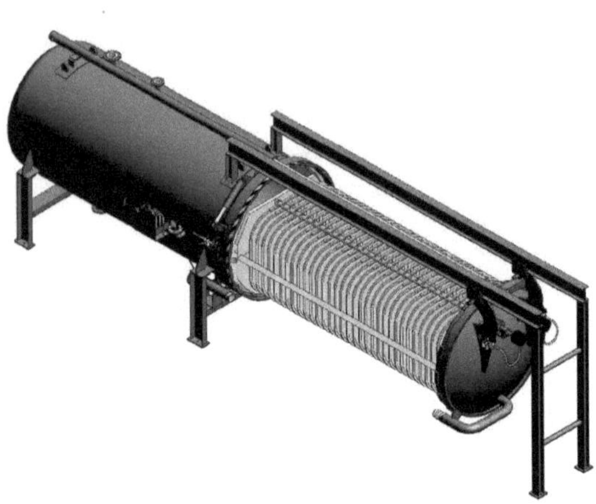

Figure 9.3 Horizontal pressure leaf filter when opened.
Source: Image courtesy of Mahle Industrial Filtration.

Neutralization Combined with Dewaxing

It has been known for a long time that wax crystals can be separated from an oil by washing the oil with an aqueous detergent solution. This has led to a patent (Gibble and Rhee, 1976) disclosing a method of dewaxing sunflower seed oil by forming an emulsion of the cold (15 °C) oil with a mixture of synthetic, organic surfactants and centrifuging and washing the oil. Soon thereafter, it was discovered that soaps could also be used as surfactants (Levine, 1977).

Accordingly, there are now several dewaxing processes comprising centrifugal separation steps. The Harburger Freudenberger Combined Degumming and Dewaxing Process (Speck, 2012) starts with crude or water degummed oil. The first step is the degumming of this oil at 70 °C. The gums that are not separated from the oil are then cooled to 25 °C in heat exchangers to start the crystallization of the wax. The crystallization is completed in stirred tanks in which the oil with gums is cooled to 6 °C over a period of 12 hours.

Then the oil is heated to 18 °C to reduce its viscosity and passed to a centrifugal separator that removes the gum phase containing the wax. Filter aid is added and the last traces of wax are removed by filtration. The degummed and dewaxed oil can then be bleached and physically refined to yield salad oil. This process has the advantage that the major part of the wax is removed by centrifuge so that the filter aid requirement and the filter capacity are reduced.

The GEA cold refining process (Zeldenrüst, 2013) is quite similar to the process previously outlined. It starts with an acid treatment at an elevated temperature and cooling of the oil, but instead of relying on the gums to entrain the wax crystals, the GEA process adds so much caustic that the oil is completely neutralized. The wax crystals are entrained by the soapstock. Then, the mixture is separated by a centrifuge at a low temperature. In this process, the oil is much more viscous than at normal degumming or neutralization temperatures. This means that a much larger separator is needed because the viscosity of sunflower oil at 18 °C is about ten times higher than at 95 °C, where it equals 8.80 (mPa.s) (Fasina et al., 2006). No filter is included in the GEA process.

It is therefore not surprising that with the so-called wet winterization process (Zeldenrüst, 2013), it is preferable to separate the gums or soapstock at an elevated temperature. Then a small amount of lye is added; the oil from which the soapstock has been removed is cooled on residual soap in the oils to concentrate the wax crystals in the aqueous phase, and then separates a less viscous, soapy water containing wax crystals at low temperature. However, as remarked by Denise (1987), centrifugal wax separation is insufficient. For clear oil with good cold stability, he writes, filtration is unavoidable.

Vacuum Stripping

The final purification step for neutralized, bleached oil is the deodorization process, and for degummed, bleached oil it is the physical refining or steam refining process. They use the same kind of equipment and process conditions, such as oil temperature, vacuum, and stripping medium, but they differ in the amount of distillate, which is fairly small in the case of deodorization but is much larger in case of the steam refining process because steam refining removes all the FFA present in the feed. The literature regularly describes the vacuum stripping process as a "steam distillation process." This is not correct because liquid water is also present in the latter and, as a result, it has one degree of freedom less than the vacuum stripping process, where both the temperature and the pressure can be freely chosen. Accordingly, the design of the scrubber, where the vapors leaving the vacuum stripping equipment are condensed, should take this difference into account.

The Physical Chemistry of Vacuum Stripping

Although the next section will list several chemical reactions that can take place during the vacuum stripping process, it is basically a physical process that is governed by the laws of Dalton and Raoult. These laws allow the amount of stripping medium to be correlated with other process parameters, an equipment parameter, and the levels of volatile impurities before and after the stripping treatment. The first such correlation was published by Bailey in 1941, but several other authors, including Deffense (1995), have pointed out that the simplification resulting from the assumption that the vapor pressure of the volatiles is quite small in comparison with the system pressure is only justified in the case of the deodorization process. For the physical refining process, this assumption can significantly overestimate the amount of steam required. Accordingly, the full Bailey equation, as derived for the batch deodorization process, reads:

$$S = \frac{PO}{EP_v} \ln \frac{V_s}{V_e} + \frac{P}{EP_v}(V_s - V_e) - (V_s - V_e) \qquad \text{(Eq. 9.1)}$$

In this equation:

S = amount of stripping steam used (mole)
P = total pressure in the system (pressure units)
O = amount of oil being processed (mole)
E = vaporization efficiency (–)
P_v = vapor pressure of pure volatile compound (same pressure units)
V_s = amount of volatiles before deodorization (mole)
V_e = amount of volatiles after deodorization (mole)

If only the first right-hand term of Eq. 9.1 is taken into account, this shows that the amount of stripping medium is proportional to the amount of oil, which is only to be expected. In the first term, the stripping medium usage is inversely proportional to the vapor pressure of the volatile compound at the stripping temperature. This means that volatile compounds with lower vapor pressure require less steam for their removal, which again is only to be expected. The stripping medium usage is also proportional to the system pressure. Because the volume of the stripping medium is inversely proportional to the pressure, for a desired stripping effect, the required stripping medium volume does not depend on the system pressure. The stripping medium volume is very large in comparison with the oil volume. When oil is deodorized at normal deodorization conditions (T = 240–260 °C; P = 5 hPa), the use of 0.5% (w/w) of steam leads to a volume that is some 4,000 times the oil volume. Another observation that can be made on reading Eq. 9.1 is that the removal of all volatiles ($V_e = 0$) requires an infinite amount of stripping medium, which means, in practice, that there will always be some volatile material left in the oil.

If oil contains a fair amount of FFA, this component will exert a significant partial pressure. It can be so high that when the oil is exposed to vacuum, some of this FFA will flash off spontaneously until the residual concentration leads to a partial vapor pressure that equals the system pressure; this flashing process requires no steam at all, which illustrates that the left-hand term of the Bailey equation is only part of the story. It requires further terms to deal with high FFA oils.

Such high FFA oils are often physically refined by allowing them to trickle down a packed column that is fed stripping medium from the bottom. For this countercurrent stripping process, I derived the following equation correlating the steam usage with process and product parameters in 1999:

$$V_e - V_e \frac{(PO)^n (P_v S - PO)}{(P_s S)^{n+1} - (PO)^{n+1}} \qquad \text{(Eq. 9.2)}$$

In this equation, the variables have the same meaning as in Eq. 9.1, and the additional variable n stands for the number of theoretical transfer units. Such a unit is defined as the deodorization column segment that ensures that the oil leaving this segment at the bottom is in physical equilibrium with the vapor leaving this segment at the top. Its actual height depends on the construction of the packed column, the residence time of the oil, and similar factors. Like the vaporization efficiency E that is part of Eq. 9.1, the variable n is an equipment parameter.

So, just as in the Bailey equation, the steam usage S is proportional to the amount of oil O being processed. To see this, just multiply O with a random number. Then the only way to ensure equality is to multiply S with the same number. Just as in the Bailey equation, the steam usage S is inversely proportional to the vapor pressure P_v of the pure compound at stripping temperature, because these factors occur as a product. In addition, just as in the Bailey equation, there will always be volatiles left in the oil that has been stripped in a counter-current column.

For both the batch and continuous systems, it follows that the larger the volume of the vapor leaving the oil, the larger the fraction of volatiles removed. In theory, this means that physical refining will remove more tocopherols than mere deodorization because the vapors leaving the oil include more FFA. Therefore, physically refined oils could run the risk of being less stable by having a lower residual antioxidant content than deodorized oils. However, in practice, the effect is marginal because of the large difference in molecular weight of water and FFA, which differ by a factor of 15. This means that even during physical refining, the volume is predominantly generated by water vapor.

The final aspect to be discussed in this section is the effect of temperature on the deodorization process. Decreasing the temperature lowers the vapor pressure of the volatiles. Consequently, more stripping medium is needed to attain the same residual fraction. But how do vapor pressures change when the temperature is changed? The answer to this question lies in Clapeyron's law:

$$\ln P = \frac{Q_{L \to G}}{RT} + C \qquad \text{(Eq. 9.3)}$$

in which:

P = vapor pressure of the pure compound
$Q_{L \to G}$ = molar heat of evaporation (J mol^{-1})
R = gas constant (8.314 J mol^{-1} K^{-1})
T = absolute temperature (K)
C = constant, the magnitude of which depends on the units used for P

If we look at the pressures P_1 and P_2 of a compound at temperatures T_1 and T_2, we can express a difference as:

$$\ln P_1 - \ln P_2 = -\frac{Q}{R}\left(\frac{1}{T_1} - \frac{T}{T_2}\right) = Q\frac{T_1 - T_2}{RT_1T_2} \quad \text{(Eq. 9.4)}$$

In other words, the ratio of the two pressures is proportional to the heat of evaporation. Because compounds that boil at a higher temperature have a higher heat of evaporation, it means that lowering the deodorization temperature has more of an effect on higher boiling tocopherols than on lower boiling fatty acids. So when we want to keep tocopherols in the oil, we should lower the deodorization temperature.

Chemical Reactions during Vacuum Stripping

During the deodorization process, the oil is heated to a temperature of 200 to 240 °C or even higher, and this temperature is maintained for a period of time (Gunstone et al., 2007). This holding period is an essential part of the process because during this period, flavor precursors are thermally broken down so that afterward, when the oil is used and heated, its agricultural origin becomes less evident. Accordingly, packed columns, in which the oil flows downward and the stripping medium flows upward in a counter-current system that ensures rapid FFA removal, have to be combined with deodorization trays that provide the oil with some residence time. Sparging steam supplied to these trays then ensures the removal of the volatile and often malodorous compounds already present in the crude oil or generated by the thermal breakdown of the flavor precursors.

Another reaction taking place during the vacuum stripping process is the so-called heat bleaching of the oil. Carotenoid coloring compounds are heat sensitive, and the oil gets lighter when heated. This is especially important for palm oil and much less so for sunflower seed oil.

There is some evidence that the steam used to strip the oil also causes some oil to be hydrolyzed, but this evidence is far from conclusive. However, it would be so easy to demonstrate. Just prepare a mixture of a neutral lauric oil and an oil containing free, long chain fatty acids, deodorize this mixture under normal industrial conditions, collect the distillate, and determine its fatty acid composition. If there is no hydrolysis, the lauric oil is not hydrolyzed and no lauric acid is liberated and removed by steam stripping. Therefore, the presence of lauric acid in the scrubber condensate would demonstrate the occurrence of hydrolysis and the amount of the lauric acid would quantify this hydrolysis. Sadly enough, the experiment has yet to be carried out.

A chemical reaction that has to be avoided or minimized is the formation of *trans* isomers. Refiners should aim at a *trans* fatty acid (TFA) content of less than 1.0% for oils like sunflower seed oil. Because linolenic acid is more prone to lead to TFA formation, soybean oil and rapeseed oil should contain less than 1.5% TFA. Since the rate of TFA formation is strongly temperature dependent, avoiding too high a temperature for too long a period of time will ensure that the TFA target is met.

Finally, the elevated temperature during the vacuum stripping process will ensure that any peroxides that have not been decomposed during the bleaching step will decompose thermally during the vacuum stripping process. Deodorized oil should have an almost negligible peroxide value.

Vacuum Stripping Process Types

The oldest type of process is the batch process. Oil is introduced into a batch deodorizer and heated. Vacuum is applied and stripping medium is sparged through the oil to remove air and later, when the deodorization temperature has been reached, the malodorous compounds are removed (the removal, as pointed out when explaining the physical chemistry of vacuum stripping, is only partial). After a set period of time, the oil is cooled, citric acid may be added to stabilize the oil, the vacuum is broken with nitrogen, and the deodorized oil is pumped to storage. The batch process hardly allows of heat recovery, which makes it expensive; it is therefore only used for small lots or specialty oils because it effectively avoids cross-contamination between batches.

In the continuous cross-flow deodorizer, the oil cascades downward from tray to tray and each tray is directly connected to the vacuum system. Successive trays perform the successive process steps of the batch process. Accordingly, the top tray may be used to deaerate the incoming oil, but in industrial practice, this treatment is normally carried out outside the deodorizer vessel. In the next tray, the oil is heated to deodorization temperature using high pressure steam, then it is deodorized in lower trays, and, in the final tray, the oil is cooled before being pumped out of the deodorizer vessel. This cooling can involve some heat exchange in a separate tray with the oil that has already been deaerated and is then preheated before final heating with high-pressure steam. Formerly, heat transfer fluids like Dowtherm® were used as liquid or in gaseous form but for presumed food safety reasons, they were abandoned in favor of high-pressure steam. A directly fired heater has been reported in the literature (Kuroda and Young, 1989) and despite its apparently satisfactory performance, it has not been adopted by industry.

The continuous deodorizer can be turned into a semi-continuous, cross-flow deodorizer by introducing a valve system that allows the tray contents to be treated for a certain period of time and only to move downward to the next tray/treatment when this has been emptied into the tray below. Incoming oil can be preheated by outgoing oil and heat exchange between trays is possible using one or two thermosiphons (De-Greyt, 2014). These measures decrease the heat demand by 45% if one thermosiphon is used and by 67% when two thermosiphons are used. This type of process is used to avoid cross-contamination between successive batches and accommodate frequent changes from one oil blend to another with a different composition. This type of equipment is suitable for a margarine plant producing a wide range of products.

A very effective type of vacuum stripping process is the counter-current process. It is fast and requires less stripping medium than the cross-flow type of process (Dijkstra, 1999). It uses a packed column with oil trickling downward and the steam moving upward. In addition to suffering from cross-contamination on type changes, it has a further drawback: If the equipment has a leak and air leaks into the vessel, this air may move through the column where it causes the oil to polymerize so that in due course (i.e., quite soon), the packing has to be replaced. Moreover, the *trans* content of the oil will have increased to currently unacceptable levels.

As previously mentioned, oil requires a holding period at elevated temperature, so the packed column has to be combined with deodorization trays. These trays are sparged with steam that is then removed from the system through the packed column. Consequently, the pressure in the trays is not as low as when they were directly connected to the vacuum system, but if the pressure drop over the column is considered to be too high, this can be remedied by a small booster pump (Kellens and de Suraÿ, 2005).

For sunflower seed oil, the last type of process will often be the preferred one. The packed column ensures rapid FFA removal with a minimum amount of steam, which also results in low steam or energy usage in the volatile condensation systems. The continuous nature of the process allows for extensive heat recovery.

Vapor Condensation

Over the years, the vapor condensation process has evolved from spraying the vapors with water in the steam ejector battery to condensing the vapors on low-temperature surface condensers. The first led to very dirty water and, besides, the water temperature determined what vacuum level could be reached. The latter uses electricity instead of steam and almost totally eliminates the aqueous effluent. There have been processes in between that condense the vapors leaving the deodorizer in scrubbers and use chilled water to reduce the motive steam usage of the condensers but current new installations often use the so-called ice-condensers as illustrated in Figure 9.4.

With such systems, vapors leaving the deodorizer are condensed in surface condensers that operate at a very low temperature (–30 °C). The low temperature maintains a low pressure in the deodorizer (<1.5 mbar). Commercially available systems consist of two or more freeze condensers with horizontally or vertically orientated straight tubes that operate alternately and switch between freezing and thawing, a refrigeration plant for the generation of the cold refrigerant, and a vessel for defrosting and cleaning of the tubes after a certain period of freezing.

In the field of vacuum stripping there are some further developments, such as the use of two scrubbers in series, one for the high boiling vapor components like tocopherols and phytosterols and the other for the FFA. These are primarily for soybean oil and palm oil; for sunflower seed oil they are not of particular interest.

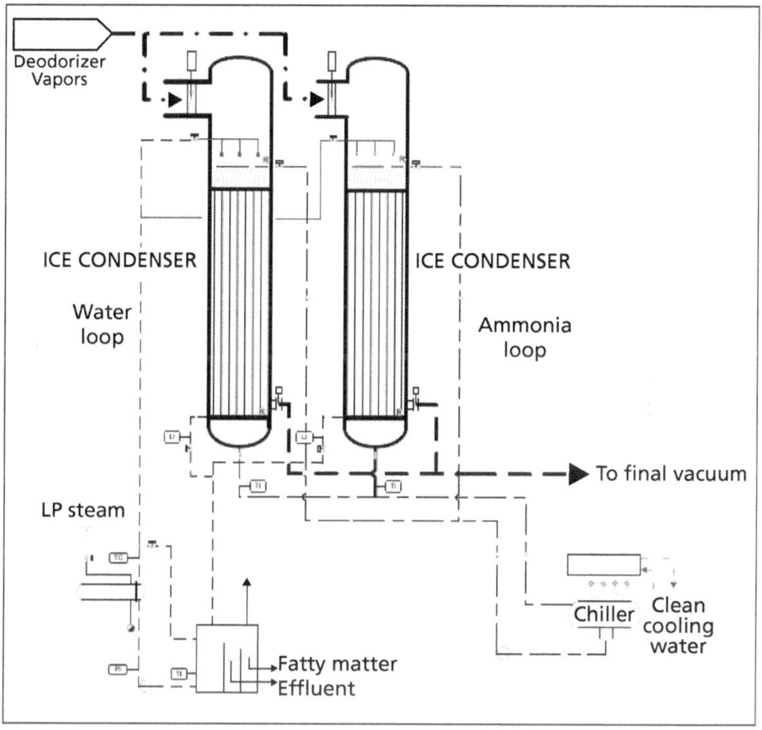

Figure 9.4 Process flow diagram of an ice-condensing vacuum system.
Source: Reprinted from De Greyt (2013), courtesy of Desmet Ballestra.

Quality Assurance

The object of quality assurance is to maintain a certain level of final product quality at minimal cost. This means that the sum of the various cost factors should be minimized. These cost factors are:

- Plant supervision and process investigation
- Product analysis
- Downgrading faulty material
- Customer complaints

How much the first two factors will cost can easily be estimated and budgeted, but budgeting customer complaints is far more difficult because, first, hopefully they don't occur too frequently and, second, their magnitude can vary considerably. If somebody buys a bottle of sunflower oil in a shop and has the impression that it developed an unpleasant off-taste rather soon, he or she may buy another brand next time but is unlikely to ask for compensation. If sunflower seed oil is sold to a customer who uses the oil to make mayonnaise and discovers that the product is unacceptably unstable because the residual wax content of the oil is too high, this customer can claim a substantial compensation. Because these complaints are so unpredictable, plant management tends to avoid them by focusing on the other three cost factors and thereby accepts that the sum is no longer minimal.

Downgrading may be necessary when sunflower oil has inadvertently been cross-contaminated with another oil. Then it can no longer be sold as sunflower seed oil, but only as vegetable oil. If it is known how much of which oil has been admixed, it may be possible to arrive at an arrangement with a customer, especially if a rebate is offered. This rebate is then the cost of downgrading. It may also be possible to reprocess the oil rather than downgrading it. Then the reprocessing cost has to be taken as this quality control cost factor.

Refiners tend to be very keen on having their products analyzed and may well spend more on analyses than necessary. If analytical certificates have been promised to a customer, the analyses listed in these certificates have to be carried out. But if the raw material met its specification and processing was perfectly normal, the final product should therefore be within specification and this should not require analytical verification.

In the long run, it is cheaper for a refiner to observe the process closely and maintain the desired process conditions so that the products meet the specifications. More emphasis on process supervision and investigation will cost in the beginning because a database of process particulars will have to be built up. Once this is available, it can be used to cut down on the product analysis so that the total cost of quality assurance is lowered.

Instead of concentrating on the analysis of the final product, intermediate analyses that verify that a process has achieved what it set out to do should be encouraged, such as:

- Residual phosphorus and iron content after degumming
- Residual FFA after neutralization
- Color after bleaching
- Cold stability after dewaxing
- Bland taste after vacuum stripping

By carrying out these intermediate analyses, they no longer have to be carried out on the final product. Besides, these analyses need not be carried out on each batch but only on batches in which the process is suspected of having deviated from what is normal and on one in every so many batches to establish a baseline and ensure that the process as a whole is not drifting away from the norm. In addition, special attention should be given to the process when a new lot of process adjuvants is being used.

Discussion

Which refining process sequence should be chosen under which circumstances? That depends on local circumstances and crude oil properties. Was the oil extracted from dehulled seeds or not or, in other words, should a centrifugal wax separation be included or not? Has the oil already been water degummed or, in other words, are there so many gums that oil recovery by using lipase enzymes should be considered?

When the oils have not been water degummed and therefore have a high phosphatide content, enzymatic hydrolysis of the gums may be justified by the resulting higher oil yield. This can be done after the gums have been separated from the oil by a centrifuge and then mixing the phospholipase with the gums in accordance with Kellens and De Greyt (2010). Or, it can be realized by cooling the oil to about 60 °C while the gums are still present, then mixing enzymes into the oil, allowing these to catalyze phosphatide hydrolysis, and heating the oil to facilitate separation. If the oil contains little wax, it can be bleached and cooled to crystallize the waxes that can then be removed by filtration and finally physically refined.

If the oil originates from non-hulled seeds and contains an appreciable amount of waxes, it is probably better to cool the oil after the enzymes have had time to cause gum hydrolysis and remove wax crystals together with the residual gums and gum hydrolysis products. Then a small amount of filter aid can be added to the oil from which most waxes have been removed; the filter aid and residual waxes can then be removed by filtration. After bleaching and physical refining, the oil should be fully refined at minimal cost. As far as I am aware, this process sequence has not been tested in industrial practice.

Table 9.G Refining Process Sequence Choices

	Low Wax Content Because of Seed Dehulling	**High Wax Content Because Seeds Were not Dehulled**
Low phosphatide content because oil has been water degummed	Acid refining, separation of the gums, bleaching, dewaxing by filtration, physical refining	Acid refining, cooling to crystallize the waxes, separation of gums and waxes, filtration, physical refining
High phosphatide content because oil has not yet been degummed	Acid refining, separation of the gums, bleaching, dewaxing by filtration, physical refining, enzymatic gum hydrolysis	Acid refining, cooling to crystallize the waxes, separation of gums and waxes, filtration, physical refining, enzymatic gum hydrolysis

References

Asbeck, L. S.; Segers, J. C. Dewaxing of Dried Oil. U.S. Patent 5,066,504, 1991.

Bailey, A. E. Steam Deodorization of Edible Fats and Oils. Theory and Practice. *Ind. Eng. Chem.* **1941,** *33,* 404–408.

Carelli, A. A.; Frizzera, L. M.; Forbito, P. R.; Crapiste, G. H. Wax Composition of Sunflower Seed Oils. *J. Am. Oil Chem. Soc.* **2002,** *79,* 763–768.

Cleenewerck, B.; Dijkstra, A. J. The Total Degumming Process: Theory and Industrial Application in Refining and Hydrogenation. *Fat Sci. Technol.* **1992,** *94,* 317–322.

Cowan, W. D. Enzyme Processing. In *Edible Oil Processing;* Hamm, W., Hamilton, R. J., Calliauw, G. H., Eds.; John Wiley & Sons: Chichester, UK, 2013; pp 197–221.

Dayton, C. L. G.; Rosswurm, E. M.; Galhardo, F. Enzymatic Degumming Utilizing a Mixture of PLA and PLC Phospholipases. PCT Patent Application WO 2008/094847, 2008.

Deffense, E. M. J. Theorie und Praxis der physikalischen Raffination. *Fette Seifen Anstrichm.* **1995,** *97,* 481–484.

De Greyt, W. F. J. Edible Oil Refining: Current and Future Technologies. In *Edible Oil Processing;* Hamm, W., Hamilton, R. J., Calliauw, G. H., Eds.; John Wiley & Sons: Chichester, UK, 2013; pp 127–151.

De Greyt, W. F. J.; Kellens, M. J. Refining Practice. In *Edible Oil Processing;* Hamm, W., Hamilton, R. J., Eds.; Sheffield Academic Press: Sheffield, UK, 2000; pp 79–128.

DeGreyt, W. F. J. Edible Oil Processing: Deodorization. AOCS Lipid Library Online (http://lipidlibrary.aocs.org/processing/deodorization/index.htm). Accessed Jan. 27, 2014.

Denise, J. Le Décirage des Huiles de Tournesol. *Rev. Franç. Corps Gras* **1987,** *34,* 133–138.

Dijkstra, A. J. Stripping Medium Requirements in Continuous Countercurrent Deodorization. *J. Am. Oil Chem. Soc.* **1999,** *76,* 989–993.

Dijkstra, A. J. Enzymatic Degumming. *Eur. J. Lipid Sci. Technol.* **2010,** *112,* 1178–1188.

Dijkstra, A. J. *Edible Oil Processing from a Patent Perspective;* Springer: New York, Heidelberg: Dordrecht, London, 2013a.

Dijkstra, A. J. The Purification of Edible Oils and Fats. *Lipid Techn.* **2013b,** *25,* 271–273.

Dijkstra, A. J.; Van Opstal, M. Process for Producing Degummed Vegetable Oils and Gums of High Phosphatidic Acid Content. U.S. Patent 4,698,185, 1987.

Dijkstra, A. J.; Van Opstal, M. The Total Degumming Process. *J. Am. Oil Chem. Soc.* **1989,** *66,* 1002–1009.

Doert, M.; Jaworska, K.; Moersel, J.-T.; Kroh, L. W. Synergistic Effect of Lecithins and Tocopherols: Lecithin-based Regeneration of α-Tocopherol. *Eur. Food Res. Technol.* **2012,** *235,* 915–928.

Duijn, G. van; Dekker, G. den. Oil Processing Design Basics. In *Edible Oil Processing;* Hamm, W., Hamilton, R. J., Calliauw, G. H., Eds.; John Wiley & Sons: Chichester, UK, 2013; pp 267–310.

Erickson, D. R. Degumming and Lecithin Processing and Utilization. In *Practical Handbook of Soybean Processing and Utilization;* Erickson, D. R., Ed. AOCS Press and United Soybean Board: Urbana, IL, and St. Louis, MO, 1995a; pp 174–183.

Erickson, D. R. Neutralization. In *Practical Handbook of Soybean Processing and Utilization;* Erickson, D. R., Ed.; AOCS Press and United Soybean Board: Urbana, IL, and St. Louis, MO, 1995b; pp 184–202.

Fasina, O. O.; Hallman, H.; Craig-Schmidt, M.; Clements, C. Predicting Temperature-dependent Viscosity of Vegetable Oils from Fatty Acid Composition. *J. Am. Oil Chem. Soc.* **2006,** *83,* 899–903.

Gibble, W. P.; Rhee, J. S. Dewaxing of Vegetable Oils. U.S. Patent 3,994,943, 1976.

Gibon, V. Palm Oil and Palm Kernel Oil Refining and Fractionation Technology. In *Palm Oil: Production, Processing, Characterization, and Uses;* Lai, O.-M., Tan, C.-P., Akoh, C. C., Eds.; AOCS Press: Urbana, IL, 2012; pp 329–375.

Gordon, R.; Gorodnitsky, I.; Grichko, V. Method for Cavitation-assisted Refining, Degumming and Dewaxing of Oil and Fat. PCT Patent Application WO 2010/098783, 2010.

Gracian, J.; Arévalo, G. Composición de las Ceras del Aceite de Girasol. *Grasas Aceites* **1981,** *32,* 1–5.

Guillaumin, R.; Drouhin, N. Composition des Cires de l'Huile de Tournesol. *Rev. Franç. Corps Gras* **1966,** *13,* 21–28.

Gunstone, F. D., Harwood, J. L., Dijkstra, A. J., Eds. *The Lipid Handbook;* Taylor & Francis: Boca Raton, FL, 2007.

Hitchman, T. Purifine® PLC: Industrial Application in Oil Degumming and Refining. *Inform* **2009,** *20,* 469–471. http://www.iomsa.org/leads/OMGsep09.pdf.

Hvolby, A. Removal of Nonhydratable Phospholipids from Soybean Oil. *J. Am. Oil Chem. Soc.* **1971,** *48,* 503–509.

Jalalpoor, M. Staggered Filtration System and Method for Using the Same for Processing Fluids such as Oils. PCT Patent Application WO 2008/025552, 2008.

Jalalpoor, M. Purification of Fatty Materials such as Oils. PCT Patent Application WO 2009/068277, 2009.

Kellens, M. J.; De Greyt, W. F. J. Oil Recuperation Process. U.S. Patent 7,696,369, 2010.

Kellens, M. J.; de Suraÿ, D. Method and Apparatus for Vacuum Stripping. European Patent 1 505 145, 2005.

Kuroda, Z.; Young, C. An Edible Oil Deodorizer with a Direct-fired Heater. *J. Am. Oil Chem. Soc.* **1989,** *66,* 1781–1783.

Levine, L. Dewaxing Process for Vegetable Oils. U.S. Patent 4,035,402, 1977.

Mag, T. K. Canola Oil Processing in Canada. *J. Am. Oil Chem. Soc.* **1983,** *60,* 380–384.

Martini, S.; Carelli, A.A.; Lee, J. Effect of Addition of Waxes on the Crystallization Behavior of Anhydrous Milk Fat. *J. Am. Oil Chem. Soc.* **2008,** *85,* 1097–1104.

Morrison, W. H., III. Sunflower Lecithin. *J. Am. Oil Chem. Soc.* **1981,** *58,* 902–903.

Morrison, W. H., III. A Rapid, Effective Method for the Reduction of Waxes in Sunflower Seed Oil. *J. Am. Oil Chem. Soc.* **1982,** *59,* 519–520.

O'Brien, R. D. Liquids Oil Technology. In *Introduction to Fats and Oils Technology;* O'Brien, R. G., Farr, W. E., Wan, P. J., Eds.; AOCS Press, Urbana, IL, 2000; pp 463–485.

O'Brien, R. D. Soybean Oil Purification. In *Soybeans, Chemistry, Production, Processing, and Utilization;* Johnson, L. A., White, P. J., Galloway, R., Eds.; AOCS Press: Urbana, IL, 2008; pp 377–408.

Przybylski, R.; Eskin, N. A. M. Oil Composition and Properties. In *Canola, Chemistry, Production, Processing, and Utilization;* Daun, J. K., Eskin, N. A. M., Hickling, D., Eds.; AOCS Press: Urbana, IL, 2011; pp 189–227.

Segers, J. C. Superdegumming, a New Degumming Process and Its Effect on the Effluent Problems of Edible Oil Refining. *Fette Seifen Anstrichm.* **1982,** *84,* 543–546.

Segers, J. C.; Sande, R. L. K. M. van de. Degumming—Theory and Practice. In *Edible Fats and Oils Processing: Basic Principles and Modern Practices;* Erickson, D. R., Ed.; AOCS Press: Urbana, IL, 1990; pp 88–93.

Sein, A.; Bueno, A.; Smits, W.; Walsh, D. *Phospholipase Composition for Optimal Enzymatic Oil Degumming: Application to Rapeseed Oil.* 11th EuroFedLipid Congress, Antalya, Turkey, 2013, p 34.

Sen Gupta, A. K. Neuere Entwicklungen auf dem Gebiet der Raffination der Speiseöle. *Fette Seifen Anstrichm.* **1986,** *88,* 79–86.

Sindhu Kanya, T. C.; Jaganmohan Roa, L.; Shamanthaka Sastry, M. C. Characterization of Wax Esters, Free Fatty Alcohols and Free Fatty Acids of Crude Wax from Sunflower Seed Oil Refineries, 2004. http://www.aseanfood.info/Articles/11017522.pdf.

Smiles, A.; Kakuda, Y.; MacDonald, B. E. Effect of Degumming Reagents on the Recovery and Nature of Lecithins from Crude Canola, Soybean and Sunflower Oils. *J. Am. Oil Chem. Soc.* **1988,** *65,* 1151–1155.

Soomro, R. K.; Sherazi, S. T. H. Extraction and Characterization of Seed Oil Waxes by Using Chromatographic Techniques. *International Journal of Industrial Chemistry* 2013, DOI:10.1186/2228-5547-4-9.

Speck, R. The Combined Degumming/Dewaxing Process. In *9th Practical Short Course: Newest Trends in Soft Processing and Downstream Technology*, Krakow, 2012.

Sullivan, F. E. *Process for Refining Fatty Oils.* U.S. Patent 4,089,880, 1978.

Transfeld, P.; Schneider, M.; Börner, G. *Method and Apparatus for Adsorptive Cleaning of Substances.* U.S. Patent 5,753,103, 1998.

Van Opstal, M.; Dewulf, F.; Cleenewerck, B.; Weber, E. *Process for the Continuous Removal of a Gum Phase from Triglyceride Oil.* U.S. Patent 4,927,544, 1990.

Wang, T. Minor Constituents and Phytochemicals of Soybeans. In *Soybeans, Chemistry, Production, Processing and Utilization;* Johnson, L. A., White, P. J., Galloway, R., Eds.; AOCS Press: Urbana, IL, 2008; pp 297–329.

Warner, K.; Vick, B.; Kleingartner, L.; Isaak, R.; Doroff, K. *Composition of Sunflower, NUSUN (Mid-oleic Sunflower) and High Oleic Sunflower Oils,* 2002. http://www.sunflowernsa.com/uploads/resources/51/warner_.pdf.

Welsh, W. A.; Bogdanor, J. M. *Modified Caustic Refining of Glyceride Oils for Removal of Soaps and Phospholipids.* U.S. Patent 5,231,201, 1993.

Zeldenrüst, R. Comparison Cold Refining, Wet Winterization and Dry Winterization. In *AOCS Practical Short Course on Fundamentals of Edible Oils Refining and Considerations in Sunflower Oil Processing,* Kiev, Ukraine, 2013.

Zou, Y.; Jiang, Y.; Yang, T.; Hu, P.; Xu, X. Minor Constituents of Palm Oil: Characterization, Processing and Application. In *Palm Oil: Production, Processing, Characterization, and Uses;* Lai, O.-M., Tan, C.-P., Akoh, C. C., Eds.; AOCS Press: Urbana, IL, 2012; pp 471–526.

10

Sunflower Oil and Lipids Biosynthesis

Monica Venegas-Calerón, Manuel A. Troncoso-Ponce, and Enrique Martínez-Force ■
Instituto de la Grasa (Consejo Superior de Investigaciones Científicas), Sevilla, Spain

Introduction

Although de novo fatty acid and lipids biosynthetic pathways are not restricted to specific tissues or organs but found in every cell of the plant, this chapter will center on the synthesis carried out in developing sunflower seeds. Sucrose is the main source of carbon provided by the mother plant to the developing sunflower embryos (Pollock and Farrar, 1996). Sunflower heterotrophic seeds, photosynthetically inactive during their development, are unable to use light energy to produce adenosine triphosphate (ATP) and NADPH, relying completely on the maternal supply from photosynthetic tissues. The photosynthates, in the form of sucrose, are hydrolyzed and the hexoses and hexoses-phosphate resulting from the enzymatic cleavage enter into the glycolysis pathway (Figure 10.1) generating carbon, reducing equivalents, and energy for de novo fatty acid synthesis (White et al., 2000).

The amount of oil in seeds of different species varies from 1% to as much as 60% of seed dry weight. The oilseed storage reserves provide the necessary energy for the germination and establishment of the subsequent seedling. In sunflower, triacylglycerols (TAGs) represent a major component of seed oils. TAG is a compact molecule for energy and carbon storage that has an important function to support seed germination. In addition to seeds, other tissues such as senescing leaves, floret tapetosomes, and pollen grains have been described for TAGs accumulation (Troncoso-Ponce et al., 2013).

The Source of Precursors for the Synthesis of Oil

From Sucrose to Hexose-Phosphates

In sunflower, most of the carbon accumulated in seeds in the form of oil comes from carbon fixed by photosynthetic tissues during the preanthesis period (Hall et al., 1990), and the main route of carbon exportation is the phloem transport of sucrose from the upper leaves to the capitulum (Alkio et al., 2002). These two observations point out the importance of sucrose metabolism during seed development and the concomitant deposition of reserve lipids. In sink tissues, previous to its utilization to provide energy and carbon skeleton, sucrose has to be transported to the cytosol and metabolized by the action of sucrose degrading enzymes.

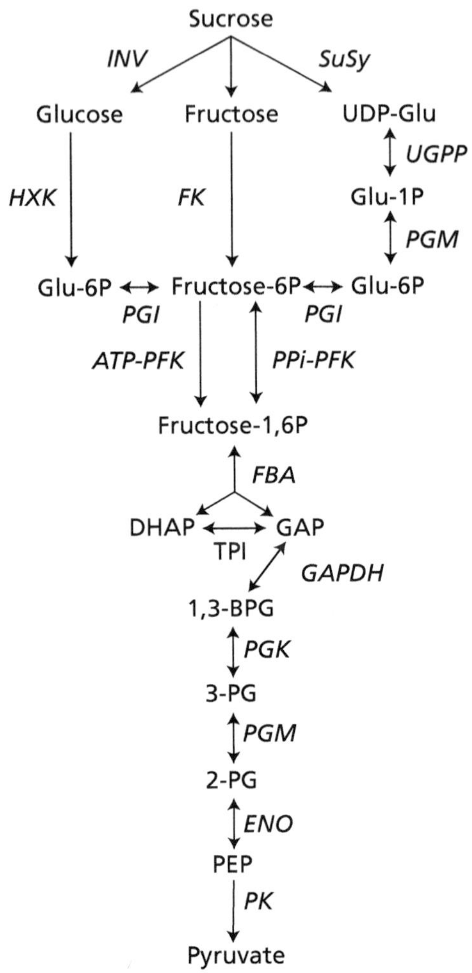

Figure 10.1 Glycolysis pathway scheme from sucrose to triose-phosphates.

Metabolites

1:3-BPG: 1:3-bis-phosphoglycerate
3-PG: 3-phosphoglycerate
DHAP: dihydroxyacetone phosphate
Fructose-1:6P: fructose-1:6-phosphate
Fructose-6P: fructose 6-phosphate
GAP: glyceraldehyde-3-phosphate
Glu-6P: glucose-6-phosphate
PEP: phosphoenolpyruvate

Enzymes

ATP-PFK: ATP-dependent phosphofructokinase
ENO: Enolase
FBA: Fructose 1:6-biphosphate aldolase
FK: Fructokinase
GAPDH: NAD-dependent glyceraldehyde-3-phosphate dehydrogenase
HXK: Hexokinase
INV: Invertase
PGI: Phosphoglucose isomerase
PGK: Phosphoglycerate kinase
PGM: Phosphoglycerate mutase
PGM: Phosphoglycerate mutase
PPi-PFK: Pyrophosphate-dependent phosphofructokinase
SuSy: Sucrose synthase
TPI: Triosephosphate isomerase
UGPP: UDP-glucose pyrophosphorylase

Soluble carbohydrate levels and enzymatic activities related to sucrose cleavage have been studied in cotyledons, hypocotyls, and developing seeds of sunflower (Kutschera and Heiderich, 2002; Luthra et al., 1991; Pfeiffer and Kutschera, 1996; Troncoso-Ponce et al., 2009, 2010a). The evolution of the content of sucrose and the two hexoses, glucose and fructose, has been measured in seeds, showing a decrease in carbohydrate levels during the period of oil synthesis, with sucrose always constituting the principal sugar. Different periods of seed formation could be described based on the kinetic of carbohydrate consumption and physiological changes during seed maturation. An initial phase is characterized by high carbohydrate content that corresponds to a high mitotic index and seed growth. Next, there is a middle phase, in which high hexoses consumption is associated with the synthesis of storage products, such as lipids. Finally, a period of low carbohydrate content is found; the reduction in sugar levels is probably due, among other factors, to a reduction of maternal exportation of photoassimilates.

Sucrose unloaded from the phloem must be degraded into hexoses by invertase (INV, EC 3.2.1.26) and/or by sucrose synthase (SuSy, EC 2.4.1.13) before its incorporation into glycolysis (Figure 10.1). Invertase irreversibly hydrolyzes sucrose into glucose and fructose, while SuSy cleaves sucrose producing UDP-glucose and fructose. Sucrose hydrolysis products enter the glycolysis pathway producing carbon, reducing equivalents, and energy for de novo fatty acid synthesis (Plaxton and Podestá, 2006). Both expression levels and enzymatic activities of these two enzymes have been studied previously in different oilseeds (Hill at al., 2003; Tomlinson et al., 2004; Troncoso-Ponce et al., 2011b). In spite of the multiple isoforms belonging to INV and SuSy families, however, only the activities of neutral and acid INV have been studied so far in developing sunflower seeds (Luthra et al., 1991; Troncoso-Ponce et al., 2009, 2010). The two activities show a similar decreasing pattern during seed formation with a maximum expression peak at the beginning of the seed formation. The products of the enzymatic sucrose degradation become part of the hexoses and hexose-phosphate pools that initiate the glycolysis pathway (Figure 10.1).

Compartmentation of Glycolytic Enzymes

Plant glycolysis is compartmentalized and a complete glycolytic pathway is present in the cytosol and in the chloroplast. Although in some chloroplasts an incomplete glycolysis has been described (Plaxton, 1996; Troncoso-Ponce et al., 2009), in other species a full set of glycolytic expressed sequence tags (ESTs) has been found, including transcripts that encode plastidial isoenzymes that catalyze a complete glycolysis pathway from hexoses phosphorylation to pyruvate synthesis (Troncoso-Ponce at al., 2011). The biochemical communication between the cytosol and the plastids is made possible due to the presence of different transporters that allow the exchange of a variety of molecules between pathways located in both sides of the plastidial membrane (Fischer and Weber, 2002). This process is particularly important in heterotrophic plastids, as in the case of sunflower seeds, as a result of their incapacity to use light energy to generate reducing equivalents and ATP; therefore, these non-green organelles have a complete dependence on cytosolic supplies (Pleite et al., 2005).

Upper Glycolysis

As previously mentioned, hexoses and hexose-phosphates are the initiators of glycolysis. Hexokinase (HXK, EC 2.7.1.1) and fructokinase (FK, EC 2.7.1.4) are two enzymes that catalyze the irreversible phosphorylation of hexoses to hexose-phosphates and are involved, together with phosphoglucose isomerase (PGI, EC 5.3.1.9), in the metabolism of glucose-6-phosphate (G6P) (Figure 10.1), an important point of connection between the cytosolic and plastidial pathways (Alonso et al., 2007; Pleite et al., 2005). These enzymes, together with other glycolytic activities, have been studied in sunflower seed crude extract and during embryo development (Troncoso-Ponce et al., 2009); in addition some of these genes have been recently cloned and their protein products characterized.

The HXK enzyme can use different hexoses as substrates: glucose, fructose, and mannose, as well as several nucleoside triphosphates, ATP, UTP, GTP, and CTP. The main product of the reaction is G6P, which is exported to the plastids and incorporated into the oxidative pentose phosphate pathway (OPPP) in developing sunflower seeds, supplying the NADPH required to the intraplastidial de novo fatty acid synthesis (Alonso et al., 2007; Pleite et al., 2005). Following a strategy involving the use of degenerated primers designed from conserved protein regions, the principal sunflower HXK isoform from seeds was cloned and characterized (Troncoso-Ponce et al., 2011). This protein belongs to the membrane-associated HXK type, a group of HXKs characterized by the presence of an N-terminal hydrophobic membrane anchor domain (Claeyssen and Rivoal, 2007). The expression levels of this transcript were around three times higher in nonphotosynthetic tissues and showed a decreasing pattern in developing seeds with a maximum during the early stages when glycolytic precursors—hexoses—are also at the highest levels (Troncoso-Ponce et al., 2009). Crude extract HXK activity in developing embryos shows a similar evolution, with a maximum reached at the beginning of seed formation. All these observations together reflect the participation of the HXK activity, and particularly this isoform, in nonphotosynthetic carbon metabolism in developing sunflower seeds. A protein version where the initial 24 N-terminal amino acid residues were removed was heterologously expressed, purified, and characterized. The purified protein had a specific activity of 19.3 U/mg protein, the highest specific activity reported for a plant HXK. The enzymatic behaviour of the truncated protein was studied, temperature and pH effect measured, possible inhibitors tested, and kinetic parameters determined using different hexoses and several phosphate donors at three pH values. Glucose was the preferred substrate, although the activity towards fructose increases at acidic pH values. Moreover, the enzyme's sensitivity to temperature and acidic conditions, together with the inhibitor effect of ADP, may play a regulatory function over the protein that could modify the flux of carbon availability. In sunflower seeds, low oxygen condition is a common situation (Rolletschek et al., 2007) that leads to cytoplasmatic acidosis and to an increase of ADP; both phenomena modify the activity of this enzyme and could indirectly affect the anabolic cellular pathways, including plastidial fatty acid synthesis.

Fructose, the hexose produced by both INV and SuSy, can also be phosphorylated by FK, an enzyme that has been found up-regulated in high oil sunflower seeds when two near-isogenic lines with different oil content were compared in a proteomic study conducted by Hajduch et al. (2007).

The glycolytic intermediate G6P can be imported to the plastid and enter in the OPPP, the main source of NADPH for lipid biosynthesis. This metabolite is the substrate and the product of PGI. The cytosolic isoform from developing sunflower seeds has been cloned and characterized. A full length cDNA encoding a cytosolic PGI (cPGI) was amplified and the recombinant protein resulting from its heterologous expression biochemically characterized. A polyclonal immune serum was raised against the purified protein, allowing the study of protein levels and the immunolocalization of the protein in seed tissues, pointing to the procambial ring as the location inside the embryo where highest levels of the protein were found. The metabolic flux map in developing sunflower embryos points out the important role played by G6P and the function of the upper glycolysis feeding the plastidial OPPP (Alonso et al., 2007). Furthermore, Pleite et al. (2005) provided evidence that G6P synthesized in the cytosol is transported into the plastid where its metabolization by the OPPP generates reducing equivalents for the de novo synthesis of fatty acid. The reversible reaction catalyzed by cPGI is a central point in the metabolism of this hexose-phosphate. The specific activities of the purified protein were 1,436 and 1,011 U/mg protein when the reaction was initiated with G6P and fructose-6-phosphate (F6P), respectively, although the comparison of catalytic efficiencies (V_{max}/K_m) obtained with the two substrates revealed that the enzyme showed a higher catalytic efficiency with F6P than with G6P. Sunflower cPGI has shown a pH-dependent substrate inhibition by G6P and is also inhibited by 6-P-gluconate (6PG) and glyceraldehyde-3-phosphate (GAP), proving the connection established by this enzyme between glycolysis (G6P and GAP) and the OPPP (6PG) and the more complex level of regulation than previously described. Also, a possible post-translational regulation mechanism of the enzyme during seed maturation has been hypothesized as a consequence of the discrepancies found between the expression levels of cPGI transcripts, the protein accumulation pattern, and the activity level during the seed development (Troncoso-Ponce et al., 2010).

Middle Glycolysis

Based on the analysis of labeling experiments in sunflower embryo culture conducted by Alonso et al. (2007), between 91% and 95% of the carbon for fatty acid synthesis is derived from triose phosphate (TP) produced from hexose that is imported into the plastid. Unfortunately, this labeling analysis could not distinguish between glycolytic reactions localized in both sides of the plastidial membrane after TP synthesis. Therefore, part of these carbon skeletons could be exported back to the cytosol and converted to phosphoenolpyruvate (PEP), which could be transported again through the plastid membrane to serve as a carbon precursor to anabolic plastidial processes. This hypothesis is based on the parallel glycolytic activities measured in both the cytosol and within plastids.

Phosphorylation of F6P to fructose-1,6-bisphosphate (F1,6P) is catalyzed irreversibly by ATP-dependent phosphofructokinase (ATP-PFK, EC 2.7.1.11) and reversibly by pyrophosphate-dependent phosphofructokinase (PPi-PFP, EC 2.7.1.90) (Figure 10.1). The last one is a strictly cytosolic enzyme, whereas ATP-PFK is also present in plastids (Plaxton, 1996). Both activities have been measured during sunflower seed development showing an opposite pattern, and although the PPi-dependent activity decreases in comparison to the increment of the ATP-dependent one, the former enzyme had higher activity during the period of oil accumulation. Preference for PPi over ATP could be related to an increase of hypoxia in the seed and has been proposed as a strategy for efficient use of oxygen in embryos (Baud and Graham, 2006). Similar conclusions have been drawn from the transcriptome analysis of different oilseeds, where EST levels of PPi-PFP were higher than either cytosolic or plastidial ATP-PFK (Troncoso-Ponce et al., 2011).

Fructose 1,6-biphosphate aldolase (FBA, EC 4.1.2.13) is an enzyme involved not only in glycolysis (Figure 10.1) and gluconeogenesis in the cytoplasm, but also in the Calvin cycle in plastids. The enzyme catalyzes the reversible cleavage of F1,6P to dihydroxyacetone-phosphate (DHAP) and glyceraldehyde 3-phosphate (GAP). This activity, in conjunction with the next glycolytic enzyme, triosephosphate isomerase (TPI, EC 5.3.1.1), takes part directly in the TP metabolism. Both activities increase during the period of oil deposition in sunflower seeds. In a comparative transcriptional analysis in leaves and immature sunflower embryos, six putative isoforms of the enzyme were detected. Two of these putative isoforms were up-regulated in developing embryos (Hewezi et al., 2006).

The highest glycolytic activity measured in sunflower seed corresponds to the reversible isomerization of DHAP and GAP catalyzed by TPI (Figure 10.1). DAHP is exchanged between the cytosol and plastids and after its conversion in GAP that is metabolized down to pyruvate (Pyr). DHAP is also converted to glycerol-3-phosphate (G3P), an important metabolite for TAG biosynthesis.

Lower Glycolysis

The NAD-dependent glyceraldehyde-3-phosphate dehydrogenase (EC 1.2.1.12) activity increases in conjunction with the oil content in developing sunflower seeds (Troncoso-Ponce et al., 2009) and is one of the most abundant relative proteins found in a proteomic study of castor been (Houston et al., 2009). This enzyme catalyzes the conversion of GAP to 1,3-bis-phosphoglycerate (1,3-PG) (Figure 10.1).

Phosphoglycerate kinase (PGK, EC 2.7.2.3) catalyzes the reversible transfer of phosphate from the C1-position of 1,3-DPG to ADP to form 3-phosphoglycerate (3-PGA) and ATP (Figure 10.1). In developing sunflower seeds, PGK is one of the glycolytic activities that increases in conjunction with oil accumulation during the period of storage lipid synthesis (Troncoso-Ponce et al., 2009). Moreover, the study of the glycolytic enzymatic activities in developing seeds from standard and low oil content sunflower lines has pointed to PGK and enolase (ENO, EC 4.2.1.11) as two of the activities probably implicated in the differences in fat content between these lines (Troncoso-Ponce et al., 2010). Three different phosphoglycerate kinase isoforms, two cytosolic (cPGK1 and cPGK2) and one plastidial (pPGK), have been cloned and characterized from developing sunflower seeds (Troncoso-Ponce, 2012). Expression levels of these three transcripts in different tissues have been measured. cPGK1 is the highest expressed isoform during embryo development, accounting for around 80% of the three genes. The expression pattern of transcripts that encode cPGK1 is similar to most of the genes related to carbon and lipid metabolism in oilseeds, and decreases during seed maturity. Meanwhile, the plastidial isoform is the highest-expressed PGK isoenzyme in photosynthetically active tissues, cotyledons, and leaves, probably reflecting its role in the Calvin cycle. The cDNAs were expressed in *Escherichia coli* and the corresponding proteins were purified to electrophoretic homogeneity using immobilized metal ion affinity chromatography and were biochemically characterized. Despite the high level of identity between sequences, the HacPGK1 isoform showed strong differences in terms of specific activity, temperature stability, and pH sensitivity in comparison to cPGK2 and pPGK. The inhibitor effect of ADP and AMP was tested, and both decrease the enzymatic PGK activity; therefore, sunflower seed PGKs could be sensitive to variations in the ATP/ADP ratio or the adenylate energy charge. This feature could be important to regulation of carbon metabolism in physiological conditions where ADP concentration is elevated, for example, during seed hypoxia conditions. A polyclonal immune serum was raised against the purified cPGK1 isoform, which showed cross-immunoreactivity with the other PGK isoforms. This serum allowed the localization of high expression levels of PGK isoenzymes in embryo tissues. The spatial distribution of PGK isoforms overlaps with the localization of cPGI, supporting the presence inside sunflower embryos of tissues where carbon metabolism is more active.

Phosphoglycerate mutase (PGM; EC 5.4.2.1) catalyzes the interconversion of 3-PGA and 2-phosphoglycerate (2-PGA) (Figure 10.1). These enzymes are divided into two evolutionarily unrelated groups in function of their dependence on the cofactor 2.3-biphosphoglycerate, which were denominated dPGM (dependent) and iPGM (independent) respectively. The dPGM enzymes are predominant in vertebrates, certain fungi, and bacteria, while higher plants contain both dPGM and iPGM enzymes. In Arabidopsis, however, these enzymes have been found in different locations, iPGM in cytosol and dPGM within plastids. In developing sunflower seeds, iPGM activity has been determined throughout the active period of fatty acid synthesis, decreasing its activity during this stage of oil deposition.

ENO, the enzyme that converts 2-PGA to PEP, has three different isoenzymes in Arabidopsis seeds, two cytosolic and one plastidial, although one of the cytosolic isoforms is inactive. Due to the central role played by PEP as a source of carbon during fatty acid synthesis, Arabidopsis mutant plants lacking the plastidial ENO isoform were characterized. These mutants displayed the same seed oil content as WT plants, which mean that the main source of carbon that supports fatty acid synthesis is located in the cytosol (Andriotis et al., 2010). In sunflower seeds, the crude extract enolase activity, which mostly corresponded to the cytosolic enzyme, increases during embryo development.

Pyruvate kinase (PK, EC 2.7.1.40) occurs as both cytosolic and plastidic isoforms and catalyzes the ADP-dependent conversion of PEP to Pyr while producing ATP. In Arabidopsis seeds, disruption of the main plastidic isoform implies a 60% reduction of seed oil content (Andre et al., 2007). This result supports the importance of this plastidial enzyme for fatty acid synthesis in developing Arabidopsis embryos. In sunflower seeds, the cytosolic and plastidial activity was measured during seed oil deposition.

Cytosolic and Plastidial Connection: Source of Carbon for Oil Biosynthesis

Different attempts to decipher the complexity of reactions that lead from sucrose to TAG have been carried out. The physiological differences that oil-rich seeds present, such as their ability to photosynthesize (green versus non-green seeds); the oil storage tissue (embryo versus cotyledon); and the nature and content of their TAG add another layer of difficulty to the efforts made by a multitude of laboratories that have been working on the regulation of oil synthesis and deposition for the last decades. Due to the economical relevance of rich oil crops, researchers have not only focused on the model plant *Arabidopsis thaliana*, but also extend their studies to agriculturally important species. The wide spectrum of approaches used for this propose ranges from individual enzymatic characterization, analysis of metabolic fluxes, and reverse genetics, to more recent transcriptome characterization of diverse oilseeds.

In Arabidopsis seeds, more than 150 carbon metabolism–related genes have been found. These genes are involved in multiple metabolic pathways, sucrose metabolism, sugar transporters, glycolysis, plastid transporters, pentose phosphate pathway, and the Calvin cycle. The number of genes involved in the generation of energy, reduction equivalents, and carbon for plastidial fatty acids synthesis reflects the importance of the process and the resources that the developing seeds mobilize to secure their oil reserves, which are essential for germination and posterior seedling development. Independent of the physiological differences previously described, most of the transcripts involved in sucrose hydrolysis, glycolysis, and plastidial transporters follow the same pattern of expression as the fatty acids core synthesis genes, decreasing during seed development, showing the co-expression and possible co-regulation of these pathways.

Due to the elevated number of isoforms and to the flexibility of the pathways (duplication of reactions in different subcellular locations), it is difficult to find regulatory points by using reverse genetics. Cernac and Benning (2004) identified the first transcription factor, WRINKLED-1, that regulates key glycolytic genes and the expression of fatty acid biosynthetic enzymes. Seeds lacking this transcriptional regulator contain a reduced amount of lipid in the mature seeds. Plastidial Pyr is the final destination of the glycolytic carbon. When plastids lack the PK activity, oil synthesis and deposition are dramatically affected. However, due to the presence of parallel glycolytic reactions in the cytosol and within plastids, and to the presence of membrane transporters, it is still unclear how sucrose-derived metabolites are distributed throughout the whole pathway. In Arabidopsis, embryo plastids have the capacity to carry out a completed glycolytic pathway. Nevertheless, in seeds lacking plastidial ENO and dPGM, a reduction in oil synthesis was not found; therefore, the cytosolic glycolysis provides enough carbon, in the form of PEP, to support plastidial oil production (Andriotis et al., 2010).

Sunflower seeds are heterotrophic tissues that rely on the maternal delivery of sucrose. The sucrose degradation products, mainly hexoses and hexose-phosphates, initiate the glycolytic pathway, a catabolic route that ends with the conversion of PEP into Pyr. The former metabolite can be found in the cytosol or within plastids, although because fatty acid synthesis takes place in plastids, the pool of Pyr for fatty acid synthesis has to be located within plastids. Metabolic exchanges between cytosolic and plastidial glycolysis occur early in the pathway. Experiments of plastid radiolabelling (Pleite et al., 2005) and analysis of metabolic fluxes in developing embryos (Alonso et al., 2007) point out G6P as an important connection between both glycolytic pathways. G6P is imported by plastids and through the OPPP generates the NADPH necessary for fatty acid biosynthesis. TP is another important point of communication between cytosolic and plastidial carbon metabolism, allowing the exchanges of these metabolites between both locations. The comparison between activities found in the cytosol and within plastids allowed the identification of the cytosolic pathways as the predominant. Additionally, some of the lower plastidial glycolytic reactions were extremely low. Therefore, most of the carbon is converted to cytosolic PEP, which is imported into the plastids and converted into Pyr that serves as the main source of carbon for fatty acid synthesis.

Fatty Acid Biosynthesis

The crucial precursor for de novo fatty acid biosynthesis is acetyl-CoA. Its biosynthesis can be achieved in a number of ways, of which three main pathways have been identified (Nikolau et al., 2000). First, pyruvate is formed from carbohydrates via glycolysis, and then the oxidative decarboxylation of pyruvate inside the plastid by the pyruvate dehydrogenase complex produces CO_2 and acetyl-CoA. Another route for the synthesis of plastidial pyruvate (and thence acetyl-CoA) is through decarboxylation of imported malate by the plastidial NADP-dependent malic enzyme. In the third pathway, acetyl-CoA is formed from pyruvate in the mitochondria and is hydrolyzed to acetate, which leaves the mitochondria to reach the plastid, where it is activated to acetyl-CoA by the acetyl-CoA synthetase (Weselake et al., 2009). The relative contribution of these pathways to fatty acid biosynthesis has not been studied so far in sunflower. De novo fatty acid biosynthesis is a process that occurs in plant plastids and needs the combined activity of two multi-enzyme complexes, acetyl-CoA carboxylase and fatty acid synthase.

Acetyl-CoA Carboxylase

Acetyl-CoA carboxylase (ACCase, EC 6.4.1.2) catalyzes the formation of malonyl-CoA from acetyl-CoA, the first committed reaction in the biosynthesis of fatty acids, and is considered to be an important regulatory enzyme for fatty acid synthesis in many tissues (Sasaki and Nagano, 2004). This reaction, which uses bicarbonate as substrate, involves the participation of biotin and is driven by the hydrolysis of ATP. The reaction takes place in two steps: (1) biotin is carboxylated by biotin carboxylase, and (2) carbon dioxide is transferred to acetyl-CoA by carboxyltransferase.

In higher plants, there are two isoenzymes located in the cytosol and the plastid. In dicotyledons, such as sunflower, the cytosolic isoform consists of a single multifunctional polypeptide; all three domains are part of a single polypeptide. This is called the *eukaryotic form* because it is the type of complex present in animals and yeast (Harwood, 1996). In contrast, the plastidial isoform is a multienzyme complex consisting of four separate proteins: biotin carboxylase, biotin carboxyl carrier protein (BCCP), and the carboxyltransferase, which is a heterodimer (α- and β-subunits). Since this multienzyme complex is similar to that in bacteria or cyanobacteria, it is called the prokaryotic form (Harwood, 1996). Three of the genes for the plastidial prokaryotic form of ACCase (biotin carboxylase, BCCP, and α-subunit of carboxyltransferase) are nuclear encoded. The fourth gene coding for the β-subunit of carboxyltransferase is plastid located (Harwood, 2005). In sunflower only, the sequence for the ACCase carboxyltransferase β-subunit has been identified during sequencing of the plastid genome (GenBank YP588126) (Timme et al., 2007). Biochemical evidence indicates that plastid ACCase is responsible for providing most, if not all, of the malonyl-CoA required for de novo biosynthesis of fatty acids (Sasaki and Nagano, 2004).

Fatty Acid Synthase

Fatty acids are synthesized from malonyl-ACP (acyl carrier protein), which is formed by transacylation of malonate from the malonyl-CoA generated by ACCase to activated holo-ACP, in a reaction catalyzed by malonyl-CoA:ACP transacylase (EC.2.3.1.39). Fatty acids are subsequently produced by an easily dissociable multi subunit complex consisting of mono functional enzymes, which is referred to as *fatty acid synthase* (FAS). As shown in Figure 10.2, the enzymatic complex includes six enzymes as well as the ACP that binds the intermediate acyl chains (Harwood, 1996): β-ketoacyl-ACP synthases I, II, and III (KASI, EC.2.3.1.41; KASII, EC.2.3.1.179; and KASIII EC.2.3.1.180, respectively), β-ketoacyl-ACP reductase (KAR, EC.1.1.1.100), β-hydroxyacyl-ACP dehydratase (HAAD, EC.4.2.1.0), and enoyl-ACP reductase (ENR, EC.1.3.1.9).

ACPs are small acidic proteins that carry the nascent acyl chains during the synthesis of 16- and 18-carbon acyl groups. All higher plants express several different ACP isoforms in a tissue-specific manner (Bonaventure and Ohlrogge, 2002). This may reflect the need of plant cells to tightly regulate the amount and the cellular destination of fatty acids produced in plastids (Ohlrogge and Browse, 1995) and to match the supply of fatty acids to different tissue and environmental demands. In sunflower, three different genes coding for ACP isoforms have been cloned: *Ha*ACP1 (GenBank HM013970), *Ha*ACP2 (GenBank HM013971), and *Ha*ACP3 (GenBank HM013972) (Aznar-Moreno et al., 2010).

The cycle of elongation starts with the condensation of malonyl-ACP with acetyl-CoA by the action of KASIII to produce acetoacetyl-ACP. This compound is subsequently reduced by KAR, an enzyme usually dependent on NADPH, to yield the β-hydroxyacyl derivative. The latter compound is then dehydrated by HAAD and finally reduced by ENR, resulting in a four carbon (butyryl) acyl-ACP derivative, which can be submitted to a further chain elongation cycle. The next condensations are catalyzed by KASI, which, unlike KASIII, uses acyl-ACPs as the primer substrates instead of acetyl-CoA. This enzyme is involved in most of the condensations necessary to produce the acyl chains, so that the final product of the cycles in which it takes part is palmitoyl-ACP (16:0-ACP). Finally, this compound can be elongated to stearoyl-ACP (18:0-ACP) in a further cycle involving the participation of the last

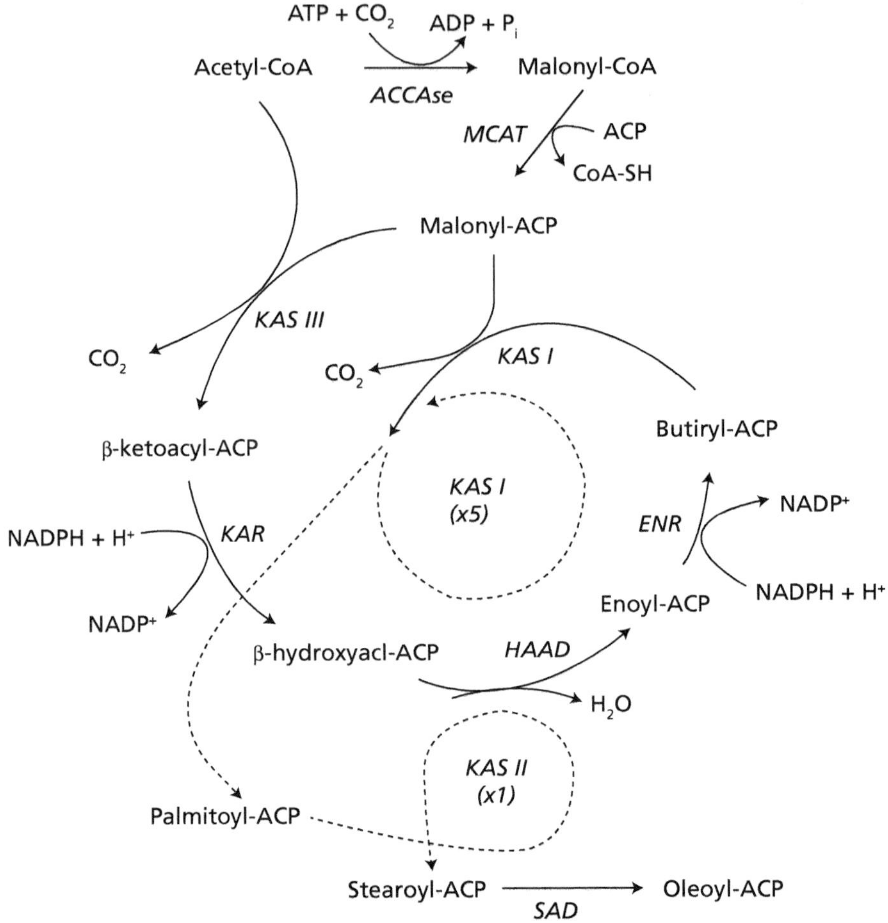

Figure 10.2 De novo fatty acid biosynthesis in plants.

ACCase: acetyl-CoA carboxylase
ACP: acyl carrier protein
HAAD: β-hydroxyacyl-ACP dehydrase
ENR: enoyl-ACP reductase
KAR: β-ketoacyl-ACP reductase
KAS: β-ketoacyl-ACP synthase
MCAT: malonyl-CoA:ACP transacylase
SAD: stearoyl-ACP desaturase

condensing enzyme, KASII (Figure 10.2). This step is especially relevant because it will determine the C16/C18 ratio of the resulting pool of fatty acids.

Most of the FAS genes involved in the biosynthesis of fatty acids in sunflower developing seeds have been isolated, cloned, and sequenced: three β-ketoacyl-ACP synthases (*Ha*KASI, GenBank EF177175; *Ha*KASII, GenBank DQ835562; *Ha*KASIII, GenBank EF514400) (González-Mellado et al., 2006a, 2006b, 2010, respectively), two β-ketoacyl-ACP reductases (*Ha*KAR1, GenBank HM021135; *Ha*KAR2, GenBank HM021136) (González-Thuillier et al., 2010b), two β-hydroxyacyl-ACP dehydratases (*Ha*HAAD1, GenBank HM044767; *Ha*HAAD2, GenBank GU595454) (González-Thuillier et al., 2010a), and two enoyl-ACP reductase isoforms (*Ha*ENR1, GenBank HM021137; *Ha*ENR2, GenBank HM021138) (González-Thuillier et al., 2010c).

Among the β-ketoacyl-ACP synthases, only *Ha*KASIII has been extensively characterized (González-Mellado et al., 2010), although it is known that the high palmitic trait in sunflower is mostly due to lower *Ha*KASII activity (Martínez-Force et al., 1999). The expression levels of *Ha*KASIII gene, examined by quantitative reverse transcription polymerase chain reaction (Q-RT-PCR), revealed higher levels in developing seeds storing oil than in leaves, stems, roots, or seedling cotyledons. Its heterologous expression in *Escherichia coli* altered the fatty acid content and composition, implying an interaction of *Ha*KASIII with the bacterial FAS complex. In vitro assays, in which purified recombinant *Ha*KASIII was added to a reconstituted *E. coli* FAS system lacking condensation activity (Heath and Rock, 1995; Yasuno et al., 2004), clearly revealed that *Ha*KASIII can use ACP acceptor substrates with 8 carbons and potentially also with 10 carbons, and is not limited to using C2-CoA in the initial condensation step. This observation, combined with the fact that purified recombinant *Cuphea wrightii* KASIII was shown to use C2-, C3-, and C4-CoAs plus C2- and C4-ACPs (Abbadi et al., 2000), implies that substrate specificity of plant KASIIIs is variable and not limited to the initial condensation step in the fatty acid biosynthesis (González-Mellado et al., 2010).

The analysis of the expression levels of the rest of the sunflower FAS proteins, reductases, and dehydratases, and their initial biochemical characterization, point to fatty acid biosynthesis as a highly regulated process (González-Thuillier, 2013).

Acyl-ACP Desaturases and Thioesterases

Although the final products of FAS are mainly C16 or C18 saturated acyl-ACPs, sunflower oil and most vegetable oils are enriched in unsaturated fatty acids, like oleic (C18:1), linoleic (C18:2), α-linolenic (C18:3), or palmitoleic (C16:1) acids.

The final products of these elongation cycles, 16:0-ACP and 18:0-ACP (Harwood, 1996), can be desaturated by the action of Δ9 stearoyl-ACP desaturase (SAD, EC 1.14.19.2), a soluble dimeric protein that is also present in the plastid stroma, to produce the corresponding acyl-ACP derivatives of palmitoleate and oleate (Shanklin and Cahoon, 1998) (Figure 10.2). Oleate is the main product of plastidial fatty acid synthesis, whereas palmitoleate is found only in trace amounts in most plant tissues. It is abundant in the seed endosperm of some tropical species like *Macadamia integrifolia* and *Doxantha unguis-cati* or cat's claw (Cahoon et al., 1998). However, although common sunflower and high-oleic lines presented only traces (0.4–0.7%) of n-7 fatty acids (palmitoleate and its derivatives), different high palmitic sunflower mutants produce oils showing contents of n-7 fatty acids ranging between 12% (Álvarez-Ortega et al., 1997) and 24% of the total oil fatty acids (Salas et al., 2004).

The desaturation reaction involves the transfer of electrons from reduced ferredoxin and the aliphatic acyl chain to a molecule of oxygen, which is also the last acceptor of hydrogen, yielding water as a by-product of the reaction (Griffiths et al., 1998). The structure of these enzymes is well known, and a form of SAD has been crystallized and studied by RX diffraction (Lidqvist et al., 1996). The gene coding for this protein has been cloned and sequenced from various plant species, including sunflower (Hongtrakul et al., 1998a; Serrano-Vega et al., 2003). Two genes encoding for SAD enzymes are expressed at high levels in the sunflower kernel, *SAD6* and *SAD17* (Genbank U91339 and U91340, respectively) (Hongtrakul et al., 1998a), which display 89% homology and are expressed at similar levels during oil synthesis in sunflower seeds.

Sunflower mutant lines deficient in SAD activity display a high stearic acid phenotype (Cantisán et al., 2000; Salas et al., 2008). In particular, Pérez-Vich et al. (2002) found this trait in the high stearic line CAS-3 cosegregates with one of the stearoyl-ACP desaturase locus, *SAD17*. However, in the CAS-14 high stearic line, where the expression of this trait is strongly influenced by temperature during seed maturation and is not uniform along the seed (Fernandez-Moya et al., 2002, 2003), the phenotype was not associated with any of the SAD locus but with a third gene (Pérez-Vich et al., 2006), probably encoding a transcription factor that regulates the expression of both *SAD6* and *SAD17* genes (Salas et al., 2008).

Although this pathway is made of discrete elements, it has been proposed that fatty acid biosynthesis is carried out by a multienzymatic complex, or *metabolon*, expediting the dynamic channeling of the metabolic flux through both the elongation and the desaturation steps (Martínez-Force and Garcés, 2002; Mekhedov et al., 2000; Roughan and Ohlrogge, 1996).

The process of de novo fatty acid synthesis inside the plastid finishes with the release of the acyl moieties from the ACP derivatives by the action of plastid acyl-ACP thioesterases (TE, EC 3.1.2.14) or their transfer into complex lipids by acyltransferases inside the plastid. In plants, there are two gene families of acyl-ACP thioesterases: *FatA*, which displays high specificity toward oleoyl-ACP and is responsible for most of the acyl-ACP hydrolysis in plants, and *FatB*, specialized in the release of saturated or shorter saturated acyl-ACP (Jones et al., 1995). Sunflower contains at least one gene of each thioesterase family in its genome: *HaFatA* (GenBank AY078350) (Serrano-Vega et al., 2005) and *HaFatB* (GenBank AF036565) (Slabaugh et al.,

1997). During sunflower seed development, both *FatA* and *FatB* thioesterases have been shown to be expressed through biochemical competence assays and by PCR amplification of thioesterase mRNA (Martínez-Force et al., 2000; Serrano-Vega et al., 2005). Recombinant *HaFatA* protein, purified after heterologous expression in *Escherichia coli*, has been biochemically characterized, showing that sunflower *HaFatA* cDNA encodes a functional thioesterase with preference for monounsaturated acyl-ACPs, being most efficient (kcat/Km) in catalyzing oleoyl-ACP (Serrano-Vega et al., 2005).

Acyl-ACP thioesterases have not yet been crystallized, although sunflower *FatA* (Serrano-Vega et al., 2005) and Arabidopsis FatB (Mayer and Shanklin, 2005) structures have been modeled on the basis of their similarity to *Escherichia coli* 4-hydroxybenzoyl-CoA thioesterase. Accordingly, it was hypothesized that the active site of these enzymes displays a hot-dog fold pattern, with a catalytic triad of amino acids similar to that in papain (Mayer and Shanklin, 2005; Serrano-Vega et al., 2005).

Long-Chain Acyl-CoA Synthetases

The free fatty acids released by thioesterases can easily move through cellular compartments, but they can only be detected at very low amounts in developing seeds because they are quickly converted by long chain acyl-CoA synthetases (LACS, EC 6.2.1.3) to their corresponding acyl-CoAs (Schnurr et al., 2002). Long chain acyl-CoAs are central metabolites in many biosynthetic and degradative pathways. So, they are involved in beta oxidation, acyl turnover of membrane lipids, and in many reactions involving the exchange or transference of a fatty acid moiety. LACS enzymes are AMP-binding proteins that are part of the superfamily of carboxyl-CoA ligases, collectively called *acyl activating enzymes* (Shockey and Browse, 2011). In their reaction, one molecule of free fatty acid is esterified to free coenzyme A, with the consumption of one equivalent of ATP. These enzymes are membrane-bound and plants usually contain several forms of LACSs with specific locations that participated in different metabolic processes. Thus, Shockey et al. (2002) reported the presence of nine LACS genes in Arabidopsis, *AtLACS1* through *AtLACS9*, which have been extensively studied in order to understand their role in the metabolism of this plant (Fulda et al., 2002, 2004; Jessen et al., 2011; Lu et al., 2009b; Schnurr et al., 2004). In this regard, oilseeds support a strong metabolic flux into TAG, which means a continuous supply of acyl-CoAs from the plastidial synthesis to feed the acyltransferases of the endoplasmic reticulum, which is responsible for the incorporation of the fatty acids into the three positions of the glycerol backbone. In this regard, the *AtLACS9* gene, expressed at high levels in seeds, and its encoded protein, located in the membrane of plastids, was initially the main candidate responsible for this process. However, reverse genomics experiments based on the study of an *AtLACS9* knockout mutant of Arabidopsis indicated that although these seeds displayed an important reduction of the plastidial LACS, they displayed a normal phenotype, which means that several LACS forms are involved in this process (Schnurr et al., 2002). Thus, a more recent work indicated that the form *AtLACS1* located in the endoplasmic reticulum is also involved in TAG synthesis (Zhao et al., 2010), indicating that several LACS forms are probably involved in this process. In sunflower, two genes, *HaLACS1* (GenBank HM490305) and *HaLACS2* (GenBank HM490306), which were cloned from developing seed mRNA and showed LACS activity when heterologously expressed in bacteria, have been described (Aznar-Moreno et al., 2013a). The proteins coded by these genes were homologous, respectively, to *AtLACS9* and *AtLACS8.* The subcellular location of these proteins was confirmed using green fluorescent protein (GFP) tags on transformed tobacco cells BY-2, with *HaLACS1* associated to the outer chloroplast membrane and *HaLACS2* with the endoplasmic reticulum (ER) membrane.

The synthesized acyl-CoAs are important intermediates in the biosynthesis of glycerolipids because they act as acyl donors in: (1) the de novo synthesis of TAGs, (2) the acyl editing of phospholipids, and (3) the modification of acyl moieties (e.g., their elongation to form very long chain fatty acids) (Bao et al., 1998; Bates et al., 2009). The size and composition of the acyl-CoA pools in tissues from wild type and mutant sunflower plants have been characterized by analyzing their corresponding acyl-etheno-CoA derivatives through high-performance liquid chromatography with fluorescence detection (Aznar-Moreno et al., 2013b). This study shows that mutations altering fatty acid metabolism in sunflower seeds provoke alterations in both the size and the composition of the acyl-CoA pools in developing seeds. These modifications were in good agreement with the changes that took place in the profile of TAG accumulation found in these mutant lines. Proportions of 18:2-CoA were lower than those found in the oils synthesized by the seed, and saturated fatty acids were in exchange at higher levels, which indicated that the former fatty acid was quickly channelled into TAGs, whereas there were restrictions for the latter ones (Aznar-Moreno et al., 2013b).

Extraplastidial Fatty Acids Modification

Normal sunflower oil is naturally rich in linoleic acid (55–70%) and consequently poor in oleic acid (20–25%). Indeed, classic varieties are qualified as low oleic. The formation of linoleic acid, the most abundant polyunsaturated fatty acid in vegetable oils, mainly occurs outside the plastid. Thus, it is produced in the ER by a reaction catalyzed by an oleate desaturase, FAD2, ($\Delta 12$ fatty acid desaturase, EC 1.3.1.35). In this case, the substrate oleoyl-moiety is esterified in a membrane phospholipid, generally at the *sn*-2 position of the phosphatidylcholine (Harwood, 1996; Sperling and Heinz, 2001). The reaction involves the concomitant reduction of molecular oxygen to water and requires NADH, NADH-cyt b5 reductase, and cyt b5 as the electron donor system (Shanklin and Cahoon, 1998). The reaction has been extensively studied in sunflower (Garcés and Mancha, 1991; Hongtrakul et al., 1998b; Kabbaj et al., 1996) and the genes encoding for the enzyme have been cloned and sequenced (Hongtrakul et al., 1998a, 1998b; Martínez-Rivas et al., 2001). Three *FAD2* genes (*FAD2-1*, GenBank AF251842; *FAD2-2*, GenBank AF251843; and *FAD2-3*, GenBank AF251844) have been identified in sunflower (Hongtrakul et al., 1998b; Martínez-Rivas et al., 2001). *FAD2-1* is seed specific and strongly expressed in developing seeds (Hongtrakul et al., 1998b), whereas *FAD2-2* and *FAD2-3* are weakly expressed in developing seeds (Martínez-Rivas et al., 2001). Of the three family members, *FAD2-1* seems to play the key role in linoleic acid synthesis in developing sunflower seeds.

In sunflower seeds, the linoleic acid content is strongly influenced by environmental temperature (Harris et al., 1978; Lajara et al., 1990). This is due not only to the fact that temperature affects the de novo oleate biosynthesis and its mobilization from preformed TAGs (Garcés et al., 1994; García-Díaz et al., 2002), but also to the regulation of *FAD2* activity by two different mechanisms: the low thermal stability of the enzyme and the availability of oxygen determined by temperature (García-Díaz et al., 2002).

Sunflower lines showing the high oleic phenotype coming from the original Soldatov's Pervenets (Soldatov, 1976) mutant genetic background show a marked decrease in *FAD2* activity due to a *FAD2-1* gene duplication leading to its genetic silencing (Hongtrakul et al., 1998b; Lacombe and Bervillé, 2001; Lacombe et al., 2009; Schuppert et al., 2006). Just recently, a new patent for high oleic sunflower lines obtained by chemical mutagenesis and selection by tilling of the *FAD2-1* structural gene has been published (León et al., 2013).

Further desaturation producing α-linolenic acid may occur by the action of a microsomal linoleate desaturase, FAD3 (Δ15 fatty acid desaturase, EC 1.14.99.0). Although this fatty acid is highly present and has major relevance in photosynthetic tissues, sunflower oil is characterized by very low amounts (0.01–0.2%) of α-linolenic acid. No gene has been identified for *FAD3* in sunflower; only the plastidial/chloroplastic linoleate desaturase *FAD7* has been described (Venegas-Calerón et al., 2006).

Common sunflower oil contains three main saturated very long chain fatty acids (VLCFAs): arachidic (20:0), behenic (22:0), and smaller amounts of lignoceric (24:0). These fatty acids account for 1–3% of the total fatty acids in sunflower seeds (Gunstone et al., 1992). Their content is augmented in high saturated sunflower lines developed by breeding and mutagenesis (Álvarez-Ortega et al., 1997). These fatty acids are synthesized in the ER from pre-existing acyl-CoAs exported from plastids and malonyl-CoA by the action of a fully dissociated enzyme complex called *fatty acid elongase* (Cassagne et al., 1994). Cycles of fatty acid elongation in the ER are analogous to intraplastidial de novo fatty acid biosynthesis and involve four enzymatic reactions.

The first enzyme of this complex is the 3-ketoacyl-CoA synthase (KCS, EC 2.3.1.119), which condenses the previously mentioned substrates to yield a β-ketoacyl-CoA with the concomitant release of CO_2. The β-ketoacyl-CoAs produced by KCS are reduced to their hydroxyl derivatives by the action of the β-ketoacyl-CoA dehydrogenases (EC 1.1.1.35, EC 1.1.1.36). The β-hydroxy-CoA products are then dehydrated and further reduced by the successive actions of a dehydratase (EC 4.2.1.17) and a reductase (EC 1.3.1.38) to yield the final 2C-elongated acyl-CoA (Lessire et al., 1998). Some of these latter enzymes have already been described and their cofactor dependence has been investigated (Barrett and Harwood, 1998; Domergue et al., 2000). Biochemical characterization in sunflower has shown that VLCFAs are synthesized from 18:0-CoA by these membrane-bound fatty acid elongases, and their biosynthesis is mainly dependent on NADPH equivalents (Salas et al., 2005). Two condensing enzymes appear to be responsible for the synthesis of VLCFAs in sunflower kernels: β-ketoacyl-CoA synthase-I (KCS-I) and β-ketoacyl-CoA synthase-II (KCS-II). Both of these enzymes were resolved by ion exchange chromatography and each displays different substrate specificities. While KCS-I displays a preference for 20:0-CoA, 18:0-CoA was more efficiently elongated by KCS-II. Both enzymes have different sensitivities to pH and Triton X-100, and their kinetic properties indicate that both are strongly inhibited by the presence of their substrates, which makes it important to consider the size of the cytosolic acyl-CoA pool when attempting to estimate their activity in vivo (Salas et al., 2005). Two different KCS genes mainly expressed in developing sunflower genes have been cloned (*HaKCS1*, GenBank EU442581; *HaKCS2*, EU496864) (Gónzalez-Mellado et al., 2008), but further studies are needed to correlate these genes with the biochemical activities already described.

Complex Lipid Synthesis: Formation of Triacylglycerols

Sunflower storage lipid synthesis predominantly occurs in developing seeds over a relatively short time period of about two weeks. The biosynthesis of TAGs is a complex process consisting of three well-differentiated phases: (1) de novo synthesis of fatty acids in the plastid, (2) formation of the acyl-CoA pool in the cytosol, and (3) TAG assembling in the ER of developing seed tissues. Synthesized TAGs are then sequestered into densely packed oil bodies (oleosomes) until their utilization (Kim et al., 2002).

TAG assembly in the ER is intimately associated with membrane metabolism and involves the action of several membrane-bound acyltransferase enzymes (Weselake, 2005). Until recently, based on biochemical research over last century, TAG biosynthesis was viewed as a sequential linear process. Acyl-CoAs were sequentially assembled onto the *sn*-1, *sn*-2, and *sn*-3 positions of the glycerol backbone, known as the Kennedy pathway, the acyl-CoA-dependent pathway, or the glycerol phosphate pathway. However, in the last decade, new activities involved in other alternative pathways also operating in the TAG biosynthesis have been discovered and characterized (Figure 10.3). Therefore, multiple interconnected pathways for TAG synthesis have actually been elucidated in plants.

Traditional Kennedy Pathway

In the traditional view, G3P serves as the glycerol backbone for TAG assembly. G3P is formed by the reduction of DHAP through the action of *sn*-glycerol-3-phosphate dehydrogenase (GPDH; EC 1.1.1.8). First, *sn*-glycerol-3-phosphate acyltransferase (EC 2.3.1.15) catalyzes the acyl-CoA-dependent acylation of G3P at the *sn*-1 position to produce lysophosphatidic acid (LPA). A second acyl-CoA-dependent acylation to generate phosphatidic acid (PA) is catalyzed by a lysophosphatidic acid acyltransferase (LPAAT; EC 2.3.1.51) in the *sn*-2. Afterward, phosphatidic acid phosphatase (PA-PEC 3.1.3.4) removes the phosphate group at the *sn*-3 position of the PA to form the 1,2-*sn*-diacylglycerol (DAG). These steps are common to the synthesis of membrane

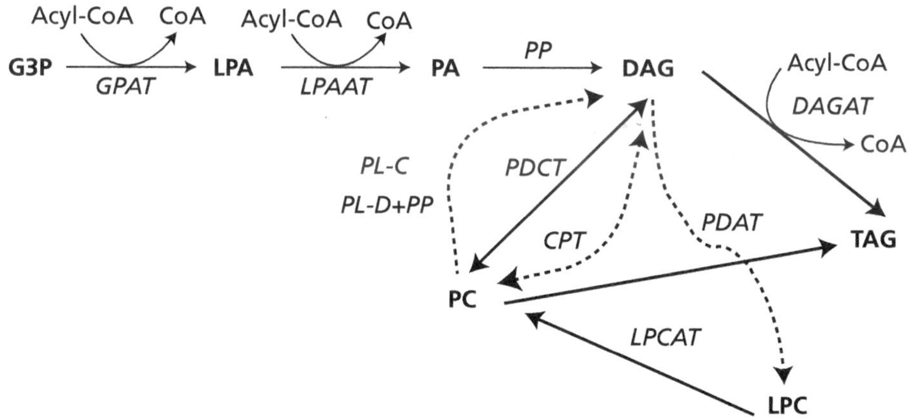

Figure 10.3 Assembly of fatty acids onto the glycerol backbone.

CPT: Diacylglycerol Cholinephosphotransferase
DAG: diacylglycerol
DAGAT: diacylglycerol acyltransferase
G3P: glycerol 3-phosphate
GPAT: glycerol 3-phosphate acyltransferase
LPA: lysophosphatidate
LPAAT: lysophosphatidate acyltransferase
LPCAT: lysophosphatidylcholine acyltransferase
LPC: lysophosphatidylcholine
PA: phosphatidate
PC: phosphatidylcholine
PDAT: phospholipid diacylglycerol acyltransferase
PDCT: Phosphatidylcholine:diacylglycerol cholinephosphotransferase
PP: phosphatidate phosphohydrolase
TAG: triacylglycerol

lipids (Figure 10.3). The final acylation at *sn*-3 of the DAG catalyzed by a diacylglycerol acyltransferase (DGAT; 2.3.1.20) is the only enzymatic reaction exclusively committed to TAG biosynthesis (Cao and Huang, 1986, 1987).

Sunflower oil contains TAG molecular species highly enriched in linoleate; therefore, characterization of sunflower acyltransferases is of special interest. Hence, the whole TAG synthesis pathway has been studied in microsomal fractions from developing seeds using several radio-labeled substrates (Griffiths et al., 1988; Mancha and Sánchez, 1981; Rochester and Bishop, 1984; Stymne and Stobart, 1984; Triki et al., 1999; Wiberg et al., 1994).

The first acylation of G3P at the *sn*-1 hydroxyl position with an activated fatty acid is catalyzed by GPAT forming LPA (Figure 10.3). In eukaryotic cells, multiple GPAT isoforms are presents (Athenstaedt and Daum, 1999; Dirks and Sul, 1997). Three plant types of GPAT localized in plastids, in the ER and mitochondria, have been described (Frentzen, 1990; Murata and Tasaka, 1997). GPAT enzymes in the chloroplast, cytoplasm, and mitochondria are different proteins and they differ also in their substrate specificities. Of the three known types, only the ER-bound GPAT seems to be involved in the Kennedy pathway activity based on similarity to yeast GPAT and other acyltransferases (Gidda et al., 2009).

To date, an eight-member gene encoding ER-bound GPAT family in Arabidopsis has been described (Beisson et al., 2007; Zheng et al., 2003). However, further characterization of five Arabidopsis genes has been reported, suggesting that these play a role in cutin and suberin synthesis instead of in seed oil synthesis in Arabidopsis (Beisson et al., 2007; Li et al., 2007; Li-Beisson et al., 2009). Therefore, the specific GPAT gene important for TAG synthesis remains to be identified to date. However, the closest Arabidopsis homolog (At5g60620) of the discovered microsomal GPAT important for TAG production in mice and humans (Cao et al., 2006) provides an obvious candidate for further study.

An asymmetrical distribution of saturated fatty acids, mainly in high stearic sunflower lines between position *sn*-1 and *sn*-3 of TAGs, has been described (Martinez-Force et al., 2004). Reported sunflower GPAT kinetic data (Fraser et al., 2003; Ruiz-Lopez et al., 2010) have shown the highest activity for linoleoyl-CoA, with palmitoyl-CoA and oleoyl-CoA also yielding significant rates. However, the low activity levels for stearoyl-CoA may indicate that this activity could be the one responsible for the asymmetric distribution (Ruiz-Lopez et al., 2010). Furthermore, these results are consistent with the main presence of linoleoyl-, palmitoyl-, and oleoyl-CoA at the *sn*-1 position of TAG in sunflower oil.

GPAT in most plants shows a particular preference for saturated fatty acids such as palmitate. For further characterization of sunflower GPAT, partial purification has been reported (Ruiz-Lopez et al., 2010), in which a putative microsomal GPAT seems to be responsible for the main activity in developing sunflower seeds consistent with previously reported data from different plants (Eccleston and Harwood, 1995; Manaf and Harwood, 2000).

In plants and other organisms, 2-lysophosphatidic acid acyltransferase (LPAAT; EC 2.3.1.51) is responsible for the second acylation of LPA to produce phosphatidic acid, a key intermediate in the biosynthesis of membrane phospholipids and storage lipids (Figure 10.3). These LPAAT activities have been found in the microsomes, plasma membrane (Bursten et al., 1991), mitochondria (Zborowski and Wojtczak, 1969), and plastids (Bourgis et al., 1999; Kim and Huang, 2004; Yu et al., 2004).

In Arabidopsis, several genes have been speculated to encode LPATs. Five LPAAT genes (*AtLPAT1–AtLPAT5*) have been identified based on sequence similarity to cloned and characterized LPAATs from other organisms (Kim and Huang, 2004). *AtL-PAT1* gene encodes the lone ubiquitous prokaryote-like LPAAT of the plastid (Kim and Huang, 2004). The plastid LPAAT is more active in vitro on 16:0- than 18:1-CoA in Brassica and Arabidopsis (Bourgis et al., 1999; Kim and Huang, 2004; Yu et al., 2004). Similarly, LPAAT activity has been confirmed for microsomal *AtLPAAT2* and *AtLPAAT3*, homologs to yeast SLC1-1 (Nagiec et al., 1993). Arabidopsis LPAAT2 is a ubiquitous ER-localized LPAAT, whereas LPAAT3 is predominantly expressed in pollen (Kim et al., 2005). *AtLPAAT2* acts at a metabolic pivotal point, being responsible for the biosynthesis of diverse glycerolipids. At this time, the identities of two other related genes, putative *AtLPAAT4* and *AtLPAAT5,* remain to be elucidated.

Microsomal LPAAT has the highest substrate stringency of the three Kennedy pathway acyl-transferases. ER-located LPAATs from several plant species exhibit a substrate preference for 18:1- over 16:0-CoA and they are also able to incorporated 18:2-CoA (Brown et al., 2002; Ohlrogge and Browse, 1995). Such discrimination against saturated acyl groups in the *sn*-2 position of TAG is relatively common among major oilseed crops. Some others species in which the seed oils have uncommon acyl moieties at the *sn*-2 position possess additional seed-specific LPAATs that are reactive toward the respective uncommon acyl-CoAs (Brown et al., 1995; Cao et al., 1990; Frentzen, 1998; Hanke et al., 1995; Knutzon et al., 1995; Laurant and Huang, 1992; Oo and Huang, 1989). Therefore, LPAAT also plays a role in the de novo synthesis of phosphatidylcholine (PC) and this may be reflected in the acyl-donor specificity of LPAATs containing high levels of polyunsaturated fatty acids. Thus, two classes of genes encoding microsomal LPAATs exist, class A and B (Frentzen, 1998). The microsomal LPAAT class A possesses substrate preferences for C18:1-CoA that are ubiquitously expressed in the plant. In contrast, individual members of the class B LPAATs display preferences for distinct, unusual saturated or unsaturated acyl groups and are normally expressed in storage organs (Maisonneuve et al., 2010).

The substrate selectivity of seed acyl-CoA:lysophosphatidate acyltransferase (LPAAT) is an important factor in determining the final acyl composition of plant oils (Bernerth and Frentzen, 1990). To evaluate the potential contribution of LPAAT isoenzymes to TAG biosynthesis in sunflower, the LPAAT activity has been studied in developing seed fractions (Brown et al., 2002). The total LPAAT activity in sunflower microsomes and fat washed pads was similar, with the specific activity being higher in the former. There was essentially no difference between the selectivity ratios obtained for both fractions, suggesting it is the same LPAATs being assayed in each fraction. Sunflower LPAAT is able to incorporate both 18:1 and 18:2 acyl groups at significant rates, exhibiting a slight substrate preference for 18:2-CoA over 18:1-CoA, while 16:0 and 18:0 are selected against. In this study, it has been demonstrated that the acyl group at the *sn*-1 of the glycerol backbone does not have a significant influence on what is incorporated at the *sn*-2 position (Brown et al., 2002).

LPAAT enzymatic assays have confirmed the ability of LPAAT to incorporate certain acyl groups, resulting in TAGs containing high levels of linoleate and oleic acid at the *sn*-2 position of TAG from sunflower, up to 58% and 37%, respectively (Ruíz-López, 2006). Moreover, stearoyl-CoA is not incorporated in TAG, due to the fact that LPAAT and the rest of the microsomal enzymes involved in TAG biosynthesis in sunflower seeds have very low specify for this fatty acid.

After the second acylation, the dephosphorylation of the resultant PA is catalyzed by PP to form DAG (Figure 10.3). In other eukaryotic systems, two classes of PP enzymes have been identified as Mg^{2+} dependent and Mg^{2+} independent (reviewed in Carman and Han, 2006). The latter class, also referred to as *PP2 enzymes,* is involved in lipid signaling. In contrast, the former class (referred to as *PP1 enzymes*) appear to play a role in the synthesis of lipids.

DAG represents an important branch point between nonpolar and membrane lipid biosynthesis. DGAT is responsible for transferring an acyl group from acyl coenzyme A to the *sn*-3 of 1,2 diacylglycerol (DAG) to form TAG (Figure 10.3). The enzyme catalyzes the final acylation reaction in the TAG biosynthesis via the Kennedy pathway; it is the only reaction not involved in membrane biosynthesis. From database DAGAT gene sequences on, two nonhomologous DGAT families have been described, DAGAT1 and DAGAT2. Both enzymes are bound to ER membrane (Kroon et al., 2006; Lardizabal et al., 2001; Liu et al., 2012; Shockey et al., 2006; Zhang et al., 2009). In addition, a new cytoplasmic type, DAGAT3, has been characterized (Hernández et al., 2012; Saha et al., 2006). DAGAT1 and DAGAT2 enzymes may play a crucial role in TAG accumulation despite differences in substrate preferences.

Previous studies in Arabidopsis and maize have established that DGAT is a key activity in determining the levels of TAGs in seed oils (Cahoon et al., 2007; Napier, 2007; Venegas-Calerón et al., 2010). Moreover, DGAT activity assays have shown the lowest specific activity of the Kennedy pathway enzymes and are regarded as a "bottleneck" in TAG synthesis (Venegas-Calerón et al., 2010). In higher plants, DGAT enzyme showed a broad acyl-CoA specificity (Banas et al., 2013), although the enzymes from different plant species differed markedly in acyl selectivity (Wiberg et al., 1994). Therefore, the fatty acid incorporation at the *sn*-3 position of the glycerol backbone may depend on the fatty acid pool available at the site of synthesis.

Selectivity of DGAT plays a crucial role in the selective channelling of uncommon acyl groups into TAG in oil seeds.

In order to elucidate the mechanism responsible for TAG accumulation in sunflower seed, DGAT activity has been studied too. According to TAG composition, microsomal sunflower DGAT utilized 18:2-CoA better than 18:1-CoA. The higher specialization of sunflower DGAT toward linoleic acid could explain its oil composition. Interestingly, acyl-CoAs and DAG with 10:0 groups served as efficient substrates for the DGAT from sunflower seeds. Although sunflower seed does not accumulate medium-chain acyl groups, 10:0-CoA is selectively used over 18:1-CoA by sunflower DGAT.

At present, although three genes that codified for DGAT activity have been described in sunflower, *DGAT1A* (GenBank: EU543809.1) (Fernández-Garcia et al., 2008), *DGAT1B* (GenBank: EU232721.1), and *DGAT2* (GenBank: EF563992.1), only the sunflower *DAGT1A* gene has been characterized (Sun et al., 2011). Overexpression of *HaDGAT1A* in yeast results in oil increased deposition (1.76-fold). Looking carefully at the fatty acids composition, it was noticed how the percentage of saturated fatty acids decreased, while the percentage of unsaturated fatty acids increased among total composition.

An Alternative Route

An acyl-CoA independent mechanism of TAG synthesis has been recently identified in yeast and plants that provides another opportunity to move fatty acids from PC to TAG (Banas et al., 2000; Ghosal et al., 2007). In this second pathway, acyl groups can be channelled into PC via the activity of lysophosphatidylcholine acyltransferase (LPCAT), and then subsequently transferred from the acyl donor, PC (or phosphatidyl ethanolamine), to DAG via the acyl-exchange activity of acyl-CoA independent phospholipid:diaylglycerol acyltransferase (PDAT, EC 2.3.1.158) (Figure 10.3).

PDAT transfers an acyl group from the *sn*-2 position of phospholipids (e.g., phosphatidylcholine, phosphatidylethanolamine) directly to the *sn*-3 position of DAG, yielding TAG and lyso-phospholipids (Banas et al., 2000; Dahlqvist et al., 2000; Oelkers et al., 2000; Stahl et al., 2004). In oilseeds, substrate specificity of the PDAT enzyme depends on the head group of the acyl donor, the acyl group transferred, and the acyl chains of the acceptor DAG (Dahlqvist et al., 2000). Furthermore, PDAT acyl group specificity varies between oilseed plants, where a large number of unusual fatty acids can be found. These unusual fatty acids are rarely found in the membranes. Instead, they are mainly incorporated into TAG. Therefore, oilseeds possess a mechanism for the efficient and selective transfer of these acyl groups from PC to TAG. This reaction is catalyzed by the phospholipid:diacylglycerol acyltransferase.

Independent of sunflower oil composition, it has been probed that PDAT activity from microsomal fractions is able to incorporate unusual fatty acids not present in its oil, such as oleic, ricinoleic, or vernolic, from the *sn*-2 position of PC into DAG (Dahlqvist et al., 2000). A recent work has demonstrated that the PDAT activity in sunflower membranes from developing seeds has shown that PE with 18:2 was the best acyl donor, followed by PE with 18:1, PC with 18:2, PC with 18:1, and PC with 18:0 at the *sn*-2 position (Banas et al., 2013)

In summary, two types of enzymes are involved in the last step of TAG synthesis in plants: DGAT and PDAT (Zhang et al., 2009). The relative contribution of DGAT and PDAT in seed TAG synthesis can vary drastically between even evolutionarily closely related species, like sunflower and safflower (Banas et al., 2013). This study has shown that, in sunflower, DAGAT activity is dominating over PDAT activity in all stages of seed development.

New Activities Involved

PC acyl editing and phosphocholine head-group exchange between PC and DAG control the majority of acyl fluxes through PC to provide PUFA for TAG synthesis (Bates et al., 2012). In recent years, the discovery of other new enzymes affecting TAG accumulation has challenged the TAG biosynthesis pathway. These new activities appear to play significant roles in incorporating acyl residues from PC into TAG by PC acyl editing, phosphocholine head-group exchange, or membrane metabolism (Bates et al., 2012).

The main PC-to-DAG conversion in Arabidopsis is catalyzed by a recently identified phosphatidylcholine:diacylglycerol cholinephosphotransferase (PDCT, EC 2.7.8.), transferring the complete phosphocholine head group from PC to DAG (Lu et al., 2009a) (Figure 10.3). PDCT seems to provide a major route for desaturated acyl chains to be returned to the DAG pool (Bates et al., 2012).

Other reactions affecting TAG accumulation through PC-derived DAG involve the rapidly reverse reaction of CDP-choline:1,2-*sn*-diacylglycerol cholinephosphotransferase (Slack et al., 1983, 1985). The DAG moiety of PC can be liberated by hydrolysis, thus becoming available for TAG final bioassembly by DGAT. Phospholipases C and D (followed by PAP) also catalyze the removal of just the choline group from PC to form PA, which can then be dephosphorylated by 3-*sn*-phosphatidate phosphohydrolase (PAP) to form DAG. In both of these head-group removals (PDCT and PLC), the acyl groups at positions *sn*-1 and *sn*-2 of PC retain their positioning on the TAG backbone (Figure 10.3).

In addition, other activities can potentially influence the movement of fatty acids into TAG. The phospholipase A2 may hydrolyze acyl groups at the *sn*-2 position of PC and may produce liberated fatty acids for subsequent activation by the action of LACS (Stahl et al., 1998).

Finally, it is important to underscore the acyl exchange between the fatty acid at the *sn*-2 position of PC with the acyl-CoA pool. This reversible reaction is catalyzed by acyl-CoA:lysophosphatidylcholine acyltransferase (LPCAT; EC 2.3.1.23). LPCAT activity was first described in rat liver microsomes (Lands, 1960), where it plays a major role in remodelling PC. In oilseeds, it is important in regulating oil quality by controlling the exchange of fatty acids through PC, making these acyl chains available for further metabolic pathways. Sunflower LPCAT has been partially purified by Fraser and Stobart (2000).

Future Prospects

In sunflower seed, the dominant lipid class is TAG, with about 90–97% of the total lipids. Linoleic acid is the dominant fatty acid with about 60% of the total fatty acids. Oleic acid is the second most abundant, with around 20% of the fatty acids among TAG, with fluctuations of fatty acids content between oleic and linoleic acid at high temperature (Pérez-Vich et al., 2002). Palmitic and stearic acid amounts represent about 13% of total fatty acid content. All other TAG fatty acids are detected in trace amounts (Osorio et al., 1995). A number of factors contribute to the final stereospecific composition of storage TAG. The relative importance of each of these will vary according to the plant species. In certain species the final composition of storage TAG will be determined by acyltransferase substrate selectivity, and in others it will be determined by availability of substrate and alternative lipid remodelling pathways.

In conclusion, according to Griffiths et al. (1998), in sunflower, saturated acyl groups are selectively utilized for acylating position *sn*-1 of glycerol-3-phosphate and are selected against in the exchange between acyl-CoA and position *sn*-2 of PC. Oleate enters position *sn*-2 of PC by acyl exchange between acyl-CoA and PC. Also, the diacylglycerol–phosphatidylcholine interconversion may be of greater importance in increasing the level of linoleate in the TAG. Overall, the TAG synthesis model highlights a complex network of enzymatic reactions mediating the movement of acyl groups between PC, acyl-CoA, DAG, PA, and TAG and contributes to the final stereospecific composition of storage TAG. Although over the last decade new activities have been characterized, the regulation flux through pathways is not completely understood. New uncharacterized activities and/or regulator genes presumably operate in addition to recently discovered pathways involved in the TAG biosynthesis, and these represent an important area for future research.

References

Abbadi, A.; Brummel, M.; Schutt, B. S.; Slabaugh, M. B.; Schuch, R.; Spener, F. Reaction Mechanism of Recombinant 3-Oxoacyl-(acylcarrier-protein) Synthase III from *Cuphea wrightii* Embryo, a Fatty Acid Synthase Type II Condensing Enzyme. *Biochem. J.* **2000**, *345*, 153–160.

Alkio, M.; Diepenbrock, W.; Grimm, E. Evidence for Sectorial Photoassimilate Supply in the Capitulum of Sunflower (*Helianthus annuus*). *New Phytol.* **2002**, *156*, 445–456.

Alonso, A. P.; Goffman, F. D.; Ohlrogge, J. B.; Shachar-Hill, Y. Carbon Conversion Efficiency and Central Metabolic Fluxes in Developing Sunflower (*Helianthus annuus* L.) Embryos. *Plant J.* **2007**, *52*, 296–308.

Álvarez-Ortega, R.; Cantisán, S.; Martínez-Force, E.; Garcés, R. Characterization of Polar and Nonpolar Lipid Classes from Highly Saturated Fatty Acid Sunflower Mutants. *Lipids* **1997**, *32*, 833–837.

Andre, C.; Froehlich, J. E.; Moll, M. R.; Benning, C. A Heteromeric Plastidic Pyruvate Kinase Complex Involved in Seed Oil Biosynthesis in Arabidopsis. *Plant Cell* **2007**, *9*, 2006–2022.

Andriotis, V. M.; Kruger, N. J.; Pike, M. J.; Smith, A. M. Plastidial Glycolysis in Developing Arabidopsis Embryos. *New Phytol.* **2010**, *185*, 649–662.

Athenstaedt, K; Daum, G. Phosphatidic Acid, a Key Intermediate in Lipid Metabolim. *Eur. J. Biohem.* **1999**, *266*, 1–16.

Aznar-Moreno, J. A.; Venegas-Calerón, M.; Salas, J. J.; Garcés, R.; Martínez-Force, E. cDNA Cloning, Expression Levels and Biochemical Characterization of ACP (Acyl Carrier Protein) Isoforms from Sunflower (*Helianthus annuus* L.). GenBank sequence, unpublished work, 2010.

Aznar-Moreno, J. A.; Venegas-Calerón, M.; Martínez-Force, E.; Garcés, R.; Mullen, R. T.; Satinder K. G.; Salas, J. J. Sunflower (*Helianthus annuus*) Long-Chain Acyl-coenzyme A Synthetases Expressed at High Levels in Developing Seeds. *Physiologia Plantarum* **2013a**, DOI:10.1111/ppl.12107.

Aznar-Moreno, J. A.; Martínez-Force, E.; Venegas-Calerón, M.; Garcés, R.; Salas, J. J. Changes in Acyl-coenzyme A Pools in Sunflower Seeds with Modified Fatty Acid Composition. *Phytochemistry* **2013b**, *87*, 39–50.

Banas, A.; Dahlqvist, A.; Stahl, U.; Lenman, M.; Stymne, S. The Involvement of Phospholipid:diacylglycerol Acyltransferases in Triacylglycerol Production. *Biochem. Soc. Trans.* **2000**, *28*, 703–705.

Banas, W.; Sanchez-Garcia, A.; Banas, A.; Stymne, S. Activities of Acyl-CoA:diacylglycerol Acyltransferase (DGAT) and Phospholipid:diacylglycerol Acyltransferase (PDAT) in Microsomal Preparations of Developing Sunflower and Safflower Seeds. *Planta* **2013**, *237*, 1627–1636.

Bao, X.; Pollard, M.; Ohlrogge, J. The Biosynthesis of Erucic Acid in Developing Embryos of *Brassica rapa*. *Plant Physiol.* **1998**, *118*, 183–190.

Barrett, P. B.; Harwood, J. L. Characterization of Fatty Acid Elongase Enzymes from Germinating Pea Seeds. *Phytochemistry* **1998**, *48*, 1295–1304.

Bates, P.; Durrett, T. P.; Ohlrogge, J.; Pollard, M. Analysis of Acyl Fluxes through Multiple Pathways of Triacylglycerol Synthesis in Developing Soybean Embryos. *Plant Physiol.* **2009**, *150*, 55–72.

Bates, P. D.; Fatihi, A.; Snapp, A. R.; Carlsson, A. S.; Browse, J.; Lu, C. Acyl Editing and Head-group Exchange Are the Major Mechanisms That Direct Polyunsaturated Fatty Acid Flux into Triacylglycerols. *Plant Physiol.* **2012**, *160*, 1530–1539.

Baud, S.; Graham, I. A. A Spatiotemporal Analysis of Enzymatic Activities Associated with Carbon Metabolism in Wild-type and Mutant Embryos of Arabidopsis Using in situ Histochemistry. *Plant J.* **2006**, *46*, 155–169.

Beisson, F.; Li, Y. H.; Bonaventure, G.; Pollard, M.; Ohlrogge, J. B. The Acyltransferase GPAT5 Is Required for the Synthesis of Suberin in Seed Coat and Root of Arabidopsis. *Plant Cell* **2007**, *19*, 351–368.

Bernerth, R.; Frentzen, M. Utilization of Erucoyl-CoA by Acyltransferases from Developing Seeds of *Brassica napus* (L.) Involved in Triacylglycerol Biosynthesis. *Plant Sci.* **1990**, *67*, 21–28.

Bonaventure, G.; Ohlrogge, J. B. Differential Regulation of mRNA Levels of Acyl Carrier Protein Isoforms in Arabidopsis. *Plant Physiol.* **2002**, *128*, 223–235.

Bourgis, F.; Kader, J. C.; Barret, P.; Renard, M.; Robinson, D.; Robinson, C.; Delseny, M.; Roscoe, T. J. A Plastidial Lysophosphatidic Acid Acyltransferase from Oilseed Rape. *Plant Physiol.* **1999**, *120*, 913–921.

Brown, A. P.; Brough, C. L.; Kroon, J. T. M.; Slabas, A. R. Identification of a cDNA That Encodes a 1-Acyl-sn-glycerol-3-phosphate Acyltransferase from *Limnanthes douglasii*. *Plant Mol. Biol.* **1995**, *29*, 267–278.

Brown, A. P.; Slabas, A. R.; Denton, H. Substrate Selectivity of Plant and Microbial Lysophosphatidic Acid Acyltransferases. *Phytochemistry* **2002**, *61*, 493–501.

Bursten, S. L.; Harris, W. E.; Bomsztyk, K.; Lovett, D. Interleukin-1 Rapidly Stimulates Lysophosphatidate Acyltransferase and Phosphatidate Phosphohydrolase Activities in Human Mesangial Cells. *J. Biol. Chem.* **1991**, *266*, 20732–20743.

Cahoon, E. B.; Salehuzzaman, S.; Shanklin, J.; Browse, J. A Determinant of Substrate Specificity Predicted from the Acyl–acyl Carrier Protein Desaturase of Developing Cat's Claw Seed. *Plant Physiol.* **1998**, *117*, 593–598.

Cahoon, E. B.; Shockey, J. M.; Dietrich, C. R.; Gidda, S. K.; Mullen, R. T.; Dyer, J. M. Engineering Oilseeds for Sustainable Production of Industrial and Nutritional Feedstocks: Solving Bottlenecks in Fatty Acid Flux. *Curr. Opin. Plant Biol.* **2007**, *10*, 236–244.

Cantisán, S.; Martínez-Force, E.; Garcés, R. Enzymatic Studies of High Stearic Acid Sunflower Seed Mutants. *Plant Physiol. Biochem.* **2000**, *38*, 377–382.

Cao, Y. Z.; Huang, A. H. C. Diacylglycerol Acyltransferase in Maturing Oil Seeds of Maize and Other Species. *Plant Physiol.* **1986**, *82*, 813–820.

Cao, Y. Z.; Huang, A. H. C. Acyl Coenzyme A Preference of Diacylglycerol Acyltransferase from the Maturing Seeds of Cuphea, Maize, Rapeseed, and Canola. *Plant Physiol.* **1987**, *84*, 762–765.

Cao, J. S.; Li, J. L.; Li, D. M.; Tobin, J. F.; Gimeno, R. E. Molecular Identification of Microsomal Acyl-CoA:glycerol-3-phosphate Acyltransferase, a Key Enzyme in de novo Triacylglycerol Synthesis. *Proc. Natl. Acad. Sci. USA* **2006**, *103*, 19695–19700.

Cao, Y. Z.; Oo, K. C.; Huang, A. H. C. Lysophosphatidate Acyltransferase in the Microsomes from Maturing Seeds of Meadowfoam (*Limnanthes alba*). *Plant Physiol.* **1990**, *94*, 1199–1206.

Carman, G. M.; Han, G. S. Roles of Phosphatidate Phosphatase Enzymes in Lipid Metabolism. *Trends Biochem. Sci.* **2006**, *31*, 694–699.

Cassagne, C.; Lessire, R.; Bessoule, J. J.; Moreou, P.; Créach, A.; Schneider, F.; Sturbois, B. Biosynthesis of Very Long Chain Fatty Acids in Higher Plants. *Prog. Lipid Res.* **1994**, *33*, 55–69.

Cernac, A.; Benning, C., WRINKLED1 Encodes an AP2/EREB Domain Protein Involved in the Control of Storage Compound Biosynthesis in Arabidopsis. *Plant J.* **2004**, *40*, 575–585.

Claeyssen, E.; Rivoal, J. Isozymes of Plant Hexokinase: Occurrence, Properties and Functions. *Phytochemistry* **2007**, *68*, 709–731.

Dahlqvist, A.; Ståhl, U.; Lenman, M.; Banas, A.; Lee, M.; Sandagar, L.; Ronne, H.; Stymne, S. Phospholipid:diacylglycerol Acyltransferase: An Enzyme That Catalyzes the Acyl-CoA-independent Formation of Triacylglycerol in Yeast and Plants. *Proc. Natl. Acad. Sci. USA* **2000**, *97*, 6487–6492.

Dirks, L. K.; Sul, H. S. Mammalian Mitochondrial Glycerol-3-phosphate Acyltransferase. *Biochim. Biophys. Acta* **1997**, *1348*, 17–26.

Domergue, F.; Chevalier, S.; Créach, A.; Cassagne, C.; Lessire, R. Purification of the Acyl-CoA Elongase Complex from Developing Rapeseed and Characterization of the 3-Ketoacyl-CoA Synthase and the 3-Hydroxyacyl-CoA Dehydratase. *Lipids* **2000**, *35*, 487–494.

Eccleston, V. S.; Harwood, J. L. Solubilization, Partial Purification and Properties of Acyl-CoA glycerol-3-phosphate Acyltransferase from Avocado (*Persea americana*) Fruit Mesocarp. *Biochim. Biophys. Acta* **1995**, *1257*, 1–10.

Fernández-Garcia, J. D.; Garcés, R.; Martínez-Force, E. Cloning and Biochemical Characterization of Diacylglycerol Acyltransferases from Sunflower (*Helianthus annuus* L.). GenBank sequence, unpublished work, 2008.

Fernández-Moya, V.; Martínez-Force, E.; Garcés, R. Temperature Effect on a High Stearic Acid Sunflower Mutant. *Phytochemistry* **2002**, *59*, 33–37.

Fernández-Moya, V.; Martínez-Force, E.; Garcés, R. Temperature-related Non-homogeneous Fatty Acid Desaturation in Sunflower (*Helianthus annuus* L.) Seeds. *Planta* **2003**, *216*, 834–840.

Fischer, K.; Weber, A., Transport of Carbon in Non-green Plastids. *Trends Plant Sci.* **2002**, *7*, 345–351.

Fraser, T.; Stobart, K. Partial Purification and Photoaffinity Labelling of Sunflower Acyl-CoA:lysophosphatidylcholine Acyltransferase. *Biochem. Soc. Trans.* **2000**, *28*, 715–718.

Fraser, T.; Chatrattanakunchai, S.; Waters, A. D.; Stobart, K. Kinetic Analysis of Microsomal Sn-glycerol-3-phosphate Acyltransferase from the Developing Seeds of Sunflower and *Mortierella alpine*. In *Advanced Research on Plant Lipids,* Proceedings of the 15th International Symposium on Plant Lipids, Okazaki, Japan, May 12–17, 2002; Murata, N., Yamada, M., Nishida, I., Okuyma, H., Sekiya, J., Hajime, W., Eds.; Springer: Dordrecht, The Netherlands, 2003.

Frentzen, M. Comparison of Certain Properties of Membrane-bound and Solubilized Acyltransferase Activities of Plant Microsomes. *Plant Sci.* **1990**, *69*, 39–48.

Frentzen, M. Acyltransferases from basic science to modified seed oils. *Fett/Lipid* **1998**, *100*, 161–166.

Fulda, M.; Shockey, J.; Werber, M.; Wolter, F. P.; Heinz, E. Two Long-chain Acyl-CoA Synthetases from *Arabidopsis thaliana* Involved in Peroxisomal Fatty Acid β-oxidation. *Plant J.* **2002**, *32*, 93–103.

Fulda, M. S.; Schnurr, J. M.; Abbadi, A.; Heinz, E.; Browse, J. A. Peroxisomal Acyl-CoA Synthetase Activity Is Essential for Seedling Development in *Arabidopsis thaliana. Plant Cell* **2004**, *16*, 394–405.

Garcés, R.; Mancha, M. In Vitro Oleate Desaturase in Developing Sunflower Seeds. *Phytochemistry* **1991**, *30*, 2127–2130.

Garcés, R.; Sarmiento, C.; Mancha, M. Oleate from Triacylglycerols Is Desaturated in Cold-induced Developing Sunflower (*Helianthus annuus* L.) Seeds. *Planta* **1994**, *193*, 473–477.

García-Díaz, M. T.; Martínez-Rivas, J. M.; Mancha, M. Temperature and Oxygen Regulation of Oleate Desaturation in Developing Sunflower (*Helianthus annuus*) Seeds. *Physiol. Plant* **2002**, *114*, 13–20.

Ghosal, A.; Banas, A.; Ståhl, U.; Dahlqvist, A.; Lindqvist, Y.; Stymne, S. Saccharomyces Cerevisiae Phospholipid:diacylglycerol Acyl Transferase (PDAT) Devoid of Its Membrane Anchor Region Is a Soluble and Active Enzyme Retaining Its Substrate Specificities. *Biochim. Biophys. Acta* **2007**, *1771*, 1457–1463.

Gidda, S. K.; Shockey, J. M.; Rothstein, S. J.; Dyer, J. M.; Mullen, R. T. *Arabidopsis thaliana* GPAT8 and GPAT9 Are Localized to the ER and Possess Distinct ER Retrieval Signals: Functional Divergence of the Dilysine ER Retrieval Motif in Plant Cells. *Plant Physiol. Biochem.* **2009**, *47*, 867–879

González-Mellado, D.; Garcés, R.; Martínez-Force, E. Cloning and Characterization of Beta-keto-acyl-ACP Synthase I from Developing Sunflower (*Helianthus annuus* L.) Seeds. GenBank sequence, unpublished work, 2006a.

González-Mellado, D.; Garcés, R.; Martínez-Force, E. Cloning and Characterization of Beta-keto-acyl-ACP Synthase II from Developing Sunflower (*Helianthus annuus* L.) Seeds. GenBank sequence, unpublished work, 2006b.

González-Mellado, D.; Garcés, R.; Martinez-Force, E. Cloning and Characterization of Beta-ketoacyl-CoA Synthases from Developing Sunflower (*Helianthus annuus* L.) Seeds. GenBank sequence, unpublished work, 2006c.

González-Mellado, D.; von Wettstein-Knowles, P.; Garcés, R.; Martínez-Force, E. The Role of β-ketoacyl-acyl Carrier Protein Synthase III in the Condensation Steps of Fatty Acid Biosynthesis in Sunflower. *Planta* **2010**, *231*, 1277–1289.

González-Thuillier, I.; Venegas-Calerón, M.; Garcés, R.; Martínez-Force, E. Cloning and Functional Characterization of Beta-hydroxyacyl-ACP Dehydratase Isoforms from *Helianthus annuus* L. GenBank sequence, unpublished work, 2010a.

González-Thuillier, I.; Venegas-Calerón, M.; Garcés, R.; Martínez-Force, E. Cloning and Functional Characterization of Beta-ketoacyl-ACP Reductase Isoforms from *Helianthus annuus* L. GenBank sequence, unpublished work, 2010b.

González-Thuillier, I.; Venegas-Calerón, M.; Garcés, R.; Martínez-Force, E. Cloning and Functional Characterization of Enoyl-ACP Reductase Isoforms from *Helianthus annuus* L. GenBank sequence, unpublished work, 2010c.

González-Thuillier, I. Deepening in the Study of Genes Related to the Fatty Acids Intraplastidial Synthesis in Developing Sunflower Seeds. Ph.D. Thesis, University of Seville, Spain, May 2013.

Griffiths, G.; Stymne, S.; Stobart, A. K. The Utilisation of Fatty-acid Substrates in Triacylglycerol Biosynthesis by Tissue-Slices of Developing Safflower (*Carthamus tinctorius* L.) and Sunflower (*Helianthus annuus* L.) Cotyledons. *Planta* **1988**, *173*, 309–316.

Griffiths, G.; Griffiths, W. T.; Stobart, A. K. Delta-9 Desaturase Activity in Developing Cotyledons of Sunflower (*Helianthus annuus* L.) *Phytochemistry* **1998**, *48*, 261–267.

Gunstone, F. D.; Harwood, J. L.; Padley, F. B. *The Lipid Handbook*; Chapman and Hall: London, 1992; pp 101–102.

Hajduch, M.; Casteel, J. E.; Tang, S.; Hearne, L. B.; Knapp, S.; Thelen, J. J. Proteomic Analysis of Near-isogenic Sunflower Varieties Differing in Seed Oil Traits. *J. Proteome Res.* **2007**, *6*, 3232–3241.

Hall, A. J.; Whitfield, D. M.; Connor, D. J. Contribution of Pre-anthesis Assimilates to Grain-filling in Irrigated and Water-stressed Sunflower Crops. II. Estimates from a Carbon budgeB. *Field Crops Res.* **1990**, *24*, 273–294.

Hanke, C.; Wolter, F. P.; Coleman, J.; Peterek, G.; Frentzen, M. A Plant Acyltransferase Involved in Triacylglycerols Biosynthesis Complements an *Escherichia coli* Sn-1-acylglycerol-3-phosphate Acyltransferase Mutant. *Eur. J. Biochem.* **1995**, *232*, 806–810.

Harris, H. M.; McWilliam, J. R.; Mason, W. K. Influence of Temperature on Oil Content and Composition of Sunflower Seed. *Aust. J. Agric. Res.* **1978**, *29*, 1203–1212.

Harwood, J. L. Recent Advances in the Biosynthesis of Plant Fatty Acids. *Biochim. Biophys. Acta* **1996**, *1301*, 7–56.

Harwood, J. L. Fatty Acid Biosynthesis. In *Plant Lipids: Biology, Utilisation and Manipulation*; Murphy, D. J., Ed.; Blackwell Publishing: Oxford, UK, 2005; pp 27–66.

Heath, R. J.; Rock, C. O. Enoyl-acyl Carrier Protein Reductase (*fabI*) Plays a Determinant Role in Competing Cycles of Fatty Acid Elongation in *Escherichia coli*. *J. Biol. Chem.* **1995**, *270*, 26538–26542.

Hernández, M. L.; Whitehead, L.; He, Z.; Gazda, V.; Gilday, A.; Kozhevnikova, E.; Vaistij, F. E.; Larson, T. R.; Graham, I. A. A Cytosolic Acyltransferase Contributes to Triacylglycerol Synthesis in Sucrose-rescued Arabidopsis Seed Oil Catabolism Mutants. *Plant Physiol.* **2012**, *160*, 215–225.

Hewezi, T.; Petitprez, M.; Gentzbittel, L. Primary Metabolic Pathways and Signal Transduction in Sunflower (*Helianthus annuus* L.): Comparison of Transcriptional Profiling in Leaves and Immature Embryos Using cDNA Microarrays. *Planta* **2006**, *223*, 948–964.

Hill, L. M.; Morley-Smith, E. R.; Rawsthorne, S. Metabolism of Sugars in the Endosperm of Developing Seeds of Oilseed Rape. *Plant Physiol.* **2003**, *131*, 228–236.

Hongtrakul, V.; Slabaugh, M. B.; Knapp, S. J. DFLP, SSCP, and SSR Markers for Delta 9-Stearoyl-acyl Carrier Protein Desaturases Strongly Expressed in Developing Seeds of Sunflower: Intron Lengths Are Polymorphic among Elite Inbred Lines. *Mol. Breed.* **1998a**, *4*, 195–203.

Hongtrakul, V.; Slabaugh, M. B.; Knapp, S. J. A Seed Specific Δ12 Oleate Desaturase Gene Is Duplicated, Rearranged, and Weakly Expressed in High Oleic Acid Sunflower Lines. *Crop. Sci.* **1998b**, *38*, 1245–1249.

Houston, N. L.; Hajduch, M.; Thelen, J. J. Quantitative Proteomics of Seed Filling in Castor: Comparison with Soybean and Rapeseed Reveals Differences between Photosynthetic and Nonphotosynthetic Seed Metabolism. *Plant Physiol.* **2009**, *151*, 857–868.

Jessen, D.; Olbrich, A.; Knüfer, J.; Krüger, A.; Hoppert, M.; Polle, A.; Fulda, M. Combined Activity of LACS1 and LACS4 Is Required for Proper Pollen Coat Formation in Arabidopsis. *Plant J.* **2011**, *68*, 715–726.

Jones, A.; Davies, H. M.; Voelker, T. A. Palmitoyl-acyl Carrier Protein (ACP) Thioesterase and the Evolutionary Origin of Plant Acyl-ACP Thioesterases. *Plant Cell* **1995**, *7*, 359–371.

Kabbaj, A.; Vervoort, V.; Abbot, A. G.; Tersac, M.; Bervillé, A. Expression of Stearate, Oleate and Linoleate-desaturase Genes in Sunflower with Normal and High Oleic Contents. *Helia* **1996**, *19*, 1–17.

Kim, H. U.; Huang, A. H. C. Plastid Lysophosphatidyl Acyltransferase Is Essential for Embryo Development in Arabidopsis. *Plant Physiol.* **2004**, *134*, 1206–1216.

Kim, H. U.; Hsieh, K.; Ratnayake, C.; Huang, A. H. C. A Novel Group of Oleosins Is Present Inside the Pollen of Arabidopsis. *J. Biol. Chem.* **2002**, *277*, 22677–22684.

Kim, H. U.; Li, Y. B.; Huang, A. H. C. Ubiquitous and Endoplasmic Reticulum-Located Lysophosphatidyl Acyltransferase, LPAT2, Is Essential for Female but Not Male Gametophyte Development in Arabidopsis. *Plant Cell* **2005**, *17*, 1073–1089.

Knutzon, D. S.; Lardizabal, K. D.; Nelsen, J. S.; Bleibaum, J. L.; Davies, H. M.; Metz, J. C. Cloning of a Coconut Endosperm cDNA Encoding a 1-Acyl-sn-glycerol-3-phosphate Acyltransferase That Accepts Medium-chain-length Substrates. *Plant Physiol.* **1995**, *109*, 999–1006.

Kroon, J. T. M.; Wei, W. X.; Simon, W. J.; Slabas, A. R. Identification and Functional Expression of Type 2 Acyl-CoA:diacylglycerol acyltransferase (DGAT2) in Developing Castor Bean Seeds Which Has High Homology to the Major Triglyceride Biosynthetic Enzyme of Fungi and Animals. *Phytochemistry* **2006**, *67*, 2541–2549.

Kutschera, U.; Heiderich, A. Sucrose Metabolism and Cellulose Biosynthesis in Sunflower Hypocotyls. *Physiol. Plant.* **2002**, *114*, 372–379.

Lacombe, S.; Bervillé, A. A Dominant Mutation for High Oleic Acid Content in Sunflower (*Helianthus annuus* L.) Seed Oil Is Genetically Linked to a Single Oleate-Desaturase RFLP Locus. *Mol. Breed.* **2001**, *8*, 129–137.

Lacombe, S.; Souyris, I.; Bervillé, A. J. An Insertion of Oleate Desaturase Homologous Sequence Silences via siRNA the Functional Gene Leading to High Oleic Acid Content in Sunflower Seed Oil. *Mol. Gen. Genom.* **2009**, *281*, 43–54.

Lajara, J. R.; Díaz, U.; Quidiello, D. R. Definite Influence of Location and Climatic Conditions on the Fatty Acid Composition of Sunflower Seed Oil. *J. Am. Oil Chem. Soc.* **1990**, *67*, 618–623.

Lands, W. E. M. Metabolism of Glycerolipids. II. The Enzymatic Acylation of Lysolecithin. *J. Biol. Chem.* **1960**, *235*, 2233–2237.

Lardizabal, K. D.; Mai, J. T.; Wagner, N. W.; Wyrick, A.; Voelker, T.; Hawkins, D. J. DGAT2 Is a New Diacylglycerol Acyltransferase Gene Family—Purification, Cloning, and Expression in Insect Cells of Two Polypeptides from *Mortiella ramaniana* with Diacylglycerol Acyltransferases activitA. *J. Biol. Chem.* **2001**, *276*, 38862–38869.

Laurant, P.; Huang, A. H. C. Organ and Development Specific Acyl CoA Lysophosphatidate Acyltransferase in Palm and Meadowfoam. *Plant Physiol.* **1992**, *99*, 1711–1715.

León, A.; Zambelli, A. D.; Reid, R. J.; Morata, M. M.; Kaspar, M.; Martínez-Force, E.; Garcés, R.; Salas, J. J.; Venegas-Calerón, M. Isolated Mutated Nucleotide Sequences That Encode a Modified Oleate Desaturase Sunflower Protein, Modified Protein, Methods and Uses. Patent No. WO2013/004280, 2013.

Lessire, R.; Domergue, F.; Spinner, C.; Lucet-Levannier, K.; Lellouche, J. P.; Mioskowski, C.; Cassagne, C. Dehydration of 3-Hydroxy Icosanoyl-CoA and Reduction of (E)-2,3-icosenoyl-CoA Are Required for Elongation by Leek Microsomal Elongase(s). *Plant Physiol. Biochem.* **1998**, *36*, 205–211.

Li, Y.; Beisson, F.; Koo, A. J. K.; Molina, I.; Pollard, M.; Ohlrogge, J. Identification of Acyltransferases Required for Cutin Biosynthesis and Production of Cutin with Suberin-like Monomers. *Proc. Natl. Acad. Sci. USA* **2007**, *104*, 18339–18344.

Li-Beisson, Y.; Pollard, M.; Sauveplane, V.; Pinot, F.; Ohlrogge, J. B.; Beisson, F. Nanoridges That Characterize the Surface Morphology of Flowers Require the Synthesis of Cutin Polyester. *Proc Natl Acad Sci USA* **2009**, *106*, 22008–22013.

Lidqvist, Y.; Huang, W.; Schneider, G.; Shanklin, J. Crystal Structure of Δ9 Stearoyl-acyl Carrier Protein Desaturase from Castor Seed and Its Relationship to Other Di-iron proteins. *EMBO J.* **1996**, *15*, 4081–4092.

Liu, Q.; Siloto, R. M.; Lehner, R.; Stone, S. J.; Weselake, R. J. Acyl-CoA:diacylglycerol Acyltransferase: Molecular Biology, Biochemistry and Biotechnology. *Prog. Lipid Res.* **2012**, *51*, 350–377.

Lu, C.; Xin, Z.; Ren, Z.; Miquel, M.; Browse, J. An Enzyme Regulating Triacylglycerol Composition Is Encoded by the ROD1 Gene of Arabidopsis. *Proc. Natl. Acad. Sci. USA* **2009a**, *106*, 18837–18842.

Lu, S.; Song, T.; Kosma, D.; Parsons, E.; Rowland, O.; Jenks, M. Arabidopsis CER8 encodes LONG-CHAIN ACYL-COA SYNTHETASE 1 (LACS1) That Has Overlapping Functions with LACS2 in Plant Wax and Cutin Synthesis. *Plant J.* **2009b**, *59*, 553–564.

Luthra, R.; Munshi, S. K.; Sukhija, P. S. Relationship of Carbohydrate Metabolism with Lipid Biosynthesis in Developing Sunflower (*Helianthus annuus* L.) seeds. *J. Plant Physiol.* **1991**, *137*, 312–318.

Maisonneuve, S.; Bessoule, J. J.; Lessire, R.; Delseny, M.; Roscoe, T. J. Expression of Rapeseed Microsomal Lysophosphatidic Acid Acyltransferase Isozymes Enhances Seed Oil Content in Arabidopsis. *Plant Physiol.* **2010**, *152*, 670–684.

Manaf, A. M.; Harwood, J. L. Purification and Characterisation of Acyl-CoA, Glycerol 3-Phosphate Acyltransferase from Oil Palm (*Elaeis guineensis*) Tissues. *Planta* **2000**, *210*, 318–328.

Mancha, M.; Sánchez, J. Incorporation of Free Fatty Acids into Acylthioesters and Lipids of Developing Sunflower Seeds. *Phytochemistry* **1981**, *20*, 2139–2142.

Martínez-Force, E.; Garcés, R. Dynamic Channelling during de Novo Fatty Acid Biosynthesis in *Helianthus annuus* Seeds. *Plant Physiol. Biochem.* **2002**, *40*, 383–391.

Martínez-Force, E.; Álvarez-Ortega, R.; Garcés, R. Enzymatic Characterisation of High-palmitic Acid Sunflower (*Helianthus annuus* L.) Mutants. *Planta* **1999**, *207*, 533–538.

Martínez Force, E.; Cantisan, S.; Serrano-Vega, M. J.; Garcés, R. Acyl-acyl Carrier Protein Thioesterase Activity from Sunflower (*Helianthus annuus* L.) Seeds. *Planta* **2000,** *211*, 673–678.

Martínez-Force, E.; Ruiz-Lopez, N.; Garcés, R. The Determination of the Asymmetrical Stereochemical Distribution of Fatty Acids in Triacylglycerides. *Anal. Biochem.* **2004**, *334*, 175–182.

Martínez-Rivas, J. M.; Sperling, P.; Lühs, W.; Heinz, E. Spatial and Temporal Regulation of Three Different Microsomal Oleate Desaturase Genes (*FAD2*) from Normal-type and High-oleic Varieties of Sunflower (*Helianthus annuus* L.). *Mol. Breed.* **2001**, *8*, 159–168.

Mayer, K. M.; Shanklin, J. A Structural Model of the Plant Acyl-acyl Carrier Protein Thioesterase FatB Comprises Two Helix/4-stranded Sheet Domains, the N-terminal Domain Containing Residues That Affect Specificity and the C-terminal Domain Containing Catalytic Residues. *J. Biol. Chem.* **2005**, *280*, 3621–3627.

Mekhedov, S.; Martínez de Ilárduya, O.; Ohlrogge, J. Toward a Functional Catalog of the Plant Genome. A Survey of Genes for Lipid Biosynthesis. *Plant Physiol.* **2000**, *122*, 389–401.

Murata, N.; Tasaka, Y. Glycerol-3-phosphate Acyltransferase in Plants. *Biochim. Biophys. Acta* **1997**, *1348*, 10–16.

Nagiec, M. M.; Wells, G. B.; Lester, R. L.; Dickson, R. C. A Suppressor Gene That Enables *Saccharomyces cerevisiae* to Grow without Making Sphingolipids Encodes a Protein That Resembles an *Escherichia coli* Fatty Acyltransferase. *J. Biol. Chem.* **1993**, *268*, 22156–22163.

Napier, J. A. The Production of Unusual Fatty Acids in Transgenic Plants. *Annu. Rev. Plant Biol.* **2007**, *58*, 295–319.

Nikolau, B. J.; Oliver, D. J.; Schnable, P. S.; Wurtele, E. S. Molecular Biology of Acetyl-CoA Metabolism. In *Recent Advances in the Biochemistry of Plant Lipids*; Harwood, J. L., Quinn, P. J., Eds.; Portland Press: London, 2000; pp 591–593.

Oelkers, P.; Tinkelenberg, A.; Erdeniz, N.; Cromley, D.; Billheimer, J. T.; Sturley, S. L. A Lecithin Cholesterol Acyltransferase-like Gene Mediates Diacylglycerol Esterification in Yeast. *J. Biol. Chem.* **2000**, *275*, 15609–15612.

Ohlrogge, J.; Browse, J. Lipid Biosynthesis. *Plant Cell* **1995**, *7*, 957–970.

Oo, K. C.; Huang, A. H. C. Lysophosphatidyl Acyltransferase Activities in the Microsomes from Palm Endosperm, Maize Scutellum, and Rapeseed Cotyledon of Maturing Seeds. *Plant Physiol.* **1989**, *91*, 1288–1295.

Osorio, J.; Fernández-Martínez, J.; Mancha, M.; Garcés, R. Mutant Sunflowers with High Concentration of Saturated Fatty Acids in the Oil. *Crop Science* **1995**, *35*, 739–742.

Pérez-Vich, B.; Fernández-Martínez, J. M.; Grodona, M.; Knapp, S. J.; Berry, S. T. Stearoyl-ACP and Oleoyl-PC Desaturase Genes Cosegregate with Quantitative Trait Loci Underlying High Stearic and High Oleic Acid Mutant Phenotypes in Sunflower. *Theor. Appl. Genet.* **2002**, *104*, 338–349.

Pérez-Vich, B.; León, A. J.; Grondona, M.; Velasco, L.; Fernández-Martínez, J. M. Molecular Analysis of the High Stearic Acid Content in Sunflower Mutant CAS-14. *Theor. Appl. Genet.* **2006**, *112*, 867–875.

Pfeiffer, I.; Kutschera, U. Sucrose Metabolism and Lipid Mobilization during Light-induced Expansion of Sunflower Cotyledons. *J. Plant Physiol.* **1996**, *147*, 553–558.

Plaxton, W. C. The Organization and Regulation of Plant Glycolysis. *Annu. Rev. Plant Physiol. Plant Mol. Biol.* **1996**, *47*, 185–214.

Plaxton, W. C.; Podestá, F. E. The Functional Organization and Control of Plant Respiration. *Crit. Rev. Plant Sci.* **2006**, *25*, 159–198.

Pleite, R.; Pike, M. J.; Garcés, R.; Martínez-Force, E.; Rawsthorne, S. The Sources of Carbon and Reducing Power for Fatty Acid Synthesis in the Heterotrophic Plastids of Developing Sunflower (*Helianthus annuus* L.) Embryos. *J. Exp. Bot.* **2005**, *56*, 1297–1303.

Pollock, C. J.; Farrar, J. F. Source-sink Relations: The Role of Sucrose. In *Photosynthesis and the Environment;* Baker, N. R., Ed.; Kluwer Academic Publishers: Dordrecht, The Netherlands, 1996; pp 261–279.

Rochester, C. P.; Bishop, D. G. The Role of Lysophosphatidylcholine in Lipid Synthesis by Developing Sunflower (*Helianthus annuus* L.) Seed Microsomes. *Archives of Biochemistry and Biophysics* **1984**, *232*, 249–258.

Rolletschek, H.; Borisjuk, L.; Sánchez-García, A.; Gotor, C.; Romero, L. C.; Martínez-Rivas, J. M.; Mancha, M. Temperature-dependent Endogenous Oxygen Concentration Regulates Microsomal Oleate Desaturase in Developing Sunflower Seeds. *J. Exp. Bot.* **2007**, *58*, 3171–3181.

Roughan, P. G.; Ohlrogge, J. B. Evidence That Isolated Chloroplasts Contain and Integrated Lipid-synthesizing Assembly That Channels Acetate into Long-chain Fatty Acids. *Plant Physiol.* **1996**, *110*, 1239–1247.

Ruiz-López, N. Caracterización del Sistema de Aciltransferencias en Semillas de Girasol (*Helianthus annuus* L.). Clonación de una Ferredoxina Heterotrófica. Doctoral thesis, University of Seville, 2006.

Ruiz-López, N.; Garcés, R.; Harwood, J. L.; Martínez-Force, E. Characterization and Partial Purification of Acyl-CoA:glycerol 3-Phosphate Acyltransferase from Sunflower (*Helianthus annuus* L.) Developing Seeds. *Plant Physiol. Biochem.* **2010**, *48*, 73–80.

Saha, S.; Enugutti, B.; Rajakumari, S.; Rajasekharan, R. Cytosolic Triacylglycerol Biosynthetic Pathway in Oilseeds: Molecular Cloning and Expression of Peanut Cytosolic Diacylglycerol Acyl-transferase. *Plant Physiol.* **2006**, *141*, 1533–1543.

Salas, J. J.; Martínez-Force, E.; Garcés, R. Biochemical Characterization of a High-palmitoleic Acid *Helianthus annuus* Mutant. *Plant Physiol. Biochem.* **2004**, *42*, 373–381.

Salas, J. J.; Martínez-Force, E.; Garcés, R. Very Long Chain Fatty Acid Synthesis in Sunflower Kernels. *J. Agric. Food Chem.* **2005**, *53*, 2710–2716.

Salas, J. J.; Youssar, L.; Martínez-Force, E.; Garcés, R. The Biochemical Characterization of a High-stearic Acid Sunflower Mutant Reveals the Coordinated Regulation of Stearoyl-acyl Carrier Protein Desaturases. *Plant Physiol. Biochem.* **2008**, *46*, 109–116.

Sasaki, Y.; Nagano, Y. Plant Acetyl-CoA Carboxylase: Structure, Biosynthesis, Regulation, and Gene Manipulation for Plant Breeding. *Biosci. Biotechnol. Biochem.* **2004**, *68*, 1175–1184.

Schnurr, J. A.; Shockey, J. M.; De Boer, G. J.; Browse, J. A. Fatty Acid Export from the Chloroplast Molecular Characterization of a Major Plastidial Acyl-coenzyme A Synthetase from Arabidopsis. *Plant Physiol.* **2002**, *129*, 1700–1709.

Schnurr, J. A.; Shockey, J. M.; Browse, J. A. The Acyl-CoA Synthetase Encoded by LACS2 Is Essential for Normal Cuticle Development in Arabidopsis. *Plant Cell* **2004**, *16*, 629–642.

Schuppert, G. F.; Tang, S. X.; Slabaugh, M. B.; Knapp, S. J. The Sunflower High-oleic mutant *Ol* Carries Tandem Repeats of FAD2–1, a Seed-specific Oleoyl-phosphatidyl Choline Desaturase. *Mol. Breed.* **2006**, *17*, 241–256.

Serrano-Vega, M. J.; Venegas-Calerón, M.; Garcés, R.; Martínez-Force, E. Cloning and Expression of Fatty Acids Biosynthesis Key Enzymes from Sunflower (*Helianthus annuus* L.) in *Escherichia coli*. *J. Chromatogr. B* **2003**, *786*, 221–228.

Serrano-Vega, M. J.; Garcés, R.; Martínez-Force, E. Cloning, Characterization and Structural Model of a FatA-type Thioesterase from Sunflower Seeds (*Helianthus annuus* L.). *Planta* **2005**, *221*, 868–880.

Shanklin, J.; Cahoon, E. B. Desaturation and Related Modification of Fatty Acids. *Annu. Rev. Plant Physiol. Plant Mol. Biol.* **1998**, *49*, 611–641.

Shockey, J.; Browse, J. Genome-level and Biochemical Diversity of the Acyl-activating Enzyme Superfamily in Plants. *Plant J.* **2011**, *66*, 143–160.

Shockey, J. M.; Fulda, M. S.; Browse, J. A. Arabidopsis Contains Nine Long-chain Acyl-coenzyme A Synthetase Genes That Participate in Fatty Acid and Glycerolipid Metabolism. *Plant Physiol.* **2002**, *129*, 1710–1722.

Shockey, J. M.; Gidda, S. K.; Chapital, D. C.; Kuan, J. C.; Dhanoa, P. K.; Bland, J. M.; Rothstein, S. J.; Mullen, R. T.; Dyer, J. M. Tung Tree GDAT1 and DGAT2 Have Non Redundant Functions in Triacylglycerol Biosynthesis and Are Localized to Different Subdomains of the Endoplasmic Reticulum. *Plant Cell* **2006**, *18*, 2294–2313.

Slabaugh, M. B.; Katemgam, S.; Hongtrakul, V.; Knapp, S. J. Sunflower FatB Thioesterase cDNA. GenBank sequence, unpublished work, 1997.

Slack, C. R.; Campbell, L. C.; Browse, J. A.; Roughan, P. G. Some Evidence for the Reversibility of Cholinephosphotransferase-catalyzed Reaction in Developing Linseed Cotyledons in Vivo. *Biochim. Biophys. Acta* **1983**, *754*, 10–20.

Slack, C. R.; Roughan, P. G.; Browse, J. A.; Gardiner, S. E. Some Properties of Cholinephosphotransferase from Developing Safflower Cotyledons. *Biochim. Biophys. Acta* **1985**, *833*, 438–448.

Soldatov, K. I. Chemical Mutagenesis in Sunflower Breeding. In *Proceedings of the Seventh International Sunflower Conference*, Krasnodar, USSR, 1976; International Sunflower Association: Vlaardingen, The Netherlands, 1976; pp 352–357.

Sperling, P.; Heinz, E. Desaturases Fused to Their Electron Donor. *Eur. J. Lipid Sci. Technol.* **2001**, *103*, 158–180.

Ståhl, U.; Ek, B.; Stymne, S. Purification and Characterization of a Low-Molecular-Weight Phospholipase A2 from Developing Seeds of Elm. *Plant Physiology* **1998**, *117*, 197–205.

Stahl, U.; Carlsson, A. S.; Lenman, M.; Dahlqvist, A.; Huang, B. Q.; Banas, W.; Banas, A.; Stymne, S. Cloning and Functional Characterization of a Phospholipid:Diacylglycerol Acyltransferase from Arabidopsis. *Plant Physiol.* **2004**, *135*, 1324–1335.

Stymne, S.; Stobart, A. K. The Biosynthesis of Triacylglycerols in Microsomal Preparations of Developing Cotyledons of Sunflower (*Helianthus annuus* L.). *Biochem. J.* **1984**, *220*, 481–488.

Sun, L.; Ouyanga, C.; Koua, S.; Wanga, S.; Yaoa, Y.; Penga, T.; Xua, Y.; Tanga, L.; Chena, F. Cloning and Characterization of a cDNA Encoding Type 1 Diacylglycerol Acyltransferase from Sunflower (*Helianthus annuus* L.). *Z. Naturforsch. C.* **2011**, *100*, 63–72.

Timme, R. E.; Kuehl, J. V.; Boore, J. L.; Jansen, R. K. A Comparative Analysis of the Lactuca and *Helianthus* (*Asteraceae*) Plastid Genomes: Identification of Divergent Regions and Categorization of Shared Repeats. *Am. J. Bot.* **2007**, *94*, 302–312.

Tomlinson, K. L.; McHugh, S.; Labbe, H.; Grainger, J. L.; James, L. E.; Pomeroy, K. M.; Mullin, J. W.; Miller, S. S.; Dennis, D. T.; Miki, B. L. Evidence That the Hexose-to-sucrose Ratio Does Not Control the Switch to Storage Product Accumulation in Oilseeds: Analysis of Tobacco Seed Development and Effects of Overexpressing Apoplastic Invertase. *J. Exp. Bot.* **2004**, *55*, 2291–2303.

Triki, S.; Demandre, C.; Mazliak, P. Biosynthesis of Triacylglycerols by Developing Sunflower Seed Microsomes. *Phytochemistry* **1999**, *52*, 55–62.

Troncoso-Ponce, M. A.; Kruger, N. J.; Ratcliffe, G.; Garcés, R.; Martínez-Force, E. Characterization of Glycolytic Initial Metabolites and Enzyme Activities in Developing Sunflower (*Helianthus annuus* L.) Seeds. *Phytochemistry* **2009**, *70*, 1117–1122.

Troncoso-Ponce, M. A.; Garcés, R.; Martínez-Force, E. Glycolytic Enzymatic Activities in Developing Seeds Involved in the Differences between Standard and Low Oil Content Sunflowers (*Helianthus annuus* L.). *Plant Physiol. Biochem.* **2010a**, *48*, 961–965.

Troncoso-Ponce, M. A.; Rivoal, J.; Cejudo, F. J.; Dorion, S.; Garcés, R.; Martínez-Force, E. Cloning, Biochemical Characterisation, Tissue Localisation and Possible Post-translational Regulatory Mechanism of the Cytosolic Phosphoglucose Isomerase from Developing Sunflower Seeds. *Planta* **2010b**, *232*, 845–859.

Troncoso-Ponce, M. A.; Kilaru, A.; Cao, X.; Durrett, T. P.; Fan, J.; Jensen, J. K.; Thrower, N. A.; Pauly, M.; Wilkerson, C.; Ohlrogge, J. B. Comparative Deep Transcriptional Profiling of Four Developing Oilseeds. *Plant J.* **2011a**, *68*, 1014–1027.

Troncoso-Ponce, M. A.; Rivoal, J.; Dorion, S.; Moisan, M. C.; Garcés, R.; Martínez-Force, E. Cloning, Biochemical Characterization and Expression of a Sunflower (*Helianthus annuus* L.) Hexokinase Associated with Seed Storage Compounds Accumulation. *J. Plant Physiol.* **2011b**, *168*, 299–308.

Troncoso-Ponce, M. A.; Rivoal, J.; Venegas-Calerón, M.; Dorion, S.; Sánchez, R.; Cejudo, F. J.; Garcés, R.; Martínez-Force, E. Molecular Cloning and Biochemical Characterization of Three Phosphoglycerate Kinase Isoforms from Developing Sunflower (*Helianthus annuus* L.) seeds. *Phytochemistry* **2012**, *79*, 27–38.

Troncoso-Ponce, M. A.; Cao, X.; Yang, X.; Ohlrogge, J. B. Lipid Turnover during Senescence. *Plant Sci.* **2013**, *205–206*, 13–19.

Venegas-Calerón, M.; Muro-Pastor, A. M.; Garcés, R.; Martinez-Force, E. Functional Characterization of a Plastidial Omega-3 Desaturase from Sunflower (*Helianthus annuus*) in Cyanobacteria. *Plant Physiol. Biochem.* **2006**, *44*, 517–525.

Venegas-Calerón, M.; Sayanova, O.; Napier, J. A. An Alternative to Fish Oils: Metabolic Engineering of Oil-seed Crops to Produce Omega-3 Long Chain Polyunsaturated Fatty Acids. *Prog. Lipid Res.* **2010**, *49*, 108–119.

Weselake, R. J. Storage lipids. In *Plant Lipids: Biology, Utilization and Manipulation*; Murphy, D. J., Ed.; Blackwell: Oxford, UK, 2005; pp 162–221.

Weselake, R. J.; Taylor, D. C.; Rahman, M. H.; Shah, S.; Laroche, A.; McVetty, P. B. E.; Harwood, J. L. Increasing the Flow of Carbon into Seed Oil. *Biotechnol. Adv.* **2009**, *27*, 866–878.

White, J. A.; Todd, J.; Newman, T.; Focks, N.; Girke, T.; de Ilárduya, O. M.; Jaworski, J. G.; Ohlrogge, J. B.; Benning, C. A New Set of Arabidopsis Expressed Sequence Tags from Developing Seeds. The Metabolic Pathway from Carbohydrates to Seed Oil. *Plant Physiol.* **2000**, *124*, 1582–1594.

Wiberg, E.; Tillberg, E.; Stymne, S. Substrates of Diacylglycerol Acyltransferase in Microsomes from Developing Oil Seeds. *Phytochemistry* **1994**, *36*, 573–577.

Yasuno, R., von Wettstein-Knowles, P., Wada, H. Identification and Molecular Characterization of the β-ketoacyl-[acyl carrier protein] Synthase Component of the Arabidopsis Mitochondrial Fatty Acid Synthase. *J. Biol. Chem.* **2004**, *279*, 8242–8251.

Yu, B.; Wakao, S.; Fan, J.; Benning, C. Loss of Plastic Lysophosphatidic Acid Acyl-transferase Cause Embryo-lethality in Arabidopsis. *Plant Cell Physiol.* **2004**, *45*, 503–510.

Zborowski, J.; Wojtczak, L. Phospholipid Synthesis in Rat Liver Mitochondria. *Biochim. Biophys. Acta* **1969**, *187*, 73–84.

Zhang, M.; Fan, J.; Taylor, D. C.; Ohlrogge, J. B. DGAT1 and PDAT1 Acyltransferases Have Overlapping Functions in Arabidopsis Triacylglycerol Biosynthesis and Are Essential for Normal Pollen and Seed Development. *Plant Cell* **2009**, *21*, 3885–3901.

Zhao, L.; Katavic, V.; Li, F.; Haughn, G.W.; Kunst, L. Insertional Mutant Analysis Reveals That Long-chain Acyl-CoA Synthetase 1 (LACS1), but Not LACS8, Functionally Overlaps with LACS9 in Arabidopsis Seed Oil Biosynthesis. *Plant J.* **2010**, *64*, 1048–1058.

Zheng, Z. F.; Xia, Q.; Dauk, M.; Shen, W. Y.; Selvaraj, G.; Zou, J. T. Arabidopsis AtGPAT1, a Member of the Membrane-bound Glycerol-3-phosphate Acyltransferase Gene Family, Is Essential for Tapetum Differentiation and Male Fertility. *Plant Cell* **2003**, *15*, 1872–1887.

11

Sunflower Oil Minor Constituents

Leonardo Velasco ■ *Instituto de Agricultura Sostenible (IAS-CSIC), Córdoba, Spain*

M. Victoria Ruiz-Méndez ■ *Instituto de la Grasa (IG-CSIC), Sevilla, Spain*

Introduction

Crude sunflower oil consists predominantly of triacylglycerols (>98%) along with a small fraction of minor constituents such as diacylglycerols, phospholipids, glycolipids, free fatty acids, aliphatic alcohols, waxes, hydrocarbons, phenolic compounds, carotenoids, chlorophyll, trace metals, tocopherols, and phytosterols (Grompone, 2005; Sánchez-Muniz and Cuesta, 2003). It may also contain pesticides resulting from agricultural practices (Gunstone, 2005). The largest fraction of minor constituents of sunflower oil constitutes the so-called unsaponifiable fraction, which includes those minor constituents that cannot be saponified by caustic treatment but are soluble in organic solvents that fat and oils are soluble in (AOCS, 2011). The unsaponifiable matter of crude sunflower oil mainly contains tocopherols, phytosterols, alcohols, hydrocarbons, and phenolics which usually comprise 0.5–1.5% of crude oil weight. The Codex Alimentarius Commission recommends a maximum unsaponifiable matter of 15 g kg^{-1} for crude sunflower oil (Codex Alimentarius, 2013). Table 11.A shows the composition of crude regular sunflower oil minor constituents.

Some minor constituents such as tocopherols and phytosterols have a major positive impact on nutritional and/or technological properties of the oil (Fernández-Martínez et al., 2009). Others, however, have a negative effect on oil quality and stability. Phospholipids, free fatty acids, and waxes are the main components within the latter group (Kamal-Eldin, 2005). Accordingly, oil extraction procedures should maximize extraction of desired minor constituents, whereas oil refining should minimize losses of desired minor constituents and maximize losses of undesired constituents (Anderson, 2005). Some of the latter (e.g., phospholipids) are co-products with important industrial uses (Salas et al., 2006).

Tocopherols

Tocopherols are a group of lipid-soluble constituents composed of a polar moiety derived from tyrosine, the chromanol ring, and a hydrophobic phytyl-derived side chain (Figure 11.1). They occur in nature in four forms, α-, β-, γ-, and δ-tocopherol, which differ in the number and position of methyl substituents on the chromanol ring (Mène-Saffrané and DellaPenna, 2010). Tocopherols are the main compounds

Table 11.A Minor Constituents in Crude Regular Sunflower Oil

Component	mg kg^{-1} Oil	Reference
Diacylglycerols	10,000–18,900	Sánchez-Muniz and Cuesta (2003)
Phospholipids	6000–12,000	Carelli et al. (2002b)
Free fatty acids	3700–4500	Sánchez-Muniz and Cuesta (2003)
Phytosterols	2400–4600	Grompone (2005)
Hydrocarbons	1000	Bastic et al. (1978)
Tocopherols	403–935	Grompone (2005)
Waxes	200–3500	Carelli et al. (2002a)
Aliphatic alcohols	59–63	Sánchez-Muniz and Cuesta (2003)
Trace metals	45–95	Gupta (2002)
Carotenoids	6.5–15.3	Premovic et al. (2010)
Phenolic compounds	4.8–16.4	Perretti et al. (2004)
Chlorophyll	0–1	Premovic et al. (2010)

with antioxidant activity in oilseeds, as well as in their extracted seed oils, where they have a major effect on their nutritional and technological properties. From a nutritional perspective, tocopherols, together with other related compounds such as tocotrienols (collectively known as tocochromanols), are compounds with vitamin E activity. They prevent oxidative, proliferative, metabolic, and inflammatory damage at the cellular level and the subsequent development of various chronic diseases (Eitenmiller and Lee, 2004; Galli and Azzi, 2010). From a technological perspective, tocopherols protect the oil from oxidative deterioration (Shahidi and Zhong, 2010).

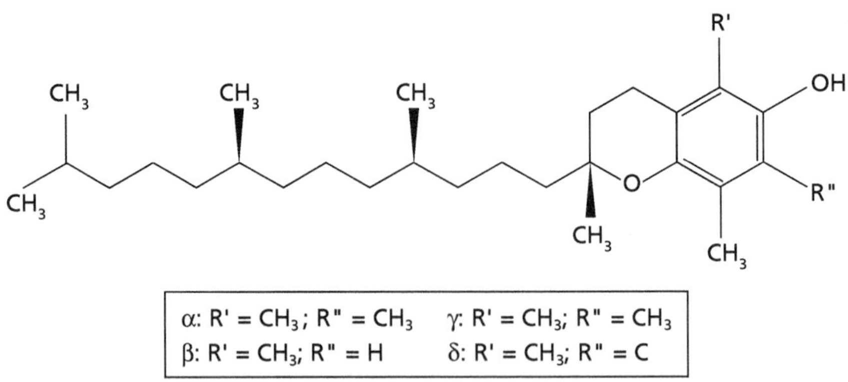

Figure 11.1 Chemical structure of tocopherols.

Crude sunflower oil has a relatively high tocopherol content predominantly made up of α-tocopherol, which makes sunflower oil the richest source of vitamin E amongst commercially important vegetable oils (Table 11.B) and one of the richest food sources (U.S. Department of Agriculture, 2012). Other important seed oils contain predominantly γ-tocopherol (soybean, rapeseed) or balanced levels of α- and γ-tocopherol (groundnut, cottonseed). Safflower is the only seed oil, together with sunflower, with predominant levels of α-tocopherol. Olive oil has a similar tocochromanol profile, though the total content is much lower than that in sunflower oil (Table 11.B).

Variation for Tocopherol Content and Profile in Sunflower

Triacylglycerols and tocopherols accumulate in the cells of sunflower seeds as the result of separate biosynthetic pathways. Because tocopherols are fat-soluble compounds, they become part of the extracted oil. Accordingly, tocopherol content in crude sunflower oil depends on the tocopherol content of the seeds, the oil content of the seeds, and the oil extraction system (Velasco et al., 2010). The oil and the tocopherol contents of the seeds are the result of genetic and environmental factors, as well as of their interactions (Marquard, 1990; Velasco et al., 2010). Therefore, understanding variation for total tocopherol content in sunflower oil requires the analysis of many explanatory variables. This explains the huge variation for tocopherol content in crude sunflower oil reported in the literature, with extreme values varying from 389 to 1873 mg kg^{-1} (Dolde et al., 1999; Velasco et al., 2002; Nolasco et al., 2004).

Tocopherol content in the oil is mainly determined by the seed oil content rather than by the seed tocopherol content (Nolasco et al., 2004). Therefore, selection for increased biosynthetic efficiency in the tocopherol pathway requires conducting selection for the tocopherol content in the seeds rather than in the extracted oil (Velasco et al., 2010). The latter authors reported a breeding program in which germplasm with increased tocopherol content in the seeds was developed and evaluated over several environments.

Despite the fact that tocopherol composition in regular sunflower seed and oil is dominated by α-tocopherol, an impressive variation for this trait has been obtained by conventional selection and mutagenesis. It includes germplasm with mid and high β-tocopherol content, high γ-tocopherol content, and mid and high δ-tocopherol contents (Table 11.C). No similar variation has been obtained thus far in other oil crops.

Genetics of Tocopherols

Tocopherols are synthesized by photosynthetic organisms, including plants and some algae and cyanobacteria. In plants, tocopherols are synthesized and localized in plastids. Tocopherol biosynthesis starts with the condensation of homogentisic acid

Table 11.B Total Annual World Production (× 10^6 t), Average Total Tocochromanol Content (Total-T; mg kg^{-1}), Concentration of Individual Tocochromanols (%), and Vitamin E Values Given as Milligram α-Tocopherol Equivalents (mg α-TE) in the Most Important Vegetable Oils at a World Scale

Oil	Production[a]	Total-T[b]	α-T[c]	β-T	γ-T	δ-T	α-T3	β-T3	γ-T3	δ-T3	PC-8	mg α-TE[d]
Palm	45.5	767	32.7	0.0	0.5	0.0	19.8	0.0	38.9	8.1	0.0	297.0
Soybean	39.1	1141	8.9	0.8	63.6	26.7	0.0	0.0	0.0	0.0	0.0	181.2
Rapeseed	21.7	643	29.9	0.0	67.3	1.7	0.0	0.0	0.0	0.0	1.1	235.4
Sunflower	13.2	675	93.8	3.3	2.8	0.1	0.0	0.0	0.0	0.0	0.0	645.9
Palm kernel	5.8	34	38.2	0.0	0.0	0.0	61.8	0.0	0.0	0.0	0.0	19.3
Groundnut	5.5	357	44.0	2.2	49.6	4.2	0.0	0.0	0.0	0.0	0.0	178.9
Cottonseed	5.0	777	49.7	1.0	49.0	0.3	0.0	0.0	0.0	0.0	0.0	428.1
Coconut	3.5	35	11.4	0.0	8.6	8.6	14.3	2.9	48.6	5.7	0.0	5.9
Olive, virgin	2.8	153	91.5	2.0	6.5	0.0	0.0	0.0	0.0	0.0	0.0	142.5
Maize	2.3	925	22.4	0.6	65.6	2.8	3.7	0.0	4.3	0.0	0.0	281.4
Sesame	1.1	469	1.3	0.6	95.1	3.0	0.0	0.0	0.0	0.0	0.0	52.2
Linseed	0.5	553	0.7	0.0	73.6	0.0	0.0	0.0	0.0	0.0	25.7	44.7
Safflower	0.1	498	90.6	1.4	6.6	1.4	0.0	0.0	0.0	0.0	0.0	457.9

[a]Average data from 2008 to 2012. Source: FAOSTAT (2013).

[b]Sources: Gunstone and Harwood (2007); Eitenmiller and Lee (2004); Kamal-Eldin (2005); Velasco et al. (2004a); Zou et al. (2012).

[c]T = tocopherol; T3 = tocotrienol; PC-8 = plastochromanol-8.

[d]Calculated following Eitenmiller and Lee (2004).

Table 11.C Sunflower Germplasm with Modified Tocopherol (T) Composition (% Total Tocopherols)[a]

Tocopherol Profile	α-T	β-T	γ-T	δ-T
Regular sunflower oil	94	3	3	0
Mid β-tocopherol[b,c,d,e,f]	50	50	0	0
High β-tocopherol[f,g,h]	25	75	0	0
High γ-tocopherol[b,c,d,e,f]	1	0	98	1
Mid δ-tocopherol[f,g,i]	3	0	39	58
High δ-tocopherol[f,g]	1	1	24	74

[a]Average values. Data in different publications may differ due to environmental effects.
[b]Demurin (1993).
[c]Demurin et al. (1996).
[d]Velasco and Fernández-Martínez (2003).
[e]Velasco et al. (2004b).
[f]Velasco and Fernández-Martínez (2012).
[g]Velasco et al. (2004c).
[h]Tang et al. (2006).
[i]Hass et al. (2006).

and phytyl diphosphate to form 2-methyl-6-phytyl-1,4-benzoquinone (MPBQ), in a reaction catalyzed by homogentisic acid phytyl transferase (HPT) (Figure 11.2). Subsequent methylation of the latter by a 2-methyl-6-phytyl-1,4-benzoquinone/2-methyl-6-solanyl-1,4-benzoquinone methyltransferase (MPBQ/MSBQ-MT) yields 2,3-dimethyl-6-phytyl-1,4-benzoquinone (DMPBQ). MPBQ and DMPBQ are cyclized by tocopherol cyclase (TC) into δ- and γ-tocopherol, respectively, which are in turn converted into β- and α-tocopherol, respectively, in a methylation step catalyzed by γ-tocopherol methyltransferase (γ-TMT) (Mène-Saffrané and DellaPenna, 2010). In sunflower, the most active branch of the pathway is that leading to the accumulation of α-tocopherol. The alternative tocopherol profiles consisting of mid and high β-tocopherol content, high γ-tocopherol content, and mid and high δ-tocopherol content are the result of mutations at two enzymes of the pathway: MPBQ/MSBQ-MT and γ-TMT (Hass et al., 2006; Tang et al., 2006).

The sunflower genome contains several MPBQ/MSBQ-MT and γ-TMT loci. Classical genetic studies demonstrated that mid β-tocopherol content around 50% was produced by mutated alleles at a single locus named *Tph1* or *MT-1* (Demurin et al., 1996; Vera-Ruiz et al., 2006), whereas high β-tocopherol content over 65% was produced by mutated alleles at two loci, *MT-1* and *MT-2*. Both loci *MT-1* and *MT-2* were identified as paralogs of the MPBQ/MSBQ-MT (Tang et al., 2006). On the other hand, high γ-tocopherol content results from recessive mutations at a single

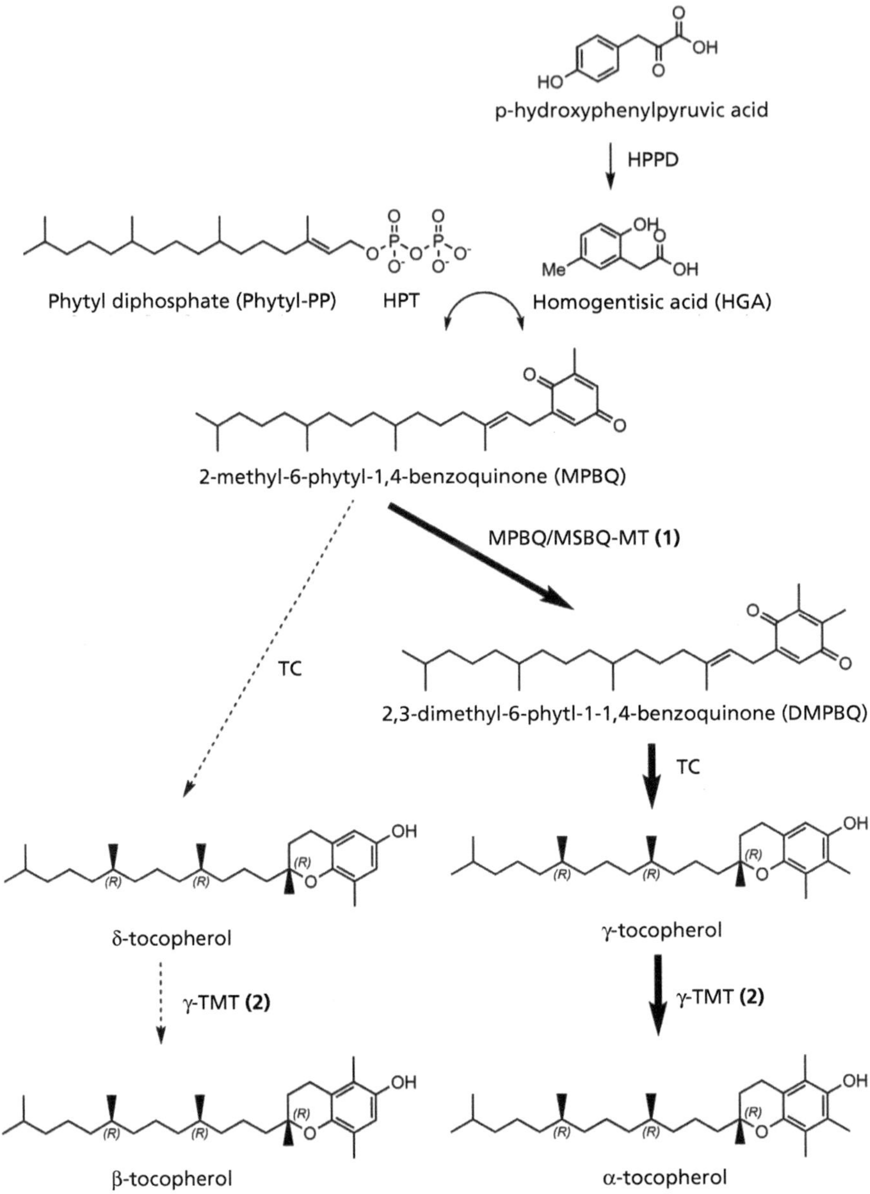

Figure 11.2 Schematic representation of tocopherol biosynthesis. Thicker arrows indicate the main branch of the pathway in sunflower. HPPD: p-hydroxyphenyl- pyruvate dioxygenase; HPT: homogentisic acid phytyl transferase; MPBQ/MSBQ-MT: 2-methyl-6-phytyl-1,4-benzoquinone/2-methyl-6-solanyl-1,4-benzoquinone methyltransferase; TC: tocopherol cyclase; γ-TMT: γ-tocopherol methyl transferase; (1) and (2) indicate the steps at which mutations for tocopherol profile have been identified in sunflower.

Tph2 locus (Demurin et al., 1996; García-Moreno et al., 2006), which was identified as a γ-TMT locus. Two different mutated alleles at this locus, *tph2* and *tph2*[a], have been identified. The latter was influenced by modifying genes, which were also identified as duplicated loci of γ-TMT, resulting in intermediate γ-tocopherol levels in the range from 5% to 85% (García-Moreno et al., 2012). Epistatic interaction of mutated alleles at the three loci *MT-1*, *MT-2*, and *Tph2* results in increased δ-tocopherol content (Hass et al., 2006). The genetic differences underlying the two different levels of δ-tocopherol (Table 11.C) have not been elucidated yet. Although some of the modified tocopherol profiles in sunflower have been found to be influenced by environmental conditions, all of them are consistently expressed across environments (Velasco and Fernández-Martínez, 2012).

Genetic analysis of a line with increased total tocopherol content concluded that the trait is highly heritable and controlled by a reduced number of genes (Del Moral et al., 2011). This was confirmed at the molecular level through the identification of seven quantitative trait loci (QTL); one of them, at linkage group (LG) 10, had a major effect because it explained more than half of the phenotypic variation for increased tocopherol content (Del Moral et al., 2012). The biosynthetic step associated with this major QTL has not been identified. Haddadi et al. (2011) followed a candidate gene analysis approach on a collection of sunflower mutants showing variation for total tocopherol content. They found a significant association between total tocopherol content and an HPT locus. Similarly, Haddadi et al. (2012) identified five QTL associated with total tocopherol content in a population of recombinant inbred lines as well as four candidate genes that co-localized with some of the QTL. Two of the candidate genes were genes underlying two enzymes involved in the tocopherol biosynthetic pathway, γ-TMT and p-hydroxyphenyl-pyruvate dioxygenase (HPPD) (Figure 11.2) (Haddadi et al., 2012).

Influence of Tocopherols on the Nutritional and Technological Properties of the Oil

Both the total amount of tocopherols present in the oil and their composition are important determinants of the nutritional and technological properties of the oil. α-Tocopherol is the tocopherol form with higher biological value as vitamin E. If a vitamin E value of 1.0 is assigned to α-tocopherol, it is generally accepted that the vitamin E values of the other tocopherols are 0.5 for β-tocopherol, 0.1 for γ-tocopherol, and 0.01 for δ-tocopherol (Eitenmiller and Lee, 2004). Regular sunflower oil contains high tocopherol content with high concentration of α-tocopherol, which makes it the commercial vegetable oil with the highest vitamin E value (Table 11.B). However, there is increasing evidence suggesting that the role of other tocopherol forms, such as γ-tocopherol, in health and disease prevention is being underestimated (Wagner et al., 2004; Ju et al., 2010).

From a technological perspective, tocopherols protect the oil from oxidative reactions that occur mainly at the double bonds of unsaturated fatty acids (Barrera-Arellano et al., 2002). The influence of the type of tocopherol on oil stability has been traditionally studied by adding tocopherols in different concentrations to purified oils (i.e., oils stripped of their natural tocopherols). The results revealed that the relative efficacy of the tocopherols depends on their concentration, as well as on oil operating temperature. At moderate temperatures, α-tocopherol usually acts as a better antioxidant than γ-tocopherol when tocopherol concentrations are low, but the opposite occurs at high tocopherol concentrations (Fuster et al., 1998; Lampi and Kamal-Eldin, 1998; Lampi et al., 1999; Seppanen et al., 2010). At high temperatures, γ- and δ-tocopherol inhibit polymerization reactions more efficiently than α-tocopherol (Barrera-Arellano et al., 1999; Lampi and Kamal-Eldin, 1998; Seppanen et al., 2010; Warner and Moser, 2009).

The availability of sunflower germplasm that naturally contains different compositions of tocopherols has enabled the study of the comparison of technological properties of oils with natural and added tocopherols. Warner et al. (2008) compared mid-oleic and high-oleic sunflower oils with different increased levels of γ-tocopherol or both γ- and δ-tocopherol, concluding that oil stability improved substantially after partial or nearly total replacement of α-tocopherol by γ- and δ-tocopherol. Marmesat et al. (2008) compared high-palmitic, high-oleic sunflower oils containing predominant levels of either α- or γ-tocopherols. The authors found higher effectiveness against lipid oxidation of γ-tocopherol as compared to α-tocopherol. The tests were carried out at high temperature to accelerate oil deterioration (120 °C) as well as to simulate the conditions applied in the frying process (180 °C). An interesting conclusion of both studies was that there is no need to completely remove α-tocopherol to improve the oxidative stability of the oil. The authors found that sunflower oil with 50% α-tocopherol and 50% γ-/δ-tocopherols has similarly enhanced oxidative stability as sunflower oil with 95% γ-/δ-tocopherols, while still being a good source of vitamin E (Marmesat et al; 2008; Warner et al., 2008).

Effect of Extraction Methods on Tocopherols

The composition of minor constituents and the oxidative stability of crude sunflower oils are strongly affected by the extraction process. Sunflower oil is obtained by mechanical and/or chemical extraction after seed preparation because in seeds and oleaginous fruits with a high fat content (>20%), such as sunflower seed, direct extraction with solvents is not profitable. The standard process is shown in Figure 11.3. It typically consists of prepressing in special presses termed *expellers* that operate in a continuous mode. The

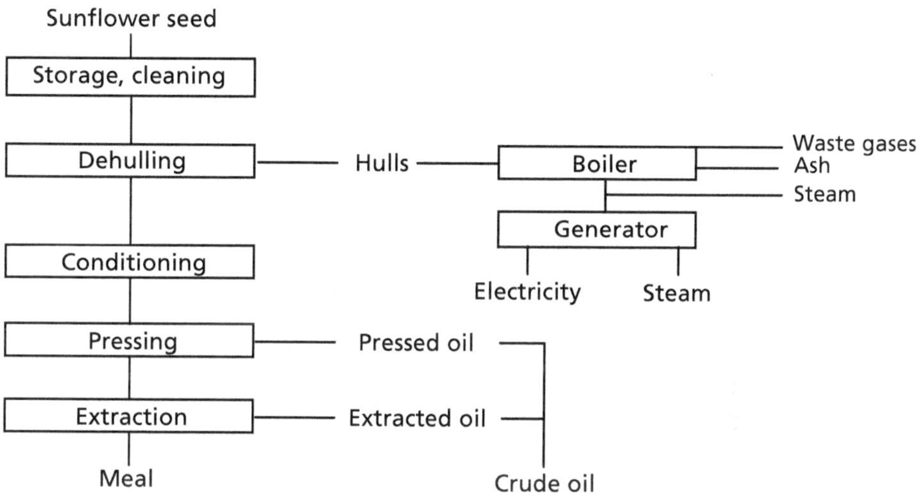

Figure 11.3 Flowchart of sunflower seed crushing.

solid matter from prepressing has a relatively high fat content (approximately 15–18%) and it is subjected to solvent extraction to recover the retained oil (Brevedan et al., 2000). The prepressed and solvent extracted oils are generally blended before storage and/or refining. The full-press is rarely used in large-scale operations, but it is more popular in small scale oilseed processing facilities (Fils, 2000).

In general, the prepressed oil is of superior quality to that obtained with solvents (Rass et al., 2008; Van Hoed et al., 2010). Tasan et al. (2011) evaluated the effects of industrial oilseed extraction methods on the quality and stability characteristics of crude sunflower oil. Some of their results are shown in Table 11.D. They found that total and individual tocopherol contents seemed to be partially affected by different extraction methods, and there was a statistically significant decline in α-, β- and γ-tocopherol levels for solvent-extracted and full-pressed oils as compared to the prepressed oil.

Total tocopherol content in the prepressed oil averaged 689.4 mg kg^{-1}, compared to 634.1 mg kg^{-1} in the solvent-extracted oil and 507.2 mg kg^{-1} in the full-pressed oil. The reduction of approximately 26.4% of total tocopherol content in the full-press process as compared to the prepress process was attributed to the longer contact time of the crude oil with air and to exposure to overheating or overpressure.

For vegetable oil extraction, the supercritical extraction method has proven to be a very interesting alternative to conventional extraction methods. Sunflower oils obtained using supercritical carbon dioxide and subcritical propane as solvents

Table 11.D Average Total Phytosterol Content (Total Phyt; mg kg^{-1}) and Concentration of Major Phytosterols (%) in the Most Important Vegetable Oils at a World Scale

Oil	Total Phyt[a]	Brass[b]	Camp[c]	Stig[d]	Sito[e]	D5Av[f]	D7Stig[g]
Palm	832	—	22.7	8.3	66.8	2.2	—
Soybean	3400	—	20.2	19.4	54.3	3.4	1.5
Rapeseed	7099	11.0	26.5	0.3	57.0	2.3	2.6
Sunflower	4436	—	8.2	7.9	59.7	2.9	14.4
Palm kernel	1310	—	9.2	10.7	70.2	6.1	0.8
Groundnut	2443	—	15.4	8.3	66.0	7.9	—
Cottonseed	4723	—	7.2	1.1	87.4	4.2	—
Coconut	975	—	11.7	13.4	52.9	16.7	—
Olive, virgin	1686	—	2.8	0.9	78.0	17.4	—
Maize	11,120	—	20.8	6.6	67.3	2.9	0.5
Sesame	5977	—	16.6	8.2	61.8	8.7	1.0
Linseed	4710	—	25.5	8.1	48.0	15.0	—
Safflower	4125	—	12.2	6.6	52.0	2.2	17.7

[a]Sources: Fernández-Cuesta et al. (2012a); Gunstone and Harwood (2007); Gül and Seker (2006); Kamal-Eldin (2005); Phillips et al. (2002); Piironen et al. (2000); Velasco et al. (2013); Verleyen et al. (2002a); Vlahakis and Hazebroek (2000); Zou et al. (2012).
[b]Brass = brassicasterol.
[c]Camp = campesterol.
[d]Stig = stigmasterol.
[e]Sito = β-sitosterol.
[f]D5Av = Δ^5-avenasterol.
[g]D7Stig = Δ^7-stigmastenol.

showed higher α-tocopherol content and oxidative stability then oils obtained using hexane as solvent (Nimet et al., 2011).

Changes in Tocopherols during Processing and Storage

Crude oils are refined to remove unwanted minor components that make oils unappealing to consumers. During refining the goal is to minimize damage to the neutral oil and loss of valuable minor constituents. The components to be removed are all glyceridic and nonglyceridic compounds that are detrimental to flavor, color, stability, and nutritional safety of the refined oil. These compounds are primarily phospholipids, free fatty acids, pigments, volatiles, and contaminants. Refining can be conducted either by means of an alkali reaction (chemical refining) or by distillation (physical refining). The major steps involved in the physical and chemical refining processes and the main components removed are shown in Figure 11.4 (Kovari, 2004).

According to several authors, total tocopherol content significantly decreases during neutralization, bleaching, and deodorization steps (Alpaslan et al., 2001; Naz et al., 2011; Tasan and Demirci, 2005). The effect of degumming on tocopherols is practically negligible (Brevedan et al., 2000), whereas the reduction in total tocopherol content during the combined degumming/neutralizing step is attributed to the longer contact time of the oil with air and alkali (Naz et al., 2011; Suliman et al., 2013a; Tasan and Demirci, 2005).

Deodorization of fats and oils usually consists of steam distillation at elevated temperature under vacuum, although nitrogen has also been used. The purpose of this step is to remove volatile compounds (mainly ketones and aldehydes) for improving the oil taste and odor, as well as to eliminate free fatty acids (in the case of physical refining) or residual free fatty acids from neutralized and bleached oils (in the case of chemical refining). The deodorization conditions also contribute to removing contaminants, extending shelf life, improving taste stability, and reducing oil color due to the breakdown of the remaining carotenes at high temperature.

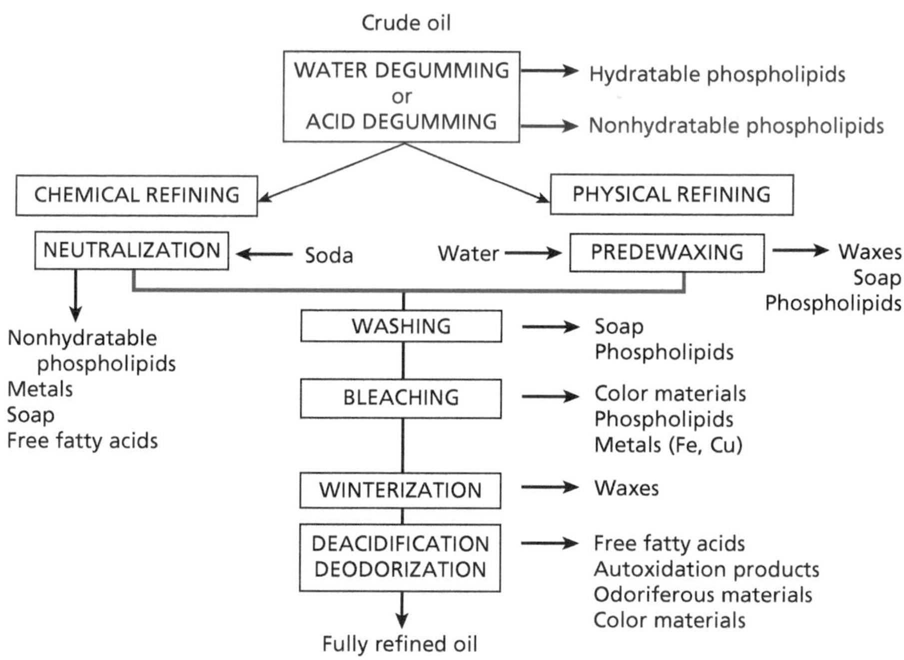

Figure 11.4 Physical and chemical refining processes.
Source: Adapted from Kovari (2004).

Physical refining reduces the loss of neutral oil, minimizes pollution, and enables recovery of high-quality free fatty acids. Nevertheless, not all the oils can be physically refined and, as a general rule, the lesser the phosphorous content and the polyunsaturated fatty acids in the oil, the easier it is for the oil to be physically refined.

Martincic et al. (2008) used response surface methodology to optimize an industrial-scale process for high-oleic sunflower oil deodorization. The authors concluded that optimization of the deodorization process requires a suitable compromise between often mutually opposing demands dictated by different oil quality indicators. Thus, the production of high-oleic sunflower oil with top-quality organoleptic and nutritional values (high tocopherol and phytosterol contents and low free and *trans* fatty acid contents) and high oxidative stability demands deodorization temperatures in the range between 220 and 235 °C and a total sparge steam above 2.0% (wt/wt in oil). The loss of tocopherols can be caused either by a thermal breakdown at temperatures higher than 240 °C, by oxidation reactions, or by chemical reactions such as the formation of tocopheryl esters (Verleyen et al., 2001). New developments in deodorization technology are driven by the continuous need for more efficient processes (lower operating costs, higher refined oil yields, and better valorization of side streams) and the increased retention of the nutritional quality of vegetable oils (De Greyt, 2012; Suliman et al., 2013b).

Romanic et al. (2009) studied the oxidative stability and tocopherol content of refined sunflower oil during long-term storage in different types of commercial packages. The results showed significant influence of packaging type on the oxidative stability of the oil. The largest changes were observed in oils packed in polyethylene canisters (volume = 5 L), where the peroxide values after 12 months were 9.34 and 10.46 mmol kg^{-1}, while the lowest peroxide values were found in Tetra brick packaging (1.05 and 1.67 mmol kg^{-1}). The highest increase in specific absorbencies was also noticed for oil packed in polyethylene canisters. The loss of tocopherol during storage was associated with the degree of oil oxidation, in such a way that tocopherol losses were 6.52% and 8.14% in oils packed in Tetra brick compared to 14.18% and 17.88% in oils packed in polyethylene canisters. The results of the Schaal-oven test confirmed the influence of packaging type and conditions on the oxidative stability of sunflower oil during storage.

Phytosterols

Phytosterols or plant sterols belong to the broad group of steroids with great physiological significance in plants and animals. In plants, this group includes important compounds such as triterpenoids, of great relevance in chemical defense against pathogens and herbivores (Flores-Sánchez et al., 2002); brassinosteroids, which are plant hormones essential for normal growth and development (Bishop and Yokota, 2001); phytosterols; and numerous secondary plant metabolites such as glycoalkaloids and saponins (Hartmann, 1998).

Plant sterols have a similar structure to mammalian cholesterol. They comprise a tetracyclic cyclopenta[a]phenanthrene structure and an alkyl side chain with 8–10 carbons (Hartmann, 1998). Figure 11.5 shows the close structural similarity between cholesterol and campesterol, one of the most important phytosterols. Phytosterol nomenclature in the literature is confusing. There have been attempts at standardization through IUPAC-IUB recommendations, but they have been only partially adopted and the trivial names are more frequently used. We will also use the trivial names in this chapter. The correspondence between trivial and systematic names for the main phytosterols can be found in Moss (1989) and Moreau et al. (2002).

The most common phytosterols are 4-desmethylsterols with a 28- or 29-carbon skeleton (Moreau et al., 2002). Most phytosterols have a double bond between C5 and C6 in the tetracyclic structure and thus are called Δ^5-sterols. Other groups of common

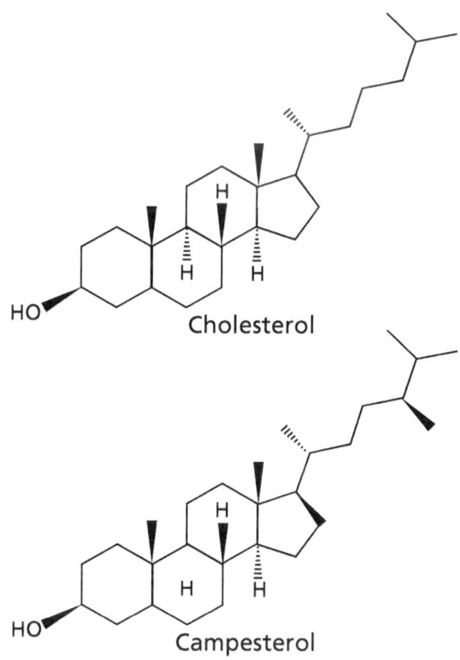

Figure 11.5 Chemical structure of mammalian cholesterol and plant campesterol.

phytosterols are Δ^7-sterols, with a double bond between C7 and C8, and $\Delta^{5,22}$-sterols with double bonds between C5 and C6 and also between C22 and C23. Saturated phytosterols with no double bonds in their structure are called *phytostanols,* but they are less common than phytosterols (Moreau et al., 2002). The main phytosterols in plants are β-sitosterol (C29, Δ^5), stigmasterol (C29, $\Delta^{5,22}$), and campesterol (C28, Δ^5). Most sterols in plants exist as free sterols, fatty acid esters, steryl glycosides, and acylated steryl glycosides (Phillips et al., 2005). Free sterols are components of plant cell membranes, where they play an essential role in the regulation of membrane fluidity and permeability (Benveniste, 2004). It has been suggested that the role of esterified sterols might be associated with the control of the amount of free sterols in cell membranes (Silvestro et al., 2013). Sterols are also precursors of other steroid compounds with a high physiological activity, such as brassinosteroids (Bishop and Yokota, 2001).

Vegetable oils are the richest natural sources of phytosterols, followed by cereal grains and nuts (Piironen et al., 2000). Because of their structural similarity with cholesterol, dietary phytosterols reduce intestinal absorption of cholesterol and contribute to reduce serum cholesterol levels (Plat and Mensink, 2005). Phytosterols are primarily synthesized from mevalonic acid through the mevalonate pathway that produces terpenes and steroids (Bishop and Yokota, 2001). Squalene, which is biosynthesized at the initial steps of the pathway, is an intermediate compound in the biosynthesis of phytosterols (Bishop and Yokota, 2001). An alternative pathway goes through 2-C-methyl-D-erythritol-4-phosphate instead of mevalonic acid (Nes, 2011). Figure 11.6 (p. 312) shows the biosynthetic pathway of the main phytosterols in sunflower. The pathway comprises two main branches from 24-methylene lophenol. The first one leads to the biosynthesis of β-sitosterol and stigmasterol, and the second one to the biosynthesis of campesterol. The first branch has Δ^7-stigmastenol as one of the intermediate products. This phytosterol is not present or is present in small amounts in vegetable oils, but it accumulates in relatively high concentrations in sunflower and safflower oils. Table 11.E shows the phytosterol content and composition in the most important vegetables oils. There are large differences among vegetable oils mainly for the total phytosterol content. In relation to the profile, β-sitosterol is predominant in all of them, with some differences in the concentrations of the other phytosterols. Rapeseed/canola contains relatively high concentration of brassicasterol, which is synthesized by desaturation of campesterol by sterol-C_{22}-desaturase. This phytosterol is absent or present in very small amounts in the other major vegetable oils. It is important to note that vegetable oils contain small amounts of cholesterol as well (Itoh et al., 1973), usually no more than 30 mg kg^{-1} (Kochhar, 1983) though higher values of around 50 mg kg^{-1} have been reported (Behrman and Gopalan, 2005). Sunflower oil contains low cholesterol levels. Behrman and Gopalan (2005) reported 15 mg kg^{-1}, whereas Schwartz et al. (2008) found no cholesterol in sunflower oil.

Table 11.E Estimated Contents of Different Types of Phytosterols in Sunflower Seeds[a]

Phytosterol Class	Seed Content (mg kg^{-1})
4,4-dimethylsterols	213 (7.1)
4-monomethylsterols	171 (5.7)
Free and esterified desmethylsterols	1994 (66.0)
Glycosides and acylated glycosides of desmethylsterols	641 (21.2)

[a]Percentages are given in parentheses.

Sources: Aguirre et al. (2012); Merah et al. (2012); Roche et al. (2010b).

Variation for Phytosterol Content and Profile in Sunflower

Several studies have reported variation for phytosterol content in sunflower oil and sunflower seeds. Most reports on sunflower phytosterols include exclusively free and esterified desmethylsterols. Therefore, total phytosterol content in this section will only refer to both groups of phytosterols, unless otherwise indicated. As previously mentioned for tocopherols, understanding genetic variation associated with biosynthetic steps of the phytosterol pathway requires evaluating germplasm for seed rather than for oil phytosterol content, in order to avoid a distorting influence of variation for oil content in the seeds. At the crude oil level, Piironen et al. (2000) reported a variation for total phytosterol content from 3740 to 7250 mg kg^{-1}. A greater range of variation, from 1250 to 7650 mg kg^{-1}, was reported by Ayerdi-Gotor et al. (2007) in the evaluation of a broad set of sunflower hybrids and inbred lines.

The analysis of phytosterols in seeds requires adapting the conventional methods used for oils. Several authors have developed procedures for the analysis of phytosterols in sunflower seeds or seed kernels. The methods basically consist of extraction of the unsaponifiable lipid fraction directly from seeds, derivatization of sterols, and GLC analysis (Fernández-Cuesta et al., 2012a; Merah et al., 2012; Roche et al., 2010a). Using these methods, Fernández-Cuesta et al. (2012a) reported a range of variation between 1426 and 4710 mg kg^{-1} kernel in 87 sunflower inbred lines, whereas Fernández-Cuesta et al. (2012b) found a variation between 2179 and 3555 mg kg^{-1} kernel in 12 confectionery landraces. Velasco et al. (2013) reported average phytosterol contents from 2954 to 3810 mg kg^{-1} kernel for 22 commercial hybrids grown under six environments. Fernández-Cuesta et al. (2014) evaluated selected accessions from a large germplasm collection under several environments and identified a maximum average phytosterol content of 3682 mg kg^{-1} seed kernel. The same authors reported a breeding line with high average kernel phytosterol content of 4957 mg kg^{-1} under the same environments. In wild *Helianthus* spp., Fernández-Cuesta et al.

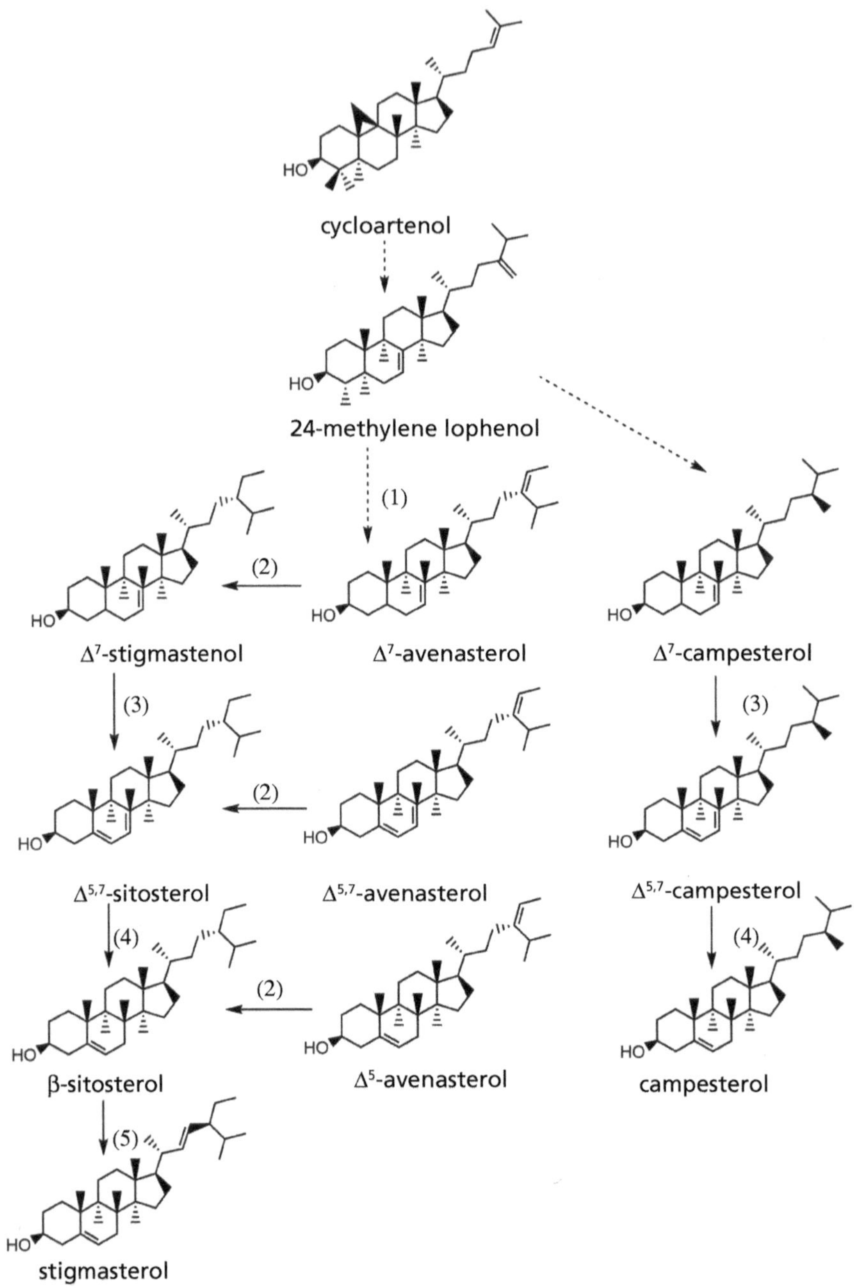

Figure 11.6 Schematic representation of the biosynthesis of the main phytosterols in sunflower. Dashed arrows indicate several enzymatic steps. The main phytosterols in sunflower are highlighted in bold. (1) Sterol methyltransferase II; (2) Δ^{24}-sterol-Δ^{24}-reductase; (3) Δ^{7}-sterol-C_5-desaturase; (4) $\Delta^{5,7}$-sterol-Δ^{7}-reductase; (5) sterol-C_{22}-desaturase.

(2011) reported a variation from 1017 to 4308 mg kg^{-1} kernel. Other authors analyzed whole seeds instead of seed kernels. Merah et al. (2012) reported a variation from 1523 to 2490 mg kg^{-1} seed in a population of recombinant inbred lines. Piao et al. (2012) analyzed a set of 100 germplasm accessions and found differences attributable to seed size. Small seeds had higher phytosterol content (2624 mg kg^{-1} seed) than medium-sized (2293 mg kg^{-1} seed) and large seeds (2003 mg kg^{-1} seed). A negative correlation between seed size and total phytosterol content was also reported by Fernández-Cuesta et al. (2012b, 2014).

The main phytosterols in regular sunflower oil are β-sitosterol (60% of total desmethylsterols), Δ^7-stigmastenol (14%), campesterol (8%), and stigmasterol (8%) (Table 11.E), although large variation for phytosterol composition has been identified in sunflower germplasm. Vlahakis and Hazebroek (2000) reported maximum concentrations of 20.4% campesterol, 17.9% stigmasterol, and 81.5% β-sitosterol. Verleyen et al. (2002a) reported germplasm with 15.1% Δ^5-avenasterol. Fernández-Cuesta et al. (2014) identified germplasm accessions with maximum levels of 18.0% campesterol, 16.1% stigmasterol, and 30.0% Δ^7-stigmastenol. In addition, maximum values of 23.9% for stigmasterol, 20.5% for Δ^5-avenasterol, and 10.6% for Δ^7-avenasterol have been reported in wild *Helianthus* spp. (Fernández-Cuesta et al., 2011).

Apart from free and esterified desmethylsterols, sunflower seeds also contain desmethylsterols in the form of steryl glycosides and acylated steryl glycosides. Their content in seeds is relatively high. In the analysis of 10 sunflower germplasms, Aguirre et al. (2012) found that the average content of free and esterified desmethylsterols in the seeds was 1994 mg kg^{-1}, whereas the average content of steryl glycosides and acylated steryl glycosides was 641 mg kg^{-1} (24%). However, both steryl glycosides and acylated steryl glycosides are poorly extracted with hexane, resulting in a relatively low content in crude sunflower oil. Aguirre et al. (2012) found an average content of free and esterified desmethylsterols in 10 hexane-extracted oils of 6552 mg kg^{-1}, whereas the average content of steryl glycosides and acylated steryl glycosides was only 176 mg kg^{-1} (3%). Steryl glycosides are undesired compounds for biodiesel production, as they form precipitates in biodiesel that may cause filter plugging (Moreau et al., 2008). Sunflower seeds and oils also contain relatively high levels of methylsterols and dimethylsterols. According to Kamal-Eldin (2005), sunflower oil contains 1110 mg $kg^{-1}$of methylsterols (4-monomethylsterols) and 680 mg kg^{-1} of 4,4-dimethylsterols, compared to 3920 mg kg^{-1} of desmethylsterols. In the seeds, Roche et al. (2010b) found an average content of 162 mg kg^{-1} of methylsterols and 173 mg kg^{-1} of dimethylsterols, compared to 1940 mg $kg^{-1}$of desmethylsterols. Merah et al. (2012) reported ranges of variation from 73 to 414 mg kg^{-1}seed for methyl sterols and from 120 to 551 mg kg^{-1} seed for dimethylsterols, compared to 1523 to 2491 mg kg^{-1} seed for desmethylsterols in a population of recombinant inbred lines. The main methylsterols in sunflower are gramisterol, citrostadienol, and obtusifoliol, whereas the main dimethylsterols are cycloartenol, 24-methylene-cycloartenol, and cyclobranol (Kamal-Eldin, 2005). A summary of the main classes of phytosterols in sunflower seeds is presented in Table 11.F.

Genetics of Phytosterols

Several studies have focused on the mode of inheritance and the influence of environmental conditions on phytosterol content and composition in sunflower. Merah et al. (2012) studied the inheritance of phytosterols in a population of recombinant inbred lines, reporting broad sense heritabilities of 0.72 for total seed phytosterol content, 0.68 for total content of desmethylsterols, 0.75 for total content of methylsterols, 0.46 for total content of dimethylsterols, and from 0.65 to 0.90 for the content of individual desmethylsterols (mg kg^{-1} seed), for example, 0.69 for β-sitosterol, 0.82 for Δ^7-stigmastenol, 0.90 for campesterol, and 0.65 for stigmasterol. For kernel phytosterol content, Velasco et al. (2013) reported broad sense heritabilities of 0.85 for total phytosterol content and from 0.95 to 0.97 for the concentration (% of total phytosterols) of campesterol, stigmasterol, and Δ^7-stigmastenol. Temperature during seed development has great influence on phytosterol content and composition, with higher temperatures inducing phytosterol accumulation (Roche et al., 2006, 2010b). This was explained as a plant reaction to compensate the effect of high temperatures on cellular membranes due to the role of phytosterols in the regulation of membrane fluidity and permeability. Also, higher temperatures led to an increase in the β-sitosterol to campesterol ratio (Roche et al., 2010b). Water stress also enhanced phytosterol accumulation (Roche et al., 2006).

Molecular marker studies have identified several QTL associated with seed phytosterols contents. Merah et al. (2012) found QTL consistently associated with seed phytosterols at LG 1, 4, and 7, while Haddadi et al. (2012) identified four QTL at LG 1, 2, 16, and 17. Merah et al. (2012) conducted candidate gene analysis, reporting four candidate genes encoding enzymatic antioxidants, and one candidate gene of the phytosterol pathway at LG 17, sterol methyltransferase II (SMT2), which catalyzes methylation of 24-methilene lophenol to 24-ethilidene lophenol and consequently directs the phytosterol biosynthesis to stigmasterol instead of campesterol (Figure 11.6).

Influence of Phytosterols on the Nutritional and Technological Properties of the Oil

Phytosterols are important nutritional constituents of vegetable oils. Because they are structurally related to cholesterol, phytosterols compete with cholesterol for absorption in the intestine, contributing to reduce serum cholesterol levels (Plat and Mensink,

Table 11.F Some Quality and Stability Characteristics of Crude Sunflower Oils[a]

Characteristics		Crude Sunflower Oil Groups			
		Pre-pressed	Solvent-extracted[b]	Mixed[c]	Full-pressed
Iron (Fe, mg kg^{-1})		0.61 ± 0.22[a]	0.80 ± 0.20[c]	0.70 ± 0.14[b]	0.71 ± 0.12[b]
Phosphorus (P, mg kg^{-1})		62.13 ± 20.82[a]	92.88 ± 22.80[c]	78.00 ± 23.29[b]	76.50 ± 2.12[b]
Free fatty acids (% oleic acid)		0.71 ± 0.19[a]	0.86 ± 0.12[a]	0.79 ± 0.13[a]	1.05 ± 0.10[b]
Peroxide value (meqO_2/kg)		8.96 ± 1.31[a]	9.20 ± 2.21[a]	9.18 ± 1.15[a]	12.10 ± 0.85[b]
Rancimat induction time (h, at 120 °C)		2.49 ± 0.25[c]	2.05 ± 0.26[b]	2.29 ± 0.17[bc]	1.83 ± 0.18[a]
Lovibond color[d]	Red	4.09 ± 0.63[a]	4.71 ± 1.02[b]	4.59 ± 0.63[ab]	5.10 ± 0.14[c]
	Yellow	38.00 ± 12.28[a]	41.63 ± 10.72[a]	39.75 ± 8.22[a]	50.00 ± 0.35[b]
α-Tocopherol (mg kg^{-1})		641.8 ± 68.4[c]	603.5 ± 42.9[b]	611.3 ± 53.5[b]	484.1 ± 63.7[a]
β-Tocopherol (mg kg^{-1})		26.7 ± 5.3[c]	17.0 ± 2.8[b]	21.6 ± 5.1[b]	13.8 ± 2.7[a]
γ-Tocopherol (mg kg^{-1})		20.9 ± 3.9[c]	13.6 ± 4.2[b]	18.4 ± 4.0[c]	9.3 ±2.4[a]
γ-Tocopherol (mg kg^{-1})		—	—	—	—
Total tocopherol (mg kg^{-1})		689.4 ± 56.8[c]	634.1 ± 46.2[b]	651.3 ± 59.7[b]	507.2 ± 61.5[a]

[a]Data from Tasan et al. (2011), given as mean ± standard deviation (SD). Means in the same row with different letters (a–c) show significant differences ($P < 0.01$, LSD multiple range test, except for color values).

[b]With hexane.

[c]Mixed equal amounts of the pre-pressed and solvent-extracted oils, taken from processing lines.

[d]5¼ inch cell.

2005). This effect has stimulated their use in a wide spectrum of phytosterol-fortified foods (Zawistowski, 2010). Due to the poor solubility and bioavailability of phytosterols, they are used as phytostanol esters after hydrogenation and esterification with fatty acids (García-Llatas and Rodríguez-Estrada, 2011). Margarine spreads were the first commercial application of phytosterol-enriched food, but the variety of food matrices has expanded beyond margarine spreads to include mayonnaise, salad dressings, dairy and nondairy beverages, chocolates, meats, cheeses, and baked goods (MacKay and Jones, 2011). In addition to their cholesterol lowering properties, there are studies suggesting that phytosterols may possess anti-cancer, anti-atherosclerosis, anti-inflammation, and antioxidative properties as well (Berger et al., 2004). The U.S. National Cholesterol Education Program recommends a phytosterol supplementation of 2 g per day to reduce low-density lipoprotein ("bad cholesterol") in blood and subsequently the risk of coronary heart disease (National Cholesterol Education Program, 2002). However, Germany's Federal Institute for Risk Assessment found indications that stanols and sterols could be a problem for those who do not have high hypercholesterolemia (BfR, 2011). For this reason, the European Union will require products that contain plant sterols and stanols to include warnings that their use is "not intended for people who do not need to control their blood cholesterol level," beginning in February 2014 (European Comission, 2013).

From a technological perspective, phytosterols increase thermal stability of oils with high levels of polyunsaturated fatty acids (Winkler and Warner, 2008a). This effect has been found to be more pronounced for phytosterols such as Δ^5-avenasterol that contain an ethylidene group in the side chain, as they protect the oil against polymerization and retard the loss of tocopherols (Gordon and Magos, 1983; Rossell, 2001). However, another study showed that the protective action of phytosterols against oil oxidation was more related to the degree of phytosterol unsaturation than to the presence of an ethylidene group (Winkler and Warner, 2008b). Even though there is a sufficient body of evidence supporting the positive impact of phytosterols on oil thermal stability, more research is still required to fully understand the influence of the phytosterol structure on oil stability.

Effect of Extraction Methods on Phytosterols

The total content and profile of phytosterols in the oil depends not only on genetic and environmental factors of sunflower, but also on technological processes of oil extraction from the seeds and its refining.

Oils extracted by press or by solvent (Figure 11.3) differ for both the total amount of phytosterols and the ratio of free to esterified sterols, with no significant differences in the composition of individual phytosterols. Solvent-extracted oils have higher total phytosterol content but lower content of esterified phytosterols than pressed oils because the solubility of these minor components depends on the oil, which acts as solvent. Most of the sterols present in sunflower seeds become part of the oil fraction after extraction, but around one-third of the total content remains in the defatted flour as sterol glycosides (Aguirre et al., 2012).

Changes of Phytosterols during Processing and Storage

Phytosterols are partially removed during the refining process previously described (Figure 11.4). The extent of phytosterol loss depends largely on the conditions applied in the various stages of the process. Most authors agree that the largest decrease in the total sterols is logged during neutralization and deodorization, although a slight decrease in the total phytosterol content may also occur during the bleaching process (Ferrari et al., 1996; Ortega-García et al., 2006; Verleyen et al., 2002b). This loss may vary between 10% and 70% of total sterols in crude oil, depending on the conditions used, although the percentages of individual phytosterols are not modified during the refining process (Ferrari et al., 1996; Naz et al., 2011).

In the vegetable oil refining process, steradienes are formed during the bleaching process on the surface of the acid activated bleaching earth or during the deodorization process, promoted by the elevated temperature. Steradienes are steroidal hydrocarbons formed by a sterol dehydration reaction. The most abundant steradiene in vegetable oils is 3,5-stigmastadiene, derived from the dehydration of β-sitosterol (Verleyen et al., 2002c). Besides steradienes, other sterol transformation products are also formed during the refining process. Among these are disteryl ethers, which are sterol condensation products, and oxidized sterols (Verleyen et al., 2002b). In addition, a considerable increase in the ratio of esterified to free sterol has been observed during the refining process (Ortega-García et al., 2006; Sciancalepore, 1981; Verleyen et al., 2002c).

Like cholesterol, phytosterols can also undergo oxidation and produce a series of oxidation products. In the case of β-sitosterol, the main oxidation products are 7β-hydroxysitosterol, 7α-hydroxysitosterol, 5α-hydroxysitosterol, 7-ketositosterol, and epoxysitosterol. Oxidation products could be generated through autoxidation at elevated temperatures, such as those in deep frying processes (Dutta and Appelqvist, 1997). The latter authors studied sterol oxides produced during storage of potato chips fried with regular and high-oleic sunflower oils and palm oil, concluding higher content of phytosterol oxides when sunflower oils were used. No quantitative or qualitative differences were observed between regular and high-oleic sunflower oils. Although the adverse effect of cholesterol oxidized products has been well documented, information on the health effects of phytosterol oxidized products is scarce (Hovenkamp et al., 2008).

Tests performed on the effect of aging, by heating oil at 50 °C for several weeks and at 100 °C for 1 h, did not show any significant variation in the phytosterol content. In contrast, heating at 200 °C for 1 h led to a 50–60% decrease of phytosterol content in the oil (Thanh et al., 2006).

Other Minor Constituents

Squalene is one of the most biologically active constituents of vegetable oils. Chemically, squalene is a naturally occurring terpenoid hydrocarbon produced by all plants and animals, including humans (Figure 11.7). In plants, squalene is synthesized as a biochemical intermediate the initial steps of the phytosterol biosynthetic pathway. Similarly, in animals it is an intermediate in the cholesterol pathway. In humans, squalene is one of the main components of skin surface lipids, which makes it of great value in the cosmetic industry (Huang et al., 2009). It is also widely used for preparation of vaccines and drug delivery emulsions in the pharmaceutical industry (Fox, 2009). Shark liver oil has traditionally been the main source of squalene. Within commercially important vegetable oils, virgin olive oil is the richest source of squalene, with typical levels between 200 and 7500 mg kg^{-1} (Boskou, 2009). In crude sunflower oil, squalene content is typically between 150 and 200 mg kg^{-1} (Grompone, 2005), though higher levels of up to 271 mg kg^{-1} have been reported (Kalogeropoulos and Andrikopoulos, 2004). Squalene has a great impact on the nutritional properties of vegetable oils, as it has been suggested as a compound possessing anticarcenogenic properties (Sotiroudis and Kyrtopoulos, 2008). Squalene is very stable both during oil storage as well as in technological processes. However, inconsistent results have been reported on its antioxidant efficacy (Amarowicz, 2009). Studies on variability for squalene content in sunflower germplasm are very scarce and practically no breeding efforts to elevate squalene content in sunflower oil have been conducted. Merah et al. (2012) identified large variation (10 to 202 mg kg^{-1}) in a population of recombinant inbred lines, as well as a moderate heredability of the trait. Further research is required to evaluate whether sunflower germplasm contains sufficient variability for squalene content to encourage breeding efforts on this trait.

Carotenoids and chlorophylls are the two most common pigments present in vegetable oils (Table 11.A). Carotenoids are tetraterpenoids, that is, they are produced from eight isoprene molecules and contain 40 carbon atoms. They are classified into two types: xanthophylls, which contain oxygen, and carotenes, which are purely hy-

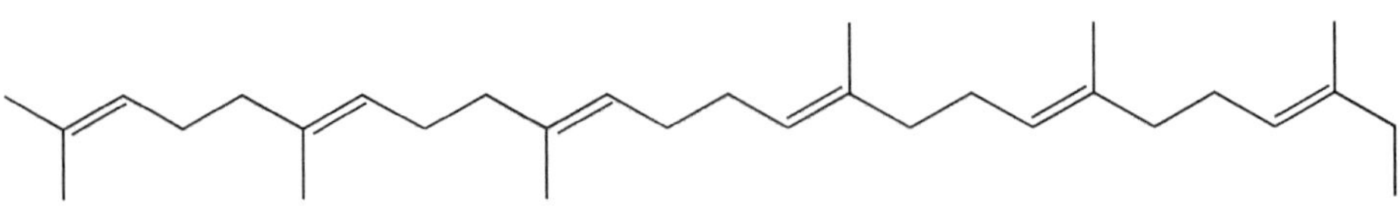

Figure 11.7 Chemical structure of squalene.

drocarbons containing no oxygen. Xanthophylls are the major carotenoids in sunflower oil, with dihydroxy carotenoids (mostly lutein; Figure 11.8) ranging from 76 to 81% of total carotenoids (Rade et al., 2004). Addition of β-carotene to sunflower oil increases its oxidation stability at room temperature and under daylight due to a synergistic effect with tocopherols (Yanishlieva et al., 2001). Important losses of carotenoids occur during oil refining. The maximum reduction occurs occurring during the bleaching process, in which around 77% of carotenoids are removed. In the subsequent deodorization step, carotenoids are further removed to a final reduction of around 85% (Rade et al., 2004). Franke et al. (2010) found carotenoids in cold pressed sunflower oil but not in refined oils. The high proportion of carotenoid removal during refining has been attributed to their thermolability (Ouyang et al., 1980).

Chlorophylls are common sensitizers that act as promoter of photo-oxidation in oil systems. Although crude sunflower oil has a low level of chlorophyll pigments, they are removed during bleaching step in the refining process.

Crude sunflower oil contains high levels of phospholipids. Phospholipids are amphipathic molecules that constitute the major components of biological membranes. They consist of a diacylglycerol with fatty acids esterified at positions *sn*-1 and *sn*-2 of the glycerol backbone, a phosphate group at the *sn*-3 position, and a variable hydrophilic head group linked to the phosphate group. Phospholipids have strong antioxidant effects due to their synergistic action with tocopherols, their metal scavenging activity, and also to their catalytic activity to decompose hydroperoxides (Carelli et al., 1997; Smouse, 1995). However, phospholipids show higher susceptibility to oxidation than triacylglycerols because they possess a higher degree of fatty acid unsaturation to maintain the fluidity of membranes (Shahidi and Zhong, 2010). For this reason, they impart unpleasant odor and taste to the oil during storage, reducing its shelf life (Desai et al., 2002). The major phospholipid groups in sunflower oil are phosphatidylcholines, phosphatidylethanolamines, phosphatidilinositols, and phosphatidic acids (Gupta, 2002). Phospholipids are removed during the degumming phase of refining as a mixture of polar lipids known as *lecithin*. Because of their emulsifying properties, lecithins have high commercial value for an extensive number of food and nonfood applications (Szuhaj, 2005).

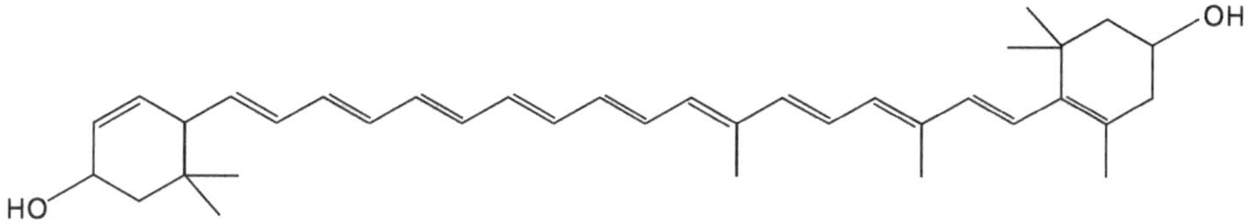

Figure 11.8 Chemical structure of lutein.

Sunflower seeds contain around 30% of hull and the hulls are characterized by high wax content, up to 3% (Carelli et al., 2002a). Accordingly, seeds are partially dehulled before oil extraction to hull levels usually below 15% (Grompone, 2005). Depending on the final hull content, the wax content in the oil may range between 200 and 3500 mg kg^{-1} (Carelli et al., 2002a). Chemically, waxes are esters of fatty acids and long chain alcohols. Waxes in crude sunflower oil generally have carbon atom numbers between 36 and 50. The content and composition of the waxes in sunflower oil are dependent on the extraction method, with higher amounts of waxes being extracted by solvent than by press extraction (Carelli et al., 2002b). Waxes are removed in the dewaxing or winterization refining step, which is a common practice in the sunflower oil industry (Figures 11.6 and 11.7). Wax concentration increases with storage conditions (time and temperature). Martini and Añón (2005) found that wax composition of sunflower seeds was approximately constant during storage for C35–C39 and showed significant differences for C40–C48 components. Waxes with higher numbers of carbon atoms cause more turbidity in crude sunflower oil due to their higher melting point, which influences the processing conditions for oil refining (Martini and Añón, 2005).

Conclusion

Important advances have been made in recent years in the characterization of the variability for minor constituents in sunflower seeds and oils, their influence on the nutritional and technological properties of the oil, and the impact of extraction and processing operations on the main minor constituents. These studies have been paralleled by breeding efforts to develop novel sunflower germplasms with modified levels of minor constituents, mainly tocopherols and phytosterols. For tocopherols, sunflower germplasm with increased levels of total tocopherol content and germplasm with modified tocopherol profiles, including increased levels of β-, γ-, and/or δ-tocopherol, have been developed by several research groups by conventional breeding. Studies at the lab and pilot plant scales have demonstrated that partially replacing α-tocopherol, predominant in regular sunflower oil, with some of the novel tocopherol profiles leads to significant enhancement of the thermoxidative stability of sunflower oil. The genetics of the novel tocopherol traits have been studied in detail and molecular tools are already available to support trait introgression into elite sunflower lines. Advances in developing novel variability for tocopherol content and composition in sunflower are unparalleled in other oilseed crops. Breeding research on seed phytosterols has recently been started and promising results on the feasibility of developing sunflower cultivars with enhanced phytosterol levels have already been reported. Major innovations in this field are expected in the near term.

Due to the existing variation for fatty acid and tocopherol profiles and the influence of both groups of traits and their synergistic interactions on the nutritional and technological properties of the oil, sunflower is currently in a privileged position among major oilseeds. Different combinations between both groups of compounds can produce oils with optimal properties for specific market niches. Recent advances in phytosterol research suggest that novel phytosterol traits will probably be available soon. Studies on the characterization of the properties of the novel sunflower oils are still insufficient and they should be encouraged. Also, further advances in the development of molecular breeding tools are required to help breeding companies transform genetic discoveries into innovative commercial sunflower oils.

References

Aguirre, M. R.; Ruiz-Méndez, M. V.; Velasco, L.; Dobarganes, M. C. Free Sterols and Steryl Glycosides in Sunflower Seeds with High Phytosterol Contents. *Eur. J. Lipid Sci. Technol.* **2012,** *114*, 1212–1216.

Alpaslan, M.; Tepe, S.; Simsek, O. Effect of Refining Processes on the Total and Individual Tocopherol Content in Sunflower Oil. *Int. J. Food Sci. Technol.* **2001,** *36*, 737–739.

Amarowicz, R. Squalene: A Natural Antioxidant? *Eur. J. Lipid Sci. Technol.* **2009,** *111*, 411–412.

Anderson, D. A Primer on Oils Processing Technology. In *Edible Oil and Fat Products: Processing Technologies*; Shahidi, F., Ed.; Bailey's Industrial Oil and Fat Products, 6th ed., Vol. 5; John Wiley & Sons: Hoboken, NJ, 2005; pp 1–56.

AOCS. Official Method Ca 6a-40. In *Official Methods and Recommended Practices of the American Oil Chemists' Society*, 6th ed.; AOCS Press: Champaign, IL, 2011 (second printing).

Ayerdi-Gotor, A.; Farkas, E.; Berger, M.; Labalette, F.; Centis, S.; Dayde, J.; Calmon, A. Determination of Tocopherols and Phytosterols in Sunflower Seeds by NIR Spectrometry. *Eur. J. Lipid Sci. Technol.* **2007,** *109*, 525–530.

Barrera-Arellano, D.; Ruiz-Méndez, V.; Márquez-Ruiz, G.; Dobarganes, C. Loss of Tocopherols and Formation of Degradation Compounds in Triacylglycerol Model Systems Heated at High Temperature. *J. Sci. Food Agric.* **1999,** *79*, 1923–1928.

Barrera-Arellano, D.; Ruiz-Méndez, V.; Velasco, J.; Márquez-Ruiz, G.; Dobarganes, C. Loss of Tocopherols and Formation of Degradation Compounds at Frying Temperatures in Oils Differing in Degree of Unsaturation and Natural Antioxidant Content. *J. Sci. Food Agric.* **2002,** *82*, 1696–1702.

Bastic, M.; Bastic, L.; Jovanovic, J. A.; Spiteller, G. Hydrocarbons and Other Weakly Unsaponifiables in Some Vegetable Oils. *J. Am. Oil Chem. Soc.* **1978,** *55*, 886–891.

Behrman, E. J.; Gopalan, V. Cholesterol and Plants. *J. Chem. Educ.* **2005,** *82*, 1791–1793.

Benveniste, P. Biosynthesis and Accumulation of Sterols. *Annu. Rev. Plant Biol.* **2004,** *55*, 429–457.

Berger, A.; Jones, P. J. H.; Abumweis, S. S. Plant Sterols: Factors Affecting Their Efficacy and Safety as Functional Food Ingredients. *Lipids Health Dis.* **2004,** *3*, 5.

BfR, Bundesinstitut für Risikobewertung. Addition of Plant Sterols and Stanols to Food: Assessment of a New Study from the Netherlands. BfR opinion 006/2012, December 2011. http://www.bfr.bund.de/cm/349/addition-of-plant-sterols-and-stanols-to-food-assessment-of-a-new-study-from-the-netherlands.pdf (accessed August 18, 2013).

Bishop, G. J.; Yokota, T. Plant Steroid Hormones, Brassinosteroids: Current Highlights of Molecular Aspects on Their Synthesis/Metabolism, Transport, Perception and Response. *Plant Cell Physiol.* **2001,** *42*, 114–120.

Boskou, D. Other Important Minor Constituents. In *Olive Oil: Minor Constituents and Health*; Boskou, D., Ed.; CRC Press: Boca Raton, FL, 2009; pp 45–54.

Brevedan, M. I. V.; Carelli, A. A.; Crapiste, G. H. Changes in Composition and Quality of Sunflower Oils during Extraction and Degumming. *Grasas Aceites* **2000,** *51*, 417–423.

Carelli, A. A.; Brevedan, M. I. V.; Crapiste, G. H. Quantitative Determination of Phospholipids in Sunflower Oil. *J. Am. Oil Chem. Soc.* **1997,** *74*, 511–514.

Carelli, A. A.; Frizzera, L. M.; Forbito, P. R.; Crapiste, G. H. Wax Composition of Sunflower Seed Oils. *J. Am. Oil Chem. Soc.* **2002a,** *79*, 763–768.

Carelli, A. A.; Ceci, L. N.; Crapiste, G. H. Phosphorus-to-Phospholipid Conversion Factors for Crude and Degummed Sunflower Oils. *J. Am. Oil Chem. Soc.* **2002b,** *79*, 1177–1180.

Codex Alimentarius. *Standard for Named Vegetable Oils. CODEX-STAN 210-1999* (revision 3, amendment 3); WHO/FAO: Rome, Italy, 2013.

De Greyt, W. F. J. Current and Future Technologies for the Sustainable and Cost-Efficient Production of High Quality Food Oils. *Eur. J. Lipid Sci. Technol.* **2012,** *114,* 1126–1139.

Del Moral, L.; Pérez-Vich, B.; Fernández-Martínez, J. M.; Velasco, L. Inheritance of Increased Seed Tocopherol Content in Sunflower Line IAST-413. *Plant Breeding* **2011,** *130*, 540–543.

Del Moral, L.; Fernández-Martínez, J. M.; Velasco, L.; Pérez-Vich, B. Quantitative Trait Loci for Seed Tocopherol Content in Sunflower. *Crop Sci.* **2012,** *52*, 786–794.

Demurin, Y. Genetic Variability of Tocopherol Composition in Sunflower Seeds. *Helia* **1993,** *18*, 59–62.

Demurin, Y.; Škorić, D.; Karlovic, D. Genetic Variability of Tocopherol Composition in Sunflower Seeds as a Basis of Breeding for Improved Oil Quality. *Plant Breeding* **1996,** *115*, 33–36.

Desai, N. C.; Mehta, M. H.; Dave, A. M.; Mehta, J. N. Degumming of Vegetable Oil by Membrane Technology. *Indian J. Chem. Technol.* **2002,** *9*, 529–534.

Dolde, D.; Vlahakis, C.; Hazebroek, J. Tocopherols in Breeding Lines and Effects of Planting Location, Fatty Acid Composition, and Temperature during Development. *J. Am. Oil Chem. Soc.* **1999,** *76*, 349–355.

Dutta, P. C.; Appelqvist, L. A. Studies on Phytosterol Oxides. I: Effect of Storage on the Content in Potato Chips Prepared in Different Vegetable Oils. *J. Am. Oil Chem. Soc.* **1997**, *74*, 647–657.

Eitenmiller, R.; Lee, J. *Vitamin E. Food Chemistry, Composition, and Analysis*; Marcel Dekker: New York, 2004.

European Commission. Commission Regulation (EU) No 718/2013 of 25 July 2013 Amending Regulation (EC) No 608/2004 Concerning the Labeling of Foods and Food

Ingredients with Added Phytosterols, Phytosterol Esters, Phytostanols and/or Phytostanol Esters. *Official Journal of the European Union,* **2013,** *L 201,* 49–50.

FAOSTAT. Statistical Database of the Food and Agriculture Organization (FAO) of the United Nations. http://faostat3.fao.org/home/index.html (accessed August 30, 2013).

Fernández-Cuesta, A.; Velasco, L.; Fernández-Martínez, J. M. Phytosterols in the Seeds of Wild Sunflower Species. *Helia* **2011,** *55,* 31–38.

Fernández-Cuesta, A.; Aguirre-González, M. R.; Ruiz-Méndez, M. V.; Velasco, L. Validation of a Method for the Analysis of Phytosterols in Sunflower Seeds. *Eur. J. Lipid Sci. Technol.* **2012a,** *114,* 325–331.

Fernández-Cuesta, A.; Nabloussi, A.; Fernández-Martínez, J. M.; Velasco, L. Tocopherols and Phytosterols in Sunflower Seeds for the Human Food Market. *Grasas Aceites* **2012b,** *63,* 321–327.

Fernández-Cuesta, A.; Jan, C. C.; Fernández-Martínez, J. M.; Velasco, L. Variability for Seed Phytosterols in Sunflower Germplasm. *Crop Sci.* **2014,** *54,* 190–197.

Fernández-Martínez, J. M.; Pérez-Vich, B.; Velasco, L. Sunflower. In *Oil Crops*; Vollmann, J., Rajcan, I., Eds.; Springer: New York, 2009; pp 155–232.

Ferrari, R. A.; Schutte, E.; Esteves, W.; Brühl, L.; Mukherjee, K. D. Minor Constituents of Vegetable Oils during Industrial Processing. *J. Am. Oil Chem. Soc.* **1996,** *73,* 587–591.

Fils, J. M. The Production of Oils. In *Edible Oil Processing*; Hamm, W., Hamilton, R. J., Eds.; Sheffield Academy Press: Sheffield, UK, 2000; pp 47–78.

Flores-Sánchez, I. J.; Ortega-López, J.; Montes-Horcasitas, M. D. C.; Ramos-Valdivia, A. C. Biosynthesis of Sterols and Triterpenes in Cell Suspension Cultures of *Uncaria tomentosa. Plant Cell Physiol.* **2002,** *43,* 1502–1509.

Fox, C. B. Squalene Emulsions for Parenteral Vaccine and Drug Delivery. *Molecules* **2009,** *14,* 3286–3312.

Franke, S.; Frölich, K.; Werner, S.; Böhm, V.; Schöne, F. Analysis of Carotenoids and Vitamin E in Selected Oilseeds, Press Cakes and Oils. *Eur. J. Lipid Sci. Technol.* **2010,** *112,* 1122–1129.

Fuster, M. D.; Lampi, A. M.; Hopia, A.; Kamal-Eldin, A. Effects of α- and γ-Tocopherols on the Autooxidation of Purified Sunflower Triacylglycerols. *Lipids* **1998,** *33,* 715–722.

Galli, F.; Azzi, A. Present Trends in Vitamin E Research. *Biofactors* **2010,** *36,* 33–42.

García-Llatas, G.; Rodríguez-Estrada, M. T. Current and New Insights on Phytosterol Oxides in Plant Sterol-Enriched Food. *Chem. Phys. Lipids* **2011,** *164,* 607–624.

García-Moreno, M. J.; Vera-Ruíz, E. M.; Fernández-Martínez, J. M.; Velasco, L.; Pérez-Vich, B. Genetic and Molecular Analysis of High Gamma-Tocopherol Content in Sunflower. *Crop Sci.* **2006,** *46,* 2015–2021.

García-Moreno, M. J.; Fernández-Martínez, J. M.; Velasco, L.; Pérez-Vich, B. Genetic Basis of Unstable Expression of High Gamma-Tocopherol Content in Sunflower Seeds. *BMC Plant Biol.* **2012,** *12,* 71.

Gordon, M. H.; Magos, P. The Effect of Sterols on the Oxidation of Edible Oils. *Food Chem.* **1983,** *10,* 141–147.

Grompone, M. A. Sunflower Oil. In *Edible Oil and Fat Products: Chemistry, Properties, and Health Effects*; Shahidi, F., Ed.; Bailey's Industrial Oil and Fat Products, 6th ed., Vol. 1; John Wiley & Sons: Hoboken, NJ, 2005; pp 655–730.

Gül, M. K.; Seker, M. Comparative Analysis of Phytosterol Components from Rapeseed (*Brassica napus* L.) and Olive (*Olea europaea* L.) Varieties. *Eur. J. Lipid Sci. Technol.* **2006,** *108,* 759–765.

Gunstone, F. D. Vegetable Oils. In *Edible Oil and Fat Products: Chemistry, Properties, and Health Effects*; Shahidi, F., Ed.; Bailey's Industrial Oil and Fat Products, 6th ed., Vol. 1; John Wiley & Sons: Hoboken, NJ, 2005; pp 213–267.

Gunstone, F. D.; Harwood, J. L. Occurrence and Characterisation of Oils and Fats. In *The Lipid Handbook,* 3rd ed.; Gunstone, F. D.; Harwood, J. L.; Dijkstra, A. J., Eds.; CRC Press: Boca Raton, FL, 2007; pp 37–141.

Gupta, M. K. Sunflower Oil. In *Vegetable Oils in Food Technology: Composition, Properties and Uses*; Gunstone, F. D., Ed.; Wiley-Blackwell: Oxford, U.K., 2002; pp 128–156.

Haddadi, P.; Yazdi-Samadi, B.; Berger, M.; Naghavi, M. R.; Calmon, A.; Sarrafi, A. Genetic Variability of Seed-Quality Traits in Gamma-Induced Mutants of Sunflower (*Helianthus annuus* L.) under Water-Stressed Condition. *Euphytica* **2011,** *178,* 247–259.

Haddadi, P.; Ebrahimi, A.; Langlade, N. B.; Yazdi-Samadi, B.; Berger, M.; Calmon, A.; Naghavi, M. R.; Vincourt, P.; Sarrafi, A. Genetic Dissection of Tocopherol and Phytosterol in Recombinant Inbred Lines of Sunflower through Quantitative Trait Locus Analysis and the Candidate Gene Approach. *Mol. Breeding* **2012,** *29,* 717–729.

Hartmann, M. A. Plant Sterols and the Membrane Environment. *Trends Plant Sci.* **1998,** *3,* 170–175.

Hass, C. G.; Tang, S.; Leonard, S.; Traber, M. G.; Miller, J. F.; Knapp, S. J. Three Non-Allelic Epistatically Interacting Methyltransferase Mutations Produce Novel Tocopherol (Vitamin E) Profiles in Sunflower. *Theor. Appl. Genet.* **2006,** *113,* 767–782.

Hovenkamp, E.; Demonty, I.; Plat, J.; Lütjohann, D.; Mensink, R. P.; Trautwein, E. A. Biological Effects of Oxidized Phytosterols: A Review of the Current Knowledge. *Prog. Lipid Res.* **2008,** *47,* 37–49.

Huang, Z. R.; Lin, Y. K.; Fang, J. Y. Biological and Pharmacological Activities of Squalene and Related Compounds: Potential Uses in Cosmetic Dermatology. *Molecules* **2009,** *14,* 540–554.

Itoh, T.; Tamura, T.; Matsumoto, T. Sterol Composition of 19 Vegetable Oils. *J. Am. Oil Chem. Soc.* **1973,** *50,* 122–125.

Ju, J.; Picinich, S. C.; Yang, Z.; Zhao, Y.; Suh, N.; Kong, A. N.; Yang, C. S. Cancer-Preventive Activities of Tocopherols and Tocotrienols. *Carcinogenesis* **2010,** *31,* 533–542.

Kalogeropoulos, N.; Andrikopoulos, N. K. Squalene in Oils and Fats from Domestic and Commercial Fryings of Potatoes. *Int. J. Food Sci. Nutr.* **2004,** *55,* 125–129.

Kamal-Eldin, A. Minor Components of Fats and Oils. In *Edible Oil and Fat Products: Specialty Oils and Oil Products*; Shahidi, F., Ed.; Bailey's Industrial Oil and Fat Products, 6th ed., Vol. 3; John Wiley & Sons: Hoboken, NJ, 2005; pp 319–359.

Kochhar, S. P. Influence of Processing on Sterols of Edible Vegetable Oils. *Prog. Lipid Res.* **1983,** *22,* 161–188.

Kovari, K. Recent Developments, New Trends in Seed Crushing and Oil Refining. *OCL—Oleagineux, Corps Gras, Lipides.* **2004,** *11,* 381–387.

Lampi, A. M.; Kamal-Eldin, A. Effect of α- and γ-Tocopherol on Thermal Polymerisation of Purified High-Oleic Sunflower Triacylglycerols. *J. Am. Oil Chem. Soc.* **1998,** *75,* 1699–1703.

Lampi, A. M.; Kataja, L.; Kamal-Eldin, A.; Piironen, V. Antioxidant Activities of α- and γ-Tocopherol in the Oxidation of Rapeseed Oil Triacylglycerols. *J. Am. Oil Chem. Soc.* **1999,** *76*, 749–755.

MacKay, D. S.; Jones, P. J. H. Phytosterols in Human Nutrition: Type, Formulation, Delivery, and Physiological Function. *Eur. J. Lipid Sci. Technol.* **2011,** *113*, 1427–1432.

Marmesat, S.; Velasco, L.; Ruiz-Méndez, M. V.; Fernández-Martínez, J. M.; Dobarganes, C. Thermostability of Genetically Modified Sunflower Oils Differing in Fatty Acid and Tocopherol Compositions. *Eur. J. Lipid Sci. Technol.* **2008,** *110*, 776–782.

Marquard, R. Untersuchungen über dem Einfluss von Sorte und Standort auf den Tocopherolgehalt verschiedener Pflanzenöle. *Fat Sci. Technol.* **1990,** *92*, 452–455.

Martincic, V.; Golob, J.; De Greyt, W.; Verhé, R.; Knez, S.; Van Hoed, V.; Zilnik, L. F.; Potočnik, K.; Rizner Hras, A.; Vila Ayala, J. Optimization of Industrial-Scale Deodorization of High-Oleic Sunflower Oil via Response Surface Methodology. *Eur. J. Lipid Sci. Technol.* **2008,** *110*, 245–253.

Martini, S.; Añón, M. C. Storage of Sunflower Seeds: Variation on the Wax Content of the Oil. *Eur. J. Lipid Sci. Technol.* **2005,** *107*, 74–79.

Mène-Saffrané, L.; DellaPenna, D. Biosynthesis, Regulation and Functions of Tocochromanols in Plants. *Plant Physiol. Biochem.* **2010,** *48*, 301–309.

Merah, O.; Langlade, N.; Alignan, M.; Roche, J.; Pouilly, N.; Lippi, Y.; Vear, F.; Cerny, M.; Bouniols, A.; Mouloungui, Z.; et al. Genetic Analysis of Phytosterol Content in Sunflower Seeds. *Theor. Appl. Genet.* **2012,** *125*, 1589–1601.

Moreau, R. A.; Whitaker, B. D.; Hicks, K. B. Phytosterols, Phytostanols, and Their Conjugates in Foods: Structural Diversity, Quantitative Analysis, and Health-Promoting Uses. *Prog. Lipid Res.* **2002,** *41*, 457–500.

Moreau, R. A.; Scott, K. M.; Haas, M. J. The Identification and Quantification of Steryl Glucosides in Precipitates from Commercial Biodiesel. *J. Am. Oil Chem. Soc.* **2008,** *85*, 761–770.

Moss, G. P. Nomenclature of Steroids. Recommendations 1989. *Pure Appl. Chem.* **1989,** *61*, 1783–1822.

National Cholesterol Education Program. *Third Report of the National Cholesterol Education Program (NCEP) Expert Panel on Detection, Evaluation, and Treatment of High Blood Cholesterol in Adults (Adult Treatment Panel III);* National Cholesterol Education Program, National Heart, Lung, and Blood Institute, National Institutes of Health, NIH Publication No. 02-5215, September 2002. http://www.nhlbi.nih.gov/guidelines/cholesterol/atp3full.pdf (accessed August 18, 2013).

Naz, S.; Sherazi, S. T. H.; Talpur, F. N. Changes of Total Tocopherol and Tocopherol Species during Sunflower Oil Processing. *J. Am. Oil Chem. Soc.* **2011,** *88*, 127–132.

Nes, W. D. Biosynthesis of Cholesterol and Other Sterols. *Chem. Rev.* **2011,** *111*, 6423–6451.

Nimet, G.; Da Silva, E. A.; Palú, F.; Dariva, C.; Freitas, L. D. S.; Neto, A. M. Extraction of Sunflower (*Heliantus annuus* L.) Oil with Supercritical CO_2 and Subcritical Propane: Experimental and Modelling. *Chem. Eng. J.* **2011**, *168*, 262–268.

Nolasco, S. M.; Aguirrezábal, L. A. N.; Crapiste, G. H. Tocopherol Oil Concentration in Field-Grown Sunflower Is Accounted for by Oil Weight per Seed. *J. Am. Oil Chem. Soc.* **2004,** *81*, 1045–1051.

Ortega-García, J.; Gámez-Meza, N.; Noriega-Rodriguez, J. A.; Dennis-Quiñonez, O.; García-Galindo, H. S.; Angulo-Guerrero, J. O.; Medina-Juárez, L. A. Refining of High Oleic Safflower Oil: Effect on the Sterols and Tocopherols Content. *Eur. Food Res. Technol.* **2006,** *223,* 775–779.

Ouyang, J. M.; Daun, H.; Chang, S. S.; Ho, C. T. Formation of Carbonyl Compounds from β-Carotene during Palm Oil Deodorization. *J. Food Sci.* **1980,** *45,* 1214–1217.

Perretti, G.; Finottia, E.; Adamuccio, S.; Della Sera, R.; Montanari, L. Composition of Organic and Conventionally Produced Sunflower Seed Oil. *J. Am. Oil Chem. Soc.* **2004,** *81,* 1119–1123.

Phillips, H. M.; Ruggio, D. M.; Toivo, J. I.; Swank, M. A.; Simpkins, A. H. Free and Esterified Sterol Composition of Edible Oils and Fats. *J. Food Comp. Anal.* **2002,** *15,* 123–142.

Phillips, K. M.; Ruggio, D. M.; Ashraf-Khorassani, M. Analysis of Steryl Glucosides in Foods and Dietary Supplements by Solid-Phase Extraction and Gas Chromatography. *J. Food Lipids* **2005,** *12,* 124–140.

Piao, X.; Jang, Y. S.; Kim, H. S. Tocopherol and Phytosterol Variations in Sunflower (*Helianthus annuus* L.) Germplasms. *J. Biobased Mater. Bioenergy* **2012,** *6,* 319–322.

Piironen, V.; Lindsay, D. G.; Miettinen, T. A.; Toivo, J.; Lampi, A. M. Plant Sterols: Biosynthesis, Biological Function and Their Importance to Human Nutrition. *J. Sci. Food Agric.* **2000,** *80,* 939–966.

Plat, J.; Mensink, R. P. Plant Stanol and Sterol Esters in the Control of Blood Cholesterol Levels: Mechanism and Safety Aspects. *Am. J. Cardiol.* **2005,** *96,* 15–22.

Premovic, T. D.; Dimic, E. B.; Takaci, A. A.; Romanic, R. S. Influence of Impurities and Hull Content in Material for Pressing on Sensory Quality Cold-Pressed Sunflower Oil. *Acta Periodica Technologica* **2010,** *41,* 69–76.

Rade, D.; Mokrovcak, Z.; Struceli, D.; Skevin, D.; Nederal, S. The Effect of Processing Conditions on the Nontriacylglycerol Constituents of Sunflower Oil. *Acta Alimentaria* **2004,** *33,* 7–18.

Rass, M.; Schein, C.; Matthäus, B. Virgin Sunflower Oil. *Eur. J. Lipid Sci. Technol.* **2008,** *110,* 618–624.

Roche, J.; Bouniols, A.; Mouloungui, Z.; Barranco, T.; Cerny, M. Management of Environmental Crop Conditions to Produce Useful Sunflower Oil Components. *Eur. J. Lipid Sci. Technol.* **2006,** *108,* 287–297.

Roche, J.; Alignan, M.; Bouniols, A.; Cerny, M.; Mouloungui, Z.; Mehrah, O. Sterol Concentration and Distribution in Sunflower Seeds (*Helianthus annuus* L.) during Seed Development. *Food Chem.* **2010a,** *119,* 1451–1456.

Roche, J.; Alignan, M.; Bouniols, A.; Cerny, M.; Mouloungui, Z.; Vear, F.; Mehrah, O. Sterol Content in Sunflower Seeds (*Helianthus annuus* L.) as Affected by Genotypes and Environmental Conditions. *Food Chem.* **2010b,** *121,* 990–995.

Romanic, R.; Dimic, E.; Lazic, V.; Vujasinovic, V. Oxidative Stability and Tocopherol Content of Refined Sunflower Oil during Long-Term Storage in Different Commercial Packagings. *Acta Alimentaria* **2009,** *38,* 319–327.

Rossell, J. B. *Frying: Improving Quality.* CRC Press: Boca Raton, FL, 2001.

Salas, J. J.; Martínez-Force, E.; Garcés, R. Accumulation of Phospholipids and Glycolipids in Seed Kernels of Different Sunflower Mutants (*Helianthus annuus*). *J. Am. Oil Chem. Soc.* **2006,** *83,* 539–545.

Sánchez-Muniz, F. J.; Cuesta, C. Sunflower Oil. In *Encyclopedia of Food Sciences and Nutrition. Second Edition*; Caballero, B., Ed.; Academic Press: New York, 2003; pp 5672–5680.

Schwartz, H.; Ollalainen, V.; Piironen, V.; Lampi, A. M. Tocopherol, Tocotrienol and Plant Sterol Contents of Vegetable Oils and Industrial Fats. *J. Plant Comp. Anal.* **2008,** *21,* 152–161.

Sciancalepore, V. The Influence of Processing on the Content and Composition of Free and Esterified Sterols in Sunflower Seed Oil. *Oli Grassi Deriv.* **1981**, *17,* 11–12.

Seppanen, C. M.; Song, Q.; Csallany, A. S. The Antioxidant Functions of Tocopherol and Tocotrienol Homologues in Oils, Fats, and Food Systems. *J. Am. Oil Chem. Soc.* **2010,** *87,* 469–481.

Shahidi, F., Zhong, Y. Lipid Oxidation and Improving the Oxidative Stability. *Chem. Soc. Rev.* **2010,** *39,* 4067–4079.

Silvestro, D.; Andersen, T. G.; Schaller, H.; Jansen, P. E. Plant Sterol Metabolism. Δ^7-Sterol-C_5-Desaturase (STE1/DWARF7), $\Delta^{5,7}$-sterol-Δ^7-Reductase (DWARF5) and Δ^{24}-Sterol-Δ^{24}-Reductase (DIMINUTO/DWARF1) Show Multiple Subcellular Localizations in *Arabidopsis thaliana* (Heynh) L. *Plos One* **2013,** *8,* e56429.

Smouse, T. H. Factors Affecting Oil Quality and Stability. In *Methods to Assess Quality and Stability of Oils and Fat-Containing Foods*; Warner, K., Eskin, N. A. M., Eds.; AOCS Press: Urbana, IL, 1995; pp 17–36.

Sotiroudis, T. G.; Kyrtopoulos, S. A. Anticarcinogenic Compounds of Olive Oil and Related Biomarkers. *Eur. J. Nutr.* **2008,** *47* (Suppl. 2), 69–72.

Suliman, T. E. M. A.; Jiang, J.; Liu, Y. Chemical Refining of Sunflower Oil: Effect on Oil Stability, Total Tocopherol, Free Fatty Acids and Colour. *Int. J. Engineering Sci.* **2013a,** *5,* 449–454.

Suliman, T. E.; Meng, Z.; Li, J. W.; Jiang, J.; Liu, Y. Optimisation of Sunflower Oil Deodorising: Balance between Oil Stability and Other Quality Attributes. *Int. J. Food Sci. Tech.* **2013b,** *48,* 1822–1827.

Szuhaj, B. F. Lecithins. In *Edible Oil and Fat Products: Processing Technologies*; Shahidi, F., Ed.; Bailey's Industrial Oil and Fat Products, 6th ed., Vol. 5; John Wiley & Sons: Hoboken, NJ, 2005; pp 361–456.

Tang, S.; Hass, C. G.; Knapp, S. J. Ty3/Gypsy-like Retrotransposon Knockout of a 2-Methyl-6-Phytyl-1,4-Benzoquinone Methyltransferase is Non-Lethal, Uncovers a Cryptic Paralogous Mutation, and Produces Novel Tocopherol (Vitamin E) Profiles in Sunflower. *Theor. Appl. Genet.* **2006,** *113,* 783–799.

Tasan, M.; Demirci, M. Total and Individual Tocopherol Contents of Sunflower Oil at Different Steps of Refining. *Eur. Food Res. Technol.* **2005**, *220,* 251–254.

Tasan, M.; Gecgel, U.; Demirci, M. Effects of Storage and Industrial Oilseed Extraction Methods on the Quality and Stability Characteristics of Crude Sunflower Oil (*Helianthus annuus* L.). *Grasas Aceites* **2011,** *62,* 389–398.

Thanh, T. T.; Vergnes, M. F.; Kaloustian, J.; El-Moselhy, T. F.; Amiot-Carlin, M. J.; Portugal, H. Effect of Storage and Heating on Phytosterol Concentrations in Vegetable Oils Determined by GC/MS. *J. Sci. Food Agric.* **2006,** *86*, 220–225.

U.S. Department of Agriculture. *USDA National Nutrient Database for Standard Reference;* Release 25, 2012. http://www.ars.usda.gov/ba/bhnrc/ndl (accessed August 30, 2013).

Van Hoed, V.; Ben Ali, C.; Slah, M.; Verhé, R. Quality Differences between Pre-Pressed and Solvent Extracted Rapeseed Oil. *Eur. J. Lipid Sci. Technol.* **2010,** *112,* 1241–1247.

Velasco, L.; Fernández-Martínez, J. M. Identification and Genetic Characterization of New Sources of Beta- and Gamma-Tocopherol in Sunflower Germplasm. *Helia* **2003,** *38,* 17–24.

Velasco, L.; Fernández-Martínez, J. M. Environmental Stability of Contrasting Seed Tocopherol Profiles in Sunflower. *Crop Sci.* **2012,** *52*, 2446–2452.

Velasco, L.; Fernández-Martínez, J. M.; García-Ruíz, R.; Domínguez, J. Genetic and Environmental Variation for Tocopherol Content and Composition in Sunflower Commercial Hybrids. *J. Agr. Sci.* **2002,** *139*, 425–429.

Velasco, L.; Pérez-Vich, B.; Fernández-Martínez, J. M. Grain Quality in Oil Crops. In *Handbook of Seed Physiology: Applications to Agriculture*; Benech-Arnold, R. L., Sánchez, R. A., Eds.; Food Products Press: Binghamton, NY, 2004a; pp 389–427.

Velasco, L.; Domínguez, J.; Fernández-Martínez, J. M. Registration of T589 and T2100 Sunflower Germplasms with Modified Tocopherol Profiles. *Crop Sci.* **2004b,** *44*, 362–363.

Velasco, L.; Pérez-Vich, B.; Fernández-Martínez, J. M. Novel Variation for Tocopherol Profile in a Sunflower Created by Mutagenesis and Recombination. *Plant Breeding* **2004c,** *123,* 490–492.

Velasco, L.; Del Moral, L.; Pérez-Vich, B.; Fernández-Martínez, J. M. Selection for Contrasting Seed Tocopherol Content in Sunflower Seeds. *J. Agr. Sci.* **2010,** *148*, 393–400.

Velasco, L.; Fernández-Cuesta, A.; García-Ruiz, J. R.; Fernández-Martínez, J. M.; Domínguez-Giménez, J. Genetic Variation and Genotype X Environment Interactions for Seed Phytosterols in Sunflower. *Crop Sci.* **2013.** DOI: 10.2135/cropsci2012.10.0593.

Vera-Ruiz, E. M.; Velasco, L.; Leon, A. J.; Fernández-Martínez, J. M.; Pérez-Vich, B. Genetic Mapping of the Tph1 Gene Controlling Beta-Tocopherol Accumulation in Sunflower Seeds. *Mol. Breeding* **2006,** *17*, 291–296.

Verleyen, T.; Kamal-Eldin, A.; Dobarganes, C.; Verhé, R.; Dewettinck, K.; Huyghebaert, A. Modeling of α-Tocopherol Loss and Oxidation Products Formed during Thermoxidation in Triolein and Tripalmitin Mixtures. *Lipids* **2001,** *36*, 719–726.

Verleyen, T.; Forcades, M.; Verhe, R.; Dewettinck, K.; Huyghebaert, A.; De Greyt, W. Analysis of Free and Esterified Sterols in Vegetable Oils. *J. Am. Oil Chem. Soc.* **2002a,** *79*, 117–122.

Verleyen, T.; Sosinska, U.; Ioannidou, S.; Verhe, R.; Dewettinck, K.; Huyghebaert, A.; DeGreyt, W. Influence of the Vegetable Oil Refining Process on Free and Esterified Sterols. *J. Am. Oil Chem. Soc.* **2002b,** *79*, 947–953.

Verleyen, T.; Szulczewska, A.; Verhe, R.; Dewettinck, K.; Huyghebaert, A.; De Greyt, W. Comparison of Steradiene Analysis between GC and HPLC. *Food Chem.* **2002c,** *78,* 267–272.

Vlahakis, C.; Hazebroek, J. Phytosterol Accumulation in Canola, Sunflower, and Soybean Oils: Effects of Genetics, Planting Location, and Temperature. *J. Am. Oil Chem. Soc.* **2000,** *77*, 49–53.

Wagner, K. H.; Kamal-Eldin, A.; Elmadfa, I. Gamma-Tocopherol—An Underestimated Vitamin? *Ann. Nutr. Metab.* **2004,** *48*, 169–188.

Warner, K.; Moser, J. Frying Stability of Purified Mid-Oleic Sunflower Oil Triacylglycerols with Added Pure Tocopherols and Tocopherol Mixtures. *J. Am. Oil Chem. Soc.* **2009,** *86*, 1199–1207.

Warner, K.; Miller, J. F.; Demurin, Y. Oxidative Stability of Crude Mid-Oleic Sunflower Oils from Seeds with High γ- and δ-Tocopherols Levels. *J. Am. Oil Chem. Soc.* **2008,** *85*, 529–533.

Winkler, J. K.; Warner, K. The Effect of Phytosterol Concentration on Oxidative Stability and Thermal Polymerization of Heated Oils. *Eur. J. Lipid Sci. Technol.* **2008a,** *110*, 455–464.

Winkler, J. K.; Warner, K. Effect of Phytosterol Structure on Thermal Polymerization of Heated Soybean Oil. *Eur. J. Lipid Sci. Technol.* **2008b,** *110*, 1068–1077.

Yanishlieva, N. V.; Raneva, V. G.; Marinova, E. M. β-Carotene in Sunflower Oil Oxidation. *Grasas Aceites* **2001,** *52,* 10–16.

Zawistowski, J. Tangible Health Benefits of Phytosterol Functional Foods. In *Food Product Development;* Smith, J., Chartes, E., Eds.; Wiley-Blackwell: Chichester, UK, 2010; pp 362–387.

Zou, Y.; Jiang; Y.; Yang, T.; Hu, P.; Xu, X. Minor Constituents of Palm Oil: Characterization, Processing, and Application. In *Palm Oil. Production, Processing, Characterization, and Uses*; Lai, O. M., Tan, C. P., Akoh, C. C., Eds.; AOCS Press: Urbana, IL, 2012; pp 471–526.

12

Sunflower Proteins

Sergio González-Pérez ■ *Instituto de Recursos Naturales y Agrobiología de Salamanca, Consejo Superior de Investigaciones Cientificas (IRNASA-CSIC), Salamanca, Spain*

Introduction

Sunflower ranks with soybean, rapeseed, cottonseed, and peanut as one of the five most important annual oilseed crops in terms of tonnage (FAO, 2012). Sunflower seeds are mainly used for their oil, but as with other oilseeds, the meal left behind after oil extraction is a valuable product because of its high protein content. Sunflower meal (SFM) comprises about 40% protein when hulled seeds are mechanically extracted, about 50% when an organic solvent is used for oil extraction (Robertson, 1972), and 53–66% for similar defatted meal prepared from dehulled seeds (Bau et al., 1983).

Soybean proteins have been consumed for thousands of years (2800 B.C.E.) and have been the subject of extensive research (Murphy, 2008); therefore, they are the main oilseed proteins used as a functional ingredient in food applications. However, other seeds, such as pea, chickpea, lupin, rapeseed, or sunflower are also suitable sources for high-quality protein applications. Sunflower seeds are particularly attractive due to their high protein content and extensive availability (FAO, 2012). Compared to other vegetable protein sources, sunflower seeds contain low or no antinutritional factors (e.g., protease inhibitors, cyanogens, goitrogenic factors, lectins, etc.) (Gassmann, 1983; González-Pérez and Arellano, 2009) and, except for lysine present at low concentration, their amino acid composition conforms with the Food and Agriculture Organization outline of human requirements (Gassmann, 1983; Raymond et al., 1991). In addition, sunflower protein consists largely of albumins and globulins (70–85%) and, consequently, has a high intrinsic solubility (González-Pérez et al., 2004).

Nowadays the main market outlet for sunflower protein relies on animal feed because the growth of sunflower as a protein source for human consumption has been limited by two factors:

1. Sunflower protein preparations and SFM, obtained as a byproduct from the sunflower oil industry, have functional properties that are generally less favorable than those of various soybean protein preparations (Gassmann, 1983; González-Pérez and Vereijken, 2007). The lower reported functionality seems to lie more in the protein denaturation, as a result of the heating steps during the oil extraction process (which may be tailored), than in the inherent properties of the proteins themselves (González-Pérez and Vereijken, 2007).

2. The phenolic compounds present in seeds, in particular chlorogenic acid (CGA), can crosslink with proteins during extraction, resulting in discoloration of the final protein preparation and alteration of its functionality and bioavailability.

Plant proteins are sustainable and versatile alternatives to animal proteins as functional ingredients in food and nonfood formulations. Nonetheless, a successful substitution of animal proteins requires technological innovations. To attain these innovations in an effective and efficient way, insights into the relation between protein structures and their functional properties are of key importance (González-Pérez and Arellano, 2009). This chapter provides information about sunflower proteins, principally seed proteins, including composition, structure, stability, and biochemical and molecular characterization. We focus on the characterization of the two major protein fractions—albumins and globulins—although oleosins and other minor proteins are also described. In addition, we provide information on the seed morphology, structure, and composition, particularly on protein subcellular location. Regarding seed composition, emphasis is added on phenolic compounds and their interaction with proteins. Moreover, we also describe how different factors (genotype, agronomic management, breeding, environment, plant tissue, and growth stage) affect the final protein content. In Chapter 13, the functional features of sunflower protein are extensively described.

Sunflower Seed: Morphology, Composition, and Subcellular Protein Location

The initial step in protein isolation, preparation of a protein extract, entails rupture of the tissues where the proteins are compartmentalized. Therefore, a good understanding of the raw material, physical structure, chemical composition, and organization of components at macroscopic and molecular level is important for designing efficient protein extraction processes. Indeed, the extraction efficiency of proteins depends largely on the solubility of the seed components (protein, carbohydrates, pigments, fibers, etc.) in the extraction medium as a function of temperature, pH, and ionic strength (González-Pérez and Arellano, 2009). Moreover, information about subcellular protein location allows alternative protein isolation approaches, such as including a preceding isolation step of protein storage organelles (Durante et al., 1989; Schnarrenberger et al., 1972). This approach has been extensively used with numerous plant seeds, but in most cases this is done for preparative and analytical purposes rather than for industrial scale application.

Morphology and Subcellular Structures

The seed is the major fraction of the plant harvested from sunflowers. The sunflower seed, achene, or nutlike fruit is an indehiscent fruit containing a single seed (Vaughan, 1970). The achene consists of kernel and pericarp:

1. The kernel (often called dehulled seed or meats by oil millers) consists of endosperm, embryo, and seed coat. The sunflower endosperm consists of two layers of aleurone cells coalesced with the seed coat (Vaughan, 1970). The seed coat or spermoderm is the outermost tissue and closely envelops the seed (Winton and Moeller, 1916). The embryo consists of two cotyledons attached to a protruding radical (Seiler, 1997; Vaughan, 1970).
2. The pericarp (testa, hull, or shell) derives from the ovary wall which is the wall of the fruit (Esau, 1960; Winton and Moeller, 1916). The epidermal cells have thick, outer walls covered with a thin cuticle (Knowles, 1978).

The hull portion of the seeds varies considerably in sunflower germplasm, from 10% to 60% (Fernández-Martínez et al., 2010), and is greater than that of equivalent structures in other oilseed crops (Connor and Hall, 1997). In oilseed cultivars, the sunflower kernel represents more than 75% of the total seed weight (Fernández-Martínez et al., 2010). The dimensions of sunflower seeds vary from 7 to 25 mm long and from 4 to 13 mm wide. The weight of an individual seed varies from 40 to 400 mg (Knowles, 1978).

There are three major groups of varieties of cultivated sunflower: (1) those for oil production (oilseed types), (2) non-oilseed or confectionery-type consumed as snacks and pet foods, (3) and those used as ornamentals. More than 90% of the world sunflower production is devoted to oil extraction (Miller and Fick, 1997; Salunkhe et al., 1992). The oilseed varieties have small black seeds with thin hulls that adhere to the kernel and high oil content (about 50%). Non-oilseed sunflower has larger seeds, generally striped, with relatively thick hulls and lower oil content (about 30%) (Knowles, 1978). Ornamental cultivars are by far the minor portion of sunflower production (Fernández-Martínez et al., 2010).

Protein Subcellular Location

Sunflower proteins, as in other seeds, accumulate mostly in protein storage vacuoles of embryo and endosperm cells, and as protein bodies directly assembled within the endoplasmic reticulum (Bolyakina and Raikhman, 1999; Molina et al., 2006; Shoar-Ghafari and Vintéjoux, 1997). In the past both terms, protein storage vacuoles and protein bodies, were considered synonyms—but protein storage vacuoles is nowadays used to differentiate vacuoles containing storage proteins from protein bodies originating from the endoplasmic reticulum (Herman and Larkins, 1999). Protein bodies vary in composition among seeds. Some contain no inclusions in their matrix; others may include globoids (membrane-bound structures) in which insoluble salts of phytic acid are localized, protein crystalloids, protein carbohydrate bodies, or crystals of calcium oxalate (Esau, 1960). The protein bodies of sunflower seed include globoids rich in phosphorus that probably represent phytate storage sites (Saio et al., 1977). Jiang et al. (2001) proposed that plant seeds store digestive enzymes in globoids located inside the protein storage vacuole. This membrane-bound structure would protect storage proteins against proteolysis in early plant development, but would be broken during germination allowing their degradation. Thus, this globoid would become a degradative vacuole within the storage vacuole (Jiang et al., 2001).

Saio et al. (1977) show some illustrative photographs of protein bodies taken by a scanning electron microscope (SEM) and a transmission electron microscope (TEM). The diameter of protein bodies measured with SEM ranged between 0.5 and 10 μm and the higher frequency was found at about 2 μm in diameter (Gallant and Saio, 1979; Saio et al., 1977). As germination proceeds, protein bodies lose their smooth spherical shape and become indented and pitted. Protein dissolves in situ and protein bodies coalesce and are replaced by vacuoles that may be fused into one large protein vacuole. Next, the storage proteins are degraded to amino acids by proteinases and peptidases (Allen and Arnott, 1982).

Composition

Table 12.A illustrates the composition of the kernel (dehulled) and the whole seed. Lipids and proteins are the most important constituents. Lipids are the major components of the sunflower seed, of which neutral triacylglycerides constitute the major lipid class. Other triacylglycerides include phospholipids and glycolipids, which constitute less than 4% of the total lipids (Salunkhe et al., 1992).

The protein content of sunflower kernels ranges between 20% and 40%. About 87–99% of the nitrogen in the sunflower seeds consists of the nitrogen in intact proteins. The remaining 1–13% originates from peptides, amino acids, or other nitrogenous substances (Gassmann, 1983; Schwenke and Raab, 1973).

The total carbohydrate content of the kernels was determined to range between 4 and 18%. One particular characteristic of sunflower seed carbohydrates is the very little starch content (0.42%) compared to monocotyledonous seeds (Bau et al., 1983). The ethanol-soluble sugars were reported to be 4.4–6.3% of the kernel weight in 10 sunflower varieties (Pomenta and Burns, 1971). Analysis of the molar sugar composition in the kernels revealed that sunflower contains high amounts of glucose (46%), followed by arabinose (16%), uronic acids (14%), and galactose (11%) (González-Pérez et al., 2002). The hulls largely consist of lignin, pentosans, and cellulosic material (Robertson and Burns, 1975). The protein content of the hulls is low, ranging between 36 and 61 g/kg dry weight (Dreher and Padmanaban, 1983; Earle et al., 1968).

Sunflower seeds also contain a substantial amount of minerals. However, they are often complexed with phytic acid, therefore, minerals in sunflower seeds are biologically not fully available (Salunkhe et al., 1992).

Mikolajczak et al. (1970) and Sabir et al. (1974) provided a detailed description of at least 10 phenolic compounds isolated from SFM. The proportion and quantity of the phenolic compounds is markedly affected by the sunflower variety (Dorrell,

Table 12.A Reported Composition Ranges of Sunflower Seeds on Dry Basis

Component	Dehulled seed[a-n] (%)	Whole seed [a, e, i, k, l, m] (%)
Protein	20.4–40.0	10.0–27.1
Peptides, amino acids, and other nonprotein nitrogen	1–13	—
Carbohydrates	4–8	18–26
Lipids	47–65	34–55
Fatty acids		—
Palmitic	5–7	—
Stearic	2–6	—
Arachidic	0.0–0.3	—
Oleic	15–37	—
Linoleic	51–73	—
Linolenic	< 0.3	—
Tocopherol	0.07	—
Carotenoids	0.01–0.02	—
Vitamin B1	0.002	—
Chlorogenic acid (CGA)	0.5–2.8	1.1–4.5
Quinic acid (QA)	0.12–0.25	—
Caffeic acid (CA)	0.05–0.29	—
Total minerals	3–4	2–4
Potasium	0.67–0.75	—
Phosphorus	0.60–0.94	—
Sulfur	0.26–0.32	—
Magnesium	0.35–0.41	—
Calcium	0.08–010	—
Sodium	0.02	—

[a]Data from Earle et al. (1968)
[b]Gheyasuddin et al. (1970)
[c]Schwenke and Raab (1973)
[d]Sabir et al. (1974)
[e]Dorrel (1976)
[f]Bau et al. (1983)
[g]Berot and Briffaud (1983)
[h]Gassmann (1983)
[i]Madhusudhan et al. (1986)
[j]Salunkhe et al. (1992)
[k]Pedrosa et al. (2000)
[l]Robertson (1975)
[m]González-Pérez et al. (2002)
[n]Weisz et al. (2009)

Source: Adapted from González-Pérez (2003).

1974; Pedrosa et al., 2000). Chlorogenic acid (CGA), caffeic acid (CA), and quinic acid (QA) are the main phenolic compounds in dehulled seeds at concentrations of 0.5–2.8%, 0.05–0.29%, and 0.12–0.25% (dry matter basis), respectively (Figure 12.1). The presence of phenolic compounds, even at small quantities, is one of the

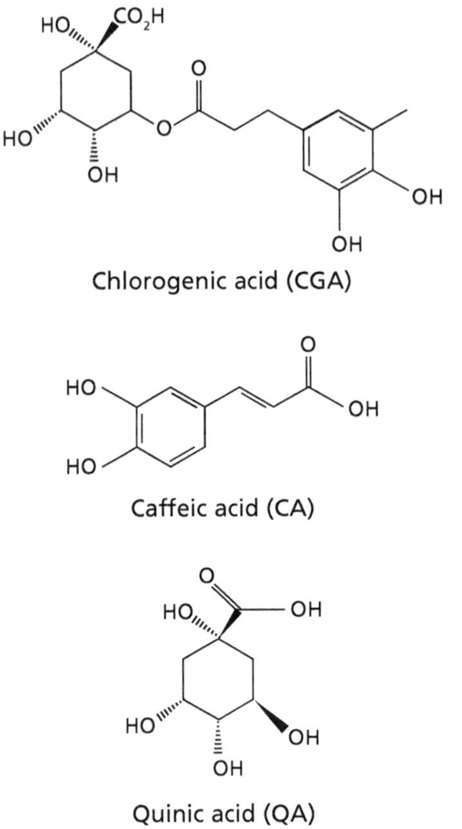

Figure 12.1 Line-angle formulas of chlorogenic acid (CGA), caffeic acid (CA), and quinic acid (QA).

main bottlenecks for the commercial use of sunflower seeds as a source of protein. Therefore, this point will be addressed in detail in the next section.

Phenolic Compounds of Sunflower Seed

Phenolic compounds are the largest group of plant constituents capable of interacting with proteins, affecting flavor and color in many food products (Maga and Katz, 1978). Actually, phenolic compounds are the major group of secondary metabolites in food produced from plants (about 6000 molecules identified) (Kroll et al., 2003). They are divided into four major biochemical groups: (1) benzoic acids, (2) cinnamic acids, (3) flavonoids, and (4) terpenoids. They can generally be found in two forms, free or esterified to other molecules such as sugars or acids (Anderson, 1985; Kroll et al., 2003).

Osborne and Campbell (1897) attributed the dark color of the protein concentrates they prepared from sunflower seeds to a compound named as helianthotannic acid at the time. Some years later this compound was identified as CGA by Gorter (1909), and then Rudkin and Nelson (1947) determined its structure as an ester of QA and CA.

Protein–Phenolic Compound Interactions. The protein-phenol binding may be either non-covalent (by hydrogen bonding and ionic and hydrophobic interactions), or covalent after oxidation of phenolic compounds to quinones (Saeed and Cheryan, 1989). The phenolic compounds can undergo oxidation under alkaline conditions or via enzymatic reactions catalyzed by polyphenol oxidase (González-Pérez et al., 2002; Pierpoint, 1969; Prigent, 2005; Prigent et al., 2007). The quinones formed after oxidation are highly reactive and spontaneously form covalent bonds with functional groups on proteins such as amines, thiols, thioethers, indole, imidazole, and disulphide groups (Loomis, 1974; Venktesh and Prakash, 1993b).

Some authors found that in sunflower protein preparations, CGA appears primarily in the form of protein-CGA complexes or covalently bound to proteins. It has been reported that CGA binds to proteins either nonpreferentially (Rahma and Rao, 1981a; 1981b) or preferentially with low molecular weight (MW) proteins (Kabirullah and Wills, 1983; Prasad, 1990; Sabir et al., 1973; Sabir et al., 1974a; Venktesh and Prakash, 1993b) or high MW proteins (Sastry and Rao, 1990). Other authors have claimed that CGA is largely present as free CGA (Durante et al., 1989; González-Pérez et al., 2002). Noncovalent interactions of both CA and QA with helianthinin (major sunflower protein) have also been described (Suryaprakash et al., 2000).

The different observations in relation to protein–CGA interaction are likely due to the variations in protein extraction technique and analytical methods used in the latter studies. However, in some cases, these divergences may be a result of different interpretation of the experimental data (González-Pérez et al., 2002). Nevertheless, there is unanimity about the irreversible quinone-protein binding. This interaction can be avoided by removing most of the CGA, or impairing its oxidation before protein extraction.

Effects of Phenolic Compounds on Protein Preparations. The protein–phenolic compounds interaction is observed as a green-brown coloration in sunflower protein preparations. This interaction can also result in reduced protein digestibility, modification of protein storage life, and alteration of its organoleptic properties (Kroll et al., 2000, 2001, 2003; Rawel et al., 2002a, 2002b; Sastry and Rao, 1990). The reduced protein digestibility arises from the resistance of condensation products to proteolytic enzymes (Muzquiz et al., 2004).

Moreover, the binding of phenolic compounds severely decreases protein solubility (about 10–80%) (Kroll et al., 2000, 2001, 2003; Prigent, 2005; Prigent et al., 2007, 2009; Rawel et al., 2002a, 2002b). Covalent interaction with CGA reduced the solubility of globular proteins and modified its isoelectric point (Kroll et al., 2001; Prigent et al., 2007; Rawel et al., 2002a). Noncovalent binding to simple (cathecol and pyrogallol) (Neucere et al., 1978) and to higher carbon-containing (flavonoids, tannins, etc.) (Prigent et al., 2009; Siebert, 2006) phenolic compounds was also reported to result in reduced protein solubility. However, there are other publications reporting that solubility of globular proteins was not affected by noncovalent interactions, even at high CGA–protein ratios (Prigent et al., 2003). The differences in experimental findings discussed earlier are due to the experimental conditions (e.g., pH, ionic strength, buffer, concentration, etc.) and the protein and phenolic compound used in those studies. The length of the phenolic molecule affects its protein binding. For example, CGA shows the highest affinity for protein binding while QA and CA show the lowest and medium affinity, respectively (Prakash and Rao, 1986). The affinity for the protein binding also depends on the number and location of the hydroxy groups on the aromatic ring (McManus et al., 1985) and the degree of polymerization of the phenolic compound (Prigent et al., 2009). Likewise, protein features such as amino acid sequence and composition, hydrophobicity, or presence of exposed functional groups on the protein structure largely affect the protein–phenol interaction. For example, proteins with proline-rich domains particularly favor interaction with phenolic compounds (Siebert, 2006; Siebert et al., 1996).

In addition to reduced protein solubility, earlier studies reported that the cross-linking of phenolic compounds to proteins modified the secondary and tertiary structure of helianthinin, but did not affect the quaternary structure (Prakash and Rao, 1986). However, more recent publications have reported modifications in protein quaternary structure in addition to secondary and tertiary structure (Prigent et al., 2007; Rawel et al., 2002b; Suryaprakash et al., 2000).

Removal of Phenolic Compounds. Removal of phenolic compounds is one of the main issues concerning the production of sunflower protein products (Gassmann, 1983; González-Pérez and Vereijken, 2007; Milic et al., 1968; Sastry and Subramanian, 1984). Numerous approaches have been reported in which the extraction of sunflower protein is preceded by, combined with, or followed by unit operations that separate phenols from the seeds. The separation process may involve the following steps: (1) extraction of seeds with mixtures of organic solvents and water (Assogna et al., 1979; Cater et al., 1972; González-Pérez et al., 2002; Mikolajczak et al., 1970; Pomenta and Burns, 1971; Prasad, 1990; Regitano d'Arce et al., 1994a, 1994b, 1994c; Saeed and Cheryan, 1988; Sanchez and Burgos, 1995; Sodini and Canella, 1977; Venktesh and Prakash, 1993a, 1993b); (2) extraction with aqueous solutions of acids, salts, and/or reducing agents (Hagenmaier, 1974; O'Connor, 1971; Pearce, 1984; Rahma and Rao, 1981c; Sastry and Rao, 1990; Sastry and Subramanian, 1984); (3) membrane filtration (O'Connor, 1971); (4) co-precipitation or complex formation with compounds other than proteins such as aluminum salts, polyvinylpyrrolidone (PVP), resins, etc. (Bau and Debry, 1980; González-Pérez, 2003; Nuzzolo et al., 1980; Petit et al., 1979); and (5) combinations thereof (Bau et al., 1983; Fan et al., 1976; Gheyasuddin, 1970; Gheyasuddin et al., 1970a; Rahma and Rao, 1979, 1981c; Raymond et al., 1984; Sosulski et al., 1972).

The efficiency of the method to render low CGA preparations with light color is generally compromised with protein alteration or/and protein losses. Often these two negative effects are exclusive; that is, the method results in a high protein yield but leads to partial denaturation, or the method minimizes protein denaturation while increasing protein losses during extraction. However, some approaches may include both negative trends. For example, removal of phenolic compounds at extreme pH values (pH <3.0) leads to low protein yields and protein denaturation. The use of organic solvents seems to be more satisfactory, even though some protein denaturation may occur (Prasad, 1990; Tranchino et al., 1983; Vermeesch et al., 1987). Organic solvent–water mixtures achieve better phenolic compound removal and lower protein denaturation than pure organic solvents (Rahma and Rao, 1981c; Sripad and Rao, 1987b). Aqueous mixtures (50–60% v/v) of methanol, ethanol, and 2-propanol resulted in significant minor protein losses and a better CGA removal compared with propanol and isobutanol (Berot and Briffaud, 1983). Extensive protein denaturation by butanol (Rahma and Rao, 1981c; Venktesh and Prakash, 1993a, 1993b) and acetone (Sanchez and Burgos, 1995, 1997) has been reported. Ethanol–water mixtures rendered protein preparations with a reduced protein solubility (Fan et al., 1976; Regitano d'Arce et al., 1994a, 1994b, 1994c) or low phenolic compound extractability (Cater et al., 1972; Saeed and Cheryan, 1988). In agreement with the latter publications, the CGA solubility in methanol, ethanol, and water was 15.2, 6.2, and 0.6 g/100 mL of solvent at 20 °C, respectively (Sabir et al., 1974a). Several researchers pointed out the high efficiency of methanol–water mixtures for phenolic compound extraction, with limited protein denaturation as well as low protein losses (Berot and Briffaud, 1983; González-Pérez et al., 2002; Mikolajczak et al., 1970; Sripad and Rao, 1987b; Weisz et al., 2009). However, it is advisable to use a low ionic strength (I) solvent mixture because it has been reported that at $I \geq$ 500 mM albumin solubility is high (Kortt and Caldwell, 1990a; Pandya et al., 2000) and as a result it is co-extracted with phenolic compounds leading to protein losses.

Breeding for sunflower seeds with low CGA would be an effective way to elude expensive and damaging or inefficient removal processes. Sunflower germplasm is a useful source to identify varieties with low antinutritive compounds, such as CGA (Fernández-Martínez et al., 2010). Dorrell (1976) found a lower CGA content in wild varieties (2.1% average) compared with cultivated sunflower (2.8% average). Sunflower varieties with low CGA levels generally contain low oil content, which is an undesirable trait for the oil industry (Dorrell, 1974; Ruso et al., 2000). However, the latter could be an advantage for protein production because oil content and protein content are generally negatively correlated (see "Environmental Factors and Agronomic Practices" and "Breeding for Protein Content" in this chapter). Finally, environmental factors and plant maturity should be taken in consideration because they strongly affect the content of phenolic compounds (Dorrell, 1974).

It must be noted that contrary to the negative effect generally reported for sunflower proteins, nowadays the effect of phenolic compounds is often considered to be beneficial for human nutrition and some food products (Baur and Sinclair, 2006; Burton-Freeman, 2010; Jang et al., 1997; Kroll et al., 2003; Smoliga et al., 2011). The antioxidant activity and protective role of phenolic compounds against oxidative damage have been reported (Kroll et al., 2003; Pedrosa et al., 2000). Sunflower phenolic compounds can also be used as antioxidants in edible oils (Weisz et al., 2009). Moreover, although in exceptional cases, phenolic compounds have even been reported to improve protein functionality (Sarker et al., 1995; Strauss and Gibson, 2004). However, the detrimental effect of phenolic compounds on sunflower protein preparations is the general and undoubtedly the main concern.

Factors Affecting Protein Content and Composition

Besides the knowledge of the raw material to design the protein extraction process, other important factors to consider are connected to the state and protein content of the protein source. Among them we find the genotype, environmental factors, agronomic practices, plant tissue, and development stage.

Genotypic Variations

Considerably large genotypic and phenotypic variations occur in sunflower at protein levels. Genotype effects on protein concentration are largely ascribed to the hull-to-kernel ratio, but deviations from this general observation are also described. Numerous publications have shown the different protein concentrations in the seed among different varieties (Miller and Fick, 1997; Nenova and Drumeva, 2012; Raymond et al., 1995). Protein content in sunflower seed of wild species has been reported to be considerably higher than that of cultivated types (Seiler, 2007; Seiler and Rieseberg, 1997). Contrary to proteins, the values for oil content found in cultivated sunflower are generally higher than the typical oil content in wild species (Fernández-Martínez et al., 2010; Miller and Fick, 1997; Ruso et al., 2000). Several authors have also reported significant differences in the amino acid composition among sunflower species (Nenova and Drumeva, 2012; Raymond et al., 1991; Robinson, 1975; Sosulski and Sarwar, 1973; Tkachuk and Irvine, 1969).

This significant genetic difference allows using electrophoretic patterns of sunflower proteins for differentiating genotypes and determining genetic purity of inbred lines. The use of various types of gel electrophoresis for identifying plant varieties and distinguishing between them is well established (Cooke, 1995). Storage seed protein and isozyme (catalytically active proteins) variants have been some of the most widely employed molecular genetic markers (Bretting and Widrlechner, 1995; Righetti and Bosisio, 1981). Marcone (1999a) describes this approach as especially useful in the analysis and exploitation of landraces (cultivated genotypes evolved by natural adaptation without breeding).

There are numerous studies based on one- and two-dimensional electrophoretic techniques to detect quantitative and qualitative differences among the protein fractions and polypeptide composition of domesticated and wild sunflower species (Aksyonov, 2005a, 2005b; Anisimova and Gavrilyuk, 1990; Anisimova et al., 1991a, 1991b, 1995, 2003, 2004; Duca et al., 2008; Nenova and Drumeva, 2012; Nikolić et al., 2008; Pallavi et al., 2010; Quillet et al., 1992; Raymond et al., 1994, 1995; Žilić et al., 2010). Although proteomic approaches are more suitable for determining changes in protein expression, they have also being used in sunflower to identify gene products underlying complex traits (Pérez-Vich and Berry, 2010). For example, Hajduch et al. (2007) reported 77 protein spots (on two-dimensional gel electrophoresis) differentially expressed when comparing two sunflower cultivars with significant differences in oil and protein content. Further identification of these proteins revealed the main metabolic processes (glycolysis and amino acid metabolism) involved in low and high oil sunflower lines.

Environmental Factors and Agronomic Practices

Genetic variability is not the single cause of protein variation. The final result, as in all crops, depends on many traits and varies greatly with environment. For example, soil fertility and fertilization, water stress, solar irradiation, diseases or time of planting largely affect growth, yield, seed composition, and quality (Connor and Hall, 1997; Connor and Sadras, 1992; Connor et al., 1985, 1993; Decau et al., 1972; Fereres et al., 1986; Gimenez and Fereres, 1986; Gimenez et al., 1994; Lencrerot et al., 1973, 1977; Puech et al., 1975; Unger, 1982, 1983). Thus, the variation in protein amount and composition in sunflower seeds is often due more to growth conditions than to genetic variability (Škorić, 1992).

Environmental Factors

Water stress is one of the major uncontrollable environmental variables limiting growth and development and subsequently seed yield. Sunflower is considered to be well adapted to drought; however, its water stress tolerance is limited by genotype (Fereres et al., 1986; Gimenez and Fereres, 1986). A proteomic study on two sunflower genotypes, displaying different levels of tolerance to drought (well watered and drought stressed), revealed some qualitative and quantitative changes on two-dimensional protein maps (Castillejo et al., 2008): a general decrease in the number of protein spots corresponding to enzymes of photosynthesis and carbohydrate metabolism was observed only in the less-drought-tolerant genotype, suggesting inhibition of the energetic metabolism. Borodulina and Suprunova (1976) demonstrated that growing sunflower under drought stress increases the globulin fraction, modifies the amino acid composition, and reduces protein quality (Borodulina and Suprunova, 1976).

Jasso de Rodríguez et al. (2002) described a dramatic effect on total biomass, seed yield, and soluble protein concentration in seeds and leaves by water deficiency at different plant growth stages. Quartacci and Navari-Izzo (1992) also reported a reduction in protein concentration associated with water stress. This effect is particularly relevant to oil and protein content during the seed filling period (Unger, 1986). Some authors reported a higher number of empty hulls (no kernel) after brief exposures to drought or shade after anthesis (phase with fully open flowers) (Connor and Hall, 1997).

Other environmental variables altering growth, development, seed yield, and composition are sunflower diseases. Sunflower is a host for over three dozen pathogenic organisms (fungal, bacterial, and viral pathogens) (Gulya et al., 1997). A number of plants can produce, in response to pathogen infection, new proteins that lead to numerous metabolic changes (Jung et al., 1993). Amino acid changes during sunflower infection have also been reported (Dulermo et al., 2009). Glutamine and arginine concentration decreased by 70% and glutamate and asparagine concentration decreased by 50% and 40%, respectively, 48 hours after infection.

Agronomic Practices

The final yield on protein levels is also modulated by agronomic practices, such as the use of fertilizers, manure, or compost; planting date and depth; plant population; weed control; and crop sequence and rotation.

Among soil management practices, fertilization—particularly nitrogen application—has been extensively studied. Nitrogen is the macronutrient required in the greatest quantity by plants for amino acid and protein production. Nitrogen deficiency affects the development and growth of leaves, florets, and seeds of sunflower plants. Therefore, protein, oil, and total biomass of sunflower are affected by nitrogen application (Blamey and Chapman, 1981; Decau et al., 1972; Lencrerot et al., 1977; Steer et al., 1986; Zubillaga et al., 2002). Steer et al. (1986) described a positive correlation between nitrogen content of plants and nitrogen availability, and reduced nitrogen content with higher plant population density. The nitrogen supply affects the nitrogen content of sunflower seeds in all developmental phases (Steer et al., 1984), and markedly during the seed filling period (Lencrerot et al., 1977). As mentioned above, the hull is composed mainly of tissues stemming from the ovary wall that develops before anthesis (Vaughan, 1970), and hull development slows down prior to the initiation of the embryo and cotyledon development. Therefore, when there is limited nitrogen supply during seed filling, the weight of the kernel decreases significantly, resulting in lower kernel/hull ratios and lower protein yields.

The concentration of all amino acids increases as the concentration of nitrogen increases in the seed. However, arginine and, to a lesser extent, glutamate increase proportionally more than the total amino acids, whereas lysine increases to a lesser extent than the total amino acids (Steer et al., 1984). The latter authors did not find any difference in seed protein composition among plants grown on different amounts of nitrogen supplies, suggesting that the differences in the amino acids were due mainly to the changes in free amino acids. Based on the studies above, controlled nitrogen application appears to be a mechanism that can be used to modulate the protein/amino acid content of the seeds.

Along with nitrogen, potassium and phosphorus are the primary macronutrients for plants. The major role of potassium is to regulate the osmotic balance of cells for the maintenance of turgor and cell extension. Potassium is essential for the formation and translocation of sugars and starches and for synthesis of proteins. However, it has been reported that potassium fertilization causes slight reduction in seed protein content—1% reduction up to 300 kg K_2O per hectare (Ahmad et al., 1999; Chaudhry and Mushtaq, 1999; Rollier et al., 1975).

Phosphorus is required at smaller amounts than either potassium or nitrogen. This macronutrient is one of the main molecules affecting photosynthesis, transport of nutrients, and energy transfer metabolic processes. Phosphorus is also a main component of plant structure at a cellular level (as phospholipid). Sunflower plants uptake phosphorus during all developmental phases, including seed filling (Jacob and Lawlor, 1991; Zubillaga et al., 2002).

Blamey and Chapman (1981) described the effects of phosphorus application on protein, oil, and energy yields of sunflower. Protein level in the seed was increased by nitrogen (0–180 Kg/ha) but reduced by phosphorus (0–60 Kg/ha) application; the opposite effects were generally observed with the oil content. Combined nitrogen and phosphorus application markedly increased the protein and oil yields, indicating that phosphorus is necessary for maximum response to applied nitrogen. Zubillaga et al. (2002) also found that phosphorus application produced greater and more consistent effects on crop performance by promoting a more efficient use of the supplied nitrogen. Other studies (Jacob and Lawlor, 1991) have shown that phosphate deficiency resulted in decreased protein and chlorophyll content of sunflower leaves due to mesophyll limitation of photosynthesis.

Increases in seed protein content in response to environmental factors or agronomic practices are often accompanied by reductions in seed oil content, although there is significant variation of this trend between genotypes (Blanchet, 1987; Blanchet and Merrien, 1982; Connor and Sadras, 1992; Merrien et al., 1983; Steer et al., 1984).

In conclusion, the genotype, environmental factors, and agronomic practices during cultivation largely influence the protein content in sunflower plants and seeds. It is important to not only deal with all these parameters individually and in detail, but also to manage knowledge about their interaction, since any factor, including those that we may be tempted to underestimate, could play an important role. For example, some authors have reported considerable variations in seed proteins based on the cultivation location, despite a strict control of many other agronomic conditions (Dorrell, 1974; Radić et al., 2009).

Plant Parts and Growth Stage

The protein content and composition also depend on the growth stage of the plant, as well as on the part (tissue) under analysis. Most of the proteins described in this chapter occur in the seed. However, some studies focus on the protein content of the whole plant or of its different parts. In sunflower plants, nitrogen concentrations in stems, leaves, and seeds were reported to be around 20, 30, and 40 g/kg, respectively (Connor and Hall, 1997). Protein concentration of the whole plant has been extensively analyzed by Seiler and collaborators (Adams and Seiler, 1984; Seiler, 1984, 1986, 2007; Seiler et al., 1991), who described variation from 38 g/kg in *H. tuberosus* to a maximum of 201 g/kg in *H. grosseserratus.* Cultivated sunflower showed a protein concentration above 204 g/kg. Seiler also reported a general protein concentration lessening as plants matured (i.e., lower during fruiting than in vegetative and flowering stages on 27 wild sunflower species), with a higher protein concentration in leaves than in stems (Seiler, 1984, 1986). Protein synthesis is one of the main activities of the green part of the plant, and therefore leaf protein concentrate could be used as an industrial protein source.

In the leaves of six cultivars, the protein concentration, but not the protein composition, was reported to be a function of the growth stage, as deduced from the increasing intensity of the electrophoresis bands and from the nearly constant protein band pattern (Jasso de Rodríguez et al., 2002). A proteomic analysis of sunflower leaves provided by da Silva et al. (2010) displayed 244 protein spots on two-dimensional gel electrophoresis. Tandem mass spectrometry allowed the identification of 22 proteins. One of the identified proteins is the photosynthesis enzyme, rubisco, which represents about 50% of the soluble leaf protein (Ono and Watanabe, 1997). Regarding the amino acid composition of sunflower leaves, different patterns were observed for free and bound amino acids in sunflower leaves throughout all developmental stages (Roy et al., 2013). In both free and bound forms, however, the amino acid content was the highest in mature leaves 3.61 ± 0.442, followed by young and senescent leaves, 2.88 ± 0.378 and 2.287 ± 0.268 μg/mg ash-free leaf, respectively.

Seed Development

After anthesis, developing sunflower seed represents a major sink for photosynthates. The upper leaves of the plant canopy export about 80% of their photoassimilates toward the flower head during seed filling (Goffner et al., 1988a, 1988b). Sucrose and nitrogen compounds pass from the apoplast of the seed coat into the growing embryo (Wolswinkel, 1987). An additional contribution to the accumulation of storage proteins during early seed development is the protein translocation from leaves and stems (Blanchet et al., 1986, Goffner et al., 1988a).

The conversion rate from free compounds (sugars, free amino acids, organic acids) to further metabolized storage compounds (lipids and proteins) varies according to seed age (Goffner et al., 1988a, 1988b). Protein deposition in seeds also varies based on their location on the flower head: peripheral seeds fill sooner. Although these differences come mainly from physiological age disparities, some authors suggested some metabolic phenomena as the origin of seed filling variation (Goffner et al., 1988a, 1988b; Lencrerot et al., 1977).

The pattern of protein deposition in seeds, in contrast to oil deposition, proceeds in line with seed growth, hence the protein concentration in the seed (dry matter basis) remains rather constant over time (Goffner et al., 1988a). Connor et al. (1993) suggested a ceiling for the total amount of oil that can be stored in the seed that does not occur for proteins. Moreover, protein synthesis begins at an earlier stage during seed filling compared with lipid synthesis (Goffner et al., 1988a). An exception to this rule is the biogenesis and accumulation of oleosins (structural proteins found in plant oil bodies) closely connected with lipid deposition in the seed. However, a later publication also detected some oleosins in very young developing embryos, before the initiation of storage protein synthesis (Mazhar et al., 1998). Mazhar et al. (1998) described two discrete groups of oleosin transcripts, one group expressing before and the other one after the storage protein synthesis.

The various seed storage protein fractions are stored together in the protein bodies (Shewry et al., 1995). These proteins are detected in sunflower embryos by 5–7 days post-fertilization, expressed during seed development as polypeptide precursors, and stored in protein bodies as matured proteins after post-translational modifications (Allen et al., 1987; Kortt et al., 1991). Some studies on seed development (Alba et al., 1980, 1981) claimed a fairly constant protein concentration in seeds during ripening, while the amino acid composition undergoes variations due to an albumin diminution in support of globulins.

Sunflower Seed Proteins

From a physiological point of view, seeds are (storage) sink organs, whereas leaves are the providing source of photoassimilates through photosynthesis (Connor and Hall, 1997). Therefore, most sunflower proteins are located in the seed, wherein about 60% of the mobilized plant nitrogen concentrates (Gachon, 1972).

Sunflower seed proteins are divided into three groups: (1) proteins involved in metabolism, (2) structural proteins (ribosomal, chromosomal, and membrane proteins), and (3) storage proteins, which include about 85% of the total sunflower protein content (Fukushima, 1991; Shewry et al., 1995). Although, there is not a clear definition for storage protein, this terminology is commonly used for proteins that represent more than 5% of the total protein fraction (Derbyshire et al., 1976). Another generally assumed characteristic for storage proteins is to be mainly generated during seed development and stored in the seed to serve as a nitrogen source during germination. Storage proteins are also characterized as a group of multimeric proteins without enzymatic activity and occurring in protein bodies (Derbyshire et al., 1976; Fukushima, 1991; Mandal and Mandal, 2000).

Sunflower seeds comprise two major groups of storage proteins, helianthinin, which belongs to a widely distributed legumin-like family of globulins, and sunflower albumins (SFAs) (Figure 12.2) (Anisimova et al., 1995; Dalgalarrondo et al., 1984; González-Pérez et al., 2004; Miflin, 1986; Prakash and Rao, 1986; Youle and Huang, 1981). Helianthinin is generally considered to be the major group of sunflower storage proteins, but as we will discuss below, the amounts reported vary greatly. In addition to helianthinin and SFAs, sunflower seeds also comprise oleosins, which are located at the outer surface of the oil bodies preventing their coalescence. An approximate overview of the relative concentrations of proteins present in sunflower seeds was visualized by densitometric analyses of electrophoresis gels (Raymond et al., 1995). The globulin content varied from 50% to 70%, whereas the albumin fraction varied between 17% and 34%. Oleosins depicted the greatest variability, ranging between 10% and 26%.

Seed Protein Fractionation: Osborne and Sedimentation Coefficients

Plant proteins were initially fractionated according to the classical method of Osborne (Osborne, 1924), which is based on the sequential extraction of proteins in water (albumins), dilute salt solutions (globulins), alcohol–aqueous mixtures (prolamins), and dilute acids or alkalis (glutelins). Although the current protein knowledge may lead to more systematic and scientific classifications (conserved domains, structure, sequence, etc.) (Boeckmann et al., 2003; Marchler-Bauer et al., 2005; Wu et al., 2003), Osborne classification is widely applied to seed proteins, where these protein fractions still have a high level of biological and functional significance (e.g., in food processing) (Shewry et al., 2006). However, we should consider that this classification depends on the conditions used for meal preparation or/and seed pretreatment, and the way the fractionation is performed (e.g., time of extraction, liquid-to-seed ratio, proportion of alcohol, salt concentration, etc.). Subsequently, the amounts of the sunflower protein fractions significantly diverge throughout literature. Globulins constitute most of the sunflower proteins, from about 40% to 90%, and albumins account for about 10–30% of the total proteins (González-Pérez and Vereijken, 2007). The minor fraction glutelins, and specifically prolamins, have received little attention

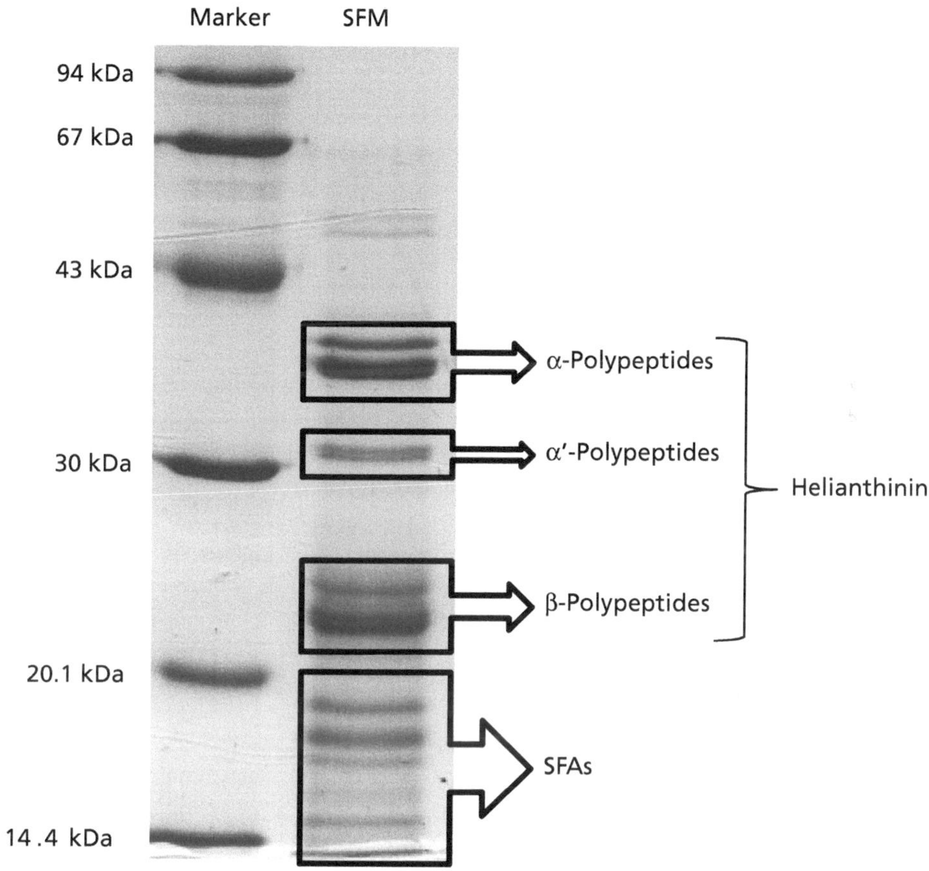

Figure 12.2 SDS-PAGE patterns of sunflower meal (SFM) under reducing conditions. Lane 1: molecular weight marker; lane 2: SFM. α', α, and β indicate acidic and basic polypeptides of helianthinin, respectively. SFAs are sunflower albumins.

(Balasaraswathi and Sadasivam, 1997; Prasad, 1988; Sammour et al., 1995; Shewry et al., 1995). The latter publications just report information about protein content, but generally do not include any further biochemical characterization.

Sunflower seed proteins are also classified according to their Svedberg sedimentation coefficient (S) that quantifies the rate at which a protein or particle settles under analytical ultracentrifugation. Although this nomenclature is commonly used throughout literature, it is confusing because the association state of proteins, and consequently their sedimentation coefficient, is largely affected by conditions such as the type of buffer, pH, ionic strength, protein concentration, and temperature.

Literature data, therefore, show extensive variation in the proportions of the different protein fractions according to the sedimentation coefficient. Furthermore, the variations can be explained by genetic and environmental factors that may affect protein composition and therefore the sedimentation coefficients (Salunkhe et al., 1992). Table 12.B gives an overview of the distribution of proteins over the different fractions according to ultracentrifugation analyses. The literature describes the presence of four fractions: 10–13S and 1–4S, the major ones; and 15–18S and 6–9S, which are present in minor amounts. The 10–13S fraction corresponds to the legumin-like oligomeric form of helianthinin, and in most analyses this fraction shows a sedimentation coefficient of about 11S. The 6–9S fraction generally has sedimentation coefficients around 7S. Contrary to soybean, sunflower globulins do not contain any genetically independent 7S vicilin-like constituent (Anisimova and Gavrilyuk, 1990; Gassmann, 1983; Lakemond, 2001; Youle and Huang, 1981). Nevertheless, various amounts of proteins with a sedimentation coefficient of 7S have been detected in sunflower seeds (Table 12.B). These 7S constituents are dissociation products of the 11S globulins, a dissociation that also occurs in soy glycinin (Gassmann, 1983; González-Pérez et al., 2004; Lakemond et al., 2000b; Schwenke et al., 1974, Schwenke et al., 1979). The latter information is confirmed by the equal SDS-PAGE pattern of the 11S and 7S isolated forms of sunflower globulins under both reducing and nonreducing conditions, as well as by the identical amino acid composition (González-Pérez et al., 2004). A small amount of a high molecular weight protein fraction with a 15–18S sedimentation coefficient is also observed (González-Pérez et al., 2004; Joubert, 1955; Madhusudhan et al., 1986; Molina et al., 2004; Rahma and Rao, 1979; Schwenke et al., 1979; Sripad and Rao, 1987a, 1987c; Venktesh and Prakash, 1993b). This fraction, which is an aggregate of 11S or/and 7S constituents, has also been reported for other oilseeds and legumes (Guéguen et al., 1988; Prakash and Rao, 1986). Finally, the 1–4S fraction corresponds to the SFAs (typically with 2S as sedimentation coefficient) and may include small amounts of a dissociated form of helianthinin (3S). Publications that analyze the globulin fraction at acid pH values show sedimentation coefficients of 1–4S corresponding to the 3S form of helianthinin (Table 12.B). In fact, the 3S form is the monomeric helianthinin molecule, the 7S form is a trimer, the 11S form is a hexamer, and the 15–18S form comprises a larger number of helianthinin subunits (González-Pérez et al., 2004; Joubert, 1955; Kabirullah and Wills, 1983; Molina et al., 2004; Rahma and Rao, 1979; Sripad and Rao, 1987a, 1987c).

Helianthinin

Helianthinin belongs to the cupin superfamily that was identified by Dunwell in 1998 (Dunwell, 1998; Dunwell and Gane, 1998). The characteristic cupin domain comprises two conserved motifs, each corresponding to two β-strands, separated by a

Table 12.B Protein Composition of Sunflower Protein According to the Sedimentation Coefficient Classification

	Fractions			
Reference	**1–4 S**	**6–9 S**	**10–13 S**	**≥ 15 S**
Joubert (1955)	Major component	Present	Major component	Present
Sabir et al. (1973	Present	Present	Major component	Not present
Schwenke et al. (1974, 1975a, 1975b, 1979) (globulin fraction)	Only acidic pH	Present	Major component	Present
Rahma and Rao (1979)	20%	10%	70%	Present
Rahma and Rao (1981a) (globulin fraction)	Not present	Not present	pH 8.3	Not present
Youle and Huang (1981)	62%	Not present	38%	Not present
Kabirullah and Wills (1983)	Present	Major component	Major component	Present
Madhusudhan et al. (1986)	Present	Present	Present	Present
Sripad and Rao (1987)	Present	Present	Present	Present
Sripad and Rao (1987) (globulin fraction)	Only acidic pH	Only acidic pH	Present	Not present
Sastry and Rao (1990) (globulin fraction)	Only acidic pH	Only acidic pH	Major component	Not present
Venktesh and Prakash (1993)	30%	5%	60%	2%
Molina et al. (2004) (globulin fraction)	Only acidic pH	At pH > 9.0 or I > 1M	At pH 6.0–9.0	Only at low ionic strength (I < 0.1 M)

Source: Adapted from González-Pérez and Vereijken (2007).

less conserved region composed of another two β-strands with an intervening variable loop (Dunwell et al., 2004). The cupin superfamily is one of the most functionally diverse families described so far, comprising both enzymatic and nonenzymatic members (Aravind and Koonin, 1999a, 1999b), and it is widely distributed in the three kingdoms of life: *Archaea, Eubacteria,* and *Eukaryota* (Khuri et al., 2001). Among cupin proteins we find single-domain cupins and two-domain bicupins. The largest family of bicupins comprises the legumin-like 11S and vicilin-like 7S globulins. The so-called 11S seed proteins are not glycosylated and form hexameric structures (Shotwell et al., 1988). Members of the 11S family include pea and broad bean legumins, rape cruciferin, rice glutelins, cotton β-globulin, soy glycinin, pumpkin 11S globulin, oat globulin, sunflower helianthinin, and so forth. The termed 7S seed proteins form trimeric structures and include globulins such as pea vicilin, common bean β-phaseolin, soybean β-conglycinin, or jack bean canavalin.

Molecular Characterization

Helianthinin is generally reported as a globular oligomeric (6 subunits) protein with a molecular weight of 300–350 kDa (Sabir et al., 1973; Schwenke et al., 1979). Besides this band, minor globulin bands of 190 kDa and 440 kDa were observed by native gel electrophoresis (PAGE) (Dalgalarrondo et al., 1984; Durante et al., 1989). These fractions, 190 kDa and 440 kDa, most probably correspond to the previously described 6–9S and 15–18S fractions, respectively. Three main helianthinin subunit groups are described: A (56–57 kDa), B (55.0 kDa), and C (51–52 kDa). These are coded by at least three loci: Hel A, Hel B, and Hel C (Raymond et al., 1991; Raymond et al., 1995; Anisimova et al., 2004; Duca et al., 2008). The isoelectric point of these subunits ranges between pH 5.0 and pH 6.8 (Durante et al., 1989; Raymond et al., 1995, González-Pérez et al., 2004), with C subunits in a more acidic range (5.4–5.7) and A and B subunits mostly between pH 6.3–6.8, except for the major B subunit (5.7–6.0).

As reported for other legumin-like seed proteins, each subunit comprises a large acidic (α) and a small basic (β) polypeptide linked by a highly conserved disulphide bond (between cysteine 103 and 312) (Vonder Haar et al., 1988). These polypeptides are post-translationally generated from larger precursor polypeptides as follows: (1) The precursor polypeptides gather into trimers in the endoplasmic reticulum; (2) the trimers are transported to the vacuole wherein they are cleaved into disulphide linked α- and β-polypeptides; and (3) the trimers finally assemble into hexamers, being stored in protein bodies (Vonder Haar et al., 1988).

The molecular weight of the basic and acidic polypeptides range from about 20 to 27 kDa and 30 to 44 kDa, respectively (Figure 12.2) (Dalgalarrondo et al., 1984, 1985; Raymond et al., 1991). The electrophoretic pattern of helianthinin revealed the presence of 19–26 polypeptide bands (Duca et al., 2008). Raymond et al. (1991,

1995) described three groups of polypeptides by two-dimensional electrophoresis: (1) 8–39 kDa focusing into a 5.6–5.8 pH range (α1), or in a pH 6.3–7.0 range (α2, α3); (2) 32 kDa focusing in a 5.2–5.4 pH range (α'1, α'2, α'3); and (3) β and β' polypeptides (22–24 kDa) focusing into a more basic pH range (6.8–8.0).

In addition to subunit heterogeneity due to various polypeptide chains within a single genotype, there is also diversity on the SDS-PAGE patterns of helianthinin among different varieties (Anisimova et al., 1991a, 1991b; Raymond et al., 1994; 1995; Duca et al., 2008). Duca et al. (2008) pointed out that most genotype dependent variability occurs in the C subunits, whereas A and B subunits exhibit a rather conserved pattern. Hence, the helianthinin C subunits' SDS-PAGE pattern is an interesting breeding selection criterion for genotype characterization.

The unique available gene sequence of one sunflower globulin subunit (Helianthinin G3 or HAG3) indicates that this particular subunit consists of a 285 amino acid acidic chain (32.64 kDa) and a 188 amino acid basic chain (20.98 kDa) (UniProtKB, 2013a; Vonder Haar et al., 1988). Sequence analysis (cDNA and genomic DNA) revealed that helianthinin is encoded by a small gene family that entails at least two divergent subfamilies (Raymond et al., 1991; Vonder Haar et al., 1988).

Solubility

A protein is soluble under specific conditions (pH, ionic strength, etc.) when it does not sediment by moderate centrifugal forces (Morrissey et al., 1985), and subsequently goes into solution or colloidal dispersion. In most functional applications, solubility is a key property and largely affects the field of application and market competitiveness of protein formulations.

The solubility of helianthinin and sunflower protein preparations is minimal between pH 3.0 and 6.0 and strongly depends on pH and ionic strength (Gheyasuddin et al., 1970b; Mattil, 1971; Prakash and Rao, 1986; Rossi et al., 1985; Schwenke et al., 1977; Sosulski and Fleming, 1977; Vermeesch et al., 1987). At low ionic strength (I = 30 mM), helianthinin shows a bell shaped curve with an isoelectric point at pH 4.0–5.5, and high solubility (>70%) at pH 3.0 (González-Pérez et al., 2004; Schwenke et al., 1978) (Figure 12.3a).

At higher ionic strengths (I > 150 mM) helianthinin solubility drastically decreases at pH <5.0 (González-Pérez et al., 2004) and becomes virtually insoluble at I = 250 mM (Schwenke et al. 1977; (González-Pérez et al., 2004) (Figure 12.3). The decreased solubility of helianthinin at pH 3.0 is attributed to protein dissociation and denaturation (González-Pérez et al., 2004; Molina et al., 2004).

Calcium has been reported to precipitate helianthinin due to a cation specific effect rather than to the associated ionic strength change (González-Pérez, 2003; Schwenke et al., 1978). Schwenke et al. (1978) observed a decrease in the solubility of helianthinin at Ca^{2+}concentrations from 10 to 150 mM, over a large pH range (4.5

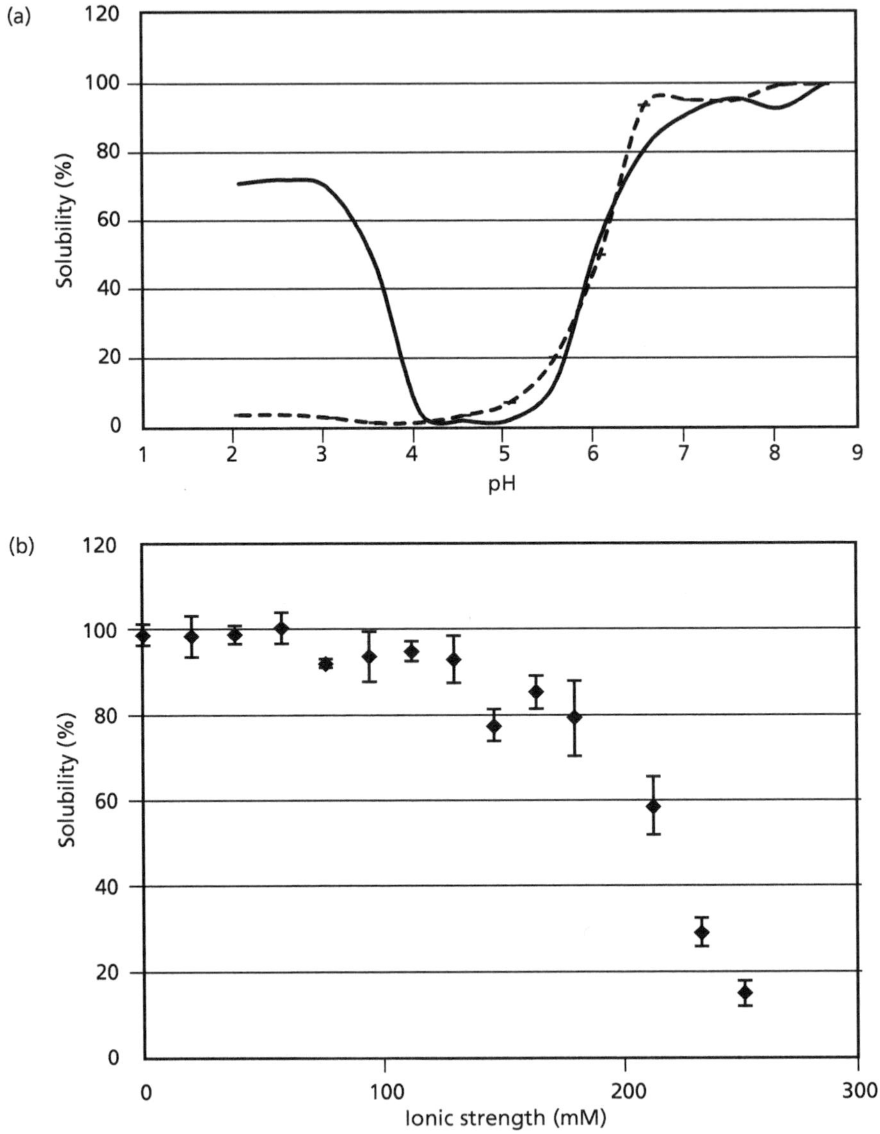

Figure 12.3 (a) pH-dependent solubility profiles of helianthinin at I = 30 mM (solid line) and 250 mM (dashed line); (b) helianthinin solubility versus ionic strength at pH 3.0.

Source: González-Pérez (2003).

to 10.0). This precipitation seems to be rather selective: at pH 8.0, only helianthinin (15S, 11S, and 7S forms) was precipitated by the addition of Ca (30 and 50 mM), whereas SFAs remained in solution (González-Pérez, 2003). Probably, some of the carboxylic groups on the helianthinin subunits are oriented in such a way that they favor the establishment of calcium bridges between them, causing aggregation and precipitation of helianthinin. This explanation agrees with the absence of calcium-induced precipitation at acidic pH, where most of the carboxylic groups are protonated. This calcium-specific precipitation has also been described for soybean proteins (Yuan et al., 2002).

Structure

In addition to solubility, information on structural stability of sunflower proteins (particularly helianthinin as major fraction) is necessary for the isolation of useful sunflower protein preparations and further functional applications. Therefore, it is of the utmost importance to know which conditions may affect structure, solubility, and conformational stability of proteins.

The net stability of the folded (native) state of a protein depends on an intricate balance between diverse interactions in the folded state, the higher conformational disorder of the unfolded state, and interactions with the solvent. The folded state is generally disrupted by environmental conditions often occurring during food and nonfood processing (e.g., extreme pH values, pressure and temperature, organic solvents, etc.) (González-Pérez and Arellano, 2009).

Secondary and Tertiary Structure. Far-UV circular dichroism (CD) studies of helianthinin indicated that the secondary structure of helianthinin is predominantly β sheet (20–30%) and random coil (40–60%) at pH 6.0–8.5, I = 0.5–1.0 M (Rahma and Rao, 1981b; Schwenke et al., 1975a; Sripad and Rao, 1987a; Suryaprakash and Prakash, 2000). The amount of α-helix structure predicted was generally lower than 10%. There are some divergences in the secondary structure predictions among authors, such as likely overestimations of α-helix content (González-Pérez et al., 2004) and high β-turn (Suryaprakash and Prakash, 2000). These divergences are derived, aside from modifications in the experimental conditions, from the use of different secondary structure prediction algorithms and wavelength ranges, as corroborated by their similar spectra (see "Comparison of Helianthinin to other Legumin-like Proteins" in this chapter).

Regarding the tertiary structure, the near-UV CD spectra of helianthinin at pH 7.0–8.5 (I = 30–500 mM) showed a maximum at 285 nm and a shoulder at 292 nm, due principally to tryptophan and some tyrosine contributions (González-Pérez et al., 2004; Sripad and Rao, 1987a). The intensity of the spectra, particularly of the 292 nm shoulder, was noticeably lower at I = 30 mM (compared to I = 250 mM), which generally points at a minor destabilization of the protein structure (Vuilleumier et al., 1993).

Severe structural variations were observed at pH 3.0: the far-UV CD spectrum displayed a negative shift (about 10 nm) in the zero crossing, and subsequently larger negative ellipticity (González-Pérez et al., 2004; Sripad and Rao, 1987a). The estimation of secondary structure elements revealed an increase of the nonstructured protein. Concurrently, the near-UV CD spectrum showed a drastic decline of insensitivity over the full range, as compared to pH 7.0–8.5. Moreover, when the pH decreased from 7.0 to 1.0, the fluorescence spectrum was quenched and red shifted in λ max (Sripad and Rao, 1987a). Similar results in the fluorescence spectrum are observed when the pH increases from 8.0 to 12.5 (Sripad and Rao, 1987c). All these results point to full unfolding (denaturation) at extreme pH values.

The near- and far-UV CD spectrum of helianthinin samples after heat treatment (105–110 °C) at pH 7.0 were similar to those at pH 3.0, pointing out an extensive denaturation and subsequent increase of the non-structured protein (González-Pérez et al., 2004).

Subunit Arrangement at Neutral and Moderate pH Values. At conditions usually used for protein isolation (neutral to moderate alkaline pH values), most helianthinin is in the 11S form (González-Pérez et al., 2004; Molina et al., 2004; Schwenke et al., 1979). Quaternary structure studies by electron microscopy, small angle X-ray scattering, and dynamic light scattering indicate that helianthinin (11S form) consists of an arrangement of six spherical subunits into an 11 nm trigonal antiprism in the dihedral point group symmetry 32 (Plietz et al., 1983, 1986; Reichelt et al., 1980; Schwenke et al., 1979; Zirwer et al., 1986). As previously stated, helianthinin may also occur in the 15–18S, 7S, and 3S form under some experimental conditions (pH, ionic strength, temperature, and protein concentration) (González-Pérez et al., 2004; Molina et al., 2004) (Figure 12.4).

According to Molina et al. (2004), helianthinin is present as 11S at pH 6.3–9.7 (I = 540 mM). At lower ionic strength (I < 300 mM), the 11S form of helianthinin was reported to dissociate into the 7S form (pH 7.0–8.0) (Schwenke et al., 1974, 1975a, 1975b). Later, the presence of 11S and 7S was reported at near neutral pH in a wide ionic strength range (30–1250 mM) (González-Pérez et al., 2004), with a larger proportion of the 11S form at high ionic strength. At ionic strengths between 30 and 250 mM, the dissociation of the 11S form into the 7S form gradually augmented when the pH was increased from 5.8 to 9.0 (González-Pérez et al., 2004). The dissociation was, however, lower at the highest ionic strength (I = 250mM).

González-Perez et al. (2004) reported a fairly slow dissociation (50%) of the 11S form into the 7S form after one week of storage (pH 7.0, 30 mM, 4 °C, 0.01% azide). Although no association of the 7S into the 11S form was observed, the low

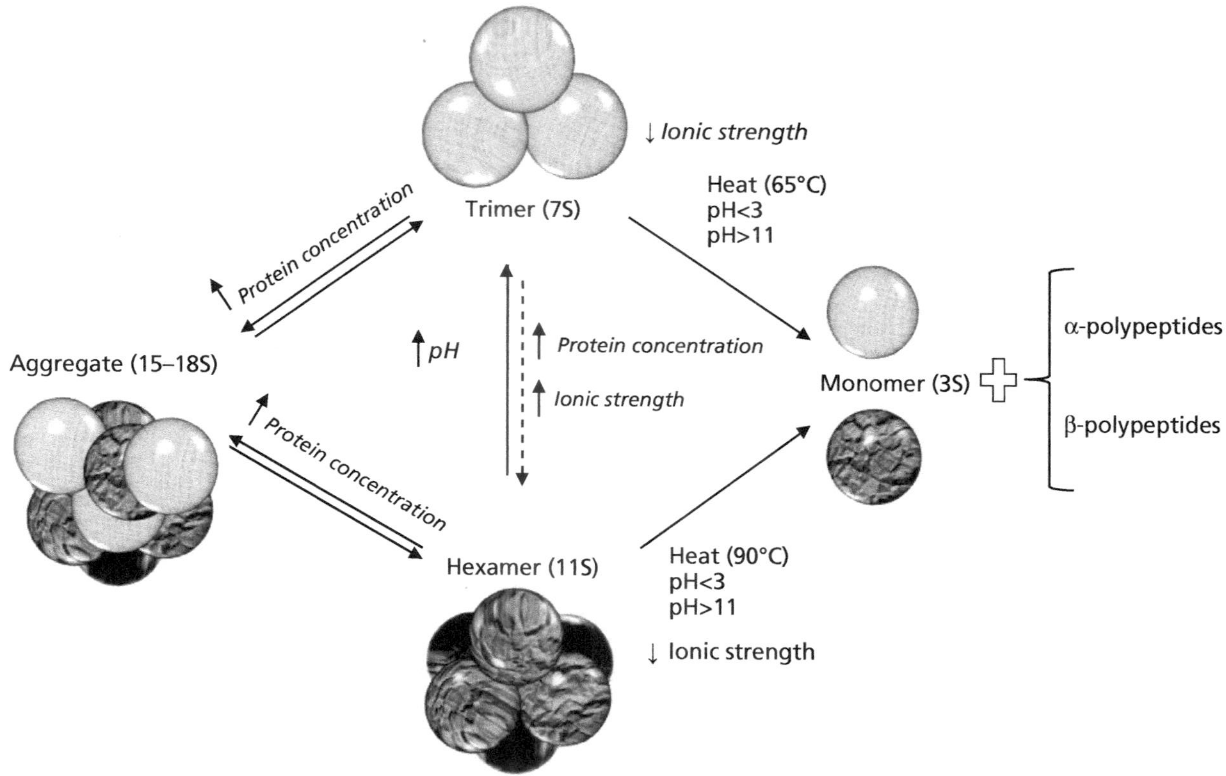

Figure 12.4 Schematic model for the quaternary structure of helianthinin at various conditions (pH, ionic strength, protein concentration, and temperature); ↑= favored by increase; ↓ = favored by reduction.

Source: Adapted from González-Pérez (2003).

rate of the process leaves open the possibility of an equilibrium, instead of an irreversible process. Schwenke et al. (1979) suggested an 11S/7S equilibrium associated with a partially irreversible dissociation of the 11S form into the 7S form of helianthinin. Other reading would be a proteolytic cleavage by endogenous seed proteases (Kortt and Caldwell, 1990b). However, the latter authors did not observe disruption of the 11S form despite the nicking of the acidic polypeptide chains.

The presence of high molecular weight aggregates (15–18S) of helianthinin is generally favored at a high concentrations of this protein, and these aggregates become the major fraction at moderate alkaline pH values (pH 8.5–9.0) and low ionic strength ($I < 100$ mM) (González-Pérez et al., 2004; Molina et al., 2004). Under very specific conditions (pH 5.0, $I = 540$ mM), at which helianthinin is marginally soluble, Molina et al. (2004) described an equilibrium between 7S, 11S, and 15S forms.

Subunit Arrangement at High Temperatures and Extreme pH Values. Exposing sunflower protein to high temperatures and extreme pH values generally results in protein denaturation that, at a molecular level, entails unfolding but not changes in its covalent structure, with the likely exception of breaking and reshuffling of disulphide bonds. Protein unfolding is not always a drawback and it may improve protein functionality (German and Phillips, 1991; German et al., 1982; González-Pérez et al., 2005a, 2005c; Martin, 2003; Nir et al., 1994; Schwenke, 1988; Schwenke et al., 1981; Van Koningsveld, 2001; Wagner and Guéguen, 1995, 1999a, 1999b). Namely, protein denaturation is a must for gel formation, generally achieved by heating. However, the positive effects of protein unfolding are usually counteracted by protein losses due to precipitation (González-Pérez and Arellano, 2009).

Extreme pH Values

Under more extreme conditions (pH > 11.0–12.0 or pH < 3.0), the 7S and 11S forms of helianthinin dissociate into 3S monomers (González-Pérez et al., 2004; Molina et al., 2004; Schwenke et al., 1975a, 1975b; Sripad and Rao, 1987a, 1987c) and denature (González-Pérez et al., 2004; Molina et al., 2004). Remarkably, this dissociation to monomers is already observed (Schwenke et al., 1975a) at pH 9.0 and ionic strengths lower than 300 mM. Conversely, dissociation to monomers, at pH 3.0, is inhibited at a higher ionic strength ($I = 1000$ mM). Actually, most authors reported a general lower degree of dissociation with increasing ionic strength (González-Pérez et al., 2004; Schwenke et al., 1974) and particularly at acidic and alkaline pH (3.0 > pH > 9.0) (Molina et al., 2004; Schwenke et al., 1975a, 1975b). For example, at pH 11.0, 11S helianthinin may dissociate to 7S ($I = 540$ mM) or to 3S ($I < 30$ mM) depending on the ionic strength (Molina et al., 2004; Sripad and Rao, 1987c). Small amounts of the monomers are, however, also present under nondenaturing conditions (González-Pérez et al., 2004).

Some authors reported reassociation of the monomers, below pH 2.5, into an aggregated form of helianthinin with a sedimentation coefficient of 11S (Prakash and Rao, 1986; Sripad and Rao, 1987a). This aggregated form does not restore the native conformation but shows a lower amount of nonstructured protein than at pH 3.0 (as 3S form). This reaggregation/partial refolding phenomenon has been described for other seed globulins such as soy glycinin, sesame α-globulin, mustard 11S, and arachin (Sripad and Rao, 1987a).

High Temperature

As previously mentioned for structural (secondary and tertiary level) changes at high temperature, the subunit arrangement after heating also resembles that at acidic conditions (i.e., dissociation to 3S and denaturation), but is noticeably more severe (González-Pérez et al., 2004; Molina et al., 2004). At low sample moisture (0–30% of the total weight), sunflower proteins show an endothermic peak at 189 to 120 °C on differential scanning calorimetry (DSC), respectively. Lower denaturation temperatures are reported for sunflower protein in solutions/dispersions. At high protein concentration (80–250 mg/mL) and high ionic strength (I = 200–500 mM), the DSC analysis of a sunflower isolate (González-Pérez et al., 2002) and that of isolated helianthinin (Molina et al., 2004) displayed a single endothermic transition at 100–105 °C. However, lower protein concentration and lower ionic strength (1–4 mg/mL and I = 10 mM) resulted in a stepwise thermal unfolding at 65 °C and 90 °C (González-Pérez et al., 2004). The lower endothermic transition corresponded to the 7S form of helianthinin, and the higher endothermic transition to the 11S form. Soy glycinin can also display independent thermal transitions for the 11S and 7S forms (Danilenko et al., 1987; German et al., 1982; Lakemond, 2001; Lakemond et al., 2000a; Schwenke et al., 1987). These results indicate that at high protein concentrations, helianthinin is mostly in the 11S form. In addition, high ionic strengths are reported to shift the 11S/7S equilibrium to the 11S form (González-Pérez et al., 2004; Schwenke et al., 1974, 1975a, 1975b). Indeed, Molina et al. (2004) showed an increase of about 10 °C in the denaturation temperature of helianthinin by increasing the ionic strength from 50 to 2000 mM. Thus, a higher ionic strength results in higher thermal stability, enthalpy, and greater cooperativity of the transition. The latter thermodynamic term relates to the ability of the polypeptide chain to unfold and is controlled by its chain topology-distribution of local and global regions of the protein structure (Shank et al., 2010).

Similar results are reported for other storage globulins such as soy glycinin and β-conglycinin (Hermansson, 1986; Kim et al., 2004; Sorgentini and Wagner, 1999), fava bean globulin (Arntfield et al., 1986), oat globulin (Harwalkar and Ma, 1987), broad bean legumin (Zheng et al., 1993), amaranth globulin (Castellani et al., 1998), red bean globulin (Meng and Ma, 2001), and flaxseed globulin (Li-Chan and Ma, 2002).

Thermal denaturation typically results in dissociation of the 11S and 7S forms of helianthinin and formation of aggregates (González-Pérez et al., 2004; Molina et al., 2004). Thermal aggregation is already reported to occur at 80 °C, surprisingly with little or no dissociation and denaturation of the hexameric form of helianthinin (Molina et al., 2004; Rahma and Rao, 1981a). At secondary structure level, thermal aggregation resulted in more β-sheet content. The formation of β-sheet after heating is generally associated with protein aggregation (Berecz et al., 2010; Choi and Ma, 2005; Ellepola et al., 2005), observed as an abrupt exothermic transition with DSC (González-Pérez et al., 2004).

Molina et al. (2004) reported the formation of SDS-resistant aggregates after thermal treatments (90 and 100 °C), with basic polypeptides at a higher proportion than acidic polypeptides (that remain in solution), as also reported for soy glycinin (German et al., 1982; Petruccelli and Anon, 1995). This result indicates cleavage of disulphide bonds in response to heating.

Some authors suggest that the dissociation, unfolding, and aggregation of helianthinin occurs sequentially in that order (González-Pérez et al., 2004; Molina et al., 2004). The latter has also been observed in the presence of chaotropic molecules such as urea, guanidine hydrochloride, and guanidine thiocyanate (Rahma and Rao, 1981a; Suryaprakash and Prakash, 2000).

Due to the high denaturation temperature of helianthinin, partial deamidation may be expected during heat denaturation. Important deamidation levels (10%) were reported after heating a sunflower protein isolate at 90 °C (Claughton and Pearce, 1989).

Based on the previous information on helianthinin dissociation and thermal denaturation, González-Pérez et al. (2004) proposed the hypothesis that the 7S and 11S forms of helianthinin are built from structurally different subunits: the 11S form unfolding at 90 °C and the 7S at 65 °C. This hypothesis is strengthened by the presence of an increased amount of nonstructured protein in the 7S form together with the destabilization of its tertiary structure.

Comparison of Helianthinin to Other Legumin-like Proteins. Several studies and reviews extensively compare and describe the genetic, biochemical, structural, and biophysical characteristics of seed globulins (Derbyshire et al., 1976; Fukushima, 1991; Mandal and Mandal, 2000; Marcone, 1999a, 1999b; Marcone et al., 1998a, 1998b; Prakash and Rao, 1986, 1988). This knowledge is of extensive relevance from an industrial point of view in order to find functional substitutes for soybean protein according to their particular properties. Indeed, oilseed and legume protein products have often been compared to well-characterized soybean protein isolates and concentrates, especially in relation to functional properties (González-Pérez and Vereijken, 2007; Martinez, 1979).

The 10–12S protein fraction is also a major fraction in legumes (such as pea, lupin, and soybean) and oilseeds (such as groundnut, sesame, rapeseed, safflower, poppy seed, and linseed) (Prakash and Rao, 1986). Marcone et al. (1998a, 1998b; Marcone 1999a, 1999b) characterized 21 legumin-like seed globulins from both monocotyledonous and dicotyledonous plants pointing out their narrow molecular weight range (300–370 kDa). In spite of the large diversity among polypeptides and amino acid sequences of these globulins, they share conserved regions in sequence and/or structure. The most conserved sequences are those essential for adequate post-translational modifications and intracellular transport (Vonder Haar et al., 1988). A common pattern of these proteins is that of the cleavage site between the acidic and basic subunits, and a proximal cysteine residue involved in the disulphide bond: N-G-x-[DE](2)-x-[LIVMF]-C-[ST]-x(11,12)-[PAG]-D (ExPASy, 2013: http://prosite.expasy.org/PS00305).

Similarities have been found in quaternary and secondary structures, showing generally low levels of α-helices and high levels of β-sheet structures. A lower homology is generally observed in the tertiary structure of seed globulins.

Additional homologies described in the previously cited publications (Derbyshire et al., 1976; Fukushima, 1991; Mandal and Mandal, 2000; Marcone, 1999a, 1999b; Marcone et al., 1998a, 1998b; Prakash and Rao, 1986, 1988) include:

- Amino acid composition: rich in acidic and aromatic amino acids and low content in lysine- and sulfur-containing amino acids.
- Nearly the same values for several hydrophobicity parameters, such as "frequency of nonpolar side chains" and "Bigelow's average hydrophobicity." Hydrophobicity is reported to be low when compared with ovoalbumin or ribonuclease.
- Low susceptibility to hydrolysis by proteolytic enzymes compares with other proteins, such as casein.
- Similar association–dissociation behavior modulated by ionic strength, pH, concentration, and temperature.
- High thermal stability, enthalpy, and cooperativity of the thermal transition.
- Similar solubility pattern, strongly affected by pH and ionic strength, generally showing a bell shaped curve with a minimum at pH 3.0–6.0.
- Other features, such as intrinsic viscosity, biological significance, biosynthesis scheme, fluorescence analysis, quaternary structure, subunit composition, and stability against chaotropic agents (Urea, SDS, etc.).

For example, comparison of helianthinin (González-Pérez et al., 2004; Rahma and Rao, 1981b; Sripad and Rao, 1987a; Suryaprakash and Prakash, 2000) and soy glycinin (Kim et al., 2004; Lakemond et al., 2000b) shows a high physicochemical behavior resemblance. Besides the similarities described in sections above, both 11S proteins are stabilized by NaCl and irreversibly dissociate and denature at acid pH (pH ≤ 3.0). They also share some structural homologies by circular dichroism. The far-UV CD spectra of both proteins at neutral pH are very similar: they exhibit a minimum around 208–212 nm, a zero crossing around 200 nm, and a shoulder at

224–228 nm, which is representative of low proportion of α-helix, and high β-sheet and random coil structure.

The near-UV CD spectrum of helianthinin and glycinin also shares some common trends: positive ellipticity between 260 and 300 nm, at neutral pH, with a maximum at 285 and a shoulder at 292 nm. Because CD spectra, particularly the near-UV, is a protein fingerprint (Pain, 2004), a similar tertiary folding of these two proteins is suggested.

Sunflower Albumins

Albumin seed proteins, with typical sedimentation coefficients of 2S, are widely distributed in both mono- and dicotyledonous plants accounting for 10–60% of the total seed proteins (Moreno and Clemente, 2008; Youle and Huang, 1981). Conversely to globulins, 2S albumins include high amounts of sulfur-containing amino acids: cysteine (6–13%) and methionine (1.6–3.8%) (Youle and Huang, 1981). 2S albumins belong to the prolamin superfamily, characterized by a highly conserved sequence of eight cysteine residues located in an α-helical globular domain. This family also includes cereal prolamins, cereal protease inhibitors of α-amylase/trypsin, indolines, lipid transfer proteins (LTPs), the soyabean hydrophobic protein, and cereal α-globulins (AllFam, 2013; Egorov et al., 1996; Moreno and Clemente, 2008; Shewry et al., 1995).

Molecular Characterization

In common with other storage proteins, 2S albumins show a high level of polymorphism and are encoded by a multigene family resulting in abundant isoforms. Because of their higher interspecific polymorphism, Raymond et al. (1995) proposed SFAs as better molecular markers of phenotypic divergence than helianthinin. The mature 2S albumins are generally composed of a large (8–10 kDa) and small (3–4 kDa) polypeptide linked by two disulphide bonds. However, SFAs are a distinct exception to this structure pattern, consisting of a single polypeptide chain (Allen et al., 1987; Anisimova et al., 1995; Egorov et al., 1996; Kortt et al., 1991; Shewry and Pandya, 1999) as noticed by the lower electrophoretic mobility under reducing conditions due to the less packed structure after the rupture of the disulphide bonds (Allen et al., 1987; Raymond et al., 1991). This particular feature also occurs in peanut 2S albumins (Lehmann et al., 2006).

SFAs make up about 10–30% of the total seed proteins (Baudet and Mosse, 1977; González-Pérez and Vereijken, 2007; González-Pérez et al., 2005b; Mazhar et al., 1998; Raymond et al., 1995, Schwenke and Raab, 1973; Sosulski and Bakal, 1969). Several individual SFA proteins (8 to 13) have been separated by reversed-phase high-performance liquid chromatography (RP-HPLC), SDS-PAGE (one and two dimensional), and isoelectric focusing (Anisimova et al., 1995, 2002; Guéguen et al., 1996; Kortt and Caldwell, 1990a; Serre et al., 2001). The levels at which these components are present vary widely between genotypes (Anisimova et al., 1995, 2002; Serre et al., 2001).

SFAs are alkaline proteins (average isoelectric point of about 8.8) with molecular weights ranging from 10 to 18 kDa (Figure 12.2) (Anisimova et al., 1995, 2002; Guéguen et al., 1996; Kortt and Caldwell, 1990a; Popineau et al., 1998; Prakash and Rao, 1986). Two types of SFAs are distinguished: methionine-rich and methionine-poor albumins (Guéguen et al., 1996; Pandya et al., 2000; Popineau et al., 1998).

HAG5 and SFA8

The amino acid sequences of two sunflower albumins are currently available: HAG5 and SFA8.

The methionine-poor 2S albumin, called HAG5 (Allen et al., 1987), consists of 134 amino acids, has an average mass of 15,78 kDa, and a theoretical isoelectric point of 8.69 (UniProtKB, 2013b). The methionine-poor 2S albumin is coded by a small gene family comprised of at least two divergent genes. The central biological role of the SFAs as storage protein is to be a reserve of nitrogen, carbon, and sulfur during germination; however, other functions as antifungal activity for HAG5 (Regente and de la Canal, 2001) and radish 2S albumins (Terras et al., 1992) have been reported.

The methionine-rich 2S protein SFA8 which was sequenced by Kortt et al. (1991), consisting of 103 amino acids with an average mass of 12.15 kDa and a theoretical pI of 5.91 (UniProtKB, 2013c). The name of this protein is based on its order of elution on RP-HPLC (Kortt and Caldwell, 1990a) and the protein contains an unusually high proportion of hydrophobic residues, as well as a large amount of sulfur-containing amino acids: 16 methionines and 8 cysteines. Because of its rich amino acid composition, SFA8 has been expressed in seeds of numerous plants, with the aim of improving their nutritive value. Among them we find fescue (Wang et al., 2001), rice (Hagan et al., 2003; Islam et al., 2005), alfalfa (Tabe et al., 1995), lupins (Molvig et al., 1997; Tabe and Droux, 2002), clover (Khan et al., 1996), and trifolium (Christiansen et al., 2000). Although some improvements on the total seed methionine were reported for these transgenic plants, generally, with the exception of lupin, the level was still below the optimal level required for human and animal feed (Ufaz and Galili, 2008).

Molecular modeling studies predict that SFA8 has a compact structure with hydrophobic residues clustered to form a large hydrophobic face. This particular hydrophobic surface patch has been proposed to largely contribute to the highly active emulsification properties reported for this protein (Pandya et al., 2000).

The three-dimensional structure of SFA8 was determined by high-resolution nuclear magnetic resonance (NMR), corroborating the predicted compact structure in

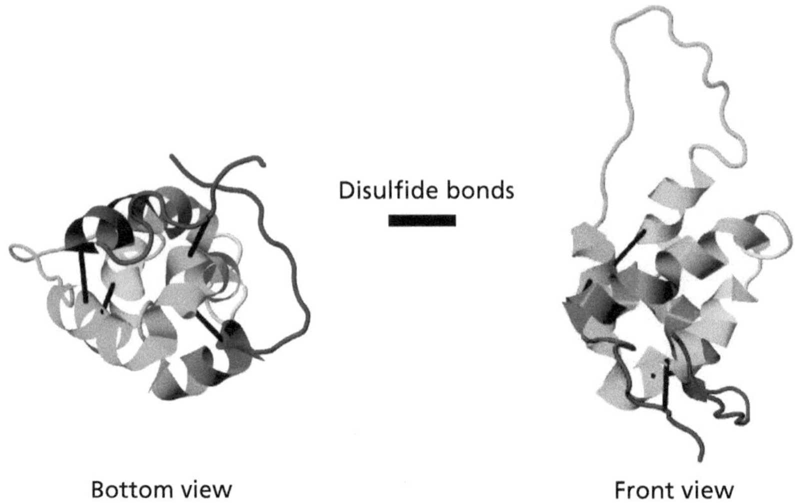

Figure 12.5 Protein structure of sunflower SFA8 (PDB: 1S6D).

bundles of five amphipathic α-helices arranged in a right-handed superhelix stabilized by four disulfide bonds (Figure 12.5) (Pantoja-Uceda et al., 2004b). Connecting two of the five helices (numbered II and IV) there is a solvent exposed fragment known as *hypervariable loop*. Similar three-dimensional structures have been described for 2S albumins from rapeseed (Pantoja-Uceda et al., 2004a) and castor bean (Pantoja-Uceda et al., 2003). SFA8 shows no tryptic, chymotryptic, or α-amylase inhibitory activity regardless of a putative reactive site similar to that of barley trypsin inhibitor (Kortt et al., 1991).

SFA8, together with a protein called SFA7 (eluting earlier on RP-HPLC), may account for about 10–20% of the total sunflower albumins (Anisimova et al., 2002). However, the final amounts and proportions of SFA7 and SFA8 are determined by genetics factors and the sulfur availability in the seed. These proteins (SFA7 and SFA8) are closely related, having similar masses (equal mobility on SDS-PAGE) and amino acid compositions, equal isoelectric points, and identical N-terminal amino acid sequences (Anisimova et al., 2002; Burnett et al., 2002; Kortt and Caldwell, 1990a). Moreover, these proteins are encoded by alleles in a different locus than the methionine-poor albumins.

Allergenicity of SFAs

The 2S albumins comprise a major group of plant food allergens (Monsalve et al., 2004; Shewry et al., 2006). Allergenic and sensitivity (intolerance) reactions to 2S albumins suffered by workers from baking and oil extraction industry (cotton, castor and rape seeds) have been documented since late 1920s (Alcocer et al., 2002; Murtagh et al., 2003). The hypervariable loop, structurally common to the methionine-rich 2S albumins, has been pointed to as an important antigenic region. Sunflower seed contains several IgE-binding sites to proteins, particularly to the methionine-rich SFA8 (Kelly and Hefle, 2000; Kelly et al., 2000). Despite these results, the allergenicity of SFA8 has not been well described because of the low prevalence of sunflower seed allergy, even in countries such as Spain where there is high consumption of sunflower seeds as snacks (Besler et al., 2001). Consequently, it is still doubtful that SFAs are the cause of the allergy of the few individuals who are sensitive to sunflower seeds (Murtagh et al., 2003).

Solubility and Structure

SFAs have been reported to be soluble at moderate ionic strength (30–250 mM) and wide pH range (2.0–8.5) (González-Pérez et al., 2005b). As deduced from the protein extraction method, SFAs are also largely soluble at alkaline pH (7.8) at high ionic strengths (500–1000 mM) (Burnett et al., 2002; Kortt and Caldwell, 1990a).

Far-UV CD studies showed that SFAs have a considerable number of α-helices, in agreement with the α-helical bundle structures that characterize the prolamin superfamily. The secondary structure of the total SFAs fraction consists of similar amounts (30–32%) of α-helical, β-sheet, and random coil elements (González-Pérez et al., 2005b; Pandya et al., 1999). The methionine-poor 2S albumin fraction has been reported to contain lower α-helical and β-sheet elements, 27% and 23%, respectively (Berecz et al., 2010). Most of the structural studies focused on SFA8 (Alcocer et al., 2002; Berecz et al., 2010; Burnett et al., 2002; Murtagh et al., 2002, 2003; Pandya et al., 1999, 2000). This methionine-rich protein also contains a high amount of α-helices, between 30% and 40%, in addition to β-sheet (about 10%), β-turn (about 20%), and random coil elements.

Conformational Stability

The conformational stability of SFA has been studied by DSC and various spectroscopic techniques (far- and near-UV CD, NMR, and fluorescence). Native and recombinant SFAs, particularly SFA8, were found to be a group of proteins with an outstanding stability against thermal and extreme pH conditions, as well as against proteolytic enzymes and chaotropic agents (guanidinium salts) (Alcocer et al., 2002; Berecz et al., 2010, 2013; González-Pérez et al., 2005b; Murtagh et al., 2002, 2003; Pandya et al., 1999).

The total SFA fraction is stable over a wide pH range (from 3.0 to 9.0, I = 30 mM), as revealed by the almost superimposable far-UV CD spectra. DSC thermograms and far-UV CD spectra showed denaturation temperatures far above 100 °C (up to 118 °C) for the same pH range (González-Pérez et al., 2005b). Moreover, the change in far-UV CD ellipticity upon heating (110 °C), associated with thermal unfolding, was partially reversible at pH 7.0 (Figure 12.6). The isolated methionine-poor 2S albumins were also highly stable to heating, with minor and reversible structural changes at 80 °C (Berecz et al., 2010).

DSC tests showed that the denaturation temperatures of native and recombinant SFA8 are above 100 °C at pH 7.0 (Berecz et al., 2010). In addition, they displayed minor and largely reversible changes on secondary and tertiary structures (far- and near-UV CD) up to 95 °C (pH 7) (Alcocer et al., 2002; Murtagh et al., 2002, 2003; Pandya et al., 1999). Native and recombinant SFA8 were also stable at a wide range of pH variation and the far-UV CD and fluorescence emission spectra did not vary over the pH ranges 2.0–7.0 and 2.0–10.0, respectively (Alcocer et al., 2002; Murtagh et al., 2002, 2003; Pandya et al., 1999). Moreover, SFA8 was shown to be very stable to chaotropic agents, with no changes observed at secondary structure levels up to 4.5 M guanidinium chloride, and it maintains a considerable amount of secondary structure at 6 M (Alcocer et al., 2002; Murtagh et al., 2003). A lower structural stability for SFA8 was reported in the presence of high concentration of guanidinium thiocyanate, being stable up to 1.5 M and completely denatured at 4 M, as revealed by one-dimensional NMR spectra.

In addition, SFA8 is highly stable against proteolysis, as revealed by simulated gastric fluid (SGF) studies. Both native and recombinant forms of this protein did not show any cleavage after 30 minutes of SGF treatment (Murtagh et al., 2002, 2003). It has been reported that 2S albumins are some of the most stable proteins to diges-

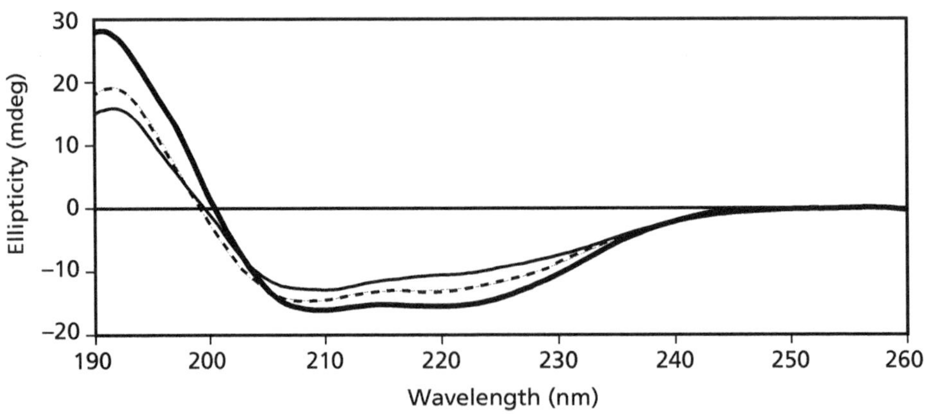

Figure 12.6 Far-UV CD spectra of SFAs (pH 7; *I* =30 mM) recorded at 20 °C (thick line), 110 °C (thin line), and 20 °C after heating up to 110 °C (dashed line).
Source: González-Pérez (2003).

tion by SGF (Murtagh et al., 2002), which may explain their attributed allergenicity because they can reach the circulatory system as native proteins.

The high stability of SFAs is a common feature of 2S albumins, which is largely explained by their reported compact structure mainly arising from the four conserved intramolecular disulfide bonds. Protein stability largely depends on the bonds involved in their tertiary structure, in which disulfide bonds together with electrostatic and hydrophobic interactions play a central role. The electrostatic interactions are, indeed, the second stabilizing interaction in 2S albumins (Pantoja-Uceda et al., 2003, 2004a, 2004b). Numerous publications have confirmed the high stability of 2S albumins from several seeds such as peanut (Lehmann et al., 2006), rapeseed (Folawiyo and Owusu Apenten, 1997; Krzyzaniak et al., 1998; Muren et al., 1996), Brazil nut (Alcocer et al., 2002; Moreno et al., 2005b; Murtagh et al., 2002, 2003), soybean (Lin et al., 2004), or sesame (Moreno et al., 2005a).

Oleosins

Seeds of many plant species store lipids as energy reserves in small (0.6–2.5 μm) intracellular organelles called *oil bodies* (lipid bodies, oleosomes, or spherosomes) (Huang, 1992; Yatsu and Jacks, 1972). Oil bodies are synthesized during seed maturation in the endoplasmatic reticulum by a budding process (Huang, 1992, 1996). These organelles mainly consist of a triacylglycerol (TAG) matrix surrounded by a phospholipid monolayer embedded with proteins known as *oleosins.* They stabilize the oil body and may provide recognition signals for lipase binding during seed germination (Huang, 1994).

Huang (1992) estimated that oleosins represent about 1–4% of the oil body mass and about 2–8% of the sunflower seed protein. Most probably, oleosins are present in much smaller amounts in sunflower protein preparations (removed during defatting) due to their high hydrophobicity. However, we can hardly corroborate this data about oleosin content in the literature because sunflower protein publications generally do not include oleosins in the protein composition of their defatted meals, concentrates, and isolates. Alexander et al. (2002) claimed that the difficulty to purify sufficient quantities for detailed analysis (such as structural analysis) was due to their hydrophobic nature (Alexander et al., 2002). Indeed, oleosin isolation generally starts with oil body isolation, removing most of the water soluble proteins.

Oleosins are alkaline proteins with molecular weight of about 15–26 kDa, depending on the isoforms and species in which they occur (Huang, 1992, 1994). They are unique to the oil bodies and are not present in other organelles, as indicated by subcellular fractionation and immunocytochemistry (Huang, 1992).

Oleosins are encoded by members of a small gene family that are only active during seed maturation (Huang, 1992). These genes share three structural domains: N-terminal, central hydrophobic, and C-terminal (Huang, 1992; Lacey, 1998). The highly conserved central hydrophobic domain of about 68–74 residues is inserted into the TAG matrix, being flanked by the less conserved N- and C-terminal amphipathic domains (about 50–70 and 55–98 residues, respectively) (Huang, 1992). Oleosins have been suggested to lie on the surface of the oil body, with positively charged residues facing the oil body (specifically phosphate groups) and negatively charged residues facing the cytosol (Huang, 1994; Lacey, 1998). Evidence for the residues' orientation is deduced by the fact that the flanking domains are digested by proteinases, but not the central hydrophobic domain (Lacey, 1998).

Two major sunflower oleosins with molecular weight of 19–19.5 kDa and 20–20.6 kDa, and a minor one with 15.5 kDa, were described in more detail by Millichip et al. (1996) and Lacey (1998). Regarding sunflower oleosin structure, Millichip et al. (1996) determined a 50–60% of α-helix content, based on data from far-UV CD spectroscopy. Alexander et al. (2002) estimated a lower amount of α-helix structures (30–50%) and about 60% α-helix for the hydrophobic domain. These authors proposed a model in which the hydrophobic domain forms an α-helical hairpin structure, being the two helices separated by a turn region. Li et al. (1993) studied the secondary structure of the N-terminal domain by far-UV CD spectroscopy, reporting structural estimations of 10% α-helix, 20–30% β-strands, 8% β-turns, and 60% random coil. Structural studies specifically targeting the sunflower oleosins' C-terminal domain are lacking.

Recently, oleosins have received increased attention due to their natural protection against coalescence. Oleosins act as emulsifiers at the oil–water interface (Li et al., 1993). Protein formulations for formation and stabilization of emulsion are of great interest for pharmaceutical and food applications (e.g., for encapsulation of liposoluble drugs).

Other Minor Proteins

In addition to storage proteins, sunflower, like other crops, contains many other seed proteins in lower concentrations, but with specific functions (metabolic and structural). Proteomic studies of sunflower seed proteins resulted in the identification and assignments of 24 (de Sousa Barbosa et al., 2013) and 44 (Hajduch et al., 2007) protein spots on two-dimensional gels. The number of identified sunflower proteins is oddly significant, considering the few entries (85) on the *Helianthus annuus* L. protein database of National Center for Biotechnology Information (NCBI) for Reference Sequence. This low number of sequenced proteins is a demonstration of the limited information about sunflower proteins compared to other crops, such as soybean with 842 entries for *Glycine max* L. in NCBI for Reference Sequence.

As previously stated, sunflower seeds have generally been reported to contain low or no antinutritional compounds as protease inhibitors; they are, however, present, likewise in legume seeds. Protease inhibitors play an important role in plant protection, forming part of a broad-spectrum defense response that also includes the production of pathogenesis-related proteins (Ryan, 1990).

A screening experiment on protease inhibitors in sunflower seeds identified six groups of protease inhibitors with apparent molecular weights of 1.5–6.0 kDa (Konarev et al., 2000). These proteins included trypsin inhibitors, as well as subtilisin and chymotrypsin inhibitors. The trypsin inhibitors have low molecular weight (1.5–2.5 kDa) and high pI (9.0–10.0). The structure of one major trypsin inhibitor, called SFTI-1 (sunflower trypsin inhibitor-1), which structurally mimics the inhibitory arm of Bowman-Birk protease inhibitors (Mylne et al., 2011), was determined in complex with trypsin (Luckett et al., 1999). It has a cyclic structure of 14 amino acids (molecular weight = 1.51 kDa) characterized by a head-to-tail cyclic backbone that is cross-braced by a single disulfide bridge and a network of hydrogen bonds (Colgrave et al., 2010; Korsinczky et al., 2004; Mylne et al., 2011). All these properties make SFTI-1 an extremely stable molecule. SFTI-1 is of great interest to chemists and the pharmaceutical industry (Mylne et al., 2011). Native SFTI-1 exhibits a potent inhibitory activity toward matriptase, a transmembrane serine protease involved in various types of cancer formation (Yuan et al., 2011). A modified SFTI-1 blocks KLK4/prostase, a trypsin-like serine protease that causes prostate cancer (Swedberg et al., 2009). In addition, SFTI-1 has been characterized as the strongest protease (peptide) inhibitor of trypsin (Ki = 0.1–1 nM) (Long et al., 2001; Luckett et al., 1999). The pharmaceutical industry also finds attractive the prospect of using SFTI-1 as protein drug scaffold (Daly et al., 2006; Korsinczky et al., 2004).

Another group of sunflower proteins belonging to the prolamin superfamily is the lipid transfer proteins (LTPs). They are low molecular weight (7–10 kDa) alkaline proteins comprising 90–95 amino acids and characterized as one of the prolamin superfamily proteins by the presence of eight conserved cysteine residues (Kader, 1996, 1997). X-ray crystallography and NMR approaches predicted a structure consisting of four α-helices connected by four disulfide bridges (Kader, 1997), quite similar to that of SFA8. Molecular modeling of a sunflower LTP also predicted the presence of an internal hydrophobic cavity (262 ångström), which is able to hold lipid molecules in vitro (Luckett et al., 2001), although this function has not been demonstrated in vivo (Gonorazky et al., 2005). Luckett et al. (2001) estimated 60% of α-helix elements and no structural modifications in a pH range from 4.0 to 10.0. The high conformational stability of LTPs against denaturation (thermal and SGF) has been recently confirmed (Berecz et al., 2010, 2013), although these authors reported a lower amount of α-helical structure.

An LTP sunflower seed protein of 10 kDa with antifungal properties has been isolated and characterized, suggesting its contribution to the innate immunity (Pagnussat et al., 2009; Regente and de la Canal, 2000; Regente and de la Canal, 2001). However, another study suggested a function related to the mobilization of lipid reserves (Gonorazky et al., 2005). Numerous functions have been proposed for LTPs, such as being involved in wax and cutin assembly, cell wall formation, the mentioned antimicrobial role, pollen tube adhesion, mobilization of lipids, and plant signaling (Pagnussat et al., 2009). The latter authors suggest that the ambiguous function attribution is due to the low sequence homology, and subsequently poor putative function assignment, of this multigenic group of proteins.

Thiamin-binding protein (TBP) is found in many types of plant seeds (Watanabe et al., 1998, 2002). TBP retains thiamin (vitamin B1) in dormant seeds and supplies nitrogen and thiamin during germination. Therefore, Watanabe et al. (1998) considered TBP to be a storage protein, namely a legumin-like globulin, although its molecular weight (230 kDa) is below the expected one (300–350 kDa) for an 11S globulin (Watanabe, 1998). TBP is present as an oligomer composed of six subunits consisting of some combination of 29-, 25-, 24-, 22-, and 21-kDa polypeptides linked by disulfide bonds. The molecular mass and the amino acid composition of these polypeptides are similar to those of helianthinin (Raymond et al., 1995; Watanabe et al., 1998, 2002).

Catalases are cell protective antioxidant enzymes that degrade two molecules of hydrogen peroxide to water and oxygen (Chelikani et al., 2004; Willekens et al., 1995). Catalases were isolated from sunflower seeds, particularly from the perixosomal cores of cotyledons with molecular weight of 55 and 59 kDa (Kleff et al., 1997). The latter study revealed that catalases are the main proteins in peroxisomal cores of sunflower cotyledons. The authors also proposed a tetrameric heme-containing structure based on the enzymatic catalase activity found at the 250 kDa protein fraction obtained by rate zonal centrifugation which allows protein separation according to size, shape, and density.

On the NCBI database we can find information about other sunflower minor proteins, such as ribosomal proteins, RNA polymerases, chloroplast proteins (enzymes, structural and photosystem I and II proteins), NADH dehydrogenases, cytochromes, proteases, synthases, or maturases. Some of the information from this database, as well as from other sources, is generated from a molecular biology approach (Almoguera and Jordano, 1992; Binet et al., 1991; Martínez-Force et al., 2000; Timme et al., 2007; Waksman et al., 1987) and based on DNA (genomic and complementary) studies rather than the biochemical and biophysical characterization of isolated proteins.

Breeding for Protein Content

The first scientific sunflower breeding study was published in 1912 by V. S. Pustovoit; this study was focused on local varieties (Panchenko, 1966; Škorić, 1992). Commercial breeding started in most of the producing countries between 1920 and 1950 (Paniego et al., 2007). Most efforts in breeding have been directed to increase oil content at the expense of protein, while maintaining low levels of hull (Jiménez et al., 1985). Indeed, one of the most important breeding successes that facilitated the spreading of sunflower as worldwide crop was a radical increase of the oil content of sunflower seeds (from 30% to 50%) (Fernández-Martínez et al., 2010).

Breeding for protein content and amino acid profile enhancement of sunflower seeds has received considerable attention particularly in regions where soybean and rapeseed are not the main crops (Fick and Miller, 1997). Generally, SFM obtained as a byproduct from the sunflower oil industry has a lower protein and lysine content than soybean meal, but higher methionine content (Norton, 1989). Accordingly, selection to increase lysine content in seeds is an important breeding goal; however. The main downside of the breeding method is its poor selectivity because increases in one amino acid are expected to result in increases in other amino acids (Škorić, 1992). Various authors showed extensive variability for lysine content among sunflower varieties (Christov et al., 1993; Ivanov, 1975; Nenova and Drumeva, 2012; Sosulski and Sarwar, 1973).

A primary step in any breeding program is the identification of genetic variability for the traits of interest. A large assortment of sunflower genetic resources has been used to search for variation in agronomic and seed quality traits (Fernández-Martínez et al., 2010). In breeding programs targeting protein content and composition, one of the most relevant parameters is the variation in the relative proportion of protein, hull, and oil in the seed (Jiménez et al., 1985). The latter research group evaluated a sunflower world collection of 430 entries. Later, Seiler (1992) expanded this evaluation for a germplasm collection of 2000 wild species. Wild sunflower species possess a wider variability for desirable economic and agronomic traits such as stress tolerance (salinity, drought, etc.); disease resistance; and oil, protein, and biomass production (Miller et al., 1992). This variability is useful in breeding programs to increase the protein concentration in domesticated sunflower species (Laferriere, 1986). Moreover, wild species cross readily with sunflower cultivars, although seed dormancy may impair successful crossbreeding of some wild germplasm entries (Snow et al., 1998). Sunflower seed dormancy plays an important role in both commercial and wild species. Sunflower may have dormancy in the embryo, seed coat, or pericarp, and often have a combination of the three. In most wild sunflower species, dormancy is primarily controlled by the seed coat and pericarp and can last for several years (Brunick, 2008).

Besides the genetic variability of traits of interest, a breeding program targeting sunflower as a protein crop must also evaluate the correlations between the protein content and other seed characteristics, as well as the influence of the environment. As previously mentioned, environmental variables play a major role in determining seed composition and could mask the breeding results. Regarding characteristic correlations, protein content has been correlated with seed yield, seed weight, dry matter mass, oil and hull content, kernel yield, lysine content, and other qualitative and quantitative characteristics (Fick and Miller, 1997; Hu et al., 2010; Jiménez et al., 1985; Miller et al., 1992; Radić et al., 2009). The most important correlation is that of oil content, because of higher price and main industrial use of sunflower oil. Most authors reported a negative correlation between the two traits, as selection of high protein content seeds usually results in lower oil concentration (Alba and Greco, 1981; Alba et al., 1980; Jiménez et al., 1985; López Pereira et al., 1999; Rojas and Fernández-Martínez, 1998; Seiler, 2007; Tavoljanskiy et al., 2004). Different results have been observed on the level and strength of the established correlations. Some authors did not confirm the existence of this antagonism and described the possibility of breeding to improve protein content of sunflower kernels while maintaining acceptable oil concentration (Bedov, 1985; Ivanov and Stoyanova, 1978; Radić et al., 2009; Seiler, 2007). Another important negative correlation to be considered when breeding for protein is that reported between lysine and protein content, likely due to modification in the albumin–globulin ratio (Borodulina and Suprunova, 1976; Girault et al., 1970). Protein fractions differ in their amino acid composition; albumins having more lysine and methionine than globulins (Salunkhe, 1992). Thus, a modification in the albumin–globulin ratio will result in a variation in the lysine content. A negative correlation with protein concentration for lysine, methionine, and cystine has also been observed in legumes and cereals (Gottschalk and Müller, 1983; Monti and Grillo, 1983; Rharrabti et al., 2001).

World collections are the main sources of genetic variability for developing improved cultivars. The largest world collections are located in the USA and Russia. The sunflower germplasm collection in the National Plant Germplasm System (NPGS) of the United States is held at the North Central Regional Plant Introduction Station (NCRPIS) in Ames, Iowa. The N.I. Vavilov All-Union Scientific Research Institute of Plant Industry (VIR) is maintained at St. Petersburg (Russia). The mission of these collections is to conserve genetically diverse crop germplasm and to manage and promote the use of germplasm for research, crop improvement, and product development (Vollmann and Rajcan, 2010). Other large collections of wild and cultivated sunflower are maintained in France (Station d'amélioration et santé des plantes, Clermont-Ferrand), Romania (Research Institute for Cereals and Industrial Crops, Fundulea), Serbia (Field and Vegetable Crops, Novi Sad), Russia (Research Institute of Oil-Bearing Plants, Krasnodar), Bulgaria (Institute for Wheat and Sunflower "Dobroudja," General Toshevo), and Italy (Eurisco, Rome).

Conclusion

The worldwide demand for proteins is increasing, both of animal and plant origin. However, animal proteins are expensive in terms of both market price and environmental impact. In addition, consumer confidence in animal products has decreased due to food safety problems related to diseases such as bovine spongiform encephalopathy, dioxin contamination, and to the use of animal hormones. Sunflower seeds are an economically interesting agricultural product because of their chemical composition. Among the various sunflower components other than oil, sunflower protein is the most promising compound with regard to its industrial use. Sunflower seeds are particularly interesting in view of their widespread availability, familiarity to both farmers and processors because of their use as an oil source, and with respect to the seed proteins, their nutritional value, high protein content, and functionality. However, the main bottlenecks are protein denaturation during oil production and the presence of high amounts of phenolic compounds. The numerous publications on sunflower protein isolation clearly indicate the difficulties that occur during recovery of a high quality protein from sunflower. It is clear that the behavior of proteins used in food applications largely depends on their processing history, which will have a direct effect on the physico-chemical, and thus the functional performance, of the proteins. Many lab scale methods that claim to render quality protein preparations (with low phenolic compounds and minimum protein denaturation) have been published. However, so far, food applications of sunflower proteins are still hampered by the lack of an inexpensive nondenaturing large scale method for their recovery. Cooperation between fundamental researchers, technology-oriented institutes, and industry is necessary to make use of methods developed and results from fundamental studies on the molecular structure, physic-chemical properties and functional properties of proteins. This cooperative approach would favor the implementation of large scale preparation methods optimized toward increased protein functionality as it has been done for soy (e.g., adaptation of desolventizing process).

Summarizing, sunflower proteins are promising as a new food protein ingredient, but more applied research using actual food systems is necessary to effectively assess sunflower proteins as a potentially widely used food ingredient.

References

Adams, R. P.; Seiler, G. J. Whole-Plant Utilization of Sunflowers. *Biomass* **1984,** *4,* 69–80.

Ahmad, R.; Saeed, M.; Ullah, E.; Mahmood, T. Effect of Potassium on Protein, Oil and Fatty Acid Contents in Two Autumn Planted Sunflower Hybrids. *Int. J. Agric. Biol.* **1999,** *1,* 325–327.

Aksyonov, I. V. Protein Markers Specify of Sunflower Inbred Lines. *Helia* **2005a,** *28,* 49–53.

Aksyonov, I. V. Use of Albumin Markers for Defining Genetic Purity of Sunflower Parent Lines and Hybrids. *Helia* **2005b,** *28,* 43–48.

Alba, A.; Greco, I. Problems in Simultaneous Breeding for Oil and Protein Content in Sunflower. In *Production and Utilization of Protein in Oilseed Crops;* Bunting E.S., Ed.; Springer Series in World Crops: Production, Utilization and Description; Springer: Netherlands, 1981, 5, pp 43–49.

Alba, A.; Greco, I.; Fiorentino, S.; Lener, M. Preliminary Results on Improvement of the Protein Content of Sunflower Seeds. *Riv. Ital. Sostanze Gr.* **1980,** *57,* 567–571.

Alcocer, M. J. C.; Murtagh, G. J.; Bailey, K.; Dumoulin, M.; Sarabia Meseguer, A.; Parker, M. J.; Archer, D. B. The Disulphide Mapping, Folding and Characterisation of Recombinant Ber e 1, an Allergenic Protein, and SFA8, Two Sulphur-Rich 2S Plant Albumins. *J. Mol. Biol.* **2002,** *324,* 165–175.

Alexander, L. G.; Sessions, R. B.; Clarke, A. R.; Tatham, A. S.; Shewry, P. R.; Napier, J. A. Characterization and Modelling of the Hydrophobic Domain of a Sunflower Oleosin. *Planta* **2002,** *214,* 546–551.

Allen, R.; Arnott, H. Protein Bodies in Dormant, Imbibed and Germinated Sunflower Cotyledons. *Food Microstruct.* **1982,** *1,* 63–73.

Allen, R. D.; Cohen, E. A.; Haar, R. A. V.; Adams, C. A.; Ma, D. P.; Nessler, C. L.; Thomas, T. L. Sequence and Expression of a Gene Encoding an Albumin Storage Protein in Sunflower. *Mol. Genet. Genomics* **1987,** *210,* 211–218.

AllFam. AF050: Prolamin Superfamily. http://www.meduniwien.ac.at/allergens/allfam/factsheet.php?allfam_id=AF050 (September 2013).

Almoguera, C.; Jordano, J. Developmental and Environmental Concurrent Expression of Sunflower Dry-Seed-Stored Low-Molecular-Weight Heat-Shock Protein and Lea mRNAs. *Plant Mol. Biol.* **1992,** *19,* 781–792.

Anderson, P. A. Interactions between Proteins and Constituents That Affect Protein Quality. In *Digestibility and Amino Acid Availability in Cereals and Oilseeds;* Finley, J. W., Hopkins, D. T., Eds.; American Association of Cereal Chemists: St. Paul, MN, 1985; pp 31–46.

Anisimova, I.; Konarev, A. V.; Gavrilova, V.; Rozhkova, V.; Fido, R.; Tatham, A.; Shewry, P. Polymorphism and Inheritance of Methionine-Rich 2S Albumins in Sunflower. *Euphytica* **2003,** *129,* 99–107.

Anisimova, I.; Gavrilova, V.; Loskutov, A.; Rozhkova, V.; Tolmachev, V. Polymorphism and Inheritance of Seed Storage Protein in Sunflower. *Russ. J. Genet.* **2004,** *40,* 995–1002.

Anisimova, I. N.; Gavrilyuk, I. P. Heterogeneity and Polymorphism of 11S Globulin in Sunflower Seeds. *Sov. Genet.* **1990,** *25,* 811–815.

Anisimova, I. N.; Gavriljuk, I. P.; Konarev, V. G. Identification of Sunflower Lines and Varieties by Helianthinin Electrophoresis. *Plant Var. Seeds* **1991a,** *4,* 133–141.

Anisimova, I. N.; Loskutov, A. V.; Borovkova, I. G. Identification of Sunflower Lines by Electrophoresis of Helianthinine and Isozymes. *Sov. Agr. Sci.* **1991b,** *6,* 11–13.

Anisimova, I. N.; Fido, R. J.; Tatham, A. S.; Shewry, P. R. Genotypic Variation and Polymorphism of 2S Albumins of Sunflower. *Euphytica* **1995,** *83,* 15–23.

Anisimova, I. N.; Konarev, A. V.; Gavrilova, V. A.; Rozhkova, V. T.; Fido, R. F.; Tatham, A. S.; Shewry, P. R. Polymorphism and Inheritance of Methionine-Rich 2S Albumins in Sunflower. *Euphytica* **2002,** *129,* 99–107.

Aravind, L.; Koonin, E. V. DNA-Binding Proteins and Evolution of Transcription Regulation in the Archaea. *Nucleic Acids Res.* **1999a,** *27,* 4658–4670.

Aravind, L.; Koonin, E. V. Gleaning Non-Trivial Structural, Functional and Evolutionary Information about Proteins by Iterative Database Searches. *J. Mol. Biol.* **1999b,** *287,* 1023–1040.

Arntfield, S.; Murray, E.; Ismond, M. Effect of Salt on the Thermal Stability of Storage Proteins from Fababean (*Vicia faba*). *J. Food Sci.* **1986,** *51,* 371–377.

Assogna, A.; Patricelli, A.; Sodini, G.; Emmi, E. Factors Affecting the Extraction of Chlorogenic Acid from Defatted Sunflower Meal. *Food Sci. Technol.* **1979,** *12,* 262–265.

Balasaraswathi, R.; Sadasivam, S. Changes in Oil, Sugars and Nitrogenous Components during Germination of Sunflower Seeds, *Helianthus annuus. Plant Food Hum. Nutr.* **1997,** *51,* 71–77.

Bau, H. M.; Debry, G. Colorless Sunflower Protein Products: Chemical and Nutritional Evaluation of the Presence of Phenolic Compounds. *J. Food Technol.* **1980,** *15,* 207–215.

Bau, H. M.; Mohtadi Nia, D. J.; Mejean, L.; Debry, G. Preparation of Colorless Sunflower Protein Products: Effect of Processing on Physicochemical and Nutritional Properties. *Am. Oil Chem. Soc.* **1983,** *60,* 1141–1148.

Baudet, J.; Mosse, J. Fractionation of Sunflower Seed Proteins. *Am. Oil Chem. Soc.* **1977,** *54,* A82–A86.

Baur, J. A.; Sinclair, D. A. Therapeutic Potential of Resveratrol: The *In Vivo* Evidence. *Nature Reviews Drug Discovery* **2006,** *5,* 493–506.

Bedov, S. A Study of Combining Ability for Oil and Protein Contents in Seed of Different Sunflower Inbreds. In *Proceedings of 11th International Sunflower Conference,* Mar del Plata, Argentine, March 6–10, 1985; Argentine Sunflower Association (ASAGIR), 1985; pp 675–682.

Berecz, B.; Mills, E. C.; Tamás, L. S.; Láng, F.; Shewry, P. R.; Mackie, A. R. Structural Stability and Surface Activity of Sunflower 2S Albumins and Nonspecific Lipid Transfer Protein. *J. Agric. Food Chem.* **2010,** *58,* 6490–6497.

Berecz, B.; Clare Mills, E.; Parádi, I.; Láng, F.; Tamás, L.; Shewry, P. R.; Mackie, A. R. Stability of Sunflower 2S Albumins and LTP to Physiologically Relevant In Vitro Gastrointestinal Digestion. *Food Chem.* **2013,** *138,* 2374–2381.

Berot, S.; Briffaud, J. Parameters for Obtaining Concentrates from Rapeseed and Sunflower Meal. *Plant Food Hum. Nutr.* **1983,** *33,* 237–242.

Besler, M.; Hefle, S. L.; Jensen-Jarolim, E. Sunflower Seed (*Helianthus annuus*). Internet Symposium on Food Allergens 2001, 3, 103–114.

Binet, M.-N.; Weil, J.-H.; Tessier, L.-H. Structure and Expression of Sunflower Ubiquitin Genes. *Plant Mol. Biol.* **1991,** *17,* 395–407.

Blamey, F.; Chapman, J. Protein, Oil, and Energy Yields of Sunflower as Affected by N and P Fertilization. *Agron. J.* **1981,** *73,* 583–587.

Blanchet, R. Main Agronomical and Physiological Characters Determining the Production of Sunflower Seeds. *Rev. Fr. Corps Gras.* **1987,** *34,* 75–80.

Blanchet, R.; Merrien, A. Influence of Water Supply on Assimilation, Yield Components and Oil-Protein Production of Sunflower. In *Proceedings of the Sunflower Workshop, EEC Plant Protein Programme,* Athens, Greece, November 23–24, 1982; Plant Breeding Institute, University of Bari, 1982; pp 185–201.

Blanchet, R.; Gelfi, N.; Dufourcq, D.; Hernandez, M.; Decau, J. Nitrogen Nutrition and Yield Production in Sunflowers (*Helianthus annuus* L.). *Agrochimica* **1986,** *30,* 480–489.

Boeckmann, B.; Bairoch, A.; Apweiler, R.; Blatter, M.-C.; Estreicher, A.; Gasteiger, E.; Martin, M. J.; Michoud, K.; O'Donovan, C.; Phan, I.; et al. The SWISS-PROT Protein Knowledgebase and Its Supplement TrEMBL in 2003. *Nucleic Acids Res.* **2003,** *31,* 365–370.

Bolyakina, Y. P.; Raikhman, L. Structural Aspects of Protein Accumulation in Developing Sunflower Seeds. *Russ. J. Plant Physiol.* **1999,** *46,* 340–348.

Borodulina, A. A.; Suprunova, L. V. Change of Sunflower Seed Protein Complex Qualitative Composition Depending on Genotype and Ecological factors. In *Proceedings of the 7th International Sunflower Conference,* Krasnodar, Russia, 27 June–3 July, 1976; pp 173–175.

Bretting, P.; Widrlechner, M. P. Genetic Markers and Horticultural Germplasm Management. *Hort Science* **1995,** *30,* 1349–1356.

Brunick, R. L. Seed Dormancy in Domesticated and Wild Sunflowers (*Helianthus annuus* L.): Types, Longevity and QTL Discovery. Ph.D. Thesis, Oregon State University, 2008.

Burnett, G. R.; Rigby, N. M.; Mills, E. N. C.; Belton, P. S.; Fido, R. J.; Tatham, A. S.; Shewry, P. R. Characterization of the Emulsification Properties of 2S Albumins from Sunflower Seed. *J. Colloid Interface Sci.* **2002,** *247,* 177–185.

Burton-Freeman, B. Postprandial Metabolic Events and Fruit-Derived Phenolics: A Review of the Science. *British Journal of Nutrition* **2010,** *104,* S1–S14.

Castellani, O. F.; Martínez, E. N.; Añón, M. C. Structural Modifications of an Amaranth Globulin Induced by pH and NaCl. *J. Agric. Food Chem.* **1998,** *46,* 4846–4853.

Castillejo, M.; Maldonado, A. M.; Ogueta, S.; Jorrín, J. V. Proteomic Analysis of Responses to Drought Stress in Sunflower (*Helianthus annuus*) Leaves by 2DE Gel Electrophoresis and Mass Spectrometry. *Open Proteomics Journal* **2008,** *1,* 59–71.

Cater, C. M.; Gheyasuddin, S.; Mattil, K. F. The Effect of Chlorogenic, Quinic, and Caffeic Acids on the Solubility and Color of Protein Isolates, Especially from Sunflower Seed. *Cereal Chem.* **1972,** *49,* 508–514.

Chaudhry, A. U.; Mushtaq, M. Optimization of Potassium in Sunflower. *Pakistan J. Biol. Sci.* **1999,** *2,* 887–888.

Chelikani, P.; Fita, I.; Loewen, P. Diversity of Structures and Properties among Catalases. *Cell. Mol. Life Sci.* **2004,** *61,* 192–208.

Choi, S.-M.; Ma, C.-Y. Conformational Study of Globulin from Common Buckwheat (*Fagopyrum esculentum* Moench) by Fourier Transform Infrared Spectroscopy and Differential Scanning Calorimetry. *J. Agric. Food Chem.* **2005,** *53,* 8046–8053.

Christiansen, P.; Gibson, J. M.; Moore, A.; Pedersen, C.; Tabe, L.; Larkin, P. J. Transgenic Trifolium Repens with Foliage Accumulating the High Sulphur Protein, Sunflower Seed Albumin. *Transgenic Res.* **2000,** *9,* 103–113.

Christov, M.; Ivanova, I.; Ivanov, P. Some Characteristics of the Helianthus Species in the Dobroudja Collection. I, Protein content and amino acid composition in proteins. *Helia* **1993,** *16,* 63–70.

Claughton, S. M.; Pearce, R. J. Preparation and Properties of Acid-Modified Sunflower Protein Isolate. *J. Food Sci.* **1989,** *54,* 357–361.

Colgrave, M. L.; Korsinczky, M. J.; Clark, R. J.; Foley, F.; Craik, D. J. Sunflower Trypsin Inhibitor-1, Proteolytic Studies on a Trypsin Inhibitor Peptide and Its Analogs. *Peptide Science* **2010,** *94,* 665–672.

Connor, D.; Sadras, V. Physiology of Yield Expression in Sunflower. *Field Crop. Res.* **1992,** *30,* 333–389.

Connor, D.; Hall, A.; Sadras, V. Effect of Nitrogen Content on the Photosynthetic Characteristics of Sunflower Leaves. *Funct. Plant Biol.* **1993,** *20,* 251–263.

Connor, D. J.; Hall, A. Sunflower Physiology. In *Sunflower Technology and Production;* Schneiter, A. A., Ed.; Agronomy Monograph 35; American Society of Agronomy: Madison, WI, 1997; pp 113–182.

Connor, D. J.; Jones, T. R.; Palta, J. A. Response of Sunflower to Strategies of Irrigation I. Growth, Yield and the Efficiency of Water-Use. *Field Crop. Res.* **1985,** *10,* 15–36.

Cooke, R. J. Gel Electrophoresis for the Identification of Plant Varieties. *J. Chromatogr. A* **1995,** *698,* 281–299.

Dalgalarrondo, M.; Raymond, J.; Azanza, J. L. Sunflower Seed Proteins: Characterization and Subunit Composition of the Globulin Fraction. *J. Exp. Bot.* **1984,** *35,* 1618–1628.

Dalgalarrondo, M.; Raymond, J.; Azanza, J. L. Sunflower Seed Protein: Size and Charge Heterogeneity in Subunits of the Globulin Fraction. *Biochimie* **1985,** *67,* 629–632.

Daly, N. L.; Chen, Y.-K.; Foley, F. M.; Bansal, P. S.; Bharathi, R.; Clark, R. J.; Sommerhoff, C. P.; Craik, D. J. The Absolute Structural Requirement for a Proline in the P3′-Position of Bowman-Birk Protease Inhibitors Is Surmounted in the Minimized SFTI-1 Scaffold. *J. Biol. Chem.* **2006,** *281,* 23668–23675.

Danilenko, A. N.; Bikbov, T. M.; Grinberg, N. V.; Leont'ev, A. L.; Burova, T. V.; Surikov, V. V.; Borisov, Y. A.; Tolstoguzov, V. B. Effect of pH on the Thermal Stability of 11S-Globulin of *Glycine max* Seeds as Indicated by Differential Scanning Microcalorimetry. *Biophysics* **1987,** *32,* 434–439.

da Silva, M. A.; Garcia, J. S.; de Souza, G. H.; Eberlin, M. N.; Gozzo, F. C.; Arruda, M. A. Evaluation of Sample Preparation Protocols for Proteomic Analysis of Sunflower Leaves. *Talanta* **2010,** *80,* 1545–1551.

Decau, J.; Pujol, B.; Lencrerot, P. Essai d'Application au Champ de Données Physiologiques Relative à la Influence de l'Alimentation Azotée sur la Comopsition Protéique de la Graine du Tornesol (*Helianthus annuus*). *C. R. Seances Acad. Sci. D* **1972,** *275,* 1231–1234.

Derbyshire, E.; Wright, D. J.; Boulter, D. Legumin and Vicilin, Storage Proteins of Legume Seeds. *Phytochemistry* **1976,** *15,* 3–24.

de Sousa Barbosa, H.; de Souza, D. L. Q.; Koolen, H. H. F.; Gozzo, F. C.; Arruda, M. A. Z. Sample Preparation Focusing on Plant Proteomics: Extraction, Evaluation and Identification of Proteins from Sunflower Seeds. *Analytical Methods* **2013,** *5,* 116–123.

Dorrell, D. G. Genetic and Environmental Modification of the Chlorogenic Acid Content of Sunflower Seeds. In *Proceedings of the 6th International Sunflower Conference,* Bucharest, Romania, July 22–24, 1974; Alex. Viorel Vranceanu: Bucarest, 1974; pp 325–328.

Dorrell, D. G. Chlorogenic Acid Content of Meal from Cultivated and Wild Sunflowers. *Crop Sci.* **1976,** *16,* 422–424.

Dreher, M.; Padmanaban, G. Sunflower Hull Flour as a Potential Dietary Fiber Supplement. *J. Food Sci.* **1983,** *48,* 1463–1465.

Duca, M.; Levitchi, A.; Anisimova, I. Polymorphism of Helianthinin Polypeptides from Various Sunflower Genotypes. Biotechnol. *Biotechnol. Equip.* **2008,** *22,* 897–900.

Dulermo, T.; Bligny, R.; Gout, E.; Cotton, P. Amino Acid Changes during Sunflower Infection by the Necrotrophic Fungus B. Cinerea. *Plant Signaling and Behavior* **2009,** *4,* 859–861.

Dunwell, J. M. Cupins: A New Superfamily of Functionally Diverse Proteins That Include Germins and Plant Storage Proteins. Biotechnol. *Genet. Eng. Rev.* **1998,** *15,* 1–32.

Dunwell, J. M.; Gane, P. J. Microbial Relatives of Seed Storage Proteins: Conservation of Motifs in a Functionally Diverse Superfamily of Enzymes. *J. Mol. Evol.* **1998,** *46,* 147–154.

Dunwell, J. M.; Purvis, A.; Khuri, S. Cupins: The Most Functionally Diverse Protein Superfamily? *Phytochemistry* **2004,** *65,* 7–17.

Durante, M.; Bernardi, R.; Lupi, M. C.; Sabelli, P. Characterization of *Helianthus annuus* L. Storage Proteins. *J. Agric. Food Chem.* **1989,** *37,* 852–855.

Earle, F.; Vanetten, C.; Clark, T.; Wolff, I. Compositional Data on Sunflower Seed. *Am. Oil Chem. Soc.* **1968,** *45,* 876–879.

Egorov, T. A.; Odintsova, T. J.; Musolyamov, A. K.; Fido, R.; Tatham, A. S.; Shewry, P. R. Disulphide Structure of a Sunflower Seed Albumin: Conserved and Variant Disulphide Bonds in the Cereal Prolamin Superfamily. *FEBS Lett.* **1996,** *396,* 285–288.

Ellepola, S. W.; Choi, S. M.; Ma, C. Y. Conformational Study of Globulin from Rice (*Oryza Sativa*) Seeds by Fourier-Transform Infrared Spectroscopy. *Int. J. Biol. Macromol.* **2005,** 37, 12–20.

Esau, K. Anatomy of seed plants. *Soil Sci.* **1960,** *90,* 149.

ExPASy. 11S Seed Storage. http://prosite.expasy.org/PS00305 (September 2013).

Fan, T. Y.; Sosulski, F. W.; Hamon, N. W. New Techniques for Preparation of Improved Sunflower Protein Concentrates. *Cereal Chem.* **1976,** *53,* 118–125.

FAO. Food Outlook: Global Market Analysis. http://www.fao.org/docrep/016/al993e/al993e00.pdf (December 2012).

Fereres, E.; Gimenez, C.; Fernandez, J. Genetic Variability in Sunflower Cultivars under Drought. I. Yield Relationships. *Crop Pasture Sci.* **1986,** *37,* 573–582.

Fernández-Martínez, J. M.; Pérez-Vich, B.; Velasco, L. Sunflower. In *Oil Crops: Handbook of Plant Breeding 4;* Springer: New York, 2010; pp 155–232.

Fick, G. N.; Miller, J. F. Sunflower Breeding. In *Sunflower Technology and Production;* Schneiter, A. A., Ed.; Agronomy Monograph 35; American Society of Agronomy: Madison, WI, 1997; pp 395–439.

Folawiyo, Y. L.; Owusu Apenten, R. K. The Effect of Heat- and Acid-Treatment on the Structure of Rapeseed Albumin (Napin). *Food Chem.* **1997,** *58,* 237–243.

Fukushima, D. Structures of Plant Storage Proteins and Their Functions. *Food Rev. Int.* **1991,** *7,* 353–381.

Gachon, L. Cinétique de l'Absorption des Éléments Nutritifs Majeurs Chez le Tournesol. *Ann. Agron.* **1972,** *23,* 547–566.

Gallant, D. J.; Saio, K. Study of Sunflower Protein Bodies Using Scanning Electron-Microscopy and Energy-Dispersive Microanalysis. *Biol. Cell.* **1979,** *35,* 8a.

Gassmann, B. Preparation and Application of Vegetable Proteins, Especially Proteins from Sunflower Seed, for Human Consumption. An Approach. *Mol. Nutr. Food Res.* **1983,** *27,* 351–369.

German, B.; Damodaran, S.; Kinsella, J. E. Thermal Dissociation and Association Behavior of Soy Proteins. *J. Agric. Food Chem.* **1982,** *30,* 807–811.

German, J. B.; Phillips, L. Protein Interactions in Foams. In *Protein Functionality in Food Systems;* Hettiarachy, N. S., Ziegler, G. R., Eds.; IFT Basic Symposium Series: Chicago, 1991; pp 181–208.

Gheyasuddin, S. Studies of the Proteins and the Polyphenols of Sunflower Seed (*Helianthus annuus*). *Dissertation Abstracts International. Section B. Physical Sci. Tech.* **1970,** *31,* 2451–2452.

Gheyasuddin, S.; Cater, C. M.; Mattil, K. F. Preparation of a Colourless Sunflower Protein Isolate. *Food Technol.* **1970a,** *24,* 242–243.

Gheyasuddin, S.; Cater, C. M.; Mattil, K. F. Effect of Several Variables on the Extractability of Sunflower Seed Proteins. *J. Food Sci.* **1970b,** *35,* 453–456.

Gimenez, C.; Fereres, E. Genetic Variability in Sunflower Cultivars under Drought. II. Growth and Water Relations. *Crop Pasture Sci.* **1986,** *37,* 583–597.

Gimenez, C.; Connor, D. J.; Rueda, F. Canopy Development, Photosynthesis and Radiation-Use Efficiency in Sunflower in Response to Nitrogen. *Field Crop. Res.* **1994,** *38,* 15–27.

Girault, A.; Baudet, J.; Mosse, J. Study of the Sunflower Seed Protein in View of Improving Their Lysine Content. In *Improving Plant Protein by Nuclear Techniques;* International Atomic Energy Agency: Vienna, Austria, 1970; pp 275–285.

Goffner, D.; Cazalis, R.; Dusert, C. P.; Calmes, J.; Cavalie, G. C-14 Photoassimilate Partitioning in Developing Sunflower Seeds. *J. Exp. Bot.* **1988a,** *39,* 1411–1420.

Goffner, D.; This, P.; Raynal, M.; Chartier, Y.; Delseny, M. Identification and Accumulation of Sunflower (*Helianthus annuus*) Seed Storage Proteins: Effects of ABA and Osmotica. *Plant Physiol. Biochem.* **1988b,** *26,* 232–233.

Gonorazky, A. G.; Regente, M. C.; de la Canal, L. Stress Induction and Antimicrobial Properties of a Lipid Transfer Protein in Germinating Sunflower Seeds. *J. Plant Physiol.* **2005,** *162,* 618–624.

González-Pérez, S. Physico-Chemical and Functional Properties of Sunflower Proteins. Ph.D. Thesis, Wageningen Universiteit, 2003.

González-Pérez, S.; Vereijken, J. M. Sunflower Proteins: Overview of Their Physicochemical, Structural and Functional Properties. *J. Sci. Food Agric.* **2007,** *87,* 2173–2191.

González-Pérez, S.; Arellano, J. B. Vegetable Protein Isolates. In *Handbook of Hydrocolloids;* Phillips, G.O., Williams, P. A., Eds.; Woodhead Publishing Series in Food Science, Technology and Nutrition No 173; Woodhead Publishing: Cambridge, UK, 2009; pp 383–419.

González-Pérez, S.; Merck, K. B.; Vereijken, J. M.; Van Koningsveld, G. A.; Gruppen, H.; Voragen, A. G. J. Isolation and Characterization of Undenatured Chlorogenic Acid-Free Sunflower (*Helianthus annuus*) Proteins. *J. Agric. Food Chem.* **2002,** *50,* 1713–1719.

González-Pérez, S.; Vereijken, J. M.; Merck, K. B.; van Koningsveld, G. A.; Gruppen, H.; Voragen, A. G. Conformational States of Sunflower (*Helianthus annuus*) Helianthinin: Effect of Heat and pH. *J. Agric. Food Chem.* **2004,** *52,* 6770–6778.

González-Pérez, S.; van Konignsveld, G. A.; Vereijken, J. M.; Merck, K. B.; Gruppen, H.; Voragen, A. G. Emulsion Properties of Sunflower (*Helianthus annuus*) Proteins. *J. Agric. Food Chem.* **2005a,** 53, 2261–2267.

González-Pérez, S.; Vereijken, J. M.; Koningsveld, G. A.; Gruppen, H.; Voragen, A. G. Physicochemical Properties of 2S Albumins and the Corresponding Protein Isolate from Sunflower (*Helianthus annuus*). *J. Food Sci.* **2005b,** *70,* C98–C103.

González-Pérez, S.; Vereijken, J. M.; van Koningsveld, G. A.; Gruppen, H.; Voragen, A. G. Formation and Stability of Foams Made with Sunflower (*Helianthus annuus*) Proteins. *J. Agric. Food Chem.* **2005c,** 53, 6469–6476.

Gorter, K. Identity of Helianthic Acid with Chlorogenic Acid. *Arch. Pharm.* **1909,** *247,* 436–438.

Gottschalk, W.; Müller, H. P. *Seed Proteins: Biochemistry, Genetics, Nutritive Value;* Martinus Nijhoff: The Hague, The Netherlands, 1983.

Guéguen, J.; Chevalier, M.; Barbot, J.; Schaeffer, F. Dissociation and Aggregation of Pea Legumin Induced by pH and Ionic Strength. *J. Sci. Food Agric.* **1988,** *44,* 167–182.

Guéguen, J.; Popineau, Y.; Anisimova, I. N.; Fido, R. J.; Shewry, P. R.; Tatham, A. S. Functionality of the 2S Albumin Seed Storage Proteins from Sunflower (*Helianthus annuus* L.). *J. Agric. Food Chem.* **1996,** *44,* 1184–1189.

Gulya, T.; Rashid, K. Y.; Masirevic, S. M. Sunflower Diseases. In *Sunflower Technology and Production;* Schneiter, A. A., Ed.; Agronomy Monograph 35; American Society of Agronomy: Madison, WI, 1997; pp 263–379.

Hagan, N.; Upadhyaya, N.; Tabe, L.; Higgins, T. The Redistribution of Protein Sulfur in Transgenic Rice Expressing a Gene for a Foreign, Sulfur-Rich Protein. *Plant J.* **2003,** *34,* 1–11.

Hagenmaier, R. D. Aqueous Processing of Full-Fat Sunflower Seeds: Yields of Oil and Protein. *J. Am. Oil Chem. Soc.* **1974,** *51,* 470–471.

Hajduch, M.; Casteel, J. E.; Tang, S.; Hearne, L. B.; Knapp, S.; Thelen, J. J. Proteomic Analysis of Near-Isogenic Sunflower Varieties Differing in Seed Oil Traits. *J. Proteome Res.* **2007,** *6,* 3232–3241.

Harwalkar, V.; Ma, C. Study of Thermal Properties of Oat Globulin by Differential Scanning Calorimetry. *J. Food Sci.* **1987,** *52,* 394–398.

Herman, E. M.; Larkins, B. A. Protein Storage Bodies and Vacuoles. *Plant Cell* **1999,** *11,* 601–613.

Hermansson, A. Soy Protein Gelation. *J. Am. Oil Chem. Soc.* **1986,** *63,* 658–666.

Hu, J.; Seiler, G.; Kole, C. *Genetics, Genomics and Breeding of Sunflower;* Science Publishers: Enfield, NH, 2010.

Huang, A. H. Oil Bodies and Oleosins in Seeds. *Annu. Rev. Plant Biol.* **1992,** *43,* 177–200.

Huang, A. H. Structure of Plant Seed Oil Bodies. *Curr. Opin. Struct. Biol.* **1994,** *4,* 493–498.

Huang, A. H. C. Oleosins and Oil Bodies in Seeds and Other Organs. *Plant Physiol.* **1996,** *110,* 1055–1061.

Islam, N.; Upadhyaya, N. M.; Campbell, P. M.; Akhurst, R.; Hagan, N.; Higgins, T. J. Decreased Accumulation of Glutelin Types in Rice Grains Constitutively Expressing a Sunflower Seed Albumin Gene. *Phytochemistry* **2005,** *66,* 2534–2539.

Ivanov, P. Variation of the Protein, Lysine and Chlorogenic Acid in Some Sunflower Inbred Lines. *Rastenievud Nauk* **1975,** *10,* 23–27.

Ivanov, P.; Stoyanova, Y. Results from Sunflower Breeding Directed to Obtaining Gene Materials of High Protein Content in the Kernel. In *Proceedings of the 8th International Sun-*

flower Conference, Minneapolis, Minnesota, July 23–27, 1978; Australian Sunflower Association: Toowoomba, 1978; pp 23–27.

Jacob, J.; Lawlor, D. Stomatal and Mesophyll Limitations of Photosynthesis in Phosphate Deficient Sunflower, Maize and Wheat Plants. *J. Exp. Bot.* **1991,** *42,* 1003–1011.

Jang, M.; Cai, L.; Udeani, G. O.; Slowing, K. V.; Thomas, C. F.; Beecher, C. W.; Fong, H. H.; Farnsworth, N. R.; Kinghorn, A. D.; Mehta, R. G.; Moon, R. C.; et al. Cancer Chemopreventive Activity of Resveratrol, a Natural Product Derived from Grapes. *Science* **1997,** *275,* 218–220.

Jasso de Rodríguez, D.; Romero-García, J.; Rodríguez-García, R.; Angulo-Sanchez, J. Characterization of Proteins from Sunflower Leaves and Seeds: Relationship of Biomass and Seed Yield. In *Trends in New Crops and New Uses;* Janick, J., Whipkey, A., Eds.; ASHS Press: Alexandria, VA, 2002; pp 143–149.

Jiang, L.; Phillips, T. E.; Hamm, C. A.; Drozdowicz, Y. M.; Rea, P. A.; Maeshima, M.; Rogers, S. W.; Rogers, J. C. The Protein Storage Vacuole: A Unique Compound Organelle. *J. Cell Biol.* **2001,** *155,* 991–1002.

Jiménez, A.; Fernández-Martínez, J. M.; Domínguez, J.; Gimeno, V.; Alcántara, M. Considerations in Breeding for Protein Yield in Sunflower. In *Eucarpia Section Oil Protein Crops;* Cordoba, Spain, 1985; pp 46–60.

Joubert, F. J. Sunflower Seed Proteins. *Biochim Biophys Acta.* **1955,** *16,* 520–523.

Jung, J.-L.; Fritig, B.; Hahne, G. Sunflower (*Helianthus annuus* L.) Pathogenesis-Related Proteins. Induction by Aspirin (Acetylsalicylic Acid) and Characterization. *Plant Physiol.* **1993,** *101,* 873–880.

Kabirullah, M.; Wills, R. B. H. Characterization of Sunflower Protein. *J. Agric. Food Chem.* **1983,** *31,* 953–956.

Kader, J.-C. Lipid-Transfer Proteins in Plants. *Annu. Rev. Plant Biol.* **1996,** *47,* 627–654.

Kader, J.-C. Lipid-Transfer Proteins: A Puzzling Family of Plant Proteins. *Trends Plant Sci.* **1997,** *2,* 66–70.

Kelly, J.; Hefle, S. 2S Methionine-Rich Protein (SSA) from Sunflower Seed Isan IgE-Binding Protein. *Allergy* **2000,** *55,* 556–560.

Kelly, J. D.; Hlywka, J. J.; Hefle, S. L. Identification of Sunflower Seed IgE-Binding Proteins. Int. Arch. *Allergy Immunol.* **2000,** *121,* 19–24.

Khan, M. R. I.; Ceriotti, A.; Tabe, L.; Aryan, A.; McNabb, W.; Moore, A.; Craig, S.; Spencer, D.; Higgins, T. J. V. Accumulation of a Sulphur-Rich Seed Albumin from Sunflower in the Leaves of Transgenic Subterranean Clover (*Trifolium subterraneum* L.). *Transgenic Res.* **1996,** *5,* 179–185.

Khuri, S.; Bakker, F. T.; Dunwell, J. M. Phylogeny, Function, and Evolution of the Cupins, a Structurally Conserved, Functionally Diverse Superfamily of Proteins. *Mol. Biol. Evol.* **2001,** *18,* 593–605.

Kim, K. S.; Kim, S.; Yang, H. J.; Kwon, D. Y. Changes of Glycinin Conformation Due to pH, Heat and Salt Determined by Differential Scanning Calorimetry and Circular Dichroism. *Int. J. Food Sci. Technol.* **2004,** *39,* 385–393.

Kleff, S.; Sander, S.; Mielke, G.; Eising, R. The Predominant Protein in Peroxisomal Cores of Sunflower Cotyledons Is a Catalase That Differs in Primary Structure from the Catalase in the Peroxisomal Matrix. *Eur J. Biochem.* **1997,** *245,* 402–410.

Knowles, P. F. Morphology and Anatomy. In *Sunflower Science and Technology;* Carter, J. F., Ed.; Agronomy Monograph 19; American Society of Agronomy: Madison, WI, 1978; pp 55–85.

Konarev, A. V.; Anisimova, I.; Gavrilova, V.; Rozhkova, V.; Fido, R.; Tatham, A.; Shewry, P. Novel Proteinase Inhibitors in Seeds of Sunflower (*Helianthus annuus* L.): Polymorphism, Inheritance and Properties. *Theor. Appl. Genet.* **2000,** *100,* 82–88.

Korsinczky, M. L.; Schirra, H. J.; Craik, D. J. Sunflower Trypsin Inhibitor-1. *Curr. Protein Pept. Sci.* **2004,** *5,* 351–364.

Kortt, A. A; Caldwell, J. B. Low Molecular Weight Albumins from Sunflower Seed: Identification of a Methionine-Rich Albumin. *Phytochemistry* **1990a,** *29,* 2805–2810.

Kortt, A. A.; Caldwell, J. B. Sunflower 11S Globulin, Susceptibility to Proteolytic Cleavage of the Subunits of Native Helianthinin during Isolation: HPLC Fractionation of the Subunits. *Phytochemistry* **1990b,** *29,* 1389–1396.

Kortt, A. A.; Caldwell, J. B.; Lilley, G. G.; Higgins, T. J. Amino Acid and cDNA Sequences of a Methionine-Rich 2S Protein from Sunflower Seed (*Helianthus annuus* L.). *Eur J. Biochem.* **1991,** *195,* 329–334.

Kroll, J.; Rawel, H. M.; Seidelmann, N. Physicochemical Properties and Susceptibility to Proteolytic Digestion of Myoglobin-Phenol Derivatives. *J. Agric. Food Chem.* **2000,** *48,* 1580–1587.

Kroll, J.; Rawel, H. M.; Rohn, S.; Czajka, D. Interactions of Glycinin with Plant Phenols: Influence on Chemical Properties and Proteolytic Degradation of the Proteins. *Mol. Nutr. Food Res.* **2001,** *45,* 388–389.

Kroll, J.; Rawel, H. M.; Rohn, S. Reactions of Plant Phenolics with Food Proteins and Enzymes under Special Consideration of Covalent Bonds. *Food Sci. Technol. Res.* **2003,** *9,* 205–218.

Krzyzaniak, A.; Burova, T.; Haertle, T.; Barciszewski, J. The Structure and Properties of Napin-Seed Storage Protein from Rape (*Brassica napus* L.). *Mol. Nutr. Food Res.* **1998,** *42,* 201–204.

Lacey, D. J. Secondary Structure of Oleosins in Oil Bodies Isolated from Seeds of Safflower (*Carthamus tinctorius* L.) and Sunflower (*Helianthus annuus* L.). *Biochem. J.* **1998,** *334,* 469–477.

Laferriere, J. Interspecific Hybridization in Sunflowers: An Illustration of the Importance of Wild Genetic Resources in Plant Breeding. *Outlook Agric.* **1986,** *15,* 104–109.

Lakemond, C. M. M. *Heat Denaturation of Soy Glycinin: Structural Characteristics in Relation to Aggregation and Gel Formation.* Ph.D. Thesis, Wageningen University, Netherlands, 2001.

Lakemond, C. M. M.; de Jongh, H. H. J.; Hessing, M.; Gruppen, H.; Voragen, A. G. J. Heat Denaturation of Soy Glycinin: Influence of pH and Ionic Strength on Molecular Structure. *J. Agric. Food Chem.* **2000a,** *48,* 1991–1995.

Lakemond, C. M. M.; de Jongh, H. H. J.; Hessing, M.; Gruppen, H.; Voragen, A. G. J. Soy Glycinin: Influence of pH and Ionic Strength on Solubility and Molecular Structure at Ambient Temperatures. *J. Agric. Food Chem.* **2000b,** *48,* 1985–1990.

Lehmann, K.; Schweimer, K.; Reese, G.; Randow, S.; Suhr, M.; Becker, W.-M.; Vieths, S.; Rösch, P. Structure and Stability of 2S Albumin-Type Peanut Allergens: Implications for the Severity of Peanut Allergic Reactions. *Biochem. J.* **2006,** *395,* 463–472.

Lencrerot, P.; Decau, J.; Cochard, B.; Marty, J. R.; Pujol, B. Contribution to Investigation on Effects of Planting and Growth Conditions of Sunflower with Regard to Quantitative Yield and Oleoprotein Composition of Its Seed. *Ann. Agron.* **1973,** *24,* 339–358.

Lencrerot, P.; Decau, J.; Puech, J. Oil and Protein-Content in Sunflower Achene (*Helianthus annuus* L.) Variations during Grain Filling, with Nitrogen Supply and Achenes Location in Inflorescence. *C R Acad Sci Hebd Seances Acad Sci D.* **1977,** *284,* 907–910.

Li, M.; Keddie, J.; Smith, L.; Clark, D.; Murphy, D. Expression and Characterization of the N-Terminal Domain of an Oleosin Protein from Sunflower. *J. Biol. Chem.* **1993,** *268,* 17504–17512.

Li-Chan, E.; Ma, C.-Y. Thermal Analysis of Flaxseed (*Linum usitatissimum*) Proteins by Differential Scanning Calorimetry. *Food Chem.* **2002,** *77,* 495–502.

Lin, J.; Fido, R.; Shewry, P.; Archer, D. B.; Alcocer, M. J. The Expression and Processing of Two Recombinant 2S Albumins from Soybean (*Glycine max*) in the Yeast *Pichia pastoris. BBA-Proteins Proteomics* **2004,** *1698,* 203–212.

Long, Y.-Q.; Lee, S.-L.; Lin, C.-Y.; Enyedy, I.; Wang, S.; Li, P.; Dickson, R.; Roller, P. Synthesis and Evaluation of the Sunflower Derived Trypsin Inhibitor as a Potent Inhibitor of the Type II Transmembrane Serine Protease, Matriptase. *Bioorg. Med. Chem. Lett.* **2001,** *11,* 2515–2519.

Loomis, W. D. Overcoming Problems of Phenolics and Quinones in the Isolation of Plant Enzymes and Organelles. *Methods Enzymol.* **1974,** *31,* 528–544.

López Pereira, M.; Sadras, V.; Trápani, N. Genetic Improvement of Sunflower in Argentina between 1930 and 1995. I. Yield and Its Components. *Field Crop. Res.* **1999,** *62,* 157–166.

Luckett, S.; Garcia, R. S.; Barker, J.; Konarev, A. V.; Shewry, P.; Clarke, A.; Brady, R. High-Resolution Structure of a Potent, Cyclic Proteinase Inhibitor from Sunflower Seeds. *J. Mol. Biol.* **1999,** *290,* 525–533.

Luckett, S.; Sessions, R. B.; Michaelson, L.; Naldrett, M. J.; Clarke, A. R.; Fido, R.; Tatham, A. S.; Shewry, P. R. Amino Acid Sequence and Molecular Modelling of a Lipid Transfer Protein from Sunflower (*Helianthus annuus* L.) Seeds. *Protein Pept. Lett.* **2001,** *8,* 241–248.

Madhusudhan, K. T.; Sastry, M. C. S.; Srinivas, H. Effect of Roasting on the Physico-Chemical Properties of Sunflower Proteins. *Food Sci. Technol.* **1986,** *19,* 292–296.

Maga, J. A.; Katz, I. Simple Phenol and Phenolic Compounds in Food Flavor. *Crit. Rev. Food Sci. Nutr.* **1978,** *10,* 323–372.

Mandal, S.; Mandal, R. K. Seed Storage Proteins and Approaches for Improvement of Their Nutritional Quality by Genetic Engineering. *Curr. Sci.* **2000,** *79,* 576–589.

Marchler-Bauer, A.; Anderson, J. B.; Cherukuri, P. F.; DeWeese-Scott, C.; Geer, L. Y.; Gwadz, M.; He, S.; Hurwitz, D. I.; Jackson, J. D.; Ke, Z.; et al. CDD: A Conserved Domain Database for Protein Classification. *Nucleic Acids Res.* **2005,** *33,* D192–D196.

Marcone, M. F. Biochemical and Biophysical Properties of Plant Storage Proteins: A Current Understanding with Emphasis on 11S Seed Globulins. *Food Res. Int.* **1999a,** *32,* 79–92.

Marcone, M. F. A Comparative Study of Thermally Induced Conformational Changes Occurring in Food-Grade Seed Proteins. *Ital. J. Food Sci.* **1999b,** *11,* 193–206.

Marcone, M. F.; Kakuda, Y.; Yada, R. Y. Salt-Soluble Seed Globulins of Dicotyledonous and Monocotyledonous Plants. II. Structural Characterization. *Food Chem.* **1998a,** *63,* 265–274.

Marcone, M. F.; Kakuda, Y.; Yada, R. Y. Salt-Soluble Seed Globulins of Various Dicotyledonous and Monocotyledonous Plants. I. Isolation/Purification and Characterization. *Food Chem.* **1998b,** *62,* 27–47.

Martin, A. H. Mechanical and Conformational Aspects of Protein Layers on Water. Ph.D. Thesis, Wageningen University, Netherlands, 2003.

Martinez, W. H. Functionality of Vegetable Proteins Other Than Soy. *J. Am. Oil Chem. Soc.* **1979,** *56,* 280–284.

Martínez-Force, E.; Cantisan, S.; Serrano-Vega, M. J.; Garces, R. Acyl-Acyl Carrier Protein Thioesterase Activity from Sunflower (*Helianthus annuus* L.) Seeds. *Planta* **2000,** *211,* 673–678.

Mattil, K. F. The Functional Requirements of Proteins for Foods. *J. Am. Oil Chem. Soc.* **1971,** *48,* 477–480.

Mazhar, H.; Quayle, R.; Fido, R. J.; Stobart, A. K.; Napier, J. A.; Shewry, P. R. Synthesis of Storage Reserves in Developing Seeds of Sunflower. *Phytochemistry* **1998,** *48,* 429–432.

McManus, J. P.; Davis, K. G.; Beart, J. E.; Gaffney, S. H.; Lilley, T. H.; Haslam, E. Polyphenol Interactions. Part 1. Introduction; Some Observations on the Reversible Complexation of Polyphenols with Proteins and Polysaccharides. *J. Chem. Soc., Perkin Trans.* **1985,** *2,* 1429–1438.

Meng, G. T.; Ma, C. Y. Thermal Properties of *Phaseolus angularis* (Red Bean) Globulin. *Food Chem.* **2001,** *73,* 453–460.

Merrien, A.; Blanchet, M.; Cavalie, M.; Cruiziat, P.; Morizet, J.; Durrieu, M.; Planchon, M.; Ribeyre, C.; Manichon, H. Physiologie de la Formation du Rendement Chez le Tournesol. *Info Tech Cetiom* **1983,** *83,* 72.

Miflin, B. J. *Plant Molecular and Cell Biology;* Oxford University Press: Oxford, U.K., 1986.

Mikolajczak, K. L.; Smith, C. R., Jr.; Wolff, I. A. Phenolic and Sugar Components of Armavirec Variety Sunflower (*Helianthus annuus*) Seed Meal. *J. Agric. Food Chem.* **1970,** *18,* 27–32.

Milic, B.; Stojanovic, S.; Vucurevic, N.; Turcic, M. Chorogenic and Quinic Acids in Sunflower Meal. *J. Sci. Food Agric.* **1968,** *19,* 108–113.

Miller, J. F.; Fick, G. N. The Genetics of Sunflower. In *Sunflower Technology and Production;* Schneiter, A. A., Ed.; Agronomy Monograph 35; American Society of Agronomy: Madison, WI, 1997; pp 441–496.

Miller, J. F.; Seiler, G. J.; Jan, C. C. Introduced Germplasm Use in Sunflower Inbred and Hybrid Development. In *Use of Plant Introductions in Cultivar Development: Proceedings of a Symposium, Part 2;* Shands, H. L., Weisner, L. E., Eds.; Crop Science Society of America: Madison, WI, 1992; pp 151–166.

Millichip, M.; Tatham, A. S.; Jackson, F.; Griffiths, G.; Shewry, P. R.; Stobart, A. K. Purification and Characterization of Oil-Bodies (Oleosomes) and Oil-Body Boundary Proteins (Oleosins) from the Developing Cotyledons of Sunflower (*Helianthus annuus* L.). *Biochem. J.* **1996,** *314,* 333.

Molina, M. I.; Petruccelli, S.; Anon, M. C. Effect of pH and Ionic Strength Modifications on Thermal Denaturation of the 11S Globulin of Sunflower (*Helianthus annuus*). *J. Agric. Food Chem.* **2004,** *52,* 6023–6029.

Molina, M. I.; Otegui, M.; Petruccelli, S. Sunflower Storage Proteins Are Transported in Dense Vesicles That Contain Proteins Homologous to the Pumpkin Vacuolar Sorting Receptor PV 72. *Electron. J. Biotechnol.* **2006,** *9,* 326–330.
Molvig, L.; Tabe, L. M.; Eggum, B. O.; Moore, A. E.; Craig, S.; Spencer, D.; Higgins, T. J. Enhanced Methionine Levels and Increased Nutritive Value of Seeds of Transgenic Lupins (*Lupinus angustifolius* L.) Expressing a Sunflower Seed Albumin Gene. *Proc. Natl. Acad. Sci. USA* **1997,** *94,* 8393–8398.
Monsalve, R. I.; Villalba, M.; Rico, M.; Shewry, P. R.; Rodríguez, R. The 2S Albumin Proteins. In *Plant Food Allergens;* Mills, E. N. C., Shewry, P. R., Eds.; Blackwell Science: Oxford, UK, 2004; pp 42–56.
Monti, L. M.; Grillo, S. Legume Seed Improvement for Protein Content and Quality. *Plant Food Hum. Nutr.* **1983,** *32,* 253–266.
Moreno, F. J.; Clemente, A. 2S Albumin Storage Proteins: What Makes Them Food Allergens? *Open Biochem J.* **2008,** *2,* 16.
Moreno, F. J.; Maldonado, B. M.; Wellner, N.; Mills, E. Thermostability and In Vitro Digestibility of a Purified Major Allergen 2S Albumin (Ses i 1) from White Sesame Seeds (*Sesamum indicum* L.). *BBA-Proteins Proteomics* **2005a,** *1752,* 142–153.
Moreno, F. J.; Mellon, F. A.; Wickham, M. S.; Bottrill, A. R.; Mills, E. Stability of the Major Allergen Brazil Nut 2S Albumin (Ber e 1) to Physiologically Relevant In Vitro Gastrointestinal Digestion. *FEBS J.* **2005b,** *272,* 341–352.
Morrissey, P. A.; Mulvihill, D. M.; O'Neill, E. M. *Functional Properties of Muscle Proteins;* Elsevier Applied Science Publishers: Barking, England, 1985.
Muren, E.; Ek, B.; Bjork, I.; Rask, L. Structural Comparison of the Precursor and the Mature Form of Napin, the 2S Storage Protein in *Brassica napus. Eur J. Biochem.* **1996,** *242,* 214–219.
Murphy, P. A. Soybean Proteins. In *Soybeans: Chemistry, Production, Processing, and Utilization;* Johnson, L.A., White, P. J., Galloway, R., Eds.; AOCS Publishing: Urbana, IL, 2008; pp 229–267.
Murtagh, G. J.; Dumoulin, M.; Archer, D. B.; Alcocer, M. J. Stability of Recombinant 2S Albumin Allergens In Vitro. *Biochem. Soc. Trans.* **2002,** *30,* 913–915.
Murtagh, G. J.; Archer, D. B.; Dumoulin, M.; Ridout, S.; Matthews, S.; Arshad, S. H.; Alcocer, M. J. C. In Vitro Stability and Immunoreactivity of the Native and Recombinant Plant Food 2S Albumins Ber e 1 and SFA-8. *Clin. Exp. Allergy* **2003,** *33,* 1147–1152.
Muzquiz, M.; Hill, G. D.; Burbano, C.; Cuadrado, C.; Pedrosa, M. M. *Recent Advances of Research in Antinutritional Factors in Legume Seeds and Oilseeds;* Wageningen Academic Publishers: Wageningen, The Netherlands, 2004.
Mylne, J. S.; Colgrave, M. L.; Daly, N. L.; Chanson, A. H.; Elliott, A. G.; McCallum, E. J.; Jones, A.; Craik, D. J. Albumins and Their Processing Machinery Are Hijacked for Cyclic Peptides in Sunflower. *Nat. Chem. Biol.* **2011,** *7,* 257–259.
Nenova, N.; Drumeva, M. Investigation on Protein Content and Amino Acid Composition in the Kernels of Some Sunflower Lines. *Helia* **2012,** *35,* 41–46.
Neucere, N. J.; Jacks, T. J.; Sumrell, G. Interactions of Globular Protein with Simple Polyphenols. *J. Agric. Food Chem.* **1978,** *26,* 214–216.

Nikolić, Z.; Vujaković, M.; Jevtić, A. Genetic Purity of Sunflower Hybrids Determined on the Basis of Isozymes and Seed Storage Proteins. *Helia* **2008,** *31,* 47–54.

Nir, I.; Feldman, Y.; Aserin, A.; Garti, N. Surface Properties and Emulsification Behavior of Denatured Soy Proteins. *J. Food Sci.* **1994,** *59,* 606–610.

Norton, G. Nature and Biosynthesis of Storage Proteins. In *Oil Crops of the World;* Robbelen, G., Downey, R. K., Ashri, A., Eds.; McGraw Hill: New York, 1989; pp 165–191.

Nuzzolo, C.; Vignola, R.; Groggia, A. Method for Preparing a Proteinic Isolate from Sunflower Seed Meal Using Aluminum Salts. U.S. Patent 4,212,799, Jul. 15, 1980.

O'Connor, D. E. Preparing Light-Coloured Protein Isolate from Sunflower Meal by Alkali Extraction under an Inert Gas Blanket Followed by Membrane Ultrafiltration. U.S. Patent 3,622,556, Nov. 23, 1971.

Ono, K.; Watanabe, A. Levels of Endogenous Sugars, Transcripts of rbcS and rbcL, and of RuBisCO Protein in Senescing Sunflower Leaves. *Plant Cell Physiol.* **1997,** *38,* 1032–1038.

Osborne, T. B. *The Vegetable Proteins;* Longmans Green: London, 1924.

Osborne, T. B.; Campbell, G. F. The Proteids of the Sunflower Seed. *J. Am. Chem. Soc.* **1897,** *19,* 487–494.

Pagnussat, L. A.; Lombardo, C.; Regente, M.; Pinedo, M.; Martín, M.; de la Canal, L. Unexpected Localization of a Lipid Transfer Protein in Germinating Sunflower Seeds. *J. Plant Physiol.* **2009,** *166,* 797–806.

Pain, R. Current Protocols in Protein Science (2004). 7.6.1–7.6.4. Wiley Online Library. John Wiley & Sons, Inc. http://onlinelibrary.wiley.com/doi/10.1002/0471140864.ps0706s05/figures.

Pallavi, H. M.; Ramegowda, K. B.; Vishwanath, K.; Shadakshari, Y. G. Evaluation and Confirmation of Hybridity on the Basis of Seed Proteins and Isozymes in Sunflower Hybrids. *Research J. Agr. Sci.* **2010,** *1,* 189–192.

Panchenko, A. Y. Sunflower Production and Breeding in the USSR. In *Proceedings of the 2nd International Sunflower Conference,* Morden, Manitoba, August 17–18, 1966. Research Branch, Canada Department of Agriculture: Morden, Canada, 1966; pp 15–29.

Pandya, M. J.; Williams, P. B.; Dempsey, C. E.; Shewry, P. R.; Clarke, A. R. Direct Kinetic Evidence for Folding via a Highly Compact, Misfolded State. *J. Biol. Chem.* **1999,** *274,* 26828–26837.

Pandya, M. J.; Sessions, R. B.; Williams, P. B.; Dempsey, C. E.; Tatham, A. S.; Shewry, P. R.; Clarke, A. R. Structural Characterization of a Methionine-Rich, Emulsifying Protein from Sunflower Seed. *Proteins: Struct., Funct., Genet.* **2000,** *38,* 341–349.

Paniego, N.; Heinz, R.; Fernandez, P.; Talia, P.; Nishinakamasu, V.; Hopp, H. E. Sunflower. In *Oilseeds;* Kole, C., Ed.; Genome Mapping and Molecular Breeding in Plants. Oilseeds 2; Springer: Berlin, Germany, 2007; pp 153–177.

Pantoja-Uceda, D.; Bruix, M.; Giménez-Gallego, G.; Rico, M.; Santoro, J. Solution Structure of RicC3, a 2S Albumin Storage Protein from *Ricinus communis. Biochemistry* **2003,** *42,* 13839–13847.

Pantoja-Uceda, D.; Palomares, O.; Bruix, M.; Villalba, M.; Rodríguez, R.; Rico, M.; Santoro, J. Solution Structure and Stability against Digestion of rproBnIb, a Recombinant 2S Albumin from Rapeseed: Relationship to its Allergenic Properties. *Biochemistry* **2004a,** *43,* 16036–16045.

Pantoja-Uceda, D.; Shewry, P. R.; Bruix, M.; Tatham, A. S.; Santoro, J.; Rico, M. Solution Structure of a Methionine-Rich 2S Albumin from Sunflower Seeds: Relationship to its Allergenic and Emulsifying Properties. *Biochemistry* **2004b,** *43,* 6976–6986.

Pearce, R. J. Preparation of Protein Isolate from Sunflower Seed. U.S. Patent 4,435,319, 1984.

Pedrosa, M. M.; Muzquiz, M.; García-Vallejo, C.; Burbano, C.; Cuadrado, C.; Ayet, G.; Robredo, L. M. Determination of Caffeic and Chlorogenic Acids and Their Derivatives in Different Sunflower Seeds. *J. Sci. Food Agric.* **2000,** *80,* 459–464.

Pérez-Vich, B.; Berry, S. T. Molecular Breeding. In *Genetics, Genomics and Breeding of Sunflower;* Hu, J., Seiler, G., Kole, C., Eds.; Science Publishers: Boca Ratón, FL, 2010; pp 222–252.

Petit, L.; Davin, A.; Guéguen, J. Purified Sunflower Seed Protein Isolates. U.S. Patent 4,174,313, Nov. 13, 1979.

Petruccelli, S.; Anon, M. Thermal Aggregation of Soy Protein Isolates. *J. Agric. Food Chem.* **1995,** *43,* 3035–3041.

Pierpoint, W. S. O-Quinones Formed in Plant Extracts; Their Reactions with Amino Acids and Peptides. *Biochem. J.* **1969,** *112,* 609–616.

Plietz, P.; Damaschun, G.; Muller, J. J.; Schwenke, K. D. The Structure of 11-S Globulins from Sunflower and Rape Seed. A Small-Angle X-ray Scattering Study. *Eur J. Biochem.* **1983,** *130,* 315–320.

Plietz, P.; Damaschun, G.; Schwenke, K. D.; Schlesier, B. Small-Angle X-ray Scattering Investigations on Seed Proteins. *Mol. Nutr. Food Res.* **1986,** *30,* 245–250.

Pomenta, J. V.; Burns, E. E. Factors Affecting Chlorogenic, Quinic and Caffeic Acid Levels in Sunflower Kernels. *J. Food Sci.* **1971,** *36,* 490–492.

Popineau, Y.; Tatham, A. S.; Shewry, P. R.; Marion, D.; Guéguen, J. 2S Sunflower Albumins: Functional Properties of Native and Modified Proteins. In *Plant Proteins from European Crops: Food and Non-Food Applications;* Guéguen, J., Popineau, Y., Eds.; Springer: Berlin, Germany, 1998; pp 131–135.

Prakash, V.; Rao, M. S. N. Structural Similarities among the High Molecular Weight Protein Fractions of Oilseeds. *J. Biosci.* **1988,** *13,* 171–180.

Prakash, V.; Rao, M. S. N. Physicochemical Properties of Oilseed Proteins. *Crit. Rev. Biochem.* **1986,** *20,* 265–363.

Prasad, D. T. Studies on the Interaction of Sunflower Albumins with Chlorogenic Acid. *J. Agric. Food Chem.* **1988,** *36,* 450–452.

Prasad, D. T. Proteins of the Phenolic Extracted Sunflower Meal: I. Simple Method for Removal of Polyphenolic Components and Characteristics of Salt Soluble Proteins. *Food Sci. Technol.* **1990,** *23,* 229–235.

Prigent, S. Interactions of Phenolic Compounds with Globular Proteins and Their Effects on Food-Related Functional Properties. Ph.D. Thesis, Wageningen Universiteit, Netherlands, 2005.

Prigent, S.; Voragen, A.; Van Koningsveld, G.; Baron, A.; Renard, C.; Gruppen, H. Interactions between Globular Proteins and Procyanidins of Different Degrees of Polymerization. *J. Dairy Sci.* **2009,** *92,* 5843–5853.

Prigent, S. V.; Voragen, A. G.; Visser, A. J.; van Koningsveld, G.A .; Gruppen, H. Covalent Interactions between Proteins and Oxidation Products of Caffeoylquinic Acid (Chlorogenic Acid). *J. Sci. Food Agric.* **2007,** *87,* 2502–2510.

Prigent, S. V. E.; Gruppen, H.; Visser, A. J. W. G.; Van Koningsveld, G. A.; de Jongh, G. A. H.; Voragen, A. G. J. Effects of Non-Covalent Interactions with 5-O-Caffeoylquinic Acid (Chlorogenic Acid) on the Heat Denaturation and Solubility of Globular Proteins. *J. Agric. Food Chem.* **2003,** *51,* 5088–5095.

Puech, J.; Lencrerot, P.; Decau, J. Effect of Reduction of Light Intensity on Total Photosynthesis of Sunflower—Repercussions on Oil and Protein Production of Seed. *C R Seances Acad. Sci. D.* **1975,** *281,* 387–390.

Quartacci, M.; Navari-Izzo, F. Water Stress and Free Radical Mediated Changes in Sunflower Seedlings. *J. Plant Physiol.* **1992,** *139,* 621–625.

Quillet, M.; Vear, F.; Branlard, G. The Use of Isozyme Polymorphism for Identification of Sunflower (*Helianthus annuus*) Inbred Lines. *J. Genet. Breed.* **1992,** *46,* 295–304.

Radić, V.; Vujaković, M.; Marjanović-Jeromela, A.; Mrđa, J.; Miklič, V.; Dušanić, N.; Balalić, I. Interdependence of Sunflower Seed Quality Parameters. *Helia* **2009,** *32,* 157–163.

Rahma, E.; Rao, M. S. N. Effect of Denaturants on the Oligomeric Structure of the 11 S Protein of Sunflower (*Helianthus Annum*). *J. Biosci.* **1981a,** *3,* 407–416.

Rahma, E. H.; Rao, M. S. N. Characterization of Sunflower Proteins. *J. Food Sci.* **1979,** *44,* 579–582.

Rahma, E. H.; Rao, M. S. N. Isolation and Characterization of the Major Protein Fraction of Sunflower Seeds. *J. Agric. Food Chem.* **1981b,** *29,* 518–521.

Rahma, E. H.; Rao, M. S. N. Removal of Polyphenols from Sunflower Meal by Various Solvents: Effects on Functional Properties. *J. Food Sci.* **1981c,** *46,* 1521–1522.

Rawel, H. M.; Czajka, D.; Rohn, S.; Kroll, J. Interactions of Different Phenolic Acids and Flavonoids with Soy Proteins. *Int. J. Biol. Macromol.* **2002a,** *30,* 137–150.

Rawel, H. M.; Rohn, S.; Kruse, H. P.; Kroll, J. Structural Changes Induced in Bovine Serum Albumin by Covalent Attachment of Chlorogenic Acid. *Food Chem.* **2002b,** *78,* 443–455.

Raymond, J.; Dalgalarrondo, M.; Azanda, J. L.; Ducastaing, A. Preparation of Protein Isolates from Sunflower Cakes. *Rev. Fr. Corps Gras.* **1984,** *31,* 233–242.

Raymond, J.; Inquello, V.; Azanza, J. L. The Seed Proteins of Sunflower: Comparative Studies of Cultivars. *Phytochemistry* **1991,** *30,* 2849–2856.

Raymond, J.; Mimouni, B.; Azanza, J. L. Variability in the 11S Globulin Fraction of Seed Storage Protein of Helianthus (Asteraceae). *Plant Syst. Evol.* **1994,** *193,* 69–79.

Raymond, J.; Robin, J. M.; Azanza, J. L. 11 S Seed Storage Proteins from Helianthus Species (Compositae): Biochemical, Size and Charge Heterogeneity. *Plant Syst. Evol.* **1995,** *198,* 195–208.

Regente, M.; de la Canal, L. Do Sunflower 2S Albumins Play a Role in Resistance to Fungi? *Plant Physiol. Biochem.* **2001,** *39,* 407–413.

Regente, M. C.; de la Canal, L. Purification, Characterization and Antifungal Properties of a Lipid-Transfer Protein from Sunflower (*Helianthus annuus*) Seeds. *Physiol. Plant.* **2000,** *110,* 158–163.

Regitano d'Arce, M. A.; Assis, R. D. P.; Lima, U. D. A. Functional Properties of Sunflower Seed Meal Obtained by Ethanol Extraction. *Arch. Latinoam. Nutr.* **1994a,** *44,* 29–32.

Regitano d'Arce, M. A.; Gutierrez, E. M.; Lima, U. D. A. Sunflower Seed Protein Concentrates and Isolates Obtention from Ethanol Oil Extraction Meals (Technical Note). *Arch. Latinoam. Nutr.* **1994b,** *44,* 33–35.
Regitano d'Arce, M. A. B.; Gutierrez, E. M. R.; Gunther, P. A. S.; Lima, U. A. Kinetics of Extraction of Chlorogenic Acid from Sunflower Cakes Using Ethanol. *Ciencia Tecnol. Aliment.* **1994c,** *14,* 149–159.
Reichelt, R.; Schwenke, K. D.; Konig, T.; Pahtz, W.; Wangermann, G. Electron Microscopic Studies for Estimation of the Quaternary Structure of the 11S Globulin (Helianthinin) from Sunflower Seed (*Helianthus annuus* L.). *Biochem. Physio. Pfl.* **1980,** *175,* 653–663.
Rharrabti, Y.; Elhani, S.; Martos-Nunez, V.; del Moral, L. F. G. Protein and Lysine Content, Grain Yield, and Other Technological Traits in Durum Wheat under Mediterranean Conditions. *J. Agric. Food Chem.* **2001,** *49,* 3802–3807.
Righetti, P. G.; Bosisio, A. B. Applications of Isoelectric Focusing to the Analysis of Plant and Food Proteins. *Electrophoresis* **1981,** *2,* 65–75.
Robertson, J. A. Sunflower: America's Neglected Crop. *J. Am. Oil Chem. Soc.* **1972,** *49,* 239–244.
Robertson, J. A.; Burns, E. E. Use of Sunflower Seed in Food Products. *Crit. Rev. Food. Sci. Nutr.* **1975,** *6,* 201–240.
Robinson, R. Amino Acid and Elemental Composition of Sunflower and Pumpkin Seeds. *Agron. J.* **1975,** *67,* 541–544.
Rojas, P.; Fernández-Martínez, J. Combining Ability of Oil and Protein among Six Sunflower Lines. In *International Symposium on Breeding of Protein and Oil Crops,* Pontevedra, Spain, April 1–4, 1998; Misión Biológica de Galicia: Pontevedra, Spain, 1998; pp 117–118.
Rollier, M.; Trocme, S.; Boniface, R. Observations Sur la Fertilisation Phospho-Potassique du Tournesol. *Inf. Techn.* CETIOM No 47. (1975).
Rossi, M.; Pagliarini, E.; Peri, C. Emulsifying and Foaming Properties of Sunflower Protein Derivatives. *Lebensmittel-Wissenschaft Und-Technologie-Food Science and Technology* **1985,** *18,* 293–299.
Roy, N.; Laskar, S.; Barik, A. Amino Acids through Developmental Stages of Sunflower Leaves. *Acta Bot. Croat.* **2013,** *72,* 23–33.
Rudkin, G. O.; Nelson, J. M. Chlorogenic Acid and Respiration of Sweet Potatoes. *J. Am. Chem. Soc.* **1947,** *69,* 1470–1475.
Ruso, J.; Rojas, P.; Fernández-Martínez, J. Evaluation of Sunflower Species for Chlorogenic Acid Content and Other Seeds Traits. In *Proceedings of the 22nd Sunflower Research Workshop;* January 18–19, 2000. National Sunflower Association: Bismarck, ND, 2000; pp 156–161.
Ryan, C. A. Protease Inhibitors in Plants: Genes for Improving Defenses against Insects and Pathogens. *Annu. Rev. Phytopathol.* **1990,** *28,* 425–449.
Sabir, M. A.; Sosulski, F. W.; MacKenzie, S. L. Gel Chromatography of Sunflower Proteins. *J. Agric. Food Chem.* **1973,** *21,* 988–993.
Sabir, M. A.; Sosulski, F. W.; Finlayson, A. J. Chlorogenic Acid–Protein Interactions in Sunflower. *J. Agric. Food Chem.* **1974a,** *22,* 575–578.
Sabir, M. A.; Sosulski, F. W.; Kernan, J. A. Phenolic Constituents in Sunflower Flour. *J. Agric. Food Chem.* **1974b,** *22,* 572–574.
Saeed, M.; Cheryan, M. Sunflower Protein Concentrates and Isolates Low in Polyphenols and Phytate. *J. Food Sci.* **1988,** *53,* 1127–1131.

Saeed, M.; Cheryan, M. Chlorogenic Acid Interactions with Sunflower Proteins. *J. Agric. Food Chem.* **1989,** *37,* 1270–1274.
Saio, K.; Gallant, D.; Petit, L. Electron Microscope Research on Sunflower Protein Bodies. *Cereal Chem.* **1977,** *54,* 1171–1181.
Salunkhe, D. K.; Chavan, J. K.; Adsule, R. N.; Kadam, S. S. Sunflower. In *World Oilseeds: Chemistry, Technology and Utilization;* Van Nostrand Reinhold: New York, 1992; pp 97–139.
Sammour, R. H.; El-Shourbagy, M. N.; Abo-Shady, A. M.; Abasary, A. M. Characterization of Sunflower Seed Proteins by Electrophoretic Techniques. *Arab Gulf J. Sci. Res.* **1995,** *13,* 591–604.
Sanchez, A. C.; Burgos, J. Thermal Gelation of Sunflower Proteins. In *Food Macromolecules and Colloids;* Dickinson, E., Lorient, D., Eds.; Special Publication; Royal Society of Chemistry: Cambridge, UK, 1995; pp 426–430.
Sanchez, A. C.; Burgos, J. Gelation of Sunflower Globulin Hydrolysates: Rheological and Calorimetric Studies. *J. Agric. Food Chem.* **1997,** *45,* 2407–2412.
Sarker, D. K.; Wilde, P. J.; Clark, D. C. Control of Surfactant-Induced Destabilization of Foams through Polyphenol-Mediated Protein-Protein Interactions. *J. Agric. Food Chem.* **1995,** *43,* 295–300.
Sastry, M. C. S.; Subramanian, N. Preliminary Studies on Processing of Sunflower Seed to Obtain Edible Protein Concentrates. *Am. Oil Chem. Soc.* **1984,** *61,* 1039–1042.
Sastry, M. C. S.; Rao, M. S. N. Binding of Chlorogenic Acid by the Isolated Polyphenol-Free 11S Protein of Sunflower (*Helianthus annuus*) Seed. *J. Agric. Food Chem.* **1990,** *38,* 2103–2110.
Schnarrenberger, C.; Oeser, A.; Tolbert, N. Isolation of Protein Bodies on Sucrose Gradients. *Planta* **1972,** *104,* 185–194.
Schwenke, K. D. Improving of Functional Properties of Oilseed and Legume Proteins by Structural Changes. *Acta Aliment. Pol.* **1988,** *14,* 15–23.
Schwenke, K. D.; Raab, B. Über Samenproteine 1. Mitt. Fraktionenverteilung der Proteine aus Sonnenblumensamen. *Mol. Nutr. Food Res.* **1973,** *17,* 373–379.
Schwenke, K. D.; Schultz, M.; Linow, K. J.; Uhlig, J.; Franzke, C. Ueber Samenproteine. 4. Isolierung der Globulin-Hauptkomponente aus Sonnenblumensamen. *Mol. Nutr. Food Res.* **1974,** *18,* 709–719.
Schwenke, K. D.; Schultz, M.; Linow, K. J. Isolierung und Charakterisierung des 11-S-Globulins aus Sonnenblumensamen (*Helianthus annuus* L.). *Mol. Nutr. Food Res.* **1975a,** *19,* 817–822.
Schwenke, K. D.; Schultz, M.; Linow, K. J. Ueber Samenproteine. 5. Dissoziationsverhalten des 11-S-Globulins aus Sonnenblumensamen. *Mol. Nutr. Food Res.* **1975b,** *19,* 425–432.
Schwenke, K. D.; Sologub, L. P.; Braudo, E. E.; Tolstoguzov, W. B. Ueber Samenproteine. 9. Einfluß von Neutralsalzen auf die Löslichkeit des 11-S-Globulins aus Sonnenblumensamen. *Nahrung* **1977,** *21,* 491–503.
Schwenke, K. D.; Robowsky, K. D.; Augustat, D. Über Samenproteins. 8. Mitt. Einfluß von Elektrolytgehalt und pH-Wert auf die Löslichkeit von Globulinen aus dem Samen von Sonnenblumen (*Helianthus annuus* L.) und aus Ackerbohnen (*Vicia faba* L.). *Mol. Nutr. Food Res.* **1978,** *22,* 425–437.

Schwenke, K. D.; Paehtz, W.; Linow, K. J.; Raab, B.; Schultz, M. On Seed Proteins. XI. Purification, Chemical Composition, and Some Physico-Chemical Properties of the 11 S Globulin (Helianthinin) in Sunflower Seed. *Mol. Nutr. Food Res.* **1979,** *23,* 241–254.

Schwenke, K. D.; Prahl, L.; Rauschal, E.; Gwiazda, S.; Dabrowski, K.; Rutkowski, A. Functional Properties of Plant Proteins. II. Selected Physico-Chemical Properties of Native and Denatured Isolates from Faba Beans, Soybeans, and Sunflower Seed. *Mol. Nutr. Food Res.* **1981,** *25,* 59–69.

Schwenke, K. D.; Grinberg, V.; Danilenko, A. N.; Burova, T. V.; Tolstoguzov, V. B. On the Acceptability of the Two-State Model of Protein Unfolding to the 11 S Globulins from Plant Seeds. *Mol. Nutr. Food Res.* **1987,** *31,* 183–184.

Seiler, G. J. Protein and Mineral Concentrations of Selected Wild Sunflower Species. *Agron. J.* **1984,** *76,* 289–294.

Seiler, G. J. Forage Quality of Selected Wild Sunflower Species. *Agron. J.* **1986,** *78,* 1059–1064.

Seiler, G. J. Utilization of Wild Sunflower Species for the Improvement of Cultivated Sunflower. *Field Crop. Res.* **1992,** *30,* 195–230.

Seiler, G. J. Anatomy and Morphology of Sunflower. In *Sunflower Technology and Production;* Schneiter, A. A., Ed.; Agronomy Monograph 35; American Society of Agronomy: Madison, WI, 1997; pp 67–111.

Seiler, G. J. The Potential of Wild Sunflower Species for Industrial Uses. *Helia* **2007,** *30,* 175–198.

Seiler, G. J.; Rieseberg, L. Systematics, Origin and Germplasm Resources of the Wild and Domesticated Sunflower. In *Sunflower Technology and Production;* Schneiter, A. A., Ed.; Agronomy Monograph 35; American Society of Agronomy: Madison, WI, 1997; pp 21–66.

Seiler, G. J.; Carr, M. E.; Bagby, M. O. Renewable Resources from Wild Sunflowers (*Helianthus* spp., Asteraceae). *Econ. Bot.* **1991,** *45,* 4–15.

Serre, M.; Feingold, S.; Salaberry, T.; Leon, A.; Berry, S. The Genetic Map Position of the Locus Encoding the 2S Albumin Seed Storage Proteins in Cultivated Sunflower (*Helianthus annuus* L.). *Euphytica* **2001,** *121,* 273–278.

Shank, E. A.; Cecconi, C.; Dill, J. W.; Marquisse, S.; Bustamente, C. The Folding Cooperativity of a Protein Is Controlled by Its Chain Topology. *Nature,* **2010,** *465,* 637–640.

Shewry, P. R.; Pandya, M. J. The 2S Albumin Storage Proteins. In *Seed Proteins;* Shewry, P. R., Casey, R., Eds.; Springer Netherlands: Dordrecht, The Netherlands, 1999; pp 563–586.

Shewry, P. R.; Napier, J. A.; Tatham, A. S. Seed Storage Proteins: Structures and Biosynthesis. *Plant Cell* **1995,** *7,* 945–956.

Shewry, P. R.; Jenkins, J.; Beaudoin, F.; Mills, E. C. Plant Protein Families as a Basis for Predicting the Allergenicity of Food Proteins. In *Allergy Matters: New Approaches to Allergy Prevention and Management;* Glissen, L. J. W. J., Wichers, H. J., Savelkoul, H. F. J., Bogers, R. J., Eds.; Wageningen UR Frontis Series 10; Springer: New York, 2006; pp 39–44.

Shoar-Ghafari, A.; Vintéjoux, C. Caractères et Réactivité des Grains Protéiques et de Leurs Globoïdes dans Différents Tissus des Cotylédons de Graines d'*Helianthus annuus* L. (Asteraceae). *Acta Bot. Gall.* **1997,** *144,* 95–106.

Shotwell, M. A.; Afonso, C.; Davies, E.; Chesnut, R. S.; Larkins, B. A. Molecular Characterization of Oat Seed Globulins. *Plant Physiol.* **1988,** *87,* 698–704.

Siebert, K. J. Haze Formation in Beverages. *LWT-Food Science and Technology* **2006,** *39,* 987–994.

Siebert, K. J.; Carrasco, A.; Lynn, P. Y. Formation of Protein-Polyphenol Haze in Beverages. *J. Agric. Food Chem.* **1996,** *44,* 1997–2005.

Škorić, D. Achievements and Future Directions of Sunflower Breeding. *Field Crop. Res.* **1992,** *30,* 231–270.

Smoliga, J. M.; Baur, J. A.; Hausenblas, H. A. Resveratrol and Health—A Comprehensive Review of Human Clinical Trials. *Mol Nutr Food Res.* **2011,** *55,* 1129–1141.

Snow, A.; Moran-Palma, P.; Rieseberg, L.; Wszelaki, A.; Seiler, G. Fecundity, Phenology, and Seed Dormancy of F1 Wild-Crop Hybrids in Sunflower (*Helianthus annuus*, Asteraceae). *Am. J. Bot.* **1998,** *85,* 794–801.

Sodini, G.; Canella, M. Acidic Butanol Removal of Color-Forming Phenols from Sunflower Meal. *J. Agric. Food Chem.* **1977,** *25,* 822–825.

Sorgentini, D. A.; Wagner, J. R. Comparative Study of Structural Characteristics and Thermal Behavior of Whey and Isolate Soybean Proteins. *J. Food Biochem.* **1999,** *23,* 489–507.

Sosulski, F.; Sarwar, G. Amino Acid Composition of Oilseed Meals and Protein Isolates. *Can. Inst. Food Sci. Technol. J.* **1973,** *6,* 1–5.

Sosulski, F.; Fleming, S. E. Chemical, Functional, and Nutritional Properties of Sunflower Protein Products. *Am. Oil Chem. Soc.* **1977,** *54,* A100–A104.

Sosulski, F. W.; Bakal, A. Isolated Proteins from Rapeseed, Flax and Sunflower Meals. *Can. Inst. Food Technol. J.* **1969,** *2,* 28–32.

Sosulski, F. W.; McCleary, C. W.; Soliman, F. S. Diffusion Extraction of Chlorogenic Acid from Sunflower Kernels. *J. Food Sci.* **1972,** *37,* 253–256.

Sripad, G.; Rao, M. S. N. Effect of Acid pH on the 11S Protein of Sunflower Seed. *J. Agric. Food Chem.* **1987a,** *35,* 668–672.

Sripad, G.; Rao, M. S. N. Effect of Methods to Remove Polyphenols from Sunflower Meal on the Physicochemical Properties of the Proteins. J. *Agric. Food Chem.* **1987b,** *35,* 962–967.

Sripad, G.; Rao. M. S. N. Effect of Alkaline pH on Sunflower 11S Protein. *J. Biosci.* **1987c,** *11,* 351–360.

Steer, B.; Hocking, P.; Kortt, A.; Roxburgh, C. Nitrogen Nutrition of Sunflower (*Helianthus annuus* L.): Yield Components, the Timing of Their Establishment and Seed Characteristics in Response to Nitrogen Supply. *Field Crop. Res.* **1984,** *9,* 219–236.

Steer, B.; Coaldrake, P.; Pearson, C.; Canty, C. Effects of Nitrogen Supply and Population Density on Plant Development and Yield Components of Irrigated Sunflower (*Helianthus annuus* L.). *Field Crop. Res.* **1986,** *13,* 99–115.

Strauss, G.; Gibson, S. M. Plant Phenolics as Cross-Linkers of Gelatin Gels and Gelatin-based Coacervates for Use as Food Ingredients. *Food Hydrocolloids* **2004,** *18,* 81–89.

Suryaprakash, P.; Prakash, V. Unfolding of Multimeric Proteins in Presence of Denaturants. A Case Study of Helianthinin from *Helianthus annuus* L. *Mol. Nutr. Food Res.* **2000,** *44,* 178–183.

Suryaprakash, P.; Kumar Ramesh, P.; Prakash, V. Thermodynamics of Interaction of Caffeic Acid and Quinic Acid with Multisubunit Proteins. *Int. J. Biol. Macromol.* **2000,** *27,* 219–228.

Swedberg, J. E.; Nigon, L. V.; Reid, J. C.; de Veer, S. J.; Walpole, C. M.; Stephens, C. R.; Walsh, T. P.; Takayama, T. K.; Hooper, J. D.; Clements, J. A.; et al. Substrate-Guided Design of a Potent and Selective Kallikrein-Related Peptidase Inhibitor for Kallikrein 4. *Chem. Biol.* **2009,** *16,* 633–643.

Tabe, L.; Wardley-Richardson, T.; Ceriotti, A.; Aryan, A.; McNabb, W.; Moore, A.; Higgins, T. A Biotechnological Approach to Improving the Nutritive Value of Alfalfa. *J. Anim. Sci.* **1995,** *73,* 2752–2759.

Tabe, L. M.; Droux, M. Limits to Sulfur Accumulation in Transgenic Lupin Seeds Expressing a Foreign Sulfur-Rich Protein. *Plant Physiol.* **2002,** *128,* 1137–1148.

Tavoljanskiy, N. P.; Chiryaev, P. V.; Scherstyuk, S. V.; Altinnikova, V. I.; Tikhomirov, V. T. Development of Original Material for Sunflower Breeding for Seed Characteristics, Oil and Protein Quality in the Conditions of Central-Chernozem Region. *Helia* **2004,** *27,* 117–122.

Terras, F.; Schoofs, H.; De Bolle, M.; Van Leuven, F.; Rees, S. B.; Vanderleyden, J.; Cammue, B.; Broekaert, W. F. Analysis of Two Novel Classes of Plant Antifungal Proteins from Radish (*Raphanus sativus* L.) Seeds. *J. Biol. Chem.* **1992,** *267,* 15301–15309.

Timme, R. E.; Kuehl, J. V.; Boore, J. L.; Jansen, R. K. A Comparative Analysis of the Lactuca and Helianthus (Asteraceae) Plastid Genomes: Identification of Divergent Regions and Categorization of Shared Repeats. *Am. J. Bot.* **2007,** *94,* 302–312.

Tkachuk, R.; Irvine, G. N. Amino Acid Compositions of Cereals and Oilseed Meals. *Cereal Chem.* **1969,** *46,* 206–218.

Tranchino, L.; Costantino, R.; Sodini, G. Food Grade Oilseed Protein Processing: Sunflower and Rapeseed. *Plant Food Hum. Nutr.* **1983,** *32,* 305–334.

Ufaz, S.; Galili, G. Improving the Content of Essential Amino Acids in Crop Plants: Goals and Opportunities. *Plant Physiol.* **2008,** *147,* 954–961.

Unger, P. W. Time and Frequency of Irrigation Effects on Sunflower Production and Water Use. *Soil Sci Soc Am J.* **1982,** *46,* 1072–1076.

Unger, P. W. Irrigation Effect on Sunflower Growth, Development, and Water Use. *Field Crop. Res.* **1983,** *7,* 181–194.

Unger, P. W. Growth and Development of Irrigated Sunflower in the Texas High Plains. *Agron. J.* **1986,** *78,* 507–515.

UniProtKB. P19084 (11S3_HELAN). http://www.uniprot.org/uniprot/P19084 (September 2013a).

UniProtKB. P15461 (2SS5_HELAN) http://www.uniprot.org/uniprot/P15461 (September 2013b).

UniProtKB. P23110 (2SS8_HELAN) http://www.uniprot.org/uniprot/P23110 (September 2013c).

Van Koningsveld, G. A. Physico-Chemical and Functional Properties of Potato Proteins. Ph.D. Thesis, Wageningen University, Netherlands, 2001.

Vaughan, J. G. The Structure and Utilization of Oil Seeds; Chapman and Hall: London, 1970.

Venktesh, A.; Prakash, V. Functional Properties of the Total Proteins of Sunflower (*Helianthus annuus* L.) Seed: Effect of Physical and Chemical Treatments. *J. Agric. Food Chem.* **1993a,** *41,* 18–23.

Venktesh, A.; Prakash, V. Low Molecular Weight Proteins from Sunflower (*Helianthus annuus* L.) Seed: Effect of Acidic Butanol Treatment on the Physicochemical Properties. *J. Agric. Food Chem.* **1993b,** *41,* 193–198.

Vermeesch, G.; Briffaud, J.; Joyeux, J. Sunflower Proteins in Human Food. *Rev. Fr. Corps Gras.* **1987,** *34,* 333–344.

Vollmann, J.; Rajcan, I. Oil Crop Breeding and Genetics. In *Oil Crops; Handbook of Plant Breeding 4;* Springer: New York, 2010; pp 1–30.

Vonder Haar, R. A.; Allen, R. D.; Cohen, E. A.; Nessler, C. L.; Thomas, T. L. Organization of the Sunflower 11S Storage Protein Gene Family. *Gene* **1988,** *74,* 433–443.

Vuilleumier, S.; Sancho, J.; Loewenthal, R.; Fersht, A. R. Circular Dichroism Studies of Barnase and Its Mutants: Characterization of the Contribution of Aromatic Side Chains. *Biochemistry* **1993,** *32,* 10303–10313.

Wagner, J. R.; Guéguen, J. Effects of Dissociation, Deamidation, and Reducing Treatment on Structural and Surface Active Properties of Soy Glycinin. *J. Agric. Food Chem.* **1995,** *43,* 1993–2000.

Wagner, J.R.; Guéguen, J. Surface Functional Properties of Native, Acid-Treated and Reduced Soy Glycinin. 2. Emulsifying Properties. *J. Agric. Food Chem.* **1999a,** *47,* 2181–2187.

Wagner, J.R.; Guéguen, J. Surface Functional Properties of Native, Acid-Treated and Reduced Soy Glycinin. 1. Foaming Properties. *J. Agric. Food Chem.* **1999b,** *47,* 2173–2180.

Waksman, G.; Lebrun, M.; Freyssinet, G. Nucleotide Sequence of a Gene Encoding Sunflower Ribulose-1, 5-Bisphosphate Carboxylase/Oxygenase Small Subunit (rbcs). *Nucleic Acids Res.* **1987,** *15,* 7181.

Wang, Z. Y.; Ye, X. D.; Nagel, J.; Potrykus, I.; Spangenberg, G. Expression of a Sulphur-Rich Sunflower Albumin Gene in Transgenic Tall Fescue (*Festuca arundinacea* Schreb.) *Plants. Plant Cell Reports* **2001,** *20,* 213–219.

Watanabe, K.; Chikushi, K.; Adachi, T.; Shimizu, M.; Yoshida, T.; Mitsunaga, T. Thiamin-Binding Protein from Sunflower Seeds. *J. Nutr. Sci. Vitaminol.* **1998,** *44,* 665–672.

Watanabe, K.; Konishi, A.; Mitsunaga, T. Molecular Characteristics of Thiamin-Binding Protein from Sunflower Seeds. *Plant Physiol. Biochem.* **2002,** *40,* 417–421.

Weisz, G. M.; Kammerer, D. R.; Carle, R. Identification and Quantification of Phenolic Compounds from Sunflower (*Helianthus annuus* L.) Kernels and Shells by HPLC-DAD/ESI-MSn. *Food Chem.* **2009,** *115,* 758–765.

Willekens, H.; Inzé, D.; Van Montagu, M.; Van Camp, W. Catalases in Plants. *Mol. Breed.* **1995,** *1,* 207–228.

Winton, A. L.; Moeller, J. *The Microscopy of Vegetable Foods;* John Wiley & Sons: New York, 1916.

Wolswinkel, P. Assimilate Transport in Developing Seeds of Sunflower (*Helianthus annuus* L.). *J. Plant Physiol.* **1987,** *127,* 1–10.

Wu, C. H.; Huang, H.; Yeh, L.-S. L.; Barker, W. C. Protein Family Classification and Functional Annotation. *Comput. Biol. Chem.* **2003,** *27,* 37–47.

Yatsu, L.; Jacks, T. Spherosome Membranes Half Unit-Membranes. *Plant Physiol.* **1972,** *49,* 937–943.

Youle, R. J.; Huang, A. H. C. Occurrence of Low Molecular Weight and High Cysteine Containing Albumin Storage Proteins in Oilseeds of Diverse Species. *Am. J. Bot.* **1981,** *68,* 44–48.

Yuan, C.; Chen, L.; Meehan, E. J.; Daly, N.; Craik, D. J.; Huang, M.; Ngo, J. C. Structure of Catalytic Domain of Matriptase in Complex with Sunflower Trypsin Inhibitor-1. *BMC Struct. Biol.* **2011,** *11,* 1–10.

Yuan, Y. J.; Velev, O. D.; Chen, K.; Campbell, B. E.; Kaler, E. W.; Lenhoff, A. M. Effect of pH and Ca^{2+}-Induced Associations of Soybean Proteins. *J. Agric. Food Chem.* **2002,** *50,* 4953–4958.

Zheng, B.-A.; Matsumura, Y.; Mori, T. Relationship between the Thermal Denaturation and Gelling Properties of Legumin from Broad Beans. *Biosci. Biotechnol. Biochem.* **1993,** *57,* 1087–1090.

Žilić, S.; Barać, M.; Pešić, M.; Crevar, M.; Stanojević, S.; Nišavić, A.; Saratlić, G.; Tolimir, M. Characterization of Sunflower Seed and Kernel Proteins. *Helia* **2010,** *33,* 103–113.

Zirwer, D.; Gast, K.; Plietz, P.; Schlesier, B.; Schwenke, K. D. Dynamic Light Scattering Studies of Globulins from Plant Seeds. *Mol. Nutr. Food Res.* **1986,** *30,* 251–255.

Zubillaga, M.; Aristi, J.; Lavado, R. Effect of Phosphorus and Nitrogen Fertilization on Sunflower (*Helianthus annuus* L.) Nitrogen Uptake and Yield. *J. Agron. Crop Sci.* **2002,** *188,* 267–274.

13

Utilization of Sunflower Proteins

Justo Pedroche ■ *Instituto de la Grasa, Consejo Superior de Investigaciones Científicas (CSIC), Sevilla, Spain*

Introduction

Production of Sunflower Meal

Agro-industrial plants generate a high number of by-products rich in proteins, including defatted meals from oilseeds after oil extraction processes; broken legume seeds after harvesting, transport, and/or handling; cereal by-products as rice, oat, or wheat bran; as well as other by-products originated from the manufacture of elaborated foods (potato juice, tomato paste, etc.). Although oil-bearing tree fruits (e.g., olive, coconut, and palm trees) provide the highest oil yields (Gunstone, 2002), oilseeds are the largest source of vegetable oils.

Sunflower seeds are the world's seventh-biggest oilseed crop according to seed production (Table 13.B) but the fourth largest one in oil production, after palm, soybean, and rapeseed (Table 13.B). According to these values, the yield of sunflower oil extraction is the second largest, very close to palm kernel and rapeseed oil extraction. At the same time, these values mean that around 60% of raw materials can be obtained as defatted meal. Ukraine, the Russian Federation, and Argentina are the main world producers of sunflower seeds and sunflower by-products, as well as suppliers to the global market (Figure 13.1, p. 397). These three countries produce over half of the sunflower seeds in the world, which is why they are often referred as the "sunflower triangle." The change in the market situation in one country has a great influence on the market in the other two countries, thereby affecting the world market of sunflower seeds and sunflower by-products (Agribusiness Handbook, 2010). In 2012, the production of sunflower seed gradually increased (Table 13.C., p. 397) and consequently, the production of sunflower oil and meal increased.

There are two types of sunflowers, oilseed and non-oilseed (or confectionery), which are nevertheless of the same species. Oilseed sunflower seeds, constituting the major part of world production, are characterized by solid black hulls that are firmly attached to the seed. They are used in the crushing industry for oil production and for wild and domestic bird feeding. Meal resulting from their crushing is mainly used for livestock feeding. The industry has bred high oleic acid oilseed sunflowers that have a fatty acid profile similar to canola and olive oil. The market share for this variety is relatively small, but it is increasing year after year. Confectionary sunflower seeds are characterized by larger, thick, striped, loosely attached hulls that

Table 13.A Seed Production of the Main Worldwide Oil Crops

Raw Material	2012
Palm fruit	249,528
Soybeans	241,841
Rapeseed	65,058
Coconuts	60,049
Cottonseed	47,191
Groundnuts, with shell	41,186
Sunflower seed	37,449
Olives	16,555
Palm kernels	14,657
Sesame seed	4,036
Linseed	2,054
Safflower seed	833

Note: The values are expressed in 1000 t.
Source: © FAO Statistics Division (2014).

Table 13.B Oil Production and Yield of the Main Worldwide Oil Crops

Raw Material	2012	Yield (%)
Oil, palm	50,199	20.12
Oil, soybean	41,538	17.18
Oil, rapeseed	23,570	36.23
Oil, sunflower	14,947	39.91
Oil, palm kernel	6,045	41.24
Oil, cottonseed	5,301	11.23
Oil, groundnut	5,170	12.55
Oil, olive virgin	3,320	20.02
Oil, coconut (copra)	3,304	5.503
Oil, sesame	1,277	31.64
Oil, linseed	544	26.49
Oil, safflower	144	17.29

Note: The values are expressed in 1000 t.
Source: © FAO Statistics Division (2014).

lend themselves to a relatively complete hulling process. These seeds are used for the human food market as roasted snack foods with shell, or dehulled seeds for the baking industry.

As discussed in previous chapters, conventional industrial oil extraction from oilseeds is commonly undertaken through mechanical pressing with a hydraulic or single expeller press, followed by solvent extraction to defat whole seeds and partially hulled seeds. Thus, the final product of sunflower defatted meal has been exposed to different physical (crushing, heating) and chemical (extraction in hydrophobic environmental with organic dissolvent) processes that will affect the final nutritional quality of the sunflower meals.

Chemical and Physical Properties of Sunflower Proteins

Chemical Composition of Sunflower Meal

The composition of the sunflower meal after oil extraction will depend mainly on the seed variety and the extraction method (Table 13.D, p. 398). Protein and crude fiber are the main compounds in sunflower meal ranging around 20–60% and 5–34%, respectively. The content of residual oil varies depending on the extraction process, with values less than 3% when solvent extraction is applied and more than 5% in meals

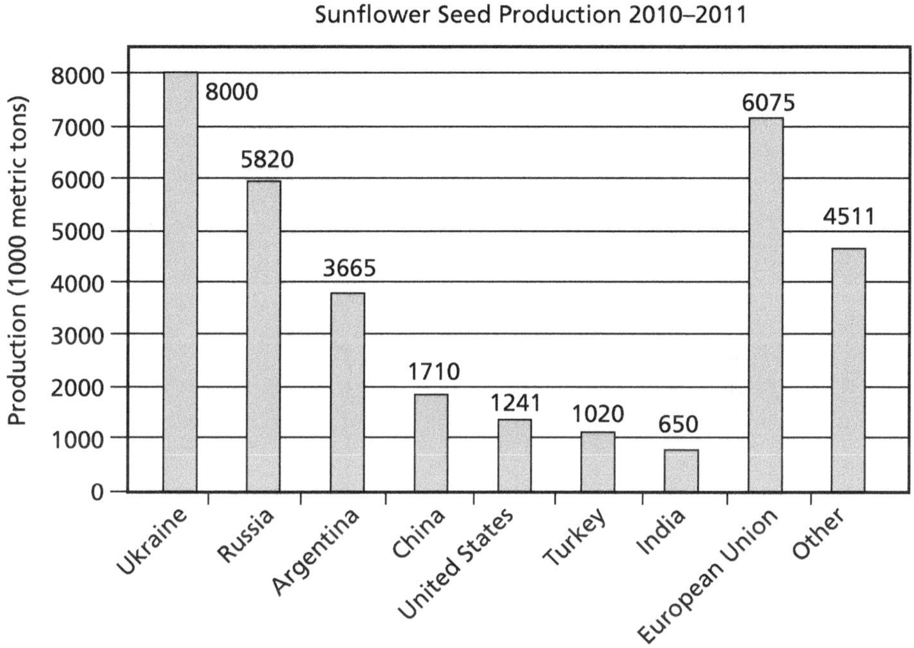

Figure 13.1 Main worldwide producers of sunflower seed.
Source: www.sunflowernsa.com/stats (2013).

obtained by mechanical extraction. Phenols, mainly chlorogenic acid, are important from a nutritional point of view with contents in sunflower meals around 2–6%. Ash and moisture don't seem to vary independently of the process carried out.

Furthermore, the amino acid composition of sunflower protein complies with the Food and Agricultural Organization (FAO) pattern, with the exception of lysine (Table 13.E, p. 400). In this regard, it must be considered that lysine is especially

Table 13.C Worldwide Production of Sunflower Seed, Oil, and Meal

Production	2007/08	2008/09	2009/10	2010/11	2011/12 Revised	2012/13 Forecast
Sunflower seed	29,259	34,753	32,171	33,572	39,473	36,310
Sunflower oil	10,200	12,871	12,543	12,418	15,038	13,812
Sunflower meal	11,607	14,295	14,018	14,128	16,978	15,639

Note: The values are expressed in 1000 t.
Source: www.sunflowernsa.com/stats (2013).

Table 13.D Literature Overview on Chemical Composition of Sunflower Meal

	Processing	Protein	Lipids	Fiber	Ash	Moisture	Phenols
Prevot et al. (1973)		28.10	—	—	—	—	—
Sodini and Canella (1977)	Hulled seeds; solvent extraction	60.75	1.47	4.33	7.25	10.0	2.59[a]
Peri and Rossi (1980)	—	21.30	—	—	—	—	—
Berot and Briffaud (1983)	—	50.05	4.85	7.10	7.35	—	—
Boni et al. (1987)	—	29.00	1.15	28.90	6.10	8.00	—
Sripad and Narasinga-Rao (1987)	—	50.80	—	—	—	—	—
Kianto (1989)	Expeller extraction	41.00	7.60	13.00	6.80	7.00	—
Kianto (1989)	Solvent extraction	46.80	2.90	11.00	7.70	7.00	—
Raymond et al. (1991)	—	29.80–36.20	—	—	—	—	
Parrado et al. (1993)	Solvent extraction	27.70	—	—	5.75	—	6.07
Economides, (1998)	—	34.00	1.06	26.00	6.71	8.50	—
Villamide and San Juan (1998)	Partial hulling; press and solvent extraction	31.46–41.75	0.73–3.61	19.23–30.13	6.92–9.01		—
Anderson (2002)	Unhulled	28.00	1.50	24.00	6.20	10.00	—
Anderson (2002)	Partially hulled	34.00	0.80	21.00	5.90	10.00	—
Anderson (2002)	Hulled	41.00	0.50	14.00	5.90	10.00	

González-Pérez et al. (2002)	Solvent extraction	55.00	4.00	—	7.00	—	2.50[a]
Irshaid et al. (2003)	—	31.20	1.50	32.10	7.10	7.30	—
Poncet et al. (2003)	—	37.20	1.80	—	—	—	—
Orliac et al. (2003)	—	34.40	1.00	—	7.60	10.00	5.70
Miñones-Conde et al. (2005)	Expeller and solvent extraction	33.30	—	—	—	—	—
Carellos et al. (2005)	—	27.50	3.08	25.90	—	7.30	—
Mushtaq et al. (2006)	—	32.00	3.54	20.50	1.39	—	—
Ramachandran et al. (2007)	—	34.10	—	13.20	6.60	9.00	—
Abdel-Hakim et al. (2008)	—	34.00	5.60	14.20	6.90	7.60	—
Geneau-Sbartaï et al. (2008)	—	32.00	2.00	—	7.10	10.40	—
Salgado et al. (2012a)	—	31.70	1.00	56.60[b]	8.00	—	2.70
Salgado et al. (2012b)	—	31.70	—	—	8.00	11.00	2.70
Salgado et al. (2012c)	—	31.70	0.90	33.50	8.00	11.00	2.70
Heuzé et al. (2012)	Mechanically extracted	27.90	13.80	26.20	5.70	8.20	—
Heuzé et al. (2012)	Non-hulled; solvent extraction	31.30	2.30	29.10	7.00	9.10	—

[a]Measured as chlorogenic acid.

[b]Carbohydrates + fibers.

Table 13.E Amino Acid Composition of Sunflower Meal

	Sunflower Meal	FAO[a]
Aspartic acid[b]	8.91±0.70	—
Glutamic acid[c]	18.96±2.36	—
Serine	4.22±0.39	—
Histidine	2.76±1.11	1.90
Glycine	5.84±0.71	—
Threonine	3.82±0.62	3.40
Arginine	9.19±2.19	—
Alanine	4.50±0.50	—
Proline	4.26±0.38	—
Tyrosine	2.23±0.47	—
Valine	4.99±0.84	3.50
Methionine	2.25±0.29	2.50[d]
Cysteine	1.61±0.25	—
Isoleucine	3.95±0.52	2.80
Triptophane	1.38±0.13	—
Leucine	6.42±0.70	6.60
Phenylalanine	4.57±0.28	6.30[e]
Lysine	3.59±0.17	5.80

Note: Values are represented as the average ± standard deviation from the references: Heuzé et al. (2012); Kianto (1989); Liebowitz and Ruckenstein (1986); Molina-Alcaide et al. (2003); Mossé and Baudet (1972); Sodini and Canella (1977); Sripad and Narasinga Rao (1987); Villamide and San Juan (1998).

[a]FAO energy and protein requirements (1985).

[b]Aspartic acid + asparagine.

[c]Glutamic acid + glutamine.

[d]Methionine + cysteine.

[e]Phenylalanine + tyrosine.

sensitive, reacting with carbonyl groups of oxidized lipids and reducing sugars through the ε-NH_2 group of this amino acid (Adams et al., 2011; Friedman, 1999). These Maillard reactions also lead to darkening of the protein material (Hofmann, 1998; Rizzi, 2008) and affect its sensory attributes (Eric et al., 2013) and certain functional properties of proteins due to the formation of protein-polyphenol complexes (Easa et al., 1996). Thus, the final content of lysine can be also influenced for the processes of oil extraction and storage conditions.

In any case, although lysine deficiency is a significant drawback from a nutritional point of view, this recommendation refers to the total intake in a balanced and varied diet, which indicates that sunflower meal amino acid composition would be part of a broad range of foods. Lysine of sunflower products could be completed with other protein resources. Thus, sunflower meal protein would contribute more than 50–60% of the value required by the FAO for this limiting amino acid. In this regard, proteins from defatted meals could be considered a valuable alternative for food ingredients because, as previously noted, these meals are low in antinutritional compounds and devoid of toxic substances.

Functional Properties of Sunflower Proteins

Proteins are the basic functional components of processed food products and thus determine textural, sensory, and nutritional properties. Functional properties have been defined as "those physical and chemical properties that influence the behavior of proteins in food systems during processing, storage, cooking and consumption" (Kinsella, 1976). The functional behavior of proteins in food is influenced by the physicochemical properties of the proteins such as their size, shape, amino acid composition and sequence, net charge, charge distribution, hydrophobicity, hydrophilicity, type of structures, and molecular flexibility/rigidity in response to external environment factors such as pH, temperature, salt concentration, or interaction with other food constituents (Damodaran, 1997).

Functional properties may be classified according to the mechanism of action on three main groups: (1) properties related to hydration (absorption of water/oil, solubility, thickening, wettability); (2) properties related to the protein structure and rheological characteristics (viscosity, elasticity, adhesiveness, aggregation, and gelling), and (3) properties related to the protein surface (emulsifying and foaming activities, formation of protein-lipid films, whippability) (Kinsella, 1979; Moure et al., 2006). As previously mentioned, these properties vary with pH, temperature, protein concentration, protein fraction, previous treatment, ionic strength, and dielectric constant of the medium, and can be affected by interactions with other macromolecules in the medium and physical, chemical, and enzymatic processing (Kinsella, 1979).

Functional Properties Related to Hydration Mechanisms

The hydrophilicity/hydrophobicity balance, which depends on the amino acid composition, particularly at the protein surface, influences the protein solubility. Higher solubility is related to the presence of a low number of hydrophobic residues, the elevated charge, and the electrostatic repulsion and ionic hydration occurring at pH above and below the isoelectric pH. Denaturation affects protein solubility due to alterations in the hydrophobicity/hydrophilicity ratio of the surface. A highly soluble protein is required in order to obtain optimum functionality required in gelation, emulsifying activity, foaming, and lipoxygenase activity (Riaz, 2006). Soluble protein preparations are easier to incorporate in food systems, unlike those with low solubility indices, which have limited functional properties and more limited food uses.

The native structure of proteins is a consequence of the interactions of amino acids with water, and some functional properties can be interpreted as a result of the protein–water interactions that are thermodynamically favorable (wettability, swelling, water retention, and solubility) or unfavorable (foaming, emulsification) (Moure et al., 2006). The interactions of protein with water are important in relation to dispersibility, water absorption and binding, swelling, viscosity, gelation, and surfactant properties because these properties influence the important functions of proteins in meat, bakery, and beverage preparations (Moure et al., 2006). The ease of dispersibility or wettability is important in food formulations and is affected by surface polarity, topography, texture and area, and by the size and microstructure of the protein particles (Kinsella, 1979).

In the case of sunflower proteins, helianthinin is a globulin protein that has been found to be the major sunflower proteins, ranging from about 40–90% of its total weight. As such, helianthinin is the main protein fraction that governs the solubility of the sunflower meal, concentrates, and isolates. González-Pérez et al. (2005) studied the solubility of helianthinin, albumins, and proteins isolated from sunflower at different pHs and ionic strengths (Figure 13.2). At low ionic strength ($I = 30$ mmol L^{-1}), helianthinin shows a bell-shaped curve with a minimum at pH 4–5.5, its isoelectric pH. At higher ionic strength ($I = 250$ mmol L^{-1}), helianthinin is almost insoluble at pH < 5. However, albumins were found to remain soluble independently of pH and ionic strength. Kabirullah and Wills (1983) obtained protein isolates based on their solubility in alkali, water, and sodium chloride (Figure 13.2). The protein solubility curves of these isolates are in agreement with what was previously explained about helianthinin solubility.

In this sense, it has been reported that the interaction of protein with water is influenced by the physicochemical characteristics of the proteins such as size, shape, amino acid composition, and net charge.

Hydrolysis of proteins is the breakdown of protein into smaller peptides and free amino acids, depending on the extension of the hydrolytic process. Consequently, increased solubility of the hydrolyzed protein is usually due to a decrease in the molecular weight of protein and, subsequently, to an increase in the number of small peptides and the corresponding increase in the ionizable amino and carboxyl groups (Creusot and Gruppen, 2008; Creusot et al., 2006). However, the hydrolysis processes needs to be controlled to improve solubility. Different enzymes can produce different sample responses to hydrolysis treatments. Hydrolysates can expose hydrophobic peptides, which increase peptide attractions/aggregation, causing a solubility

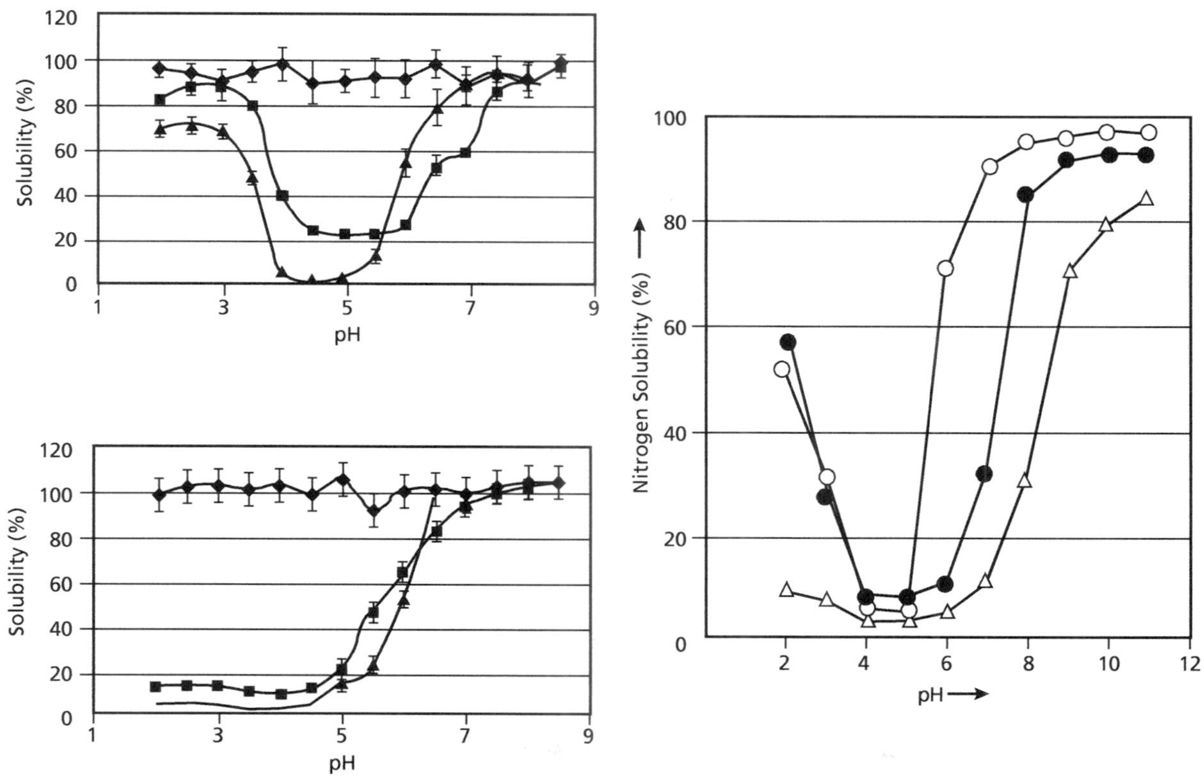

Figure 13.2 pH-Dependent solubility profiles of helianthinin (▲), sunflower albumins (◆), and sunflower isolate (■) at I = 30 mM (a) and I = 250 mM (b) (from González-Pérez et al., 2004, 2005) and protein solubility of sunflower protein isolate at various pH levels: alkali isolate (●), water-isolate (○), and salt isolate (△) (from Kabirullah and Wills, 1983).

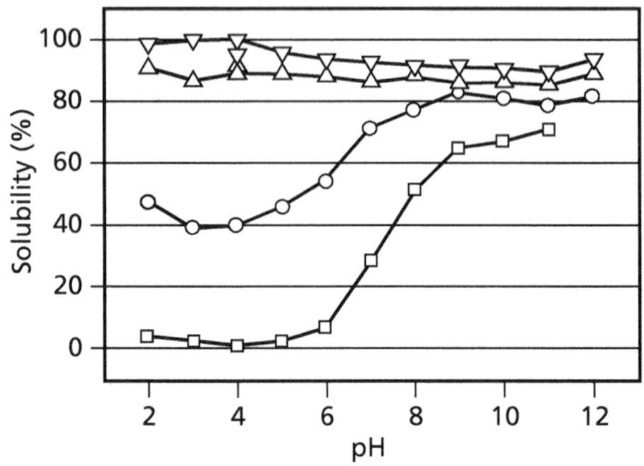

Figure 13.3 Solubility at different pH values of sunflower protein isolate (□) and protein hydrolysates at 5.62% (○), 23.2% (△), and 46.4% (▽) of degrees of hydrolysis.

Source: Miñones-Conde et al. (2005).

decrease (Paraman et al., 2007). Sunflower protein hydrolysates have been shown to have higher solubility than native proteins (Miñones-Conde et al., 2005; Parrado et al., 1993; Sánchez and Burgos, 1996; Villanueva et al., 1999). As a higher degree of hydrolysis is reached, a higher disappearance of the isoelectric point is produced (Figure 13.3). A high solubility (>80–90 %) independent of the pH can be observed in sunflower protein hydrolysates with a degree of hydrolysis higher than 25–30% (Miñones-Conde et al., 2005; Villanueva et al., 1999).

Functional Properties Related with Protein Structure and Rheology

Solubility, hydrodynamic properties, hydrophobicity, and microstructure have been reported to play an important role in the rheological properties of proteins (Krause et al., 2001, 2002). Due to the increased interactions of hydrated proteins shown in water absorption and swelling, viscosity increases exponentially with protein concentration (Kinsella, 1979).

Protein gels are three-dimensional matrices or networks of intertwined, partially associated polypeptides with entrapped water, and they are characterized by a relatively high viscosity, plasticity, and elasticity (Kinsella, 1979). The ability of protein to form gels and provide a structural matrix for holding water, flavors, sugars, and food ingredients is useful in food applications and in new product development, providing an added dimension to protein functionality. Gelling is important in comminuted sausage products and is the basis of many Asian textured foods (e.g., tofu). Factors known to affect gelation include pH, ionic strength, reducing agents, urea, temperature, the presence of nonprotein components, and the mechanical forces applied to the system (Sathe, 2002). Properties of the gel are determined by the interactions between the solvent and the molecular net, resulting in transparent or coagulant gels. Coagulant gels are formed by proteins containing nonpolar residues (Shimada and Mastsushita, 1980), whereas protein containing hydrophilic amino acids form transparent gels (Moure et al., 2006).

Sunflower proteins display a low capacity to form gels. Gel formation involves several steps such as denaturation, aggregation, and network formation (Fleming and Sosulski, 1975). The high denaturation temperature of sunflower proteins (proteins denaturation temperature is around 100 °C), amino acid composition (albumins rich in sulfur amino acids are not very abundant in sunflower meals, concentrates, or isolates), and a solubility very dependent of pH could be the reasons for this poor functional property.

However, Sánchez and Burgos (1996, 1997) have observed that good gelation is obtained in trypsinized sunflower protein solutions. They concluded that the main effects in this thermogelation seemed to be due to hydrophobic interactions. Hydrogen bonds seemed to be involved in stabilizing the residual gel structure when gels are cooled. Some native disulphide bridges helped to stabilize the gel structure, but electrostatic interactions did not seem important.

In the last two decades, interesting studies on the possibility of forming gels or biodegradable and renewable films have been conducted, looking at the ability of such proteins to form highly stable networks. The major reasons for this research are the environmental consequences of the intensive use of synthetic plastics, the petroleum price increase, and the new and emerging applications for biodegradable materials. Thus, new synthetic biodegradable polymers such as polylactic acid have appeared recently. However, these synthetic biodegradable polymers have a major drawback: They degrade in compost conditions, so their elimination will need an efficient classification and the development of more composting units. Moreover, natural biopolymer processing has many advantages, such as the use of renewable carbon sources, low energetic cost, and biodegradability, and some drawbacks related to moisture: special processing conditions and water sensitivity of formed materials (Rouilly et al., 2006). Parameters such as film thickness, density, water content, water vapor permeability, mechanical properties, glass transition temperature, and type of interactions involved in the protein network formation (which in all cases were mainly hydrogen bonds, hydrophobic interactions, and disulfide bonds) are the main factors to take in account for this purpose. Most protein-based films are made by casting. This involves the evaporation of solvents from protein solutions, resulting in the formation of a cross-linked protein network. Although this process has the advantage of producing simple protein films that can be subjected to numerous tests (mechanical, water resistance, barrier properties), it also has the major disadvantage that it is difficult to scale-up for industrial purposes (Guerrero et al., 2010). For this reason, a number of authors now shape proteins via a dry method, such as thermo-molding (Cunningham et al., 2000; Foulk and Bunn, 2001; Sun et al., 2008). In particular, sunflower proteins have been shown to possess adequate properties for use in the preparation of films by casting (Ayhllon-Meixueiro et al., 2000) and by thermo-pressing (Orliac et al., 2003; Salgado et al., 2008), and also for the production of materials by injection molding as a combined effect of temperature and pressure (Rouilly et al., 2006).

Functional Properties Related to Protein Surface

Emulsions and foams are two phase systems commonly found in food systems for which formations are significantly affected by protein surface activity (Kinsella, 1979; Moure, et al., 2006). Emulsions are generated by mixing two immiscible liquids (e.g., oil and water). The liquids are immiscible because of their relative polarities. When a liquid of low polarity, such as fat, is mixed with water, strong driving forces limit the contact between the two liquids resulting in phase separation (Jideani, 2011).

To stabilize these food dispersions (emulsions and foams), emulsifiers (lipids, phospholipids, proteins, some polysaccharides, etc.) must be placed at the interface so they can form a film around droplets or bubbles. Proteins and low-molecular weight emulsifiers (LMWEs) (i.e., lipids and phospholipids) are often used simultaneously in food formulations, and knowledge of their interfacial physical–chemical characteristics, such as surface activity, structure, interactions, stability, dynamic phenomena and interfacial rheological properties, as well as the kinetics of the film formation at fluid interfaces, is important for the optimization of dispersion formation, stability, and texture (Miñones-Conde et al., 2006). Mixtures of different emulsifiers (LMWEs and proteins) used in commercial food formulations often exhibit properties superior to those of the individual emulsifier alone due to synergistic interactions between emulsifier molecules (Bos et al., 1997; Dickinson, 2001; Rodríguez-Patino et al., 2003). Interactions between molecules of emulsifiers could affect not only the film structural characteristics (Rodríguez-Patino, et al., 2001) but also dynamic phenomena in mixed monolayers (Carrera et al., 2005). Properties of good foaming proteins include: (1) solubility in the aqueous phase and rapid adsorption during shipping and bubbling; (2) concentration at the air–water interface and unfolding to form cohesive layers of protein around air droplets with reduction of surface tension; and (3) sufficient viscosity and mechanical strength to prevent rupture and coalescence (Hettiarachchy and Ziegler, 1994; Kinsella, 1979).

An overview of the foam and emulsion properties of sunflower protein products, as determined in several studies, is given in Tables 13.F and 13.G (p. 411), respectively.

In light of the results obtained by different authors, highly diverging values are found in terms of formation and stability of foam and emulsion in meals, protein concentrates, and isolates. This diversity of results will probably come from compositional

Table 13.F Literature Overview on Foam Properties of Sunflower Protein

Reference	Material	Conditions and Evaluated Parameters	Main Results and Conclusions
Lawhon et al. (1972)	Meal	Dispersions (8–12%); pH 4–6; foam expansion in presence of sugar	Similar foam expansion to soy meal
Lin et al., 1974	Flour; isolate; concentrate	Dispersions (3%); foam capacity and stability	Isolate with similar foam capacity and stability to soy isolate; flour and concentrate with better foam capacity and stability than soy flour and concentrate
Huffman et al. (1975)	Meal	Dispersions (1–10%); pH 2–11; temperature 15–85 °C; foam expansion and stability	Best foam expansion at pH 4 and best foam stability at pH 9; decrease in foam expansion above 55 °C
Wu et al. (1976)	Meal	Dispersions (3%); pH 7; foam stability	Re-extraction of meal with several solvents (benzene, chloroform, petroleum ether, chloroform/methanol, ethanol/ether/water) did not affect foam stability; methanol washing
Canella et al. (1977)	Meal; isolate; concentrate	Dispersions (3%); foam expansion and stability	Foam expansion and stability were highest for isolate and lowest for meal; poorer properties than soy proteins
Canella (1978	Flour; concentrate	Dispersions (1–12%); pH 1–12; temperature 10–80 °C; foam expansion and stability	Sunflower flour had highest foam expansion (pH 7–10) and lowest stability, followed by sunflower concentrate and soy concentrate; maximum foam stability of flour at pH 6–8 and minimum at pH 2–5
Rahma and Rao (1981)	Meal	Dispersions (1%); foam capacity and stability	Aqueous ethanol decreased foam capacity; acidic n-butanol increased foam capacity; other solvents decreased foam capacity
Rossi and Germondari (1982)	Meal	Dispersions (5%); pH 2–7; foam capacity and stability	Foam capacity and stability high at pH > 5; increased foam volume and reduced foam stability with increasing ionic strength; better foam properties than soy meal
Canella et al. (1985)	Meal; albumins	Dispersions (8%); pH 2–10; foam expansion and stability	Albumins resulted in voluminous foams (maximum at pH 7.7) and relatively stable foams (pH 6–10)

Continued

Table 13.F *Continued*

Reference	Material	Conditions and Evaluated Parameters	Main Results and Conclusions
Khalil et al. (1985)	Seed; flour; isolate	Dispersions (0.3 mg nitrogen mL^{-1}); pH 3–7; foam capacity and stability	Isolate (pH 7) had lower foam capacity but higher stability than flour; foam capacity of flour maximum at pH 7 and minimum at pH 4; foam stability of flour maximum at isoelectric pH (4–5) and minimum at pH 7; heating decreased foam capacity; soybean proteins had higher foam capacity
Khalil et al. (1985)	Isolate	Dispersions (3%); pH 7; foam capacity and viscosity	Lower foam capacity than soybean isolate; heating decreased expansion and viscosity of foam
Raymond et al. (1985)	Isolate	Dispersions (0.5-5.0%); pH 1.5–0.5; temperature 15–60 °C; foam expansion and stability	Maximum foam expansion at pH 7.5; highest stability between pH 6.5 and 10.5; no effect of temperature; sunflower isolate had higher foam expansion and stability than soy isolate
Rossi et al. (1985)	Meal; concentrate	Dispersions (4%); pH 2–9; foam capacity and stability	High foam capacity and stability far from isoelectric point (pH 4.5–5); improved foam properties by salt addition (pH 4–6.5); heat denaturation did not affect or slightly improved foam expansion and stability
Kabirullah and Wills (1988)	Flour; isolate	Dispersions (0.5–2.5%); pH 4–10; foam expansion and stability	Similar foam properties for soy and sunflower isolates, but lower for sunflower flour; best foam expansion and stability of isolates at pH 7–10; foam stability decreased at pH > 6 for meal
Claughton and Pearce (1989)	Isolate	Dispersions (5%); pH 2.5–8; foam expansion and stability	Strong linear correlation between solubility and foam expansion, but not with foam stability; protein denaturation by acidification improved foam expansion and stability
Booma and Prakash (1990)	Helianthinin flour	Dispersions (8.7 and 1%); pH 6; foam capacity and stability	Helianthinin hardly stabilised foams
Venktesh and Prakash (1993)	Meal	Dispersions (3%); foam volume and stability	Acidic butanol and heating generally decreased foam volume and stability
Wastyn et al. (1993)	Isolate; concentrate	Dispersions in water; foam capacity and stability	Lack of foam capacity and stability

Guéguen et al. (1996)	Albumins	Dispersions (0.2 and 0.3%); pH 7.8 and 8; foam capacity and stability	Foams with little or no stability
Popineau et al. (1998)	Albumins	Dispersions (0.1%); pH 7; foam capacity and stability	Poor foam formation and stabilization; disulfide bond reduction resulted in dense foams with moderate stability
Pawar et al. (2001)	Meal; isolate; concentrate	Dispersions (1%); foam capacity and stability	Foam capacity lower for meal; foam stability higher for meal and isolate than for concentrate; increased foam capacity after extraction with acidic *n*-butanol
González-Pérez et al. (2005)	Isolate; helianthinin; albumins	Protein solutions (0.05%); pH 3–8; foam volume and stability; effects of heat and acid denaturation	Sunflower albumins formed foams with high volumes but little stability; helianthinin produced low foam volumes especially at alkaline and neutral pH, but with high foam stability; helianthinin protein denaturation resulted in improved foam properties
Miñones-Conde et al. (2005)	Isolate; hydrolysates with different degree of hydrolysis (DH: 5.6%, 23.3%, and 46.3%)	Protein solutions (0.01–1 × 10–6% w/w) Temperature 20 °C, pH 7, and I = 0.05 M; Surface pressure, Adsorption efficiency and superficial activity	Low solubility of SPI in acidic aqueous solutions is overcome for its hydrolysates; the solubility of sunflower proteins hydrolysates increases with the degree of hydrolysis; superficial activity is lower for the protein isolate than for the hydrolysates and decreases at the highest degree of hydrolysis
Miñones-Conde et al. (2006)	Isolate; hydrolysates with different degree of hydrolysis (DH: 5.6%, 23.3%, and 46.3%)	Protein solutions (0.96–1.24 mg/mL) Temperature 20 °C, pH 7, and I = 0.05 M; mixed with DPPC; Surface pressure (π)-area (A) isotherms	The structural characteristics of the mixed monolayers depend on the interfacial composition and degree of hydrolysis; at surface pressures lower than the equilibrium surface pressure of hydrolysates ($_{\pi e}$SPI hydrolysate), both DPPC and protein are present in the mixed monolayer; at higher surface pressures (at $\pi >$ πeSPI hydrolysate), collapsed protein residues may be displaced from the interface by DPPC molecules

Continued

Table 13.F *Continued*

Reference	Material	Conditions and Evaluated Parameters	Main Results and Conclusions
Martinez et al. (2005)	Isolate; protein hydrolysates (DH: 1.5% and 9.8%)	Proteins (3% wt) and polysaccharide (PS) (0.25% or 0.5 wt%) solutions; PS used were xanthan (X), γ and κ-carrageenan (γC and κC), guar (G) and locust bean gum (LB); foam overrun and foam stability	A small degree of hydrolysis (DH: 1.5%) enhanced both foam overrun and stability; an increase of DH to 9.8% did not further improve foaming properties; overrun of foams was decreased in the presence of all the PS; increasing PS concentration the detrimental effect could be partially reverted
Rodríguez-Patino et al. (2007)	Isolate; hydrolysates with different degree of hydrolysis (DH: 5.6%, 23.3%, and 46.3%)	Protein solutions (0.01–1 × 10–6 % w/w); temperature 20 °C; pH 7; I = 0.05 M; surface tension and dilatational; foam capacity and stability; foam maximum density	Superficial activity is lower for isolate than for the hydrolysates, and decreases at the higher DH; foaming characteristics are poor for isolate, but are improved for SPI hydrolysates; foam capacity and stability are optimum for a low DH (3–5%).
Karayannidou et al. (2007)	Isolates; limited protein hydrolysates	Dispersions or 1.0% (w/v) pH value 5.5 or 7.0; 0.1 % (w/v) XG; 0.25, 0.5 M NaCl solutions; foam ability and stability	Limited proteolysis isolate had better emulsifying and foaming properties than had the native isolates; NaCl enhanced foam ability while the addition of xanthan gum improved foam stability
Miñones-Conde and Rodríguez-Patino (2007)	Isolate; hydrolysates with different degree of hydrolysis (DH: 5.6%, 23.3%, and 46.3%)	Protein solutions (0.1–1 × 10–5 w/w) Temperature 20 °C, pH 7, and I = 0.05 M; surface pressure, adsorption time	Optimum functionality of sunflower protein hydrolysates occurs at low degrees of hydrolysis and high protein concentrations in solution

Table 13.G Literature Overview on Emulsion Properties of Sunflower Protein

Reference	Material	Conditions and Evaluated Parameters	Main Results and Conclusions
Lin et al. (1974)	Flour; isolate; concentrate	Dispersions (5.5%); emulsion capacity	Emulsion capacity of flour superior to that of concentrates and isolates of sunflower and to that of soy flour isolates and concentrates
Huffman et al. (1975)	Meal	Dispersions (6%); pH 5.2, 7, and 10.8; emulsion capacity	Highest emulsion capacity at pH 7
Wu et al. (1976)	Meal	Dispersions; pH 7; emulsion capacity	Re-extraction of meal with several solvents (benzene, chloroform, methanol, petroleum ether, chloroform/methanol, ethanol/ether/water) did not affect emulsion capacity
Canella et al. (1977)	Flour; concentrate; isolate	Dispersions (5.5%); emulsion capacity	Emulsion capacity of sunflower isolate higher than for meal but smaller than for concentrate; emulsion capacity of isolate better than for soy isolate; emulsion capacity of concentrate poorer than for soy concentrate
Rahma and Rao (1981)	Meal	Dispersions (8%); emulsion capacity	Aqueous ethanol treatment decreased emulsion capacity
Schwenke et al. (1981)	Isolate	Dispersions (0.5%); pH 5 and 7; emulsion activity, capacity, and stability	Emulsion activity, capacity, and stability higher at pH 7 than at pH 5; protein denaturation (pH 2, 24 h) had no effect on emulsion activity and stability but decreased emulsion capacity; better emulsion properties than soy proteins
Rossi and Germondari (1982)	Meal	Dispersions (7%); pH 5.2, 7 and 10.8; emulsion activity and stability	Emulsion activity and stability high at pH > 5; minimum emulsion activity close to pH 3.75; better emulsion properties than soy meal

Continued

Table 13.G *Continued*

Reference	Material	Conditions and Evaluated Parameters	Main Results and Conclusions
Canella et al. (1985)	Meal; albumins	Dispersions (1%) for emulsion capacity; dispersions (7%) for emulsion activity and stability	Low emulsion activity and stability of helianthinin compared with albumins and sunflower meal; thermal denaturation of helianthinin improved emulsion activity and stability; emulsion capacity lower for albumins and helianthinin than for meal
Khalil et al. (1985)	Isolate	Dispersions (10%); pH 8; emulsion capacity	Lower emulsion capacity than for soy isolate; heating improved emulsion capacity
Raymond et al. (1985)	Isolate	Dispersions (0.1%); pH 3–10; emulsion capacity	Maximum emulsion capacity at pH 8; equivalent to soy isolate
Rossi et al. (1985)	Meal; concentrate	Dispersions (4%); pH 2–9; emulsion activity and stability	Emulsion activity maximum (meal and concentrate) between pH 6.5 and 7.5 and minimum at pH 4.5–5; high emulsion stability of concentrate and independent of pH; emulsion properties improved close to isoelectric point by salt addition; heat denaturation reduced emulsion properties
Brueckner et al. (1986)	Concentrate; isolate	Dispersions in water; emulsion activity, capacity, and stability	Emulsion activity and stability of sunflower concentrates similar to those of soy concentrates, but higher emulsion capacity; emulsion activity, capacity, and stability much better for soy isolates
Booma and Prakash (1990)	Helianthinin flour	Dispersions (8.7 and 4%); pH 6; emulsion capacity	Emulsion capacity of helianthinin double that of sunflower flour
Venktesh and Prakash (1993)	Meal	Dispersions (5%); pH 7; emulsion activity, capacity, and stability	Higher emulsion stability in water than in presence of NaCl (1 mol L^{-1}); increased emulsion capacity and stability after heating in presence of salt
Wastyn et al. (1993	Isolate; concentrate	Dispersions in water; emulsion capacity	Good emulsion capacity

Guéguen et al. (1996)	Albumins	Dispersions (0.05 and 0.1%); pH 8; creaming flocculation and resistance to coalescence	Stable emulsions; different emulsion stabilization activities between several albumins; a methionine-rich albumin (SFA 8) was most active in emulsion stabilisation
Popineau et al. (1998)	Albumins	Dispersions (0.1%); pH 7; creaming flocculation and resistance to coalescence	Resistance to coalescence was much higher with methionine-rich albumins than with methionine-poor albumins; disulfide bond reduction resulted in very stable emulsions
Pawar et al. (2001)	Meal; isolate; concentrate	Dispersions (8%); emulsion activity, capacity, and stability	Increased emulsion activity, capacity, and stability for protein products with small amounts of phytate and phenolic compound
Burnett et al. (2002)	Albumins	Dispersions (up to 0.5%); pH 7; droplet size, surface tension and surface dilation viscosity	Stable emulsions with SFA 8 and SFA 7 (methionine-rich proteins); less hydrophobic sunflower albumin proteins (lipid transfer proteins, SF-LTP) gave unstable emulsions
González-Pérez et al. (2005)	Isolate; helianthinin; albumins	Protein solutions (0.2–0.5%); pH 3–8; emulsion formation and stability; effects of heat and acid denaturation	Sunflower albumins, helianthinin, and isolate formed stable emulsions, with exception of SFAs at alkaline and neutral pH values; helianthinin protein denaturation resulted in improved emulsion properties
Karayannidou et al. (2007)	Isolates, limited protein hydrolysates	Dispersions or 1.0% (w/v) pH value 5.5 or 7; emulsion preparation and stability	Limited proteolysis isolate had better emulsifying properties than had the native isolates

(intrinsic parameters) and operational (extrinsic parameters) differences. Thus, different protein contents of the samples used as well as the presence of other nonprotein minor compounds found in the inherent composition of sunflower, such as polysaccharides or chlorogenic acid, are known to affect to the functional properties of foods. Likewise, the methods of oil extraction and the processing of meal to obtain protein concentrates and isolates will influence in the final composition of these protein products, affecting protein conformation and the amino acid residues depending on the severity of the conditions used. Furthermore, not only intrinsic parameters but also parameters extrinsic to the composition of sunflower meals, concentrates, and isolates, such as the choice of pH and ionic strength, affect the solubility of sunflower proteins, which govern the first step in the formation of foams and emulsions by altering the inclusion of these proteins in the interphase of air–water or oil–water, respectively.

In any case, a series of conclusions or key aspects could be drawn up for a successful formation of foams and emulsions and their subsequent stability by different protein products of sunflower. As previously discussed, solubility seems critical in the formation of disperse phases. Thus, factors that can improve the solubility will have a positive effect on this purpose. Claughton and Pearce (1989) observed that a greater solubility of isolates was well correlated with foaming capacity but not with foam stability. Three-step mechanisms are observed in foam and emulsion formation: adsorption, aggregation and subsequent reorganization, and refolding of the proteins. The first two steps, adsorption and aggregation, involve the location of amphipathic components, such as proteins, at the interface, producing the correspondent foam or emulsion. The subsequent reorganization, interaction, and refolding of the proteins and other compounds are responsible for the different grades of stability of dispersed phases formed. Thus, good adsorption and formation of an adsorbed layer in the interface is not of good stability.

Likewise, dispersions carried out far from the isoelectric point of sunflower protein (pH 4.0–5.0) seem to give better results for the formation of foams (Canella, 1978; Claughton and Pearce, 1989; Kabirullah and Wills, 1988; Khalil et al., 1985; Raymond et al., 1985; Rossi and Germondari, 1982; Rossi et al., 1985) and emulsions (Huffman et al., 1975; Raymond et al., 1985; Rossi and Germondari, 1982; Rossi et al., 1985; Schwenke et al., 1981). However, foam stability did not show the same pattern as foam formation at pH values far from pI. This effect could be due to several reasons: (1) the minimum foamability of protein solutions observed near pI might be a consequence of the protein precipitation observed, i.e., there is not enough protein capable of adsorbing at the surfaces; (2) hydrophobic interactions are reduced among the protein molecules at pH values far from the pI due to electrostatic repulsion, which could facilitate the flexibility of the protein molecules and enable proteins to diffuse more rapidly to the interfaces.

Similarly, protein denaturation could help to expose possible hydrophobic groups hidden in the original conformation of the protein, improving of foaming capacity. In so doing, the amino acid composition and the conformation of the sunflower proteins (globulins and albumins) contribute quite considerably to the formation and stability of foams and emulsions. Thus, it seems that amino acid composition and conformation of sunflower albumins play a greater role in the formation of foams and emulsions than helianthinin, which seems to be involved in stabilization parameters. However, several authors have shown albumins, mainly albumins with methionine rich residues in its amino acid composition, to form stable emulsions (Burnett et al., 2002; Guéguen et al., 1996; Popineau et al., 1998), reconfirming the importance of certain amino acids in the formation and/or stabilization of emulsions.

Protein hydrolysis has been shown as a modification of proteins that can improve the functional properties of proteins by (1) increasing the protein solubility, and (2) exposing hydrophobic groups that may be cloistered in the interior of the protein conformation and that are essential in functional aspects, such as oil absorption and formation of foams and emulsions. As shown in the results from several authors, sunflower protein hydrolysates with different degree of hydrolysis contribute in broken peptide bonds contribute in a different way in the formation and stabilization of foams and emulsions. As commented previously, when higher degrees of hydrolysis are reached, higher protein solubility is obtained. This involves very fast adsorption to the interface, but more difficulty in forming a stable absorbed layer and, subsequently, more difficulty forming stable foams and emulsions (Figure 13.4). Thus, a limited degree of hydrolysis usually improves the emulsifying and foaming capacities, whereas excessive hydrolysis often causes loss of some of these functionalities (Barać et al., 2011; Pedroche et al., 2004; Yust et al., 2010; Zhao et al., 2011). Some studies confirm that the hydrolysis of sunflower native proteins confers a positive effect to produce stable foams, but in a limited way without exceeding hydrolysis levels of 5–8% (Karayannidou et al., 2007; Martínez et al., 2005; Miñones-Conde et al. 2005, 2006; Miñones-Conde and Rodríguez-Patino 2007; Rodríguez-Patino et al., 2007).

Nevertheless, extensive proteolysis reduces the incidence of allergenic factors and can result in the formation of small peptides with biological (antioxidant, antihypertensive, antithrombotic, anti-inflammatory) activities. For this reason, proteolysis could be considered as an interesting approach in further studies to compensate for the lack of functionality of sunflower proteins.

Processing

Factors Affecting the Food Applications of Sunflower Meal

The suitability of sunflower meal proteins for food applications will depend mainly on three factors: antinutritional components, amino acid composition, and protein denaturation occurring during oil extraction.

Different studies about antinutritional factors in sunflower seeds and meals have been carried out regarding the content of several compounds such as chlorogenic acid,

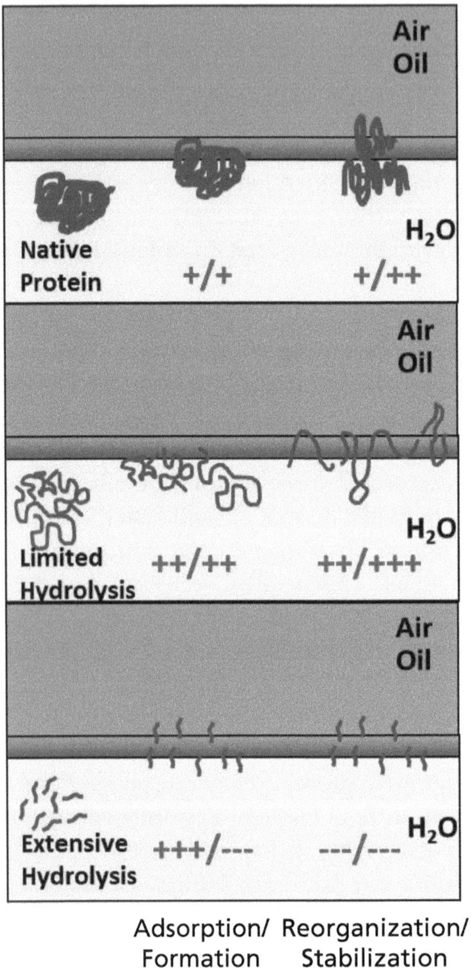

Figure 13.4 Effect to the degree of hydrolysis to the formation and stabilization of foams and emulsions.

saponin, phytic acid, trypsin inhibitors, and even fiber, which can be found at high levels in sunflower seeds and meals.

The comparatively high content of phenolic compounds in Helianthus (2–7% in defatted flour) (Table 13.D), mainly chlorogenic and caffeic acids, has been the main restriction for sunflower protein application in food commodities so far. Conventional processes of protein extraction yield dark-colored products with decreased nutritional and functional quality. They affect the quality of sunflower protein in several ways, such as reducing digestibility, altering organoleptic properties, prolonging or shortening storage life and stability, and adversely altering the functional properties and behavior of sunflower protein in food systems (Bau et al., 1983; González-Pérez and Vereijken, 2007; Kroll et al., 2003; Moure et al., 2006; Sastry and Rao 1990). The interaction may become irreversible when, under alkaline conditions, phenolic compounds autocatalytically oxidize to quinones and react with functional protein groups, forming covalent bonds with reactive groups such as amines, thiols, thioethers, indole, imidazole, and disulfide groups (Ozdal et al., 2013; Rawel et al., 2002a; Saeed and Cheryan, 1989; Świeca et al., 2013).

Many methods have been proposed for removing phenolic compounds from sunflower seeds and meals. Of all of the methods described, the most promising ones with respect to the efficiency of chlorogenic acid extraction are those that extract phenolic compounds with mixtures of organic solvents and water (De Leonardis et al., 2005; González-Pérez et al., 2002). However, disadvantages of these methods are not only that organic solvent–water mixtures are known to cause protein denaturation, which may result in lower solubility and protein recovery, but also that they can alter the digestibility and availability of essential amino acids, involving higher processing costs for solvent management and safety precautions (Grinberg et al., 1998; Rawel et al., 2002b; Sripad and Rao, 1987; Suryaprakash et al., 2000).

On the other hand, all phenolic compounds exhibit at least one aromatic OH group imparting antioxidant activity. Therefore, they are supposed to protect plants from oxidative damage as a consequence of biotic or abiotic stress factors (Robards and Antolovich, 1997). In addition, phenolics are widely associated with antimicrobial (Daglia, 2012), antidiarrheal (Nwidu et al., 2011), antiulcer (Hisam et al., 2012), anti-inflammatory (Lopresti et al., 2012), antiviral, and anticarcinogenic activities (Bishayee et al., 2013). Moreover they are potential protectors against cardiovascular diseases (Quiñones et al., 2013), a further health-promoting effect, which is possibly associated with the prevention of oxidative stress (Havsteen, 2002; Kammerer et al., 2007).

Although for many years research studies aimed to improve the quality of the products enriched in sunflower proteins by removing phenolic compounds (González-Pérez and Vereijken, 2007; Karayannidou et al., 2007; Pickardt et al., 2009; Sripad and Narasinga Rao, 1987), recently, a tendency to keep or even add these compounds in view of their beneficial action has been set (Raskin et al., 2002; Salgado et al., 2012a, 2012b).

In addition, the recovery of phenolic compounds from food processing by-products for their use as functional ingredients also deserves increasing efforts in sustainable food production. Besides food applications, these bioactive compounds also bear a large potential in the pharmaceutical and cosmetics industries as demonstrated by recently launched products such as mouthwashes and herbal cosmetics (Petti and Crispian, 2009; Saraf and Kaur, 2010; Soares de Moura, 2010).

With regard to antinutritional compounds in sunflower meals, Ingale and Shrivastava (2011) investigated several antinutritional factors in the LSF-11 and LSF-8 varieties of sunflower seeds. Antinutritional factors such as cyanogenic glucoside, tannin, oxalate, and hemagglutinin activity contents ranged around 4.026–4.175, 0.623–0.651, 0.098–0.113 g/100g, and 1:16–1:8 (w:w), respectively, in both varieties. They concluded that sunflower seeds have no trypsin inhibitors and relatively low levels of antinutritional factors (cyanogenic glucoside, tannin, and oxalate). Besides that, authors found these varieties are nontoxic feed for rats. Their potential for nutritional exploitation in the human population is very important not only from a nutritional point of view, but also because they would not require prolonged and expensive heat-treatment in an industrial scale prior to use.

On the other hand, Della Gatta and Piergiovianni (1996) found that the addition of sunflower meal in baked goods enhanced protein content but had a detrimental effect on bread quality. The incorporation of sunflower meal produced bread with a remarkable content of trypsin inhibitors compared with the control. Consequently, proteins are less susceptible to proteolysis. This lower availability of proteins partially counteracts the improvement of the nutritional quality obtained with the incorporation of sunflower meal. The incorporation of 10% sunflower meal was found to be acceptable from a functional point of view. Likewise, Petit (2003) concluded that both flaxseed and sunflower seed are acceptable fat sources for midlactating cows (feeding up to 30% of plant seeds) in terms of milk production. However, protein concentration in milk composition was slightly higher for cows fed flaxseed (3.41%) than for those fed sunflower seeds (3.25%), despite the higher digestibility of crude protein in sunflower (62.1%) than flaxseed-based diets (56.0%). These differences in protein concentration in the cows' milk were attributed to different seed size and amino acid composition of both raw materials, rather than a trypsin inhibitor affecting digestibility results of crude protein. This smaller size, which might increase the rate of passage from the rumen, joined to the amino acid composition could increase the supply of amino acids for milk protein synthesis.

Fiber is mainly constituted of polysaccharides such as cellulose, hemicellulose, lignin, pectic substances, and so forth, and it is generally more abundant on the cover of the grains (Pérez et al., 2007). It is of great importance for its abundance in some commodities like sunflower. The antinutritional interactions of fiber with proteins occurs mainly by the crosslinking with specific amino acids, which increases the volume of nondigestible fraction compounds and affects the digestibility of certain individual amino acids, especially lysine (Baer et al., 1997; De Lange et al., 1998; Sánchez-Lozano et al., 2007). Furthermore, high levels of fiber lead to a shorter residence time of feed in the digestive system and, therefore, lower enzyme attack and nutrient digestibility in the small intestine, diminishing efficient nutrient utilization (Wenk, 2001). Nutritionists usually avoid high fiber content in poultry diets because they have low energy values. Furthermore, fiber may cause abrasion of the intestinal mucosa, decreasing its functionality, and thus increasing the amino acid requirements for the synthesis of mucosa cells (Parsons et al., 1983). However, sunflower products with high fiber content are being reevaluated for different technological applications. It has been demonstrated that broiler chickens may thrive on sunflower meal with insoluble fiber contents up to 30% with no adverse effect on growth (Kalmendal et al., 2011). In addition, these fiber-enriched products exerted some positive effects on digestion and small intestinal health (Ibrahim and El Zubeir, 1991; Kalmendal et al., 2011). Besides that, sunflower meal is also used as substrate for the production of bioethanol (Ruiz et al., 2006), biodiesel (Iriarte and Villalobos, 2013), and biogas (Fernández-Cegri et al., 2012).

Another factor to take into account in the food application of sunflower proteins is protein denaturation during oil extraction and meal processing. Denaturation involves modifications in the three-dimensional organization of protein structure. Protein denaturation may occur during seed conditioning, expelling (temperatures up to 140 °C can be reached), and desolventising/toasting because temperatures in these operations exceed the denaturation temperature of sunflower globulins such as 11S (the main protein fraction in sunflower meal and protein isolates), which is about 95 °C (González-Pérez and Vereijken, 2007; Rouilly et al., 2003). Protein denaturation results in residual sunflower meals with high contents of insoluble proteins. Likewise, association of proteins with other nonprotein compounds may occur during the preparation of the protein isolate, mainly at alkaline pHs. Alkaline processing can alter protein quality due to undesirable reactions involving (1) racemization of amino acids, and (2) formation of lysinoalanine by reactions of lysine with dehydroalanine, produced by degradation of cysteine and serine. Both of them lead to the reduction of protein digestibility and loss of essential amino acids (Moure et al., 2006).

Sunflower Protein Isolates and Hydrolysates

As previously noted, a large amount of agro-industrial waste rich in proteins is generated, which includes defatted meals from oil extractions of plant seeds. Although its primary use is currently intended for animal feed, it represents one of the reservoirs of proteins with the greatest potential for the food industry. Processes intended to obtain and improve these proteins through the production of protein isolates have been developed. The main raw materials used for preparation of plant protein isolates are from seeds rich in proteins (Table 13.H).

The elaboration of protein isolates involves the purification and concentration of this component of food to a concentration up to 80–90%. The nonprotein components that are separated and discarded during the purification steps can interact with proteins during the process, affecting the quality of the final product. These constituents include: lipids, fiber, carbohydrates, and minor components such as mineral substances, polyphenols, glucosinolates, phytates, phorbol esthers, and so forth. Sunflower seeds show a chemical composition including around 40% lipids, 30% proteins, and 25% carbohydrates, and it is a raw material with low levels of antinutritional factors, even if polyphenols (mainly chlorogenic acid) are present in a high proportion. Sunflower defatted meal obtained after oil solvent extraction (hexane) presents protein contents around 40%, depending on the variety of sunflower seeds and the process of oil extraction (Table 13.D).

Several attempts have been made to obtain sunflower protein isolates from defatted meals. They are mainly based on the principles of solubilization of protein fractions and subsequently protein precipitation at the isoelectric point. Commonly, sunflower proteins are solubilized under alkaline conditions due to the protein solubility profile, and soluble protein is recovered from the supernatant obtained after centrifugation. In a later step, solubilized proteins are precipitated, lowering the pH of the extract by isoelectric precipitation, pI (pH 4–5), where sunflower proteins exhibit a minimum solubility, and approximately 10–20% of the protein is soluble (Figure 13.2) (Karayannidou et al., 2007; Martínez et al., 2005; Miñones-Conde and Rodríguez-Patino, 2007; Miñones-Conde et al., 2005, 2006; Orliac et al., 2002, 2003; Rouilly et al., 2003; Rodríguez-Patino et al., 2007; Rouilly et al., 2001; Salgado et al., 2010; 2012a, 2012b, 2012c; Villanueva et al., 1999). In this regard, acidic or neutral extractions have been developed with the goal of preventing irreversible binding of phenolic acids to proteins (Karayannidou et al., 2007; Pawar et al., 2001; Pickardt et al., 2009, 2011; Saeed and Cheryan, 1989, Weisz et al., 2010, 2013). In this last case, due to the poor solubility of sunflower proteins in the low-acid region, optimization of protein extraction can be carried out on salts addition (Gheyasuddin et al., 1970; Pawar et al., 2001; Pickardt et al., 2009, 2011). The use of ultrafiltration techniques has also been reported for the protein recovery (González-Pérez et al., 2002, 2005; Karayannidou et al., 2007; O'Connor, 1971). This ultrafiltration step has an added benefit of removing low molecular weight constituents, including phenolic compounds. Conversely, occurrence of membrane fouling could cause severe reduction in membrane performance that includes loss of productivity and selectivity.

Besides that, different procedures or modifications have been developed during the production of sunflower protein isolates or recovery of sunflower protein fractions, most of them with the goal of reducing or removing the polyphenols content from the final sunflower protein isolates. Addition of Na_2S0_3 (0.01 M), dioxane (4% [v/v]), or urea (8 M) to the solution buffer in the extraction process prevented the

Table 13.H Chemical Composition of Different Raw Materials

	Proteins (%)	Lipids (%)	Carbohydrates (%)	Moisture (%)	Protein Efficiency Ratio	Limiting Amino Acid	Antinutritional Factors
Soybean	40	20	15	7	2.3	Met	TI[a], hemagglutinins, α-galactosides
Cotton	53	30	10	6	2.3	Lys, Met	Gossypol
Peanut	25	50	20	9	1.7	Met	Aflatoxins
Sunflower	30	40	25	6	2.1	Lys	Polyphenols
Rapeseed	26	33	35	3	2.6	Equilibrated	Glucosinolates, sinapine
Lupine	40	10	45	10	0.7	Met	Alkaloids
Pea	23	2	70	12	2.0	Met	TI, phytates
Bean	30	2	60	12	1.8	Met	TI, vicine, tannins
Alfalfa	25	5	45	80	1.6	Met	Alkaloids, tannins, cumarine
Wheat	14	2	78	10	1.8	Lys	—
Oat	13	10	72	9	2.2	Lys	—
Maize	10	5	80	10	1.2	Lys, Trp	—
Rice	8	1	87	12	1.7	Lys	—
Potato	9	0	73	78	2.0	Lys, Thr	—

[a]Trypsin inhibitors.

Source: Cheftel et al. (1985).

binding between proteins and phenols (Sastry and Narasinga-Rao, 1990). Adsorbent and ion exchange resins have been also used to remove the phenolic compounds during the purification of sunflower protein (Weisz et al., 2010). Mixtures of organic solvents and water (ethanol 95% [v/v], 2-propanol 70% [v/v], and methanol, 80% [v/v]) were used for the removal of chlorogenic acid. These treatments did not produce any denaturation of the sunflower proteins in the subsequent processes of protein extraction at alkaline pH and diafiltration (González-Pérez et al., 2002).

However, these protein isolates in their native form have two main disadvantages from a technological and nutritional point of view in applications for human consumption. On the one hand, they show a minimum of solubility at pH around 4–5. Most protein drinks available in the market are high pH (>4.5) beverages (Eckert and Riker, 2007). According to Diosady et al. (2007) and Eckert and Riker (2007), soy protein isolates have not been typically used in beverages with pH less than 4.5, because, upon protein dispersion, the protein does not dissolve and instead forms a colloidal dispersion, which results in a cloudy beverage. On the other hand, allergiesfrom food native proteins are becoming more frequent, and several authors have reported that sunflower seeds have the potential to cause severe anaphylactic reactions (Axelsson et al., 1994; Besler et al., 2001; Kelly and Hefle, 2000; Mayaud-Marret et al., 2000; Yagami, 2010; Zitouni et al., 2000). Kelly and Hefle (2000) observed that methionine-rich 2S albumin from sunflower seeds was specifically recognized by IgE from the serum of the sunflower seed–allergic subject. In this regard, other 2S albumin proteins from plant seeds, such as rapeseed, soy, castor beans, Brazil nuts, and walnuts has also been associated with food allergy (Alcocer et al., 2012; Dubois et al., 2013; Puumalainen et al., 2006; Vissers et al., 2011).

Hydrolysate-based nutritionals have several potential advantages over elemental diets. First, small peptides are better absorbed and nutritionally processed than intact protein (Ziegler et al., 1990) or free amino acids (Grimble and David, 1989; Minami et al., 1992; Siemensma et al., 1993; Silk et al., 1980). Second, hydrolysates are generally less chemically reactive than amino acid solutions. This allows improved product functionality and stability. Third, hydrolysates have the potential for improving flavor better than amino acids. Finally, nutritionally complete hydrolysates are generally less expensive than a nutritionally balanced mix of amino acids (Mahmoud and Cordle, 2000).

As previously discussed, protein hydrolysis has a direct effect on the improvement of the protein solubility and, in addition, on other functional properties such as gelation, emulsifying, and foaming capacity (Miñones-Conde et al., 2005; Parrado et al., 1993; Sánchez and Burgos, 1996; Villanueva et al., 1999). During the last decades, the physicochemical properties of colloidal dispersions have been the subject of a great number of research studies that mainly addressed the capacity and stability of foams, emulsions, or gels. There has also been research focusing on the behavior of these dispersions during their formation and reorganization in model systems.

Recently, a better understanding of processes that govern the physicochemical and structural changes in colloidal dispersions has been obtained by approaching a physiological point of view. It is not only important to set the parameters to control the formation and stabilization of foams, emulsions, or gels, but also it seems that this rheological conformation is essential in the control of the assimilation and bioavailability of food and bioactive compound formulations during the intake and passage through the gastrointestinal system.

Likewise, reduction of food allergen reactivity by hydrolysis has been studied extensively. Hydrolysate-based hypoallergenic products for infant nutrition have been available for almost 50 years (Knight, 1985). These products usually contain hydrolysates of cow's milk casein or whey proteins, as well as rice proteins. Soy protein hydrolysates have also been evaluated for hypoallergenic infant formulas.

Production of protein hydrolysates is initially based on enzymatic hydrolysis of the raw material chosen. The use and availability of food-grade commercial proteases derived from microorganisms is an advantage from the technological point of view because they are considered to be safe and low-cost components (Tavano, 2013). Traditionally, the use of products from enzymatic hydrolysis has been restricted to the field of rheological properties of food proteins. Today, the bioactivity of some peptides offers an interesting opportunity to create a personalized and targeted nutrition (Clemente, 2000). The world market for products with bioactive ingredients is continuously growing. Thus, certain proteins and, more specifically, certain peptides located in the sequence of them, have antioxidant, antimicrobial, immunomodulatory, antihypertensive, and other biological activities that could be used as preventive treatment or control in certain diseases (García et al., 2013; Hettiarachchy et al., 2012; Korhonen and Pihlanto, 2006). Consequently, it is very likely that, in the coming decades, the bioactive peptides market becomes an autonomous sector promoting new application areas related to the field of nutrition and health. Similarly, the growing demand in such foods can be explained by the increase in the cost of health services, the constant increase in life expectancy, and the desire of elder people to improve the quality of their later years.

With regard to sunflower, protein hydrolysates with certain bioactive properties prepared from its meal have been reported. These involve antioxidative (Megías et al., 2008, 2009a; Ren et al., 2010; Taha et al., 2013), antihypertensive (Megías et al., 2004, 2009a, 2009b), antimicrobial (Taha et al., 2013), hypocholesterolemic (Megías et al., 2009c), and anticariogenic (Mylne et al., 2011) activities, as well as some promising results on the treatment of patients with liver failure (Bautista et al., 2000). Megías et al. (2004) identified a peptide with the sequence Phe-Val-Asn-Pro-Gln-Ala-Gly-Ser (FVNPQAGS) that inhibits angiotensin-I converting enzyme (ACE). This peptide corresponds to a fragment of helianthinin, the 11S globulin from sunflower seeds, which is the main storage protein in sunflower. It has been purified from a final sunflower protein hydrolysate obtained by hydrolysis using pepsin and pancreatin, resembling those peptides generated during the digestion of proteins in the organism that might be readily absorbed at the intestinal epithelium and then reach the bloodstream and act peripherally.

Applications of Sunflower Proteins

Sunflower Meal Proteins in Animal Feed

The main outlet for sunflower meal is the animal feed market. Sunflower meal is a valuable protein source for various species/categories in livestock feeding. However, specific restrictions in animal feed are taken into consideration such as the maximum incorporation rates, the species concerned or r nonprotein compounds, (i.e., fiber) (Table 13.I) (OECD, 2007).

As discussed previously, the nutrient composition of sunflower meal is dependent on the oil content of the seed, extent of hull removal, efficiency of oil extraction, and the processing temperature. The fiber in sunflower meal is low in digestibility and may be a disadvantage when balancing rations for nonruminant and high producing animals.

At present, sunflower meal is mainly used for rabbits and ruminants (respective incorporation rates are 10–12% and 10–20%). Levels for laying birds (less than 5%) or pigs (less than 1%) have been noted (OECD, 2007). Hulled meal, in which the protein content has been increased nearly to 40% by an efficient hulling, can be a useful feed for broilers and fattening pigs. In these cases, the incorporation rates of previously hulled sunflower meal may then be increased to at least 10% (OECD, 2007).

Sunflower meal is more ruminally degradable (74% of crude protein) than soybean meal (66%) or canola meal (68%). Sunflower meal is high in protein, but due

Table 13.I Maximum Incorporation Rates for Sunflower Extracted Meal in Rations for Livestock Feed

Species/Category	Incorporation Level (%)
Dairy cows	30
Rearing calves	15
Cattle and bulls	40–50
Sheep, goat	No limitation
Rabbits	30
Growing-finishing pigs	5–10
Poultry	5–10

to the lack of a sufficient content of lysine, is more suitable for ruminants than non-ruminants (OECD, 2007).

Sunflower Proteins in Human Consumption

Concerning direct human consumption, sunflower seeds are either consumed plain (salted only) or as roasted sunflower nuts with a variety of flavorings added (salted, barbecue, dill pickle, bacon, as well as others), chocolate or honey covered snacks, as well as part of crackers, tofus, or bread. Sunflower seed milk can be also found in many markets.

Nevertheless, more than two-thirds of all agricultural land is devoted to growing feed for livestock, while only 8% is used to grow food for direct human consumption (LEAD-FAO, 2013).

In this regard, sunflower protein as an ingredient in food formulations through products such as defatted flour and protein isolates or hydrolysates is not widespread. This is despite the fact that, as commented previously, sunflower seeds are considered a good quality product, containing low amounts of antinutritional compounds and a balanced amino acid composition. Several studies have exposed the safety and nutritional advantages of sunflower protein products for human consumption (Gandhi et al., 2008; Gassmann, 1983; Skoric, 2009; Yañez et al., 1969). The main reasons to explain this reduced application could be, on the one hand, fiber and polyphenol contents of sunflower meals as the main handicaps for the implementation of these products in processed foods for human consumption, and on the other hand, limitations to the production of large quantities to satisfy the growing market demand make protein isolates or hydrolysates uncompetitive compared with other plant ingredients found in elaborated foods such as protein concentrates or isolates of soybean, rice, or wheat.

Trends, Recommendations, and Conclusions

The projections of land use in the developing countries (such as India, the Philippines, Brazil, and Thailand) indicate that the four main oil crops (oil palm, soybeans, sunflower seed, and rapeseed) will continue to account for a good part of future expansion of harvested area. These developing countries have so far been net exporters of oil meals. This has enabled them to maintain a positive, though declining, trade balance in terms of the value of their combined trade of oilseeds, oils, and meals. However, with the development of their livestock sector and their own meat consumption, the prospect is that their net exports of meals could turn into net imports of agricultural products to feed livestock production.

LEAD researchers have also found that the global livestock industry uses dwindling supplies of freshwater, destroys forests and grasslands, and causes soil erosion, while pollution and the runoff of fertilizer and animal waste create dead zones in coastal areas and smother coral reefs. There also is concern over increased antibiotic resistance, since livestock accounts for 50% of antibiotic use globally (LEAD-FAO, 2013).

In recent years, in many countries, public concern about the safety of foods of animal origin has heightened due to problems that have arisen due to bovine spongiform encephalopathy (BSE), dioxin contamination, outbreaks of food-borne bacterial infections, growing concern about veterinary drug residues, and microbial resistance to antibiotics. However, can be observed in Figure 13.5, the per capita protein supply from animal products is higher in those continents with a high ratio of developed countries (Europe, North America, and Oceania) than in continents with a high ratio of developing countries (Africa, Asia). In addition, meat production is projected to double by 2020 due to increased global consumption of meat and population growth (FAO, 2011; Gandhi and Zhou, 2014; Kristensen et al., 2014). Most of this increase

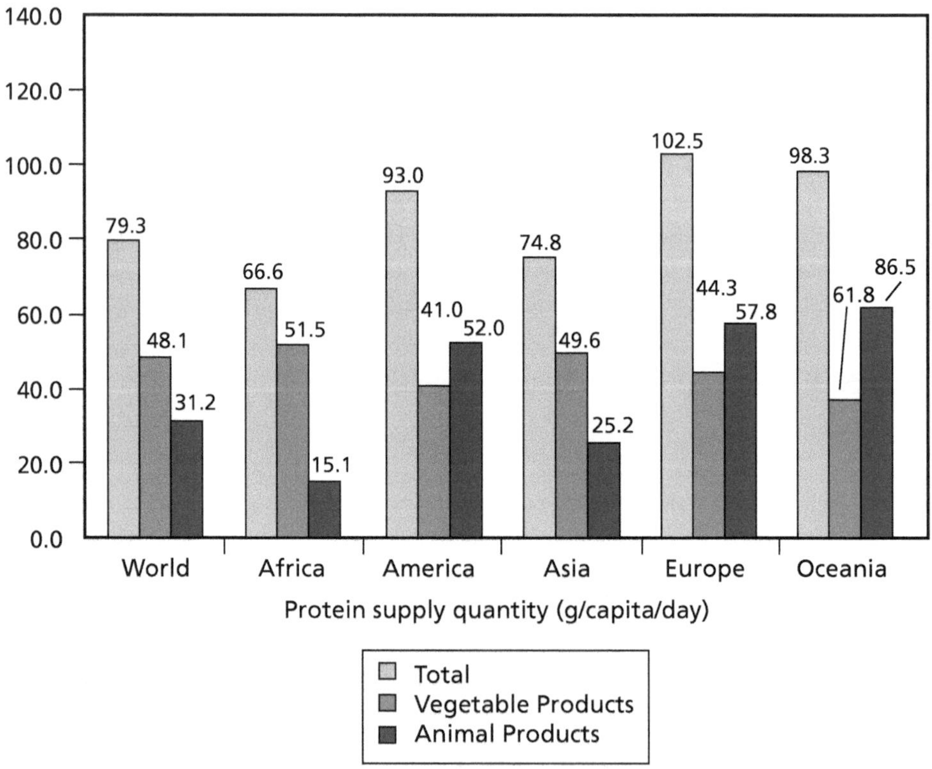

Figure 13.5 Protein supply per capita at a worldwide level.

in production will come through industrialized animal production systems. These trends will have major consequences on the global environment.

Thus, there is a concomitant rise in the need for animal feeds particularly oil cakes and meals, but at the same time, there is increased public concern about contaminants and health, and demand for safety, regulation, and traceability. Against this daunting background, the collection of good quality data relating to world market forces and the availability of proteins for the feed industry should be addressed. Innovative developments in the feed industry should be sought with a view to providing alternative sources of proteins and new amino acid technologies (FAO, 2002).

In fact, interest in the use of plant proteins is increasing. Plant proteins are economic and sustainable alternatives to animal proteins as functional ingredients in food formulations. Oilseeds are the most important source of plant protein preparations. Among these alternative sources, as has been noted previously, sunflower seeds are particularly interesting in view of their widespread availability in areas where soy is not produced or is only sparsely produced. Furthermore, sunflower is a well-known crop to both farmers and processors because of its extensive use for oil extraction. In addition, sunflower seeds have been reported to contain low amounts of antinutritional factors (such as protease inhibitors, cyanogens, and glusosinolates), and their amino acid composition (except for lysine) complies with the FAO pattern in amino acid requirements into a balanced diet.

Until recently, the focus of the industry in sunflower cultivation has been put almost exclusively on oil extraction and production. Most of the research activities carried out in recent years have clearly revealed the potential of sunflower proteins as a high value-added component for human nutrition and biofilms production. However, limitation on the production of large quantities to satisfy the growing market demands include the lack of viable bioprocesses that are transferable to industrial scale. Consequently, the main current use of this protein byproduct is in animal feed. Process development for production of sunflower-based protein concentrates, isolates, and hydrolysates tailored for specific food applications and the improvement of the functional and rheological properties of the sunflower proteins could expand their market potential.

To date, a large number and diverse range of high added-value products are also present in sunflower meals, mainly polyphenols and fiber, not considering the sunflower meal as a whole value chain of sunflower production. All of these products (protein, polyphenols, and fiber) and the beneficial activities associated with them are highly demanded by society and their valorization could lead to economic viability and sustainability of sunflower exploitation. Therefore, a new innovative industrial model is required for the use of valuable sunflower biomass, based in the concept of a biorefinery that could radically transform the already-developed single-product-oriented model into multiple high value-added products from sunflower biomass, achieving an integral valorization in the most economically profitable way.

References

Abdel-Hakim, N. F. M.; Lashin, M. E.; Al-Azab, A. A.; Nazmi, H. M. Effect of Replacing Soybean Meal Protein by Other Plant Protein Sources on Growth Performance and Economic Efficiency of Mono Sex Nile Tilapia (*Oreochromis niloticus*) Cultured in Tanks. In *Proceedings of the 8th International Symposium on Tilapia in Aquaculture,* Cairo, Egypt, 2008; pp 739–755.

Adams, A.; Kitryté, V.; Venskutonis, R.; De Kimpe, N. Model Studies on the Pattern of Volatiles Generated in Mixtures of Amino Acids, Lipid-oxidation-derived Aldehydes, and Glucose. *J. Agric. Food Chem.* **2011**, *59,* 1449–1456.

Agribusiness Handbook: Sunflower Crude and Refined Oils. FAO: Rome, Italy, 2010. http://www.fao.org/docrep/012/al375e/al375e.pdf.

Alcocer, M.; Rundqvist, L.; Larsson, G. Ber e 1 Protein: The Versatile Major Allergen from Brazil Nut. *Biotechnol. Lett.* **2012**, *34,* 597–610.

Anderson, V. *Sunflower Meal in Beef Cattle Diets.* North Dakota State University Extension Service, Fargo, ND, 2002.

Axelsson, I. G. K.; Ihre, E.; Zetterström, O. Anaphylactic Reactions to Sunflower Seed. *Allergy* **1994**, *49,* 517–520.

Ayhllon-Meixueiro, F.; Vaca-Garcia, C.; Silvestre, F. Biodegradable Films from Isolate of Sunflower (*Helianthus annuus*) Proteins. *J. Agric. Food Chem.* **2000**, *48,* 3032–3036.

Baer, D. J.; Rumpler, W. V.; Miles, C. W.; Fahey, G. C. Dietary Fiber Decreases the Metabolizable Energy Content and Nutrient Digestibility of Mixed Diets Fed to Humans. *J. Nutr.* **1997**, *127,* 579–586.

Barać, M.; Cabrilo, S.; Pešić, M.; Stanojević, S.; Pavlićević, M.; Maćej, O.; Ristić, N. Functional Properties of Pea (*Pisum sativum*, L.) Protein Isolates Modified with Chymosin. *Int. J. Mol. Sci.* **2011**, *12,* 8372–8387.

Bau, H. M.; Mohtadi-Nia, D. J.; Mejean, L.; Debry, G. Preparation of Colorless Sunflower Protein Products: Effect of Processing on Physicochemical and Nutritional Properties. *J. Am. Oil Chem. Soc.* **1983**, *60,* 1141–1148.

Bautista, J.; Corpas, R.; Cremades, O.; Hernández-Pinzón, I.; Ramos, R.; Villanueva, A.; Sánchez-Vioque, R.; Clemente, A.; Pedroche, J.; Vioque, J.; et al. Sunflower Protein Hydrolysates for Dietary Treatment of Patients with Liver Failure. *J. Am. Oil Chem. Soc.* **2000**, *77,* 121–126.

Berot, S.; Briffaud, J. Parameters for Obtaining Concentrates from Rapeseed and Sunflower Meal. *Plant Foods Hum. Nutr.* **1983**, *33,* 237–242.

Besler, M.; Hefle, S. L.; Jensen-Jarolim, E. Sunflower Seed (*Helianthus annuus*). *Internet Symposium on Food Allergens* **2001**, *3,* 103–114. http://www.food-allergens.de.

Bishayee, A.; Thoppil, R. J.; Mandal, A.; Darvesh, A. S.; Ohanyan, V.; Meszaros, J. G.; Háznagy-Radnai, E.; Hohmann, J.; Bhatia, D. Black Currant Phytoconstituents Exert Chemoprevention of Diethylnitrosamine-initiated Hepatocarcinogenesis by Suppression of the Inflammatory Response. *Mol. Carcinog.* **2013**, *52,* 304–317.

Boni, R.; Assogna, A.; Grillo, F.; Robertiello, A.; Petrucci, F.; Giacomozzi, E.; Patricelli, A. Method for Preparing Protein Hydrolysates Soluble in an Acid Environment, and the Hydrolysates Obtained. European Patent EP0 271 964 A2, 1987.

Booma, K.; Prakash, V. Functional Properties of the Flour and the Major Protein Fraction from Sesame Seed, Sunflower Seed and Safflower Seed. *Acta Alim.* **1990**, *19,* 163–176.

Bos, M. T.; Nylander, T.; Arnebrandt, T.; Clark, D. I. Protein/emulsifier Interactions. In *Food Emulsifiers and Their Applications;* Hasenhuette, G. L., Hartel, R. W., Eds.; Chapman & Hall: New York, 1997; pp 95–146.

Brueckner, J.; Mieth, G.; Muschiolik, G. Functional Properties of Plant Proteins in Selected Foods. *Nahrung* **1986**, *30,* 428–429

Burnett, G. R.; Rigby, N. M.; Mills, E. N.C.; Belton, P. S.; Fido, R. J.; Tatham, A. S.; Shewry, P. R. Characterization of the Emulsification Properties of 2S Albumins from Sunflower Seed. *J. Colloid Interface Sci.* **2002**, *247,* 177–185.

Canella, M. Whipping Properties of Sunflower Protein Dispersions. *Food Sci. Technol.* **1978**, *11,* 259–263.

Canella, M.; Castriotta, G.; Bernardi, A.; Boni, R. Functional Properties of Individual Sunflower Albumin and Globulin. *Lebensm Wiss. Technol.* **1985**, *18,* 288–292.

Canella, M.; Castriotta, G.; Sodini, G. Functional Properties of Sunflower Products after Extraction of Phenolic Pigments by Acid Butanol. *Riv. Ital. Sost. Grasse,* **1977**, *54,* 73–76.

Carellos, D. C.; Lima, J. A.F.; Fialho, E. T.; Freitas, R. T. F.; Silva, H.O.; Branco, P. A. C.; Souza, Z. A.; Vieira-Neto, J. Evaluation of Sunflower Meal on Growth and Carcass Traits of Finishing Pigs. *Cienc. Agrotec.* **2005**, *29,* 208–215.

Carrera, C.; Rodríguez Niño, M. R.; Rodríguez Patino, J. M. Dynamic Phenomena in Caseinate–Monoglyceride Mixed Films at the Air–Water Interface. *Food Hydrocol.* **2005**, *19,* 395-405.

Cheftel, J. C.; Cuq, J. L.; Lorient, D. Protéines alimentaires: Biochimie, propriétés fonctionnelles, valeur nutrionnelle, modifications chimiques. Ed. Technique et Documentation—Lavoisier, Paris, France, 1985.

Claughton, S. M.; Pearce, R. J. Preparation and Properties of Acid-modified Sunflower Protein Isolate. *J. Food Sci.* **1989**, *54,* 357–361.

Clemente, A. Enzymatic Protein Hydrolysates in Human Nutrition. *Trends Food Sci. Technol.* **2000**, *11,* 254–262.

Creusot, N.; Gruppen H. Hydrolysis of Whey Protein Isolate with Bacillus licheniformis Protease: Aggregating Capacities of Peptide Fractions. *J. Agric. Food Chem.* **2008**, *56,* 10332–10339.

Creusot, N.; Gruppen, H.; van Koningsveld, G. A.; de Kruif, C. G.; Voragen, A. G. J. Peptide–Peptide and Protein–Peptide Interactions in Mixtures of Whey Protein Isolate and Whey Protein Isolate Hydrolysates. *Int. Dairy J.* **2006**, *16,* 840–849.

Cunningham, P.; Ogale, A. A.; Dawson, P. L.; Acton J. C. Tensile Properties of Soy Protein Isolate Films Produced by a Thermal Compaction Technique *J. Food Sci.* **2000**, *65,* 668–671.

Daglia, M. Polyphenols as Antimicrobial Agents. *Curr. Opin. Biotechnol.* **2012**, *23,* 174–181.

Damodaran, S. Food Proteins: An Overview. In *Food Proteins and Their Applications;* Damodaran, S., Paraf, A., Eds; New York: Marcel Dekker, 1997; pp 1–24.

De Lange, C. F. M.; Garbet, V. M.; Gillis, D.; Patience, J. F. Digestible Energy Contents and Apparent Ileal Amino Acids Digestibilities in Regular or Partial Mechanically Dehulled Canola Meal Samples Fed to Growing Pigs. *Can. J. Anim. Sci.* **1998**, *7,* 641–648.

De Leonardis, A.; Macciola, V.; Di Domenico, N. A First Pilot Study to Produce a Food Antioxidant from Sunflower Seed Shells (*Helianthus annuus*). *Eur. J. Lipid Sci.* **2005**, *107,* 220–227.

Della Gatta, C.; Piergiovanni, A. R. Technological and Nutritional Aspects in Hyperproteic Bread Prepared with the Addition of Sunflower Meal. *Food Chem.* **1996**, *5,* 493–494.

Dickinson, E. Milk Protein Interfacial Layers and the Relationship to Emulsion Stability and Rheology. *Colloids Surf. B* **2001**, *20,* 197–210.

Diosady, L. L.; Xu, L.; Chen, B. K. Production of High-quality Protein Isolates from Oilseeds Seeds. U.S. Patent 20,070,237,877, October 2007.

Dubois, J. L.; Piccirilli, A.; Magne, J.; He, X. Detoxification of Castor Meal through Reactive Seed Crushing. *Ind. Crops Prod.* **2013**, *43,* 194–199.

Easa, A. M.; Armstrong, H. J.; Mitchell, J. R.; Hill, S. E.; Harding, S. E.; Taylor, A. J. Maillard Induced Complexes of Bovine Serum Albumin—A Dilute Solution Study. *Int. J. Biol. Macromol.* **1996**, *18,* 297–301.

Eckert, M.; Riker, P. Overcoming Challenges in Functional Beverages. *Food Technol.* **2007**, *61,* 20–26.

Economides, S. The Nutritive Value of Sunflower Meal and Its Effect on Replacing Cereal Straw in the Diets of Lactating Ewes and Goats. *Livestock Prod. Sci.* **1998**, *55,* 89–97.

Eric, K.; Raymond, L. V.; Huang, M.; Cheserek, M. J.; Hayat, K.; Savio, N. D.; Amédée, M.; Zhang X. Sensory Attributes and Antioxidant Capacity of Maillard Reaction Products Derived from Xylose, Cysteine and Sunflower Protein Hydrolysate Model System. *Food Res. Int.* **2013**, *54,* 1437–1447.

FAO. *Protein Sources for the Animal Feed Industry.* Expert Consultation and Workshop Bangkok, April 29–May 3, 2002.

FAO. *World Livestock 2011–Livestock in Food Security.* FAO: Rome, Italy, 2011.

FAO/WHO/ONU. *Energy and Protein Requirements: Report of a Joint FAO/WHO/UNU Expert Consultation.* Technical Report Series No. 724, 1985.

FAOSTAT. FAO Statistics Division, 2014. http://faostat.fao.org (accessed July 18, 2014).

Fernández-Cegrí, V.; De la Rubia, M. A.; Raposo, P.; Borja, R. Effect of Hydrothermal Pretreatment of Sunflower Oil Cake on Biomethane Potential Focusing on Fibre Composition. *Bioresource Technol.* **2012**, *123,* 424–429.

Fleming, S. E.; Sosulski, F. Gelation and Thickening Phenomena of Vegetable Protein Products. *J. Food Sci.* **1975**, *40,* 805–807.

Foulk, J. A.; Bunn, J. M. Properties of Compression Molded, Acetylated Soy Protein Films. *Ind. Crops Prod.* **2001**, *14,* 11–22.

Friedman, M. Chemistry, Biochemistry, Nutrition, and Microbiology of Lysinoalanine, Lanthionine, and Histidinoalanine in Food and Other Proteins. *J. Agric. Food Chem.* **1999**, *47,* 1295–1319.

Gandhi, A. P.; Jha, K.; Gupta, V. Studies on the Production of Defatted Sunflower Meal with Low Polyphenol and Phytate Contents and Its Nutritional Profile. *ASEAN Food J.* **2008**, *15,* 97–100.

Gandhi, V. P.; Zhou, Z. Food demand and the food security challenge with rapid economic growth in the emerging economies of India and China. *Food Res. Int.* **2014**, *63,* 108–124.

García, M. C.; Puchalska, P.; Esteve, C.; Marina, M. L. Vegetable Foods: A Cheap Source of Proteins and Peptides with Antihypertensive, Antioxidant, and Other Less Occurrence Bioactivities. *Talanta* **2013**, *106,* 328–349.

Gassmann, B. Preparation and Application of Vegetable Proteins, Especially Proteins from Sunflower Seed, for Human Consumption. An Approach. *Nahrung* **1983**, *27,* 351–369.

Geneau-Sbartaï, C.; Leyris, J.; Silvestre, F.; Rigal, L. Sunflower Cake as a Natural Composite: Composition and Plastic Properties. *J. Agric. Food Chem.* **2008**, *56,* 11198–11208.

Gheyasuddin, S.; Cater, C. M.; Mattil, K. F. Effect of Several Variables on the Extractability of Sunflower Seed Proteins. *J. Food Sci.* **1970**, *35,* 453–456.

González-Pérez, S.; Merck, K. B.; Vereijken, J. M.; van Koningsveld, G. A.; Gruppen, H.; Voragen A. G.J. Isolation and Characterization of Undenatured Chlorogenic Acid Free Sunflower (*Helianthus annuus*) Proteins. *J. Agric. Food Chem.* **2002**, *50,* 1713–1719.

González-Pérez, S.; Merck, K. B.; Vereijken, J. M.; Van Koningsveld, G. A.; Gruppen, H.; Voragen, A. G.J. Conformational States of Sunflower (*Helianthus annuus*) Helianthinin: Effect of Heat and pH. *J. Agric. Food Chem.* **2004**, *52,* 6770–6778.

González-Pérez, S.; Merck, K. B.; Vereijken, J. M.; Van Koningsveld, G. A.; Gruppen, H.; Voragen, A. G. J. Physico-chemical Properties of 2S Albumins and the Corresponding Protein Isolate from Sunflower (*Helianthus annuus*). *J. Food Sci.* **2005**, *70,* 98–103.

González-Pérez, S.; Vereijken, J. M. Sunflower Proteins: Overview of Their Physicochemical, Structural and Functional Properties. *J. Sci. Food Agric.* **2007**, *87,* 2172–2191.

Grimble, G. K.; David, B. A. Silk Peptides in Human Nutrition. *Nutr. Res. Rev.* **1989,** *2,* 87–108.

Grinberg, V. Y.; Grinberg, N. Y.; Burova, T. V.; Dalgalarrondo, J.; Haertlé, T. Ethanol-induced Conformational Transitions in Holoa-lactalbumin: Spectral and Calorimetric Studies. *Biopolymers* **1998**, *46,* 253–265.

Guéguen, J.; Popineau, Y.; Anisimova, I. N.; Fido, R. J.; Shewry, P. R.; Tatham, A. S.; Functionality of the 2S Albumin Seed Storage Proteins from Sunflower (*Helianthus annuus* L.). *J. Agric. Food Chem.* **1996**, *44,* 1184–1189.

Guerrero, P.; Retegi, A.; Gabilondo, N.; de la Caba, K. Mechanical and Thermal Properties of Soy Protein Films Processed by Casting and Compression. *J. Food Eng.* **2010**, *100,* 145–151.

Gunstone, F. D. Production and Trade of Vegetable Oils. In *Vegetable Oils in Food Technology Composition, Properties and Uses;* Gunstone, F. D., Ed.; Blackwell Publishing: Oxford, UK, 2002; pp 1–17.

Havsteen, B. H. The Biochemistry and Medical Significance of the Flavonoids. *Pharmacol. Ther.* **2002**, *96,* 67–202.

Hettiarachchy, N. S.; Ziegler, G. R. *Protein Functionality in Food Systems;* Marcel Decker: New York, 1994.

Hettiarachchy, N. S.; Satō, K.; Marshall, M. R.; Kannan. A., Eds. *Bioactive Food Proteins and Peptides: Applications in Human Health Hettiarachchy*; CRC Press Taylor & Francis Group: Boca Raton, FL, 2012.

Heuzé, V.; Tran, G.; Hassoun, P.; Lebas, F.; Lessire, M. Sunflower Meal, 2012. http://www.feedipedia.org/node/732.

Hisam, E. E. A.; Zakaria, Z. A.; Mohtaruddin, N.; Rofiee, M. S.; Hamid, H. A.; Othman, F. Antiulcer Activity of the Chloroform Extract of *Bauhinia purpurea* Leaf. *Pharmacol. Biol.* **2012**, *50,* 1498–1507.

Hofmann, T. Studies on the Relation between Molecular Weight and the Color Potency of Fractions Obtained by Thermal Treatment of Glucose/Amino Acid and Glucose/Protein Solutions by Using Ultracentrifugation and Color Dilution Techniques. *J. Agric. Food Chem.* **1998**, *465,* 3891–3895.

Huffman, V. L.; Lee, C. K.; Burns, E. E. Selected Functional Properties of Sunflower Meal (*Helianthus annuus*). *J. Food. Sci.* **1975**, *40,* 70–74.

Ibrahim, M. A.; El Zubeir, E. A. Higher Fibre Sunflower Seed Meal in Broiler Chick Diets. *Anim. Feed Sci. Technol.* **1991**, *33,* 343–347.

Ingale, S.; Shrivastava, S. K. Chemical, Nutritional and Anti-nutritional Study of New Varieties of Oil Seeds from Sunflower, Safflower and Groundnut. *Int. J. Biotechnol. App.* **2011**, *3,* 118–129.

Iriarte, A.; Villalobos, P. Greenhouse Gas Emissions and Energy Balance of Sunflower Biodiesel: Identification of its Key Factors in the Supply Chain. *Resour. Conserv. Recy.* **2013**, *73,* 46–52.

Irshaid, R. H.; Harb, M. Y.; Titi, H. H. Replacing Soybean Meal with Sunflower Seed Meal in the Ration of Awassi Ewes and Lambs. *Small Ruminant Res.* **2003**, *50,* 109–116.

Jideani, V. A. Functional Properties of Soybean Food Ingredients. In *Food Systems, Soybean Biochemistry, Chemistry and Physiology;* Tzi-Bun, N., Ed.; InTech Europe: Rijeka, Croatia, 2011; pp 345–366.

Kabirullah, M.; Wills, R. B. H. Characterization of Sunflower Protein. *J. Agric. Food Chem.* **1983**, *31,* 953–956.

Kabirullah, M.; Wills, R. B. H. Foaming Properties of Sunflower Seed Protein. *J. Food Sci. Technol.* **1988**, *25,* 16–19.

Kalmendal, R.; Elwinge, K.; Holm, L.; Tauson, R. High-fibre Sunflower Cake Affects Small Intestinal Digestion and Health in Broiler Chickens. *Br. Poult. Sci.* **2011**, *52,* 86–96.

Kammerer, D. R.; Saleh, Z. S.; Carle, R.; Stanley, R. Adsorptive Recovery of Phenolic Compounds from Apple Juice. *Eur. Food Res. Technol.* **2007**, *224,* 605–613.

Karayannidou, A.; Makri, E.; Papalamprou, E.; Doxastakis, G.; Vaintraub, I.; Lapteva, N.; Artikov, G. Limited Proteolysis as a Tool for the Improvement of the Functionality of Sunflower (*Helianthus annus* L.) Protein Isolates Produced by Seeds or Industrial By-products (Solvent Cake). *Food Chem.* **2007**, *104,* 1728–1733.

Kelly, J. D.; Hefle, S. L. 2S Methionine-rich Protein (SSA) from Sunflower Seed Is an IgE-binding Protein. *Allergy* **2000**, *55,* 556–559.

Khalil, M.; Ragab, M.; Abd el Aal, M. H. Foaming Properties of Oilseed Proteins. *Nahrung* **1985**, *29,* 201–207.

Kianto, U. Utilization of Sunflower Meal for Swine and Poultry Feeds. In *Proceedings of the World Congress on Vegetable Protein Utilization in Human Foods and Animal Feedstuffs;* Applewhite, T. H., Ed.; AOCS: Urbana, IL, 1989; pp 415–423

Kinsella, J. E. Functional Properties of Food Proteins: A Review. *Crit. Rev. Food Sci. Nutr.* **1976**, *7,* 219–280.

Kinsella, J. E. Functional Properties of Soy Proteins. *J. Am. Oil Chem. Soc.* **1979**, *56,* 242–258.

Knight, R. J. Processing and Evaluation of the Antigenicity of Protein Hydrolysates. In *Nutrition for Special Needs in Infancy: Protein Hydrolysates;* Lifshitz, F., Ed.; Marcel Dekker: New York, 1985; pp 105–115.
Korhonen, H.; Pihlanto, A. Bioactive Peptides: Production and Functionality. *Int. Dairy J.* **2006**, *16,* 945–960.
Krause, J. P.; Bagger, C.; Schwenke, K. D. Rheological Properties of Modified Lupin Proteins. *Narung/Food* **2001**, *45,* 412–415.
Krause, J. P.; Schultz, M.; Dudek, S. Effect of Extraction Conditions on Composition, Surface Activity and Rheological Properties of Protein Isolates from Flaxseed (*Linum usitativissimum L.*). *J. Sci. Food Agric.* **2002**, *82,* 970–976
Kristensen, L.; Støier, S.; Würtz, J.; Hinrichsen, L. Trends in Meat Science and Technology: The Future Looks Bright, But the Journey Will Be Long. *Meat Sci.* **2014**, 98, 322–329.
Kroll, J.; Rawel, H. M.; Rohn, S. Reactions of Plant Phenolics with Food Proteins and Enzymes under Special Consideration of Covalent Bonds. *Food Sci. Tech. Res.* **2003,** *9,* 205–218.
Lawhon, J. T.: Cater, C. M.; Mattil, K. F. A Comparative Study of the Whipping Potential of an Extract from Several Oilseed Flours. *Cereal Sci. Today* **1972**, *17,* 240–244.
LEAD-FAO. Livestock, Environment and Development Initiative. http://www.fao.org/agriculture/lead/lead/en/ (accessed May 29, 2013).
Liebowitz, Z.; Ruckenstein, C. Processing and Utilization of Sunflower Meals and Residues. In *Proceedings of the World Conference on Emerging Technologies in the Fats and Oils Industry;* Baldwin, A. R. Ed.; AOCS: Urbana, IL, 1986; pp 208–212.
Lin, M. J. Y.; Humbert, E. S.; Sosulski, F. W. Certain Functional Properties of Sunflower Meal Products. *J. Food Sci.* **1974**, *39,* 368–370.
Lopresti, A. L.; Hood, S. D.; Drummond, P. D. Multiple Antidepressant Potential Modes of Action of Curcumin: A Review of its Anti-inflammatory, Monoaminergic, Antioxidant, Immune-modulating and Neuroprotective Effects. *J. Psychopharmacol.* **2012**, *26,* 1512–1524.
Mahmoud, M. I.; Cordle, C. T. Protein Hydrolysates as Special Nutritional Ingredients. In *Novel Macromolecules in Food Systems*; Doxastakis, G., Kiosseoglou, V., Eds; Elsevier Science B. V.: Amsterdam, Holland, 2000; pp 181–215.
Martínez, K. D.; Baeza, R. I.; Millán, F.; Pilosof, A. M.R. Effect of Limited Hydrolysis of Sunflower Protein on the Interactions with Polysaccharides in Foams. *Food Hydrocol.* **2005**, *19,* 361–369.
Mayaud-Marret, C.; Malod-Panisset, A.; Bidat, E. L'allergie au tournesol et à son huile: rôles du contact, de l'ingestion et de l'inhalation. *Revue française d'allergologie et d'immunologie clinique* **2000**, *46,* 92–94.
Megías, C.; Yust, M. M.; Pedroche, J.; Lquari, H.; Girón-Calle, J.; Alaiz, M.; Millán F.; Vioque J. Purification of an Ace Inhibitory Peptide after Hydrolysis of Sunflower (*Helianthus annuus* L.) Protein Isolates. *J. Agric. Food Chem.* **2004**, *52,* 1928–1932.
Megías, C.; Pedroche, J.; Yust, M. M.; Girón-Calle, J.; Alaiz, M.; Millán F.; Vioque J. Production of Copper-chelating Peptides after Hydrolysis of Sunflower Proteins with Pepsin and Pancreatin. *LWT Food Sci. Technol.* **2008**, *41,* 1973–1977.
Megías, C.; Pedroche, J.; Yust, M. M.; Alaiz, M.; Girón-Calle, J.; Millán F.; Vioque J. Stability of Sunflower Protein Hydrolysates in Simulated Gastric and Intestinal Fluids and Caco-2 Cell Extracts. *LWT—Food Sci. Technol.* **2009a**, *42,* 1496–1500.

Megías, C.; Pedroche, J.; Yust, M. M.; Alaiz, M.; Girón-Calle, J.; Millán F.; Vioque J. Purification of Angiotensin Converting Enzyme Inhibitory Peptides from Sunflower Protein Hydrolysates by Reverse-phase Chromatography Following Affinity Purification. *LWT—Food Sci. Technol.* **2009b**, *42,* 228–232.

Megías, C.; Pedroche, J.; Yust, M. M.; Alaiz, M.; Girón-Calle, J.; Millán F.; Vioque J. Sunflower Protein Hydrolysates Reduce Cholesterol Micellar Solubility. *Plant Foods Hum. Nutr.* **2009c**, *64,* 86–89.

Minami, H.; Morse, E. L.; Adibi, S. A. Characteristics and Mechanism of Glutamine–Dipeptide Absorption in Human Intestine. *Gastroenterol.* **1992**, *103,* 3–11.

Miñones-Conde, J.; Rodríguez-Patino, J. M. The Effect of Enzymatic Treatment of a Sunflower Protein Isolate on the Rate of Adsorption at the Air–Water Interface. *J. Food Eng.* **2007**, *78,* 1001–1009.

Miñones-Conde, J.; Yust, M. M.; Pedroche, J.; Millán, F.; Rodríguez-Patino, J. M. Effect of Enzymatic Treatment of Extracted Sunflower Proteins on Solubility, Amino Acid Composition, and Surface Activity. *J. Agric. Food Chem.* **2005**, *53,* 8038–8045.

Miñones-Conde, J.; Miñones, J.; Rodríguez-Patino, J. M. Structural Characteristics of Dipalmitoylphosphatidylcholine and Hydrolysates from Sunflower Protein Isolate Mixed Monolayers Spread at the Air-Water Interface. *J. Phys. Chem. B* **2006**, *110,* 11582–11591.

Molina-Alcaide, E.; Yáñez-Ruiz, D. R.; Moumen, A.; Martín-García, A. I. Ruminal Degradability and in Vitro Intestinal Digestibility of Sunflower Meal and in Vitro Digestibility of Olive By-products Supplemented with Urea or Sunflower Meal: Comparison between Goats and Sheep. *Animal Feed Sci. Technol.* **2003**, *110,* 3–15.

Mossé, J.; Baudet, J. Biochemical Aspects of Quality and Content of Lysine in Sunflower Seed Proteins. In *Proceedings of the 5th International Sunflower Conference,* Clermont Ferrand, France, 1972; pp 437–440.

Moure, A.; Sineiro, J.; Domínguez, H.; Parajo, J. C. Functionality of Oilseed Protein Products: A Review. *Food Res. Int.* **2006**, *39,* 945–963.

Mushtaq, T.; Sarwar, M.; Ahmad, G.; Nisa, M. U.: Jamil, A. The Influence of Exogenous Multienzyme Preparation and Graded Levels of Digestible Lysine in Sunflower Meal-based Diets on the Performance of Young Broiler Chicks Two Weeks Post Hatching. *Poult. Sci.* **2006**, *85,* 2180–2185.

Mylne, J. S.; Colgrave, M. L.; Daly, N. L.; Chanson, A. H.; Elliott, A. G.; McCallum, E. J.; Jones, A.; Craik, D. J. Albumins and Their Processing Machinery Are Hijacked for Cyclic Peptides in Sunflower. *Nat. Chem. Biol.* **2011**, *7,* 257–259.

National Sunflower Association. http://www.sunflowernsa.com/stats (accessed May 29, 2013).

Nwidu, L. L.; Essien, G. E.; Nwafor, P. A.; Vilegas, W. Antidiarrheal Mechanism of *Carpolobia lutea* Leaf Fractions in Rats. *Pharmacol. Biol.* **2011**, *49,* 1249–1256.

O'Connor, D. E. Preparing Light-colored Protein Isolate from Sunflower Meal by Alkali Extraction under an Inert Gas Blanket Followed by Membrane Ultrafiltration. U.S. Patent 3,622,556, 1971.

OECD (Organisation for Economic Co-operation and Development). *Consensus Document on Compositional Considerations for New Varieties of Sunflower: Key Food and Feed Nutrients, Anti-nutrients and Toxicants.* Environment Directorate, Joint Meeting Of The Chemicals

Committee and the Working Party on Chemicals, Pesticides and Biotechnology. Series on the Safety of Novel Foods and Feeds, 2007, No. 16.

Orliac, O.; Rouilly, A.; Silvestre, F.; Rigal, L. Effects of Additives on the Mechanical Properties, Hydrophobicity and Water Uptake of Thermo-moulded Films Produced from Sunflower Protein Isolate. *Polymer*, **2002**, *43,* 5417–5425.

Orliac, O.; Rouilly, A.; Silvestre, F.; Rigal, L. Effects of Various Plasticizers on the Mechanical Properties, Water Resistance and Aging of Thermo-Moulded Films Made from Sunflower Proteins. *Ind. Crops Prod.* **2003**, *18,* 91–100.

Ozdal, T.; Capanoglu, E.; Altay, F. A Review on Protein–Phenolic Interactions and Associated Changes. *Food Res. Int.* **2013**, *51,* 954–970.

Paraman, I.; Hettiarachchy, N. S.; Schaefer, C.; Beck, M. I. Hydrophobicity, Solubility, and Emulsifying Properties of Enzyme-modified Rice Endosperm Protein. *Cereal Chem.* **2007**, *84,* 343–349.

Parrado, J.; Millan, F.; Hernandez-Pinzón, I.; Bautista, J.; Machado, A. Characterization of Enzymatic Sunflower Protein Hydrolysates. *J. Agric. Food Chem.* **1993**, *41,* 1821–1825.

Parsons, C. M.; Potter, L. M.; Brown R. D. Effect of Dietary Carbohydrate and of Intestinal Microflora on Excretion of Endogenous Amino Acids by Poultry. *Poult. Sci.* **1983**, *62,* 483–489.

Pawar, V. D.; Patil, J. N.; Sakhale, B. K.; Agarkar, B. S. Studies on Nitrogen Extractability of Defatted Sunflower Meal. *J. Food Sci. Technol.* **2001**, *38,* 217–219.

Pedroche, J.; Yust, M. M.; Lqari, H.; Girón-Calle, J.; Alaiz, M.; Vioque, J.; Millán, F. *Brassica carinata* Protein Isolates: Chemical Composition, Protein Characterization and Improvement of Functional Properties by Protein Hydrolysis. *Food Chem.* **2004**, *88,* 337–346.

Pérez, E. E.; Crapiste, G. H.; Carelli, A. A. Some Physical and Morphological Properties of Wild Sunflower Seeds. *Biosystems Eng.* **2007**, *96,* 41–45.

Peri, C.; Rossi, M. Produzione di farine proteiche di girasole. Tecnotogia di preparazione e qualith delle proteine. *Riv. It. Sost. Grasse*, **1980**, *57,* 465–469.

Petit, H. V. Digestion, Milk Production, Milk Composition, and Blood Composition of Dairy Cows Fed Formaldehyde Treated Flaxseed or Sunflower Seed. *J. Dairy Sci.* **2003**, *86,* 2637–2646.

Petti, S.; Crispian, S. Polyphenols, Oral Health and Disease: A Review. *J. Dent.* **2009**, 37, 413–423.

Pickardt, C.; Neidhart, S.; Griesbach, C.; Dube, M.; Knauf, U.; Kammerer, D. R.; Carle, R. Optimisation of Mild-acidic Protein Extraction from Defatted Sunflower (*Helianthus annus* L.) Meal. *Food Hydrocol.* **2009**, *23,* 1966–1973.

Pickardt, C.; Weisz, G. M.; Eisnera, P.; Kammererb, D. R.; Neidhart, S.; Carle, R. Processing of Low Polyphenol Protein Isolates from Residues of Sunflower Seed Oil Production. *Procedia Food Sci.* **2011**, *1,* 1417–1424.

Poncet, C.; Rémond, D.; Lepage, E.; Doreau, M. Comment mieux valoriser les protéagineux et oléagineux en alimentation des ruminants. *Fourrages* **2003**, *174,* 205–229.

Popineau, Y.; Tatham, A. S.; Shewry, P. R.; Marion, D.; Guéguen, J. 2S Sunflower Albumins: Functional Properties of Native and Modified Proteins. In *Plant Proteins from European Crops. Food and Non-food Applications;* Guéguen, J., Popineau, Y., Eds.; INRA Editions: Nantes, 1998; pp 131–135.

Prevot, A.; Blochk, C.; Defromont, C. Enrichissement en proteines des tourteaux de tournesol par voie mécanique. *Rev. Franc. Corps Gras* **1973**, *20,* 33–43.

Puumalainen, T. J.; Poikonen, S.; Kotovuori, A.; Vaali, K.; Kalkkinen, N.; Reunala, T.; Turjanmaa, K.; Palosuo, T. Napins, 2S Albumins, Are Major Allergens in Oilseed Rape and Turnip Rape. *J. Allergy Clin. Immunol.* **2006**, *117,* 426–432.

Quiñones, M.; Miguel, M.; Aleixandre, A. Beneficial Effects of Polyphenols on Cardiovascular Disease. *Pharmacol. Res.* **2013**, *68,* 125–131.

Rahma, E. H.; Rao, M. S.N. Removal of Polyphenols from Sunflower Meal by Various Solvents: Effects on Functional Properties. *J. Food Sci.* **1981**, *46,* 1521–1522.

Ramachandran, S.: Singh, S. K.: Larroche, C.: Soccol, C. R.: Pandey, A. Oil Cakes and Their Biotechnological Applications: A Review. *Bioresour. Technol.* **2007**, *98,* 2000–2009.

Raskin, I.; Ribnicky, D. M.; Komarnytsky, S.; Illic, N.; Poukev, A.; Bortsjuk, N.; Brinker, A.; Moreno, D. A.; Ripoll, C.; Yakoby, N.; et al. Plants and Human Health in the Twenty-first Century. *Trends Biotechnol.* **2002**, *20,* 522–531.

Rawel, H. M.; Czajka, D.; Rohn, S.; Kroll, J. Interactions of Different Phenolic Acids and Flavonoids with Soy Proteins. *Int. J. Biol. Macromol.* **2002a**, *30,* 137–150.

Rawel, H. M.; Rohn, S.; Kruse, H. P.; Kroll, J. Structural Changes Induced in Bovine Serum Albumin by Covalent Attachment of Chlorogenic Acid. *Food Chem.* **2002b**, *78,* 443–455.

Raymond, J.; Rakariyatham, N.; Azanza, J. L. Functional Properties of a New Protein Isolate from Sunflower Oil Cake. *Food Sci. Technol.* **1985**, *18,* 256–263.

Raymond, J.; Inquello, V.; Azanza, J. L. The Seed Proteins of Sunflower: Comparative Studies of Cultivars. *Phytochem.* **1991**, *30,* 2849–2856.

Ren, J.; Zheng, X. Q.; Liu, X. L.; Liu, H. Purification and Characterization of Antioxidant Peptide from Sunflower Protein Hydrolysate. *Food Technol. Biotechnol.* **2010**, *48,* 519–523.

Riaz, M. N. *Soy Application in Foods.* CRC Taylor and Francis: London, 2006; pp 39–226.

Rizzi, G. The Strecker Degradation of Amino Acids: Newer Avenues for Flavor Formation. *Food Rev. Int.* **2008**, *24,* 416– 435.

Robards, K.; Antolovich, M. Analytical Chemistry of Fruit Bioflavonoids. *Analyst* **1997**, *122,* 11–34.

Rodríguez-Patino, J. M.; Rodríguez Niño, M. R.; Carrera, C.; Cejudo, M. J. The Effect of pH on Monoglyceride–Caseinate Mixed Monolayers at the Air–Water Interface. *Colloid Interface Sci.* **2001**, *240,* 113–126.

Rodríguez-Patino, J. M.; Rodrıíguez Niño, M. R.; Carrera, C. Protein–Emulsifier Interactions at the Air–Water Interface. *Curr. Opin. Colloid Interface Sci.* **2003**, *8,* 387–395.

Rodríguez-Patino, J. M.; Miñones-Conde, J.; Millán, H.; Pedroche, J.; Carrera, C.; Pizones, V.; Millán, F. Interfacial and Foaming Properties of Enzyme-induced Hydrolysis of Sunflower Protein Isolate. *Food Hydrocol.* **2007**, *21,* 782–793.

Rossi, M.; Germondari, I. Production of a Food-grade Protein Meal from Defatted Sunflower. II. Functional Properties Evaluation. *Lebensm. Wiss. Technol.* **1982**, *15,* 313–316.

Rossi, M.; Pagliarini, E.; Peri, C. Emulsifying and Foaming Properties of Sunflower Protein Derivatives. *Lebensm. Wiss. Technol.* **1985**, *18,* 293–299.

Rouilly, A.; Orliac, O.; Silvestre, F.; Rigal, L. DSC Study on the Thermal Properties of Sunflower Proteins According to Their Water Content. *Polymer* **2001**, *42,* 10111–10117.

Rouilly, A.; Orliac, O.; Silvestre, F.; Rigal, L. Thermal Denaturation of Sunflower Globulins in Low Moisture Conditions. *Thermochimica Acta* **2003**, *398,* 195–201.
Rouilly, A.; Orliac, O.; Silvestre, F.; Rigal, L. New Natural Injection-moldable Composite Material from Sunflower Oil Cake. *Biores. Technol.* **2006**, *97,* 553–561.
Ruiz, E.; Cara, C.; Ballesteros, M.; Manzanares, P.; Ballesteros, I.; Castro, E. Ethanol Production from Pretreated Olive Tree Wood and Sunflower Stalks by an SSF Process. *Appl. Biochem. Biotechnol.* **2006**, *129,* 631–643.
Saeed, M.; Cheryan, M. Chlorogenic Acid Interactions with Sunflower Proteins. *J. Agric. Food Chem.* **1989**, *37,* 1270–1274.
Salgado, P. R.; Schmidt, I. C.; Molina-Ortiz, S. E.; Mauri, A. N.; Laurindo, J. B. Biodegradable Foams Based on Cassava Starch, Sunflower Proteins and Cellulose Fibers Obtained by a Baking Process. *J. Food Eng.* **2008**, *85,* 435–443.
Salgado, P. R.; Molina-Ortiz, S. E.; Petruccelli, S.; Mauri, A. N. Biodegradable Sunflower Protein Films Naturally Activated with Antioxidant Compounds. *Food Hydrocol.* **2010**, *24,* 525–533.
Salgado, P. R.; Drago, S. R.; Molina-Ortiz, S. E.; Petruccelli, S.; Andrich, O.; González, R. J.; Mauri, A. N. Production and Characterization of Sunflower (*Helianthus annuus* L.) Protein-enriched Products Obtained at Pilot Plant Scale. *LWT—Food Sci. Technol.* **2012a**, *45,* 65–72.
Salgado, P. R.; Molina-Ortiz, S. E.; Petruccelli, S.; Mauri, A. N. Functional Food Ingredients Based on Sunflower Protein Concentrates Naturally Enriched with Antioxidant Phenolic Compounds. *J. Am. Oil Chem. Soc.* **2012b**, *89,* 825–836.
Salgado, P. R.; López-Caballero, M. E.; Gómez-Guillén, M. C.; Mauri, A. N.; Montero, M. P. Exploration of the Antioxidant and Antimicrobial Capacity of Two Sunflower Protein Concentrate Films with Naturally Present Phenolic Compounds. *Food Hydrocol.* **2012c**, *29,* 374–381.
Sánchez, A. C.; Burgos, J. Thermal Gelation of Trypsin Hydrolysates of Sunflower Proteins: Effect of pH, Protein Concentration and Hydrolysis Degree. *J. Agric. Food. Chem.* **1996**, *44,* 3773–3777.
Sánchez, A. C.; Burgos, J. Factors Affecting the Gelation Properties of Hydrolyzed Sunflower Proteins. *J. Food Sci.* **1997**, *62,* 284–288.
Sánchez-Lozano, N. B.; Tomás-Vidal, A.; Martínez-Llorens, S.; Nogales-Mérida, S.; Espert, J.; Moñino, A.; Pla, M.; Jover, M. Growth and Economic Profit of Gilthead Sea Bream (*Sparus aurata* L.) Fed Sunflower Meal. *Aquaculture* **2007**, *272,* 528–534.
Saraf, S.; Kaur, C. D. Phytoconstituents as Photoprotective Novel Cosmetic Formulations. *Pharmacognosy Rev.* **2010**, *4,* 1–11.
Sastry, M. C.S.; Narasinga-Rao, M. S. Binding of Chlorogenic Acid by the Isolated Polyphenol-free 11S Protein of Sunflower (*Helianthus annuus*) Seed. *J. Agric. Food Chem.* **1990**, *38,* 2103–2110.
Sathe, S. K. Dry Bean Protein Functionality. *Crit. Rev. Biotechnol.* **2002**, *22,* 175–223.
Schwenke, K. D.; Prahl, L.; Rauschal, E.; Gwiazda, S.; Dabrowski, K.; Rutkowski, A. Functional Properties of Plant Proteins II. Selected Physico-chemical Properties of Native and Denatured Isolates from Faba Beans, Soybeans, and Sunflower Seed. *Nahrung* **1981**, *25,* 59–69.
Shimada, K.; Mastsushita, S. Relationship between Thermocoagulation of Proteins and Amino Acid Compositions. *J. Agric. Food Chem.* **1980**, *28,* 13–17.

Siemensma, A. D.; Weijer, W. J.; Bak, H. J. The Importance of Peptide Lengths in Hypoallergenic Infant Formulae. *Trends Food Sci. Technol.* **1993**, *4,* 16–21.

Silk, D. B. A.; Fairclough, D.; Clark, M. L.; Hegarty, J. E.; Marrs, T. C.; Addison, J. M.; Burston, D.; Clegg, K. M.; Matthews, D. M. Use of a Peptide Rather Than Free Amino Acid Nitrogen Source in Chemically Defined "Elemental" Diets. *J. Parenteral Enteral Nutr.* **1980**, *4,* 548–553.

Skorić, D. Possible Uses of Sunflower in Proper Human Nutrition. *Med Pregl.* **2009**, *62,* 105–110.

Soares de Moura, R. Process to Obtain Tooth-paste or Mouth-wash Oral Antiseptics, Containing Synthetic or Natural Antioxidant from Polyphenol Rich Plants, for Prevention and Treatment of Periodontal Disease. PCT International Application WO2010148464 A1 20101229, 2010.

Sodini, G.; Canella, M. Acidic Butanol Removal of Color-forming Phenols from Sunflower Meal. *J. Agric. Food Chem.* **1977**, *25,* 822–825.

Sripad, G.; Narasinga Rao, M. S. Effect of Methods to Remove Polyphenols from Sunflower Meal on the Physicochemical Properties of the Proteins. *J. Agric. Food Chem.* **1987**, *35,* 962–967.

Sun, S.; Song, Y.; Zheng, Q. Thermo-molded Wheat Gluten Plastics Plasticized with Glycerol: Effect of Molding Temperature. *Food Hydrocol.* **2008**, *22,* 1006–1013.

Suryaprakash, P.; Kumar, R. P.; Prakash, V. Thermodynamics of Interaction of Caffeic Acid and Quinic Acid with Multi-subunit Proteins. *Int. J. Biol. Macromol.* **2000**, *27,* 219–228.

Świeca, M.; Gawlik-Dzikia, U.; Dzikib, D.; Baraniaka, B.; Czyżc J. The Influence of Protein–Flavonoid Interactions on Protein Digestibility in Vitro and the Antioxidant Quality of Breads Enriched with Onion Skin. *Food Chem.* **2013**, *141,* 451–458.

Taha, F. S.; Mohamed, S. S.; Wagdy, S.M; Mohamed, G. F. Antioxidant and Antimicrobial Activities of Enzymatic Hydrolysis Products from Sunflower Protein Isolate. *World App. Sci. J.* **2013**, *21,* 651–658.

Tavano, O. L. Protein Hydrolysis Using Proteases: An Important Tool for Food Biotechnology. *J. Mol.Cat. B Enz.* **2013**, *90,* 1–11.

Venktesh, A.; Prakash, V. Functional Properties of the Total Proteins of Sunflower (*Helianthus annuus* L.) Seed: Effect of Physical and Chemical Treatments. *J. Agric. Food Chem.* **1993**, *41,* 18–23.

Villamide, M. J.; San Juan, L. D. Effect of Chemical Composition of Sunflower Seed Meal on its True Metabolizable Energy and Amino Acid Digestibility. *Poult. Sci.* **1998**, *77,* 1884–1892.

Villanueva, A.; Vioque, J.; Sánchez-Vioque, R.; Clemente, A.; Pedroche, J.; Bautista, J.; Millán F. Peptide Characteristics of Sunflower Protein Hydrolysates. *J. Am. Oil Chem. Soc.* **1999**, *76,* 1455–1460.

Vissers, Y. M.; Jansen, A. P.H.; Ruinemans-Koerts, J.; Wichers, H. J.; Savelkoul, H. F. J. IgE Component-resolved Allergen Profile and Clinical Symptoms in Soy and Peanut Allergic Patients. *Allergy* **2011**, *66,* 1125–1127.

Wastyn, M. M.; Zhang, A.; Hagen, A.; Kastner, R.; Taufratzhofer, E. Functionality of Protein Isolates and Concentrates from Pea, Sunflower, Rapeseed and Potato. In *Food Proteins:*

Structure and Functionality; Schwenke, D. K., Mothes, R., Eds.; VCH: Weinheim, Germany, 1993; pp 324–326.

Weisz, G. M.; Carle, R.; Kammerer, D. R. Sustainable Sunflower Processing-II. Recovery of Phenolic Compounds as a By-product of Sunflower Protein Extraction. *Innov. Food Sci. Emerg. Technol.* **2013**, *17,* 169–179.

Weisz, G. M.; Schneider , L.; Schweiggert, U.; Kammerer, D. R.; Carle, R. Sustainable Sunflower Processing-I. Development of a Process for the Adsorptive Decolorization of Sunflower [*Helianthus annuus* L.] Protein Extracts. *Innov. Food Sci. Emerg. Technol.* **2010**, *11,* 733–741.

Wenk, C. The Role of Dietary Fibre in the Digestive Physiology of the Pig. *Anim. Feed Sci. Tech.* **2001**, *90,* 21–33.

Wu, J. S.; Wisakowsky, E. E.; Burns, E. E. Emulsion Capacity and Foam Stability of Re-extracted Sunflower Meal (*Helianthus annuus*). *J. Food Sci.* **1976**, *41,* 965–966.

Yagami, A. Anaphylaxis to Lipid Transfer Protein from Sunflower Seeds. *Allergy* **2010**, *65,* 1340–1341.

Yañez, E.; Barja, I.; Pak, N.; Donoso, G.; Ballester, D.; Maccioni, A.; Spada, R. Mönckeberg, F.; Chichester, C. O. Fish-protein Concentrate and Sunflower Presscake Meal as Protein Sources for Human Consumption. *Amer. J. Clin. Nutr.* **1969**, *22,* 878–886.

Yust, M. M.; Pedroche, J.; Millán-Linares, M. C.; Alcaide-Hidalgo, J. M.; Millán, F. Improvement of Functional Properties of Chickpea Proteins by Hydrolysis with Immobilised Alcalase. *Food Chem.* **2010**, *122,* 1212–1217.

Zhao, G.; Liu, Y.; Zhao, M.; Ren, J.; Yang, B. Enzymatic Hydrolysis and Their Effects on Conformational and Functional Properties of Peanut Protein Isolate. *Food Chem.* **2011**, *127,* 1438–1443.

Ziegler, F.; Ollivier, J. M.; Cynober, L.; Masini, J. P.; Coudray-Lucas, C.; Levy, E.; Giboudeau, J. Efficiency of Enteral Nitrogen Support in Surgical Patients: Small Peptides vs Non-degraded Proteins. *Gut* **1990**, *31,* 1277–1283.

Zitouni, N.; Errahali, Y.; Metche, M.; Kanny, G.; Moneret-Vautrin, D. A.; Nicolas J. P.; Fremont, S. Influence of Refining Steps on Trace Allergenic Protein Content in Sunflower Oil. *J. Allergy Clin. Immunol.* **2000**, *106,* 962–967.

14

Food Uses of Sunflower Oils

Joaquín J. Salas, Miguel A. Bootello, and Rafael Garcés ■ *Instituto de la Grasa (Consejo Superior de Investigaciones Científicas), Sevilla, Spain*

Introduction

Sunflower (*Helianthus annuus*) is a plant belonging to the family of the Compositae (Asteraceae), and the genus *Helianthus.* It is an annual plant that has a large yellow inflorescence consisting of a circular head, or *capitulum,* with small flowers housing achenes that contain kernels very rich in oil. This plant was first domesticated and cultivated by natives of Central and North America, and introduced to Europe by Spanish explorers in the 16th century. Sunflower was used as an ornamental plant until the sunflower extraction industry was developed during the 18th century. At this time, the production and consumption of sunflower oil was especially popular in Russia, which led to a rapid increase of the cultivated area (Autino et al., 1993; Melgarejo, 1998). The cultivation of sunflower was extended all over the world during the 19th and the beginning of the 20th centuries, and today it is the fourth largest vegetable oil produced along with soybean, palm, and rapeseed. Hence, it is an important agricultural commodity.

Sunflower is cultivated in temperate climates (temperature range between 20 °C and 25 °C). This plant grows better in dry climates with high solar irradiation and deep soils in which it is able to develop its long pedicular system. The seeds of sunflower are produced within an achene and consist of a shell, composed mainly of lignin and cellulose material, and the kernel, which accounts for the 80% of the total weight of the seeds and is rich in oil (up to 55%). The final oil content of sunflower seeds is around 50% w/w (Autino et al., 1993; Bockisch, 1998).

Sunflower oil is extracted from sunflower seeds through a process involving the production of flakes that are later extruded to extract most of the oil. The extruded solids are then extracted with hexane. The crude sunflower oil is submitted for refining, a process in which it is neutralized, degummed, bleached, deodorized, and winterized to remove free fatty acids, phospholipids, pigments, volatile compounds, and waxes, respectively (Gupta, 2002). Sunflower oil is one of the most desirable oils in the world, and in some countries it is preferred over other vegetable oils such as soybean, cottonseed, and rapeseed oils.

High Oleic and Mid Oleic Sunflower Oils

Common sunflower lines were submitted to breeding and mutagenesis programs during the 1970s to increase the level of oleic acid, producing the first reported high oleic sunflower mutant population, the Pervenet progeny, which has been used to produce commercial seeds of high oleic sunflower all over the world. High oleic sunflower was produced by chemical mutagenesis and it was, together with canola, one of the first modified seed oils to be commercialized at large scale. The high oleic mutant first reported by Soldatov (1976) contained up to 75% of oleic acid at the expense of reduced amounts of linoleic acid (Table 14.A). This mutant line has been inbred to produce lines with even higher oleate content, reaching levels up to 90% oleic acid (Garcés et al., 2011; Warwel et al., 2001). The high oleic sunflower oil is produced and commercialized worldwide at higher prices than regular sunflower oil.

Another commercial type of sunflower is the mid oleic line, promoted by the National Sunflower Association (NSA) in the United States. This oil contained oleic acid levels between high oleic and regular sunflower, 43.1–71.8% (Table 14.A) and saturated fatty acid content below 10%. This new type of oil has been commercialized with the trademark NuSun and it is an alternative sunflower oil having advantages of the two oils previously described (Gupta, 1998; Kleingartner, 2002).

High Saturated Sunflower Oils

After the success of the production of the high oleic sunflower, new programs of mutagenesis were launched in order to produce sunflower oils with new characteristics and functionalities. The most interesting phenotypes resulting from these programs were sunflower lines displaying high levels of saturated fatty acids: the high stearic and high palmitic sunflower mutants. The high stearic mutant produced oils that con-

Table 14.A Fatty Acid Typical Composition of Sunflower Oils from Different Types of Seeds[a]

Sunflower Oil Type	16:0	16:1	16:2	18:0	18:1 n-9	18:1 n-7	18:2	20:0	22:0
Regular	6.3	0.1	—	4.6	26.7	—	61.1	0.3	0.9
Mid oleic	4.9	—	—	3.8	57.9	—	32.3	0.3	0.8
High oleic	3.8	—	—	4.1	82.1	—	8.7	0.4	0.9
High stearic	7.4	—	—	27.1	16.1	—	46.3	1.5	1.6
High stearic–high oleic (HSHO)	5.4	—	—	24.9	57.8	—	8.2	1.8	1.9
High palmitic	34.7	5.1	0.6	2.6	6.9	3.4	45.1	0.5	1.1
High palmitic–high oleic (HPHO)	31.7	7.4	—	2.0	50.5	3.9	2.7	0.5	1.3

[a]Mol% by gas chromatography.

Source: Codex-Stan 210-1999; Salas et al. (2011); Serrano-Vega et al. (2005).

tained much higher levels of stearic acid than regular sunflower, mainly at the expense of oleic acid (Table 14.A) (Fernández-Moya et al., 2005). Furthermore, the high palmitic mutant produced oils rich in palmitic acid, also at the expense of both oleic and linolenic acid (Martínez-Force et al., 1999). Sunflower oils high in saturated fatty acids are helpful for many applications that require high oxidative stability and melting point. Therefore, new lines were developed, including the high palmitic–high oleic (HPHO) (Fernández-Martínez et al., 1997) and the high stearic–high oleic (HSHO) lines (Fernández-Moya et al., 2005). These lines are of a great interest for many food applications because they are very stable at high temperatures, which makes them excellent for frying and cooking. They could also be a source of saturated fatty acids alternative to animal, hydrogenated, and tropical fats (Garcés et al., 2009). Most solid vegetable fats or butters commercialized in the world come from tropical species such as palm, coconut, cocoa, or other more exotic species like shea or kokum (Salas et al., 2009). They are used in a large variety of food formulations, such as margarines, shortenings, chocolates, confectionary, or ice creams. The use of high saturated lines of sunflower would assure a safer local supply of these saturated fats in temperate climates, as well as a lower ecological impact caused by the expansion of the cultivation of palm at the expense of the rainforests in tropical countries.

Lipid Composition

Regular and High Oleic Sunflower Oils

Regular sunflower oil is rich in linoleic acid, which accounts for 48–74% of the total fatty acids of the oil (Table 14.A). It contains low levels of saturated fatty acids, mainly palmitic acid and stearic acid, and unlike other seed oils such as soybean and rapeseed, a negligible amount of linolenic acid. The presence of some very long chain saturated fatty acids in small amounts, in the form of arachidic and behenic acids, is also remarkable. The variability of the content of linoleic acid in different sunflower oils is caused by cultivation temperature (Harris et al., 1978) and affects the nutritional value and the functionality of sunflower oils.

Fatty acids in vegetable oils are esterified to glycerol forming triacylglycerides (TAGs). The most abundant TAG species in regular sunflower oil are trilinolein (LLL) and oleoyl-dilinoleins (LLO and LOL) (1,2-dilinoleoyl-3-oleoyl glycerol and 1,3-dilinoleoyl-2-oleoyl-glycerol), with very low levels of triolein (OOO). The composition of TAGs of the high oleic sunflower oil is simple because its main component is triolein, which can account for 35–72% of total TAGs (Table 14.B). Most of the saturated fatty acids are in the *sn*-1,3 positions of the glycerol backbone and an almost undetectable amount is present in the *sn*-2 position (Table 14.C). Linoleic acid has a slightly higher occurrence at the *sn*-2 position, whereas oleic acid is slightly more abundant in the *sn*-3 position (Álvarez-Ortega et al., 1997; Boukhchina et al., 2003; Rossell et al., 1983).

High Saturated Sunflower Oils

The highly saturated sunflower lines comprise high palmitic and high stearic lines produced by mutagenesis from regular sunflower. Stearic acid content of the high stearic lines range from 20% to 30% (Table 14.A), which is up to five times more than regular sunflower. Most of the increase in stearic acid takes place at the expense of oleic acid, and this change affects the physical characteristics of the oil, increasing the solid fat content of the oil at low temperatures. This was probably caused by an activation of oleate desaturase activity to compensate for the high levels of saturated fatty acids and increase the fluidity of seed membranes (Martínez-Force et al., 1998). Another remarkable change in the composition of high stearic mutants was the higher levels of very long chain fatty acids, arachidic and behenic, which accounted for 3% of the total fatty acids, contrasting with amounts below 1% in the regular or high oleic lines.

Palmitic acid content in the high palmitic mutant can reach to higher than 30% at the expense of oleic and stearic fatty acids (Table 14.A). The levels of very long chain fatty acids are also slightly higher than those found in the regular sunflower line. However, the most remarkable aspect of the high palmitic line is the presence of n-7 fatty acids, palmitoleic (16:1), palmitolinoleic (16:2), and *cis*-vaccenic (18:1 n-7), in the oil that are not present in the other lines of sunflower.

The HSHO and HPHO lines differ from regular high saturated lines because they have high levels of oleic acid and lower levels of linoleic acid, which often remain at levels even lower than those obtained in high oleic sunflower oils. The transference of the high stearic trait to a high oleic background involves a significant decrease in the level of stearic acid, which in this mutant ranged from 16% to 25% of the total fatty acid (Table 14.A).

The triacylglycerol composition of the high saturated sunflower mutants is determined by the distribution of saturated fatty acids within TAGs synthesized by sunflower. There are very low levels of saturated fatty acids in the *sn*-2 position of sunflower TAGs (Table 14.C). This also occurs in the case of high saturated mutants; however, the distribution of these saturated fatty acids is not symmetrical in the positions *sn*-1 and *sn*-3, so the 1,3-random distribution is not accomplished by them (Martínez-Force et al., 2004). This asymmetry makes monosaturated TAGs (those of the form saturated-unsaturated-unsaturated, or SUU) more abundant than expected, whereas the content of disaturated TAGs (saturated-unsaturated-saturated, or SUS) is lower than that predicted on the basis of the fatty acid composition (Tables 14.B and 14.D). This has importance from the point of view of the oil applications because for some food applications, high levels of solid are required, and this is better provided by oils with higher contents of disaturated TAGs. Thus, the predominant TAGs in

Table 14.B Triacylglycerol Typical Composition of Regular, High Oleic, High Stearic, and High Stearic–High Oleic (HSHO) Sunflower Oils[a]

TAG Species (%)	Regular	High Oleic	High Stearic	HSHO
POP	0.3	—	—	0.5
PLP	0.7	—	0.8	—
POSt	0.5	0.4	2.1	4.5
POO	3.0	6.4	1.0	6.2
PLSt	1.2	—	6.8	0.6
POL	7.5	—	3.3	1.4
PLL	6.2	—	4.8	—
StOSt	—	0.5	3.6	10.4
StOO	2.2	11.8	4.1	37.7
StLSt	—	—	12.0	—
OOO	6.4	70.9	—	14.3
StOL	5.2	—	14.6	7.7
LOO	18.8	6.6	2.7	4.0
StLL	5.2	—	24.6	1.2
LLO	26.9	—	6.5	0.8
LLL	14.8	—	7.6	—
OOA	—	0.9	—	2.8
StOA	—	—	—	1.5
StLA	—	—	1.1	—
OLA	—	—	0.8	0.5
LLA	—	—	1.2	—
OOB	—	2.5	—	3.7
StOB	—	—	—	1.4
StLB	—	—	0.7	—
OLB	0.5	—	0.8	0.7
LLB	0.6	—	0.9	—
UUU	66.9	77.5	16.8	19.1
SUU	30.4	21.6	56.1	61.9
SUS	2.7	0.9	27.1	19.0

[a]Mol% by gas chromatography

P: palmitic acid; St: stearic acid; O: oleic acid; L: linoleic acid; A: arachidic acid; B: behenic acid; S: saturated fatty acid; U: unsaturated fatty acid.

Source: Fernández-Moya et al. (2000); Salas et al. (2011).

Table 14.C Fatty Acid Distribution in the *sn*-2 and *sn*-1,3 Positions of the Triacylglycerol of Sunflower Oils from Different Types of Seeds[a]

Sunflower Oil Type		16:0	16:1	16:2	18:0	18:1 n-9	18:1 n-7	18:2	20:0	22:0
Regular	*sn*-1,3	10.1	—	—	7.6	38.7	—	42.0	0.5	1.1
	sn-2	0.6	—	—	0.6	40.8	—	58.0	—	—
High Oleic	*sn*-1,3	4.2	—	—	6.4	84.3	—	3.1	0.5	1.5
	sn-2	0.5	—	—	0.6	96.6	—	2.3	—	—
High Stearic–High Oleic	*sn*-1,3	7.2	—	—	33.1	46.9	—	7.3	2.7	2.8
	sn-2	1.7	—	—	1.9	87.4	—	9.0	—	—
High Stearic	*sn*-1,3	10.6	—	—	38.2	15.2	—	32.3	1.9	1.8
	sn-2	0.7	—	—	1.6	32.4	—	65.3	—	—
High Palmitic–High Oleic	*sn*-1,3	43.4	9.9	—	3.1	33.3	5.3	2.4	1.0	1.6
	sn-2	1.6	3.1	—	0.8	85.9	3.8	4.8	—	—
High Palmitic	*sn*-1,3	40.9	5.6	1.8	5.0	4.1	7.1	34.0	0.5	1.0
	sn-2	1.4	2.9	0.6	1.1	6.9	1.7	85.4	—	—

[a]Mol% by gas chromatography.

Source: Álvarez-Ortega et al. (1997); Fernández-Moya et al. (2005).

high stearic sunflower oil are 1-stearoyl-2,3-dilinoleoyl-glycerol (StLL) and 1-stearoyl-2-linoleyl-3-oleyl-glycerol (StOL), followed by the disaturated TAG 1,3-distearoyl-2-linoleyl-glycerol (StLSt). In all cases, the proportion of monosaturated TAGs was higher than that estimated by calculations from fatty acid composition of the three TAG positions. In the case of the HSHO oil, the TAG composition was more simple, with 1-stearoyl-2,3-dioleoyl glycerol (StOO), OOO, and 1,3-distearoyl-2-oleoyl glycerol (StOSt) being the predominant TAG species (Table 14.B).

The TAG composition of the high palmitic sunflower line has a pattern similar to the high stearic line, with high proportions of the TAGs 1-palmityl-2,3-dilinoleyl-glycerol (PLL) and 1,3-dipalmityl-2-linoleyl glycerol (PLP) and LLL. However, the presence of n-7 fatty acids adds complexity to the composition of this oil (Table 14.D). So, some lines of sunflower with higher contents of palmitoleic acid accounted for up to 34 identified TAG species (Salas et al., 2007). Palmitoleic acid was randomly distributed in the three positions of the glycerol, whereas *cis*-vaccenic acid showed preference for the external *sn*-1 and *sn*-3 positions (Table 14.D). Furthermore, the HPHO line was rich in 1-palmityl-2,3-dioleyl-glycerol (POO), 1,3-dipalmityl-2-oleyl glycerol (POP), and 1-palmityl-2-oleyl-3-linoleoyl glycerol (POL). The triacylglycerol POP is one of the main components of palm oil and palm fractions, so this oil could be an alternative to these fats in food formulations. Regarding the thermal

Table 14.D Triacylglycerol Typical Composition of High Palmitic and High Palmitic–High Oleic (HPHO) Sunflower Oils

TAG Species (%)	High Palmitic	HPHO	TAG Species (%)	High Palmitic	HPHO
PPoP	0.2	0.9	PLA	0.2	—
POP	0.8	19.4	OOO	—	6.9
PAsP	0.2	—	StOL	0.5	—
PLP	14.5	11.5	StAsL	0.8	—
PPoL	4.0	1.5	OOAs	—	3.9
PPlL	1.6	—	LOO	—	1.3
POSt	—	2.4	POB	—	1.4
POO	—	30.3	StLL	4.9	—
POAs	—	5.6	LLO	2.7	—
PLSt	4.0	—	AsLL	5.6	—
POL	3.9	10.1	LLL	13.6	—
PAsL	6.9	—	PLB	0.9	—
PLL	28.9	—	OOB	—	1.5
PoOL	—	0.8	LLB	1.0	—
PoLL	3.7	—	UUU	26.7	12.9
PlLL	1.1	—	SUU	52.5	51.5
StOO	—	2.5	SUS	20.8	35.6

[a]Mol% by gas chromatography.

Source: Fernández-Moya et al. (2000).

behavior of sunflower oils, the melting curves obtained by differential scanning calorimetry show endothermic peaks in a wide range of temperatures according to the TAGs composition (Figure 14.1). Regular and high oleic sunflowers, richer in LLL and OOO, respectively, exhibit narrower melting intervals at lower temperatures than high saturated sunflowers.

High Saturated Sunflower Oil Fractions

The high saturated sunflower oils contain species of TAGs with a high melting point, which would be of great interest for food formulations requiring solid fats, having the advantage of coming from a well-established crop that can be cultivated in temperate climates. However, there are certain limitations in the accumulation of saturated fatty

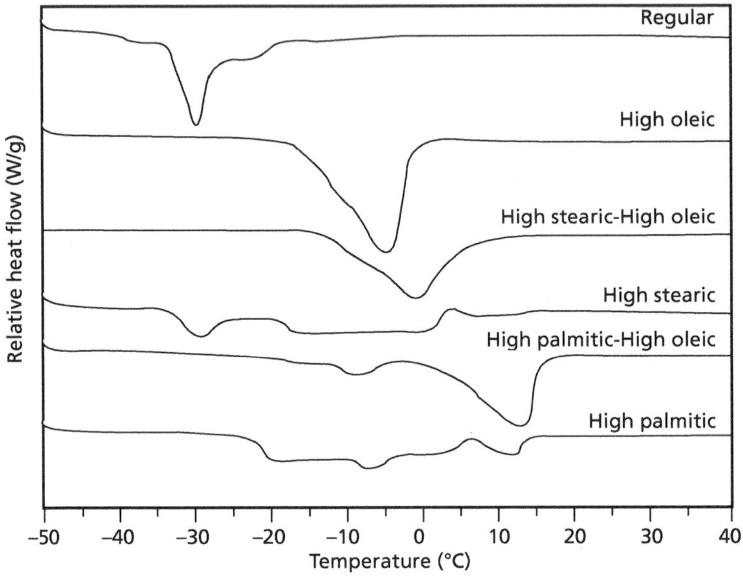

Figure 14.1 Differential scanning calorimetry melting curves of sunflower oils from different types of seeds.

Source: Author's unpublished data.

acids in sunflower, and the levels reached today are not enough to provide solid fats from these mutants. If we look at the melting curve of HSHO and HPHO sunflower oils (Figure 14.2), they display more solid content than common or high oleic sunflower. However, these levels are still far from those required for confectionary or margarine applications, which are limited by the availability of palm fractions or cocoa butter. A technical solution for this problem is the concentration of high melting point TAGs by means of oil fractionation. This process is widely used in the processing of palm, palm kernel, or shea fats and oils. It consists of the crystallization of high melting point TAGs, followed by a step of phase separation, usually consisting of high pressure filtration to produce a liquid fraction or olein and a solid fraction or stearin (Figure 14.3). In this process, only physical steps are applied, so there is no chemical transformation of the oil fatty acids (Timms, 2005).

The most interesting high saturated sunflower oil from the fractionation point of view is HSHO oil, which is commercialized under the brand Nutrisun. This oil was fractionated by dry or solvent fractionation techniques to yield stearins enriched in stearic acid for a broad range of applications. The dry fractionation process consists of a crystallization step followed by phase separation by filtration. The filter cake is then squeezed to remove the olein trapped within the crystals. This method is usually

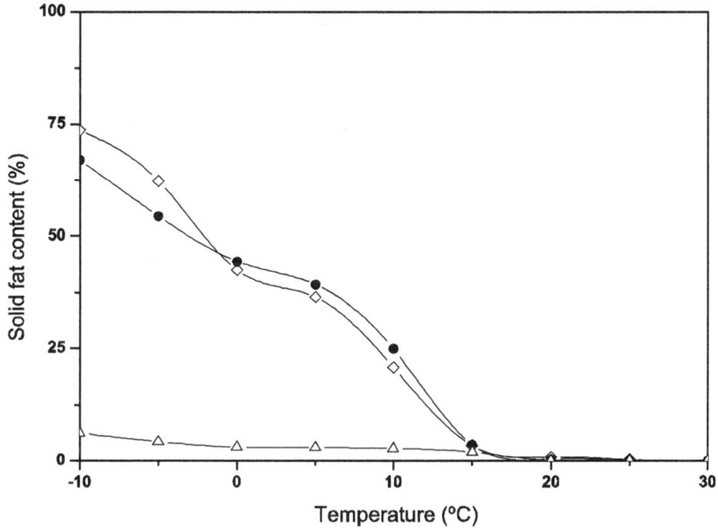

Figure 14.2 Solid fat content profiles obtained by p-NMR of regular sunflower oil (△), high stearic–high oleic sunflower oil (◇), and high palmitic–high oleic sunflower oil (●)

Source: Author's unpublished data.

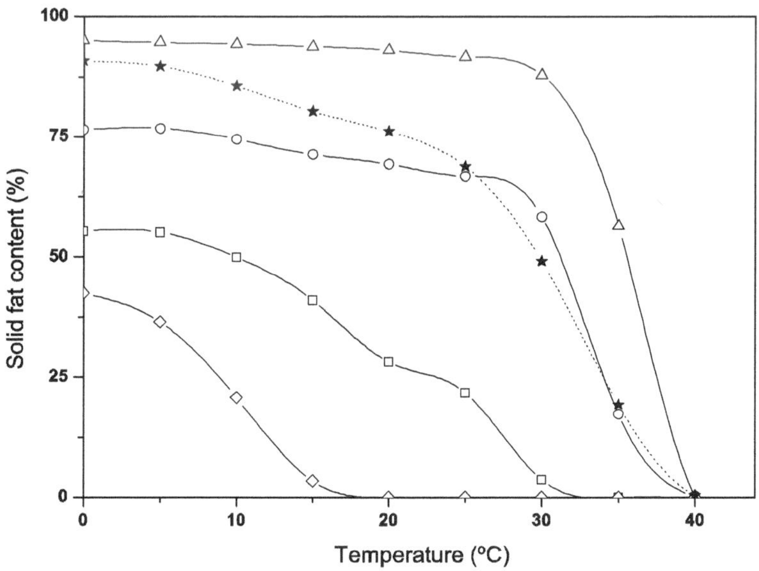

Figure 14.3 Solid fat content profiles obtained by p-NMR of high stearic–high oleic sunflower oil (◇), sunflower dry stearin (□), sunflower hard stearin 80% SUS (○), sunflower hard stearin 95% SUS (△), and Ivory Coast cocoa butter (★).

Source: Adapted from Bootello et al. (2012) and author's unpublished data.

Table 14.E Fatty Acid Composition of Sunflower Stearins, Shea Stearin, and Cocoa Butter[a]

	16:0	18:0	18:1	18:2	20:0	22:0	Total Saturated
Sunflower Dry Stearin	4.7	26.4	61.4	3.0	1.8	2.7	35.6
Sunflower Hard Stearin 80% SUS	6.0	48.9	39.0	1.0	2.9	2.2	60.0
Sunflower Hard Stearin 95% SUS	4.4	54.8	33.0	0.1	3.0	4.7	66.9
Shea Stearin	3.7	60.9	30.9	2.8	1.7	—	66.3
Ivory Coast Cocoa Butter	25.5	37.8	32.7	2.7	1.3	—	64.6

[a]Mol% by gas chromatography.

Source: Bootello et al. (2011, 2012).

applied for the fractionation of palm oil and represents an alternative that is less efficient than solvent fractionation, but involves lower costs of investment and operation. Furthermore, HSHO oil can be also processed by solvent fractionation, yielding fractions with very high levels of stearic acid. In solvent fractionation, the oil is dissolved in an organic solvent and then submitted for crystallization. The TAGs crystallized in those conditions are then filtered under vacuum and washed with fresh solvent to remove the remaining liquid phase that could be trapped in crystals. This method is more efficient and produces fractions that are enriched in saturated fatty acids. However, the use of solvents increases the costs of operation, so it is only used to produce fractions with very high added value, as in the case for shea stearin, which is used for the formulation of confectionary fats.

Table 14.F Triacylglycerol Composition of Sunflower Stearins, Shea Stearin, and Cocoa Butter[a]

	PPP	PStP	POP	PLP	PStst	POSt	POO	PLSt	POL
Sunflower Dry Stearin	—	—	0.5	—	—	5.6	5.7	—	0.5
Sunflower Hard Stearin 80% SUS	—	—	0.4	—		8.8	1.6	—	0.4
Sunflower Hard Stearin 95% SUS	—	—	0.3	—	—	8.1	0.4	—	0.4
Shea Stearin	—	—	0.6	—	—	6.2	2.0	—	—
Ivory Coast Cocoa Butter	0.2	0.6	14.0	1.2	0.9	39.2	1.6	3.1	0.3

[a]Mol% by gas chromatography.

Source: Bootello et al. (2011, 2012).

The dry fractionation of HSHO was run at temperatures around 18 °C and yielded stearins with up to 27% of stearic acid (Table 14.E), which were enriched in disaturated TAGs (30% of total TAG composition) (Table 14.F) (Bootello et al., 2011). These fractions showed important increase in their solid content at room temperature (Figure 14.3), although they were far from the solid content required for the formulation of confectionary fats, so they were called *soft stearins.* The solvent fractionation of HSHO oils produces stearins with much higher levels of saturated fatty acids, with stearic acid content ranging from 40% to 55%, and levels of disaturated TAGs as high as 95.8% (Salas et al., 2011). The melting curves of these fractions were similar to those of fats such as cocoa butter and shea stearin; therefore, they could be used for the production of confectionary fats. They were called *sunflower hard stearins.*

Sunflower Nutrition Facts

Regular and High Oleic Sunflower Oils

Regular sunflower oil, like other vegetable oils, is a component of healthy diet, and a source of unsaturated fatty acids and fat soluble vitamins. Sunflower oil is rich in linoleic acid, which is an essential n-6 polyunsaturated fatty acid. The biological effects of the n-6 fatty acids are mediated by their conversion to n-6 eicosanoids, the n-6 prostaglandins and leukotrienes, which are hormones that act at different levels in human metabolism, especially on the inflammatory response (Simopoulos, 2002). Furthermore, the intake of polyunsaturated oils has a well-known effect on the composition of human blood lipoprotein cholesterol, which has an important impact on the incidence of cardiovascular disease (Erkkilä et al., 2008). Sunflower oil induces a decrease in the plasma total lipoprotein cholesterol, acting on the low density lipoprotein. The high density lipoprotein often remains unaltered in these studies. This

StStSt	StOSt	StOO	StLSt	OOO	StOL	LOO	LLO	StOA	OOA	StOB	OOB	Total SUS
—	21.5	23.5	—	25.6		3.0	1.1	3.1	1.8	3.1	2.9	33.9
0.6	55.5	7.6	—	5.6	0.4	0.8	0.2	7.9	0.7	8.0	1.5	80.7
0.7	68.5	1.5	—	1.0	—	0.2	—	9.0	—	9.9	—	95.8
2.3	73.7	5.1	—	6.7	—	—	—	3.5	—	—	—	83.9
0.6	30.8	2.6	2.0	0.3	0.5	—	—	1.8	0.3	—	—	92.0

means that the intake of sunflower prevents one of the main factors causing arteriosclerosis, and it is recommended in a healthy diet over saturated fats such as those present in animal-derived products like butter or tropical fats such as palm or palm kernel oils or derivatives (Katan et al., 1995). The increase in the level of oleic acid makes high oleic sunflower quite interesting from the nutritional and technical points of view. Thus, the guidelines of the Food and Agricultural Organization (FAO) of the United Nations on the intake of oils and fats recommend the intake of oils with high levels of oleic acid because they are stable and reduce the levels of blood cholesterol, exerting a protective effect against arteriosclerosis. In this regard, high oleic sunflower contains the same nutrients and vitamins of regular sunflower, but has higher levels of oleic acid, which makes it a better fit in a healthy diet than many other oilseeds.

High Saturated Sunflower Oils and Fractions

High stearic sunflower oils represent a healthier alternative to other fats rich in saturated fatty acids. Thus, the intake of medium chain saturated fatty acids, typically present in palm kernel and coconut fat, increases the levels of total blood cholesterol, which induces arteriosclerosis and increases the risk of cardiovascular disease (Katan et al., 1995). A similar effect is caused by high intakes of fats rich in palmitic acid, such as palm oil or fractions. Thus, the consumption of these fatty acids should be reduced in a healthy diet. Moreover, stearic acid is the only saturated fatty acid that does not have any effect on the levels of blood cholesterol and, therefore, does not affect cardiovascular health (Elson, 1992). However, the fats rich in stearic acid are not very common and are often expensive, such as cocoa butter or stearin fractions coming from shea butter. High stearic sunflower oils could be a healthier and convenient alternative to these fats, having the advantage of being rich in tocopherols that are not present in palm or palm fractions. HSHO oils are a source of fats rich in stearic acid with very low levels of palmitic acid and very high levels of the healthy oleic acid. Saturated fatty acid remains in the *sn*-1,3 positions, and it has the possibility of being processed by fractionation to produce fats enriched in saturated fatty acids that can cover a high variety of food applications.

Uses of Sunflower Oils

Domestic Use and Retailing

Sunflower oil is broadly used for retailing and as frying oil. The refined sunflower oil has a pale yellow color and a mild flavor. It is one of the most appreciated oils for domestic use in Western countries, where it is a common ingredient of salads and cooking recipes, as a well as the choice oil for frying. Moreover, sunflower oil has a very low melting point (below 0 °C); therefore, it stays clear at refrigeration temperatures, which is an advantage for the formulation of sauces and emulsions that have to be kept cold.

The reduced amount of polyunsaturated fatty acids makes high oleic sunflower oil especially stable against oxidation. Marmesat et al. (2009) reported that the rates of oxidation of high oleic sunflower at 40 °C in the dark were much lower than those found for regular sunflower oil, which pointed to longer shelf life of the former oil. This result was in agreement with previous data (Martín-Polvillo et al., 2004) and confirmed that high oleic sunflower performed better than more unsaturated oils in most of the oxidations tests to which it was submitted.

Sunflower Oil in Frying

In industry, sunflower is mostly used as frying oil both in continuous and batch processes. It is also used in mayonnaises and oil-based dressings. Batch frying involves frying a certain amount of food product on demand. The oil is heated for relatively short periods of time and is allowed to cool down until the next batch of frying. This process causes a higher exposition of the oil to air and a higher incorporation of oxygen into the oil, which takes place when the oil is cooled down and there is infrequent oil replacement. On the contrary, continuous frying is the common practice in industry for the production of fried and par-fried foods. In this process, the oil is kept at the frying temperature while an amount of fresh oil equivalent to that taken by the fried product is fed into the fryer. Moreover, at the frying temperature, the solubility of oxygen is lower and the steam released by the food acts a protective barrier against the air. Therefore, this method of frying assures a lower oxidation rate of the oil.

The oil oxidation is accelerated by high temperatures during frying, so the stability of the oil is especially important for this application. Vegetable oil autoxidation is a well-known process that starts with the formation of hydroperoxides through a reaction of oxygen with fatty acids (Frankel, 1991). The reaction is then propagated via free radicals and termination reactions involving the formation of carbonyl compounds and triacylglycerol and fatty acid dimers or oligomers. The water present in the food products being fried is also the cause of fatty acid hydrolysis. The rate of oxidation of a given vegetable oil is a function of its composition of fatty acids and of the presence of antioxidants. The rate of reaction of molecular oxygen with the aliphatic carbon chain of fatty acids to form hydroperoxides strongly depends on the presence of double bonds. So, saturated fatty acids are the most stable ones, followed by monounsaturated and polyunsaturated. An extra double bond in the acyl chain of a fatty acid increases the rate at which it is oxidized. So, linoleic and linolenic acids are oxidized 27 times and 77 times quicker, respectively, than oleic acid, which means that the content of these fatty acids will broadly determine the stability of oils (Holman and Elmer, 1947). As it has been stated, fatty acid oxidation occurs through a free radical mediated chain reaction, so radical scavengers are very active in hampering this process. The most active antioxidants in oilseeds are tocopherols, with γ-tocopherol being the most active as an antioxidant. During the frying process, the oil goes through various alterations, but the one determining whether the oil is still suitable to be used in food preparation is the level of polar or polymerized triacylglycerides that should not exceed 25% of the total triacylglycerides of the oil.

Regular sunflower is broadly used as frying oil in many countries. This oil is rich in linoleic acid, like other oilseeds such as soybean or canola, but it has the advantage of having very low levels of linolenic acid. With regard to natural antioxidants, the level of tocopherols is lower than other seed oils and it contains lower levels of the most active α-tocopherol. Masson et al. (1997) found that regular sunflower oil performed better than soybean oil or mixes of soybean and canola in batch frying, probably because of its lower content of linolenic acid. However, the stability of sunflower oil was lower than that found for olive oil in the same frying system (Bastida and Sánchez-Muñiz, 2002). The use of sunflower oil in continuous frying was also tested by Cuesta et al. (1993). They found that the alteration of the oil after 75 frying operations of 8 minutes at 180 °C was mostly oxidative and not hydrolytic. Most of the alteration of the oil took place in the first 20 operations, with little alteration of the oil in the following operations in which the oil reached a near–steady state. In the whole trial, the oil did not reached the maximum allowed level of 25% of altered triacylglycerides. Furthermore, studies on the stability of the fried product were also carried out. These works pointed out that the level of polyunsaturated fatty acids in the oil exerted a stronger influence on stability than the level and composition of tocopherols (Masson et al., 2002).

High Oleic Sunflower Oil in Frying

The disruption of high oleic sunflower oils in the seed oil market caused increasing interest in the performance of these oils in processes requiring oils with high stability, as it is the case for frying. In this regard, Dobarganes et al. (1993) tested the frying performance of three different high oleic sunflower oils with different levels of unsaturation and compared them with those from regular sunflower and olive oil. Results indicated that the stability of the oils in frying operations depended strongly on the level of oleic acid of the oils. Thus, all high oleic sunflower oils tested displayed lower oxidation rates than regular sunflower oil and stability similar to olive oil. The most stable oil in this study was a high oleic sunflower having 80.8% of oleic acid. These results were mostly reproduced by Barrera-Arellano et al. (1997) by using the method of Rancimat to estimate the oxidative stability of sunflower oils from both varieties. In all cases, the induction times observed for high oleic oils were higher than those for common sunflower. The performance of high oleic sunflower oil was also investigated in industrial trials in which several oilseeds and fats were tested for continuous frying

(Niemela et al., 1996). In that work, high oleic sunflower was shown to be especially suitable for deep fat frying, although some disadvantages were found for prefrozen food against oils with a higher melting point. So, high oleic sunflower could partially melt if the temperature of the food rises to a point close to 0 °C during the transport or retailing, which would lead to caking once the freezing temperature was recovered.

The stability of oil at the time of frying was very dependent on the level of unsaturation. However, studies on the stability of fried foods during the process of storage indicated that the level of natural tocopherols was also essential in determining the shelf life of the processed food. Márquez-Ruiz et al. (1999b) reported that potato chips fried in high oleic sunflower oil with low levels of α-tocopherol displayed lower shelf life than those fried in regular sunflower oil enriched in that natural antioxidant. Furthermore, the oxidative stability of oils may be improved by the addition of tocopherols. So, high oleic sunflower oil was reported to have a better performance as frying oil with added γ-tocopherol (Lampi and Kamal-Eldin, 1998).

Some studies on the frying stability of mid stearic acid have been also carried out. Kiatsrichart et al. (2003) compared the stability of mid stearic with commercial canola oil in discontinuous potato frying. Both oils showed a similar pan-frying stability.

High Saturated Sunflower Oils in Frying

Several highly saturated sunflower oils were tested for thermal stability by Márquez-Ruíz et al. (1999a). In this study, the TAGs from regular, high oleic, high stearic, high palmitic, and HPHO sunflower oils were purified and submitted to thermoxidative treatment at 180 °C. This method allowed for comparison of the stability of the oil TAGs by suppressing any possible influence of the minor components. The formation of dimers and altered TAGs was monitored, and regular sunflower was the oil that oxidized at the highest rate. High oleic sunflower oil displayed records similar to high stearic and high palmitic oils, whereas the HPHO displayed by far the highest stability, producing half the number of polar compounds as the regular sunflower oil and less than two-thirds of those of the high oleic line. This result showed the great potential of the high saturated sunflower oils having a high oleic background in food applications in which high stability is required. Thus, in a later study, the rates of thermoxidation of the HPHO sunflower oil were compared with those from palm olein, a liquid fraction obtained from palm oil that is considered to be very stable for frying (Guinda et al., 2003). Results in this work showed that HPHO sunflower performed better than palm olein at frying conditions. Thus, after 10 hours of oxidation at 180 °C, HPHO oil contained less than a half of the polar compounds of the palm olein, and almost three times fewer triacylglycerol polymers, which indicated that this oil was much more stable than the palm olein, although the saturated fatty acid content of the latter was higher. The explanation of this result was in the lower content of linoleic acid of the HPHO sunflower oil and the presence of tocopherols in this oil. Industrial tests were carried out with HPHO oil, testing the batch frying of 200 kg of almonds, peanuts, and sunflower seeds during 4 hours of operation at 180 °C (Marmesat et al., 2005). The content of altered TAGs in the HPHO oil after this process was only 1.3%, which confirmed its high oxidative stability. The performance of this oil was, in general, higher than that of palm olein, and the products fried in it displayed better appearance and a longer shelf life. These works demonstrated that HPHO sunflower oil was a real alternative to palm oleins, and it displayed higher thermostability and frying performance, lower content of saturated fatty acids, and a higher nutritional value. Further research on this oil studied the effect of the tocopherol composition. The oil enriched in γ-tocopherol was more stable than the regular HPHO oil, showing a higher stability index and lower rates of formation of polar compounds and triacylglycerol polymers (García-Moreno et al., 2006). Similar results have been obtained with HSHO sunflower oils.

Interesterification of Sunflower Oil

Interesterification is a catalytic reaction that facilitates fatty acid redistribution on the glycerol backbone of triacylglycerol molecules (Gibon, 2011). This rearrangement of fatty acids leads to a modification of the triacylglycerol composition and, consequently, to a product that has different physical properties (solid fat content, melting point, or hardness) than the original stock (Sreenivasan, 1978). Hard fats, such as palm oil fractions or fully hydrogenated oils, are usually blended or interesterified with a liquid vegetable oil rich in polyunsaturated fatty acids to obtain fats with better functional properties, such as good plasticity and spreadability at room temperature (Noor Lida and Ali, 1998). Sunflower oil can be interesterified with solid fats for the production of plastic fat and structured lipids. Both regular sunflower and high oleic sunflower are good sources of trilinolein and triolein, and can be used as a substrate for the enzyme-catalyzed interesterification (Xu, 2000).

Enzymatic interesterification of palm stearin/regular sunflower oil mixtures with 1,3-specific and nonspecific lipases for the production of plastic fats has been widely studied (Ghosh and Bhattacharyya, 1997; Lai et al., 1998, 1999). According to Lai et al. (1998), the interesterification of a 40:60 palm stearin with sunflower oil blend using lipases from *Pseudomonas* and *Rhizomucor* would be the most appropriate for the preparation of a more fluid product. In general, the melting point of the interesterified products decreased as the proportion of sunflower oil increased, resulting in a 50% reduction of the solid fat index. An interesterified palm stearin/sunflower oil blend is generally softer than the noninteresterified one due to the higher formation of beta-prime crystals. This type of polymorph is the desired form in margarine and shortening to avoid problems with sandiness and the rapid development of graininess (Miskandar et al., 2005). The physicochemical properties of the blend palm stearin (or palm oil)/palm kernel olein/regular sunflower oil before and after chemical interesterification have been reported (Dian et al., 2006, 2007; Noor Lida et al., 2002). Osorio et al. (2006) investigated the operational stability of the enzyme-catalyzed interesterification of these ternary blends evaluating the biocatalyst in terms of the decrease of the solid fat content, showing that these palm-/sunflower-based interesterified products can be used as alternatives to partially hydrogenated fat.

Chemical and enzymatic interesterification of sunflower oil and fully hydrogenated oils are alternatives to *trans* fatty acid–rich hardstocks in food products. Ahmadi et al. (2008) studied the interesterification of fully hydrogenated canola oil and fully hydrogenated soybean oil (which can be considered a source of stearic acid and tristearin) with high oleic sunflower oil for the production of zero-*trans* stearic and oleic acid–rich shortenings and spreads. The resulting products exhibited plasticity and melting characteristics comparable to commercial shortening (Ahmadi and Marangoni, 2009).

The interesterification of animal fats, such as beef tallow and lard, with sunflower oil has been used to improve textural and thermal properties of these fats (Grompone, 1989). The chemical interesterification of beef tallow with regular sunflower oil was an appropriate method to obtain shortenings for use in the baking industry (Grompone, 1991). The incorporation of unsaturated fatty acids from sunflower into saturated triacylglycerides from tallow resulted in a reduction of the hardness in the interesterified product (Rodríguez et al., 2001). Segura et al. (2011) studied the enzyme-catalyzed interesterification of blends of beef tallow and high oleic sunflower oil. These authors reported that the interesterification process enhanced the plastic range of the blends, with a reduction of the solid fat content with respect to simple blending, and an increasing concentration of monosaturated-type triacylglycerol. On the other hand, the occurrence of smaller and more regular crystals in the modified tallow avoids sandy mouth feel caused by large crystals. Lard and high oleic sunflower oil have also been interesterified to produce plastic fats (Seriburi and Akoh, 1998), attempting to use this animal fat in a more general-purpose shortening with improved creaming properties. Butter stearin fraction, obtained from anhydrous milk fat, interesterified with regular sunflower oil yielded spread fats with desirable physical properties and fatty acid composition suitable for utilization in a variety of food products (Pal et al., 2001).

The production of structured lipids involves the use of *sn*-1,3-specific lipases for regio-selective enzymatic interesterification. The enzymatic reaction can take place as an ester–ester exchange (starting from a blend of oils or from a blend of oil and fatty acid esters) or by means of acidolysis (from a blend of oil and fatty acids) (Gibon, 2011). Acidolysis is the most popular technique for producing structured lipids. Regular sunflower oil and high oleic sunflower oil are suitable substrates for this reaction because these fats contain abundant linoleic and oleic acid at their *sn*-2 positions. The acidolysis of sunflower oils with a mixture of palmitic and stearic acid by means of an *sn*-1,3-specific lipase has been carried out to produce fat mixtures with high SUS triacylglycerol content (Carrín and Crapiste, 2008; Pacheco et al., 2010; Ray et al., 2013). In this process, however, it is necessary to strip out the residual fatty acid from the resulting oil and post-fractionate it to isolate the fraction suitable for cocoa butter equivalent (CBE) applications (Gibon, 2011).

Margarines and Confectionary Fats

Regular and High Oleic Sunflower Oils

Margarines consist of an emulsion of an aqueous phase containing salt and flavorings and a fatty phase, which is made with a hardstock or a solid fat mixed with a liquid oil. The components are mixed and tempered to favor the formation of small beta-prime fat crystals, which confer creaminess and good mouth feel to the final product. The fats used in margarines are usually from vegetable origin, so they are a cholesterol-free alternative to butter. However, partially hydrogenated seed oils have often been used as hardstock for margarine formulation. These chemically transformed fats contain *trans* fatty acids that are not recommended in a healthy diet because they induce arteriosclerosis. In this regard, partially hydrogenated sunflower has been used for making margarines, although this fat by itself tends to crystallize in β form, which confers sandiness to the final product. Therefore, it was usually mixed with other hydrogenated oils for a better crystallization pattern (Rivarola et al., 1987), or additives were included in the formulations to avoid undesirable phase transitions (Herrera and Márquez Rocha, 1996). In recent times, the use of partially hydrogenated oils has been reduced in margarines, and the hardstock fat usually comes from vegetable oils interesterified with palm stearins or fully hydrogenated seed oils. Interesterification of sunflower oil with fully hydrogenated soybean yielded a fat with a crystallization profile adequate for producing zero-*trans* margarine (Zeitoun et al., 1993). In any case, regular sunflower oil is often used as liquid oil in the elaboration of margarines as a source of n-6 fatty acids and α-tocopherol to enhance the nutritional properties of the final formulations.

High Saturated Sunflower Oils

The highly saturated sunflower oils are a good source of saturated fatty acids appropriate for the formulation of foods requiring solid fats. High stearic or high palmitic sunflower oils display higher melting points and solid contents than regular and high oleic oils, but they do not have enough levels of solid by themselves to be used as the sole source of fat in this kind of formulation. However, the solid fractions obtained from HSHO oil can cover a range of compositions and can be used as solid fat components in many food products. Soft sunflower stearins obtained by dry fractionation from HSHO display broad melting profiles typical of a plastic fat, and are appropriate for the production of fillings or hardstock fats for spreads or margarines (Bootello et al., 2011). These products are often formulated using lauric fats, palm fractions, or even hydrogenated oils. The use of soft sunflower stearins would mean a healthy alternative to these fats, based on fractions rich in stearic and oleic acids. These fats have also been successfully tested as bakery fats with very promising results.

Similarly, hard sunflower stearins would be a healthy alternative to cocoa butter substitutes and replacers. These hard stearins can be used as source of stearic acid in the formulation of CBEs (Bootello et al., 2012), which are the only healthy alternative to cocoa butter for the production of confectionary fats. Cocoa butter equivalents are produced by blending palm mid fraction (PMF) with fractions rich in stearic acid. They are typically prepared by fractionation of tropical butters to produce a fat with a melting profile and polymorphisms similar to cocoa butter. Blends of sunflower hard stearins and PMF produce fats that have these characteristics and are compatible with cocoa butter (Bootello et al., 2012). Moreover, these CBEs formulated with sunflower stearins display crystallization kinetics and polymorphism similar to those shown by cocoa butter or CBEs formulated with shea butter (Bootello et al., 2013). These sunflower CBEs can be used for the preparation of sweets, compound coatings, and other confectionary products as a healthy alternative to cocoa butter that can be produced locally in countries with temperate climates.

Conclusion

Sunflower produces oil rich in linoleic acid and vitamin E that is very appreciated by consumers all over the world. Thus, it has typically been one of the oils most used for retailing and for domestic consumption. In industry, regular sunflower has been used for frying, showing similar performance to other oilseeds such as soybean and canola; for the production of emulsions and sauces; and in margarine formulations.

In the last decades, many new lines of sunflower with modified fatty acid composition have been produced. One of the most successful was the production of mutants with very high levels of oleic acid, which kept intact the other properties of regular sunflower oil. This oil displayed a fatty acid composition similar to olive oil, the basis of the Mediterranean diet and of the healthy diet. Moreover, it is more stable than most seed oils used for frying and storage. Therefore, the cultivation and consumption of high oleic sunflower oil is steadily increasing and it has been broadly accepted by industry and consumers.

In more recent years, a second generation of modified sunflower oils has been developed. The fatty acid composition of these new lines differs much more from that in regular sunflower, displaying high levels of stearic acid in high oleic background, having a high potential for industrial applications. The high stearic–high oleic oils can be fractionated to produce fractions with high levels of solids and different melting profiles that can be used in broad variety of food formulations, including fillings, spreads, coatings, and confectionary products. Furthermore, these sunflower-derived fats, rich in stearic and oleic acids, are healthier than those that are being used now for the same products, which are mostly based on lauric, hydrogenated fats or palm fractions. The HSHO oil has also been demonstrated to be an alternative to palm oleins for industrial frying, providing higher stabilities, lower levels of saturated fatty acids, and higher vitamin contents. The use of sunflower for the production of hardstock fats for industry has the extra advantages of avoiding any use of GMO organisms, local production in countries with temperate climates, and the use of a crop that is not demanding in terms of agronomy.

References

Ahmadi, L.; Marangoni, A. G. Functionality and Physical Properties of Interesterified High Oleic Shortening Structured with Stearic Acid. *Food Chem.* **2009**, *117,* 668–673.

Ahmadi, L.; Wright, A.; Marangoni, A. G. Chemical and Enzymatic Interesterification of Tristearin/Triolein-Rich Blends: Chemical Composition, Solid Fat Content and Thermal Properties. *Eur. J. Lipid Sci. Technol.* **2008**, *110,* 1014–1024.

Álvarez-Ortega, R.; Cantisán, S.; Martínez-Force, E.; Garcés, R. Characterization of Polar and Nonpolar Seed Lipids Classes from Highly Saturated Fatty Acid Sunflower Mutants. *Lipids* **1997**, *32,* 833–837.

Autino, H.; Wnuk, F.; Vacca, P. Historia, Presente y Futuro de una Oleaginosa Atractiva: El Girasol. *Aceites y Grasas* **1993**, *3,* 21–30.

Barrera-Arellano, D.; Márquez-Ruiz, G.; Dobarganes, M. C. A Simple Procedure to Evaluate the Performance of Fats and Oils at Frying Temperatures. *Grasas y Aceites* **1997**, *48,* 231–235.

Bastida, S.; Sánchez-Muñiz, F. J. Polar Content vs. TAG Oligomers Content in the Frying-Life Assessment of Monounsaturated and Polyunsaturated Oils Used in Deep-Frying. *J. Am. Oil Chem. Soc.* **2002**, *79,* 447–451.

Bockisch, M. Vegetable Fats and Oils. In *Fats and Oils Handbook*; Bockisch, M., Ed.; AOCS Press: Urbana, IL, 1998; pp 174–344.

Bootello, M. A.; Garcés, R.; Martínez-Force, E.; Salas, J. J. Dry Fractionation and Crystallization Kinetics of High-Oleic High-Stearic Sunflower Oil. *J. Am. Oil Chem. Soc.* **2011,** *88,* 1511–1519.

Bootello, M. A.; Hartel, R. W.; Garcés, R.; Martínez-Force, E.; Salas, J. J. Evaluation of High Oleic-High Stearic Sunflower Hard Stearins for Cocoa Butter Equivalent Formulation. *Food Chem.* **2012**, *134,* 1409–1417.

Bootello, M. A.; Hartel, R. W.; Levin, M.; Martínez-Blanes, J. M.; Real, C.; Garcés, R.; Martínez-Force, E.; Salas, J. J. Studies of Isothermal Crystallization Kinetics of Sunflower Hard Stearin-based Confectionery Fats. *Food Chem.* **2013**, *139,* 184–195.

Boukhchina, S.; Gresti, J.; Kallel, H.; Bézard, J. Stereospecific Analysis of TAG from Sunflower Oil. *J. Am. Oil Chem. Soc.* **2003**, *80,* 5–8.

Carrín, M. E.; Crapiste, G. H. Enzymatic Acidolysis of Sunflower Oil with a Palmitic-Stearic Acid Mixture. *J. Food Eng.* **2008**, *84,* 243–249.

Codex Alimentarius. *Standard for Named Vegetable Oils, Codex-Stan 210-1999.*

Cuesta, C.; Sánchez-Muñiz, F. J.; Garrido-Polonio, C.; López-Varela, S.; Arroyo, R. Thermoxidative and Hydrolitic Changes in Sunflower Oil Used in Frying with a Fast Turnover of Fresh Oil. *J. Am. Oil Chem. Soc.* **1993**, *70,* 1069–1073.

Dian, N. L. H. M.; Sundram, K.; Idris, N. A. DSC Study on the Melting Properties of Palm Oil, Sunflower Oil, and Palm Kernel Olein Blends before and after Chemical Interesterification. *J. Am. Oil Chem. Soc.* **2006**, *83,* 739–745.

Dian, N. L. H. M.; Sundram, K.; Idris, N. A. Effect of Chemical Interesterification on Triaclyglycerol and Solid Fat Contents of Palm Stearin, Sunflower Oil and Palm Kernel Olein Blends. *Eur. J. Lipid Sci. Technol.* **2007**, *109,* 147–156.

Dobarganes, M. C.; Márquez-Ruiz, G.; Pérez-Camino, M. C. Thermal Stability and Frying Performance of Genetically Modified Sunflower Seed (*Helianthus annuus* L.) Oils. *J. Agric. Food Chem.* **1993**, *41,* 678–681.

Elson, C. E. Tropical Oils: Nutritional and Scientific Issues. *Crit. Rev. Food Sci.* **1992**, *31,* 79–102.

Erkkilä, A.; de Mello, V. D.; Risérus, U.; Laaksonen, D. E. Dietary Fatty Acids and Cardiovascular Disease: An Epidemiological Approach. *Prog. Lipid Res.* **2008**, *47,* 172–187.

Fernández-Martínez, J. M.; Mancha, M.; Osorio, J.; Garcés, R. Sunflower Mutant Containing High Levels of Palmitic Acid in High Oleic Background. *Euphytica* **1997**, *97,* 113–116.

Fernández-Moya, V.; Martínez-Force, E.; Garcés, R. Identification of Triacylglycerol Species from High-Saturated Sunflower (*Helianthus annuus*) Mutants. *J. Agric. Food Chem.* **2000**, *48,* 764–769.

Fernández-Moya, V.; Martínez-Force, E.; Garcés, R. Oils from Improved High Stearic Acid Sunflower Seeds. *J. Agric. Food Chem.* **2005**, *53,* 5326–5330.

Frankel, E. N. Recent Advances in Lipid Oxidation (Review). *J. Sci. Food Agr.* **1991**, *54,* 495–511.

García-Moreno, M. J.; Vera-Ruiz, E. M.; Fernández-Martínez, J. M.; Velasco, L.; Pérez-Vich, B. Genetic and Molecular Analysis of High Gamma-Tocopherol Content in Sunflower. *Crop Sci.* **2006**, *46,* 2015–2021.

Garcés, R.; Martínez-Force, E.; Salas, J. J.; Venegas-Calerón, M. Current Advances in Sunflower Oil and its Applications. *Lipid Tech.* **2009**, *21,* 79–82.

Garcés, R.; Martínez-Force, E.; Salas, J. J. Vegetable Oil Basestocks for Lubricants. *Grasas y Aceites* **2011**, *62,* 21–28.

Ghosh, S.; Bhattacharyya, D. K. Utilization of High-Melting Palm Stearin in Lipase-Catalyzed Interesterification with Liquid Oils. *J. Am. Oil Chem. Soc.* **1997**, *74,* 589–592.

Gibon, V. Enzymatic Interesterification of Oils. *Lipid Tech.* **2011**, *23,* 274–277.

Grompone, M. A. Physicochemical Properties of Fractionated Beef Tallows. *J. Am. Oil Chem. Soc.* **1989**, *66,* 253–255.

Grompone, M. A. Propiedades Físicas y Químicas de las Grasas Bovinas Fraccionadas e Interesterificadas. *Grasas y Aceites* **1991**, *42,* 349–355.

Guinda, A.; Dobarganes, M. C.; Ruiz-Méndez, M. V.; Mancha, M. Chemical and Physical Properties of a Sunflower Oil with High Levels of Oleic and Palmitic Acids. *Eur. J. Lipid Sci. Technol.* **2003**, *105,* 130–137.

Gupta, M. K. NuSun: The Future Generation of Oils. *INFORM* **1998**, *9,* 1150–1154.

Gupta, M. K. Sunflower Oil. In *Vegetable Oils in Food Technology: Composition, Properties and Uses*; Gunstone, F. D., Ed.; Blackwell Publishing: Oxford, UK, 2002; pp 128–156.

Harris, H. C.; McWilliam, J. R.; Mason, W. K. Influence of Temperature on Oil Content and Composition of Sunflower Seed. *Aust. J. Agric. Res.* **1978**, *29,* 1203–1212.

Herrera, M. L.; Márquez Rocha, F. J. Effects of Sucrose Ester on the Kinetics of Polymorphic Transition in Hydrogenated Sunflower Oil. *J. Am. Oil Chem. Soc.* **1996**, *73,* 321–326.

Holman, R. T.; Elmer, O. C. The Rates of Oxidation of Unsaturated Fatty Acids and Esters. *J. Am. Oil Chem. Soc.* **1947**, *24,* 127–129.

Katan, M. B.; Zock, P. L.; Mensink, R. P. Dietary Oils, Serum Lipoproteins, and Coronary Heart Disease. *Am. J. Clin. Nutr.* **1995**, *61,* 1368–1373.

Kiatsrichart, S.; Brewer, M. S.; Cadwallader, K. R.; Artz, W. E. Pan-Frying Stability of NuSun Oil, a Mid-Oleic Sunflower Oil. *J. Am. Oil Chem. Soc.* **2003**, *80,* 479–483.

Kleingartner, L. W. NuSun Sunflower Oil: Redirection of an Industry. In *Trends in New Crops and New Uses*; Janick, J., Whipkey, A., Eds.; ASHS Press: Alexandria, VA, 2002.

Lai, O. M.; Ghazali, H. M.; Chong, C. L. Effect of Enzymatic Transesterification on the Melting Points of Palm Stearin-Sunflower Oil Mixtures. *J. Am. Oil Chem. Soc.* **1998**, *75,* 881–886.

Lai, O. M.; Ghazali, H. M.; Chong, C. L. Use of Enzymatic Transesterified Palm Stearin-Sunflower Oil Blends in the Preparation of Table Margarine Formulation. *Food Chem.* **1999**, *64,* 83–88.

Lampi, A. M.; Kamal-Eldin, A. Effect of α- and γ-Tocopherols on Thermal Polymerization of Purified High-Oleic Sunflower Triacylglycerides. *J. Am. Oil Chem. Soc.* **1998**, *75,* 1699–1703.

Marmesat, S.; Mancha, M.; Ruiz-Méndez, M. V.; Dobarganes, M. C. Performance of Sunflower Oil with High Levels of Oleic and Palmitic Acids during Industrial Frying of Almonds, Peanuts and Sunflower Seeds. *J. Am. Oil Chem. Soc.* **2005**, *82,* 505–510.

Marmesat, S.; Morales, A.; Velasco, J.; Ruiz-Méndez, M. V.; Dobarganes, M. C. Relationship between Changes in Peroxide Value and Conjugated Dienes during Oxidation of Sunflower Oils with Different Degrees of Unsaturation. *Grasas y Aceites* **2009**, *60,* 155–160.

Márquez-Ruiz, G.; Garcés, R.; León-Camacho, M.; Mancha, M. Thermoxidative Stability of Triacylglycerol from Mutant Sunflower Seeds. *J. Am. Oil Chem. Soc.* **1999a**, *76,* 1169–1174.

Márquez-Ruiz, G.; Martín-Polvillo, M.; Jorge, N.; Ruiz-Méndez, M. V.; Dobarganes, M. C. Influence of Used Frying Oil Quality and Natural Tocopherol Content on Oxidative Stability of Fried Potatoes. *J. Am. Oil Chem. Soc.* **1999b**, *76,* 421–425.

Martínez-Force, E.; Álvarez-Ortega, R.; Cantisán, S.; Garcés, R. Fatty Acid Composition in Developing High Saturated Sunflower (*Helianthus annuus*) Seeds: Maturation Changes and Temperature Effect. *J. Agric. Food Chem.* **1998**, *46,* 3577–3582.

Martínez-Force, E.; Álvarez-Ortega, R.; Garcés, R. Enzymatic Characterization of High-Palmitic Acid Sunflower (*Helianthus annuus L.*) Mutants. *Planta* **1999**, *207,* 533–538.

Martínez-Force, E.; Ruiz-López, N.; Garcés, R. The Determination of the Asymmetrical Distribution of Fatty Acids in Triacylglycerides. *Anal. Biochem.* **2004**, *334,* 175–182.

Martín-Polvillo, M.; Márquez-Ruiz, G.; Dobarganes, M.C. Oxidative Stability of Sunflower Oils Differing in Unsaturation Degree during Long-Term Storage at Room Temperature. *J. Am. Oil Chem. Soc.* **2004**, *81,* 577–583.

Masson, L.; Robert, P.; Romero, N.; Izaurieta, M.; Valenzuela, S.; Ortiz, J.; Dobarganes, M. C. Comportamiento de Aceites Poliinsaturados en la Preparación de Patatas Fritas para Consumo Inmediato: Formación de Nuevos Compuestos y Comparación de Métodos Analíticos. *Grasas y Aceites* **1997**, *48,* 273–281.

Masson, L.; Robert, P.; Dobarganes, M. C.; Urra, C.; Romero, N.; Ortiz, J.; Goicoechea, E.; Pérez, P.; Salamé, M.; Torres, R. Stability of Potato Chip Fried in Vegetables Oils with Different Degrees of Unsaturation. Effect of Ascorbyl Palmitate during Storage. *Grasas y Aceites* **2002,** *53,* 190–198.

Melgarejo, M. Girasol en Argentina. *Aceites y Grasas* **1998**, *8,* 49–52.

Miskandar, M. S.; Yaakob, C. M.; Yusoff, M. S. A.; Rahman, R. A. Quality of Margarine: Fats Selection and Processing Parameters. *Asia Pac. J. Clin. Nutr.* **2005**, *14,* 387–395.

Niemela, J. R. K.; Wester, I.; Lahtinen, R. M. Industrial Frying Trials with High Oleic Sunflower Oil. *Grasas y Aceites* **1996**, *47,* 1–4.

Noor Lida, H. M. D.; Ali, A. R. M. Physicochemical Characteristics of Palm-Based Oil Blends for the Production of Reduced Fat Spreads. *J. Am. Oil Chem. Soc.* **1998**, *75,* 1625–1631.

Noor Lida, H. M. D; Sundram, K.; Siew, W. L.; Aminah, A.; Mamot, S. TAG Composition and Solid Fat Content of Palm Oil, Sunflower Oil, and Palm Kernel Olein Blends before and after Chemical Interesterification. *J. Am. Oil Chem. Soc.* **2002**, *79,* 1137–1144.

Osorio, N. M.; da Fonseca, M. M.; Ferreira-Dias, S. Operational Stability of Thermomyces Lanuginosa Lipase during Interesterification of Fat in Continuous Packed-Bed Reactors. *Eur. J. Lipid Sci. Technol.* **2006**, *108,* 545–553.

Pacheco, C.; Crapiste, G. H.; Carrín, M. E. Lipase-Catalyzed Acidolysis of Sunflower Oil: Kinetic Behavior. *J. Food Eng.* **2010**, *98,* 492–497.

Pal, P. K.; Bhattacharyya, D. K.; Ghosh, S. Modifications of Butter Stearin by Blending and Interesterification for Better Utilization in Edible Fat Products. *J. Am. Oil Chem. Soc.* **2001**, *78,* 31–36.

Ray, J.; Nagy, Z. K.; Smith, K. W.; Bhaggan, K.; Stapley, A. G. F. Kinetic Study of the Acidolysis of High Oleic Sunflower Oil with Stearic-Palmitic Acid Mixtures Catalised by Immobilized *Rhizopus oryzae* Lipase. *Biochem. Eng. J.* **2013**, *73,* 17–28.

Rivarola, G.; Segurola, J. A.; Añón, M. C.; Calvelo, A. Crystallization of Hydrogenated Sunflower-Cottonseed Oil. *J. Am. Oil Chem. Soc.* **1987**, *64,* 1537–1543.

Rodríguez, A.; Castro, E.; Salinas, M. C.; López, R.; Miranda, M. Interesterification of Tallow and Sunflower Oil. *J. Am. Oil Chem. Soc.* **2001**, *78,* 431–436.

Rossell, J. B.; King, B.; Downes, M. J. Detection of Adulteration. *J. Am. Oil Chem. Soc.* **1983**, *60,* 333–339.

Salas, J. J.; Moreno-Pérez, A.; Martínez-Force, E.; Garcés, R. Characterization of the Glycerolipid Composition of a High-Palmitoleic Acid Sunflower Mutant. *Eur. J. Lipid Sci. Technol.* **2007**, *109,* 591–599.

Salas, J. J.; Bootello, M. A.; Martínez-Force, E.; Garcés, R. Tropical Vegetable Fats and Butters: Properties and New Alternatives. *OCL* **2009**, *16,* 254–258.

Salas, J. J.; Bootello, M. A.; Martínez-Force, E.; Garcés, R. Production of Stearate-Rich Butters by Solvent Fractionation of High Stearic-High Oleic Sunflower Oil. *Food Chem.* **2011**, *124,* 450–458.

Segura, N.; Claro da Silva, R.; Schäfer de M. Soares, F.; Gioielli, L. A.; Jachmanián, I. Valorization of Beef Tallow by Lipase-Catalyzed Interesterification with High Oleic Sunflower Oil. *J. Am. Oil Chem. Soc.* **2011**, *88,* 1945–1954.

Seriburi, V.; Akoh, C. C. Enzymatic Interesterification of Lard and High-Oleic Sunflower Oil with *Candida antarctica* Lipase to Produce Plastic Fats. *J. Am. Oil Chem. Soc.* **1998**, *75,* 1339–1345.

Serrano-Vega, M. J.; Martínez-Force, E.; Garcés, R. Lipid Characterization of Seeds Oils from High-Palmitic, and Very High-Stearic Acid Sunflowers Lines. *Lipids* **2005**, *40,* 369–374.

Simopoulos, A. P. Polyunsaturated Fatty Acids in Biology and Diseases. The Importance of the Ratio of Omega-6/Omega-3 Essential Fatty Acids. *Biomed. Pharmacother.* **2002**, *56,* 365–379.

Soldatov, K. Chemical Mutagenesis in Sunflower Breeding. In *Proceedings of the 7th International Sunflower Conference,* 7th International Sunflower Conference, Krasnodar, USSR; International Sunflower Association: Vlaardingen, The Netherlands, 1976; pp 352–357.

Sreenivasan, B. Interesterification of Fats. *J. Am. Oil Chem. Soc.* **1978**, *55,* 796–805.

Timms, R. E. Fractional Crystallization—The Fat Modification Process for the 21st Century. *Eur. J. Lipid Sci. Technol.* **2005**, *107,* 48–57.

Warwel, S.; Brüse, F.; Demes, C.; Kunz, M.; Rusch, M. K. Polymers and Surfactants on the Basis of Renewable Resources. *Chemosphere* **2001**, *43,* 39–48.

Xu, X. Production of Specific-Structured Triacylglycerols by Lipase-Catalyzed Reactions: A Review. *Eur. J. Lipid Sci. Technol.* **2000**, *102,* 287–303.

Zeitoun, M. A. M.; Neff, W. E.; List, G. R.; Mounts, T. L. Physical Properties of Interesterified Fat Blends. *J. Am. Oil Chem. Soc.* **1993**, *70,* 467–471.

15

Oxidative Stability of Sunflower Seed Oil

Nurhan Turgut Dunford ■ *Department of Biosystems and Agricultural Engineering and Robert M. Kerr Food and Agricultural Products Center, Oklahoma State University, Stillwater, OK, USA*

Introduction

Oxidative stability refers to resistance of edible oils to oxidation during processing, handling, and storage (Guillen and Cabo, 2002). Product quality and shelf life are significantly affected by the oxidative stability of the oil used in the food formulation. The reaction mechanism of oil oxidation has been reviewed extensively elsewhere (Choe and Min, 2006; Frankel and Nessim, 1998; Frankel, 2005). In summary, two types of oxidation, autoxidation and photosensitized oxidation, are responsible for edible oil oxidation. Autoxidation is driven by atmospheric triplet oxygen, $^3O_2$, which reacts with lipid radicals forming lipid peroxy radicals and hydroperoxides leading to free radical chain reactions. The primary oxidation products, lipid hydroperoxides, are relatively stable at low temperatures and in the absence of metals. However, free radical formation from fatty acids or acylglycerols accelerates in the presence of heat, metals (e.g., iron and copper), and light. Quantity of the total primary oxidation products present in edible oils is determined by the American Oil Chemists' Society (AOCS) official method Cd8-53 for peroxide value (PV) and is widely used as a measure of oil quality (AOCS, 2004). It is important to note that there have been reports indicating that the accuracy of PV determination is affected more by the rates of hydroperoxide generation and degradation than the amount of hydroperoxides present in the oil sample (Van de Voort et al., 1994). High hydroperoxide generation and degradation rates resulted in large errors in PV measurements. Primary oxidation products are readily decomposed to alkoxy radicals, and then are formed into secondary oxidation products including aldehydes, ketones, acids, esters, alcohols, lactones, ethers, furan derivatives, and short-chain hydrocarbons. The latter low–molecular weight compounds generate off-flavors and make oil undesirable or unacceptable to consumers and food processors. The amount of secondary oxidation products in edible oils is commonly measured by the AOCS official method Cd 18-90 for *p*-anisidine value (AV) (AOCS, 2004). While evaluating oil quality based on AV and PV, it is important to keep in mind that the timing of decomposition of primary oxidation products to secondary oxidation products varies with the type of oil. For example, secondary oxidation products are formed immediately after hydroperoxide formation in olive and rapeseed oils. Yet, in sunflower and safflower seed oil, secondary oxidation products are formed after the concentration of hydroperoxides is reached to a substantial level (Van de Voort et al., 1994). Hence, PV and AV should be used together to evaluate oil quality.

Photosensitized oxidation occurs in the presence of light, sensitizers (e.g., chlorophyll), and oxygen. Singlet oxygen, $^1O_2$, formed during this process readily reacts with the high electron density double bonds of unsaturated fatty acids without forming alkyl radicals and producing hydroperoxides (Choe and Min, 2006). The type and amount of secondary oxidation products generated from the decomposition of hydroperoxides formed by singlet oxygen may be different from those produced by autoxidation. For example, photosensitized oxidation produced more 2-decenal and octane and less octanal and 10-oxodecanoate (decomposition products of hydroperoxides) in oleic acid than autoxidation did (Frankel, 1985). Although substantial amounts of 2-heptenal and 2-butenal could be found in $^1O_2$-oxidized linoleic and linolenic acids, their concentrations were negligible as the same fatty acids went through autoxidation.

The oxidative stability of oils is influenced by many factors, including fatty acid composition, processing, storage and handling conditions (exposure to heat, light, sensitizers, metals, oxygen), and presence of free fatty acids, mono- and diacylglycerides, antioxidants, and oxidation products in the oil. This chapter reviews the effects of fatty acid composition, seed and oil processing conditions, and antioxidants on sunflower seed oil oxidative stability. Sunflower seed oil is widely used in frying applications by food industry. Oil stability during frying was discussed in Chapter 14 of this book.

Effect of Fatty Acid Composition on Sunflower Seed Oil Stability

Fatty acid composition determines chemical and physical properties of oils. Oils rich in polyunsaturated fatty acids (PUFAs) have much lower oxidative stability than that of the oils rich in monounsaturated fatty acids (MUFAs) and saturated fatty acids (SFAs). Oxidative stability of a fatty acid decreases with increasing numbers of double bonds on the molecule. Nevertheless, the correlation between the relative oxidation rate and the number of double bonds on a fatty acid is not linear. The relative rates of autoxidation of different unsaturated fatty acids were compared based on the oxygen absorption capacity of their methyl or ethyl esters at 37 °C (Holman and Elmer, 1947). Oleic acid (C18:1) was used as a reference with assigned oxidative stability of 1. The relative rate of oxidation of the other unsaturated fatty acids increased with increasing numbers of double bonds as follows: linoleic (C18:2) = 41, linolenic (C18:3) = 98, and arachidonic acid (C20:4) = 195. It has also been reported that the oxidative stability of edible oils is correlated with their oleic, linoleic, and linolenic acid contents (Frankel and Huang, 1994; Gunstone and Hilditch, 1946). Addition of 1–5% methyl linoleate to methyl oleate significantly decreased the oxidative stability (Gunstone and Hilditch, 1946). Flavor and oxidative stability of soybean oil could be improved through lowering the linoleic acid content by blending soybean oil with peanut oil (49–61% oleic acid) (Cowan et al., 1971).

Conventional, also referred to as regular, sunflower seed oil (SUN) is rich in linoleic acid and low in SFA (Table 15.A) (Grompone, 2011). Decreasing linoleic content and increasing oleic content in the oil are beneficial for improving oxidative stability. Indeed, many oilseeds, including sunflower seeds, have been modified through genetic engineering or traditional breeding to produce oils with a fatty acid profile suitable for different product applications and that will meet the consumer and food industry demands for oils with high nutritional value, oxidative stability, and functional properties (Grompone, 2011; Skoric et al., 2008). High oleic content oils are preferred by food industry because of their higher oxidative stability than vegetable oils with high linoleic content. Today high oleic canola, sunflower, peanut, and many other oilseeds are commercially available (Jonnala et al., 2005; Merrill et al., 2008). Codex Alimentarius provides chemical composition and properties of vegetable oils (Codex Alimentarius, 2001). The maximum and minimum palmitic, oleic, and linoleic acid contents of SUN, mid oleic (NuSun), and high oleic sunflower seed oil (HOSUN) defined by Codex are as follows: palmitic (C16:0), 5.0–7.6, 2.6–5.0, and 4.0–5.5; oleic (C18:1), 14.0–39.4, 75.0–90.7, and 43.1–71.8, and linoleic (C18:2), 48.3–74.0, 2.1–17.0, and 18.7–45.3, respectively. Typical complete fatty acid compositions of SUN, NuSun, and HOSUN from different sources are shown in Table 15.A.

Table 15.A Typical Fatty Acid Composition of Conventional and Modified Sunflower Seed Oil[a]

Fatty Acid Name	Conventional	NuSun	HOSUN-A	HOSUN-B
Myristic (C14:0)	<0.1	—	—	—
Palmitic (C16:0)	6.2	4.5	4.2	3.8
Palmitoleic (C16:1)	—	0.1	0.1	0.1
Stearic (C18:0)	4.8	3.5	3.5	3.6
Oleic (C18:1)	25.1	60.6	83.7	76.2
Linoleic (C18:2)	58.0	29.3	6.2	10.6
Arachidic (C20:0)	2.4	0.3	0.3	1.9
Linolenic (C18:3)	1.6	0.1	0.1	1.6
Eicosenoic (C20:1)	1.4	0.2	0.3	1.8
Behenic (C22:0)	0.5	0.9	1.1	0.6
Lignoceric (C24:0)	—	0.3	0.4	—

[a]Conventional sunflower seed oil and HOSUN-B are adapted from Merrill et al. (2008). NuSun and HOSUN-A are commercial oil samples received from a distributor and analyzed in the author's laboratory (unpublished data).

Oil stability index (OSI) is an automated method that determines oil stability by measuring the conductance produced when evolving volatile organic acids are collected in deionized water (Jebe et al., 1993). Stable secondary oxidation products (volatile organic acids) are produced while oil is oxidized by a stream of air that is bubbled through the oil, usually at 110 °C. Effect of heating on OSI of NuSun and HOSUN was examined in the author's research laboratory. Fatty acid composition of both oils is shown in Table 15.A. As expected, unheated HOSUN had significantly higher OSI (110 °C, 20 L/min air flow rate) than that for NuSun (Table 15.B). OSI of both oils decreased over 50% after 12 hours of heating in an open fryer (without any product in the oil) at 180 °C. Even after 48 hours of heating, HOSUN had significantly higher OSI than that of NuSun, confirming much higher stability of HOSUN than NuSun. Another study that examined seven brands of SUN collected from retail stores revealed that OSI of the oils ranged from 3.08 to 9.08 hours (Majumdar, et al., 2007). Over 90% decrease in OSI was observed for all the oils examined in the latter study after 18 hours of heating at 180 °C.

In an effort to improve thermal stability of the oil, a sunflower mutant containing high levels of oleic acid (57.7%) and palmitic acid (27.8%) and low linoleic acid (2.3%) in the seed oil (HOHP) was developed (Fernández-Martínez, et al., 1997). Oxidative stability of the HOHP sunflower oil was examined (Guinda et al., 2003). OSI at 120 °C and total polar compounds after heating the oil at 180 °C for 10 hours were determined. The test results demonstrated that HOHP sunflower oil produced half the amount of total polar compounds and fewer TAG polymers than the palm olein treated under similar conditions. OSI of the HOHP was very high at 19 hours. DSC thermograms of the HOHP oil and palm olein showed significant differences. The HOHP oil had lower solid fat contents than the palm olein above 15 °C and remained liquid at room temperature. Thermostability of the HOHP sunflower oil from seeds genetically modified for varying levels of α- and γ-tocopherols was studied

Table 15.B Effect of Heating Time (at 180 °C) on Oxidative Stability Index (OSI) of Sunflower Seed Oil[a]

	OSI (h)	
Heating Time	**NuSun**	**HOSUN**
No heating	8.93 ± 0.2	19.33 ± 0.2
12 h	4.44 ± 0.04	8.39 ± 0.35
24 h	3.7 ± 0.14	2.76 ± 0.03
48 h	0.5 ± 0	1.02 ± 0.07

[a]See Table 15.A for the fatty acid composition of oils.

by Marmesat et al. (2008). OSI (at 120 °C) of the oil rich in γ-tocopherol (897 mg/kg in refined oil, OSI of 75.2 h; 808 mg/kg in crude oil, OSI of 27.3 h) was much higher than the HOHP oil rich in α-tocopherol (936 mg/kg in crude oil, OSI of 27 h; 826 mg/kg in refined oil, OSI of 17.5 h). Furthermore, degradation and the loss of natural tocopherol in HOHP oil heated at 180 °C were significantly lower in γ-tocopherol rich HOHP as compared to α-tocopherol rich HOHP. A mixture of (50:50, v/v) α- and γ-tocopherol rich HOHP oils had similar oxidative stability as γ-tocopherol rich HOHP. The authors concluded that the mixture of oils containing α- and γ-tocopherol would satisfy both consumer demands for oils rich in α-tocopherol and the need of the processors for high oxidative stability.

Blending oils with different fatty acid compositions and flavors is a common practice in food industry to achieve desired oxidative stability and flavor profile. Oxidative stability of SUN (17% oleic) could be improved by blending it with a HOSUN variety (TRISUN®, 80% oleic and 10% linoleic) (Yodice, 1990). Active oxygen method (AOM) was used to monitor the accelerated oxidation of oil blends. During the AOM test, which is similar to the OSI test, oxygen is bubbled through an oil sample while heating it at 97.8 °C, and PV of the oil is monitored. The time required (in hours) for the PV of the oil to reach to 100 meq peroxide/kg of oil is reported as AOM value of the oil tested. AOM of the sunflower seed oil blends increased with increasing oleic acid content in the oil. Oils with 17, 51, 65, and 80% oleic acid had AOM values of 11.5, 15.5, 22, and 43.5 hours, respectively. These results clearly demonstrate the stabilizing effect of increasing oleic acid when replaced with linoleic acid in edible oils.

Thermal stability of the blends prepared with three sunflower oils with very different fatty acid composition—SUN (664 mg/h α-tocopherol and OSI of 4.8 h, at 110°C), HOSUN (283 mg/kg α-tocopherol and OSI of 18.3 h), and HOHP (747 mg/kg α-tocopherol and OSI of 34.4 h)—was examined at 180 °C (Marmesat et al., 2012). There was no significant difference in the minor components of the oils except for α-tocopherol contents. The polymer formation was clearly dependent on the degree of oil unsaturation. The order of polymer formation in each oil was as follows: HOHP < HOHP/SUN, 95:5 (OSI = 23.6 h) < HOSUN < HOSUN/SUN 80:20 (OSI = 9.8 h) < HOSUN/SUN 50:50 (OSI = 6.4 h) < HOSUN/SUN 25:75 (OSI = 5.3 h) < SUN. The total amount of polar compounds in oil increased with increasing degree of unsaturation. Total polar content in SUN was highest, 25.7%, during 10 hours of heating at 180 °C. All the other oils had total polar content lower than 25%, the maximum value allowed for human consumption in most of the European countries where regulations have been established. HOHP treated under similar conditions contained only 6.7% total polar compounds. The oil blend HOHPSO/HLSO (95:5, v/v), demonstrated excellent performance producing a low level of total polar compounds of 7%.

Blending of oils from different sources has been examined in an effort to improve the oxidative stability of biodiesel produced from seed oils (Serqueira et al., 2014). Biodiesel produced from SUN contained 6.3% palmitic, 2.4% stearic, 30.4% oleic, and 60.1% linoleic methyl esters and 12 mg/kg tertiary butylhydroquinone (TBHQ). The induction point of SUN biodiesel as measured by using official Rancimat test, 0.9 hours, did not meet the EN 14214 standards. Blending SUN methyl esters with rapeseed, corn, soybean, canola, and cottonseed biodiesel increased the induction point of the blended biodiesel. The addition of 90% corn oil methyl esters into 10% SUN methyl ester resulted in a blend that met the minimum induction point requirement by European standards (6 h). The following methyl ester blends met the minimum ASTM D6751 induction point requirement of 3 hours: 30:70% rapeseed-sunflower (3.7 h), 30:70% corn-sunflower (3 h), and 40:60% (w/w) (3 h). The authors (Serqueira et al., 2014) suggested that using oil blends rather than adding antioxidants to biodiesel would be an alternative for reducing the cost of biodiesel.

Oxidative stability of biodiesel blends containing HOSUN (82% oleic, 8% linoleic, and 0.2% linolenic) and soybean oil methyl esters (24% oleic, 56.3% linoleic, and 7.7% linolenic) was studied by Moser (2008). OSI of soybean and sunflower methyl esters were 5.0 and 6.2 hours, respectively. A binary blend of these samples, 1:3 soybean:sunflower methyl esters, resulted in improved OSI of 6.4 hours. Better OSI values were obtained with sunflower and palm methyl ester blends (9.2 h for 3:1 palm:sunflower methyl ester blend). OSI of the pure palm methyl esters was 10.3 hours. As expected, these findings were well correlated with the fatty acid composition and the saturation levels of oils used for the study.

Comparison of oxidative stability of edible oils is often based on inherent stability, which is a theoretical value defined as the sum of the relative oxidation rates multiplied by the weight percent of each constituent fatty acid in the oil (List and Erickson, 1985). Both inherent stability and iodine value (IV) reflect the degree of unsaturation of the constituent fatty acids in oil; hence, they are inversely correlated with oxidative stability. High IV and inherent stability values indicate low oxidative stability. Inherent stability of SUN was calculated as 6.8, which was lower than that for safflower oil, 7.6 (List and Erickson, 1985). Higher stability of SUN was partly due to a lower IV, 136, than safflower oil, 149. Inherent stability of HOSUN (90% oleic) was the lowest, 0.2, among the sunflower seed oils with various fatty acid profiles (Garcés et al., 2009). The oils examined in the latter study, SUN (58% linoleic and 29% oleic), high stearic/high oleic (24% stearic and 62% oleic), high stearic/high linoleic (30% stearic and 50% linoleic), high palmitic/high linoleic (32% palmitic and 46% linoleic), and high palmitic/high oleic (29% palmitic and 56% oleic) had inherent stabilities of 5.9, 0.7, 5.3, 4.7, and 0.6, respectively. Oils with high oleic acid content had significantly lower inherent stability than the oils with high linoleic.

The differences in oxidative stability of oils with various degrees of unsaturation diminish as the oxidation temperature increases (Warner et al., 1997). This is because the increase in oxidation rate induced by high temperature is greater than the difference in oxidation rates of oleic and linoleic acid (Eldin, 2006). Furthermore, the solubility of oxygen in the oil decreases by about 25% for every 10 °C increase in temperature (Robertson, 2000). The effect of temperature on oil oxidation was examined for soybean (IV =125–133), canola (IV = 114–116), and sunflower oils (IV = 139–141) deodorized under similar conditions (Warner et al., 1989). PV was used to monitor the progress of oxidation. Soybean oil was found to be the most stable, followed by canola and sunflower oil when oils were stored at 60 °C in the dark. On the other hand, stability tests carried out under both Rancimat and AOM conditions at 100 °C showed that canola oil was the most stable, followed by soybean and sunflower oils. Hence, comparisons of the stabilities of various oils require testing at different temperatures using the same oxidative stability test (Ragnarsson and Labuza, 1977).

In addition to the degree of unsaturation of the fatty acids attached to a TAG molecule, the position of the unsaturated fatty acids on the glycerol moiety might affect the oxidative stability of the oil at low temperatures (Hoshina et al., 2004). Linoleic and linolenic acids in *sn*-1,3 positions on the glycerol molecule could be more prone to oxidation, while unsaturated fatty acids in the *sn*-2 position may be relatively protected from oxidation reactions (Hoshina et al., 2004). Indeed, formation of volatile compounds during soybean oil storage at 60 °C in the dark was correlated with the concentration of linolenic acid at the *sn*-1,3 positions (Neff et al., 1992). Distribution of TAG species in sunflower seed oil with 64.4% linoleic, 23.1% oleic, 8.2% palmitic was reported as follows: 25.2% Oleic-Linoleic-Linoleic, 29.2% Linoleic-Linoleic-Linoleic, 9.9% Oleic-Linoleic-Oleic, and 6.7% Palmitic-Oleic-Linoleic. Considering that the majority of the PUFA, linoleic, in SUN (IV = 131) is in either *sn*-1 or *sn*-3 position of the glycerol molecule, SUN is expected to oxidize easily. However, the literature on the effect of the types of TAG species on oil oxidative stability is controversial. Further research is needed to better understand the effect of the position of the fatty acids on a glycerol moiety on oil oxidative stability.

Oil oxidation can be determined by monitoring oxygen depletion, PV, formation of volatile compounds, conjugated diene content, and flavor scores. Analysis of the volatile compounds accumulating in the headspace of an oil stored in a closed container is an analytical method used to evaluate oxidation. There are only a few studies examining the oxidation of oils at low temperatures because of the very slow oxidation rates at low temperatures. The headspace composition of 27 samples of sunflower oil stored in closed bottles at room temperature was examined using Fourier Transform Infrared Spectroscopy (FTIR) and Proton Nuclear Magnetic Resonance (^{1}H NMR) techniques (Guillen and Goicoechea, 2008). The headspace composition of the samples varied with the degree of oil oxidation. Hexanoic, acetic, and formic acids, followed by pentanoic, heptanoic, octanoic, propanoic, butanoic, and nonanoic acids, were the main acids detected in all the samples, but decanoic acid was present only in the most oxidized oil samples. Only 1-octen-3-ol was detected in the headspace of the nonoxidized sunflower oil. Other alcohols, 1-pentanol and 1-hexanol, were also found in the headspace of oxidized oils. Esters were detected in the samples as well. Ethylacetate was identified in the headspace of all samples kept at room temperature regardless of the degree of oxidation of the sample. Pentyl methanoate, pentyl hexanoate, and dibutyl-2-butenedioate were only present in oxidized samples in small amounts. The concentration of the latter esters increased as the level of oil oxidation increased. The presence of terpenes, sesquiterpenes, and of triterpenes in nonoxidized vegetable oils is common. These compounds are plant metabolites. The concentration of some of these compounds, α-pinene, limonene, α-campholene aldehyde, sesquiterpene, α-ylangene, junipene, calarene, (E)-farnesene, aristolen, triterpene, and kaur-16-ene, in the headspace of the sunflower oil samples was not correlated with the degree of oil oxidation. However, the headspace of the most oxidized sample contained a larger number of these compounds, in slightly higher concentrations, than the headspace of less oxidized samples, indicating that these compounds could be produced during oil oxidation at low temperatures in the presence of limited oxygen. Alkyl benzene derivatives were present in very small amounts in the nonoxidized sample. The authors (Guillen and Goicoechea, 2008) reported that the number of benzene derivatives and their concentration rose with the increasing degree of the oxidation level of the oils. The latter trend was most pronounced with toluene and other alkyl benzene derivatives. In addition to the above mentioned hydrocarbons, some polycyclic aromatic hydrocarbons (PAHs)—naphthalene, their two methyl derivatives, and some of their dimethyl derivatives, especially 1,6-dimethylnaphthalene, biphenyl, and 9H-fluorene—were also detected in the headspace of the oil samples. The European Union and some European countries (Commission Regulation EC No. 208/2005) limited the amount of the latter compounds in edible oils due to their toxicity. Although no ketones were detected in the headspace of the nonoxidized sample, the concentration of ketones increased with the degree of oil oxidation. The following ketones were identified in the headspace of the oxidized oil samples: 2-heptanone and 3-nonen-2-one are significant, followed by 3-octen-2-one and 2-nonanone. Aldehydes are the most abundant group of carbonyl derivatives formed during oil oxidation. Alkanals, alkenals, alkadienals and alkatrienals, and aromatic aldehydes were found in the headspace of the sunflower oil stored at room temperature (Guillen and Goicoechea, 2008). Four to ten carbon alkanals were present in the nonoxidized oil. The concentration of alkanals, specifically 2-octenal and 2-heptenal concentrations, increased with the increasing degree of sunflower oil oxidation. Guillén and Goicoechea (2008) is the first study reporting that sunflower oils oxidized at room temperature and in the presence of a limited amount of oxygen contained monoaromatic and polyaromatic hydrocarbons, oxygenated derivatives of alkanals, as well as of some toxic oxygenated α- and β-unsaturated aldehydes with hydroxyl-, oxo-, and epoxy-functional groups.

Lubricants derived from vegetable oils are biodegradable and nontoxic and have high solubilizing power for polar contaminants and additive molecules, very low volatility, and excellent temperature-viscosity properties (Randles, 1992). However, poor oxidative stability of vegetable oils is a problem for their use in lubricant formulations (Becker and Knorr, 1996). A combination of additives in high oleic oils blended with a synthetic fluid was examined in an effort to improve the oxidation behavior of vegetable oil derivative lubricants (Erhan et al., 2006). The oxidation stability of oils was determined as their onset temperatures (OT) using a Pressure Differential Scanning Calorimetry (PDSC). A high OT would suggest a high oxidation stability of the oil. Neat high oleic sunflower (89.1% oleic, 4.8% linoleic), safflower (80.3% oleic, 12.9% linoleic), and soybean (86.8% oleic, 1.08% linoleic, 1.76% linolenic) oils had OTs of 193.3, 177.4, and 197.7 °C, respectively. A polyalphaolefin (PAO) from BP (Naperville, IL) and two additives, zinc diamyl dithiocarbamate (ZDDC) and antimony dialkyldithiocarbamate (ADDC), were used to formulate blends. OT of the sunflower oil plus 20% PAO blend without additive was 198.5 °C, which is higher than the OT of neat sunflower oil. Sunflower oil plus 20% PAO plus additive blend had the highest OT, 264 °C, among the blends prepared using the same formula but different oils—safflower and soybean oils. ZDDC has been reported to function both as radical scavenger and hydroperoxide decomposer (Becker and Knorr, 1996) and converts the hydroperoxides formed during the oxidation process to nonradical products, thus preventing free radical chain reactions. ADDC is an antiwear and extreme pressure additive and also exhibits antioxidant properties. The combination of 2% ZDDC and 2% ADDC provided a synergistic effect. Hence, oil blends were prepared with this additive combination. The effectiveness of this additive combination was tested in sunflower, soybean, and safflower oils and their 20% PAO blends. The PDSC temperature ramping method was used to measure OT of the blends. According to the authors (Erhan et al., 2006) this method has the advantage that it can be used for determination of oxidation stability of neat base oils as well as additized oils, while the PDSC isothermal method is good only for additized oils. Once again, sunflower oil plus 20% PAO plus the additive blend had the highest OT, 264 °C, among the similar blends prepared with safflower and soybean oils. A modified PDSC isothermal method was also used to assess the oxidative stability of formulated products. In this method, the oil was heated in an air atmosphere at 200 °C and oxidation induction time (OIT) was determined as the time at which oxidation started. This method is more sensitive to differences in oxidation stability of additized oils (Sharma and Stipanovic, 2003). Similar to the findings of the previous tests, high oleic sunflower oil blend had the highest OIT among the blends examined. The oxidation stability of the vegetable oil–based lubricants was further evaluated using the rotary bomb oxidation test (RBOT), which measures the time at which a rapid pressure drop is observed due to an increase in the rate of consumption of oxygen indicating the depletion of the antioxidant and rapid oxidation of the lubricant. The RBOT time usually coincides with oxidation induction time. High oleic sunflower oil had a much higher (40%) RBOT time compared to the other oils. Addition of additives to the oil increased the oxidation stability as much as 53%. The poor oxidation stability of vegetable oils limits their use in high temperature applications like engine oils. The oven oxidation test on the samples investigated in the study (Erhan et al., 2006) was carried out at 85 °C because this temperature is suitable for applications such as drip, hydraulic, elevator, and compressor fluids for agricultural and heavy equipment. The weight loss/gain and total acid number of the samples were measured after 28 days. Neat oil samples gained some weight due to the oil oxidation in the absence of antioxidants and some oxopolymerization products were formed. Addition of PAO and additives lead to weight loss indicating that volatility was not a problem. The volatility relates to the potential loss of oil and to the environmental impact of the volatile components of the fluid. Both European and U.S. original equipment manufacturers and industry organizations include volatility in their specifications. The authors concluded that vegetable oil–based formulations, including high oleic sunflower oil based lubricant formulations, performed similar or better than the petroleum-based lubricants for use in high-temperature applications.

Effect of Antioxidants on Sunflower Seed Oil Stability

Antioxidants significantly delay or prevent oxidation reactions when they are present even at very low concentrations in oil, 100–200 mg/kg oil. Commercial food product formulations usually include antioxidants to protect quality and extend the shelf life of the products. Antioxidants are important food ingredients because they prevent rancidity and deterioration of color, flavor, texture, nutritional quality, and the safety of foods. Antioxidants are classified into primary and secondary antioxidants based on their mode of action. Primary antioxidants, such as tocopherols, break free radical chain reactions and react with peroxyl radicals and convert them to stable products. Secondary antioxidants including carotenoids, citric acid, ascorbic acid, and ascorbyl palmitate bind metal ions, scavenge oxygen, decompose hydroperoxides to nonradical products, absorb UV radiation, and deactivate singlet oxygen (Jadhav et al., 1996).

Effect of Antioxidants Naturally Present in Sunflower Seed Oil on Oxidative Stability

Most vegetable oils naturally contain tocopherols and some contain both tocopherols and tocotrienols, which are together referred to as *tocols*. Significant differences in tocopherol content were found among sunflower varieties (Škorić, 1992). Typical total tocopherol content of SUN was about 800 mg/kg oil. About 90–99% of the tocopherols in wild varieties of SUN were in the form of α-tocopherol. No δ-tocopherol was detected in wild sunflower seeds. There have been attempts to increase the oxidative stability of SUN by breeding for high tocopherol content in seeds (Demurin et al., 1996; Škorić, 1992). However, this approach has limitations. It appears that there is an optimal level of tocopherols for maximum protection. For example, increasing α-tocopherol content over 700–900 mg/kg does not improve sunflower seed oil oxidative stability (Demurin et al., 1996). Furthermore, the tocopherol content of sunflower seed oil is shown to be negatively correlated with seed oil content (Demurin, et al., 1996). Nevertheless, it was possible to develop sunflower seeds with increased γ-, δ-, and β-tocopherol content through breeding (Demurin et al., 1996). Rancimat tests carried out at 100 °C using mechanically pressed oils from high linoleic sunflower varieties (58–63% linoleic and 24–33% oleic) showed that oxidation induction time was the highest, 8.7 hours, for the oil obtained from a new variety with 306, 212, 263, and 227 mg/kg α-, β-, γ-, and δ-tocopherol content, respectively. The induction time of the wild variety (63% linoleic and 26% oleic) that had 1764 and 50 mg/kg α- and β-tocopherol, respectively, was only 4.8 hours. Significantly higher total tocopherol content (1814 mg/kg, 97% in α form) of the wild variety than the new variety (1008 mg/kg total tocopherol, 58% linoleic and 32% oleic) did not result in higher oxidative stability. The high oxidative stability of the new variety can be explained by its lower linoleic, higher oleic, and higher β-, γ-, and δ-tocopherol contents than the wild variety. When individual tocopherols were tested for their relative antioxidant potencies in fats and oils, the following order was observed: $\delta > \gamma = \beta > \alpha$ (Olcott and Veen, 1968). A similar trend, higher oxidative stability with increasing γ- and δ-tocopherol content, was observed with HOSUN (86% oleic and 6% linoleic) varieties (Demurin et al., 1996). HOSUN, which contained 321, 20, 722, and 28 mg/kg α-, β-, γ-, and δ-tocopherol, had significantly higher induction time, 78.5 hours, than that of HOSUN, 26.5 hours, with 1440, 42, and 12 mg/kg with α-, β-, γ-, and no δ-tocopherol content, respectively.

In a later study, Warner et al. (2008) examined the oxidative stability of NuSun and HOSUN extracted from seeds bred to have increased γ- and δ-tocopherols. The storage tests were carried out at 60 °C and PV and hexanal were monitored as indicators of oxidation levels. NuSun varieties examined in the study (56–67% oleic and 23–33% linoleic) had 33–393, 12–82, 333–680, and 3–221 mg/kg α-, β-, γ- and δ-tocopherols, respectively. The levels of α-, β-, γ-, and δ-tocopherols in control SUN (56.1% oleic and 33.2% linoleic) were 909, 34, 8, and 3 mg/kg, respectively. The modified HOSUN (70–76% oleic and 15–19% linoleic) had 44–512, 8–109, 223–678, and 42–140 mg/kg of α-, β-, γ-, and δ-tocopherols, respectively. The control for the HOSUN group contained 70.1% oleic, 19.4% linoleic, and 779, 34, 10, and 2 mg/kg α-, β-, γ-, and δ-tocopherols, respectively. The PV of the samples decreased significantly as the amount of γ- and δ-tocopherols increased. A trend similar to that of the PV was observed for the hexanal contents of the oxidized oils. However, the variances in hexanal production were not as distinct as peroxide generation. HOSUN (70.1% oleic and 19.4% linoleic) with 308, 47, 522, and 121 mg/kg α-, β-, γ-, and δ-tocopherols, respectively, had the lowest PV even though it contained a moderate rather than a low amount of α-tocopherol, indicating that low α-tocopherol content is not a requisite for good oil oxidative stability (Warner et al., 2008).

Effect of Added Antioxidants on Sunflower Seed Oil Oxidative Stability

The majority of the antioxidants naturally present in edible oils is lost during refining. Synthetic or natural antioxidants are usually added into oils after processing to maintain product quality. Carelli et al. (2005) examined the effect of added tocopherol on sunflower seed oil oxidative stability. OSI of the SUN increased linearly with the increasing amount of δ-tocopherol added to the oil. This result confirms that δ-tocopherol acts as a primary antioxidant involved in free radical chain reactions (Yanishlieva and Marinova, 1992). Antioxidant activity of δ-tocopherol in SUN was similar to that of butylatedhydroxyltoluene (BHT), which is a synthetic antioxidant commonly used for stabilization of edible oils. α-Tocopherol was not very effective in increasing OSI of the oil. Studies with TAG purified from SUN showed that antioxidant activity of α-tocopherol was greatest at concentrations less than 700 mg/kg. The amount of α-tocopherol naturally present in the oil (670 mg/kg) was close to the maximum level, hence, additional α-tocopherol did not improve OSI of the oil (Carelli et al., 2005). Citric acid mainly acts as a chelating agent and binds metals in the oil. Since the metal content of the oil examined in the study was quite low (3.2 and 0.3 mg/kg iron and copper, respectively), citric acid addition did not have a significant effect on OSI of the oil (Carelli et al., 2005).

The antioxidant activity of added α- and δ-tocopherol, citric acid, ascorbic acid, and ascorbyl palmitate was investigated in SUN (19.7% oleic and 68.4% linoleic, OSI 10 h measured at 98 °C). The amount of total tocopherols naturally present in the oil was 700 mg/kg oil (Carelli et al., 2005). Rancimat (measured at 98 °C) and storage tests (at 30 °C for 35 days, 68 °C for 23 days, and 130 °C for 48 h) were carried out on the samples to monitor oxidation. The fastest increase in OSI with increasing antioxidant concentration was observed with ascorbic acid. The rate of increase in OSI slowed down between 400 and 800 mg/kg ascorbic palmitate concentrations in the oil after a significant increase up to 400 mg/kg. Ascorbic acid and ascorbyl palmitate were much less effective than the synthetic antioxidant T-Butyl Hydroquinone (TBHQ) and their performance was intermediate between propyl gallate (PG) and BHT (Carelli, et al., 1998; Carelli, et al., 2005). PV and AV measurements did not reveal any significant difference among the SUN oils containing different types of antioxidants stored at 30 °C for 35 days (Carelli et al., 2005). It is important to note that use of PV for monitoring oxidation reactions should be limited to the early stages of oxidation since the hydroperoxide concentration reaches a maximum and then decreases at advanced stages of oxidation (Crapiste et al., 1999). After storage, SUN oils containing 100 mg/kg of added ascorbyl palmitate or δ-tocopherol had significantly lower amounts of polar compounds than the oils containing the same amount of other antioxidants stored under the same conditions. The presence of a high level of polar compounds in oils indicates quality deterioration. An evaluation of PV of the SUN stored at 68 °C alluded to the following order of antioxidant effectiveness: δ-tocopherol > ascorbyl palmitate > ascorbic acid > citric acid. The only antioxidant that reduced the degradation of the naturally occurring tocopherols in SUN was δ-tocopherol (Carelli et al., 2005).

Although ascorbic acid showed a lower antioxidant activity at 68 °C than the other antioxidants examined in the study and exhibited no effect in SUN at 30 °C, it was the best antioxidant in Rancimat tests carried out at 98 °C (Carelli et al., 2005). This behavior was explained by the very low solubility of ascorbic acid in oils. During Rancimat tests, the continuous airflow facilitates mixing and emulsification, and better contact between the antioxidant and the oil is achieved. Consequently, the effectiveness of hydrophilic antioxidants is improved (Velasco, et al., 2000). When SUN with added ascorbic acid was stored at 130 °C for 24 hours, all naturally occurring tocopherols were lost and PV and concentration of polar compounds increased significantly. This result indicates that ascorbic acid deteriorates at high temperature and loses its protective effect (Carelli et al., 2005).

Phospholipids (PL) are constituents of all cell membranes and are naturally present in vegetable oils. They are usually removed from the crude oil during refining, specifically during the degumming process. Antioxidant properties of PL in oils have been reported (King et al., 1992). Several mechanisms have been proposed for the action of PL as antioxidants: (1) PL work synergistically with tocopherols (Judde et al., 2003); (2) phosphatidylcholine (PC), phosphatidylserine (PS), or phosphatidylethanolamine (PE), or the sugar moiety of phosphatidylinositol (PI), have been shown to have metal-chelating properties (Pokorný et al., 1992); (3) PL are aligned at interphases and act as an oxygen barrier between oil and air interface (Frankel, 2005); and (4) aldehydes formed during oil oxidation react with PL generating melanophosphatide complexes which inactivates hydroperoxides (Frankel, 2005; King et al., 1992). The effect of soybean lecithin containing 62% PL on oxidative stability of SUN was examined by storage tests carried out at 60 °C for 8 weeks (Pokorný et al., 1992). Oxidative stability of SUN significantly improved at PL concentrations larger than 0.2% based on oil weight. The inhibitory effect of PL was particularly high in the presence of metals, copper, iron, or manganese in the oil.

The hydroperoxide decomposing activity of nitrogen containing PL was substantially higher than that of phosphatidic acids.

The synergistic effect of commercial deoiled soybean lecithin and tocopherols in SUN (67% linoleic and 21% oleic and 656 mg/kg total tocopherol as 95.2% α-, 3.5% β-, and 1.2% γ-tocopherol) and HOSUN (10.6% linoleic and 80.2% oleic and 679 mg/kg total tocopherol as 94.6% α-, 2.8% β-, and 0.7% γ-tocopherol) was evaluated by storing the oils at 40 °C in open vials and monitoring OSI and PV of the samples (Judde et al., 2003). OSI of SUN and HOSUN increased from 5 to 6 hours and 11.9 to 22 hours, respectively, by 1% lecithin addition to the oil. Change in PV by lecithin addition was less pronounced for SUN (PV = 23.2 with no lecithin and PV = 22.4 with lecithin after 35 days of storage) than that for HOSUN (PV = 19.5 with no lecithin and PV = 8.7 with lecithin after 35 days of storage). Better stabilization of oils could be achieved with lecithin containing high PC and PE content. The antioxidant activity of lecithin was assumed to be due to the synergistic effect between nitrogen containing PC and PE with γ- and δ-tocopherols. No synergism was observed with α-tocopherols, especially when the tested oil was rich in linoleic acid.

Consumer concerns about the safety of synthetic antioxidants generated substantial interest in antioxidants from natural sources. Various plant extracts have been tested for their antioxidant properties. Ammari et al. (2012) examined the effect of *Nigella sativa* L. (Ranunculaceae) seed ethanol extract as natural antioxidant on SUN purchased in a grocery store in France. The accelerated oxidation tests carried out at 120 and 140 °C demonstrated that the seed extract at 800 mg/kg concentration had antioxidant effect similar to BHT at 200 mg/kg concentration in SUN.

Caffeic acid, coumarins, and 3,4-dihydroxybenzoic are ubiquitous in plants. Esculetin and fraxetin are hydroxycoumarins that are allowed for use in food formulations in some countries (Yanishlicva and Marinova, 1996). Oxidative stability of commercial SUN was examined in the presence of 0.01, 0.02, and 0.05% 3,4-dihydroxybenzoic acid, caffeic acid, esculetin, and fraxetin at 25 and 100 °C. Oxidative stability of SUN at 100 °C could be improved by adding either fraxetin or caffeic acid at the concentration of 500 mg/kg oil. 3,4-Dihydroxybenzoic acid did not show antioxidant activity in SUN at either temperature. Although esculetin did not demonstrate any antioxidant effect at 25 °C, 500 mg/kg, it improved the oxidative stability of SUN at 100 °C. The effectiveness of these compounds as antioxidants in SUN at 100 °C was in the following order: fraxetin > caffeic acid > esculetin. Esculetin used in the latter study was prepared by acid hydrolysis of pure esculin isolated from the ethyl alcohol extract of *Fraxinus ornus* bark. Yanishlieva and Marinova (1996) also noted that SUN was much more difficult to stabilize than the pure TAG species present in the oil.

Roman et al. (2013) examined the oxidation kinetics of SUN (26.0% oleic and 62.3% linoleic) and HOSUN (82.3% oleic and 8.8% linoleic) during heat treatment at 100, 120, and 150 °C for 6 hours in the absence or presence of α-tocopherol and a mixture of α-tocopherol and chlorogenic acid. α-Tocopherol addition significantly delayed the formation of oxidation products in both oils. A larger portion of tocopherols in HOSUN, 66.2%, than that of SUN, 14.5%, was consumed due to the antioxidant activity. Tocopherol degradation due to heat exposure, 33% reduction, was similar for both SUN and HOSUN. Although chlorogenic acid did not improve the oxidative stability of the oils, it reduced α-tocopherol degradation during heating.

The results discussed above clearly demonstrate that the effect of antioxidants in SUN varies significantly with the type of antioxidant, the temperature, and the testing method used.

Effect of Processing on Oil Stability

Edible oils are extracted from oilseeds by mechanical pressing and/or using an organic solvent such as hexane and then refined to remove undesirable oil constituents and improve final product quality and shelf life. Degumming, neutralization, bleaching, and deodorization are the common refining steps performed to produce edible grade vegetable oils. Processing may change the chemical composition of vegetable oils, subsequently affecting their oxidative stability. Indeed, various compounds that are present in minor amounts but still contribute to very important nutritional and functional attributes of the oils are inadvertently lost or removed during the refining process. For example, the refining process may result in the loss of vitamins, phytosterols, phenols, and aroma compounds naturally present in oils (Dunford, 2004). PL are removed during degumming. Removal of these compounds could have a significant effect on the oxidative stability of oils.

Brevedan et al. (2000) reported that hexane extracted sunflower oil contained higher amounts of free fatty acids (FFA) than in mechanically pressed oil, while Karaali (1985) published data contradicting the findings of the previous study. The difference in the results could be due to variations in the initial quality of the seeds and/or extraction conditions used to obtain the oil. High FFA content in the oil reduces the oxidative stability of the oil.

Significant differences in OSI of the oils obtained by mechanical pressing and hexane extraction have been reported (Brevedan et al., 2000). SUN cold pressed in the laboratory had the lowest OSI (measured at 98 °C), 10.2 hours, followed by commercially pressed, 20.2 hours, and hexane extracted oils, 29.9 hours. Considering that oils examined in the latter study did not have substantial differences in PV (0.83–3.05 meq/kg), AV (0.39–2.0), total tocopherol (696–752 mg/kg), and metal (Fe + Cu = 8.0–9.2 mg/kg) contents, the low OSI of the oil pressed in the laboratory can be attributed to its significantly lower PL content (trace amounts) than that of the commercially pressed, 7.37 g/kg, and hexane extracted SUN, 11.95 g/kg.

Indeed, the degumming process significantly reduced the OSI of the oils. OSIs for the commercially pressed (1.32 g/kg residual PL) and hexane extracted and then degummed oils (2.07 g/kg residual PL) were 12.5 and 16.3 hours, respectively. The positive effect of PL on oil oxidative stability was discussed in the previous section.

Fatty acid and TAG compositions and organoleptic properties of supercritical carbon dioxide (SC-CO_2) extracted sunflower oil is reported to be similar to those of hexane-extracted oil. However, oxidative stability of SC-CO_2 extracted SUN is shown to be lower than that of hexane extracted oil (Calvo et al., 1994). The low PL content of SC-CO_2 extracted oil leads to poor oxidative stability.

Antioxidant addition to crude oil is not a common practice since it is presumed that subsequent refining processes would remove both the oxidation products and the antioxidants present in the oil, deeming the cost of antioxidant addition untenable. Experimental data on the advantages of protecting crude oil during storage by adding antioxidants prior to refining is limited. Sherwin and Luckadoo (1970) examined the oxidative stability of sunflower seed oil stored in crude form with and without antioxidants and the refined oil produced from the stored oils. Synthetic antioxidants, BHT, TBHQ, and PG, were evaluated for their effect on oxidative stability of the samples using AOM method (Sherwin and Luckadoo, 1970). In the latter study, AOM was defined as the number of hours until PV of the oil heated at 99 °C reached to 70 meq/kg oil. TBHQ was more effective than BHA and PG in protecting crude SUN during 16 weeks storage. AOM of the crude oil prior to storage was 40 and 8 hours with and without TBHQ treatment, respectively. At the end of the 16 week storage period, AOM of the crude oils stored with and without TBHQ treatment decreased to 16 and 4 hours, respectively. Oxidative stability of the deodorized oil obtained from the crude oil stored without antioxidant addition (AOM = 8 h) was higher than that of the crude oil stored under the same conditions (4 h). Addition of 0.02% TBHQ to the deodorized oil obtained from crude oil stored without antioxidant further improved the AOM of the oil to 34 hours. These results are expected because refining/deodorization removes most of the oxidation products and improves the oxidative stability of the oil. Interestingly, deodorized oil obtained from the crude oil treated with 0.02% TBHQ prior to storage had significantly lower AOM, 6 hours, than that of the refined oils obtained from the crude oils stored without antioxidant, 8 hours, and oils treated with BHA, 8 hours, and PG, 7 hours. Sherwin and Luckadoo (1970) reported that a notable discoloration was observed in the TBHQ treated crude oil after four months of storage. Considering that it took 24 hours longer for the sensory panelists to detect rancidity in all the deodorized oils obtained from the antioxidant treated crude oils (longer shelf life) as compared to the refined oil obtained from crude oil stored without antioxidant treatment, it is possible that color discoloration affected the AOM measurements.

A study on the deodorized sunflower oils (15.7% oleic and 73.3% linoleic, IV= 140.5) containing no added antioxidants or metal chelating agents demonstrated the significant effect of bleaching on oil oxidative stability (List et al., 1972). The alkali refined (deacidified or neutralized) samples were more stable than their bleached counterparts as reflected in accelerated storage flavor scores and peroxide values determined by the 8 hour AOM. These results indicate that some of the oil components (naturally occurring metal chelators and antioxidants) removed during bleaching are beneficial for oxidative stability of SUN. Both alkali-refined and bleached oils had predominantly buttery, beany, and nutty flavors prior to storage. Rancid flavors were dominant in both oils after accelerated storage at 60 °C for 4 days. The bleached oil had a higher rancid flavor intensity score than that of the deacidified oil. In addition, paint-like flavors essentially absent in the stored neutralized oil were detected in the bleached samples by the taste panelists. Concentration of the bleaching earth also had a significant effect on oil oxidative stability (List et al., 1972). Sunflower oils treated with 6% bleaching earth in the laboratory had inferior flavor and oxidative stability than the same oil bleached with 2% earth. Even after the antioxidant treatment of the oils, PV increased at a faster rate in the bleached oil than in the alkali-refined but not the bleached sample. BHT, TBHQ, and nordihydroguaiaretic acid (NDGA) addition was ineffective and propyl gallate (PG), BHA, and 2,4,5-trihydroxybutyrophenone (THBP) were highly effective in improving flavor stability of the bleached oil. It is interesting to note that although TBHQ was the most effective antioxidant in inhibiting the increase in PV, it did not improve the flavor stability of the bleached oil. Unfortunately, AV of the samples was not reported in the study. Poor flavor stability could be due to high AV.

Commercial antioxidant mixtures containing both phenolic antioxidants and a metal scavenger improved the flavor and oxidative stabilities of refined unbleached oil. Although phenolic antioxidants alone improved oxidative stability as measured by the AOM test, flavor stability did not improve significantly for antioxidant-treated refined, unbleached samples after accelerated storage. Conversely, alkali-refined and bleached sunflower oil responded to treatment with certain phenolic antioxidants. Although iron and copper are deleterious to oil stability at concentrations of 0.1 mg/kg, metal-inactivating agents such as citric acid were effective in improving flavor stability.

Morrison (1975) also examined the effect of bleaching on the oxidative stability of oils obtained from sunflower seeds grown in the northern and southern United States. The oils differed in their linoleic and oleic contents, which were 61.5 and 26.4% and 37.9 and 50.9% for northern and southern oils, respectively. Based on AOM values, the crude northern oil was less stable than the southern oil. Bleaching the southern oil with 1% activated clay slightly decreased the initial stability as compared to the deacidified oil, while no change was observed in the stability of northern oils. When oils were bleached by using 3% activated clay initial oxidative stability of both oils improved. Bleaching with 1% activated clay decreased the storage stability of both northern and southern oils as compared to refined oils, but with 3% clay, storage stability was unaffected. Bleaching had an adverse effect on both oils by lowering their initial oxidative stability, but storage characteristics of the southern oil were better than that of the northern oil, which is probably due to the differences in their linoleic acid content.

The type of unit operations used for oil refining significantly affect the oxidative stability of the final product due to the changes occurring to minor compounds in the oil during processing. The effect of soft column deodorization (SCD) on tocopherol content of sunflower oil was compared to that of traditional deodorization (Alpaslan et al., 2001). Tocopherol loss during sunflower deodorization was reduced by 50% when traditional method (12.1% loss) was replaced with SCD (5.6% loss). Hence, SCD-refined oil is expected to have higher stability.

Ultrasound is useful in reducing processing times and increasing efficiency of many processes such as sterilization, pasteurization, extraction, emulsification, cleaning, cutting, freezing, thawing, and filtration. A study on the effect of high power ultrasound (20 kHz) on refined sunflower oil samples showed that PV of the oil increased from 5.38 meq O_2/kg oil for untreated oil to 6.33 meq O_2/kg oil for the oil sonicated at 20 kHz, 150 W for 2 minutes (Chemat et al., 2004). PV of the treated oil increased rapidly to 30 meq O_2/kg oil during storage. Headspace analysis of the sonicated samples showed that limonene and hexanal were the major components. Fishy, rancid, and metallic odors were detected in the oil even after 1 minute of sonication. The odors were attributed to the formation of (Z)-hept-2-enal and (2E,4E)-deca-2,4-dienal, respectively. Sensory tests carried out with sonicated oil samples indicated that sunflower and corn oil degraded more severely than soybean and olive oil, and the sensory properties of the peanut oil were not affected with sonication. The stability of peanut oil during sonication was attributed to its lower unsaturated fatty acid content than the other oils examined in the study. Unsaturated fatty acid content of the sunflower oil examined in the study was 85.8% (Chemat, et al., 2004). When the sonicated sunflower oil was stored at room temperature in darkness for longer than 24 hours it became cloudy. This effect was attributed to the formation of polymers following degradation of TAG. The authors (Chemat et al., 2004) argued that oil oxidation during sonication was due to cavitation, which is described as the formation, growth, and sometimes the implosion of micro bubbles created in oil when ultrasound waves propagate through it. The collapse of the bubbles leads to energy accumulation in hot spots where temperatures of above 5000 °C and pressures of about 500 MPa have been measured (Povey and Mason, 1998). Cavitation can lead to thermal oxidation. Furthermore, during sonication, free radicals are generated and the mechanical forces (shear forces) created by microstreaming and shock waves may lead to oil oxidation.

Microwave heating has been examined as a pretreatment step for oilseeds prior to extraction. PV of the hexane extracted oil from microwave-treated sunflower seeds (2450 MHz, 5–15 min) increased significantly with increasing treatment time (Anjum et al., 2006). Although AV of the oil samples did not change significantly, the formation of conjugated dienes and trienes increased with microwave heating. A significant decrease was also observed in the tocopherol content of the treated oil samples.

Interesterification is used to produce oil or fat mixtures with improved melting properties, to produce plastic fats like margarines, and to improve oxidative stability of the final product. Lipase-catalyzed acyl exchange between lard and high oleic sunflower oil has been investigated (Seriburi and Akoh, 1998). Sunflower oil (Trisun® Extra) was interesterified with lard at 55 °C for 24 hours using SP435 lipase from *Candida antarctica* to produce plastic fats. Trisun (84.7% oleic) had a considerably higher OSI value than lard. The physical blend was more stable than the interesterified blend (11.6 vs. 3.9 h), which means that interesterification lowered the oxidative stability of the lard–Trisun blend. This could be due to the loss of tocopherols during the downstream processing of the product mixture. The authors indicated that adding back antioxidants such as citric acid or BHA may increase the stability of the interesterified product.

For applications that require high oxidative stability, highly unsaturated oils with desirable flavor profile oils are hydrogenated to improve their oxidative stability. Oxidative stability of two hydrogenated sunflower oils with different fatty acid compositions was examined (Morrison and Robertson, 1978). As expected, hydrogenation process decreased linoleic acid (from 64.4 to 53.3% and from 33.4 to 30.8%) and increased oleic acid (from 26.0 to 32.1% and from 55.5 to 56.9%) contents of both oils examined in the latter study. The calculated IVs were also reduced (from 133.9 to 119.9 and from 105.6 to 102.3). *Trans* fat contents of the hydrogenated samples were 6.4 and 3.3%. Hydrogenation increased the AOM values and the overall stability for both oils. The oil with lower PUFA content had the higher initial AOM value and overall stability than the unhydrogenated oil. Inherent stability was not affected by hydrogenation.

The function of the oil bodies that are found in oilseeds is to store and protect neutral lipids during seed dormancy. They consist of a neutral lipid core surrounded by a charged coat of phospholipids, oleosin proteins, and other compounds associated with them (Frandsen et al., 2001; Huang, 1992). Oil bodies are relatively robust and stable in a range of detergents and solvents and during heating or drying (Mancuso et al., 1999; White et al., 2008). These characteristics lead to development of a commercial method for recovering recombinant proteins (Vanrooijen and Moloney, 1995). The oil bodies may have the potential as a natural plant source of food emulsions. Although stability of oil bodies in vivo is confirmed, their in vitro stability has not been established. Fisk et al. (2008) examined oxidative stability of oil bodies recovered from sunflower seeds in vitro. The oxidative stability of sunflower oil body suspensions (10%, w/w lipid) were evaluated over 8 days at three temperatures (5, 25, and 45 °C) against three comparable sunflower oil emulsions stabilized with dodecyltrimethylammonium bromide (DTAB), polyoxyethylene-sorbitan monolaurate (Tween 20), and sodium dodecyl sulfate (SDS) (17 mM). It was found that sunflower seed oil bodies were stable against oxidation at 5 °C and, when stored at elevated temperatures, hydroperoxide formation was significantly lower than in emulsions formulated with DTAB, SDS, and Tween 20. No hexanal formation was detected in the oil body emulsions, whereas hexanal was present in the headspace of the formulated emulsions. The higher oxidative stability of the oil body suspensions was explained with the presence of residual seed proteins in the continuous phase and a strongly stabilized lipid–water interface. The purity of oil bodies had an effect on the oxidation rate. The salt-washed oil body preparation showed less hydroperoxide formation than the unwashed preparations during incubation studies. The authors indicated that oil bodies can be used in industrial applications, specifically in delivery systems designed for fat-soluble vitamins.

Conclusion

Literature reviewed in this chapter clearly demonstrates that oxidative stability of sunflower oil is strongly correlated with its fatty acid composition. Today a wide range of sunflower oils with varying levels of PUFA and SFA are commercially available. High oleic sunflower oil tends to have higher stability than the high linoleic varieties. Biotechnology has been an excellent tool to improve sunflower seed and oil properties including agronomic, functional, and stability characteristics. Ongoing research will certainly lead to new and improved sunflower varieties with higher oil stability, better functionality for specific applications, and higher nutritional value.

Extraction and refining techniques used for production significantly affect oxidative stability of oils. Choosing softer processing techniques that minimize the loss of antioxidants naturally present in oils and formation of oxidation products or adding synthetic or natural antioxidants to refined oil improve their stability. Hydrogenation and interesterification could increase the stability of high linoleic type sunflower oil. However, conditions used for oil modification may affect the oxidative stability of the final product due to the loss of natural antioxidants during processing. *Trans* fatty acid formation during hydrogenation is a concern for edible oils. New processing techniques that produce very low or no *trans* fats should be further investigated.

High oleic and palmitic sunflower oil has the potential as a *trans*-free fat alternative. Sunflower oil is preferred for numerous food applications because of its favorable flavor profile. It is expected that ongoing research on agronomic properties and chemical composition of sunflower seeds and advanced processing techniques will result in sunflower-derived products with improved properties and broaden the utilization of sunflower oil and meal in food and other industrial applications.

References

Alpaslan, M.; Tepe, S.; Simsek, O. Effect of Refining Processes on the Total and Individual Tocopherol Content in Sunflower Oil. *Int. J. Food Sci. and Tech.* **2001,** *36* (7), 737–739.

Ammari, F.; Jouan-Rimbaud-Bouveresse, D.; Boughanmi, N.; Rutledge, D. N. Study of the Heat Stability of Sunflower Oil Enriched in Natural Antioxidants by Different Analytical Techniques and Front-face Fluorescence Spectroscopy Combined with Independent Components Analysis. *Talanta* **2012,** *99*, 323–329.

Anjum, F.; Anwar, F.; Jamil, A.; Iqbal, M. Microwave Roasting Effects on the Physico-chemical Composition and Oxidative Stability of Sunflower Seed Oil. *J. Am. Oil Chem. Soc.* **2006,** *83* (9), 777–784.

AOCS. *Official Methods and Recommended Practices.* American Oil Chemists' Society: Urbana, IL, 2004.

Becker, R.; Knorr, A. An Evaluation of Antioxidants for Vegetable Oils at Elevated Temperatures. *Lubrication Science* **1996,** *8* (2), 95–117.

Brevedan, M. I. V.; Carelli, A. A.; Crapiste, G. H. Changes in Composition and Quality of Sunflower Oils During Extraction and Degumming. *Grasas y Aceites* **2000,** *51* (6), 417–423.

Calvo, L.; Cocero, M. J.; Díez, J. M. Oxidative Stability of Sunflower Oil Extracted with Supercritical Carbon Dioxide. *J. Am. Oil Chem. Soc.* **1994,** *71* (11), 1251–1254.

Carelli, A. A.; Bodnariuk, P.; Crapiste, G. H. Effectiveness of Antioxidants in Sunflower Oil. In *Advances in Oils and Fats, Antioxidants, and Oilseed By-Products*; Koseoglu, S. S., Rhee, K. C., Wilson, R. F., Eds.; AOCS Press: Urbana, IL, 1998; pp 258–261.

Carelli, A. A.; Franco, I. C.; Crapiste, G. H. Effectiveness of Added Natural Antioxidants in Sunflower Oil. *Grasas y Aceites* **2005,** *56* (4), 303–310.

Chemat, F.; Grondin, I.; Costes, P.; Moutoussamy, L.; Sing, A. S. C.; Smadja, J. High Power Ultrasound Effects on Lipid Oxidation of Refined Sunflower Oil. *Ultrasonics Sonochemistry* **2004,** *11* (5), 281–285.

Choe, E.; Min, D. B. Mechanisms and Factors for Edible Oil Oxidation. *Compr. Rev. Food Sci. Food Saf.* **2006,** *5*, 169–186.

Codex Alimentarius. *Standard for Named Vegetable Oils. CODEX STAN 210-1999.* WHO/FAO: Rome, 2001; pp 11–25.

Cowan, J. C.; Moser, H.; List, G. R.; Evans, C. D. Organoleptic and Oxidative Stability of Blends of Soybean and Peanut Oils. *J. Am. Oil Chem. Soc.* **1971,** *48* (12), 835–839.

Crapiste, G. H.; Brevedan, M. I. V.; Carelli, A. A. Oxidation of Sunflower Oil during Storage. *J. Am. Oil Chem. Soc.* **1999,** *76*, 1437–1443.

Demurin, Y.; Skoric, D.; Karlovic, D. Genetic Variability of Tocopherol Composition in Sunflower Seeds as a Basis of Breeding for Improved Oil Quality. *Plant Breeding* **1996,** *115* (1), 33–36.

Dunford, N. T. Effects of Oil and Oilseed Processing Techniques on Bioactive Compounds. In *Nutritionally Enhanced Oil and Oilseed Processing*; Dunford, N. T., Dunford, H. B., Eds.; AOCS Press: Urbana, IL, 2004; pp 25–37.

Eldin, A.-K. Effect of Fatty Acids and Tocopherols on the Oxidative Stability of Vegetable Oils. *Eur. J. Lipid Sci. Technol.* **2006,** *58*, 1051–1061.

Erhan, S. Z.; Sharma, B. K.; Perez, J. M. Oxidation and Low Temperature Stability of Vegetable Oil-based Lubricants. *Ind. Crops Prod.* **2006,** *24* (3), 292–299.

Fernández-Martínez, J. M.; Mancha, M.; Osorio, J.; Garcés, R. Sunflower Mutant Containing High Levels of Palmitic Acid in High Oleic Acid Background. *Euphytica* **1997,** *97*, 113–116.

Fisk, I. D.; White, D. A.; Lad, M.; Gray, D. A. Oxidative Stability of Sunflower Oil Bodies. *Eur. J. Lipid Sci. Technol.* **2008,** *110* (10), 962–968.

Frandsen, G. I.; Mundy, J.; Tzen, J. T. C. Oil Bodies and Their Associated Proteins, Oleosin and Caleosin. *Physiol Plant* **2001,** *112*, 301–307.

Frankel, E. N. Chemistry of Autoxidation: Mechanism, Products and Flavor Significance. In *Flavor Chemistry of Fats and Oils*; Min, D. B., Smouse, T. H., Eds.; AOCS Press: Urbana, IL, 1985; pp 1–34.

Frankel, E. N. *Lipid Oxidation*, 2nd ed. The Oily Press: Dundee, Scotland, 2005.

Frankel, E. N.; Huang, S. W. Improving the Oxidative Stability of Polyunsaturated Vegetable Oils by Blending with High-oleic Sunflower Oil. *J. Am. Oil Chem. Soc.* **1994,** *71* (3), 255–259.

Frankel, E. N.; Nessim, E. Stability Methods. In *Lipid Oxidation*; The Oily Press Ltd.; Glasgow, UK, 1998; pp 99–114.

Garcés, R.; Martínez-Force, E.; Salas, J. J.; Venegas-Calerón, M. Current Advances in Sunflower Oil and Its Applications. *Lipid Technology* **2009,** *21* (4), 79–82.

Grompone, M. A. Sunflower Oil. In *Vegetable Oils in Food Technology*; Wiley-Blackwell: London, UK, 2011; pp 137–167.

Guillén, M. D.; Cabo, N. Fourier Transform Infrared Spectra Data versus Peroxide and Anisidine Values to Determine Oxidative Stability of Edible Oils. *Food Chemistry* **2002,** *77*, 503–510.

Guillén, M. D.; Goicoechea, E. Formation of Oxygenated α,β-Unsaturated Aldehydes and Other Toxic Compounds in Sunflower Oil Oxidation at Room Temperature in Closed Receptacles. *Food Chemistry* **2008,** *111* (1), 157–164.

Guinda, Á.; Dobarganes, M. C.; Ruiz-Mendez, M. V.; Mancha, M. Chemical and Physical Properties of a Sunflower Oil with High Levels of Oleic and Palmitic Acids. *Eur. J. Lipid Sci. Technol.* **2003,** *105*, 130–137.

Gunstone, F. D.; Hilditch, T. P. The Autoxidation of Methyl Oleate in Presence of Small Proportions of Methyl Linoleate. *J. Chem. Soc.* **1946**, 1022–1025.

Holman, R. T.; Elmer, O. C. The Rates of Oxidation of Unsaturated Fatty Acids and Esters. *J. Am. Chem. Soc.* **1947,** *24*, 127–129.

Hoshina, R.; Endo, Y.; Fujimoto, K. Effect of Triacylglycerol Structures on the Thermal Oxidative Stability of Edible Oil. *J. Am. Oil Chem. Soc.* **2004,** *81* (5), 461–465.

Huang, A. H. C. Oil Bodies and Oleosins in Seeds. *Annu Rev. Plant Physiol Plant Mol Biol.* **1992,** *43*, 177–200.

Jadhav, S. J.; Nimbalkar, S. S.; Kulkarni, A. D.; Madhavi, D. L. Lipid Oxidation in Biological and Food Systems. In *Food Antioxidants: Technological, Toxicological, and Health Perspectives*; Madhavi, D. L., Deshpande, S. S., Salunkhe, D. K., Eds.; Marcel Dekker: New York, 1996; pp 5–63.

Jebe, T.; Matlock, M.; Sleeter, R. Collaborative Study of the Oil Stability Index Analysis. *J. Am. Oil Chem. Soc.* **1993,** *70* (11), 1055–1061.

Jonnala, R. S.; Dunford, N. T.; Dashiell, K. E. New High Oleic Peanut Cultivars Grown in the Southwestern U.S. *J. Am. Oil. Chem. Soc.* **2005,** *85*, 125–128.

Judde, A.; Villeneuve, P.; Rossignol-Castera, A.; Guillou, A. Antioxidant Effect of Soy Lecithins on Vegetable Oil Stability and Their Synergism with Tocopherols. *J. Am. Oil Chem. Soc.* **2003,** *80* (12), 1209–1215.

Karaali, A. The Effects of Refining on the Chemical Composition of Turkish Sunflower Seed Oil. *Fette Seifen Anstrichmittel* **1985,** *87*, 112–117.

King, M. F.; Boyd, L. C.; Sheldon, B. W. Antioxidant Properties of Individual Phospholipids in a Salmon Oil Model System. *J. Am. Oil Chem. Soc.* **1992,** *69* (6), 545–551.

List, G. A.; Erickson, D. R. Oil and Fat Products. In *Bailey's Industrial Oil and Fat Products*; Applewhite, T. H., Ed.; John Wiley and Sons: New York, 1985; Vol. III, pp 275–277.

List, G. R.; Evans, C. D.; Moser, H. Flavor and Oxidative Stability of Northern-grown Sunflower Seed Oil. *J. Am. Oil Chem. Soc.* **1972,** *49* (5), 287–292.

Majumdar, T. K.; Vasudish, C. R.; Premavali, K. S.; Bawa, A. S. Stability Evaluation of Heated Sunflower Oil Using Rancimat. *J. Food Sci. Technol.* **2007,** *44* (2), 143–145.

Mancuso, J. R.; McClements, D. J.; Decker, E. A. The Effects of Surfactant Type, pH, and Chelators on the Oxidation of Salmon Oil-in-water Emulsions. *J. Agric. Food Chem.* **1999,** *47*, 4112–4116.

Marmesat, S.; Velasco, L.; Ruiz-Méndez, M. V.; Fernández-Martínez, J. M.; Dobarganes, C. Thermostability of Genetically Modified Sunflower Oils Differing in Fatty Acid and Tocopherol Compositions *Eur. J. Lipid Sci. Technol.* **2008,** *110*, 776–782.

Marmesat, S.; Morales, A.; Velasco, J.; Carmen Dobarganes, M. Influence of Fatty Acid Composition on Chemical Changes in Blends of Sunflower Oils during Thermoxidation and Frying. *Food Chemistry* **2012,** *135* (4), 2333–2339.

Merrill, L. I.; Pike, O. P.; Ogden, L. V.; Dunn, M. L. Oxidative Stability of Conventional and High-Oleic Vegetable Oils with Added Antioxidants. *J. Am. Oil Chem. Soc.* **2008,** *85*, 771–776.

Morrison, W. H., III. Effects of Refining and Bleaching on Oxidative Stability of Sunflowerseed Oil. *J. Am. Oil Chem. Soc.* **1975,** *52* (12), 522–525.

Morrison, W. H.; Robertson, J. A. Hydrogenated Sunflowerseed Oil: Oxidative Stability and Polymer Formation on Heating. *J. Am. Oil Chem. Soc.* **1978,** *55* (5), 451–453.

Moser, B. R. Influence of Blending Canola, Palm, Soybean, and Sunflower Oil Methyl Esters on Fuel Properties of Biodiesel. *Energy & Fuels* **2008,** *22* (6), 4301–4306.

Neff, W. E.; Selke, E.; Mounts, T. L.; Rinsch, W.; Frankel, E. N.; Zeitoun, M. A. M. Effect of Triacylglycerol Composition and Structures on Oxidative Stability of Oils from Selected Soybean Germplasm. *J. Am. Oil Chem. Soc.* **1992,** *69* (2), 111–118.

Olcott, H. S.; Veen, J. Comparison of Antioxidant Activities of Tocol and Its Methyl Derivatives. *Lipids* **1968,** *3* (4), 331–334.

Pokorný, J.; Réblová, Z.; Ranný, M.; Káňová, J.; Pánek, J.; Davídek, J. Natural Lecithins and Phosphorylated Acylglycerols as Inhibitors of Autoxidation of Fats and Oils. *Food / Nahrung* **1992,** *36* (5), 461–465.

Povey, M. J. W.; Mason, T. J. *Ultrasound in Food Processing*. Blackie Academic International: London, 1998.

Ragnarsson, J. O.; Labuza, T. P. Accelerated Shelf-life Testing for Oxidative Rancidity in Foods—A Review. *Food Chemistry* **1977,** *2* (4), 291–308.

Randles, S. J. Environmentally Considerate Ester Lubricants for the Automotive and Engineering Industries. *Journal of Synthetic Lubrication* **1992,** *9* (2), 145–161.

Robertson, G. L. Shelf Life of Packed Foods: Its Measurements and Prediction. In *Developing New Food Products for a Changing Marketplace*; Brody, A. L., Lord, J. B., Eds.; Technomic Publishing: Lancaster, UK, 2000; pp 329–353.

Roman, O.; Heyd, B.; Broyart, B.; Castillo, R.; Maillard, M.-N. Oxidative Reactivity of Unsaturated Fatty Acids from Sunflower, High Oleic Sunflower and Rapeseed Oils Subjected to Heat Treatment, Under Controlled Conditions. *LWT—Food Science and Technology* **2013,** *52* (1), 49–59.

Seriburi, V.; Akoh, C. Enzymatic Interesterification of Lard and High-oleic Sunflower Oil with *Candida antarctica* Lipase to Produce Plastic Fats. *J. Am. Oil Chem. Soc.* **1998,** *75* (10), 1339–1345.

Serqueira, D. S.; Fernandes, D. M.; Cunha, R. R.; Squissato, A. L.; Santos, D. Q.; Richter, E. M.; Munoz, R. A. A. Influence of Blending Soybean, Sunflower, Colza, Corn, Cottonseed, and Residual Cooking Oil Methyl Biodiesels on the Oxidation Stability. *Fuel* **2014,** *118*, 16–20.

Sharma, B. K.; Stipanovic, A. J. Development of a New Oxidation Stability Test Method for Lubricating Oils Using High-pressure Differential Scanning Calorimetry. *Thermochim. Acta* **2003,** *402* (1/2), 1–18.

Sherwin, E. R.; Luckadoo, B. M. Studies on Antioxidant Treatments of Crude Vegetable Oils. *J. Am. Oil Chem. Soc.* **1970,** *47* (1), 19–23.

Škorić, D. Achievements and Future Directions of Sunflower Breeding. *Field Crops Research* **1992,** *30* (3–4), 231–270.

Skoric, D.; Jocić, S.; Sakac, Z.; Lecic, N. Genetic Possibilities for Altering Sunflower Oil Quality to Obtain Novel Oils. *Can. J. Physiol. Pharmaco.* **2008,** *86*, 215–221.

Van de Voort, F. R.; Ismail, A. A.; Sedman, J.; Dubois, J.; Nicodemo, T. The Determination of Peroxide Value by Fourier Transform Infrared Spectroscopy. *J. Am. Oil Chem. Soc.* **1994,** *71* (9), 921–926.

Vanrooijen, G. J. H.; Moloney, M. M. Plant Seed Oil-bodies as Carriers for Foreign Proteins. *Biotech. Adv.* **1995,** *13*, 72–77.

Velasco, J.; Dobarganes, M. C.; Marquez-Ruiz, G. Application of the Accelerated Test Rancimat to Evaluate Oxidative Stability of Dried Microencapsulated Oils. *Grasas y Aceites* **2000,** *51* (4), 261–267.

Warner, K.; Frankel, E. N.; Mounts, T. L. Flavor and Oxidative Stability of Soybean, Sunflower and Low Erucic Acid Rapeseed Oils. *J. Am. Oil Chem. Soc.* **1989,** *66* (4), 558–564.

Warner, K.; Orr, P.; Glynn, M. Effect of Fatty Acid Composition of Oils on Flavor and Stability of Fried Foods. *J. Am. Oil Chem. Soc.* **1997,** *74* (4), 347–356.

Warner, K.; Miller, J.; Demurin, Y. Oxidative Stability of Crude Mid-Oleic Sunflower Oils from Seeds with High γ- and δ-Tocopherol Levels. *J. Am. Oil Chem. Soc.* **2008,** *85* (6), 529–533.

White, D. A.; Fisk, I. D.; Mitchell, J. R.; Wolf, B.; Hill, S. E.; Gray, D. A. Sunflower-seed Oil Body Emulsions: Rheology and Stability Assessment of a Natural Emulsion. *Food Hydrocolloids* **2008,** *22* (7), 1224–1232.

Yanishlieva, N. V.; Marinova, E. M. Inhibited Oxidation of Lipids. 1: Complex Estimation and Comparison of the Antioxidative Properties of Some Natural and Synthetic Antioxidants. *Fat Science and Technolology* **1992,** *94*, 374–379.

Yanishlicva, N. V. M.; Marinova, E. M. Antioxidative Effectiveness of Some Natural Antioxidants in Sunflower Oil. *Z Lebensm Unters Forch* **1996,** *203*, 220–223.

Yodice, R. Nutritional and Stability Characteristics of High Oleic Sunflower Seed Oil. *Lipid / Fett* **1990,** *92* (3), 121–126.

16

U.S. and Canada Perspectives on Sunflower Production and Processing

Larry Kleingartner ■ *Retired, National Sunflower Association, USA; Consultant and Adjunct Professor, Bismarck, ND, USA*

Introduction

Sunflower (*Helianthus annuus L.*) is a relatively new crop in North America. Although the plant is native to North America, domesticated sunflower did not become a significant crop in this continent until the mid-1970s. The wild species made a very circuitous route to Russia where the modern high oil yielding varieties were developed. Those varieties were brought to North America during the Cold War. A combination of demand for the oil and the need for a rotational crop popularized sunflower in specific North American regions. The road to farmer acceptance of sunflower has not been easy. Native pests have been numerous, federal farm programs were often detrimental to acreage growth, and the U.S. domestic market has not been eager to use a vegetable oil that was not a "least cost" oil. In the short 40-year history, tremendous changes to the North American sunflower industry have taken place. Markets were gained and lost, the fatty acid composition of the oil was changed dramatically, and the production base moved to a more arid region, leaving most processing plants out of position. Despite these challenges, the crop has endured and opportunities on the horizon point to the continuation of this crop as a viable production choice in North America.

Sunflower in North America

Commercial sunflower became a viable crop in the middle 1970s in North Dakota (ND), Minnesota, and Manitoba. The production was mostly isolated in a narrow band of highly productive soil called the Red River Valley. There had been small pockets of production in the region in earlier years, filling a demand for confection and bird food. New and improved varieties of sunflower planting seed from the former Soviet Union were imported into Canada. The seed made its way across the U.S.–Canadian border circumventing the prohibition of U.S./Soviet Union trade during the peak of the Cold War.

The late 1960s were also a time of new insight into issues of heart health. Animal fat was the cooking fat of choice, but research was consistently pointing to animal fat as a contributing factor in deaths associated with heart disease. Developed countries began making a switch to vegetable oils. In the United States, hydrogenated soybean oil was the main substitute for animal fat, while sunflower oil became one of the main substitutes in Western Europe. The Soviets were the main suppliers of sunflower seed and oil to Western Europe until internal demand, production issues, and exports to satellite nations such as Cuba severely curbed supply. Farmers in the United States and Canada quickly geared up production to supply an increasing European demand for sunflower seed that the Europeans crushed into oil and meal. At this same time, farmers in the Red River Valley were eager for alternative crops to spring wheat and barley. Both crops were produced in excess and prices were generally depressed. The concept of rotations with a broadleaf crop was gaining interest, and farmers often found a "yield bump" in barley and wheat when alternating production cycles with sunflower. Farmers enjoyed good financial returns from sunflower and acreage expanded dramatically. Planted acreage peaked in 1979 at 2,248,100 ha and was never reached again. The vast majority of the production was trucked or railed to the Great Lakes ports of Duluth-Superior, where ships were loaded for Rotterdam, Hamburg, and Lisbon.

European Union (EU) agricultural policy was redirected in the early 1980s. Farmers were given lucrative subsidies to produce oilseeds, mostly rapeseed and sunflower. Western European demand for North American sunflower seed declined dramatically, and so did plantings. Three large sunflower crush plants were built in North Dakota during the late 1970s/early 1980s, but the domestic demand for sunflower oil was small. Mexico filled part of the European void with significant imports of sunflower seed and later sunflower oil. The U.S. market for sunflower oil did not gain traction until the issue of *trans* fatty acids emerged as a health concern. But for sunflower oil to be a viable domestic choice, the fatty acid structure of the oil had to be genetically altered.

Production, Markets, and Government Policy

With the exception of one year, North Dakota has been the nation's largest producer of sunflower in the history of U.S. production. The other major state is South Dakota, while Minnesota, Kansas, Nebraska, Colorado, and Texas are generally minor contributors. Manitoba is the largest Canadian producing province, with Saskatchewan and Alberta producing smaller amounts. Yields gradually increased over time with improved hybrids, production practices, and farmer knowledge (Figure 16.1). Total production (Figure 16.2) obviously varied with planted hectares (Figure 16.3) and yield. The low production level of 925 t in 2011 was likely an anomaly based on extreme wet planting conditions that greatly minimized total area planted in North Dakota and Minnesota.

Sunflower is a rotation crop in North America. It is usually planted every third or fourth year in regions where the crop has gained acceptance. Sunflower rotations are necessary to maximize yields by reducing pest pressure such as weeds, insects, and

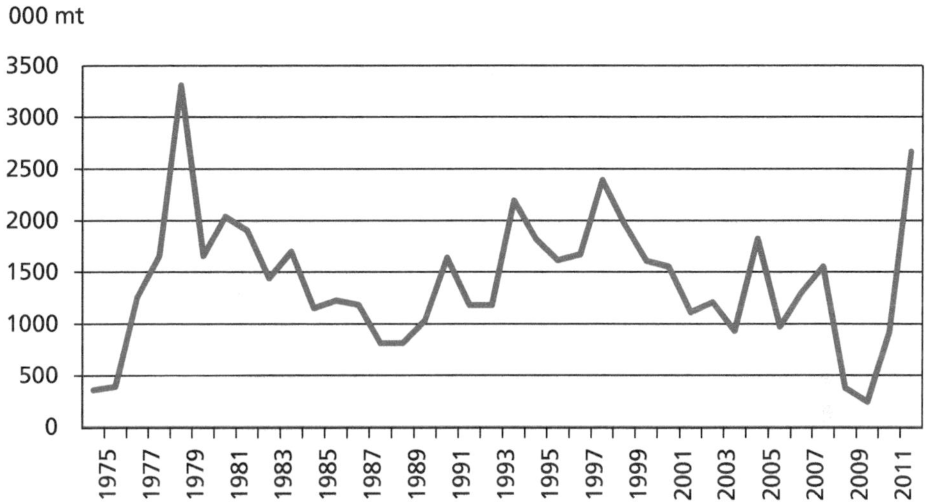

Figure 16.1 Total U.S. sunflower seed production.
Source: USDA NASS.

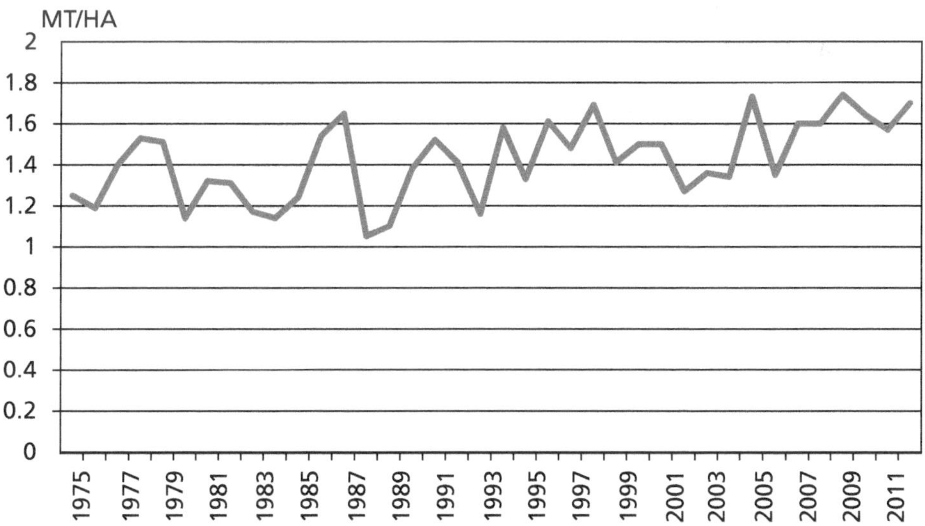

Figure 16.2 United States national average yield.
Source: USDA NASS.

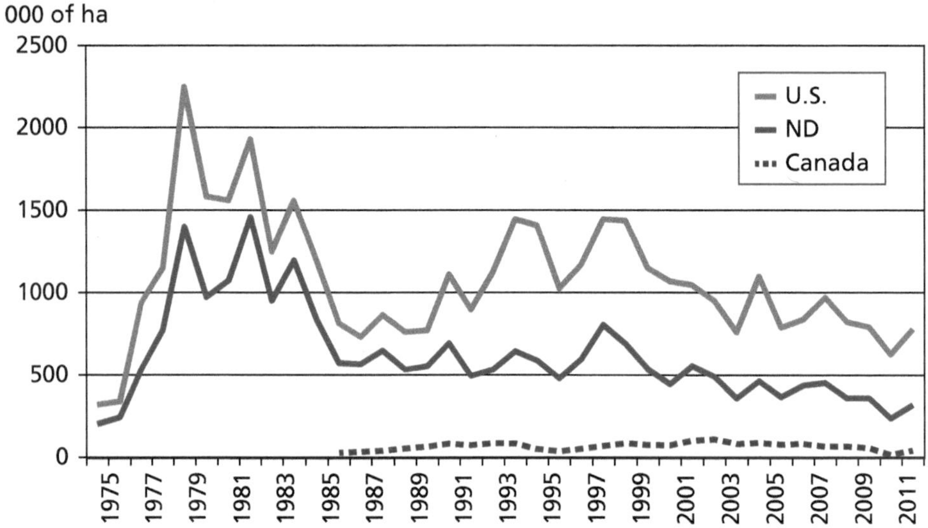

Figure 16.3 U.S. planted hectares 1975–2012 with ND and Canadian composition. Source: USDA NASS and AAFC.

disease and to maximize the use of soil moisture. Small grains such as wheat and barley are the most common rotation crops with sunflower, but farmers are finding that maize is also an excellent rotation crop with sunflower. Farmers often alter rotations annually by adding or deleting a crop. Price is a significant factor in planting decisions. Crop pests are significant factors as well. A farmer's prior year experience with a crop in terms of yield or production challenges can carry over to the next year's planting decision. Weather is a major factor. Sunflower is generally one of the last crops to be planted. There have been some years when extended rainfall late into the planting season significantly impacted the amount of sunflower acres that were planted. The opposite is true in dry years, when farmers cannot justify planting this late-season crop into extremely dry soil. Roundup Ready® technology was a significant cause of sunflower acreage decline and its shift to more arid regions. All of these factors combined have resulted in variable production levels from one year to the next.

Types of Sunflower Produced

North America is unique in the multiple types of sunflower that are grown for specific markets. The majority of acres are oil-type. In this category are three distinct fatty acid types for oil extraction (Table 16.A). These are NuSun, high oleic sunflower oil

Table 16.A Fatty Acid Composition of Five Types of Sunflower Oil

Fatty Acid	16:0	18:0	18:1	18:2	18:3	Misc[a]
Traditional	7.0	4.5	18.7	67.5	0.8	1.5
Mid Oleic (NuSun)	4.6	4.2	63	26	0.6	1.6
High Oleic (HOSO)	3.5	5.2	82	9.0	—	0.3
High Stearic/High Oleic	4	18	70	4	—	4.0
Low Saturate	1.9	0.7	94	2	0.2	1.2

[a]Misc may contain less of 0.5% of a combination of other fatty acids.

(HOSO), and traditional high linoleic. Soon to be introduced will be a low saturate/high oleic and a high stearic/high oleic. These oilseed types are the small black seed. Any of these can and are used for the bird food market. There are hulling types as well. These are unique varieties that are hulled for the seeds. Some of these hulling types have white stripes on the hull, while others are the traditional black-hulled sunflower. A hulling variety often is slightly larger with a slight separation between the hull and seed. This separation provides hulling efficiency required for quality product. The hulling types can also be used in bird food or can be crushed for oil. This depends on market dynamics.

Confection sunflower is a unique part of the North American industry. This sunflower is large with a white-and-black-striped hull. The product is used for snacking. The hybrid seed companies have developed various sizes and shapes for specific markets domestically and internationally. Industry designations include extra-long, jumbo, and giant, among other descriptors. In addition to the white-and-black-striped hulls, confection types are lighter in test weight due to the seed size. There are also HOSO and traditional linoleic hybrids in confection sunflower.

Market Structure

The initial demand for North American sunflower came from Europe. This demand was in the form of seed that was shipped to European ports and crushed in European mills. Demand for seed also came from Mexico and was mostly railed from U.S. elevator locations to Mexican mills for crushing into oil. In the early era of North American sunflower production, three crushing plants were built in North Dakota, each with a capacity of a minimum of 1,000 t per day. There was an existing plant in Red Wing, Minnesota, that primarily processed linseed and sunflower. A mill in Altona (Manitoba), Canada, also crushed sunflower for a short period. The U.S. crushing industry faced many challenges, and most of the facilities changed hands at least once. The challenges were related to inconsistent production, demand for seed from

Europe and/or Mexico, and a domestic food market that was heavily dependent on soybean oil. A large mill was added in Kansas. There is a small mill in Lamar, Colorado, that processes mostly sunflower. Most of these mills added refining capacity when the industry switched to oleic-type sunflower. Oleic-based sunflower oil was in greater demand in the domestic market and new refining capacity was required. There are no sunflower mills in Canada at the present time.

Most of the confection processing plants are located along the Red River that separates North Dakota and Minnesota. A great deal of consolidation has taken place over the years and the existing plants have sizable capacities and some also roast and bag consumer-ready product. In addition, there are two processing plants in Kansas, one plant in Colorado, and a roasting and packaging plant was recently established in northern Texas. Most of these processors have satellite facilities throughout the production region for farmer delivery and storage. There are several processing plants in Manitoba and one in Alberta. All of these processing plants have highly sophisticated sorting and cleaning machinery to ensure quality product.

Feeding wild birds is a popular backyard hobby and is a sizable business in North America. Processing wild bird food started mostly as a cottage industry, but has since matured into larger cleaning and bagging facilities. These plants are scattered throughout the production region with the largest combined capacity located in South Dakota. Demand for bird food was considered inelastic until the 2007–2009 recession. This era witnessed high commodity prices and reduced personal income, leading some consumers to either reduce or eliminate the feeding of wild birds in their backyard.

Price History and Price Discovery

Sunflower is a high oil content seed and approximately 80% of the value of sunflower comes from the oil, with the remainder from the protein meal. Historically, the Chicago soybean oil futures contract was the basic price indicator for sunflower oil and seed. A sunflower seed future's contract was developed at the Minneapolis Grain Exchange in the late 1970s, but trading was minimal and the contract was dropped. Pricing signals at the industry inception in the late 1970s came from Western European mills and not the Chicago soybean oil contract. However, as the European market declined, the domestic crush plants became a more significant factor in demand. It is at this time that the Chicago soybean oil contract played a larger role in sunflower pricing discovery. Sunflower oil, however, had a very small domestic share of the oil market until 2005. The U.S. market was dominated by inexpensive soybean oil that was always plentiful as a byproduct of the soybean protein meal demand. Traditional high linoleic sunflower oil was quite similar to the fatty acid structure of soybean oil. Like soybean oil, sunflower oil could not be used for industrial frying or baking applications in the United States without hydrogenation. The market had little interest in paying a premium for sunflower oil that performed similarly to soybean oil. Thus, the vast majority of U.S.-produced sunflower oil was exported to Mexico and North African countries. Most of the food industries of these countries were government managed and subsidized. These regions preferred sunflower oil and paid a reasonable premium over soybean oil. This market strategy worked fine until large Argentine crops of sunflower oil hit the world market.

High oleic sunflower oil (HOSO) was developed in the mid-1980s. Farm production had to be isolated and segregated in the market chain. All of this added to cost and the U.S. market was not ready to pay a premium for the oil. Part of the difficulty for HOSO acceptance was that *trans* fatty acids had not yet emerged as a health issue. Further, HOSO seed and oil were under patent and this placed an artificial limit on demand. Finally, food labeling laws were permissive, allowing a processed food manufacturer to label any potential oil that might be used. So production volumes remained low, despite a superior fatty acid composition as defined by today's market.

By the year 2000, the issue of *trans* fat as a health risk had gained a great deal of recognition. The U.S. Food and Drug Administration (FDA) announced its intention to require the labeling of trans fatty acids. This event set off a scramble within North American food manufacturers to find a suitable replacement for the market dominated by partially hydrogenated soybean oil. Oleic-based sunflower oils such as HOSO and newly developed NuSun found new domestic demand. The traditional price relationship between U.S. sunflower and soybean oil of approximately $40/t escalated to approximately $110/t. However, sunflower oil price discovery was not easily accomplished because most buyers made advanced purchases, leaving minimal oil for spot cash purchasing. A disastrous 2011 production season limited sunflower oil production and sent prices skyrocketing, requiring food companies to find alternative sources of supply. This was extremely damaging to future market demand.

Bird food is considered an opportunity market. When farmer seed supplies are ample, this market lags the price at the crush market. However, when farmer supplies are tight, the bird food market will pay a substantial premium over the price of the crush plants. It all depends on supply and demand. Demand for bird food is a national market with major retail chains supplying the product to customers. However, the northeast part of the United States continues to be a dominant region for demand. Harsh winters in that region can spur demand. Although bird food processors have consolidated and expanded their storage capacities, many are not capable of accepting large farmer deliveries over a short period of time. That can be a limiting factor because farmers with large production may want to deliver a large portion of their crop at harvest or at a later time when they are reducing their on-farm storage inventories.

The bird food market can be complementary to the crushing industry and farmer market. Crushing plants discount seed with oil content under 40% and pay a premium for seed with oil content exceeding 40%. The bird food market does not. Farmers with low oil content seed will likely find higher values in the bird food market, while farmers with high oil content seed will find higher values at crushing plants. Sometimes bird food can be a market outlet for slightly off-grade seed. The estimated annual size of the North American birdseed market for sunflower seed is between 400,000 and 500,000 t (National Sunflower Association, 2012a). Other seeds, such as millet, are used in bird food. But the preferred seed for most bird species and bird feeders is the small black oilseed.

Confection is a separate market within the sunflower industry. The planting hybrids are distinctly different from oil types, as are the seed size and the genetics. The confection industry compares its product to eating an apple as opposed to drinking the apple juice. Thus, a great deal of care is required in producing, processing, and packaging confection sunflower that is directly consumed as a whole seed. Differentiation begins with the development of hybrid planting seeds, which must have a particular seed size and color. There are specific hybrids for specific markets. For example, production that is exported to Spain, Turkey, and parts of the Middle East is generally for a larger and longer seed. Consumers in this region prefer to eat one seed at a time. North American consumers generally put a handful of seeds in their mouth, so these seeds must be slightly smaller.

Confection processors take great care in selecting farmers to grow the crop. In order to guarantee the best contract price, farmers must be vigilant in seeding rates, plant spacing, control of disease and insects, harvesting, and storage. Seed delivery that does meet specifications is rejected. There are not many market alternatives for rejected confection sunflower. The crushing plants cannot utilize these seeds due to the large size, light test weight, and low oil content. Typically, confection farmers are paid \$80 to \$120/t premium over oil type sunflower. There may be additional price premiums for large seed. It is important to note that confection sunflower acreage has declined (Figure 16.4). This is despite increasing demand in the United States and several overseas markets. Hybrid seed companies have developed hybrids that produce a greater percentage of large seed that can be used for human consumption. The percentage of large seed produced in 2008–2012 was 83% over a 20/64 sieve. This compares to 60% during the period of 1999–2002 (National Sunflower Association, 1999–2012b).

Most of these same confection processors also hull sunflower for the seeds. The hulled seeds are used in a variety of food products from bread to trail mixes in North America and around the world. These small seeds are loaded with nutrition and provide a crunchy nutty flavor. Specific hybrids developed for this market are called "hullers" by the U.S. industry. These hybrids may produce slightly larger seed when compared to oil types, and often these hybrids are lower in oil content. Hullers can, however, be used in the oil crush market. Generally, the hulling market is priced at a

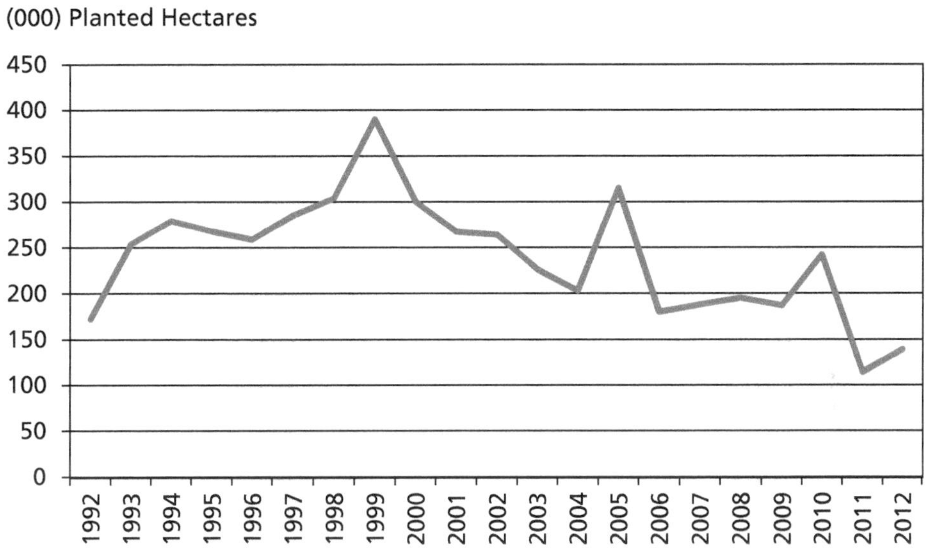

Figure 16.4 U.S. and Canada combined confection.

premium to oil types, but at a discount to confection sunflower. The size of this market has varied considerably over the last two decades. Competition from Chinese exports has reduced North American demand in key export markets such as Germany.

Government Policy

North American government policy has had significant impacts on the fortunes and misfortunes of sunflower production and markets. United States sunflower production got an initial boost due to government farm policy. In the 1970s and early 80s, farmers were prohibited from planting certain program crops (wheat, maize, barley) on all of their so-called base acres. These base acres received a government subsidy. The remaining acres were considered "free acres," and oilseeds and minor crops could be planted on those acres without a subsidy or penalty. However, that all changed with the 1985 U.S. farm bill, when agriculture was going through a very depressed era. In the 1985 farm bill, subsidies for program crops were increased considerably. Farmers could increase their base acres by not planting non-program crops like sunflower, soybeans, and other minor crops. The result of this policy was devastating to U.S. sunflower and other non-program crops. Farmers stopped rotating crops in an effort to maximize payments from the federal government. Production of sunflower and other crops dropped drastically in between 1985 and 1990.

During this same era, Argentine crushing plants were given preference via a differential taxation program where exports of oil and meal were taxed at a lower rate than exports of unprocessed seed. This provided Argentine sunflower oil processors with a price advantage in the international marketplace. The EU found such great success in their production scheme for self-sufficiency in grains and oilseeds that they provided export subsidies for surplus vegetable oils, among other farm products. As U.S. farmers were reducing their oilseed production due to planting restrictions, world market prices for vegetable oil were depressed due to export subsidies from two large production regions. The U.S. policy already included export subsidies for major program crops. Oilseed groups lobbied for similar programs for vegetable oil. A unique Sunflower Oil Assistance Program (SOAP) was created in 1989. The program allowed U.S. exporters to compete internationally with other countries subsidizing oil exports. This program came to an end in 1996, after the EU oilseed production program was found in violation of international trade rules. There was a renewed effort worldwide to reduce export subsidies of all kinds with accelerated negotiations within the World Trade Organization.

Since 1991, U.S. farm policy has consistently provided more planting flexibility for most crops. Policymakers recognized that farmers were in the best position to select which crops to plant, and thereby to manage their farms in the most efficient manner. Despite some exceptions, the U.S. government has continued to follow a "market-centered" policy. Both the U.S. and Canadian governments have developed crop insurance programs for most crops. Crop insurance has become a vital component for most farmers to guarantee against the risks of production perils and sharp price declines. The insurance program is largely based on market signals/factors, thus making the program crop neutral.

The U.S.-based biofuels programs have created new market opportunities for crops like maize. The production of ethanol is corn based in the United States and has been supported by government subsidies and market mandates. This has created an unusually high demand for maize and has pushed commodity prices much higher as crops compete for planting area throughout most of North America. Oilseed prices have been only partially influenced by biodiesel. However, potential biodiesel demand in North America does keep a base floor under the vegetable oil price. If market prices fall too low, biodiesel plants can increase demand and production. Even though no sunflower oil in North America is used in biodiesel production, the program indirectly supports sunflower seed prices when vegetable oil prices fall sharply.

Research and Education Infrastructure

Sunflower (*Helianthus annuus* L.) is native to North America. There are 50 native species of sunflower that are commonly called "wild sunflower." Most often wild sunflower has been and continues to be a noxious weed for many farmers. Historically, Native Americans cultivated certain species of wild sunflower for food purposes and contributed to genetic improvement of some of the species. Wild sunflower has been an important source of genes resistant to certain diseases. These wild plants have evolved and adapted over the centuries. The U.S. Department of Agriculture Agricultural Research Service (USDA ARS) has made a consistent effort to collect and catalog wild species. The collection is housed at the USDA ARS North Central Regional Plant Introduction Station in Ames, Iowa. The collection is an important resource to sunflower researchers around the world. There is a renewed effort and interest to tap wild sunflower species for disease and insect resistance and to enhance drought resistance (Seiler and Rieseberg, 1997).

Public research in the United States is often a combination of USDA ARS and state universities' research. Sunflower research in Canada is a combination of provincial, federal, and university research. Both Canada and the United States funded public sunflower research as the crop increased in popularity and value. However, funding at the federal and state/provincial levels has since decreased. Research at USDA ARS has largely been basic work on long-term problems and opportunities. The USDA team of scientists in the Sunflower and Plant Biology Research Unit is interdisciplinary and is located on the campus of North Dakota State University (NDSU) in Fargo, North Dakota. The research unit is a leader in world sunflower research and hosts visiting scientists and students from around the world. The scientists make improved germplasm releases. These releases are available free of charge to public and private entities for testing and incorporation into elite breeding lines. Over the last two decades, releases have been concentrated in disease and herbicide resistance and altered fatty acids, such as high oleic and low saturates. At the present time, the unit's scientists are working on disease and insect resistance, incorporating molecular tools, and developing an efficient double haploid breeding technique.

The USDA ARS sunflower researchers have developed important partnerships with other public and private research entities. Universities located in the sunflower production region of North America have staff that work on sunflower. Many of these scientists interact and closely coordinate their sunflower efforts with the USDA ARS sunflower researchers. This provides, for example, field testing of new genetic events over a broad geographic region from Texas to Manitoba. Private hybrid seed companies have extensive nurseries and often cooperate with USDA ARS scientists in field testing.

Sunflower is at a significant disadvantage to the major crops because it does not have the technology to utilize genetically modified organisms (GMO). Crops with GMO technology have been more efficient in disease, herbicide, and insect resistance breeding than non-GMO crops like sunflower (Stanley et al., 2003). The most significant barrier for sunflower GMO technology is cost. Bringing a single GMO event to market can cost $5–25 million in regulatory costs alone (Watson, 2011). At this point, sunflower producers in Europe cannot utilize GMO technology because of political resistance. It is unlikely that producers in Russia and the Ukraine would be willing or able to pay for this technologically improved seed. There would also be resistance from end users of sunflower oil and confection sunflower who simply do not want to have sunflower end products produced from GMO seed.

Despite the absence of GMO technology, the sunflower industry is advancing genetic material through genomics. Several initiatives have been developed with public and private cooperation to produce single nucleotide polymorphism (SNP) breeding platforms. One of the public initiatives recently published its SNP maps. The National Sunflower Association (NSA) in the United States has created an SNP initiative with the goal of lowering the overall costs of using this technology for members of a consortium. The Government of Canada has funded a comprehensive physical mapping project. Some of the larger hybrid seed companies have SNP capability with in-house laboratories. The vast majority of hybrid seed companies are internationally based or have partnerships with companies in other countries. Most have sophisticated research programs dedicated to increasing yield and oil content. Some have specific interests in creating unique fatty acid profiles in the oil. Many of the international companies depend on public releases from USDA ARS and other public entities that carry specific disease resistance genes. This partnership between the private and public research entities provides an excellent teamwork approach to improving hybrids. It also promotes efficiency in this era of declining resources.

The United States has another resource that is important for technology transfer to farmers. The Cooperative Extension Service consists of professionals who usually reside at a university or are associated with a university. The task of this group is to keep farmers informed of the latest research advances. This is done through classroom-style seminars and/or field trials developed for educational purposes. This applied research and teaching system was most important to the North American sunflower industry in its infancy. It continues to be vital as new-to-market crop-protection products are introduced and improved production techniques are advanced.

With the decline of sunflower production areas in Canada, the commitment to public research programs has also declined. However, there is a unique federal grant in place with the private sector to develop confection hybrids that are specific to Canadian growing conditions.

The USDA National Sclerotinia Initiative is another example of public–private partnership. This multicommodity partnership provides grants to public entities for high-risk research on this most difficult disease. Sunflower is particularly susceptible to Sclerotinia, as are many other broadleaf crops. This disease has had a substantial yield impact in North America and other sunflower producing locations. Breeding resistance is challenging because of the multigenetic nature of the disease and the multiple ways the disease can infect the plant. Good progress has been made in better understanding the disease and in identifying genes that can reduce the vulnerability of the plant to the disease.

Evolution of Fatty Acids in Sunflower Oil

North American sunflower oil has traditionally consisted of linoleic acid (18:2) at 55–70%, oleic acid (18:1) at 18–33%, and the remaining 10–12% consisting of a variety of saturated fatty acids. The U.S. food market did not need additional vegetable oils high in total polyunsaturated fatty acids. The market was already dominated by soybean oil with linoleic levels of 54% and linolenic (18:3) levels of 8% (Kamal-Edin and Andersson, 1997). Soybean oil was inexpensive, always available, and functional for most uses when hydrogenated. However, these conditions changed as the issue of *trans* fatty acids emerged as a health concern by the middle 1990s. A series of studies conducted by J.T. Judd concluded that *trans* fatty acids are detrimental to heart health (Judd et al., 1994). This changed the landscape of the U.S. vegetable oil market, with the FDA requiring the labeling of *trans* fat on food products by January of 2006 (U.S. Federal Register, 2003). The Canadian Government followed this lead, as did most European countries.

This set in motion an effort on behalf of the U.S. sunflower industry to switch the production to an oleic-type sunflower. High oleic (18:1) sunflower was in the market since 1985. The seed and oil were under patent and the oil was expensive. Food companies were not keen on paying a high premium for this oil with only one supply available and a patent in place (Hulke and Kleingartner, 2013). Plus, there was little reason to switch to HOSO when partially hydrogenated soybean oil met the needs of the domestic oil users. Lead by the members of the NSA and USDA ARS, a mid oleic sunflower was developed and called NuSun. The oleic level was established between 55% and 75%, compared to the 80%-plus oleic level in HOSO. This was one way to avert the high oleic patent issue and to take advantage of growing market demand for oil that was naturally stable without the need for hydrogenation.

Switching the entire industry from traditional linoleic sunflower was a huge effort and all entities of the industry had to be active participants. By 2005, the U.S. sunflower oil production was predominately oleic acid based (National Sunflower Association, 2005). By the 2008–2009 market year, 95% of U.S.-produced sunflower oil was consumed in North America, a huge turnaround from the mid-1990s when the vast majority of the U.S. sunflower oil was exported (Kleingartner, 2006). By 2010, there was essentially no traditional high linoleic sunflower oil produced in the United States.

As food companies scrambled to find alternatives to hydrogenated soybean oil, NuSun oil became the industry standard for natural stability. Major food processing companies extensively tested the oil and companies like Frito Lay made significant commitments to its use in a number of significant product lines.

Essentially, the levels of linoleic and oleic are reversed in NuSun/HOSO. The combined saturated fatty acids in HOSO are about 25% less than in traditional linoleic sunflower oil. The process of switching to oleic-based sunflower was expensive and time consuming. Efforts to improve yield, oil content, and disease resistance in hybrid seeds had to be delayed while researchers placed all of their emphasis on producing NuSun planting seeds that would be acceptable to farmers with an oleic level acceptable to oil users. Crushing plants had to develop a two-tier pricing system and separate storage facilities for seed and oil. Farmers had to be convinced to switch to NuSun planting seed and they had to ensure the production from those fields would be segregated from traditional sunflower. A separate pricing system created by the crushing plants provided that incentive.

By 2012, the leaders of the NSA determined that a second transition from NuSun to HOSO was needed. In essence, NuSun had literally outgrown its usefulness. Eliminating NuSun would be more efficient for hybrid seed companies in the production of planting seeds and more efficient for the overall marketplace (Lilleboe, 2012a). The expiration of the HOSO patents removed a huge barrier and oil users generally have no concerns about higher oleic levels in the oil. Most are interested in competitively priced oil that is available in consistent volumes. Another reason for eliminating NuSun has been the announcement of two additional types of high oleic sunflower that will be coming to the market soon. One is high oleic with very low saturate levels and the other is high oleic with high levels of stearic (18:0). Both products have been in field and laboratory testing for a number of years and commercialization is expected in the 2014–2015 production seasons (Kleingartner, 2010; Mullally, 2012). It is possible that sunflower production in North America could transition into these two latter fatty acid types due to their unique profiles. High oleic/low saturated fatty acid sunflower oil may be of considerable appeal to a broad market because of its saturate level under 3%. That may allow food products to state and promote "zero saturates" on the front of the product package (Mullally, 2012). Other oilseeds have now developed mid oleic and high oleic oils, reducing the market advantages that HOSO sunflower oil enjoyed just a few years ago. Moving to these newly introduced sunflower oils may keep sunflower oil competitive within the North American market, which is price conscious but is highly sensitive to *trans* and saturated levels.

Farmer Adaptation and Evolution of Production Systems

The production of sunflower in North America has encountered a host of native pests. Most of these pests have coevolved with native sunflower species. These native pests have adapted quickly to attacking commercial fields of healthy and robust sunflower, which is in contrast to the more spindly wild relatives. Researchers and farmers have adjusted to most of these pests as they have surfaced, but not without crop loss. Crop protection companies have provided important pesticides to combat most of the pests. However, there is usually a time period of experimentation and labeling regulations before a pesticide can be made available. It is in this time period when farmers have experienced minor to major crop losses.

There is an assortment of insects that attack sunflower and this is unique to North America compared to other production areas around the world. There are three common insect pests that attack the plant when the flower is in bloom. Two are Lepidopteron species and are unique to specific regions of North America. The banded sunflower moth (*Cochylis hospes* Walsingham) is most commonly found in North Dakota, Minnesota, and Manitoba. The female lays eggs most typically on the bracts just prior to bloom (stage R4) (Schneiter and Miller, 1981). A related insect is the sunflower moth (*Homoeosoma electellum* Hulst), which is found in the southern region of the United States, affecting production in Texas, Kansas, Nebraska, and Colorado. This moth lays eggs on the head of the blooming sunflower as well. The larvae of both insects burrow into the developing seeds, causing considerable damage to yield and quality. Pesticides are commonly used for both insects when adult numbers exceed thresholds. Often the sunflower moth moves with southern winds and can appear in huge numbers, often overwhelming the plants, resulting in significant losses (Randolph et al., 1972). Another common insect at bloom is the red sunflower seed weevil (*Smicronyx fulvus* LeConte), which is found most commonly in North and South Dakota. This insect is easily scouted and insecticide treatment is often warranted. The female lays eggs in the developing seed and the larvae consume a portion of the developing achene. These three main insects are of particular concern to the confection sunflower industry where seed quality is of utmost importance.

There are two stem-boring insects of significance. Both of these insects are of most concern in the southern production region from Texas to Nebraska and points in between. The sunflower stem weevil (*Cylindrocopturus adspersus* LeConte) lays eggs on the stem of the plant in the vegetative stage. Larvae burrow into the stem and tunnel throughout the stem. A plant can be overwhelmed, resulting in lodging prior to harvest (Rogers and Jones, 1979). In such cases, a portion of the field may be impacted. Drought conditions can exacerbate lodging due to the plants having thin and weak stems. The other stem-boring insect is called the dectes (*Dectes texanus* LeConte) and is also mostly found in the southern production region. This stalk-burrowing insect also causes lodging. This larva is much larger and only one is found in a stalk. The larva overwinters at the base of the plant and begins to prepare the chamber by girdling the inner walls of the stalk as the plant dries down (Michaud et al., 2009). This weakens the plant and can result in lodging. Insecticides for controlling either stem insect are generally not successful due to the long period of egg laying and the difficulty in scouting for adult egg-laying females. Recommendations to minimize lodging include lower plant populations and the use of "stay green" hybrids (Michaud et al., 2009). There are a number of additional insects that can appear periodically, resulting in damage. Most of the sunflower insect populations are cyclical and can be absent or in great abundance from one year to the next.

Diseases on sunflower are not unique to North America. Diseases, however, have contributed the most to yield decline when compared to insects (National Sunflower Association, 2012c). Sclerotinia caused by *Sclerotinia sclerotiorum* (Lib.) generally occurs when soil conditions are wet. The disease mostly affects the northern portion of the production region including North Dakota, Minnesota, and Manitoba. The disease can manifest in three separate ways: (1) basal stalk rot, (2) mid-stalk rot, and (3) head rot. All three can be devastating, with severe yield consequences. Head rot can be most severe, with yields reduced up to 100% (Masirevic and Gulya, 1992). The disease is dependent on sclerotia, which reside in the soil, a result from previous crop infections. Sclerotinia is common to most broadleaf crops; thus, the residue of sclerotia can be considerable. Basal stalk rot infection occurs via the roots of the growing sunflower plant. Both mid-stalk and head infection occur from ascospores produced by the sclerotia. This condition requires continuously wet soil. Infection occurs and manifests under cool, wet, and cloudy conditions. Thus, it is more common in the northern production region. There are no fungicide rescue treatments available in either the United States or Canada that have proven to be effective for any of the Sclerotinia infections. Long-term rotations between susceptible crops and hybrid tolerance are the "control" recommendations at this time (Markell and Harveson, 2012). A great deal of research at the USDA ARS has been ongoing since 2002 to develop genetic resistance to Sclerotinia. Integrating wild species genes from both annual and perennials resistant to the disease is making good progress, and several releases of germplasm have been made to the private sector.

A second disease of economic significance is downy mildew (caused by *Plasmopora halstedii* (Farl.) Berl. and de Toni). This is also a disease that impacts mostly the northern production region of Manitoba, North Dakota (Table 16.B), and Minnesota, when soils are excessively wet at planting and early emergence. A diseased seedling usually dies or continues to grow but remains stunted and does not produce seed. A fungicide seed treatment provided 100% protection until 1985, when the disease developed resistance to the fungicide. Since that time, the disease has been problematic in years when spring soil conditions are wet. New fungicide treatments since fungicide resistance emerged have not provided complete protection (Albourie et al., 1998). Disease-resistant genes have been identified from wild species and have been incorporated into many commercial hybrids. This is providing acceptable protection in combination with multiple fungicide seed treatments. However, the disease has a history of overwhelming resistant genes and fungicides. The disease has contributed

Table 16.B Downy Mildew (DM) Incidence, 1985–2005[a]

Description	1984	1991	1995	2001	2005
Incidence (percentage of fields)	59	16	18	43	52
Severity (percentage of plants)	1	0.05	0.3	1.8	4.9
Percent of fields with >15% DM	5	0	0	3	14

[a]Incidence and severity of sunflower downy mildew recorded during field surveys in North Dakota.

Source: Gulya (2005).

to uneven plant stands, which has been identified as a key production problem and yield-limiting factor in both the United States and Canada, according to the NSA Crop Survey (National Sunflower Association, 2012c).

Phomopsis stem canker is caused by *Diaporte helianthi/Phomopsis helianthi* and has become more evident in North America. The disease, much like Sclerotinia, is most common in the northern region including North Dakota, Minnesota, and Manitoba. The disease is also found around the world in most sunflower production regions. The soil-borne disease infects the stem leading to yield reduction and often lodging. The disease was particularly serious in 2011 in Minnesota; 90% of the surveyed fields were infected, according to the NSA Crop Survey, and 45% of the plants in those fields were diseased (National Sunflower Association, 2012c). Fungicide testing has not yielded conclusive results and there are no recommendations for fungicide use at this time (Markell et al., 2010). An aggressive research program is underway at NDSU to determine which of three identified Phomopsis species are most common in North America (Mathew et al., 2012). When this is completed, plant breeders can efficiently target their resistant breeding programs to the most abundant and virulent species.

Sunflower rust is caused by *Puccinia helianthi* and is found on native sunflower in North America. Disease pustules can often be found on the lower leaves in cultivated sunflower in the southern regions from Texas to Nebraska and can rapidly spread to the upper leaves and sunflower head under the optimal climatic conditions (Sayler, 2005). Rust is not as common in the northern production region of North America, but it can have serious yield implications when it manifests. A number of NDSU scientists have been researching fungicide control and application timing. Researchers are now confident that the disease can be controlled with fungicides (Friskop et al., 2012). In addition, aggressive breeding work is ongoing in North America using resistant genes in both oil and confection sunflower (Qi et al., 2010). Sunflower rust is common in other parts of the world and can be especially damaging in Australia (Gulya et al., 1997).

Other sunflower diseases can be found in North America, but they are minor and often not identifiable without close inspection by trained pathologists. These diseases are not of economic consequence at the present time. USDA ARS pathologists and NSA crop survey volunteers do conduct intensive annual field surveys to determine distribution and frequency of one or more of these secondary diseases. In the event one or more of these diseases becomes more widespread, research on control is quickly implemented. This is the history of past disease events.

One pest that is unique to North America is the red-winged blackbird (*Agelouis phoeniceus*). Other regions of the world also encounter bird pressure, but these are of different species. Damage in North America again varies within regions, but parts of North and South Dakota, Minnesota, and Manitoba suffer the greatest losses. This area is called the Prairie Pothole region, which has thousands of small wetlands overgrown with cattails (*Typha latifolia*). The cattails, combined with standing water, provide an excellent habitat for a number of blackbird species, especially in the fall of the year when, on migration, millions of blackbirds pass through the Prairie Pothole region. The migration coincides with the sunflower crop in the ripening stage (stages R7–R9) (Schneiter and Miller, 1981). Damage can be severe on a per field basis. The NSA Crop Survey assesses bird damage. Yield reduction over a five year average, ending in 2007, indicates average bird damage of 3.85% in North Dakota, 2.81% in South Dakota, and 3.66% in Minnesota (National Sunflower Association, 2012c). Control of bird damage is extremely difficult and recommendations are multifaceted, including eliminating cattail stands in wetlands, harassment techniques, early harvest with the aid of desiccants, and the use of repellant chemicals (Linz and Homan, 2012). The use of chemical repellants has been inconsistent because there are many variables to consider, such as timing, weather conditions, and the habits of a particular flock. The huge population of several blackbird species in North America is problematic for overwintering sites in the extreme southern region of the United States and in northern Mexico. In addition to human disease concerns, airports are most concerned about bird strikes with aircraft.

Acquiring pesticide labels in the United States and Canada has been a serious challenge. This is problematic for any new crop and has been of real consequence in gaining new herbicide labels. From 1970 to 2000, the herbicides of choice were from the dinitroaniline (DNAn) family. This group of herbicides controlled only several of the troublesome weeds in sunflower production. The most negative aspect of this family of herbicides was that the majority of the products needed to be soil incorporated via multiple tillage treatments (Blamey et al., 1997). This dried out the soil, reducing even emergence of seedlings. This limited the western movement of the crop, where soil moisture has consistently been a yield-limiting factor for all crops. Furthermore, it eliminated the possibility of reduced- or no-till systems, which had become popular with many farmers in an effort to maintain soil moisture and soil health (Lilleboe, 1996).

There are many species of weedy plants that have been problematic for sunflower producers in North America. Most are small-seeded broadleaf weeds. Kochia (*Kochia* sp.) has been and continues to be a significant weed problem. In the late 1990s, limited use of a preplant herbicide (Sulfentrazone) was made available to sunflower producers in the United States. By 2000, the product became widely available in the United States and later in Canada. This product was excellent in control of Kochia. Also during this period, two post-emergent herbicides from the imidazolinone and sulfonylurea families became available. Sunflower resistance to these two herbicides groups was incorporated into hybrids and the resistant genes came from wild sunflower species (Miller and Zollinger, 2004). With these herbicide advances, sunflower production under no-till and reduced-tillage systems became feasible, allowing farmers in more arid regions to plant the crop. However, weeds continue to be of concern, especially in the present environment where numerous weedy species are developing resistance to Glyphosate, which is broadly used as a fall and spring "burndown" prior to sunflower planting (Sayler, 2004).

History of Yield

Yield increase in U.S. sunflower has averaged 1% per year over the 37-year history of sunflower production records (USDA NASS, 2013b). It is important to note that U.S. breeders have had numerous breeding requirements other than yield over this period. Breeding for defensive traits like downy mildew, rust, and herbicide resistance has required the introgression of genes from wild species. The inheritance of these genes is often accompanied by negative yield factors (Hulke and Kleingartner, 2013). Breeders have also spent considerable time on rearranging fatty acids. Conversion to NuSun required from 5 to 10 years. During that period, breeders could not focus on overall yield enhancement. Although a great deal of progress has been accomplished incorporating defensive traits in today's hybrids, more work in this area is anticipated. Sclerotinia resistance requires multiple genes from wild species and yield drag is expected in initial introductions and in years when no disease occurs. It takes important breeding time to eliminate the "yield drag" of incorporating defensive traits from genes of wild species (Lilleboe, 2012b). The goal of reducing or eliminating diseases through breeding programs continues to be a notable goal. The yield potential in fields with no identifiable production problem is substantially greater than average yields determined by USDA National Agriculture Statistics Service (NASS). This has been confirmed in NSA crop surveys where fields designated as having "no production problem" averaged 2540 kg/ha compared to the average USDA NASS yields of 1620 kg/ha (1457 lb/ac). This data is a three-year average (2010–2012) encompassing all states in the survey (National Sunflower Association, 2012c). New tools for breeders will hopefully reduce the time required for introgressing defensive genes and maintaining high yield potential.

Movement of the Crop

The North American sunflower crop was initially established in the Red River Valley. This valley is a former lake bottom and consists of highly productive soils. It extends in a narrow band following the Red River from South Dakota, through Minnesota and North Dakota, and ending in Manitoba. The sunflower crop progressively moved from the Red River Valley to areas further west. This was largely due to lower rainfall in the western areas and sunflower diseases and insects that proliferated in the higher rainfall Red River Valley. The progression of the crop farther west was accelerated when new herbicide systems became available. Glyphosate and gluphosinate genetically engineered resistant soybeans and maize made those crops more competitive in some of the traditional sunflower-producing regions of the east central Dakotas, Minnesota, and Manitoba. For example, the area planted to sunflower in Barnes County, North Dakota (east central), was 54,788 ha in 1993 compared to 2671 ha (6,600 ac) in 2011. This contrasts with Emmons County, North Dakota (west central), with a sunflower area of 1255 ha (3,100 ac) in 1993 and 29,946 ha (74,000 ac) in 2011 (USDA NASS, 2013). Successive high rainfall years from 1999 to 2011 in the eastern Dakotas and western Minnesota and Manitoba contributed to sunflower diseases in that region, but had a yield increase affect for maize and soybeans. The sunflower crop also shifted to the south in portions of Kansas, Colorado, Nebraska, and Texas. These areas were historically wheat/fallow rotations. Adding a second crop to this rotation provided more potential revenue, provided better weed control in succeeding wheat crops, and allowed sunflower processors to broaden their dryland production areas. Irrigation is also a central part of this region's agriculture. Sunflower is providing a method to reduce water consumption through reduced irrigation. Research has shown that one to three well-timed irrigations on sunflower can provide excellent yields and reduce water consumption when compared to maize production (Aiken et al., 2012).

The combination of the federal Conservation Reserve Program (CRP) and strong world demand for food, fuel, and fiber has placed a great deal of pressure on commodity prices and competition for acres. At its peak, CRP took 34.7 million U.S. acres out of production (Soil and Water Conservation Society and Environmental Defense Fund, 2008). A U.S. federal biofuels incentive for the production of ethanol from maize raised prices and competition for planted area. According to USDA NASS (2013a), U.S. area planted to corn in 1993 was 30 million ha compared to the projected 39 million ha in 2013 (97 million ac). This combined with the wet weather cycle and Roundup Ready technology forced U.S. and Canadian sunflower acres into more arid regions and also lowered overall planted area and production.

Biodiesel made from vegetable oil in the United States and Canada has had an effect on vegetable oil prices. However, that program has not driven oilseed prices in the same manner that ethanol has with maize prices. That is largely due to the federally mandated blending of ethanol with gasoline (Loris, 2012). Biodiesel production and consumption in the United States pales in comparison to ethanol, but it is large enough that the food industry cannot allow additional vegetable oil volume to be used for fuel. This concern has kept a price floor under vegetable oil prices in North America (Kleingartner, 2008).

Expectations for the Future

The 40-year history of commercial sunflower in North America has been marked by extreme highs and lows. Area planted has varied considerably from one year to the next. New production problems appeared consistently throughout this history. However, the crop has persevered when some pundits predicted its demise. The North American industry has made many adjustments from dependency on Europe as a primary market to totally revolutionizing the industry in the development of NuSun. Adjustments continue to be made and the industry has the advantage of dealing with crises and overcoming long odds.

The key to the future of North American sunflower production is consistent increases in average yield. It has been pointed out that much of the breeding time and energy has been directed to defensive traits and fatty acid adjustments of the oil. Much work has been completed in both of these areas, although work on integrating Sclerotinia resistance continues in the future. The critical concern is keeping up with yield increases in soybeans and maize. Obviously sunflower is at a disadvantage in terms of sheer volume of both of these crops. That volume attracts significant private and public resources to those research communities. Soybeans and maize have the advantage of GMO. Those industries have also adapted the newest and latest genomic techniques. This is in contrast to the majority of sunflower breeding, which largely depends on traditional backcrossing that is time and resource consuming (Hulke and Kleingartner, 2013).

However, there is a great deal of sunflower work ongoing that will pay dividends in the near future. There are a number of private hybrid seed companies with a large international focus that are investing in new technology and breeding techniques. This includes high throughput genotyping technologies such as SNP (Bachlava et al., 2012). Whole genome sequencing is in development and a completed genome map should be released in 2014 (Lilleboe, 2010). Doubled haploid technology is being researched at the USDA ARS Sunflower and Plant Biology Research Unit in Fargo, North Dakota, and likely in one or more private laboratories around the world as well. This technology could greatly reduce the time required to bring an event to the marketplace. As discussed, the plethora of wild species brings many advantages to the sunflower breeding community.

The advances that have been made in altered fatty acids in sunflower provide a considerable advantage over other oils in the marketplace, especially in the North American market. Since much of this work is now completed, breeders can concentrate on yield enhancement. In the present regulatory environment, sunflower will not be able to utilize GMO technology. The matter of outcrossing of cultivated sunflower genes to wild species has been conclusively determined (Snow et al., 1998). This would obviate a sunflower GMO herbicide event; however, disease resistance or other quality enhancement might be acceptable to regulators. The substantial cost in passing the regulatory hurdles would be too great for one or two companies in the sunflower industry; it could only occur if the European prohibitions against GMO were to be lifted. The EU farmer market for hybrid planting seed would be necessary to justify the cost of research and regulatory approval. Sunflower farmers in Argentina, Russia, and the Ukraine would also need to be buyers of GMO sunflower seed. In order for this to occur, a sunflower GMO event would need to be of great significance in terms of disease resistance and/or yield enhancement. In the meantime, sunflower oil is the only non-GMO oil in the world of any significant volume other than palm oil. This is providing advantages in some markets. In the United States there are a number of individual states that are considering labeling requirements for food products made from GMO ingredients. If implemented, this would provide a market boost for non-GMO ingredients like sunflower oil. The bottom line is that GMO sunflower is simply not feasible at the present time.

The increasing number of weeds that are resistant to Glyphosate is of great concern among North American agronomists and farmers. These "super weeds" are of greatest concern in regions where monoculture has become the norm (Powles, 2008). Examples are successive years of soybeans on the same ground or the production of only two crops in moderate rotation, such as soybeans and maize with no rotation of herbicides. Glyphosate is an easy and inexpensive way to produce these crops. However, this important weed control tool is no longer recommended as a primary or standalone herbicide (Boerboom and Owen, 2012). Sunflower can play a role in expanded rotations of these crops in some of the traditional areas where sunflower was a prevalent crop just a few years ago. The advantage of sunflower and other non–Roundup Ready crops in a narrow rotation is the use of different herbicide families.

Of the production issues facing North American sunflower producers, only the problem of blackbird depredation has no foreseeable solution. Most of the recommendations for reducing damage are labor intensive and require consistent monitoring. Farmers in the present era are highly leveraged and mechanized. They plant large acreages spread over a large area. Attempting to monitor and move blackbirds out of a particular sunflower field is not economical. Farmers in high-risk blackbird areas simply have abandoned the crop.

The issue of climate change can also be a factor in planting decisions going forward. For much of the last 15 years, the northern sunflower production region of the Dakotas, Minnesota, and Manitoba experienced above average rainfall (National Sunflower Association 2012c). This has led to a proliferation of sunflower diseases and an expansion of crops such as soybean and maize, which generally require higher rainfall. A weather change to drier and warmer growing seasons could benefit the acreage of warm season broadleaf crops like sunflower. Similar weather patterns in the southern growing region of Kansas to Texas could reduce maize area under irrigation and shift acres to sunflower and other more drought-tolerant crops.

Looking to the future of where sunflower might fit in the United States also requires considering U.S. farm policy, government funding for research, and international trade policy. Farm policy in the United States has been inequitable for minor oilseeds and several other crops (Johnson and Monke, 2012). This was particularly egregious in the 15 years from 1980 to 1995. Since that time, U.S. farm policy has been more impartial among crops and it is likely to continue that way in the future. Federal and state government funding of agricultural research is at a critical juncture as government budgets continue to be under pressure and other funding priorities are considered. There has been considerable consolidation of both state and federal agricultural research in the past decade, and that is likely to continue. International trade issues continue to be a concern, especially during periods of surplus production. World trade agreements have resolved some issues, and there are now more safeguards in place to protect against the introduction of egregious trade distortions.

Supporting Organizations

The United States and Canada each have a supporting organization called the National Sunflower Association (NSA). Both organizations work closely together on issues of mutual interest. The U.S. NSA was created in 1980 and was instrumental in changing discriminatory federal farm policy, creating export subsidies for sunflower oil, and assisting in the development of crop insurance programs for all types of sunflower. The organization has been most active in supporting and gaining new pesticide labels. The NSA was instrumental in leading the change from linoleic to oleic sunflower. The development of the National Sclerotinia Initiative was the result of the NSA. Overall, the U.S. NSA continues to be most active in production research and recently created the NSA SNP Initiative. This initiative consists of nine private seed companies, USDA ARS, and the NSA.

The organizational structure of the U.S. NSA includes farmers, researchers, and all the related support industries. It publishes *The Sunflower* magazine, conducts an annual crop survey, and organizes the annual research forum, among many other activities. The website, www.sunflowernsa.com, is filled with historic and current information. The NSA also conducts market development programs in Mexico, Canada, Turkey, and Spain.

The NSA of Canada was organized in 1996 and also includes a broad spectrum of growers, industry leaders, and researchers. This organization has a smaller production base than its U.S. counterpart. Of real significance is the organization's project of developing confection hybrid seed that is best suited to production in northern Canada. The organization has been successful in developing crop insurance instruments and in gaining new pesticide labels. The organization participates in the annual crop survey conducted by the U.S. NSA. There is a very close working relationship between the two organizations.

References

AAFC Agriculture and Agri-Food Canada, Market Analysis Division. Internet Statistics Canada. http://www5.statcan.gc.ca/subject-sujet/theme-theme.action?pid=920&lang=eng&more=0 (accessed March 12, 2013).

Aiken, R.; Lamm, F.; Seiler, G. Sunflower Canopy and Oilseed Yield Formation under Deficit Irrigation. Internet Sunflower Research Forum. http://www.sunflowernsa.com/uploads/research/1190/aiken_canopy_paper_12.pdf (accessed Mar 20, 2013).

Albourie, J. M.; Tourvielle, J.; Tourvielle de Labrouhe, D. Resistance to Metalaxyl in Isolates of the Sunflower Pathogen *Plasmopara halstedil. Eur. J. Plant Pathol.* **1998**, *104*, 235–242.

Bachlava, E.; Taylor, C. A.; Tang, S.; Bowers, J. E.; Mandel, J. R.; Burke, J. M.; Knapp, S. J. SNP Discovery and Development of a High-Density Genotyping Array for Sunflower. *PLoS ONE* **2012,** *7*, e29814. DOI:10.1371/journal.pone.0029814.

Blamey, F. P. C.; Zollinger, R. K.; Schneiter, A. A. Sunflower Production and Culture. In *Sunflower in Technology and Production*; Schneiter, A. A., Ed.; ASA-CSSA-SSSA: Madison, WI, 1997; pp 595–670.

Boerboom, C.; Owen, M. Facts about Glyphosate Resistant Weeds. The Glyphosate, Weeds and Crop Series, GW-1. Internet Purdue Extension Service, 2012. https://www.extension.purdue.edu/extmedia/GWC/GWC-1.pdf (accessed Apr 9, 2013).

Friskop, A. J.; Gulya, T. J.; Halley, S.; Schaefer, J.; Schatz, B.; Wunsch, M.; Hendrickson, P.; Meyer, S.; Markell, S. Managing Sunflower Rust (*Puccinia helianthi*) in Early Onset Epidemics with Fungicides in North Dakota: 2011. Internet Sunflower Research Forum, 2012.

Gulya, T. J.; Rashid, K. Y.; Masirevic, S. M. Sunflower Diseases. In *Sunflower Technology and Production*; Schneiter, A. A., Ed.; ASA-CSSA-SSSA: Madison, WI, 1997; pp 263–380.

Hulke, B. S.; Kleingartner, L. W. Genetic Contributions to Yield Gain for U.S. Sunflower. In *Genetic Contribution to Yield Gains for Major U.S. Crops*; Carver, B., Diers, B., Smith J. S.; Specht, J. C., Eds.; CSSA: Madison, WI, 2013. (In print).

Johnson, R.; Monke, J. *What Is the Farm Bill?* Congressional Research Service, Government Printing Office: Washington, D.C., 2012; RS22131:6.

Judd, J. T.; Clevidence, B. A.; Muesing, R. A.; Wittes, J.; Sunkin, M. E.; Podczasy, J. J. Dietary *Trans* Fatty Acids: Effects on Plasma Lipids and Lipoproteins of Healthy Men and Women. *Am. J. Clin. Nutr.* **1994**, *59*, 861–868.

Kamal-Edin, A.; Andersson, R. A. Multivariate Study of the Correlation between Tocopherol Content and Fatty Acid Composition in Vegetable Oils. *J. Am. Oil Chem. Soc.* **1997**, *74*, 376.

Kleingartner, L. Ten Years of NuSun®. *The Sunflower* **2006**, *32*, 8–9.

Kleingartner, L. Biodiesel: What It Means for Sunflower. *The Sunflower* **2008**, *34*, 14–15.

Kleingartner, L. Another Type of Sun Oil on the Horizon. *The Sunflower* **2010**, *36*, 14–15.

Lilleboe, D. Granules in No-Till Sunflower. *The Sunflower* **1996**, *22*, 22–24.

Lilleboe, D. Sunflower Genome Holds Promise of Sustainable Ag. *The Sunflower* **2010**, *36*, 22.

Lilleboe, D. High Oleic Sunoil: Soon to Become the Standard? *The Sunflower* **2012a**, *38*, 8–10.

Lilleboe, D. Building New Hybrids: Opportunities and Challenges. *The Sunflower* **2012b**, *38*, 18–21.

Linz, G. M.; Homan, H. J. Preliminary Evaluation of 9,10 Anthraquinone Bird Repellent for Managing Blackbird Damage to Ripening Sunflower. Internet Sunflower Research Forum, 2012. http://www.sunflowernsa.com/uploads/research/1172/linz_eval.anthro_12.pdf (accessed Apr 4, 2013).

Loris, N. Two Cheers for Ethanol Subsidies Expiring—But Costly Mandate Remains. The Heritage Foundation, 2012. http://www.heritage.org/research/reports/2012/01/ethanol-subsidies-expiring-but-the-costly-mandate-remains (accessed Apr 6, 2013).

Markell, S.; Harveson, R. Tips for Management of Diseases. *The Sunflower* **2012**, *14* (2), 14y–15y.

Markell, S.; Halley, S.; Schatz, B. G.; Gulya, T. J.; Mathew, F.; Friskop, A. J.; Aberle, E.; Misek, K. Evaluation of Fungicides and Fungicide Timing for Management of Phomopsis in Sunflowers. Internet Sunflower Research Forum, 2010. http://www.sunflowernsa.com/uploads/research/448/Markell_phomopsis_Paper_10.pdf (accessed Feb. 18, 2013).

Masirevic, S.; Gulya, T. J. Sclerotinia and Phomopsis—Two Devastating Sunflower Pathogens. *Field Crop. Res.* **1992**, *30*, 271–300.

Mathew, F.; Castlebury, L.; Gulya, T.; Jordahl, J.; Heitkamp, E.; Choe, S.; Alananbeh, K.; Markell, S. Phomopsis Diversity and Pathogencity: An Update. Internet Sunflower Research Forum, 2012. http://www.sunflowernsa.com/uploads/research/1205/mathew_phomoopsis_13.pdf (accessed Feb 18, 2013).

Michaud, J. P.; Stahlman, P. W.; Jyoti, J. L.; Grant, A. K. Plant Spacing and Weed Control Affect Sunflower Stalk Insects and the Girdling Behavior of *Dectes texanus* (Coleoptera: Cerambycidae). *J. Econ. Entomol.* **2009**, *102*, 1044–1053.

Miller, J. F.; Zollinger, R. Utilization of Cross-Resistance to Create Herbicide-Resistant Sunflower Hybrids. Internet Sunflower Research Forum, 2004. http://www.sunflowernsa.com/uploads/research/136/136.pdf (accessed Feb 27, 2013).

Mullally, S. Groundbreaking Sunoil Being Introduced. *The Sunflower* **2012**, *38* (1), 26–27.

National Sunflower Association. Internet Friskup Fungicide Trials Paper. http://www.sunflowernsa.com/uploads/research/1184/friskip_fungicide.trials_paper_12.pdf (accessed Mar 12, 2013).

National Sunflower Association. Internet U.S. Crop Quality Report, 2005. http://www.sunflowernsa.com/uploads/resources/145/2005_cropquality.pdf (accessed May 3, 2013).

National Sunflower Association. Internet U.S. Crop Quality Report, 1999–2012b. http://www.sunflowernsa.com/stats/us-crop-quality (accessed May 1, 2013).

National Sunflower Association. Internet U.S. Supply & Disappearance, 2012a. http://www.sunflowernsa.com/stats/us-supply (accessed May 1, 2013).

National Sunflower Association. Internet Sunflower Crop Survey, 2012c. http://www.sunflowernsa.com/growers/yield-and-survey/sunflower-crop-survey (accessed May 1, 2013).

Powles, S. B. Evolution in Action: Glyphosate-Resistant Weeds Threaten World Crops. *Outlooks on Pest Manag.* **2008**, *19,* 256–259.

Qi, L.; Gulya, T. J.; Hulke, B. S.; Vick, B. A. New Sunflower Rust Projects at the USDA Sunflower Research Unit. Internet Sunflower Research Forum, 2010. http://www.sunflowernsa.com/uploads/research/444/qi_Rust_10.pdf (accessed Mar 2, 2013).

Randolph, N. M.; Teetes, G. L.; Baxler, M. C. Life Cycle of the Sunflower Moth under Laboratory and Field Conditions. *Ann. Entomol. Soc. Am.* **1972**, *65,* 1161–1164.

Rodgers, C. E.; Jones, O. R. Effects of Planting Date on Soil Water on Infestation of Sunflower by Larvae of [*Cylindrocopturus adspersus*]. *J. Econ. Entomol.* **1979**, *72,* 529–531.

Sayler, T. Ten Weed Control Tips for Flowers. *The Sunflower* **2004**, *31,* 10–14.

Sayler, T. It Pays to Control Rust in Confection Sunflower. *The Sunflower* **2005**, *31* (6), 20–21.

Schneiter, A. A.; Miller, J. F. Description of Sunflower Growth Stages. *Crop Sci.* **1981**, *21,* 901–903.

Seiler, G. J.; Rieseberg, L. H. Systematics, Origin, and Germplasm Resources of the Wild and Domesticated Sunflower. In *Sunflower Technology and Production*; Schneiter, A. A., Ed.; ASA-CSSA-SSSA: Madison, WI, 1997; pp 21–65.

Snow, A. A.; Rieseberg, L.; Alexander, H.; Cummings, C.; Pilson, D. *Assessment of Gene Flow and Potential Effects of Genetically Engineered Sunflowers and Wild Relatives.* International Biosafety Symposium, Braunschweig, Germany, Sept 6–10, 1998.

Soil and Water Conservation Society and Environmental Defense Fund. Internet Conservation Reserve Program (CRP) Program Assessment, 2008. http://www.swcs.org/documents/filelibrary/CRPassessmentreport_3BEFE868DA166.pdf (accessed Apr 2, 2013).

Stanley, J.; Hutchinson, K.; Godwin, I.; Gregg, P.; Jessop, R.; Katz, M.; King, K.; Prior, J.; Sindel, B. Biotechnology (GMO) Issues and Research Priorities in Natural Resource Management; Report to Land and Water Australia, product number PR030520, 2003.

U.S. Federal Register. 2003. Vol. 68 (1330) 41466.

USDA NASS. Internet Corn: Acreage by Year, US. http://www.nass.usda.gov/Charts_and_Maps/Field_Crops/cornac.asp (accessed Apr 2, 2013a)

USDA NASS. Internet Quick Stats. http://quickstats.nass.usda.gov/results/2F4810C1-956A-3B6C-A7E6-1CFF35C22D3D?pivot=short_desc (accessed Mar 23, 2013b).

Watson, E. "Ill-Defined, Irrational and Costly" GM Approval Process Must Be Reviewed. Internet Food Navigator, 2011. http://www.foodnavigator-usa.com/Regulation/Ill-defined-irrational-and-costly-GM-approval-process-must-be-reviewed-academic (accessed Mar 18, 2013).

17

South America Perspectives on Sunflower Production and Processing

Carlos E. Feoli ▪ *INTA-ASAGIR Agreement, Instituto Nacional de Tecnología Agropecuaria—Asociación Argentina de Girasol, Pergamino, Argentina*

Jorge Ingaramo ▪ *ASAGIR, Asociación Argentina de Girasol, CABA, Argentina*

Introduction

The southern portion of Latin America has always been one of the main sunflower regions across the world. Argentina was a pioneer in this crop adoption and has the highest production. After economic reforms in former Soviet Union countries, a remarkable technology adoption took place, mainly in the Ukraine and Russia, forcing Argentina out of the leading position it held until the late 20th century. Argentina has vast experience in sunflower and developed one of the best organized export-orientated value chains. Their local scientists and researchers generate most of the available crop knowledge around the world. Input and crushing industries accompany farmers scattered around eight different Argentine provinces, and they work together to maintain Argentina's significant position in world trade for sunflower oil and protein.

The first section of this chapter summarizes the history of the crop in South America since its introduction to Argentina and how it spread to neighboring countries. Agro-ecological rationality of cultivating this crop in this vast grassland region—one of the five greatest in the world—also stands out. We also describe sunflower cultivation and industrialization traditions with a focus on Argentina, and the history that established the basis for its global market leadership is described. Next we describe sunflower's presence in Argentina, its history, and local industry and production tendencies. We also go further into the topic of Argentina's participation in global markets.

We move on to address sunflower oil and byproduct price formation and structure, as well as distribution, transport, crushing, and commercialization, taking into consideration particularities of the business conditions for Argentina and land use competition with other activities. A special emphasis on price formation difficulties is made, given that sunflower products are available at different distances from ports and have at least 40% domestic consumption. There are huge challenges to be faced as a result of the existence of strongly discriminative commercial policies that restrict sunflower development: export taxes and an erratic exchange rate policy. These are mainly in place for tax collection purposes and to reduce domestic prices, and the policies have been followed since 2002. Next, we analyze the future of the Argentine sunflower complex after Black Sea production expanded. The challenges faced by the entire value chain, represented by Argentine Sunflower Association (ASAGIR), are presented. Finally, we present the main conclusions.

Sunflower in South America

Brief Regional History: Crop Introduction and Development

Sunflower is a Native American plant that has been grown since 3000 B.C. It was first introduced in Spain and France in the 16th century and in Russia in the 18th century. Russian growers were among the first to promote sunflower crops, and British farmers may have been the first to grow sunflower for industrial purposes. Russian immigrants brought sunflower seeds to Argentina in the middle of 19th century and started planting it across the country.

In Argentina, sunflower was a family subsistence crop until 1929. Many documents point out that in Jewish colonies, settled in Carlos Casares, Buenos Aires Province, and Basavilbaso in the Entre Rios Province in the middle of the 19th century, sunflower was used as an ornamental flower, as bird food, and for human consumption as a roasted and salted confection. Jewish colonies first introduced the crop to Bolivia and Paraguay. In Uruguay, it was introduced by Argentine farmers who had frequent contact with their peers from the other side of the border. Spreading north through the continent, the first references to sunflower in Brazil date back to 1924, at which time sunflower was brought by European immigration (ASAGIR, 2013; Garcia, 1988; Ingaramo et al., 2008).

Agronomic Rationale for Sunflower Cropping in South America

Several factors must be considered to explain Argentine sunflower production. Tradition shows the long and strong relationship between sunflower and Argentine farmers. Argentina's climate gives optimum conditions for crop establishment and development, while its soils give physical and chemical support to crop production. In Argentina, sunflower is grown in areas that range from 24° to 38° latitude south. It spreads along very different environments, from hot temperate climates to milder ones, and from almost dry areas (less than 500 mm rains) to very humid ones (more than 1000 mm rains). Days free from frost range from 140 to more than 200 days. Sunflower grain production crops are rain fed in Argentina, although irrigated crops are used to assist the hybrid seed industry (ASAGIR, 2013).

Average sunflower yields were far lower than 2 t/ha until the 2010 harvest; since then they have been very close to 2 t/ha. According to research and field trial results, using the best available hybrids in biotic and abiotic stress-free environments, like field comparative trials that are conducted to maximize high production techniques, potential yields in Argentina overmatch 4 t/ha (Hall et al., 2012). These potential yields can also be obtained in neighboring countries such as Bolivia, Paraguay, Brazil, and Uruguay.

Update on Sunflower Crop in Main South American Producing Countries

Despite Argentina's planted area and production reduction, this country remains the primary regional supplier, due to the limited production of the four above mentioned countries.

Argentina

Argentina has stabilized national yields during 2010, 2011, and 2012 sowing periods, with a planted area of around 1.8 million ha. Average yields were 2,015 t/ha, an impressive increase of 29% over the previous triennium, totaling a production of 3.48 million t, including 2012 estimates (MinAgri, 2013).

Bolivia

During the sowing season of 2012, sunflower planted area increased 51.7% compared to 2011, totaling 287,020 ha.

Paraguay

During the 2010/2011 season, 70,000 ha were planted with sunflower in Paraguay. Itapua state has 62% of the total production. Country yields were around 1.6 t/ha, totaling a production of 109,000 t. In the following season, yields decreased to an average of 1.5 t/ha, with an area increase of 14% up to 80,000 ha, resulting in a total production of 120,000 t (INBIO, 2013).

Brazil

The sunflower planted area in Brazil in the last years was only 0.15% of its agricultural surface, in contrast to soybean, at 19.2%. This enormous difference makes sunflower almost a niche in Brazil, in spite of some company and state effort to spread the crop over arid regions (Gazzola et al., 2013). Nevertheless, they produced a surface of over 76,000 ha with yields of 1.5 t/ha that totaled a production of 81,000 t in the 2012 season. This satisfies both Brazil's own domestic market needs and maintenance of sunflower seed stocks (Nunes, 2013).

Uruguay

Uruguay surface planted with sunflower has been low, mainly due to diseases and bird damage. In the last three years, sunflower covered about 7,000 ha, with a total production of 65,000 t meant only for refined domestic consumption (Souto, 2007).

Table 17.A Harvested Sunflower Area in South America (thousand ha)

	Harvest Season				
Country	**2008**	**2009**	**2010**	**2011**	**2012**
Argentina	2612	1967	1542	1756	1800
Brazil	111	75	71	66	76
Paraguay	181	194	168	70	80
Bolivia	213	305	235	143	217
Uruguay	44	50	10	4	6

Source: Oil World (2012).

Table 17.B Sunflower Production in South America (thousand t)

	Harvest Season				
Country	**2008**	**2009**	**2010**	**2011**	**2012**
Argentina	4650	2483	2220	3669	3470
Brazil	147	109	81	113.3	81.4
Paraguay	280	196	262	109	120
Bolivia	319	457	311	153	194
Uruguay	54	51	9	4	5

Source: Oil World (2012).

Table 17.C South American Yields (t/ha)

	Harvest Season				
Country	**2008**	**2009**	**2010**	**2011**	**2012**
Argentina	1.80	1.36	1.49	1.99	1.95
Brazil	1.30	1.50	1.10	1.30	1.50
Paraguay	1.60	1.60	1.60	1.60	1.50
Bolivia	1.50	1.50	1.30	1.10	0.90
Uruguay	1.20	1.00	0.90	1.00	0.80

Source: Oil World (2012).

Argentina—Regional Leader in Sunflower Production and Crushing: A Fully Articulated Value Chain

Following the USDA sunflower seed ranking for 2012 (Table 17.D), Argentina's leading role clearly stands out; it is ranked fourth (after all EU countries combined) with more than 3 million t, whereas Bolivia, Brazil, and Paraguay are placed 15th, 16th, and 30th, respectively. Combined, these countries have less than 14% of Argentina's production. Given this remarkable scale difference and common sunflower production evolution in all South American countries, we will mainly describe Argentina's production profile. Even though Argentina has a far greater role in sunflower production in South America, it is worth pointing out that regardless of Bolivia's significant volume differences with Argentina, sunflower has similar importance in relative terms. Even so, Bolivia hasn't yet developed a value chain, limiting its relevance to raw production. Moreover, its impact over the global market is close to none, given that it sells directly to regional commercial partners.

As for the other countries in the region, a great sunflower area expansion was expected, but they were soon discouraged. Circumstances such as lack of a mature oil industry, absence of well-adapted genetic material, and poor crop management knowledge worked against sunflower increase in the region. Even in sensitive environments where sunflower has a naturally more stable behavior, there has been no steady planting progress. For example, the Brazilian semiarid region has remarkable potential, but in spite of promotion by some companies and the state, it hasn't found its place.

Sunflower physiology is best adapted to mild cold weather conditions; sunflower performs worst in tropical weather. Despite this difference, there are no fundamental agro-climatic conditions that prevent sunflower development in other countries. The only country that has a relevant agro-climatic conditioning factor is Uruguay, which has severe bird pest damage (by *Zenaida auriculata* sp.) that can lessen yields

Table 17.D Oilseed, Sunflower Seed 2012 Ranking

Rank	Country	Production (1000 t)	Rank	Country	Production (1000 t)
1	Ukraine	9000	6	United States	1264
2	Russia	7959	15	Bolivia	250
3	EU-27	6950	16	Brazil	115
4	Argentina	3230	30	Paraguay	64

Source: USDA-FAS, Production, Supply and Distribution (PS&D) database.

randomly from 0 up to 60% (Rodriguez, 2010). In spite of any such climate differences, the main cause of such a significant sunflower volume difference between Argentina and the other countries in the region can be attributed to the fact that Argentina has a well-organized value chain and better crop-management skills, accompanied by higher yields, better technology, more expertise, and longer crop production tradition. A reasonable assumption would be that in the event of a consistent shift in relative prices compared to soybean, sunflower may extensively spread out over the region, originating from Argentina.

In Argentina, the sunflower value chain products are one of the country's main exports, in spite of the important market distortions generated by its own arbitrary commercial and exchange rate policies over the last 12 years. The introduction of export taxes with a present statutory rate of 32% over sunflower seed Free on Board (FOB) price strongly discourages farmers from planting sunflower, using technology, and other related investments. This tax also lessens the entire value chain competitiveness by reducing the whole operating scale. However, the low tax rate differential of only two percentage points, accounting for 30% of the FOB value for crushed products exports, compared with 32% for grain, doesn't encourage the oil industry to create local added value. Furthermore, refined industry does not obtain reimbursement of indirect taxes, a measure that is broadly admitted by the World Trade Organization.

In spite of the decline of sunflower planted area since the end of 1990s, production is still enough to fully satisfy the domestic market. Domestic prices are under strict control to ease consumption and export duties, reduce internal prices favoring refined industries, and meet broadening supermarket demand and consumption for livestock as sunflower meal or pellets.

Argentina's total sunflower production in grain, meal, and oil has been around 10% of global sunflower production for the business years 2010 to 2013. In spite of a strong domestic consumption (12.3 liters/inhabitant), global trade fluctuates around 14% in meal, 16.2% in oil, and a low 4.4% in untransformed sunflower seed (USDA, 2013).

ASAGIR has made a balance that, on the one hand, establishes a slight decrease in implanted area in 2012 and 2013, mainly for climatic reasons. On the other hand, for the third consecutive year, yields have been above historical average at almost 2 t/ha for the last triennium (2009–2012), which is 29% more than the previous one.

Sunflower in Argentina

Crop History in Argentina

For practical purposes, sunflower history is divided into six periods according to its evolution as a crop (ASAGIR, 2013):

Period 1—mid 19th century to 1920: Plantation and consumption for domestic purposes.

After being a plant cultivated for domestic uses, at the beginning of the 1920s, a few hectares of sunflower were planted and a small amount of its oil was exported. Russians used to plant sunflower as a cultural tradition to eat toasted grain and to feed chicken and birds.

Period 2—1930 to 1949: Extensive farming expansion from 400 ha to 1 million ha.

In 1930, crop expansion began, along with sunflower's acceptance amongst farmers and consumers, driving a growing industry. Statistics references point out that in the 1928/29 growing season, 400 ha were planted; 505,600 ha in 1939; 1,806,000 ha in 1949; and a year later 103,000 t were exported. The main reason for such a gap was a strong demand for food due to the olive oil shortage in Europe, as a consequence of the Spanish Civil War (1936–1939) and World War II (ASAGIR, 2013).

Period 3—1950 to 1960: First sunflower area contraction.

Low prices due to international context and susceptibility to diseases (such as rust and Sclerotinia, among others at the time called "the Black Pest," now identified as a complex of fungal agents, *Verticillium dahliae, Macrophomina phaeseolina, Albugo tragopogonis,* and *Alternaria helianthi*). Under those conditions, farmers experienced very poor economic results, which led to a massive reduction in planted area to 560,000 ha in 1955.

Period 4—1960 to 1975: Crop recovery.

During this period, advances over breeding lines related to disease resistance and new technologies resulted in sensitive yield increments, favoring the return of sunflower to lands that had abandoned it during the 1950s. Regarding breeding lines, during the 1960s the INTA, an Argentine governmental agricultural agency, developed breeding lines like *Manfredi INTA, Impira INTA, Guayacan INTA, Cordobes INTA,* and *Pehuen INTA* that started a new era of higher yields and rust resistance, recapturing the area lost in the previous decade. Previously existing breeding lines such as Klein, developed in 1938, had better oil content (38%), less height, and a shorter cycle compared to open-pollinated cultivars planted until then such as *Gigante de Rusia* and *Gigante Americano,* both of Russian origin. In 1972, the first sunflower hybrid G104 was introduced with low market penetration. Later, the G97 hybrid got wider adoption.

Period 5: 1975 to 1980—Beginning of hybrid adoption by farmers.

Around 1975, replacement of single breeding lines with hybrids began, with an important presence of private seed companies. This was crucial for the crop expansion favored by increments in oil content and yields. Earlier, sunflower farming was conducted with very low technology use and care, which accounted for very uncertain results. Low yields, disease susceptibility, low oil content, and cycle unevenness were the main reasons that sunflower had a low priority in farmers' planning; sunflower was usually relegated to low productivity soils or was a second crop immediately after winter cereals harvest. In some cases, poor results were compensated with oat sowed between sunflower lines during weed removal labors in advanced stages (R4–R6) of the crop, for future cattle consumption.

Period 6: 1980 to present—Crop management enhancement.

The improved results obtained with the use of hybrids motivated farmers to change sunflower crop conduction. This key element started successive technology changes. Early sowing dates, herbicide use, insect control, and improved harvest technology with fewer losses became the rule more than the exception (Feoli, 1991a, 1991b, 1991c). Fertilization continued to be limited, though. At the time, farmers usually pointed out that weed control with sunflower was poor, becoming the biggest problem, especially when available herbicide products had a limited control spectrum. Later on, new herbicides and new technologies, like Clearfield and Clearfield Plus, resulted in improved weed control. Consequently, after 2000, weed effect in sunflower crop was minimized (Sala et al., 2010). However, new challenges started to appear for the crop with the new competition of palm oil in global markets and the advent of the huge demand for soybean, which started competing for land space and subjugated sunflower, reducing productive regions and soils and forcing the crop to meet new challenges.

Among Argentinian and most South American farmers, there are different technology adoption behaviors: There are those that minimize risk and choose to make low investments on inputs, and others who have less risk aversion and who invest highly on inputs in order to optimize productive response of the best breeding lines and every step in between (Sammarro et al., 2013).

Sunflower Production Tendencies in Past and Present Days

Area, Production, and Yield in Argentina

Figures 17.1, 17.2, and 17.3 show the main tendencies in the last 40 years for planted area, yields, and total Argentine production from 1970 to 2013.

When analyzing planted area, three periods can be clearly identified. The first period, with a certain instability but also a subtle ascendant tendency, goes from 1970 to 1994. The second period, until year 2000, is characterized by maximum expansion in sunflower crop history in Argentina, which concluded with the oversupply of 1999. By that time, Argentina was by far the global market leader and price former. This almost monopolistic position ended up undermining its own business. The abundant

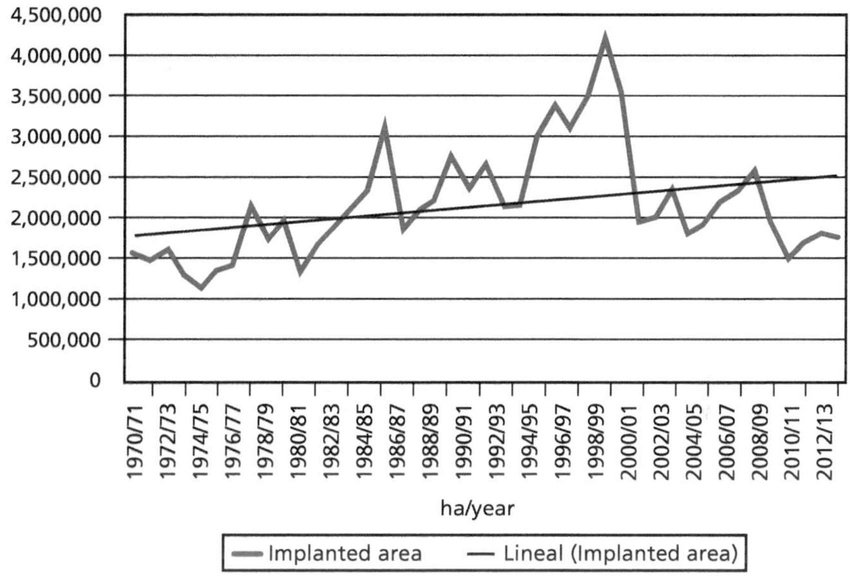

Figure 17.1 Argentina sunflower planted area, 1970–2012.
Source: Based on data Bolsa de Cereales (2013).

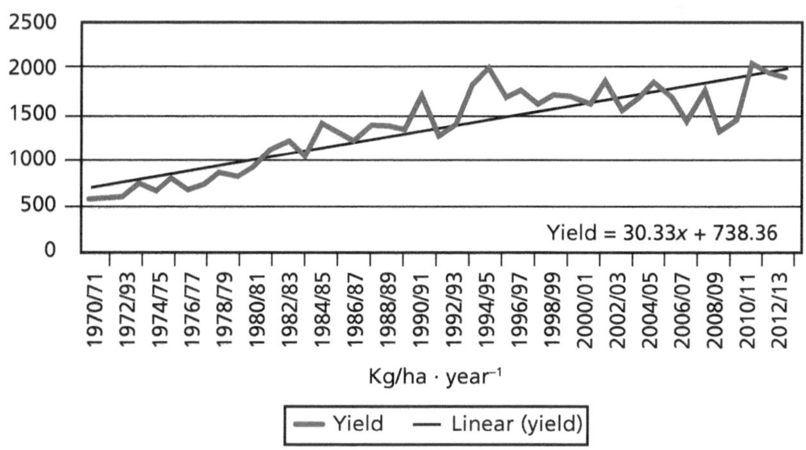

Figure 17.2 Argentina sunflower yields, 1970–2012.
Source: Based on data Bolsa de Cereales (2013).

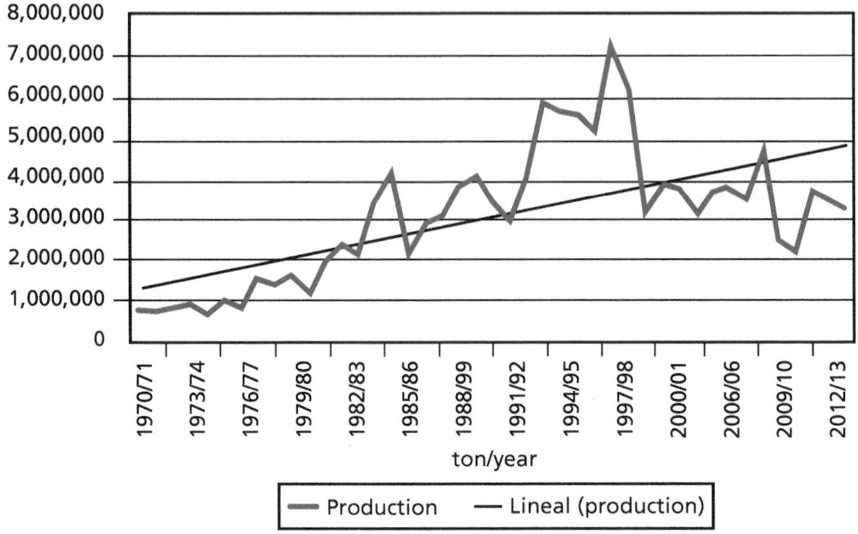

Figure 17.3 Argentina's total sunflower production, 1970–2012.

Source: Based on data Bolsa de Cereales (2013).

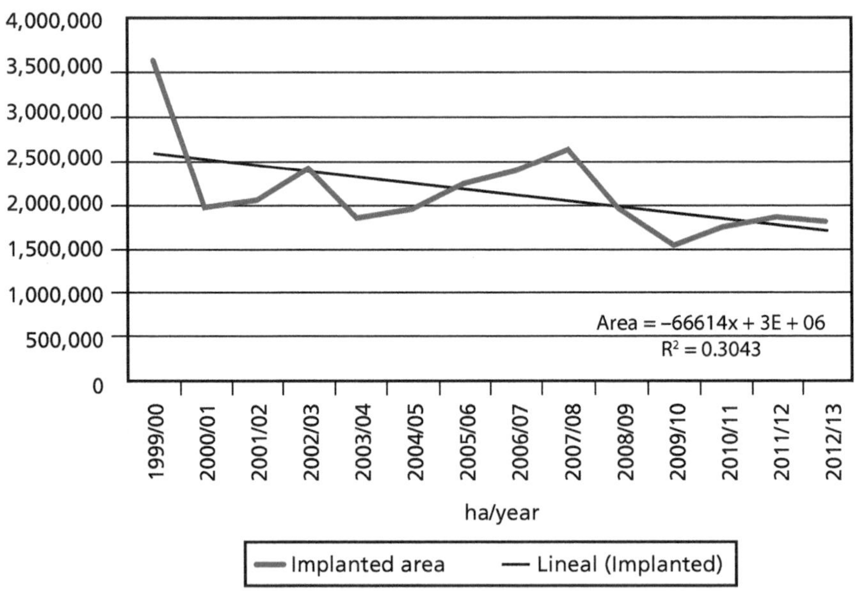

Figure 17.4 Argentina sunflower implanted area, 2000–2013.

Source: Based on data Bolsa de Cereales (2013).

production of 1999 was the cause for these pernicious prices, which came together with the eruption of palm oil from Asia. This combination led to over abundant oil of different qualities, a severe blow that still remains in some local farmer's memories.

Since 2000, there is not only less area but also a growing instability, founded mainly on the difficulties sunflower has in local competition with soybean for most productive lands, combined with the country's high export duties since 2002, and also a relative participation decrease in the global sunflower market (Figure 17.4).

In regard to yields, during the 1970s and 1980s, yields (see Figure 17.2) had clearly increased: In the 1990s, yields stagnated around 1.6 t/ha, and better crop conduction and technology allowed yields to reach the 2 t/ha mark in the 2009 to 2012 period (Figure 17.5).

During the 1987–2002 period, production is characterized by variability as a consequence of area and yield fluctuation (Figure 17.6). The strong production grow during the 1990s is associated with planted area increase, as previously mentioned.

Sunflower Crop Localization and Migration

Sunflower is grown extensively in several specific agro-ecosystems over Argentina that are distinguished by seasonal rainfall, radiation and temperature patterns, soil properties (texture, soil depth, organic matter content), the role of sunflower in the cropping

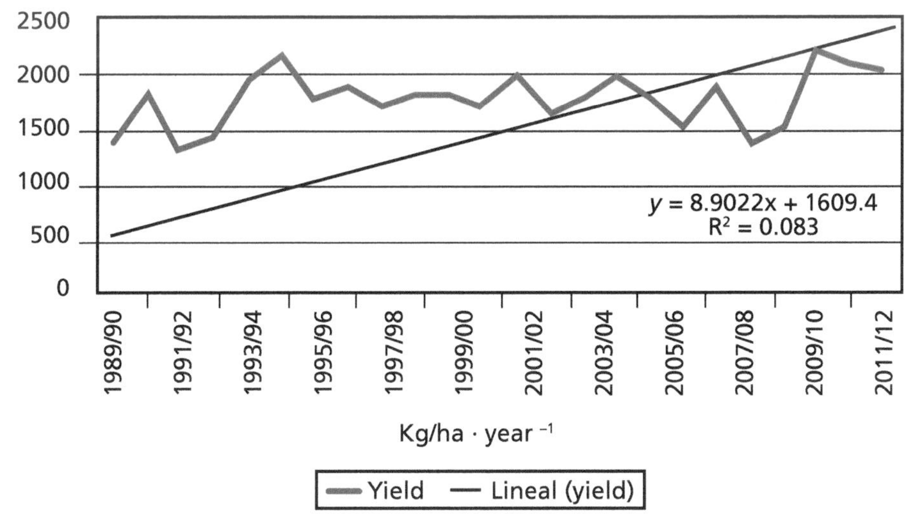

Figure 17.5 Argentina's sunflower yields, 1990–2013.

Source: Based on data Bolsa de Cereales (2013).

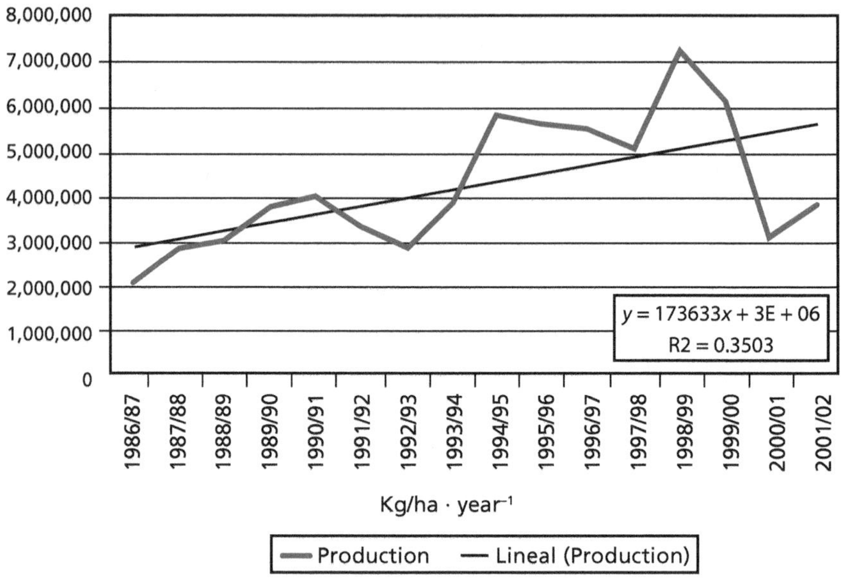

Figure 17.6 Argentina's sunflower production, 1987–2002.

Source: Based on data Bolsa de Cereales (2013).

system (sole within-season crop, lead crop of a seasonal sequence of two crops), and crop management (time of sowing). ASAGIR classified sunflower area in eight regions (Figure 17.7) with similar agro-economic conditions: southeast of Buenos Aires Province (SE), southwest of Buenos Aires Province (SO), west of Buenos Aires Province (OBA), center of Buenos Aires Province (CBA), San Luis and La Pampa Provinces (SL-LP), central Argentina (CENTRO), Entre Ríos Province (ER), and northeast Argentina (NEA) (ASAGIR, 2013).

Importantly, the SE to NW gradient across the Pampas region reflects increasing temperature and rainfall and a gradation in soil texture from thick to fine. Petrocalcic layers that limit soil depth in the SE region and annual rainfall distribution in this region are almost isohygrous, in contrast to the summer-dominant patterns for the remaining Pampean regions. Petrocalcic layers are also a feature in some soils of the SL-LP region, but these layers tend to be deeper in the soil profile than those of the SE region. Sunflower is grown as a sole crop within a season across all the Pampean regions, with sowing dates occurring later from north to south. Chapman and de la Vega (2002) have described weather (rainfall, temperature) conditions for the NEA region. Soils in the NEA region are fairly heterogeneous, but lighter and deeper soils are more frequent in the western portion of this area, and shallower and heavier

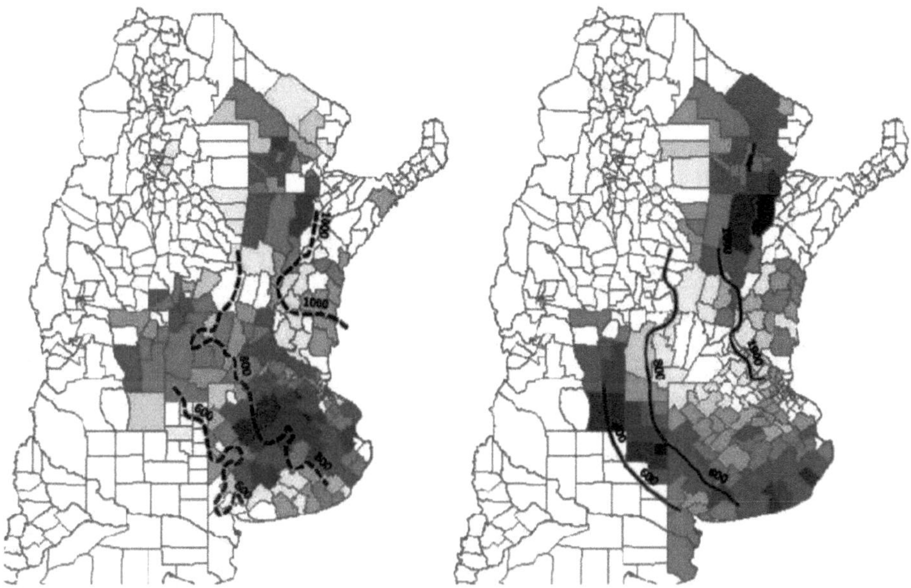

1969–70 Sowing Season
1925–65 Isohyet

2008–09 Sowing Season
1966–2001 Isohyet

Figure 17.7 Sunflower area/agriculture exploited land.

Source: CREA based on MingAgri information.

soils are found in the east (Ledesma and Zurita, 2004; Mosconi et al., 1981). In this region, crops are sown and harvested earlier than in the others (August–September vs. October; December vs. February–March), and waterlogging can be a significant constraint in high rainfall (Niño) years (Chapman and de la Vega, 2002). Provided early season rains are sufficient, sunflower can act as a lead crop for a sequence of two crops within a season in the NEA.

Farm size varies across regions. The mean farm size reported in the 2002 census (the latest available with data on a reporting district basis) was 1670 ha in the SL-LP; 600 to 770 ha in the OBA, SE, SO, and CENTRO regions; and 300 to 490 ha in the ER, CBA, and NEA regions (National Institute for Statistics and Census, 2013). Actual land surface farmed by a specific farmer may be greater due to widespread use of rented land and a continuous trend toward agriculture consolidation or bigger management units. For instance, in Buenos Aires, the number of farms dropped 38% between 2002 and 2008 (Bolsa de Cereales de Buenos Aires, 2012). This trend was similar, although minor, in other provinces where sunflower is grown.

A very technologically updated industry processes sunflower grain. Several private companies like Aceites Vegetales de la Pampa, Cargill, and Oleaginosa Moreno Hnos (crude oil industry); AGD, Gente de la Pampa, Molinos Cañuelas SA, Molinos Río de la Plata, Nidera, Bunge, and Vicentín (both refining and crude oil industry); among others, have facilities located along the Argentine sunflower area (CIARA–CEC, 2013) with high-volume capacity for grain reception and conditioning, crushing, and refinery, as well as bottling plants.

Sunflower Migration Process to Less Productive Areas

Sunflower production areas in Argentina have changed dramatically in the last 15 to 20 years, migrating from the established areas in the 1970s closer to the central Argentine region and to areas closer to the present western agriculture border: the Province of La Pampa and west of the Buenos Aires Province (Figure 17.7). The main cause was the decrease in area as a consequence of the drop in international prices and domestic prices in 1999 and 2000. Later on, however, when sunflower prices recovered, sunflower area increased up to 2.2 million ha.

This new context was the consequence of many different factors. Changes in rain pattern (Negri, 2010), isohyet of 800 mm/year, has moved to the west, broadening land suited for agriculture, the only factor that can be methodologically proved. This meant, alongside technology improvements such as no-till planting, that sunflower was available to grow in areas that were until not long ago used exclusively for cattle. Thanks to its unique physiological characteristics, sunflower has better yield stability when water is the limiting factor, due to its osmotic adjustment ability (Chimenti et al., 2002; Lisanti et al., 2010; Pearson, 2002). Among all crops, sunflower was the logical pioneer to explore arid environments.

Much of the retraction in sunflower area is a consequence of the great soybean area expansion; it went from 37,700 ha in 1970 to 18,886,634 ha in 2010. Also, since the late 1990s, soybean has had better prices and lower price risk given that it can be hedged in the Chicago Board of Trade. In addition, it had better potential yields, easier management, and lower implantation costs due to RR-glyphosate technology and its self-pollinated seed (which allows farmers to store grain to use as seed the following growing season and save the cost of buying seed). The increased competiveness of soybean over the last decades has relegated sunflower to areas where soybean can't be sowed with successful economic results. Even in the same field, sunflower goes to the lowest productive areas, choosing the highest potential environments for soybeans (Negri, 2010). In spite of the migration to less productive areas, sunflower yields have sustained the level of previous years. Ricardo Negri's (2010) research also shows that no matter what weather conditions sunflower has been under, it has presented stable results in every seeding season, granting reassurance for farmers against other crops with more diverse results. The author concludes that although probable net margins are better for soybean or maize, many companies and farmers have begun to notice, probably in reviewing their financial data, that one of the reasons that they have earned money in years of drought is due to sunflower. Its stable yields, granted by osmotic adjustment and deep root exploration, would probably continue to ensure its presence in many farmers' crop portfolios.

Sunflower Oil Uses and Opportunities for Specialties

Sunflower seeds consist of approximately 50% oil and 20% protein. The fat composition has always been of the linoleic type. However, industry demands for special characteristics, namely tolerance to high temperatures over long periods of time, have promoted development of mid-oleic and high-oleic types to replace the hydrogenation process associated with unhealthy trans fatty acids (TFAs).

Presently, the Argentine market offers the high linoleic or traditional oil type, almost 90% of total (commercially known as SFO), and high oleic type (HO) (Table 17.E). Mid-oleic type, patented in US as NuSun, is not commercialized in Argentina, although many hybrid lines available have the same qualities under given environmental conditions, such as in the NEA region. Very important byproducts are high protein pellets, low protein pellets, and regular protein pellets.

Regarding the non-oleic types, Argentina is also a strong player on confectionary sunflower, with its own business particularities. As a dressing, any of the current

Table 17.E Sunflower Oil Fatty Profiles

Sunflower Oil Type	Monounsaturated Oleic C 18:1	Polyunsaturated Linoleic—C 18:2	Saturated Stearic C 18:0	Saturated Palmitic C 16:0
		Percentage		
High oleic	77[a] > 82[b]	9	5	3
Mid oleic	55 – 76.9[a] / 65[b]	26	—	9
SFO—linoleic	> 54.9[a] / 20[b]	> 69.9	—	—
High linoleic	—	< 70	—	—
HL/S	10.2[b]	49.6[b]	—	—
HO/S	60.8[b]	8.2[b]	—	—
High palmitic	—	—	—	> 25
High stearic	5.2[b]	—	30	7

[a]According to Norma IRAM 5529 (2013).

[b]Most frequent acid content.

Source: Based on Velasco et al. (2008); Norma IRAM 5529 (2013); Universidad de Sevilla (2013); Instituto de Agricultura Sostenible (2013).

fat profiles can be used, but Argentine taste is used to the linoleic type, which has a particular flavor; this type is also very valued in Europe (Feoli, 2004; Universidad de Sevilla, 2013).

Planted Argentine High Oleic Sunflower Area Evolution

In the early 1990s, high oleic sunflower was introduced in Argentina, beginning with less than 8,000 ha. Today 250,000 ha are planted according to private estimations. (Table 17.F).

Confectionary Sunflower Business Description

Although common sunflower is much more broadly sown, the planting of confectionery sunflower, for which kernel size is the objective, is sustaining growth. It is commercialized as whole toasted seeds or as sweet or salted toasted kernels.

Morphologically speaking, the confectionary sunflower plant is similar to the conventional type; the differences lie in achene size, larger striped pericarp, and lower oil content. Breeding advances allow confectionary sunflower to have similar yields to regular sunflowers. Although there are no official statistics, average yields in the 2012/13 seeding season were around 2.1 t/ha.

It's a much smaller business compared to oil sunflower because it is a specialty and, consequently, most of the production is backed up by a supply contract in which seed size determines the premiums. Grading is similar to oil sunflower, with no premiums for oil content. A sunflower kernel is classified by its size; confection grade seeds has 550 achenes per ounce (kernel size > 8.75 mm), bakery grade seeds are smaller, and the smallest seeds are used to feed birds (ASAGIR, 2013). Global demand for confectionery sunflower has grown remarkably during these years and Argentina is the only relevant producer in the counter season, becoming the third largest exporter with more than 70,000 ha (ASAGIR, unpublished data) planted mainly in La Pampa, south of Córdoba, and northwest of the Buenos Aires Province. Europe and the Middle East are the main destinations of Argentine confectionary sunflower.

Table 17.F High Oleic Sunflower Area Evolution

Year	Implanted Area (ha)	Year	Implanted Area (ha)	Year	Implanted Area (ha)
2001	50,000	2007	300,000	2010	200–220,000
2003	90,000	2008	220,000	2011	220–230,000
2005	180,000	2009	180,000	2012	250–280,000

Source: ASAGIR (2013).

Sunflower for Biodiesel: Characteristics, Potential, and Limitations

The chemical combination of triglycerides (fats or oils) and methanol, in the presence of a catalyst, results in methyl esters (biodiesel) and glycerin. Biodiesel quality, cetane number, stability toward oxidation, fuel lifetime without residue accumulation, and solidification point, among other factors, affect engine performance. The raw material used to produce biodiesel affects its properties.

Sunflower biodiesel has remarkable properties compared to other oils: Its melting point is –12 °C and its cetane number satisfactorily complies with standards. The ideal oil for biodiesel must have oleic acid with only one double bond (a characteristic best associated to storage stability), a satisfactory cetane number (59 to 61), and a low solidification point (Cámara Argentina de Biocombustibles, 2014).

G. Pereyra Irujo et al. (2007) analyzed the quality variability of sunflower based biodiesel with a validated model that predicted results based on fatty acid profiles. Results indicated that predicted values complied with standard parameters having low variability for density, viscosity, and cetane number. Perfect biodiesel must be composed only of monounsaturated fatty profiles. This provides great future potential for high oleic sunflower that complies with this request and has an oil productivity over 1.5 t/ha. However, unsaponifiable levels reflect wax content: Sunflower oil scores negatively in this point, affecting quality for biodiesel due to a very high wax content. This requires a dewaxing process commonly known as *winterization.*

In brief, sunflower is not, at least in the short term, a probable input for biofuel. It is an expensive oil, and although high oleic varieties comply with quality demand, regular sunflower (linoleic type), which is the most vastly produced, is not very suitable for biodiesel. In addition, the wax content demands an extra process that elevates production costs.

Crushing and Trading: Argentine Sunflower Price and Market Characteristics

Crushing Sector: Companies and Characteristics

According to Hinrichsen (2013) and other qualified informants, the total crushing capacity in Argentina, mainly aimed at soybean, was 186,461 t/day in 2012 (Figure 17.8, Table 17.G).

There is a new plant under construction by Oleaginosa Moreno and Glencore, located at Timbues close to Rosario, with a soybean crushing capacity of 10,000t/day. By entering into operation, this will take the country's total crushing capacity for all crops to 196,381 t. This will continue to ratify Rosario and it surroundings as the paragon oil cluster, accounting for 77% of country's total crushing capacity. As a result, it will transform Argentina into the second crushing leader, over the United States and after China, which has the capacity to crush a remarkable 350,000 t/day.

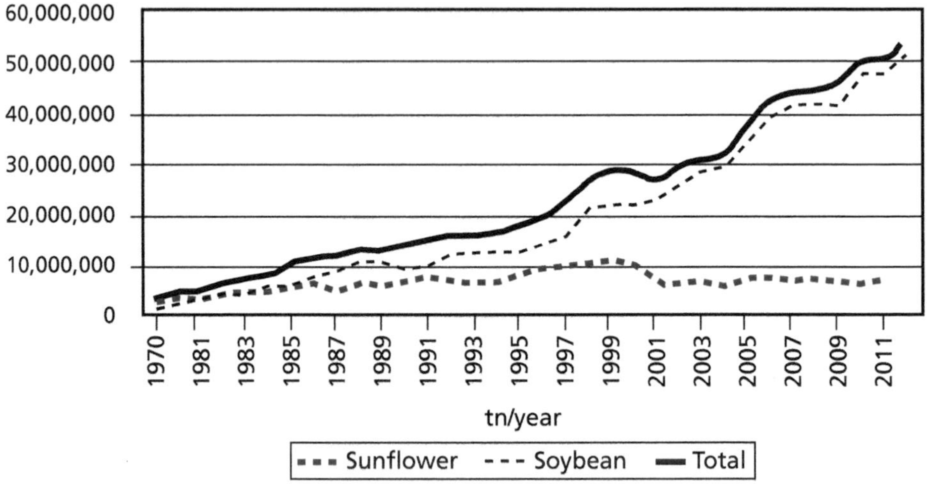

Figure 17.8 Argentina crushing evolution, 1970–2012.

Source: CIARA (2013).

Regarding sunflower crushing exclusively, by the second semester of 2011 (no investments were made to the date), Argentina had a total theoretical capacity of 30,136 t/day. Some facilities can switch between sunflower and soybean, and very few are devoted exclusively to sunflower crushing (Table 17.H).

Starting 50 years ago there has been a continuous strong evolution of the crushing industry, particularly for soybean since the end of 1990s, and a stagment for sunflower (Figure 17.9).

Table 17.G Argentina Total Crushing Capacity

Province	t/Day	%
Santa Fe	140,706	75.4
Buenos Aires	20,500	11
Córdoba	18,950	10.1
Entre Ríos	1,970	1.1
Santiago del Estero	3,000	1.6
La Pampa	400	0.2
Others	935	0.5
Total	186,461	100

Source: Hinrichsen (2013).

Table 17.H Crushing Plants: Companies, Capacity, Type of Grain, and Location

Company	Location	Grain Processed	Maximum Capacity (t/day)
Center Buenos Aires			
Oleaginosa Moreno SA (4)	Gral. Villegas	Sf Sb	2000
SEDA SA	Lezama	Sf Sb	150
Oleaginosa Moreno SA	Daireaux	Sf Sb	1800
Nidera Sociedad Anónima	Saforcada (Junín)	Sf Sb	2200
Molino Cañuelas SACIFIA	Cañuelas	Sf	1200
Germaíz SA	Baradero	Mz Sb Sf	410
Region Subtotal			***7760***
South of Buenos Aires and La Pampa			
Gensiroil SA	Bahia Blanca	Sf	500
Cargill SACI	Ing. White	Sf Sb	2200
Cargill SACI	Pto. Quequén	Sf Sb	2000
Oleaginosa Moreno Hnos. SA	Quequen	Sf	1350
Gente de La Pampa SA	Catriló	Sf	300
Aceites Vegetales La Pampa	Gnrl. Pico	Sf	100
Region Subtotal			***6450***
Northern Region (Chaco and Sta. Fe)			
Buyatti SAICA	Reconquista	Sb Sf Sff C	1456
Vicentin SAIC (Planta Ruta 12)	San Lorenzo	Mz Sb Sf	4000
Region Subtotal			***5456***
Center Region (Cordoba and Z. Nucleo)			
Bunge Argentina SA	Tancacha	Sb Mz Sff Sf	1000
Agroindustria Madero SA	Villa Madero	Sf Sb	700
Molinos Rio de la Plata SA (1)	Rosario	Sf Sff	1500
Aceitera Gral. Deheza SAICA	Gral. Deheza	Sb Sf	7000
Oleag. Gral. Cabrera OLCA SA	Gral. Cabrera	Mz Sf Sb	270
Region Subtotal			***10,470***
TOTAL Argentina's Sunflower Crushing Capacity			**30,136**

Sf: (sunflower); Sb: (soybean); Mz: (maize); C: (cotton); Sff: (safflower).

Source: Based on J.J. Hinrichsen table cited by CIARA statistics and others.

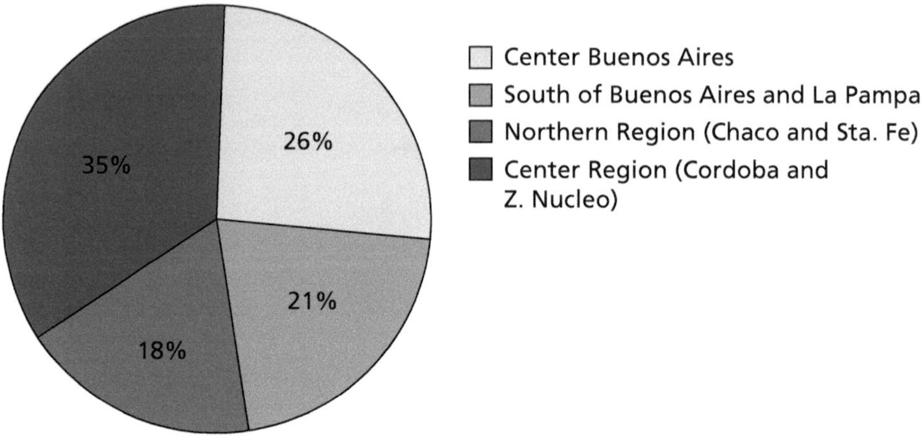

Figure 17.9 Argentina's sunflower crushing capacity distribution by regions.

Source: ASAGIR. Based on CIARA-CEC (2013).

Global Argentine Market Share for Oil and Meal

Argentina's production, domestic consumption, and exports through the triennium of 2010–2013, in global trade terms, are shown according to the following:

Production

- Argentina's total sunflower seed production averaged 3.41 million t for the mentioned triennium. The total world production was 36.7 million t; therefore, Argentina's participation in global production was 9.3%.
- Argentina's meal production averaged 1.52 million t. Total world meal production was 14.6 million t; therefore, Argentina's participation in global production was 10.5%.
- Argentina's oil production averaged 1.46 million t. Total world oil production was 13.7 million t; therefore, Argentina's participation in global production was 10.6%.

Domestic Consumption

- Domestic meal consumption was around 755,000 t. Domestic participation in local production went from 45.3% in 2011 to 51.2% in 2013, with an average of 49.5%.
- Domestic oil consumption was around 496,000 t (12.3 liters per inhabitant). Domestic participation in local production went from 29.3% in 2011 to 38.3% in 2013, with an average of 34.1%. This remarkable increase was largely due to

strict price control over the product for individual consumption in a context of high inflation.

Exports

- Grain exports averaged 74,300 t for the mentioned triennium, with no significant changes during recent years. Only 2.17% of Argentinian sunflower seed is exported untransformed. In global trade participation terms, this accounts for 4.4% of total volume.
- Meal exports grew 0.7% from 0.745 million t to 0.750 million t between 2011 and 2013, but overall world trade decreased from 16.2% to 12.9%, with an average of 13.7%.
- Oil exports decreased 8% between 2011 and 2013, from 0.978 million t to 0.900 million t. In global trade terms, this means participation decreased 31.9%, from 21.6% to 14.7%, with an average of 16.2%.

Market Morphology Sunflower Seed and Byproducts

Sunflower seed price is primarily defined by the value of its oil, which depends on the competing oils around the world and is affected by various uses. In the last 10 years, biodiesel's unprecedented production growth generated big changes in the oilseed market, enabling vegetable oil to also be affected by the crude oil market (and biofuel regulation policies). Before then, the international and Argentine sunflower market value depended only on the supply and demand of vegetable oils. Sunflower oil trading is massively marketed in bulk (unrefined) or refined, either bottled or not bottled.

Based on a USDA report, for example, the prices of oil and sunflower protein in March 2013 were $1,221/t and $299/t, respectively (USDA, 2013). In the Northern Hemisphere season 2012/2013, 92.5% of the 36.4 million t produced were crushed, resulting in a 41.2% and 43.8%, of oil and protein respectively. For CIF Rotterdam theoretical price of $649/t, we find that 79.3% corresponds to oil and 20.7% to protein.

As mentioned, there are various segments of oil prices and qualities. Sunflower oil can be equated with canola oil for its quality. Using the same price formation procedure, canola is weighted 68% on oil and 32% on protein.

The sunflower oil market represented a worldwide volume in 2012/2013 of 8.7% of the consumption among the nine major oils (13.5 million t compared to 155.6 million t) (Figure 17.10). Under stable conditions, sunflower and canola oils hold the highest price in bulk markets. In terms of global consumption measured in bulk, oils such as coconut, olive, palm kernel, and peanut all have shares inferior to 3.4%. Other mass consumption oils represent 33.7% (palm), 27.7% (soybean), 14.5% (canola), and 8.7% (sunflower), which have predominant commercialization in bulk.

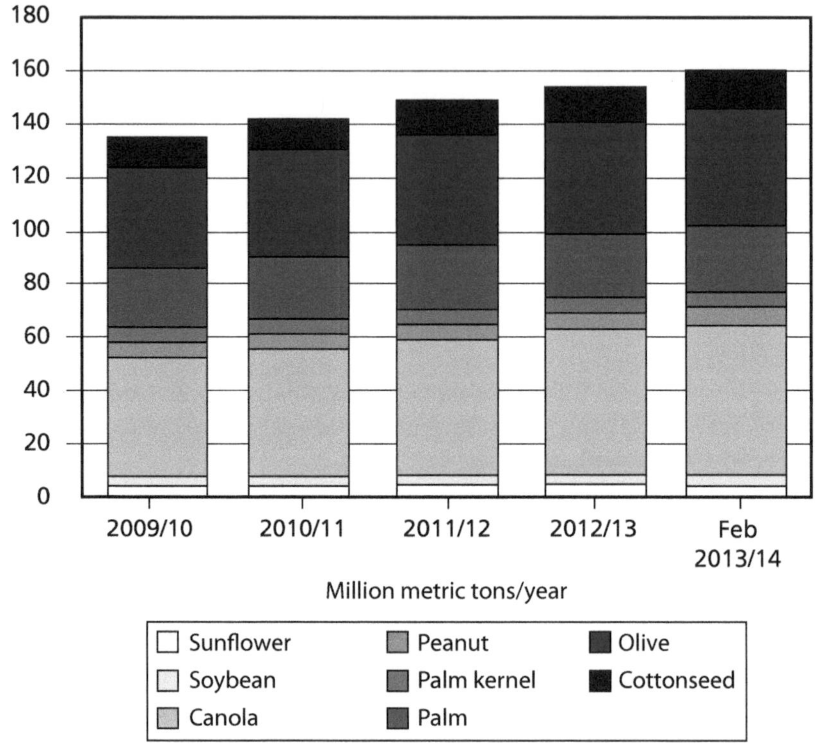

Figure 17.10 Shares of major oils in global oil consumption.

Source: Based on USDA (2014).

Meanwhile, soybean oil (with a share of 27.7%) has a lower price. But sometimes under strong global shortage in the production of soybean, as happened in 2011/12, prices become similar to those of higher quality products, such as sunflower and canola. This is influenced by demand for soybean protein and soybean oil in markets such as Asia, which have much less substitutability.

International Price Evolution in the Last 15 Years

Sunflower oil price in the context of competitiveness with the rest of the vegetable oils has been through at least three stages in recent years. Stage 1: Since the mid-90s, it held a value range in Rotterdam of $545 to $730/t (annual averages in current dollars). Until then there had been no massive harvests of palm oil from Indonesia, Malaysia, and to a lesser extent, Thailand, pushed down to the global market. In Argentina, the sunflower crop area was approximately 2 million ha in 1993/94 and it grew significantly to reach 3.9 million ha in 1998/99, increasing production from under 3.9 million t to 7.1 million t. Because Argentina was the largest exporter in the market, the increase in production hurt the exceptional prices and made them collapse from $730/t in 1997/98 to $413/t two seasons later, a decrease of more than 43%. Simultaneously, the global production of palm oil between 1998/99 and 2000/01 went from 19.3 million t to 24.3 million t, growing from 40.6% to 49.0% of the total oilseed world production.

Sunflower oil, which was traded at premiums of 21.5% and 14.6% over palm and canola oils in 1997/98, went on to have premiums of 81.1% and 15.7% in 2000/01, respectively. This shows that if the price of sunflower oil drops, it drags down canola and other lower quality oils, such as palm oil, which fell 61% in the same period. In summary, when prices are low, market tends to favor sunflower oil consumption, which contributes to the fall of palm oil. The more the supply increases the deeper price falls, and this is what happened in this period.

Stage 2: As a result of falling prices in sunflower oil in the late 1990s, local producers were deeply discouraged and the planted area declined between 1998 and 2000 from 3.9 million ha to 1.9 million ha, which means a reduction slightly less than 52%. Since then, it had kept its area between 1.8 million and 2.3 million ha, according to the crop competitiveness compared to soybean. During this period, economic reforms in countries of the former Soviet Union resulted in Argentina finally losing its leading role as sunflower oil exporter at the hands of the Ukraine and the Russian Federation. These countries were favored due to their proximity of the Black Sea markets, and they were close to the lead importers of both grain and derivatives, the European Union. There was a brief period of slight price recovery, to values of $587/t and $592/t in 2001/01 and 2002/03, respectively. The first of the cited years presented a premium for sunflower oil, 78.4% and 30.2% compared to palm and canola, respectively, while the following year it only registered premiums of 40.6% and 0.7%. This downfall in premiums in Rotterdam was due to the European Union's massive canola use for biodiesel production, which marked the start of the third stage.

Stage 3: As stated, this stage is marked by the increasing use canola oil for biodiesel in the European Union and, as a consequence, the beginning of shifting European food consumption to sunflower oil. Palm oil has grown in five years (between 2000/01 and 2005/06) from 24.3 million t to 34.8 million t in world production and 17.2 million t to 25.6 million t in world imports, reaching, in this latter case, a global share of 59.4% and always catering to the most dynamic markets of the fastest growing global middle class, mainly India and China. The impact of the use of canola for biodiesel can be seen with the following indicators:

- At the beginning of the century, canola oil consumption was 12.2 million t and 1.2 million t for food and industrial use, respectively. In the last three seasons, stated consumption was 16.3 million t and 7.3 million t, on average.

- Between 93% and 96.5% of total canola oil consumption is justified by the increase of industrial consumption, and this is surely due to the use of 5.7 million t in biofuel production.
- Between the beginning of the century and the last three seasons, world food consumption grew 33.7%, total canola oil consumption grew by 76.1%, and industrial consumption grew a staggering 608.3%, which explains the maintenance of sunflower and canola oils among the most demanded in the massive premium segment.
- Premiums paid to sunflower oil depended on macroeconomic factors in the last 10 years, such as the abundant global monetary liquidity or changes in the activity level in the lead consumer, the European Union. The average of the last 11 years, not including this season (2013), was around 40% over palm and was barely significant (3.5%) relative to canola oil. There were moments of strong growth in the palm oil supply (2004/05) during which the premium for sunflower oil reached 79.9%, then went down in 2011/12 to 21.5%, and then went back up to almost 52% in March 2013. This did not happen with canola oil, which had years of strong premium, such as the bubble of 2008 (16.2%), but also had significant discounts (17.5% in 2005/06). In the season 2011/12 they were at the same level. In March 2013, premium over regular price grew to 5.1%.

The global market is now segmented. On one hand it is supplied by palm oil, which serves Asian demand (that also uses it for biofuel, like soybean oil). On the other hand over supply has a spill over effect on the global price for sunflower oil, reducing it. On the other hand, sunflower and canola supply the European Union, where they are the most preferred oils for human consumption. Recent decrease in canola supply, as a result of the steady demand for biofuel use, is favoring sunflower demand, which focus exclusively on the human consumption market.

Science and Technology in Crop Production

Research and Products in Science & Technology (S&T) Institutions and Companies Involved

Breeding of edible crops started in Argentina at the beginning of the 20th century, but until the 1980s research related to agriculture was almost entirely conducted by public institutions, such as National Department of Agriculture and Livestock, INTA (the National Institute of Agricultural Technology), and some national universities: del Sur, Mar del Plata, and Buenos Aires. At the beginning of the 1990s, private seed and pesticide companies, in addition to the S&T carried out by public institutions, acquired a strong leadership in new technologies development, and several public–private joint ventures were boosted, benefiting farmers and consumers alike. This generated great advantages for related value chains. The history of research in other South American countries was similar or began later.

This shift didn't mean that public institutions abandoned their work; on the contrary, it was boosted by the public–private partnerships. Nowadays, several seed companies, such as ACA, Advanta Semillas, Argenetics, Dow AgroSciences, KWS, Limagrain, Nidera, Nuseed, Pannar, Pioneer, Sursem, Syngenta, and Tecnoseeds, among others, have very important breeding programs in Argentina. In fact, they provide commercial hybrid seed to producers. INTA develops research and extension programs to allow farmers to grow crops according to the best available technology (Feoli, 2002). Research work includes breeding, which is developed in joint ventures with private companies, and innovative techniques for crop management. Private consultants assist growers in decision making as well as in cooperative and other commercial organizations.

S&T Brief Development History

As in several crop species and in many cropping systems, sunflower yields obtained by farmers are often lower than those attainable using locally optimized agricultural best practices and adapted, current cultivars. Hall et al. (2012) estimate yield gaps as the difference between mean farmer yields and attainable yields. Farmer yields have been estimated using regional or national statistics and by sampling farmers' fields. Attainable and potential yields have been estimated using on-farm experiments, yield contest results, research station experiments, and breeders' trials. The results of this analysis for the sunflower growing regions of Argentina, conducted on behalf of the ASAGIR, showed: (1) Yield gaps ranged from 0.37t/ha to 1.18 t/ha across regions; (2) a country average of 0.75 t/ha, equivalent to 41% of the mean country yield of 1.85 t/ha; (3) mean yields for the top decile of comparative yield trial from 3.2 to 4.2 t/ha; and (4) the highest yields for this decile reached values of 3.9 to 4.8 t/ha. They conclude that the magnitude of yield gaps in Argentina justifies further research to reduce them.

Authors expect to find which factors affect yields, to determine whether the size of current yield gaps justify further research into yield gap reduction, and to define a framework that would allow infrequent but recurring reports of high grain yields of 4 to 5 t/ha in contrast to national yield averages of 1.7 to 1.9 t/ha.

Support for S&T

In Argentina, agricultural S&T development began with experimental farms sponsored by the National Department of Agriculture at the beginning of the 20th century. The Universities of Buenos Aires and La Plata have research programs since the beginning of the 20th century and INTA and CONICET (local research agencies) were created in 1956, consolidating diverse disciplines in sunflower research programs.

In the beginning of the 1990s, the Secretaria de Ciencia y Tecnica (SECYT), a governmental agency that promotes and sponsors science and technology work, was founded.

Sunflower Price Formation in Argentina: Commercial Institutions and Information Systems

In Argentina, sunflower is an export-oriented business; sunflower products are primarily shipped as oil and flour, and to a lesser extent seed is also exported. Given that 60% of the oil produced is exported in a normal season and that flour has a low relative value, over 80% of sunflower value is explained by the international value of its oil. Argentina is the world's second largest exporter of sunflower oil (after the Ukraine), with 14.7% of the market in the last campaign. That means that Argentina has no pricing formation capacity as before. In those growing seasons in which North Hemisphere production goes down, an eventual pricing formation capacity appears around June. There is a window of opportunity that appears when global production falls, during the months of May to August, and Argentina becomes by far the most significant exporter.

Because the United States is not a major producer of sunflower, there isn't a market like Chicago for forward prices that producers and traders can access worldwide to hedge price risk. There is just an international reference price for oils in Rotterdam that consists of collected statistical contracts between exporters and industries in different countries and industries and importers or users from the European Union. It is not a market in which risk can be discharged. It is merely based on transactions accompanied by delivery of goods, like any industrial export. The lack of an international reference price is due to the low world volume of sunflower seed or oil and its geographically scattered production and consumption.

In the absence of forward markets, price values of reference, like the CIF Rotterdam, make it almost impossible to allow negotiating local contracts between producers and exporters. In these markets, it is only possible to negotiate positions for a few months, mainly when the planted area and production can be estimated, making it impossible to arbitrate positions further into the future. This impossibility of coverage and having no midterm price orientation are mistaken by many traders as lack of transparency, when in fact it's a matter of illiquidity (in Argentina, a very low number of buyers).

In Argentina, soybean exports and industry are gathered around the city of Rosario and ports in the Paraná River; consequently, a unique reference price (Rosario Board of Trade) is possible. The geographical distribution of the sunflower crop differs from that of soybean, and as in other countries, harvest times and freight costs also vary and consequently, the farmer has no unique reference price.

Industry and exports have locations close to: (1) Domestic consumption markets, (2) ports, or (3) sunflower seed production centers. There is less industrial capacity intended for sunflower than for soybean, and it is located according each to company's logistical convenience, present or past.

Because sunflower is less of a commodity than soybean, there are even more specialties, such as high oleic, that are mainly produced by contract. Also, the industry pays premiums for quality and oil content, from a base price. Usually freight costs are negotiated with product price. This is why different prices are given to farmers depending on location, commercial agreement, and seed quality.

Typical types of grain transactions between farmers and/or middlemen with the industry are: (1) forwards, (2) available on delivery, and (3) delivery to trader and price fixed by producer at some future date as a function of spot price at that time.

Daily "spot prices" are determined by a committee of representatives from the supply, demand, traders, and public sector in arbitration chambers in grain exchange markets or trade boards of main delivery points. This committee settles on a price that reflects the average of operatives reported that are less than the ones truly made. This price is valid for transactions that take them as reference, either used among farmers and middlemen; farmers and industries or exporters; or middlemen and consumers. Its adoption as a reference price is not mandatory in any contract, and it is possible to use other reference prices. Most farmers' perceptions are that spot prices do not reflect available-on-delivery prices, given that they associate as true market value with the price paid, on a certain date, for a given batch. Spot price truly reflects the average price informed by traders.

Distances to ports from main production areas in any of Argentina's four common sunflower productive regions (NEA, CENTRE, WBA, LP) are frequently long enough to discourage production (due to increased freight costs).

In Argentina, agricultural future markets work with U.S. dollars market rates, but it is also frequent to have rates in Argentine pesos. There are three usual references that farmers and middlemen usually check before selling: (1) sunflower grain price that can be established both in Argentine pesos or U.S. dollars; (2) regional differences, for example freight cost (Table17.I); and (3) commercial conditions, like grain volume or payment time. It's important to notice that when sunflower seed is delivered, the brute value paid for grain is unknown, given that oil quantity and oil premiums are yet to be determined. Usually, an additional 7% or 10% is paid as a result of oil content.

Value Chain Organizations: The ASAGIR Model

ASAGIR is a nonprofit organization that aims to increase crop production efficiency and product and byproduct quality, and also promotes product diversification. Founded at the beginning of the 1980s, ASAGIR has organized and supported sunflower-related technical and scientific meetings, including the 11th and 18th

Table 17.1 Usual Distances for Sunflower Seed Freights

Origin	Destination	Distance (km)
Charata (NEA)	Rosario	813
Reconquista (NEA)	Rosario	504
Deheza (C)	Rosario	310
Trenque Lauquen (WBA)	Deheza	506
Trenque Lauquen (WBA)	CABA	440
Trenque Lauquen (WBA)	Bahia Blanca	317
Gral. Pico (LP)	Bahia Blanca	425

Source: Ruta (2013).

International Sunflower Conferences held at Mar del Plata in 1985 and 2012, respectively, and local conferences at regular intervals in 2002, 2003, 2005, 2007, and 2010. Currently, ASAGIR is working on programs and activities in line with the concept of the food value chain, which seeks to foster the integration of all tiers in the chain, from seed production to oil and byproduct commercialization.

Its main current activities include support for scientific research projects and market studies, the organization of workshops and scientific or technical meetings, and participation in international meetings and study groups. These activities relate to crop production, crushing industry issues, and sunflower uses and nutrition, and the results of these studies are available at the ASAGIR website (ASAGIR, 2013).

National Economic Impact of the Sunflower Value Chain in 2013

The business value of sunflower—understood as gross revenue generated by total transactions for the entire value chain: input provision, agricultural production, industrialization, and export—can be placed between $1.920 billion and $2.041 billion for 2013. Total sunflower-related exports would be $1,026 to $1.147 billion, according to the source used for the production value, given that the Ministry of Agriculture gives an additional 200,000 t (estimated by the authors based on MinAgri, 2013, and Bolsa de Cereales, 2013). Current export duties (high tax rates in force since November 2007) have an impact over FOB prices of 30% for the finished product (oil and flour pellets) and 32% for plain seed export. It is important to take notice of the low protection for local added value, measured by a tariff differential of only 2 percentage points, a very small difference compared with up to an 11% spread of some Black Sea countries, oil industry protection policies.

Sunflowers effective tax collection in export duties alone, in fiscal year 2013, ranged between $308 million and $345 million. This tax also distorts local prices for sunflower seeds, given that the local price is the FOB price less export duties, generating an additional loss for farmers of $270 million. That is also a loss for the government, given that in recent years less than 60% of sunflower is exported.

This net margin loss accounts for $342 to $350/ha, considering the added effect of export duties and its impact on the domestic price of sunflower seeds. This tax burden on production doesn't consider tax collection in the form of income tax (35%), among other national and provincial taxes. A reasonable hypothesis would be that with an export duty reduction, the total collection related to sunflower business would be higher, given that increasing collection from other taxes, sustained by production growth, will fully make up for any direct loss in export duties collection and generate more added value for the entire economy.

A more reasonable policy, as was in force in the 1990s, would be a tariff differential of 3.5% between raw material and finished product, with an export duty of zero for transformed products. This would be an adequate protection policy for the local industry and would reduce farmers' gross income difference to only $37/ha, a more encouraging fee compared to current estimate of $350/ha.

Prospective for the Argentine Sunflower Complex

Argentina's Sunflower Status in a New Global Context

Argentina is a strong exporter of oilseed and its byproducts. It has an industrial capacity that largely exceeds domestic needs, and it has been one of the quickest in terms of investment to respond to stimuli and economic incentives, not only to export oil but also to take the lead as a biodiesel producer. There is a strong demand for soybean-based biodiesel, whereas sunflower is not used given that it has a higher relative value and higher paraffin content that requires de-waxing, which is a very expensive process. At the same time, the world has a growing demand for renewable energy. The European Union has set ambitious targets for biodiesel blended with petroleum diesel. Most of the canola produced in the EU is for this purpose, whereas sunflower is oriented toward food consumption, directly or as component of other edibles.

The changing behavior of global oil markets, according to the OECD-FAO Agricultural Outlook: 2012–2021, shows tendencies for increasing demand for high quality oils, such as sunflower. Over the coming decade, world oilseed production is expected to expand by only 20% compared to the 2009–2011 average. High costs, environmental constraints, and sustained profitability of competing crops will limit growth to only about half the rate observed over the previous decade. Production growth would be based equally on continued area expansion and yield improvements.

Oilseed production and export will continue to be dominated by traditional players, but emerging exporters, such as the Ukraine and Paraguay, are expected to increasingly contribute to global export growth by 2021. Although South American soybean producers continue to dominate global meal exports, Indonesia and Malaysia will expand their share of vegetable oil exports to over 60%. Imports of oilseeds and products will be less concentrated, but China and the European Union will remain as the dominant importers.

In the rest of the world, production is set to expand as much as 34% as a result of traditional producers such as Canada, as well as a number of emerging, fast-growing countries such as Paraguay, the Ukraine, and the Russian Federation. Biodiesel use is expected to grow considerably in developed and developing countries. However, food consumption will stagnate in the developed world, while per capita annual food use in developing countries is expected to expand by 2 kg or 12% over the next 10 years, leaving it still at only about three-quarters of the level currently found in developed countries. World vegetable oil production is expected to increase about 28% in this period. Perennial oil crops such as palm and coconut oil will be the main contributors. Indonesia and Malaysia are set to remain the world's two largest producers, accounting for 20% and 14% of global oil output in 2021. Palm oil is projected to increase by 37%, accounting for one-third of global vegetable oil production. China, the EU, and the United States will expand their global production between 19% and 24%, maintaining their shares. China will grow much less in crushing capacity, relying more on imported oil than imported oilseed. Vegetable oil consumption is expected to grow by about 2% per year, less than half the rate of the last decade. The share of vegetable oil consumption used for world biodiesel production is expected to rise from 12% in 2009–2011 to 16% in 2021. It is expected that regulations set forth by biofuel policies will shape not only biofuel markets but also all agricultural commodity markets. In the developed countries, food use is expected to grow 10% over the entire outlook, with the average consumption per capita unchanged at 24 to 25 kg per year. In developing countries, the average per capita intake of vegetable oil is set to rise by about 2 kg to 18 kg per year. Argentina may further expand its export-oriented biodiesel industry, which could absorb 31% of the domestic vegetable oil output.

The projection point for oil and meal would be a slowdown in trade growth, decelerating more in developed countries. Indonesia and Malaysia's total exports are set to climb to 65% by 2021. Argentina should remain the third-largest exporter of vegetable oil with a share of 9%, as well as a top supplier of oil meals (with a share of 40%), shipping over 60% of its oil output and close to 90% of its domestic meal. On average, about 50% of vegetable oil consumption in developing Asia would come from imports. This will increase their share in global vegetable oil import to 51% by 2021 (OECD-FAO, 2012).

Under the OECD-FAO projected outlook, demand for high quality oils will rise, buoyed up by populations joining the middle class and becoming more quality-conscious consumers. Sunflower oil would lead the supply if that EU canola-based biodiesel promotion policies continue, which is the most probable scenario.

Challenges and Opportunities: Production Tendencies and Reduction of Gaps between Farmers and Attainable Yields

Recent studies by ASAGIR show the main challenges and opportunities in the context of threats, strengths, and weaknesses, usually called SWOT (a SWOT analysis is a structured planning method used to evaluate the strengths, weaknesses, opportunities, and threats involved in a project or in a business venture) for the entire value chain (Table 17.J).

Under ASAGIR SWOT analysis (ASAGIR, 2013), Argentina production may follow three possible scenarios:

1. Sunflower faces a limited technology improvement without important changes over productivity and cost reduction. As a result of handing over area to other competing crops, such as soybean, wheat, and maize, production stagnates or declines.
2. New technology adoption favors productivity and cost reduction, with a slight area increase, enabling production to reach similar levels as those obtained in the first decade of the 21st century (an average of 4.5 million t).
3. The production of scenario 2 is accompanied by quality and use differentiation, pushing higher price and export earnings. High oleic or mid-oleic oil profiles became favored by a hydrogenated oil labeling policy, granting higher consumer choosing. As for linoleic flavor, demand persists in direct consumption. Acreage and production increase, reaching new production levels.

Challenges for Argentine Sunflower Competitiveness: Economic Policy, Bird Damage, and Good Management

In order to take advantage of eventual opportunities, sunflower production must increase through better crop competiveness. To accomplish such an objective, some productive difficulties must be solved:

- Relative backwardness in technology availability and adoption, compared to other crops planted in the country: No-till planting adoption, hybrid yields, wider herbicide resistance, yield stability, earlier sowing dates, technology and management, plant coverage, nutritional diagnostics, nutritional deficit, diseases susceptibility, weed and crop management techniques
- High seed and oil export taxes and low exchange rate

Table 17.J Argentina's Sunflower Value Chain SWOT

Strengths	**Opportunities**
• Strategic crop for certain environments and rotations • High productivity germplasm available • Installed crushing industry, efficient and competitive • Variety of acidic profiles available (especially tailored for different industrial demands) • Established highly competitive R & D teams • Coordinated value chain • Production enters global market in counter season.	• Potential new uses for specific industrial demands • Probable increase in oil demand • Access to new markets • Increased production of biofuels from vegetable oils • Higher sunflower flour and pellets transformation into meat (pork, chicken, fish, etc.)
Weaknesses	**Threats**
• Low crop competitiveness compared to other crops • Lack of future markets/risk hedging markets in Argentina and lack of a unified reference price • High freight incidence in some productive regions • Important market distortion caused by export taxes • Grain contamination with insecticides, protection barriers from buying and competing countries	• Sensitive production increases of competing countries

Source: ASAGIR, 2012, unpublished.

In order to overcome these challenges, ASAGIR has agreed on the following goals:

1. Short-term goals: technology gaps
   - To reduce the technology gap by promoting adoption of available technology
   - To adjust sunflower technology to ease no-till planting adoption
   - To enhance resistance to *Plasmopara halstedii, Sclerotinia sclerotiorum, Phoma oleracea,* and *Verticillium dahliae*

- To continue with hybrid testing networks
- To develop tools and crop-management techniques to reduce the impact of plant diseases, combining genetic resistance, sowing dates, planting systems, and insecticide use
- To intensify the use of molecular markers in order to identify traits for disease resistance, drought tolerance, and oil quality, among others
- To adjust weed control and management methods, especially under no-till planting conditions
- To consolidate a germplasm bank for sunflower, in order to ease genetic resource availability in breeding programs
- To develop alternative technologies for pest control during storage in order to eliminate residue-related problems

2. Mid- and long-term goals

- To broaden sunflower genome knowledge and promote interaction with research groups in other countries
- To promote physiological studies to explain yield variability in sunflower
- To identify causes of empty seeds
- To enhance use of conjugated linoleic acid (CLA) in cattle production to increase the nutritional quality of beef
- To improve silage bag technologies to help commercialization

Conclusion

Argentina is by far the largest producer of sunflower in South America. The history of the crop, which was introduced to this country more than 80 years ago, traces the development of a huge value chain that grows from scientific and technical bases to the formation of modern exchange markets. Local farmers have adopted peak technologies, and for this reason yields are among the highest around the globe. The crushing sector developed a high volume capacity since 1998 based on an export-oriented model of investment.

The sunflower area downfall at the end of last century can be attributed to a crisis based on global oversupply of oilseeds that was fostered by the soybean boom and the incorporation of palm oil, as well as the oversupply of Argentinean sunflower. The logical result was a huge drop in prices that motivated producers to leave the crop. In 2003, canola oil, the first competitor of sunflower in markets that demanded high quality oils, had begun to be used to fulfill the EU biodiesel quota. Sunflower prices partially recovered, along with farmers' trust. Nevertheless, the 2008 crisis was a setback for oilseed price recovery, with oil-based crops suffering a much greater impact than other less elastic commodities, such as protein-based crops, especially soybean, which was sustained by the strong demand from China.

Since then, the sunflower competiveness has been compromised. Alongside the aforementioned challenges, the strong evolution of the Argentinean soybean cluster (located along Rosario city, close to the most fertile lands in the country) set the basis for the advantages of assigning this area to soybean production or wheat–soybean or wheat–maize crop sequences, relegating sunflower to distant and less productive areas. To make matters worse, other issues arose, such as oppressive export duties (30% to 32% of FOB price), elevated growth of pigeons and other bird pest populations (Canevelli, 2010), and commercial restrictions in the EU due to insecticide traces. Moreover, agriculture progress in the Ukraine and Russia, closer to the EU, forced Argentina sunflower oil commercial options to lesser markets.

The challenges that Argentina's sunflower value chain faces are many. In spite of the excellent human capacity to organize its value chain, strong efforts must be made to regain lost area, economic sustainability, and appropriate profit balance. These challenges may rely on the sunflower crop's natural advantages: adaptation and efficiency in drought conditions that competing crops can't tolerate. Nevertheless, it might be impossible to regain the virtuous cycle that allowed Argentina to become the lead supplier of meal and oil, if tax discrimination and continuous low real exchange rate persist. In addition, the Argentinean peso has been under an artificial overvaluation that has lessened the competiveness of Argentina's entire economy.

In conclusion, it can be said that international prospects for high quality oils are favorable. We believe that human consumption demand for sunflower oil, at similar value with soybean oil, will grow significantly.

Argentine sunflower expansion depends, in local terms, on crop competitiveness, growth, and the tax policy evolution. Eastern Europe's production progress and its influence over sunflower export product prices are the external conditioning factors. The only way Argentina may take advantage of eventual opportunities will be with improvements in technology availability and adoption. However, this option is presently discouraged by the previously mentioned tax policies that don't favor input-intensive use strategies that bring about higher yields. Once the best available technology has been adopted, a reduction of yield gap will follow, increasing added value and having a similar impact on the gross domestic product. Political changes that would allow the Argentine sunflower complex to regain competitiveness are significant and should be made as soon as possible. The present insufficient number of hectares may reach at least the level seen before the global crisis of 2008, and there would be opportunities for offering in bulk, as in the refined or specialties world markets.

References

ASAGIR (Asociación Argentina de Girasol) Home Page. http://www.asagir.org.ar (accessed June 11, 2013).

Bolsa de Cereales de Buenos Aires. Número estadístico 2010-2011. Buenos Aires, Argentina, 2012.

Bolsa de Cereales de Buenos Aires Home Page. http://www.bolcereales.com.ar (accessed May 8, 2013).

Cámara Argentina de Biocombustibles Home Page. http://www.carbio.com.ar (accessed January 22, 2014).

Canavelli, S. B. *Manejo del daño por aves en girasol: Estado actual de las investigaciones en Argentina.* Workshop ASAGIR 2010, Mar del Plata, Argentina, Mar 30–31, 2010.

Chapman, S. C.; de la Vega, A. J. Spatial and Seasonal Effects Confounding Interpretation of Sunflower Yields in Argentina. *Field Crops Res.* **2002,** *73,* 107–120.

Chimenti, C.A.; Pearson, J.; Hall, A. J. Osmotic Adjustments and Yield Maintenance under Drought in Sunflower. *Field Crops Res.* **2002,** *75,* 235–246.

CIARA–CEC (Argentine Chamber of Fats and Oil) Home Page. http://www.ciaracec.com.ar (accessed April 17, 2013).

Feoli, C. Evaluación de factores limitantes de la producción de girasol en el centro-oeste de Buenos Aires. In *Proceedings of Primera Reunión Nacional de Oleaginosos,* Rosario, Argentina, October 10–11, 1991a.

Feoli, C. *Los rendimientos de girasol en el Centro Oeste de la Prov. de Buenos Aires: Un análisis conceptual*; Carpeta de Producción Vegetal EEA Pergamino, Volumen X: Pergamino, BA, Argentina, 1991b.

Feoli, C. *Un estudio exploratorio de factores limitantes de la producción de girasol;* Carpeta de Producción Vegetal EEA Pergamino, Volumen X: Pergamino, BA, Argentina, 1991c.

Feoli, C. (2002) Aportes del INTA al Girasol. IDIA XXI 2 (3), 119–123.

Feoli, C. Girasol Diferenciado. *Agroclaves,* Ago, 2004, 21–22.

García, A. Evolucao da cultura do girasol no Brasil. In *Manejo del cultivo, control de plagas y enfermedades del girasol;* Molestina, C. J., Ed.; IICA: Montevideo, Uruguay, 1988; pp 19–21 (diálogo XXII).

Gazzola, A.; Gonçalves Ferreira, C. T., Jr.; Cunha, D. A.; Bortolini, E.; Dias Paiao, G.; Pestana, J.; Cruanhes D'Andréa, M. S.; Souza Oliveira, M.; In *A cultura do Girassol* [Online]; Vescove Primiano, I., Ed.; http://www.lpv.esalq.usp.br/lpv506/LPV-0506%20-%20GIRASSOL%20APOSTILaO%202012.pdf (accessed April 20, 2013).

Hall, A. J.; Feoli, C.; Ingaramo, J.; Balzarini, M. Gaps between Farmer and Attainable Yields across the Sunflower Growing Regions of Argentina. *Field Crops Res.* **2012,** *143,* 119–129.

INBIO (Instituto de Biotecnologia Agricola) Home Page. http://www.inbio.org.py (accessed April 23, 2013).

Ingaramo, J.; Feoli, C. *La importancia del complejo girasol en la Argentina*; Technical Report No. 72: El cultivo de girasol en la región semiárida pampeana; Quiroga, A., Pérez Fernández, J., Eds.; Anguil, Argentina, June 2008.

Instituto de Agricultura Sostenible Home Page. http://www.ias.csic.es/ (accessed June 12, 2013).

Hinrichsen, S. A. *La capacidad de crushing de la industria aceitera.* Informativo Semanal Bolsa de Comercio de Rosario, March 15, 2013; pp 7–8.

Ledesma, L. L.; Zurita, J. J. *Suelos del Departamento 9 de Julio, Provincia del Chaco*; Carta de suelos de la República Argentina. INTA y Ministerio de la Producción del Chaco, Buenos Aires, 2004.

Lisanti, S.; Giuliano, J.; Hall, A.; Chimenti, C. *Identificación de genotipos de girasol con diferente grado de ajuste omótico usando granos de polen.* Workshop ASAGIR 2010, Mar del Plata, Argentina, Mar 30–31, 2010.

MinAgri Statistics Home Page. http://www.siia.gov.ar (accessed May 15, 2013).

Mosconi, F. P.; Priano, L. J. J.; Hein, N. E.; Moscatelli, G.; Salazar, J. C.: Gutierrez, C. T.; Caceres, L. *Mapa de suelos de la provincia de Santa Fe.* INTA/MAG (Ministerio de Agricultura y Ganadería de Santa Fe), INTA, Buenos Aires, 1981.

National Institute for Statistics and Census Home Page. http://www.indec.mecon.ar (accessed June 4, 2013).

Negri, R. Vuelve el girasol a los mejores lotes? In *Proceedings of 5 to Congreso Argentino de Girasol: Con la mirada en la competitividad,* Buenos Aires, Argentina, June 1–2, 2010.

Norma IRAM 5529. Vegetable oils. Sunflower oil. Instituto Argentino de normalización y certificación: Buenos Aires, Argentina, 2013.

Nunes, P. Conjuntura Agrícola. Produção e consumo de óleos vegetais no Brasil, 2013. http://www.deser.org.br/documentos/doc/Produ%C3%A7%C3%A3o%20e%20consumo%20de%20%C3%B3leos%20vegetais.pdf (accessed April 20, 2013).

OECD-FAO. *Agricultural Outlook 2012–2021* [Online]; OECD Publishing and FAO: 2012. http://dx.doi.org/10.1787/agr_outlook-2012-en (accessed May 6, 2013).

Oil World. *Oil World Annual 2012, Global Analysis of all Major Oilseeds, Oils & Oilmeals Supply, Demand and Price Outlook;* Mielke, T., Ed.; Hamburg, Germany, 2012; Vol. 1.

Pearson, J. I. Efecto del nivel de expresión del ajuste osmótico sobre el número y peso de los granos en girasol expuesto a condiciones de estrés hídrico. Tesis de grado, Universidad de Buenos Aires, 2002.

Pereyra Irujo, G.; Covi, M.; Izquierdo, N.; Nolasco, S. M.; Quiróz, F.; Lúquez, J.; Aguirrezábal, L. *Calidad de aceite de girasol para la producción de biodiesel en Argentina un estudio por simulación*; INTA EEA Balcarce: Balcarce, Argentina, 2007.

Rodríguez, E. Políticas gubernamentales en el manejo del daño de aves en cultivos de girasol. In *Proceedings of 5 to Congreso Argentino de Girasol: Con la mirada en la competitividad,* Buenos Aires, Argentina, June 1–2, 2010.

Ruta 0 Home Page. http//:www.ruta0.com (accessed April 2013).

Sala, C. A.; Bulos, M.; Altieri, E. *Tecnología Clearfield en Girasol: Pasado, presente y perspectivas.* Workshop ASAGIR 2010, Mar del Plata, Argentina, Mar 30–31, 2010.

Sammarro, D. G; Brihet, J. M.; Costa, R.; Puricelli, E. Resultados Generales: Girasol. In Relevamiento de tecnologia Agricola aplicada: Campaña 2010/11; Eds. Bolsa de Cereales; Buenos Aires, 2013; pp 46–47.

Souto, G. *El cultivo de girasol en Uruguay.* Conference 1er Encuentro Nacional de Girasol, Mercedes, Uruguay, June 2007.
Universidad de Sevilla Home Page. http://www.us.es (accessed June 14, 2013).
USDA. *Oilseeds: World Market and Trade* [Online]; Circular series. March 2013, http://usda01.library.cornell.edu/usda/fas/oilseed-trade//2010s/2013/oilseed-trade-03-08-2013.pdf (accessed July 25, 2013).

18

Sunflower Production in the European Union

Leonardo Velasco and José M. Fernández-Martínez ■ *Instituto de Agricultura Sostenible (IAS-CSIC), Alameda del Obispo, Córdoba, Spain*

Juan Fernández ■ *Limagrain Ibérica, Sevilla, Spain*

Introduction

Sunflower (*Helianthus annuus* L.) was domesticated around 3000 B.C. in North America and became a staple food for Native Americans. The species was introduced in the Old World by Spanish explorers in 1510, where it was first used as ornamental plant and snack food (Putt, 1997). Large-scale use of sunflower for oil production started in Russia in the 1830s (Skoric, 2012). In Western Europe, sunflower production started to extend to Spain in the late 1960s and to France in the late 1970s. Sunflower acreage in both countries showed important fluctuations in subsequent years (see Figure 18.1), mainly attributed to the changes happening in the European Union (EU) common agricultural policy (CAP). Such fluctuations did not affect some countries, such as Hungary (see Figure 18.1), which became a member of the EU in 2004.

Current Situation of Sunflower Production in the European Union

The average sunflower acreage in the EU-28 from 2008 to 2012 was 4 million ha, which was about 16.2% of the world acreage. Seed production was 7.3 million t, representing 20.5% of the world production (see Table 18.A). The countries with the largest sunflower acreage in the EU are Romania, Spain, Bulgaria, France, and Hungary, which together account for more than 90% of the EU sunflower acreage and also close to 90% of the EU sunflower production. Crushing capacity is in accordance with seed production, with around 6.3 million t of sunflower seeds crushed per year, which represents 19.3% of the world crushing capacity (Oil World, 2013). Total sunflower oil production per year in the EU between 2008 and 2012 was 2.47 million t; this was mainly concentrated in France (0.55 million t), Spain (0.37 million t), Hungary (0.31 million t), Romania (0.27 million t), The Netherlands (0.21 million t), and Germany (0.19 million t). The Netherlands and Germany produce small

---

This chapter does not consider sunflower producing countries such as Bulgaria, Croatia, Greece, and Romania, which are included in the chapter on Eastern European zone, except when the global EU figures are calculated. Part of the information provided in this chapter was obtained from technical personnel of several seed and oil companies, who preferred to remain anonymous.

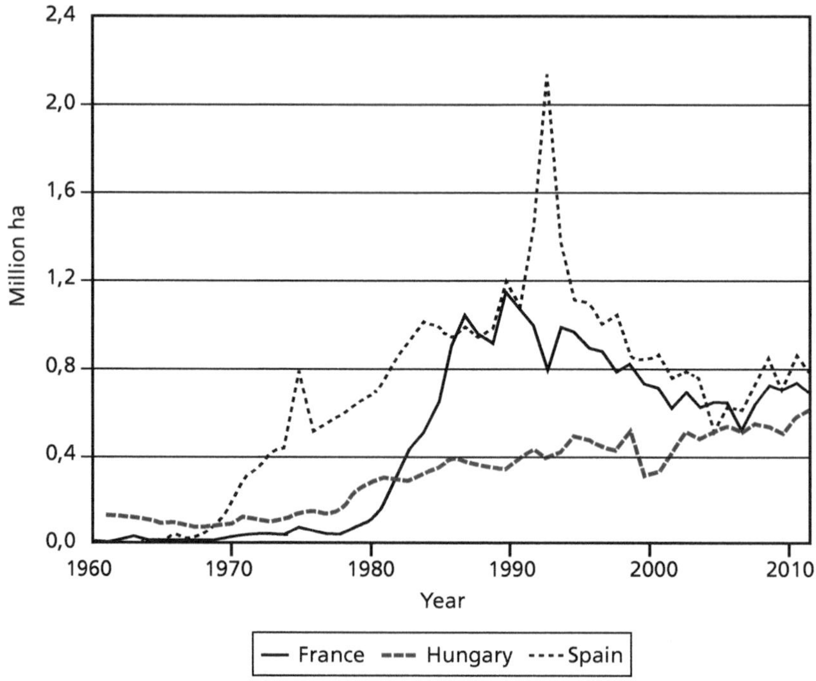

Figure 18.1 Sunflower acreage in France, Hungary, and Spain from 1961 to 2012.

amounts of seed (or no seed), so their crushing industry is based on intra-EU seed import (Oil World, 2013). The EU, mainly due to Romania and Bulgaria, is the main world exporter of sunflower seed, with 0.6 million t per year in the past five years, accounting for 35% of total world exports (Oil World, 2013).

The relative contribution of the EU to the world's sunflower production relies on a higher average seed yield recorded in the EU (1.8 t/ha), when compared to the world average (1.4 t/ha). There are large differences in sunflower yield between different EU countries, with particularly low yields in Portugal (0.6 t/ha) and Spain (1.1 t/ha), and to a lesser extent in Greece and Romania (1.5 t/ha). The case of Spain, with more acreage but about half the production achieved in France, is a paradigm of the limitations for sunflower production in some EU areas. The particularities of the main sunflower producing countries in the EU are outlined next.

Spain

Sunflower represents the largest oilseed production in Spain. It is mainly cultivated in three regions: Andalucía, Castilla y León, and Castilla La Mancha, which together account for about 95% of the total national production (Figure 18.2, p. 558), (MA-

GRAMA, 2012). Ninety percent of sunflower production is under dry land conditions, which produce low seed yields due to the need for late sowing after the freezing period and the occurrence of heat and drought stress during seed filling (Domínguez-Giménez, 1997). Average seed yield in Spain is around 1.1 t/ha (Table 18.A). There is some variation in seed yields between the three main production regions: 1.4 t/ha in Andalucía, 1.2 t/ha in Castilla y León, and 0.8 t/ha in Castilla La Mancha (MAGRAMA, 2012). Besides drought conditions during seed filling, the main problem limiting sunflower yield in Spain is the occurrence of broomrape (*Orobanche cumana* Wallr.) infestations, particularly in Andalucía and Castilla La Mancha, but spreading rapidly to new areas in recent years (Pineda-Martos et al., 2013). The large majority of hybrids cultivated in Andalucía and Castilla La Mancha are resistant to sunflower broomrape race F, which is predominant in both regions. The use of imidazolinone (IMI)-resistant hybrids, which are an alternative for controlling both broomrape and weeds, has increased in the last 10 years, representing a market share of 9% in 2013. Downy mildew (*Plasmopara halstedii*) was first identified in Spain and described in 1973 (Jimenez-Diaz, 1973). It was the main disease problem in the 1970s (Sackston, 1978), but after the introduction of genetic resistance and fungicide treatments, the

Table 18.A Average Sunflower Acreage, Seed Yield, and Seed Production in the European Union from 2008 to 2012

Country	Acreage (ha)	Yield (kg/ha)	Production (t)
Austria	25,496	2687	68,786
Bulgaria	732,635	1908	1,396,498
Croatia	31,190	2790	87,704
Czech Republic	260,90	2356	61,445
France	693,189	2425	1,680,764
Germany	25,335	2126	53,808
Greece	41,901	1450	60,527
Hungary	556,206	2293	1,277,066
Italy	113,710	2129	242,788
Poland	2784	1746	4,843
Portugal	20,902	574	12,094
Romania	882,790	1522	1,343,688
Slovakia	83,912	2218	185,605
Slovenia	285	1911	570
Spain	780,782	1113	867,680
EU	4,017,207	1828	7,343,866
World	24,737,202	1446	35,776,071

Source: FAOSTAT (2014).

problem was temporarily well solved. However, there has been a significant incidence of this disease in the last decade, particularly under favorable environmental conditions for infection, due the appearance of new more-virulent races and the pathogen's built up resistance to fungicides (García-López et al., 2008). Unlike in other markets such as the United States and France, the use of mid-oleic or high oleic acid hybrids is still very low in Spain, representing less than 15% of the total acreage in 2013.

France

Unlike Spain, the main oilseed crop in France is rapeseed, with around 1.5 million ha compared to 0.7 million ha covered by sunflower (PROLEA, 2011). Sunflower production is mainly concentrated in five regions (Figure 18.3), which together account for more than 80% of the whole production (PROLEA, 2011). The two main regions are Midi-Pyrénées and Poitou-Charentes. The main sunflower production areas are under oceanic or semi-oceanic climates, with fairly even distribution of precipitation throughout the year. This leads to high yields compared to regions with Mediterranean climates, such as Spain. However, such environmental conditions are more favorable for fungal diseases, mainly downy mildew, white rot (*Sclerotinia sclerotiorum*), phomopsis (*Diaporthe helianthi*), and phoma (*Phoma macdonaldii*) (Vear et al., 2007). Sunflower broomrape is currently not a major problem for sunflower pro-

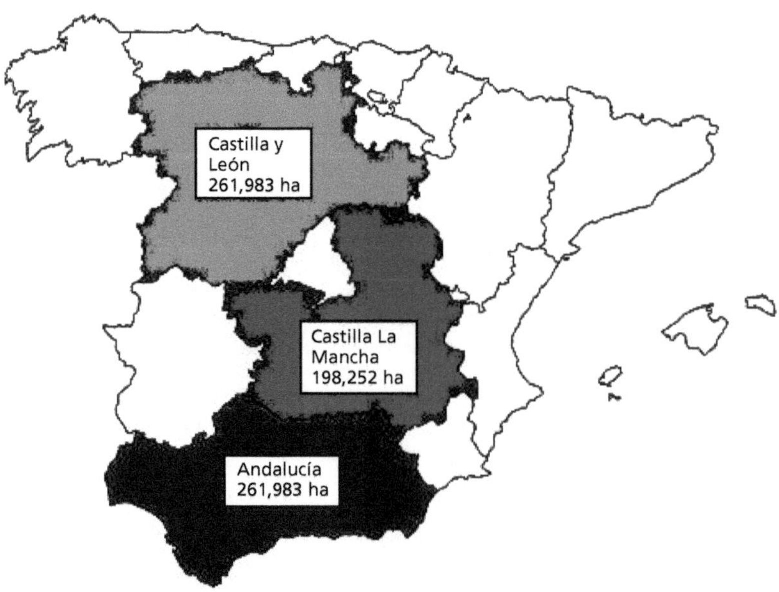

Figure 18.2 Main sunflower production regions in Spain in 2011.

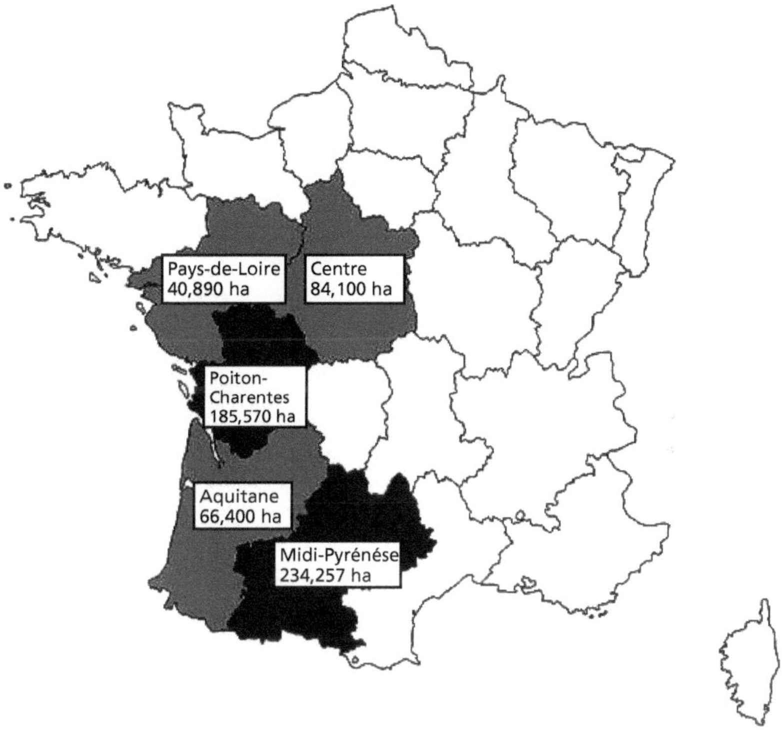

Figure 18.3 Main sunflower production regions in France in 2011.

duction in France, although the existence of field infestations has been recently reported and there is concern about this parasite spreading on a large scale (Jouffret and Lecomte, 2010). There is no information so far on the predominant races of sunflower broomrape in France. A notable characteristic of sunflower production in France is the relevance of high oleic acid hybrids, which account for more than 50% of the total sunflower production (Kleintgartner, 2011; Labalette and Raoul, 2012).

Hungary

Sunflower is the main oilseed crop in Hungary, although there is also significant production of rapeseed. In 2012, the sunflower acreage was 615,079 ha compared to 164,916 ha of rapeseed (FAOSTAT, 2014). Sunflower production is mainly concentrated in the Northern Great Plain, Southern Great Plain, and Northern Hungary regions (Figure 18.4), which together account for more than 65% of national production. Average seed yield is high, close to 2.3 t/ha (Table 18.A). High oleic

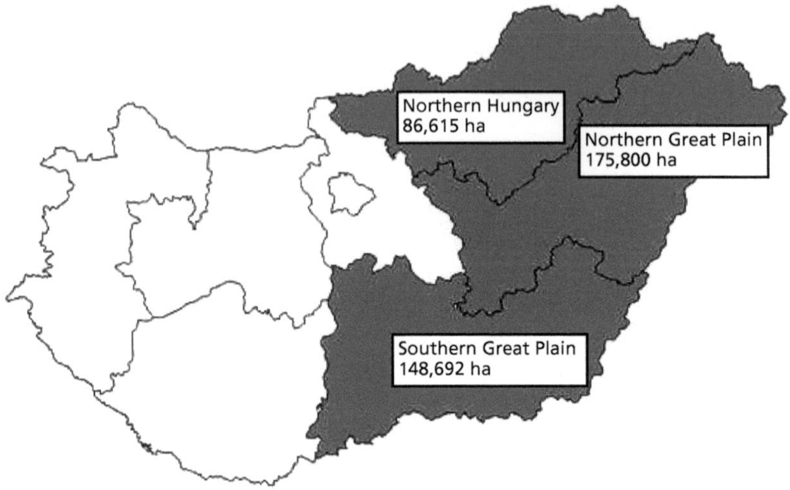

Figure 18.4 Main sunflower production regions in Hungary in 2012.

hybrids have a small market share in Hungary, generally around 5%, although there are yearly variations. Confectionery hybrids also account for around 5% of the market. One important characteristic of sunflower production in Hungary is the predominance of herbicide-tolerant hybrids, which represented around 80% of the whole market in 2013. Fungal diseases are the most limiting factors for sunflower production (Szabó, 2007). Genetic resistance to downy mildew is mandatory for all hybrids registered in Hungary. Sunflower broomrape is widely distributed, but infections are not particularly important. Its incidence is more important in the Southern Great Plain region, especially in dry years. To the best of our knowledge, there are no reports on the occurrence of races beyond race E.

Italy

Sunflower production in Italy is concentrated in three regions of the central part of this country: Marche, Toscana, and Umbria (Figure 18.5), which together account for more than 70% of the total sunflower acreage. Seed yields are high and more or less similar in these three regions: 2.30 t/ha in Marche, 2.15 t/ha in Toscana, and 2.06 t/ha in Umbria (AGRIT, 2011). High oleic acid hybrids have an important position in the market, with a share around 35%. White rot caused by Sclerotinia and downy mildew represent the main diseases, although most of the cultivated hybrids possess genetic resistance to the latter one. Broomrape infestations have not been reported. The market share for herbicide-tolerant hybrids is similar to that in France, around 17%.

Figure 18.5 Main sunflower production regions in Italy in 2011.

Types of Sunflower Hybrids

Currently, sunflower hybrids in the EU area are classified according to the following characteristics: conventional (high linoleic), high oleic, broomrape resistant, and herbicide tolerant. Confectionery hybrids have a small presence in the EU, which makes the EU (particularly Spain) a large importer of confectionery sunflower seeds (Nabloussi et al., 2011).

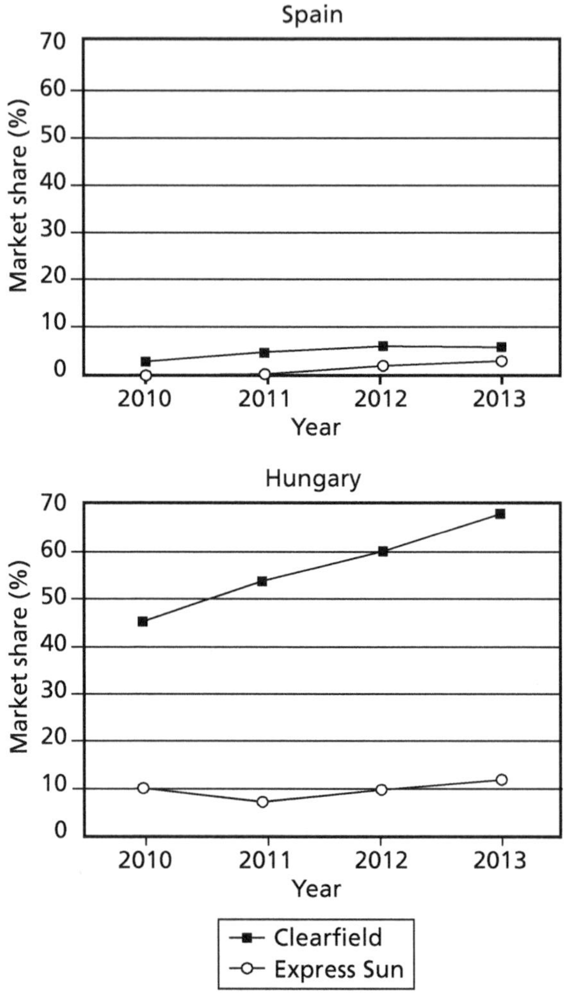

Figure 18.6 Evolution of the market share for herbicide tolerant hybrids (Clearfield and Express Sun) in Spain, France, Hungary, and Italy from 2010 to 2013.

Herbicide-tolerant hybrids are in turn divided into two different classes: tolerant to imidazolinones (IMI) and tolerant to tribenuron methyl or sulfonil urea (SU). Hybrids tolerant to IMI are marketed under the Clearfield trademark, whereas those resistant to tribenuron methyl are marketed under the Express Sun trademark. Figure 18.6 shows the evolution of both types of herbicide-tolerant hybrids in Spain, France, Hungary, and Italy from 2010 to 2013. Herbicide-tolerant hybrids are particularly important in Hungary, which has a 68% market share for Clearfield and a 12%

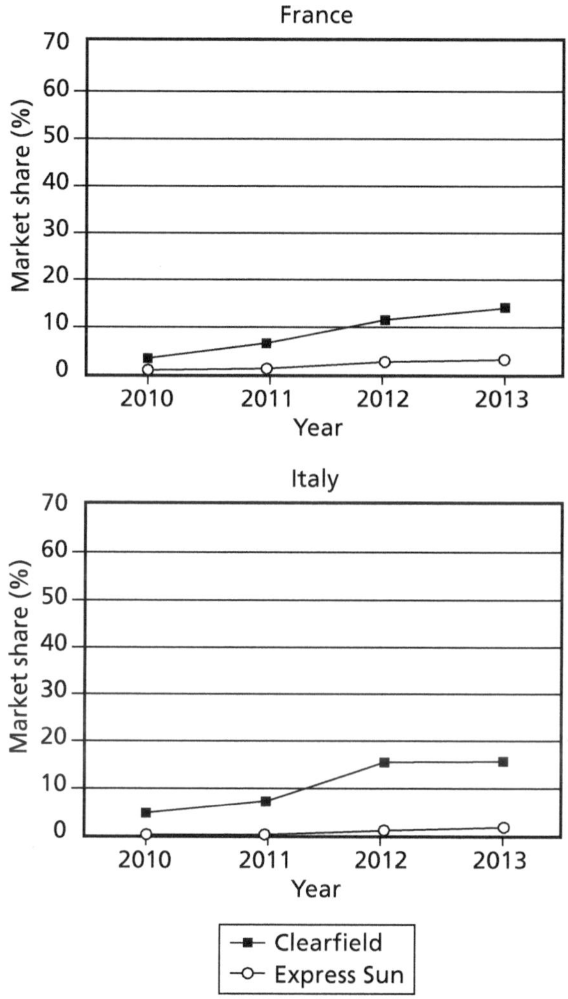

Figure 18.6 *Continued*

market share for Express Sun, together accounting for 80% of the sunflower seed marketed in this country in 2013. At a lower scale there is also a growing trend for both types of herbicide-tolerant hybrids, mainly Clearfield, in France and Italy, and to a lesser extent in Spain.

The use of broomrape-resistant hybrids is particularly important in Spain, where it is estimated that 45% of sunflower acreage is infested by broomrape race F. This percentage rises to 85% in the Guadalquivir Valley area of Andalucía (see Figure 18.2), which represents more than 40% of the Spanish sunflower production. All hybrids marketed in this area are resistant to broomrape race F or are tolerant to IMI herbicides, which allow effective broomrape control (Alonso et al., 1998). Nonetheless, Clearfield technology is having a slow penetration in the Spanish market. Another area with heavy infestation of broomrape race F is Cuenca province in Castilla La Mancha (Figure 18.2). In France, broomrape was absent until a few years ago (Jouffret and Lecomte, 2010). Currently, the broomrape infested area is not more than 3%, but heavily infested plots have been observed in recent years, particularly in the Midi Pyrénées region; this is the region with the largest sunflower acreage (Figure 18.3). Sunflower broomrape is also present in other western regions such as Poitou-Charentes and Pays de Loire. Race F has been identified in the southwestern area, which is promoting cultivation of race F–resistant hybrids throughout the whole sunflower area in western and southwestern France. Broomrape is not a major problem in Hungary, where race E is still predominant. It is mainly controlled through Clearfield technology, since Clearfield hybrids account for 68% of the total sunflower acreage in the country. No broomrape infestations have been detected in Italy so far. The development of Clearfield hybrids that also carry broomrape resistance genes, thus conferring double protection against the parasite, is a new strategy that is being currently introduced by the main sunflower seed companies. This strategy is expected to reduce the selection pressure leading to the development of new virulent races while simultaneously allowing adequate weed control.

High oleic acid hybrids are particularly important in France (>50% market share) and Italy (>30%) and to a lesser extent in Spain and Hungary (Figure 18.7). Market growth from 2010 to 2013 has been small or nonexistant. Current forecasts do not anticipate a major growth in the high oleic acid market in these countries in the short term.

Markets

The EU is the third-largest worldwide sunflower crusher, after the Ukraine and Russia. Annual sunflower crushing in the EU is around 19% of the total world sunflower oilseed production, whereas sunflower seed production is around 20.5% (FAOSTAT, 2014; Oil World, 2013). Therefore, the EU is a net exporter of sunflower seed. Annually, the EU crushes the equivalent of 84% of its internal sunflower production. Net seed export (excluding internal EU trade) equals around 4% of sunflower production. The main exporters within the EU are Bulgaria and Romania, which together account for 84% of EU exports. Most of the seed (59%) is exported to Turkey. The main importers are Spain (18% of total EU imports), followed by The Netherlands (15%), and Germany (13%). The two main crushers in the EU are France (1.3 million t; 21%) and Spain (1.1 million t; 18%). Hungary and Italy crush 0.8 and 0.4 million t

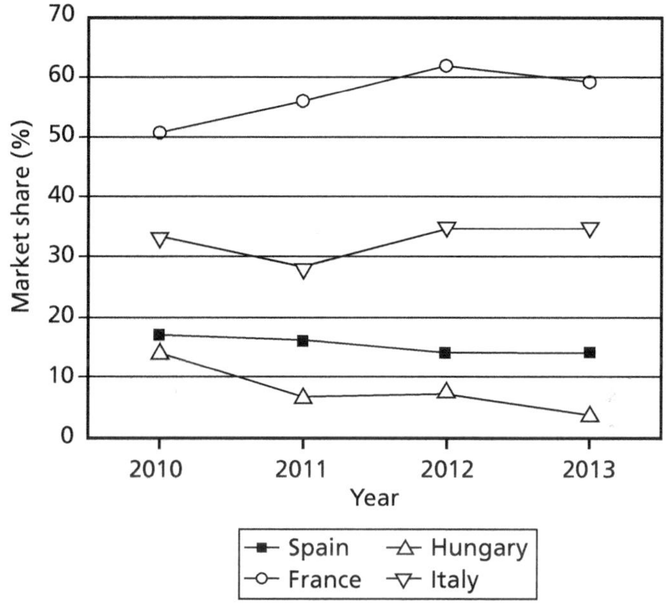

Figure 18.7 Evolution of the market share for high oleic hybrids in Spain, France, Hungary, and Italy from 2010 to 2013.

annually, respectively. It is noteworthy that The Netherlands, with no sunflower seed production, crushes 0.5 million t annually. Sunflower crushing in this country mainly takes place around the Rotterdam port (CBI, 2012).

Unlike sunflower seed, the EU is a net importer of sunflower oil and sunflower meal. Net sunflower oil imports per year were 0.8 million t in the past four years. The large majority of the oil is imported from Ukraine (62%), Russia (14%), and Argentina (10%). Annual sunflower meal imports have averaged 0.3 million t in the past four years, mainly also from Ukraine (59%), Russia (24%), and Argentina (14%) (FAOSTAT, 2014; Oil World, 2013).

Sunflower oil prices in the EU have risen considerably (Figure 18.8). Considering averages over 5-year periods, the average price has risen from $514/t in 1999 to 2003, to $908/t in 2004 to 2008, to $1135/t in 2009 to 2013 (Oil World, 2013).

European Union Common Agricultural Policy

The current European Union integrated by 28 member states traces back to the European Economic Community, created by the Treaty of Rome in 1957. That treaty,

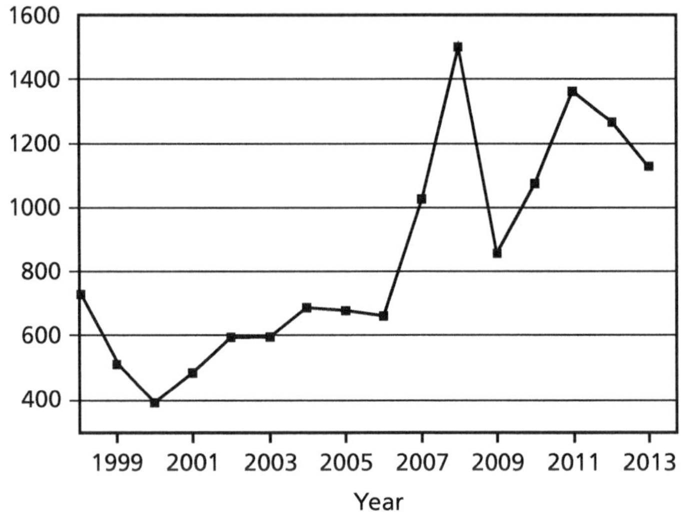

Figure 18.8 Evolution of sunflower oil prices (annual averages) in the EU from 1998 to 2013.

signed by Belgium, France, Italy, Luxembourg, The Netherlands, and West Germany, foresaw a common agricultural policy (CAP) that started to be implemented in 1962 (European Commission, 2012). The CAP was initially oriented to supporting minimum pricing for agricultural commodities as a way to increase agricultural production and enhance the standard of living of farmers. Because that policy led to unbalanced production and subsequent surpluses in several commodities, the model shifted in 1993 from market support to producer support, in the form of direct payments to farmers. This had a huge initial impact on sunflower production in some countries such as Spain, where sunflower acreage in 1993 doubled over previous years, reaching more than 2 million ha (Figure 18.1). Acreage reduction in 1994 and subsequent years was due to stabilization measures in which the payment was reduced in proportion to the excess of the maximum guaranteed area imposed by the EU/USA Blair-House Agreement (Alonso, 2008). Such payments, initially coupled with production, were decoupled in 2005 through the single farm payment (SFP) system. Receipt of the SFP was made partly dependent on the fulfilment of food safety, environmental, and animal welfare standards.

In general terms, the CAP had a significant influence on sunflower production in the two countries that entered the EU several decades after its creation: Spain (1986) and Hungary (2004). In the case of Spain, average sunflower acreage in the 5-year period previous to EU accession was 900,000 ha comparing to an average acreage of approximately 700,000 ha for the period of 2005 to 2012, after the last big reform of the CAP. Conversely, sunflower acreage in Hungary increased by 150,000 ha when comparing equivalent periods of time. This can be explained by several years of insufficient profitability of the crop shortly before EU accession (Slezák et al., 2003).

Currently, the EU is discussing a new reform of the CAP for the 2014 to 2020 period. In this reform, there will be more flexibility for member states in the budgeting and implementation of financial instruments. The main scheme, in which payments to farmers are decoupled from production, will continue; in some vulnerable sectors, however, a small percentage of the payment will be linked to production. Direct payments will be coupled with other measures to promote rural development, environmental protection, and internal convergence. They will be limited to active farmers and divided into a basic payment, a green payment associated with environmentally friendly production (30% of the direct payment), and other support measures related to areas of natural constraints, young farmers, small farms, and so forth (European Commission, 2013).

Research and Supporting Organizations

Research

Sunflower research in the EU has played a major role in the development of this crop. For example, the cytoplasmic male sterility universally used today for commercial hybrid production was discovered in France (Leclercq, 1966). France also made major achievements on genetic resistance to fungal diseases. Research groups in Romania and Spain have been at the forefront of sunflower broomrape research since the 1980s.

In France, the main institution conducting sunflower research is the National Institute for Agronomic Research (Institut National pour la Recherche Agronomique, INRA). Historically, the three main INRA sites conducting sunflower research were Montpellier, mainly focused on genetic and molecular studies, wild species conservation, characterization and utilization, and development of new sources of cytoplasmic male sterility (Lacombe and Bervillé, 2001; Serieys, 2002); Clermont-Ferrand, with major specialization in genetic resistance to fungal diseases, especially the development of durable resistance to downy mildew (Tourvieille et al., 2004; Vear et al., 2007); and Toulouse, concentrated on agronomy and sunflower physiology (Blanchet et al., 1981). In 2010, the INRA center in Toulouse was selected to host the Sunflower 2010 National Program that focused on revitalizing public sunflower research in important areas such as agronomy, ecophysiology, phytopathology, genetics, and genomics (INRA, 2013). INRA also collaborates closely in sunflower research with other institutions such as universities, for example, the School of Engineers (l'Ecole d'Ingénieurs) PURPAN in Toulouse and the National Center for Scientific

Research (Centre National de la Recherche Scientifique, CNRS). In addition to the development of basic germplasm with relevant traits, INRA has been involved in the development of parental inbred lines that were released to private seed companies and used to develop hybrid cultivars that achieved important market shares.

Sunflower research in Spain is more scattered across several institutions. The main research on genetics, breeding, agronomy, and phytopathology was initially conducted at the National Institute for Agricultural Research (Instituto Nacional de Investigaciones Agrarias, INIA), which was largely decentralized to regional governments in the early 1980s. After that, a solid group of researchers from two institutions, the Institute for Sustainable Agriculture (Instituto de Agricultura Sostenible, IAS) of the National Scientific Research Council (Consejo Superior de Investigaciones Científicas, CSIC) and the Agrarian and Fishing Technologies Research and Education Institute of Andalusia (Instituto Andaluz de Investigación y Formación Agraria, Pesquera, y Alimentaria, IFAPA) interacted at Córdoba as the main research group in Spain on the previously mentioned disciplines. Other research groups working actively on sunflower mainly belong to CSIC and universities, for example the Instituto de la Grasa (CSIC), which has played in major role in the development and characterization of fatty acid sunflower mutants (Osorio et al., 1995). In the 1970s and early 1980s, public breeding was involved both in the development of genetic materials with improved characteristics and also in developing parental lines and cultivars that were transferred to the private sector. However, from the middle 1980s onward, there was a declining involvement in cultivar development by public institutions, which switched the emphasis to the development of basic germplasm with novel traits such as modified fatty acid and tocopherol profiles (Pérez-Vich et al., 1999; Velasco et al., 2004) or new sources of broomrape resistance (Velasco et al., 2012).

Sunflower research in other countries such as Hungary and Italy has been less intense or at least less known internationally. Important research on phytopathology has been conducted at the Plant Protection Institute of Szent István University at Gödöllo, Hungary (Virányi, 2008; Virányi and Masirevic, 1991). Similarly to other eastern countries such as Romania, Bulgaria, and Serbia, a public company, the Cereal Research Institute at Szeged (Hungary), carried out development and marketing hybrid cultivars, but at the present time multinational seed companies are installed in Hungary and have achieved a remarkable market share. Sunflower research in Italy has been led by research groups of several universities such as the Università degli Studi di Pisa, the Università degli Studi della Basilicata in Potenza, and the Università di Udine. They have carried out specific research on agronomy and drought tolerance (Baldini and Vannozzi, 1999), disease resistance (Baldini et al., 2002), and oil uses (Tahmasebi Enferadi et al., 2006) and have also participated in the introduction of hybrid cultivars in Italy. Significant advances in sunflower research have been carried out in Germany, too, where a research group directed by Professor Wolfgang Friedt at the Institute of Crop Science and Plant Breeding of the Justus-Liebig Universität of Giessen pioneered molecular and biotechnological investigations in sunflower (Friedt et al., 1991; Rönicke et al., 2005). Finally, there has also been relevant basic research on this crop in the Czech Republic by the Research Institute of Crop Production of Prague-Ruzyne, where numerous studies on the genetics of important sunflower traits have been conducted (Kovácik and Skaloud, 1992).

The V.S. Pustovoit Award is the highest honor conferred to individuals working in the sunflower industry, and it is awarded by the International Sunflower Association (ISA). Several public and private researchers from current EU countries have received the V.S. Pustovoit Award: Dr. P. Leclercq (France) and Dr. A. Vranceanu (Romania) in 1980; Dr. G. Piquemal (France) and Dr. G. Teodorova (Bulgaria) in 1992; Dr. A. Kovacic (Czech Republic) in 1996; Dr. H. C. Iliescu (Romania) and Dr. A. Benvenuti (Italy) in 2000; Dr. G. P. Vannozzi (Italy), Dr. J. M. Fernández-Martínez (Spain), Dr. F. Vear (France), and Dr. Florin Stoenescu in 2004; Dr. Ferenc Viranyi (Hungary) and Dr. L. Cuk (France) in 2008; and Dr. M. Christov (Bulgaria), Dr. J. Domínguez (Spain), and André Pouzet (France) in 2012.

Supporting Organizations

The main worldwide supporting organization for sunflower is the International Sunflower Association, which has its current headquarters in France. In that country there is also a well-organized network of organizations supporting oilseeds named PROLEA (www.prolea.com), which includes the French Federation of Producers of Oil and Protein Crops (Féderation Française des Producteurs d'Oléagineux et de Protéagineux, FOP), the Technical Center for Oilseed Crops and Industrial Hemp (Centre Technique Interprofessionnel des Oléagineux Metropolitains, CETIOM), the National Professional Association of Oilseeds (Organisation Nationale Interprofessionnelle des Oléagineux, ONIDOL), and Sofiprotéol, which is the financial branch of the network. The mission of FOP is to defend the interests of producers at national and international levels. CETIOM has as main role to enhance the economic competitiveness of oilseed crops and oilseed products. ONIDOL conducts research to promote oilseed crops as well their markets and processing industries. In Spain, there is a Spanish Sunflower Association (Asociación Española del Girasol, AEG) with the main goal of promoting sunflower development in this country. Like the ISA, the AEG is not formed by growers or industries, but by individuals who do not defend particular interests other than the general interest of the sunflower crop and industry.

An important worldwide organization supporting sunflower research has been the FAO Sunflower Research Network, in which EU countries collaborated actively. It was established at a consultation meeting held in Bucharest in October 1975 (Vranceanu, 1978). The first coordination center of the network was the Research

Institute for Cereal and Industrial Crops of Fundulea, Romania, and the coordinator was Professor Viorel Vranceanu (1975–1990). After 1990, the coordination center was the Institute of Field and Vegetable Crops, Novi Sad, with Professor Dragan Skoric as coordinator. The last technical consultation meeting was held in 2005 in Novi Sad, Serbia. The main purpose of this research network was to undertake joint applied research on sunflower with division of tasks and adequate methodology, as well as exchange of experience, information, and germplasm among participating institutions. The structure of the network was a coordination center and a number of liaison centers conducting different subnetworks on different topics (disease control, plant physiology, breeding and genetics, experimentation of hybrids, biotechnology, and wild sunflower species). Research institutions of countries of the EU (France, Spain, Hungary, Czech Republic, and Germany) have participated actively as liaison centers of different subnetworks. One important issue of the FAO Research Network has been the publication of the scientific journal *Helia,* which started in 1978 as an information bulletin of the FAO and is currently an important scientific journal for sunflower researchers and industry.

Conclusion

Sunflower plays a major role in rotations with cereals in the EU. Having a deeper root system than cereals, sunflower uses part of the fertilization not taken by cereals, thus preventing leaching and subsequent groundwater contamination. Therefore, it is expected that sunflower will continue being a key crop in the EU regions where it is currently grown. The evolution of sunflower acreage will depend on several factors. The incorporation of a green direct payment in the new CAP will favor sunflower cultivation but also other groups of crops such as legumes, which will compete with sunflower in rotations with cereals in some regions. Demand for biofuels has become one of the driving forces for vegetable oil markets in recent years. Changes in the biofuel policy in the EU, as well as developments in the next generations of biofuels, will have an impact on the evolution of the demand of sunflower oil for nonfood uses (OECD-FAO, 2012). Public and private research has played and will continue to play an important role in the development of sunflower in the EU. Breeding advances will also determine the evolution of sunflower acreage and yields in the EU. In particular, resistance to fungal diseases and broomrape will continue to be a key aspect of sunflower breeding. The use of herbicide-tolerant hybrids enables growers to face the objective of advancing sowing time as a means of gaining yield, especially in the hot Mediterranean countries affected by drought conditions, although this will require breeding research to improve cold tolerance. Current forecasts predict the stabilization of a sunflower cultivation area around 4 million ha in the EU in the next 10 years, together with a slight increase in seed yield. The EU will continue to be a net importer of sunflower meal and oil, with no significant changes in meal but a significant decrease in oil net imports (FAPRI-ISU, 2012). These forecasts, however, might change if the profitability of the crop is improved in the next generations of sunflower cultivars.

References

AGRIT Statistiche Agro Ambientali. *Statistiche Agronomiche di Superficie, Resa e Produzione. Bollettino 20 Settembre 2011*; Ministero delle Politiche Agricole, Alimentari e Forestali: Rome, Italy, 2011.

Alonso, L. C. Sunflower in Spain: Past and Present Trends in an International Context. In *Proceedings of the 17th International Sunflower Conference*, Córdoba, Spain, June 8–12, 2008; Velasco, L., Ed.; Consejería de Agricultura y Pesca, Junta de Andalucía: Sevilla, Spain, 2008; pp 661–666.

Alonso, L. C.; Rodríguez-Ojeda, M. I.; Fernández-Escobar, J.; López-Calero, G. Chemical Control of Broomrape (*Orobanche cernua* Loefl.) in Sunflower (*Helianthus annuus* L.) Resistant to Imazethapyr Herbicide. *Helia* **1998,** *29,* 45–54.

Baldini, M.; Vannozzi, G. P. Yield Relationships under Drought in Sunflower Genotypes Obtained from a Wild Population and Cultivated Sunflowers in Rain-out Shelters in Large Pots and Field Experiments. *Helia* **1999,** *30,* 81–96.

Baldini, M.; Turi, M.; Vischi, M.; Vannozzi, G. P.; Olivieri, A. M. Evaluation of Genetic Variability for *Sclerotinia sclerotiorum* (Lib.) de Bary Resistance in Sunflower and Utilization of Associated Molecular Markers. *Helia* **2002,** *36,* 177–190.

Blanchet, R.; Marty, J. R.; Merrien, A.; Puech, J. Main Factors Limiting Sunflower Yield in Dry Areas. In *Production and Utilization of Protein in Oilseed Crops;* Bunting, E. S., Ed., World Crops, Vol. 5; Martinus Nijholf: The Hague, Netherlands, 1981; pp 205–223.

CBI. Market Information Database, 2012. http://www.cbi.eu (accessed April 1, 2014).

Domínguez-Giménez, J. Presente y futuro del cultivo del girasol en España. *Vida Rural* **1997,** *53,* 52–53.

European Commission. *The Common Agricultural Policy: A Story to be Continued;* Publications Office of the European Union: Luxembourg, 2012.

European Commission. *Overview of CAP Reform 2014–2020.* DG Agriculture and Rural Development: Brussels, Belgium, 2013.

FAOSTAT. Statistical Database of the Food and Agriculture Organization (FAO) of the United Nations, 2014. http://faostat3.fao.org/home/index.html (accessed March 17, 2014).

FAPRI-ISU. Food and Agricultural Policy Research Institute and Iowa State University 2012 World Agricultural Outlook, 2011. http://www.fapri.iastate.edu/outlook/2012/ (accessed April 1, 2014).

Friedt, W.; Nichterlein, K.; Dahlhoff, M.; Köhler, H.; Gürel, A. Recent Progress and Prospects of Biotechnology in Sunflower Breeding. *Lipid/Fett.* **1991,** *93,* 368–374.

García-López, J.; García-Ruiz, R.; Martínez-Aguilar, J.; Molinero-Ruiz, L. Effectiveness of the Genetic Resistance to *Plasmopara halstedii* under Natural Conditions and Diversity of the Pathogen within Sunflower Fields. In *Proceedings of the 17th International Sunflower*

Conference, Córdoba, Spain, June 8–12, 2008; Velasco, L., Ed.; Consejería de Agricultura y Pesca, Junta de Andalucía: Sevilla, Spain, 2008; pp 163–168.

INRA. *Dossier de Presse: Inauguration Pôle de Recherche Agrogénomique du Tournesol.* Institut National de la Recherche Agronomique de Toulouse Midi Pyrénées: Auzeville, France, 2013.

Jiménez-Díaz, R. M. Notas sobre la presencia en España del mildiu del girasol. *An. INIA Ser. Prot. Veg.* **1973,** *3,* 95–105.

Jouffret, P.; Lecomte, V. Orobanche sur tournesol. Un parasite en extension. *Lettre d'Informations Regionales CETIOM, Zone Sud* [Online] 2010. http://www.cetiom.fr/fileadmin/cetiom/regions/Sud/PDF/2010/oleomail/OleoMail_TO_Prophylaxie_O_cumana_sud_07092010.pdf (accessed July 2, 2013).

Kleingartner, L. W. World Sunflower Market Slow to Transition to High Oleic. *NSA Sunflower Magazine* [Online] 2011, October/November. http://www.sunflowernsa.com/magazine (accessed March 17, 2014).

Kovácik, A.; Skaloud, V. Study on Inheritance of Agronomically Important Traits of Sunflower. In *Proceedings of the 13th International Sunflower Conference,* Pisa, Italy, Sept 7–11, 1992; International Sunflower Association: Paris, France, 1992; pp 1099–1006.

Labalette, F.; Raoul, Y. Current Uses of the French Sunflower Production. In *Proceedings of the 18th International Sunflower Conference,* Mar del Plata, Argentina, Feb 27–March 1, 2012. http://www.asagir.org.ar/asagir2008/buscar_congreso.asp (accessed March 14, 2014).

Lacombe, S.; Bervillé, A. A Dominant Mutation for High Oleic Acid Content in Sunflower (*Helianthus annuus* L.) Seed Oil Is Genetically Linked to a Single Oleate-desaturase RFLP Locus. *Mol. Breed.* **2001,** *8,* 129–137.

Leclercq, P. Une sterilité mâle utilisable pour la production d'hybrides simples de tournesol. *Ann. Amelior. Plant.* **1966,** *16,* 135–144.

MAGRAMA. *Anuario de Estadística.* Ministerio de Agricultura, Alimentación y Medio Ambiente: Madrid, Spain, 2012.

Nabloussi, A.; Fernández-Cuesta, A.; El-Fechtali, M.; Fernández-Martínez, J. M.; Velasco, L. Performance and Seed Quality of Moroccan Sunflower Varieties and Spanish Landraces Used for Confectionery and Snack Food. *Helia* **2011,** *55,* 75–82.

OECD-FAO. Oilseeds and Oilseed Products. In *OECD-FAO Agricultural Outlook 2011–2020.* Organisation for Economic Co-operation and Development (OECD) Publishing: Paris, France, 2012; pp 133–246.

Oil World. *Oil World Annual Report.* ISTA Mielke GmbH: Hamburg, Germany, 2013.

Osorio, J.; Fernández-Martínez, J. M.; Mancha, M.; Garcés, R. Mutant Sunflower with High Concentration in Saturated Fatty Acid in the Oil. *Crop Sci.* **1995,** *35,* 739–742.

Pérez-Vich, B.; Garcés, R.; Fernández-Martínez, J. M. Genetic Control of High Stearic Content in the Seed Oil of Sunflower Mutant CAS-3. *Theor. Appl. Genet.* **1999,** *99,* 663–669.

Pineda-Martos, R.; Velasco, L.; Fernández-Escobar, J.; Fernández-Martínez, J. M.; Pérez-Vich, B. Genetic Diversity of Sunflower Broomrape (*Orobanche cumana*) Populations from Spain. *Weed Res.* **2013,** *53,* 279–289.

PROLEA. *De la production à la consommation. France, Europe, Monde. Statistiques des oléagineux et protéagineux 2011–2012.* Proléa Documentation: Paris, France, 2011.

Putt, E. D. Sunflower Early History. In *Sunflower Production and Technology*; Schneiter, A. A., Ed.; ASA-CSSA-SSSA: Madison, WI, 1997; pp 1–21.

Rönicke, S.; Hahn, V.; Vogler, A.; Friedt, W. Quantitative Trait Loci Analysis of Resistance to *Sclerotinia sclerotiorum* in Sunflower. *Phytopathol.* **2005,** *95*, 834–839.

Sackston, W. E. Sunflower Disease Mapping in Europe and Adjacent Mediterranean Countries. *Helia* **1978,** *1*, 21–31.

Serieys, H. *Report on the Past Activities of the FAO Working Group "Identification, Study and Utilization in Breeding Programs of New CMS Sources;"* FAO: Rome, Italy, 2002; pp 1–54.

Skoric, D. The Genetics of Sunflower. In *Sunflower Genetics and Breeding*; Skoric, D., Sakac, Z., Eds.; Serbian Academy of Sciences and Arts: Novi Sad, Serbia, 2012; pp 1–163.

Slezák, Z.; Fáró, N.; Széles, G. The Situation of the Sunflower Seed Production in Hungary before the EU Accession. *Agriculturae Conspectus Scientificus* **2003,** *68*, 81–285.

Szabó, A. The Effect of Crop Density on the Product Yield, Yield Safety and Quality of Sunflower Hybrids. PhD Dissertation, University of Debrecen, Hungary, 2007.

Tahmasebi Enferadi, S.; Rabiei, Z.; Vannozzi, G. P. Protection of Biodiesel Based on Sunflower Oil from Oxidative Degradation by Natural Antioxidants. *Helia* **2006,** *44*, 25–32.

Tourvieille, D. L.; Walser, P.; Mestries, E.; Gillot, L.; Penaud, A.; Tardin, M.C.; Pauchet, I. Sunflower Downy Mildew Resistance Gene Pyramiding Alternation and Mixture: First Results Comparing the Effects of Different Varietal Structures on Changes in the Pathogen. In *Proceedings of the 16th Int. Sunflower Conf*erence, Fargo, ND, Aug 29–Sept 2, 2004; Seiler, G. J., Ed.; International Sunflower Association: Paris, France, 2004; pp 111–116.

Vear, F.; Serre, F.; Roche, S.; Walser, P.; Tourvieille de Labrouhe, D. Recent Research on Downy Mildew Resistance Useful for Breeding Industrial-use Sunflowers. *Helia* **2007,** *46*, 45–54.

Velasco, L.; Pérez-Vich, B.; Fernández-Martínez, J. M. Novel Variation for Tocopherol Profile in a Sunflower Created by Mutagenesis and Recombination. *Plant Breeding* **2004,** *123,* 490–492.

Velasco, L.; Pérez-Vich, B.; Yassein, A. A. M.; Jan, C. C.; Fernández-Martínez, J. M. Inheritance of Resistance to Sunflower Broomrape (*Orobanche cumana* Wallr.) in an Interspecific Cross between *Helianthus annuus* and *H. debilis* subsp. *tardiflorus*. *Plant Breeding* **2012,** *131*, 220–221.

Virányi, F. Research Progress in Sunflower Diseases and Their Management. In *Proceedings of the 17th International Sunflower Conference,* Córdoba, Spain, June 8–12, 2008; Velasco, L., Ed.; Consejería de Agricultura y Pesca, Junta de Andalucía: Sevilla, Spain, **2008** pp 1–12.

Viranyi, F.; Masirevic, S. Pathogenic Races of Sunflower Downy Mildew in Europe: Present State, Problems and Prospects. *Helia* **1991,** *15*, 7–10.

Vranceanu, A. V. Organization and Objectives of the F.A.O. Research Network on Sunflower. *Helia* **1978,** *1*, 7–9.

19

Eastern Europe Perspectives on Sunflower Production and Processing

Yalcin Kaya ■ *Trakya University, Edirne, Turkey*

Igor Balalic and Vladimir Milic ■ *Institute of Field and Vegetable Crops, Novi Sad, Serbia*

Introduction

Sunflower originated in North America and most probably was domesticated around 1000 B.C. by Native Americans as an ornamental crop. It was then brought to Europe by Spanish explorers at the beginning of 15th century. However, it was not until the late 1800s, when it was introduced to Russia, that the sunflower became a food crop. In the 1860s, Russian farmers made significant improvements in the way that the sunflower was cultivated. At that time, they became the world's largest producer of sunflower seeds. In the recent years of the 21st century, sunflower became a major crop again in Eastern Europe. Sunflower is one of the main crops in the rotation system and the main oil crop in Eastern Europe in recent years (FAO 2010; Kaya et al., 2011; Škorić, 2012a, 2012b).

After the major producer, Argentina, turned to more soybean and corn in recent years, Eastern European countries, especially Ukraine and Russia, lead the production and crushing of sunflower in the world. Eastern European sunflower producers comprise the Russian Federation, Ukraine, Turkey, Bulgaria, Romania, Moldova, Serbia, and others. While sunflower is one of the main crops in the rotation system and the main oil crop in Eastern Europe, European countries share only 14% of world arable land areas, even including Russia's vast land area in addition to the Black Sea basin. Except small countries such as Moldova, Serbia, Greece, and others, Eastern European countries are in the top 10 countries in sunflower production and planted areas in the world. Because of its high importance in the region, sunflower could easily be assumed to be the typical plant of the Black Sea. Russia and Ukraine are the main sunflower producers (ranking first and second, respectively) in the world. Although sunflower crushing capacities in Russia and Ukraine have impeded exportable sunflower supplies, sunflower oil productions and exports are gradually increasing year by year. Turkey is the largest importer in Eastern Europe and has an approximate 30% market share of the world sunflower trade. Ukraine is the largest sunflower oil exporter in the world, and Russia is the third largest. Sunflower is a summer crop and is highly influenced by climatic conditions, so both production and trade are dependent on seasonal seed yields. All sunflower production in Eastern Europe is of the linoleic type, as opposed to the new trend for oleic-type sunflower in the rest of the world (Kaya et al., 2008a, 2011; World Statistics Database, 2013).

Sunflower in Eastern Europe

Regional History of Crop Introduction and Development

Eastern Europe (Balkans and the region of the Black Sea) is a considerably large region (2.4 million km^2) and is inhabited by about 220 million people (16 million of which live in coastal areas). Eastern Europe is steeped in history and culture and forms a vital trading area linking Europe with Asia (Figure 19.1). The Black Sea's surface area is 423,000 km^2, and it is largest interior sea in the world. The Black Sea region is one of the largest inland water basins in the world and its area is about one-third the size of continental Europe. Although it is entirely isolated from the world's oceans, it receives river inputs from large land areas and from the longest rivers in Europe, such as the Danube, the Dnieper, and the Don. Most of Eastern Europe has large areas of flat land that are very suitable for agricultural production (Kaya et al., 2008a) (Table 19.A).

Agriculture in Eastern Europe has vast potential. For instance, around 75% of Ukrainian territory is involved in agricultural production. Moreover, between 90% and 95% percent of Ukrainian lands are arable, and this makes up nearly one-third

Figure 19.1 Eastern European map.

of the world's richest black soils. The total arable land of Eastern Europe is two times bigger than EU-25 (about 100 million ha), but the agricultural production is mostly non-irrigated and in dry lands (Table 19.A). Turkey has the most irrigated area among Eastern European countries. Because they have large cultivated lands, farmers prefer field crops such as wheat, barley, sunflower, and so forth, which grow in rain-fed conditions, especially in countries near the upper part of the Black Sea (Russia and Ukraine). Eastern European countries have more production and planted area than EU-25 in many field and horticultural crops such as wheat, barley, corn, cotton, potatoes, sugar beets, lentils, cabbages, peppers, pumpkins, cherries, apples, apricots, and so forth (Kaya et al., 2008a).

Sunflower Production

Sunflower is a vital crop in Eastern Europe, and its harvesting area and production cover almost half of the world's totals (Table 19.B). As a spring crop, sunflower has

Table 19.A Land Use in Eastern European Countries in 2012

Countries	Total Area 1000 Ha	Agricultural Area 1000 Ha	Arable Land 1000 Ha	Irrigated Land 1000 Ha	Perm, Pasture 1000 Ha
Bulgaria	11,099	5,325	3,355	588	1,742
Serbia	5,093	3,302	3,099	163	833
Croatia	5,641	1,580	1,457	4	124
Greece	13,080	3,799	2,675	1,555	1,124
Macedonia	2,486	592	547	128	450
Montenegro	1,381	514	175	3	325
Albania	2,740	666	551	188	115
Bosnia Herzegovina	5,120	1,101	1004	3	97
Moldova	3,384	2,534	1,843	300	391
Romania	23,839	14,837	9,398	3,077	4938
Russia	1,707,540	216,651	123,465	4,600	91,351
Turkey	77,482	41,690	25,938	5,215	13,167
Ukraine	60,370	41,396	32,544	2,208	7,939
Total	**1,919,255**	**333,987**	**206,051**	**18,032**	**122,596**
%	**14.3**	**6.7**	**14.7**	**6.5**	**3.5**
World	**13,427,880**	**5,012,266**	**1,404,130**	**277,098**	**3,471,729**

Source: FAOSTAT (2013).

Table 19.B Sunflower Harvested Area and Production by Season in Eastern Europe

	Harvesting Area, 1000 Ha				Seed Production, 1000 t			
	2012	2010	2007	2003	2012	2010	2007	2003
Russia	6,500	5,800	5,250	4,875	7,993	5,820	5,250	4,870
Ukraine	5,350	4,950	3,650	4,000	8,850	8,070	4,430	4,480
Romania	1,088	810	830	1,190	1,260	1,285	570	1,505
Bulgaria	680	692	520	660	1,348	1,380	525	790
Turkey	570	500	475	425	1,100	1,020	690	560
Moldova	295	252	230	350	295	382	170	390
Serbia	167	169	155	199	366	378	295	354
Croatia	33	26	21	28	90	80	54	69
Greece	65	56	12	8	82	85	19	16
Macedonia	5	4	4	5	9	8	5	7
Albania	2	2	2	2	3	3	2	2
Bosnia Herzegovina	1	1	1	1	1	1	1	1
Montenegro	—	—	—	—	—	—	—	—
E. Europe Total	**14,756**	**13,262**	**10,320**	**11,545**	**21,397**	**18,512**	**11,325**	**12,620**
World Total	**25,451**	**23,923**	**21,305**	**22,835**	**35,976**	**33,605**	**26,430**	**26,970**
%	**58**	**55**	**48**	**51**	**60**	**55**	**43**	**47**

Sources: Oil Word Annual (2013); World Statistics Database (2013).

been influenced a lot by climatic conditions, and it is dependent on summer rainfall and temperatures during the vegetation period. For instance, in the 2007–08 season, the harvesting area was reduced by 12% and sunflower production plummeted by nearly one-third, or 3.67 million metric tons (t), due to historical drought and the highest temperatures of the last 80 years. During that season, sunflower seed production in Romania was 62% lower than estimated due to these severe drought and extreme heat conditions (Kaya et al., 2008a).

Sunflower production areas in Turkey are mostly located in the Trakya region, which is the European part of Turkey and has more than 50% of the total Turkish sunflower area. Central Anatolia, which is mostly irrigable, is an emerging sunflower production area that is rotated with sugar beet. In Ukraine, the main sunflower seed producing provinces (oblasts) are Donetsk, Dnipropetrovsk, Zaporizhya, Kharkiv, Odesa, and Kirovograd. These oblasts actually subsidize a lower domestic price for the processing enterprises in the rest of the country with large capacities to process sunflower seeds. Major regions of sunflower growing in Russia are North Caucasus, Volga, and Central Black Earth. Central Black Earth and North Caucasus are also sugar beet growing regions. Similarly, eastern parts of Bulgaria and Romania (mostly Black Sea coasts) have huge plains where mostly sunflower has been grown, alternated with wheat (Kaya et al., 2008a).

Sunflower is growing in Turkey. Planting is mostly mechanized (with pneumatic planters), fertilizer is applied, and hybrid seed is used (Kaya, 2004). However, in Ukraine, only about half of the country's large agricultural enterprises are profitable, and most farms have neither the cash nor access to credit to enable them to purchase the additional inputs (fertilizer, herbicide, etc.). Yield improvement has mainly been focused on the largest enterprises: farms over 10,000 ha in Ukraine. With the continuous changing of inefficient farms into large and successful enterprises, overall sunflower and other crop productivity will gradually increase in Ukraine. A similar situation could also be observed in Russia (Kaya et al., 2008a).

Crop Rotation Systems

Sunflower–wheat is the main rotation system in Eastern Europe due to higher mechanization use and less need for workers. Because of higher prices and lower production costs together than wheat (mostly related to less fertilizer and chemical use), sunflower seems to be one of the most profitable crops in the region. Farmers' decisions are mostly related to market price, and the profitability of sunflower has expanded significantly in recent years (Kaya et al., 2008a; USDA, 2004).

It was advised in the past, particularly in the Soviet Union, that sunflowers should not be planted in the same field more than once every 7 years. The main reason for this restriction was to prevent the establishment and spread of soil-borne fungal diseases, mainly Phomopsis and Sclerotinia. Crop rotation was the best or only option to control disease in sunflower fields because most farmers had limited access to fungicides and disease-resistant hybrids due to financial problems. Chemical control was difficult and expensive, so farmers in Russia typically relied on the use of long rotations. Frequent sunflower plantings could result in another serious problem, especially in dry regions like Volgograd oblast: infestation by broomrape parasite, which is one the most important yield-reducing biotic factors in sunflower production. Because of its deep rooting system, sunflower depletes soil fertility and moisture and removes higher amounts of nutrients from the soil than other crops in the rotation. Therefore, it needs longer rotation in cropping designs in these areas because most of the main sunflower-growing regions in Russia and Ukraine are located in the drought-risk zones of agriculture (USDA, 2004).

Because of the higher demand for sunflower seed, Russia's sunflower area has doubled and Ukraine's has tripled since 2001, and many farmers have left the longer, traditional sunflower rotations (four- to seven-year rotations), opting instead for two- to four-year crop-rotation schemes (USDA, 2004).

Sunflower–wheat is the main rotation system in the Trakya region of Turkey. However, in the Central Anatolian region, sunflower is rotated with wheat, sugar beet, and corn because all crops are irrigated due to the limited rainfall. Therefore, in this region and in countries with more rain, such as Serbia, Croatia, and Romania, corn prices play very important roles in farmers' decisions on whether to plant sunflower (Kaya et al., 2008b).

Planting and Harvest Dates

Planting sunflower as a summer crop in the Western Balkan countries, Eastern Europe, and the Black Sea countries takes place during March and April. The summer crops are mainly corn and sunflower, and their sowing time is completed in late May. Sunflower harvest begins at the end of August and continues through October.

Sunflower Processing

The application of 17% and 20% sunflower export taxes, in force since 2001 in Ukraine and Russia, respectively, have been promoting higher sunflower crushing/sunflower oil production in Eastern Europe (Table 19.C). Based on a U.S. Foreign Agricultural Service (FAS) report, Russia's oilseed crushing capacity was 14 million t

Table 19.C Sunflower Crushing and Sunflower Oil Production by Seasons in Eastern Europe

	Sunflower Crushings, 1000 t				Sunflower Oil Production, 1000 t			
	2012	**2010**	**2007**	**2003**	**2012**	**2010**	**2007**	**2003**
Russia	8,467	6,211	4,740	4,070	3,632	2,616	2,005	1,720
Ukraine	9,125	7,332	4,150	3,325	3,937	3,125	1,715	1,400
Romania	790	752	480	960	329	314	205	385
Bulgaria	475	360	400	360	199	152	170	145
Turkey	1,625	1,497	960	1,185	329	324	380	510
Moldova	216	197	140	270	91	83	60	110
Serbia	388	362	312	265	155	145	125	106
Croatia	37	42	50	50	15	17	20	20
Greece	93	78	34	—	35	30	13	19
Macedonia	6	7	6	5	3	3	3	1
Albania	2	2	2		1	1	1	1
Bosnia Herzegovina	45	48	43	12	17	19	17	5
E. Europe Total	**21,269**	**16,888**	**11,317**	**10,502**	**8,743**	**6,829**	**4,714**	**4,422**
World Total	**35,314**	**30,464**	**24,300**	**23,470**	**14,800**	**12,536**	**9,735**	**9,580**
%	**55**	**60**	**55**	**47 45**	**59**	**55**	**48**	**46**

Sources: Oil Word Annual (2013); World Statistics Database (2013).

in 2012–13, and a further increase was forecasted for 2013–14. Although oilseed processing has decreased at small plants recently, new and modern crushing plants owned by large agro-holding companies are increasing year by year in the market.

Turkey has over 5 million t of sunflower crushing potential, and the factories are working at less than 50% capacity due to a weaker competitive force compared to cheaper factories from other parts of the Black Sea region. The crushing factory investments of multinational firms like Bunge, Cargill, Glencore, ADM, Noble, and so on also play a big role in the Black Sea region. With increasing crushing capacities in both Ukraine and Russia, export remains limited only by sunflower seed quantity. Bulgaria and Romania, however, have replaced the earlier positions of Ukraine and Russia as raw sunflower seed exporters. Although almost all Black Sea countries are sunflower exporters, Turkey is the largest sunflower importing country in the world (Table 19.D).

With gradually enlarging crushing capacities, both Ukraine and Russia started to export more sunflower oil. Ukraine is now the leading worldwide sunflower oil exporter. Russian sunflower oil imports were also sharply reduced, and Russia became

Table 19.D Sunflower Trade by Seasons in Eastern Europe

	Sunflower Seed Exports, 1000 t				Sunflower Seed Imports, 1000 t			
	2012	2010	2007	2003	2012	2010	2007	2003
Russia	326	15	45	350	28	34	5	10
Ukraine	282	410	80	930	18	7	5	5
Romania	105	192	175	135	28	36	30	30
Bulgaria	150	166	175	200	6	11	5	10
Turkey	—	—	—	5	754	649	450	520
Moldova	113	108	50	100	1	4	10	—
Serbia	10	1	1	2	23	38	11	1
Croatia	43	28	18	13	1	1	25	8
Greece	—	—	—	—	2	5	11	43
Macedonia	1	—	—	1	—	3	1	—
Bosnia Herzegovina	—	—	—	—	43	39	40	13
Albania	—	—	—	—	—	—	—	—
E. Europe Total	**1030**	**920**	**544**	**1736**	**904**	**827**	**593**	**640**
World Total	**1632**	**1508**			**1730**	**2151**		
%	**63**	**61**			**52**	**38**		

Sources: APK (2013); Oil Word Annual (2013); TUIK (2013); World Statistics Database (2013).

the second-largest sunflower oil exporter in the world, surpassing Argentina in 2012 (Table 19.E).

Because Turkey is the main buyer in Eastern Europe, and the major part of Turkish imports are of Bulgarian origin, Turkey sunflower export prices influence this area of seed production. A heavy drought in 2006–07 created a huge reduction in sunflower crops in Bulgaria and other Black Sea countries in 2007–08 in turn causing delivered-at-place (DAP) prices to increase by 170%. Record prices for other oilseeds and vegetable oil also influenced the increase of sunflower prices.

Sunflower Research Systems in Eastern Europe

Great success was obtained by Eastern European public institutes on sunflower breeding. Many open-pollinated (OP) varieties have high-yield and adaptation capabilities, such as Vniimik-1646, 8931, Perodovik, Armavirski, and Pervenets; these were developed by public institutes during the 20th century and were still planted in some areas in the world. Later, sunflower hybrid breeding was also very successful

Table 19.E Sunflower Oil Trade and Their Proportions in the World (%) by Seasons in Eastern Europe

	Sunflower Oil Exports, 1000 t				Sunflower Oil Imports, 1000 t			
	2012	**2010**	**2007**	**2003**	**2012**	**2010**	**2007**	**2003**
Russia	1387	393	614	91	17	101	132	201
Ukraine	3614	2703	1.925	929	2	1	—	1
Romania	2	3	14	38	17	15	25	18
Bulgaria	37	36	27	15	3	5	20	13
Turkey	553	149	94	15	744	225	163	93
Moldova	66	42	61	54	4	1	1	1
Serbia	78	78	48	4	4	4	12	10
Croatia	6	6	7	8	32	31	24	17
Greece	—	—	4	7	22	25	56	44
Macedonia	10	9	1	1	44	40	31	25
Bosnia Herzegovina	21	20	17	—	45	33	37	34
Albania	—	—	—	—	34	35	37	35
E. Europe Total	**5774**	**3439**	**2812**	**1162**	**968**	**516**	**538**	**492**
World Total	**7256**	**4771**			**7232**	**4834**		
%	**80**	**72**			**13**	**11**		

Sources: APK (2013); FAOSTAT (2013); APK (2013); World Statistics Database (2013).

in Eastern Europe public programs. Some public sunflower breeding institutions in Eastern Europe are listed in Table 19.F.

Public breeding programs for sunflower have focused primarily on developing genetic materials that aid in resistance to diseases, insects, and herbicides; producing high oil or exchanged fatty acid content; and promotion of other important traits. Western and U.S. institutions, such as the USDA, South Dakota State University, Cordoba-Spain Research Institute, and German universities, have these kinds of goals in their breeding programs, and they distribute and share their findings and the genetic materials they have developed. Other breeding programs, such as INRA in France, have similar objectives, but they have also developed inbred lines and are selling them to private breeding companies and getting royalties (Kaya, 2010).

Some breeding programs, such as Trakya–Edirne, have both public and private sunflower breeding programs for developing lines and hybrids. However, they do not market these hybrids; rather, they sell the inbred lines and rights for producing those hybrids to private companies and collect only royalties. On the other side, some public institutes in Eastern Europe, such as Dobroudja–Bulgaria, Fundulea–Romania, Osijek–Croatia, Vniimik–Russia, Novi Sad–Serbia, and Kharkov–Ukraine, develop and market their own hybrids, some of which have very high market shares. Novi Sad, especially, has an international market share that is highly competitive with private companies (Kaya, 2010).

Vavilov, Novi Sad, and Dobroudja have made great contributions to sunflower genetic resources and have developed many interspecific hybrids and new genes with important yield traits and tolerance to biotic and abiotic stress. These institutions, along with many other universities in Eastern Europe, have contributed research on both agronomic and physiological studies such as rotation systems, planting designs, time, seed biology, plant morphology, fertilizer applications, and nutrient uptake, in addition to molecular genetics studies such as genetic characterization of yield and other traits, heritability of genes, and gene transfer.

Sunflower Production by Country in Eastern Europe

Russia

Area, Production, Yield, and Localization in the Past and Today

The main agricultural region covers the Central region in the European part of Russia and extends its eastern borders to western Siberia. Russia has about 200 million ha of agricultural land, of which roughly 120 million ha are planted in row crops (mainly grain crops, annual or perennial forage crops, sunflower, potatoes, and vegetables) or temporary fallow. The remaining lands are permanent meadow or pasture. Russia is divided into seven federal districts, but the Central, Southern, Volga, and Siberian

Table 19.F Public Sunflower Breeding Programs in Eastern Europe

The Institution Name	City	Country	Web Page
Trakya Agricultural Research Institute	Edirne	Turkey	www.ttae.gov.tr
Aegean Agricultural Research Institute	Izmir	Turkey	www.etae.gov.tr
Dobroudja Research Institute	Dobroudja	Bulgaria	www.dai-gt.org
National Agricultural Research and Development Institute (NARDI)	Fundulea	Romania	www.ricic.ro
Plant Breeding and Genetics Institute	Odessa	Ukraine	www.sgi.od.ua/
Yurjev Plant Production Institute	Kharkov	Ukraine	www.yuriev.com.ua
Institute for Field Crops	Lugansk	Ukraine	
Institute for Oilseed Crops, NAAS	Zaporizhya	Ukraine	www.oilseed.marka.net.ua
South Plant Biotechnology Center NAAS	Odessa	Ukraine	www.genome.intes.odessa.ua
Institute for Field Vegetable and Crops	Novi Sad	Serbia	www.nsseme.com
Agricultural Institute—Osijek	Osijek	Croatia	www.poljinos.hr
All Russia Research Institute Oil Crops	Krasnodar	Russia	www.vniimk.ru
N. I.Vavilov Research Institute of Plant Industry	St. Petersburg	Russia	http://www.vir.nw.ru
Don Experimental Station—L.A. Zdanova—VNIIMK	Rostov Russia		
Research Ins. of Field Crops "Selectia"	Balti, C. Esilor	Moldova	www.agriculture.md/selectia

regions cover 90–95% of Russia's grain crop areas. Sunflower is grown throughout, but mainly in the southern part of Russia. Irrigation is not common in Russia, even in dry land areas. Although the average annual rainfall in the main crop production regions of the European part of Russia is around 600 mm, it is only 400 mm during the growing season. The climate is suitable for growing both winter and spring crops. The Volga district is one of most important agricultural regions in Russia due to higher yield potential and production for many grain crops (USDA, 2013).

Russia has the largest sunflower planting areas in the world, reaching over 7 million ha in 2011 (Table 19.G). After a slight reduction to 6.1 million ha in 2012, sunflower sowing areas increased to 7.1 million ha in 2013 with production at its highest during the last 10 years (World Statistics Database, 2013). However, this increase in area happened in less-fertile areas in Russia because the more-fertile land is preferred for planting spring grains and pulses, soybeans, and fodder crops. Most sunflower farmers in these less-fertile regions cannot afford to buy hybrid seeds and expensive chemicals (USDA, 2013).

Sunflower is one of the most profitable crops in Russia due to higher price, recently consistent demand, and lower production costs relative to wheat. Oil crop planting areas have been gradually increasing in the last decade in Russia, particularly for oilseeds other than sunflower. Although sunflower production is still greatly dependent on weather as a spring crop and changes from year to year, sunflower still remains the dominant oil seed in Russia. It is expected that sunflower areas will not exceed 15% of the total arable land, but sunflower is still a consistently profitable crop and the main crop in the rotation systems by region. Despite the likely negative impact on yield resulting from increasing sunflower planting in some oblasts, Russian sunflower yield has benefited from the same growth in agriculture that has advanced corn yield. Sunflower yield has increased by about 50% since 2001, mostly due to the introduction of hybrids, instead of OP varieties.

The Southern and North Caucasus districts of Russia remain the major sunflower seed production areas (accounting for 38% of Russia's total sunflower seed crop in 2012). However, sunflower planted areas in Southern European Russia have reduced due to strong competition with other crops in recent years. This region offers many advantages for grain production, including good climate and closer proximity to key export ports. On the other hand, the Central Federal District has increased sunflower seed production by 80%, mainly because of higher yielding cultivars and better agronomical applications (30% of total Russian sunflower seed production in 2012). Due to higher sunflower demand in recent years in the Central Federal District, several new crushing plants have been erected, especially in Voronezh and Belgorod oblasts. Sunflower seed production increased about 39% in the Volga Valley Federal District, mostly due to increased planting area, although the average yield was still lower (APK Sunflower Market Report, 2013; USDA, 2004, 2013a).

Table 19.G Sunflower harvested Areas, Seed, Oil and Meal Production by Seasons in Russia

Year	Harvested Area (1000 Ha)	Seed Production (1000 t)	Oil Production (1000 t)	Meal Production (1000 t)
1992	2650	3073	894	986
1993	2760	2765	720	685
1994	3040	2553	650	618
1995	3960	4200	920	880
1996	3410	2765	650	631
1997	3290	2831	649	610
1998	3570	3000	750	730
1999	5000	4150	1240	1110
2000	4350	3915	1250	1120
2001	3420	2670	1025	920
2002	3798	3685	1365	1225
2003	4850	4850	1610	1450
2004	4650	4800	1815	1625
2005	5400	6450	2320	2081
2006	5900	6750	2465	2220
2007	5000	5650	2130	1917
2008	6000	7350	2565	2306
2009	5600	6425	2505	2253
2010	5550	5350	2082	1874
2011	7200	9627	3552	3199
2012	6500	7959	2971	2678

Sources: Oil Word Annual (2013); World Statistics Database (2013).

Crushing and Trading, Price, and Market Characteristics

Crushing. Russia's oilseed crushing capacity reached 14 million t in recent years and is expected to continue to increase in the future. Oilseeds processing is rising, especially at new and modern crushing plants built by large agro-holding companies, but the rate of smaller plants is reducing (Table 19.H). It is expected that the sunflower crop area will be the same as in previous years, so sunflower crush in the next few years could remain the same. Russia produced about 3.3 million t of sunflower oil in 2013. Sunflower seed crushing still seems a profitable and attractive business for large agro-holding companies, and they continue to increase their crushing capacities (USDA, 2013b).

Recent reports indicated that the domestic oilseed market is saturated, and if crude sunflower oil prices reduce in the future only large crushing companies would be able to survive and afford the higher sunflower production costs. Therefore, increased sunflower prices have attracted big companies to invest in the construction of large and modern crushing plants in recent years (USDA, 2012).

High yields in sunflower production directly affect oil exports from Russia. Export growth is stimulated by permanently increasing global demand for sunflower oil. Application of new technologies, such as growing hybrid varieties of sunflowers, has a good influence on industry performance. This is a major driving factor for improving quality and quantity of the crop in Russia. The record harvest of sunflower seeds in recent years stimulated an increase of export volume. The major supplier of sunflower seed oil to Russia is Ukraine, which controls 95% of the import volume in the world. Meanwhile, the major buyer of Russian sunflower seed oil is Turkey, with about 40% of export volume. According to analysts, export of sunflower seed oil from Russia is forecasted to be about 4 million t by 2020. Oil manufacturers have to pay special attention to the renovation of production equipment and updating of technologies. Today, equipment and machinery in many production facilities are worn out, which can further affect development of the oil pressing industry. More than half of the companies beyond the group of leaders use outdated equipment, and this affects product quality. About 250 companies produce sunflower seed oil in Russia, and major manufacturers are oil-extracting facilities of agricultural holdings. The largest manufacturers in Russia are Aston OJSC, Bunge CIS LLC, Efko OJSC, MEZ (Oil Pressing Plant), Yug Rusii (South of Russia) LLC, and Solnechnye Produkty (Sunny Products) GK. The largest manufacturer of sunflower seed oil in Russia today is MEZ Yug Rusi, with about 30% of domestic production volume. Its production facilities are located in the Central and Southern Federal Districts. All major manufacturers are Russian companies except Bunge CIS, which is a Russian branch of a transnational company. Major production facilities are located in the Krasnodar Territory, Rostov, Belgorod, Saratov, and the Voronezh region sunflower growing areas. These locations provide high availability of raw seeds, which reduces shipment costs.

Consumption. Sunflower seed oil is the most popular vegetable oil in Russia, where it is commonly used for cooking. It first appeared and was used about 200 years ago. Sunflower seed oil still ranks first in the oil consumption in Russia, despite increasing availability of other kinds of oil, such as olive oil, rapeseed oil, flaxseed oil, and others. All these years, the oil extracting industry has been developing and improving production technology because applications of sunflower oil are not only limited to home cooking, but are also used in the food processing industry and for nonfood purposes. Currently, 35% of all vegetable oil produced in Russia is packed for food uses; the rest of the oil goes to other industries. About 95% of packed sunflower seed oil is

Table 19.H Sunflower Seed Domestic Consumption, Seed Crushing and Exports in Russia

Year	Seed Domestic Consumption (1000 t)	Seed Crushing (1000 t)	Oil Export (1000 t)	Seed Export (1000 t)	Meal Export (1000 t)
1992	3055	2235	10	76	0
1993	2285	1800	40	450	0
1994	2075	1625	21	533	13
1995	2698	2300	25	1200	5
1996	1898	1618	35	1220	21
1997	1890	1610	35	950	10
1998	2140	1860	55	890	5
1999	3288	3000	195	847	129
2000	3201	3020	130	729	203
2001	2688	2485	41	18	97
2002	3486	3300	103	186	163
2003	4300	3900	136	311	523
2004	4812	4389	226	45	479
2005	6045	5610	616	396	831
2006	6530	5980	711	162	866
2007	5648	5160	322	37	666
2008	6917	6210	802	160	965
2009	6720	6065	504	20	660
2010	5555	5045	181	8	588
2011	9345	8600	1427	332	1681
2012	7715	7200	900	250	960

Source: Oil Word Annual (2013); World Statistics Database (2013).

sold in retail markets: large retail chains (56%), retail stores (32%), and street markets (6%). Sunflower seed oil consists of about two-thirds of the domestic food and industrial vegetable oil consumption of Russia. Total vegetable oil consumption is at about 25 L per capita per year, lower than in Europe and in the U.S. (USDA, 2013).

Trade. Growing sunflower and other oil crops was the most profitable segment of agriculture in recent years. Still, export of sunflower seeds today is insignificant due to the high export tariff. During the next few years, the export tariff will certainly decline because of Russia's entry to WTO. This can have positive influence on industry development. The current sunflower seed export duty is 20%, but not less than 30 Euro per metric ton. Russia accessed in WTO in 2012; therefore, these Russian export duties must be reduced from 20% (but not less than 35 Euro/t) to 6.5% (but not less than 9.75 Euro/t) in the next 4 years. This policy could affect domestic prices and lead to increasing exports based on demand. Russia has significant opportunities to increase vegetable oil exports by both increasing profitability of companies that work on vegetable oil and constructing new port and storage facilities, opening new directions for trade. For this goal, Russia is improving port facilities, constructing 2 million t capacity tank terminals in Novorossiysk for vegetable oil handling and the exporting of sunflower seed oil. Russia exports vegetable oil to Turkey and Egypt from the Azov-Black Sea region at smaller docks because the closer destinations mean smaller ships are used for imports. However, larger ships could be used for bigger capacity exports to farther destinations such as India and China (USDA, 2013b; World Sunflower Market Update, 2013).

Sunflower seed exports increased by approximately 200,000 t from 119,000 t in 2011 to 325,000 t in 2012. Turkey was the major destination in 2012 with 194,000 t, followed by Kazakhstan with 81,000 t and the Netherlands with 27,500 t. The total sunflower oil export of the Russian Federation was 1,387,200 t in 2012, which was double that of the previous year (607,000 t). Similarly, for seed export, Turkey was the main destination with 506,000 t in 2012. This amount was 4 times higher than that in 2011. Egypt (242,000 t), Uzbekistan (92,600 t), and Kazakhstan (67,800 t) followed Turkey. In 2012, Russia exported 1,800,800 t of sunflower seed meal. Leaders in imports of Russian sunflower seed meal were Turkey (485,100 t), Italy (355,300 t), Latvia (259,500 t), Spain (232,700 t), France (76,600 t), Israel (63,000 t), and Denmark (57,200 t) (FAOSTAT, Oil World Annual, 2013).

Trade policy of Russia is also affected by the domestic vegetable oil market because Russia produces more oilseed than domestic need and also the total production fluctuates from year to year. Vegetable oils are exported duty free, and import tariffs on vegetable oil vary due to custom regulations based on the Customs Union TR on Oils and Fats, adopted on December 9, 2011, in Russia (USDA, 2013b).

Research, Seed Science, and Technology in Sunflower

The first written record of sunflower as an oil plant dates back to 1818, but sunflower began as an oil crop in Russia in the 1830s. Sunflower selection began simultaneously with its expansion as an oil crop. At the end of the 19th century, farmers started to select the best plants in their fields in various parts of Russia, aimed at improving the available sunflower populations. The farmers, especially those in Saratov and Voronezh, systematically worked on the improvement of the production qualities of the existing forms of sunflower. As a result, many local sunflower populations were created. Scientifically grounded sunflower breeding began in 1912 with the variety development program established at the Kruglik Plant Breeding and Experiment Station.

The first significant sunflower variety developed in Russia, Saratovski 169, was grown in large areas that periodically exceeded 1 million ha (Demurin, 2003; Gontcharov, 2012; Gontcharov et al., 2006; Skoric, 1992).

At the beginning of the 20th century, big sunflower breeding centers were established in Krasnodar (VNIIMK), Rostov-on-Don (Don Breeding Station), and at Armavir Station and Saratov Station. The most important contributions for sunflower breeding were performed by Pustovoit and Zhdanov in the former Soviet Union. All-Russia Research Institute of Oil Crops (VNIIMK) in Krasnodar was set up in 1912. High-yielding varietal populations, based on individual selection and seed reserve, were developed and used through the selection cycle. At that time, the local Russian varieties had 28–33% oil content. Vasilij Stepanovic Pustovoit, who established VNIIMK, created the sunflower variety Kruglik 41, having 36–37% oil, in 1927. During the 1950s, the varieties VNIMK 8931 and Armavirski 3497, with oil content of 47%, were created. It is important to mention that an armored layer had been incorporated into the husk and sunflower moth damage was reduced (Demurin, 2003; Gontcharov, 2012).

All the varieties created through 1927 were resistant to broomrape race A, but then a new broomrape race B occurred. In 1933, new resistant varieties Armavirski 611, 762, 768 were spread on 20,000 ha in the Krasnodar region. The Russian high-yielding sunflower varieties Peredovik, Armavirski 3497, Mayak, VNIMK 8931, VNIMK 6540, Smena, and others contributed much to the expansion of sunflower to all continents (Demurin, 2003). The breeding goals were to develop three sunflower varieties: high oleic, resistant to main diseases, and suitable for use as both confectionary and oil-type hybrids with high seed yielding capacity, oil content, and seed weight (Gontcharov, 2012).

Since 1982, VNIIMK institute has focused on research activity on sunflower genetics investigating mainly variability and inheritance of the seed traits that influence oil quality. Main investigations currently focus on three main characteristics: fatty acid profiles, tocopherol composition, and free fatty acid content. Successful results have been achieved on fatty acid and tocopherol, with the first high oleic hybrid Hermes developed and released in 2009. Although many high oleic lines and prominent new hybrids were developed, there is no current demand in the Russian market for high oleic sunflower. The pleiotropic effect of the high palmitic and high oleic mutations in sunflower seed oil was observed in near-isogenic lines derived from the VK580 genetic base. The negative effect of the high palmitic mutation on plant height was proved beyond doubt with both near-isogenic lines and by the inheritance analysis (Demurin et al., 2010).

Breeding for resistance or tolerance to diseases is one of the most important goals in sunflower breeding. The most important diseases in sunflower production in Russia are new races of broomrape, downy mildew, and *Fusarium*. Broomrape has almost

100 years of history on sunflower, affecting sunflower production areas and becoming a severe threat to large sunflower areas. The key method to control broomrape is the development of resistant and tolerant hybrids. During the last century, sunflower breeders in Russia have been challenged repeatedly by broomrape; biotype A (broomrape race A) was challenged by biotype B and then by E. Recently, the E race is primarily covered in Russian territory. Many of the released Russian hybrids and OP varieties are resistant to race E, but new races of broomrape invaded large cultivated areas of Russia. After much effort finding resistance, line VK-623, which was resistant to the race F from Spain, was discovered. New promising inbred sunflower lines obtained from VK-623, which is resistant to both broomrape E and F races, were developed. However, more aggressive races have infested sunflower plants in many parts of Russia, making it necessary to search for new donors of resistance and different approaches to their use. As an alternative, Genetic Laboratory of VNIIMK started a program of developing imidazolinone-resistant analogues (IMI) of the VNIIMK elite sunflower lines in order to produce new hybrids for the Clearfield technology. The cultivated sunflower was crossed with *H. tuberosus* (used as a pollinator) to develop new resistance sources in the future. Different wild populations of *H. tuberosus* were used as pollinators for crossing with cytoplasmic male sterility (CMS) lines of cultivated sunflower (Antonova et al., 2011, 2012; Gontcharov, 2009; Gontcharov et al., 2004).

Fusarium is one of the most important diseases for many crops, but it was previously regarded as a minor pathogen in sunflower. Five *Fusarium* species were observed in the sunflower fields of the Voronezh and Belgorod regions of Russia in the 1990s; then, a special breeding program was started to develop *Fusarium*-resistant lines and hybrids in the 2000s. Both field experiments and lab methods for *Fusarium*-resistance testing in VNIIMK Institute observed higher resistant levels to the pathogen in the new breeding lines; then those lines were developed rapidly and integrated in the hybrid breeding program of VNIIMK, combining test methods with individual selection for *Fusarium* resistance (Gontcharov et al., 2006).

In the middle of the 20th century, only modern high-oil sunflower varieties were cultivated in Russia. No breeding efforts for confectionary use were performed until the end of the 20th century. After demand grew for confectionery sunflower, VNIIMK started a confectionery breeding program to develop new OP varieties. Four OP confectionery varieties—SPK, Lakomka, Borodinskiy, and Oreshek—were developed, and only Borodinskiy could be identified as a "classical" confectionary variety. The other three varieties are classified as intermediates between oil type and confectionary type. After a commercial success of confectionary OP varieties, VNIIMK started a new confectionery hybrid breeding program in 1999. The evaluation of new confectionary inbred lines and hybrids was done in 2009 and 2010. The aim of breeding efforts was to develop sunflower hybrids suitable for use as both as confectionary- and oil-type hybrids. These hybrids should have higher seed yield than high-oil-content sunflower and higher 1000-seed weight. On the basis of a 3-year trial, the most prominent hybrid was Katyusha (Gontcharov, 2012; Gontcharov and Beresneva, 2011).

Ukraine

Area, Production, Yield, and Localization in the Past and Today

Ukraine is one of the main world producers of sunflower seed and sunflower byproducts, with Russian Federation and Argentina, and is a supplier to the global market. Ukraine is ranked second in sunflower planted areas (after Russia), with 21% of total sunflower world planted area, and is the leader of sunflower seed production, producing 24% of total sunflower seed in the world in recent years. In sunflower crushing they have 26% of total world production; in sunflower oil production, 27%; and in meal production, 25%. Sunflower planted areas in Ukraine are over 5 million ha in recent years, and sunflower production reached 9 million t. Sunflower yield is about 1.6 t per ha in recent years; the record yield was in 2011, at 1.83 t per ha (Table 19.I).

Oilseed production in Ukraine seems more profitable than other grain crops. Sunflower seed production is especially attractive to Ukrainian farmers due to greater export opportunities and lower production costs compared with other crops in the rotation. The high profitability of sunflower when compared to other crops comes from strong domestic demand from the crushing industry for export and frequent need to replant winter grain areas lost due to winter kill in Ukraine. Sunflower is produced mainly by big agricultural enterprises but also in fewer amounts by smaller farmers (USDA, 2012).

Crushing and Trading, Price, and Market Characteristics

From September 2012 until August 2013, Ukrainian enterprises produced 3.4 million t of sunflower seed oil. (Table 19.J). Seed crushing capacity increased almost five times in the last 10 years. Normally, Ukraine prefers to export crude oil instead of sunflower seed, and in many years they apply a higher tax to seed exports. Ukraine could export more seeds again in the future due to the recently reduced export duty on sunflower seed (10%). The largest buyer of sunflower seeds from Ukraine is Turkey.

Trade. Sunflower seed exports are expected to be lower due to recent higher demand from the domestic crushing industry in Ukraine, but seed exports could also rise, due to expected higher demand from the global market. Ukraine had an export duty of 23% for seeds from 1999 to 2001 and of 17% from 2002 to 2007. Since 2008, this has decreased by about 1% per year. Turkey is the main import destination (203,000 t) and has a 72% share of the total sunflower seed export (282,000 t). The Nether-

Table 19.I Sunflower Harvested Areas, Seed, Oil and Meal Production By Years in Ukraine

Year	Harvested Area (1000 Ha)	Seed Production (1000 t)	Oil Production (1000 t)	Meal Production (1000 t)
1992	1630	2127	777	760
1993	1629	2075	767	787
1994	1725	1570	588	560
1995	2008	2860	943	875
1996	2026	2123	421	450
1997	2002	2308	560	550
1998	2431	2266	530	520
1999	2800	2794	840	850
2000	2842	3457	970	950
2001	2396	2251	842	832
2002	2720	3270	1150	1150
2003	3810	4252	1320	1300
2004	3427	3050	1265	1250
2005	3800	4900	1925	1880
2006	4200	5700	2050	2025
2007	3400	4200	1795	1769
2008	4500	7000	2631	2600
2009	5000	7600	2975	2947
2010	5400	8400	3327	3304
2011	5800	10500	4347	4361
2012	6000	9000	3781	3807

Sources: Oil Word Annual (2013); World Statistics Database (2013).

lands and Georgia followed Turkey as top importers in 2012 (APK Sunflower Market Report, 2013; World Statistics Database, 2013).

Sunflower seed oil exports were about 35% higher in the 2012 season compared to 2011, but this amount is almost three times higher than in the 2010 season. Higher domestic production and profitability lead to increasing sunflower oil exports. India (1,131,000 t), Egypt (483,000 t), Turkey (276,000 t), and Algeria (174,000 t) are the main destinations for Ukrainian sunflower oil export, and it seems that they are likely to retain positions in the future. France, the Netherlands, Italy, Spain, Poland, Russia, Belarus, Saudi Arabia, Syria, and Lebanon are also traditional destinations for Ukrainian sunflower exports. China and Iran could also be more important destinations soon (USDA, 2012).

Table 19.J Sunflower Seed Domestic Consumption, Seed Crushing and Exports in Ukraine

Year	Seed Domestic Consumption (1000 t)	Seed Crushing (1000 t)	Oil Export (1000 t)	Seed Export (1000 t)	Meal Export (1000 t)
1992	757	1850	120	200	0
1993	677	1825	100	200	0
1994	532	1400	150	100	0
1995	700	2300	200	450	0
1996	367	1003	209	1100	10
1997	411	1365	180	785	150
1998	357	1300	205	876	190
1999	405	2100	430	450	338
2000	417	2330	550	1020	600
2001	514	2057	308	95	597
2002	266	2800	911	331	848
2003	303	3150	978	928	1089
2004	380	3000	642	12	862
2005	417	4550	1514	220	1337
2006	380	4847	1867	336	1483
2007	413	4240	1325	75	1273
2008	465	6220	2098	767	2209
2009	505	7050	2645	353	2516
2010	535	7885	2652	444	2927
2011	540	10300	3263	282	3838
2012	555	8960	3400	250	3600

Sources: Oil Word Annual (2013); World Statistics Database (2013).

Ukrainian sunflower meal exports reached 4 million t in 2012, setting a record that was three times more than the export level in 2008. Meal exports are increasing because of higher domestic production and rising demand from Poland and Belarus. Italy, Turkey, Israel, Spain, the Netherlands, Lithuania, the UK, and Morocco are other large buyers of Ukrainian sunflower seed meal, and they are expected to maintain their positions as main buyers in the future (USDA, 2012).

Sunflower seed producers both for hybrid and OP varieties in Ukraine should compete with foreign hybrids belonging to international companies. However, farmers prefer the higher quality imported seeds that provide better seed quality and higher yield potential. Ukrainian farmers should consider many costs when deciding which seed to buy, such as increasing fertilizer and fuel costs, losses after winter killing, associated customs charges, and so forth (USDA, 2012).

Consumption. Sunflower is the main and most-consumed vegetable oil in Ukraine. Sunflower meal is second in preference after soybean meal and is mostly being exported. Sunflower meal consumption in Ukraine seems to increase mainly due to the domestic crushing market demand and the demand for cake for feeding in animal production. There is also some increase in feed production in the Ukraine that will increase consumption of sunflower meal in the upcoming years (USDA, 2012).

Research, Seed Science, and Technology

The Yuryev Plant Production Institute (UAAS) of NAAS in Kharkiv; the Institute of Oil Crops of the National Academy of Agrarian Sciences in Zaporozhye; and the Plant Breeding and Genetics Institute, National Center of Seed and Cultivar Investigation (UAAS) in Odessa are the most important institutes dealing with sunflower in Ukraine.

Main directions of research of the Yuryev Plant Production Institute (YPPI) include: development of theoretical bases for selection of crops; the formation and maintenance of the National Center for Plant Genetic Resources of Ukraine; development of new varieties and hybrids of cereals, legumes, and oilseeds; the development of highly efficient agricultural seed cultures; conducting primary and zoned elite seed varieties and hybrids; development of technology for growing field crops; and providing scientific support and innovation (Kyrychenko and Kolomatska, 2011).

The YPPI coordinates the work of development and improvement of modern technology selection processes and creates lines and hybrids of sunflower for different areas of use. They also work on the organization of scientific innovation-oriented agricultural production in the Kharkiv, Donetsk, Lugansk, Poltava, and Sumy regions. They have three experimental stations: Ustymivsky DS Plant, Donetsk, and Luhansk DSDS. YPPI is the leader in providing training of scientific personnel of NAAS on the genetics of complex traits, information support breeding, and genetic studies (effective from 1974) and, since 2013, on the formation and maintenance of the bank of plant genetic resources (Kyrychenko and Kolomatska, 2011).

For the last 10 years in Ukraine, the assortment of sunflower hybrids has been essentially changed. YPPI has 30 new cultivars; among these, 10 are oleic types, 2 are palmitic, and 18 are linoleic. In the state trials of Ukraine and Russia, there are more than 35 hybrids with different vegetative periods, grain yields, and levels of resistance to downy mildew and broomrape. A separate group of stearic and palmitic hybrids, with contents of stearin up to 16% and of palmitin up to 22%, has been initiated. In the working collection of the breeders, there are more than 100 maternal and 1490 paternal forms. The conditions for a wide recombination of the genotypes have been created to develop experimental hybrids for which they will run trials in different ecological areas in Ukraine, Russia, and other countries. Research coordination has made it possible to create the prerequisites for sunflower germplasm examination in the YPPI. Checking the combining ability of sunflower inbred lines, sterile analogues, and pollen fertility restorers has permitted transferring 10 hybrids of joint selection by PPI and SGI, and 2 hybrids—PPI + IOC and IOC + SGI—to the State Variety Test. The created hybrids are ideal for plant height (145–160 cm), head diameter (17–22 cm), grain yield in competitive trials (3.9–4.5 t/ha), oil content (48–56%), oleic acid (60–88%), and resistance to downy mildew and the major races of broomrape (*Orobanche cumana*). A study of experimental material with higher β- and γ-tocopherol content is being conducted. Lines resistant to Eurolighting and Granstar are being developed as well. Over the last 10 years, industrial seed production of more than 27 hybrids has been initiated. In 2011, YPPI provided monitoring of major industrial hybrids' growth (Yason, Oskil, Dariy, Kyi, Zorepad, Boyets, Svitoch, Kvin, Kapral, and others) on more than 15,000 ha of hybridization plot that provide the seed for 1.8–2.0 million planting units of one of the most important oil crops (Kyrychenko and Kolomatska, 2011).

A number of highly productive sunflower hybrids and varieties with resistance to the main disease pathogens have been developed by the researchers of the laboratory for breeding and genetics of sunflower and the immunologists of YPPI since 1986. Eighteen hybrids and one variety of the Institute selection were registered on the Ukrainian Plant Variety List in 2010. Six of the hybrids—Kharkovskiy 49, Svitoch, Zorro, Dariy, Yason, and Queen—are listed in the State Plant Variety Register of the Russian Federation (Kyrychenko et al., 2010).

The possibility of genetic control of saturated and unsaturated fatty acid content in sunflower oils has been proved at the YPPI. Owing to the use of mutagens, a series of breeding entries with contents of palmitin from 4% to 32%, palmetoleate from 4% to 20%, oleate from 45% to 95%, and linoleate from 45% to 85% was obtained. The results of genetic analysis indicate that oleic acid content in oils is controlled by either dominant alleles of several independent loci or recessive alleles with a considerable contribution to dispersion of the effects of dominant modifier genes. Inheritance of high oleate content by hybrids is accomplished by incomplete dominance of paternal forms with a higher degree of the trait. This creates favorable possibilities for the selection of lines and hybrids of the NuSun type (mid-oleic type). The main method used to solve this problem is creating a biological "mixer" in the hybridization of a high oleic sunflower with the lines of a common linoleic type. The advantages of this method are the possibility of acquiring a wide genetic diversity of lines for the heterotic breeding and the development of competitive hybrids of NuSun with the yield of 4.9–5.3 t/ha, oil content of 48–54%, oleic acid content of 50–65%, and high resistance to downy mildew and main races of broomrape. These are industrial hybrids such as Dariy, Bohun, Kvin, Maksymus, Oreol, and others. Hybrids of NuSun are of particular importance because they combine higher oleic acid and saturated fatty acid contents. High contents of oleic and palmitic acids are controlled by independent genetic systems. Hybrids were developed as a result of crosses between high oleic lines and lines that have a high content of palmitic acid glycerides, which combine 70% oleate with up to 20% palmitate. Oils of this type are going through trials as a raw material for the production of lubricants for two-stroke engines (as a substitute for high-cost synthetics) and as a raw material for the production of hydraulic liquid. The carriers of genetic determinants for high oleic content are the lines X526V, X1228V, and X176V and palmitate lines X1008V, X1012V, and others, which are registered in the National Center for Plant Genetic Resources of Ukraine and stored in its depository (Burlov and Burlov, 2010; Kyrychenko et al., 2011).

The Institute of Oil Crops of the National Academy of Agrarian Sciences in Zaporozhye was founded in 1989. Today, the Institute is one of the largest scientific and industrial complexes of Ukraine. The focus of their work is creating new varieties and hybrids of oilseed crops on the basis of modern genetics, breeding, and biotechnology, as well as reproduction and cultivation of elite seeds. The main objectives of the scientific work of the Institute are: high-quality seed varieties in hybrids of sunflower and other oilseeds, modern technologies for cultivation of oilseeds, biochemical and chemical analyses (oil content, fatty acid composition), and so forth. For example, the "Slavyanin" sunflower hybrid with high oleic content created by the Institute in collaboration with the All-Russian Research Institute of Oil Crops (Krasnodar) was the first registered in the catalog of the year 2000 in Ukraine.

The agricultural company Sady Ukrainy in Kharkiv is the largest Ukrainian partner of the Institute of Field and Vegetable Crops in Novi Sad (Serbia). One in every six sunflower hectares sown in the Ukraine in 2005 was planted with seeds produced by the Sady Ukrainy company. There are 57 sunflower hybrids from the Institute of Field and Vegetable Crops registered in Ukraine. Nine of them are joint with Sady Ukraine (Dragan, Imperator, Terminator, and others). The agro-industrial company Flora, in Odessa, has three joint sunflower hybrids (Sasha, Pesma, Akkcent) with the Institute of Field and Vegetable Crops (Zaytsev et al., 2006).

Romania

Area, Production, Yield, and Localization in the Past and Today

The climate and soil conditions of Romania are favorable for growing sunflower, and it is one of the most important agricultural crops, as well as the most important oil crop. Thus, sunflower ranks third in cultivated land use, after maize and wheat. Romanian sunflower areas are about 10% of the arable lands. Sunflower is considered a low input and low production-cost crop, stimulating farmers to cultivate larger surfaces but without forgetting the limits imposed by crop rotation. Sunflower-cultivated surface has continuously increased because of the importance of its seeds for oil production in a country where sunflower oil is still in the top position in human consumption (Popescu, 2012).

Sunflower production increased from 556,000 t in 1990 to 1.80 million t in 2011 (Table 19.K). Crop rotation is limiting sunflower area to less than 1 million ha. The increased sunflower production was imposed by the demand of the Romanian oil industry, which the reason to export to almost all domestic sunflower seeds. The EU's higher needs for fats and oils of vegetable origin also stimulated farmers to produce

Table 19.K Sunflower Harvested Areas, Seed, Oil and Meal Production by Years in Romania

Year	Harvested Area (1000 Ha)	Seed Production (1000 t)	Oil Production (1000 t)	Meal Production (1000 t)
1992	560	618	248	214
1993	590	696	259	395
1994	582	767	286	436
1995	715	933	336	514
1996	910	1180	403	615
1997	778	858	365	562
1998	820	970	325	490
1999	1043	1300	350	456
2000	876	721	304	402
2001	800	823	303	370
2002	860	1003	281	374
2003	1153	1506	350	464
2004	925	1558	387	514
2005	957	1340	368	465
2006	981	1526	331	407
2007	748	547	256	311
2008	808	1170	173	271
2009	791	1098	241	356
2010	810	1262	327	389
2011	960	1789	319	396
2012	1088	1260	329	410

Sources: Oil Word Annual (2013); World Statistics Database (2013).

more sunflower seeds. They are aware that sunflower is a low input and low cost crop, as long as it requires a small quantity of seeds and fertilizers per surface unit. Romania is situated among the top sunflower producers in Europe, contributing 25% to the EU-27 sunflower production (Popescu, 2012).

Crushing and Trading, Price, and Market Characteristics

The Romanian oil industry is booming and is very competitive, crushing almost 950,000 t of sunflower seed each year thanks to great investments from the large foreign companies in this field (Cargill, Bunge, etc.). Sunflower remains a top product in Romanian exports. Romania is ranked third in Europe for refined oil production. The market has developed rapidly in recent years, and the international processors (mainly from the United States) produce two-thirds of the domestic oil. Two U.S. companies (Bunge and Cargill) have a 50% market share in the Romanian market, and Romania will soon be a greater refined oil supplier in Europe. There are also local investors with significant market shares (Argus Constanţa, 20%; Agricover Buzău, 9.5%; and 10 other similar companies) that are struggling to survive in an increasingly competitive market (Ionel, 2009; Popescu, 2012).

By tradition, Romania is a net sunflower oil exporter. The sunflower oil export was 150,000 t in recent seasons, which is about three times lower than in previous seasons (Table 19.L). The main destinations of sunflower oil exports were Spain, Hungary, and Greece. In sunflower seed exports, the main destinations were Spain, Hungary, France, Italy, and Slovakia (in the EU) and Turkey. Pakistan is also another main destination, but it is not a regular importer. In sunflower oil cake exports, Spain, Hungary, France, Italy, and Germany are the main buyers. The local livestock production in Romania absorbed 160,000 t in recent seasons due to the rehabilitation of the pig industry and development of the calf- and poultry-raising sectors (Ionel, 2009; Oil World Annual, 2013; Popescu, 2012; World Statistics Database, 2013).

The income from oilseeds exports is quite stable due to previous contracting. This happens not only with sunflower oil, but also with sunflower oil cakes. The oil factories have small storage facilities that ensure their operation for one month. The oil factories have long-term contractual relations with the silos, and it seems that both parties are satisfied with this partnership. Romania has a positive trade balance from the trade of oilseeds, oils, and oil cakes. Almost all farmers sell their oilseeds for production at harvest time because they do not have adequate storage facilities. Even when they have storage facilities, they cannot maintain suitable storage conditions for a very long time. In sunflower production, about 70,000 t of seeds do not go to the market. This quantity of seeds is used by farmers for obtaining oil in small processing units. The sunflower subsidy accounted for 20% of the total sunflower price, out of which 95% is the direct payment per hectare and 5% is the diesel oil subsidy in recent seasons in Romania (Ionel, 2009; Popescu, 2012).

Table 19.L Sunflower Seed Domestic Consumption, Seed Crushing and Exports in Romania

Year	Seed Domestic Consumption (1000 t)	Seed Crushing (1000 t)	Oil Export (1000 t)	Seed Export (1000 t)	Meal Export (1000 t)
1995	893	866	80	25	100
1996	1080	1035	150	20	180
1997	973	953	90	25	181
1998	860	830	80	100	187
1999	588	860	125	290	195
2000	648	687	34	104	174
2001	685	697	41	128	186
2002	720	704	9	128	116
2003	795	876	38	366	148
2004	790	970	91	310	177
2005	733	889	83	40	234
2006	738	785	60	26	182
2007	685	602	54	47	71
2008	630	524	45	148	271
2009	700	688	92	192	356
2010	715	752	147	192	158
2011	635	765	154	463	215
2012	680	790	141	105	181

Sources: Oil Word Annual (2013); World Statistics Database (2013).

In Romania, the consumption of vegetable oil has been 24 kg per capita in recent years. Domestic consumption of sunflower oil was 272,000 t and it was more than 50% of the total domestic fats and oils consumption (512,000 t) in 2012. In Romania, animal fats are traditionally consumed during the period of winter holidays, when the pigs are slaughtered; it is in autumn that the maximum level of vegetable oil consumption is found. For the most part, sunflower oil is consumed in Romania, but other vegetable oils, which are mainly imported, are also available. In order to boost oil and margarine sales in recent years, specific advertisements emerged that were supported by processors (Ionel, 2009; Popescu, 2012).

Romanian consumers mainly prefer sunflower oil. There are several Romanian brands that are recognized at a national level and they are preferred, depending on the consumers' incomes. Both the local producers and the importers are trying to diversify their supplies and strengthen their roles on the market. The first three producers (Unilever, Orkla, and Rozal Brinkers) have 90% of the margarine sales. The competition is very high for obtaining the best raw oil prices possible for margarine production. At present, the main player on the oilseed market is the Cargill company, which owns about 10% of the total storage capacity and uses about 50% of the total useful storage capacity so it can buy more seed during the harvest and then supply more sunflower seed for liquid oil or margarine over a long period (Ionel, 2009; Popescu, 2012).

Research, Seed Science, and Technology in Sunflower

The Fundulea Institute was a pioneer in breeding hybrid sunflower, releasing the first commercially grown hybrids in 1971. Many hybrids bred at Fundulea were registered and have been cultivated in countries such as Italy, France, Spain, and China. The National Agricultural Research and Development Instititute (NARDI) in Fundulea celebrated 50 years of research on field crops in 2007. However, NARDI continues a longer tradition of agricultural research in Romania, being the main successor of the former Romanian Agronomical Research Institute (ICAR) in Bucharest.

The main directions of NARDI Fundulea are developing new varieties or hybrids of sunflower that are adapted to the Romanian environment; improving crop management practices, including for use in organic agriculture, adapted to the Romanian environment; and so forth. Breeding results for sunflower made the NARDI known worldwide, concomitantly with classical selection work resulting in many varieties that produced higher yields than old varieties. The first hybrids (Romsun 52 and Romsun 53) were released in 1971 and produced commercially, and then Romania became the first country largely growing sunflower hybrids. New hybrids were released later utilizing cytoplasmic male sterility (CMS) and pollen fertility restoration for hybrid seed production (Verzea, 2007).

Significant genetic progress was achieved between 1965 and 1982 for disease resistance to downy mildew, broomrape parasite, and others. Through genetic research, dominant genes were identified for controlling resistance to five broomrape races (A, B, C, D, and E) in Romania. NARDI identified a differential set of broomrape races containing cumulative resistant lines to the five old races having the dominant genes *Or1, Or2, Or3, Or4,* and *Or5* (Vranceanu et al., 1980). The race F was found in Romania, too. Dominant gene *Or6* was identified in the inbred line LC 1093, which confers resistance to this race (Pacureanu-Joita, 1998). Changes of the broomrape race composition in Romania have been reviewed (Pacureanu-Joita et al., 2009). Significant progress was also achieved in developing self-compatible hybrids. Thirty-two hybrids have been released since 1971, and they are growing widely not only in Romania, but also in other countries such as France, Spain, Greece, Italy, and China. A new trend has been emphasized recently to develop higher oleic acid content and herbicide (IMI and SU) resistance in Romania. The first high oleic hybrids are undergoing regional trials, and the first IMI and SU herbicide-resistant sunflower hybrids will be available soon (Pacureanu et al., 2009, 2012).

Bulgaria

Area, Production, Yield, and Localization in the Past and Today

Sunflower is the main oil crop in Bulgaria, and planting area has increased in recent years because of higher profitability and low input requirement, as well as higher world prices and exporting possibilities in Bulgaria. Due to the higher oilseed demand of the EU, sunflower prices seems to increase continuously while more rapeseed goes to biodiesel production. Due to higher adaptation capabilities to dry conditions, sunflower is suitable to Bulgaria's climate, but higher rates of disease and pests have severely limited the sunflower planted area in some years. In the past, it was not advised that sunflower be planted more than once every 5 years in a crop rotation, but now sunflower is in rotation more frequently because of higher profitability over other crops and the fact that there is no alternative crop to compensate. Sunflower is a row crop in Bulgarian agriculture, and cultivation helps maintain proper rotation. Sunflower plays a significant role in Bulgarian economy and international trade that increases year by year (USDA, 2006).

At present in Bulgaria, an average of 650,000–700,000 ha of sunflower are planted and, therefore, sunflower is the second most important crop after wheat. However, sunflower area reached a record 762,000 ha in 2012 (Table 19.M). Sunflower seed production is about 1.3–1.4 million t and seed yield is 1.8–2.0 t/ha in recent years. Sunflower planted areas are expected to increase due to attractive and higher prices for sunflower this season. Reseeding of winterkill rapeseed areas, earlier planting at the last week of March, strong export demand, growing local crush capacity, and improving the domestic market are factors promoting sunflower planting in Bulgaria. Sunflower planted areas are mostly located in fertile plains of eastern and central parts such as Plovdiv, Dobroudja, Razgrad, Rouse, and so forth.

Recently, sunflower oil production in Bulgaria has been about 150,000 t and meal production about 180,000–200,000 t. Sunflower oil is the most popular oil in Bulgaria among oilseeds (total 262,000 t in 2012); the other was rapeseed oil (15,000–20,000 t). Similarly, sunflower meal is the most preferable and produced meal among oilseed meals, followed by corn germ and rapeseed meal.

Crushing and Trading, Price, and Market Characteristics

Bulgaria is a net sunflower exporter country (Table 19.N). Although Bulgaria exports mainly sunflower seed, oil and meal are also exported. With increasing local crushing capacity in recent years, domestic crushing and exports may be higher in the future due to higher demand both in the EU and in Turkey. Turkey is the main destination for sunflower seed export, but EU countries such as the Netherlands, Portugal, Ro-

Table 19.M Sunflower Harvested Areas, Seed, Oil and Meal Production by Years in Bulgaria

Year	Harvested Area (1000 Ha)	Seed Production (1000 t)	Oil Production (1000 t)	Meal Production (1000 t)
1992	475	594	126	178
1993	469	432	149	176
1994	495	601	139	180
1995	586	766	190	216
1996	499	526	159	183
1997	452	438	155	199
1998	538	524	183	180
1999	592	605	174	178
2000	511	425	154	154
2001	389	405	108	125
2002	471	645	128	149
2003	659	788	135	153
2004	592	1078	154	171
2005	635	934	192	209
2006	750	1196	182	195
2007	602	564	156	168
2008	721	1300	79	159
2009	683	1317	116	188
2010	729	1536	168	163
2011	684	1439	156	183
2012	762	1348	148	215

Sources: Oil Word Annual (2013); World Statistics Database (2013).

mania, France, and Germany are also main destinations for import due to having no export tax.

In Bulgaria, private companies dominate the oil production sector (14 of 15 companies are private). Seventy new factories with generally small production capacities mainly processed sunflower seed. otal oilseed processing capacity of Bulgaria is about 800,000–850,000 t per year. Some of the main oilseed oil producers in Bulgaria are Papas Olio AD (Jambol), Bisser Oliva AD (Stara Zagora), Kaliakra AD (Dobrich), Niva AD (Kostinbrod), Zvezda AD (Dolna Mitropolia), Rosa AD (Popovo), Oliva EAD (Knezha), and Zlatna Trakia Pliska OOD (Shumen) (Balabanova, 2004).

Although domestic hybrids and varieties (such as San Luka, Albena, Merkuri, Musala, and from OP cultivars, Favorit) were grown in the previous 10 years, recently

Table 19.N Sunflower Seed Domestic Consumption, Seed Crushing and Exports in Bulgaria

Year	Seed Domestic Consumption (1000 t)	Seed Crushing (1000 t)	Oil Export (1000 t)	Seed Export (1000 t)	Meal Export (1000 t)
1995	315	479	61	144	116
1996	325	407	21	95	—
1997	350	441	35	70	1
1998	338	400	29	54	25
1999	345	395	13	254	27
2000	280	343	21	52	29
2001	243	277	108	103	27
2002	240	330	129	229	21
2003	305	340	133	297	43
2004	333	379	148	268	41
2005	355	465	37	350	53
2006	358	432	48	301	81
2007	333	371	43	227	24
2008	275	352	56	243	159
2009	280	415	88	388	188
2010	240	360	99	166	163
2011	240	405	108	221	183
2012	230	475	132	149	215

Sources: Oil Word Annual (2013); World Statistics Database (2013).

international seed companies are so active that many foreign hybrids are preferred and make up most of what is being sold in the seed market in Bulgaria (Balabanova, 2004).

Sunflower is grown mainly for seeds, from which the most-used vegetable oil in Bulgaria is produced. The industrial utilization of sunflower oil includes the liquid oil industry, food oil product (mayonnaises and margarines) production, and nonfood purposes such as paints, varnishes, and so forth. Vegetable oil consumption per capita is about 26 kg in recent years (Balabanova, 2004).

Research, Seed Science, and Technology in Sunflower

The purposeful breeding work on sunflower in Bulgaria began in 1941. Breeding was primarily aimed at developing sunflower varieties resistant to the parasite *Orobanche cumana*. Since 1962, Dobrudzha Agricultural Institute (DAI)–General Toshevo has become a leading center for sunflower breeding and a coordinator of the scientific research work on this crop. Investigations on the use of heterosis in breeding and the production of self-pollinated lines were initiated. Since 1985, the breeding work at DAI has been focused on the use of heterosis by developing lines and hybrids, production of seeds from parental lines, and registration of both their own and joint hybrids. One breeding success was the development of the hybrid Albena; it was introduced in France in 1986, and at its best it occupied about 40% of the total area sown with sunflower in this country. Due to its good adaptability and productivity, it became a world standard. It has been successfully grown in France, Germany, and Austria (Christov et al., 2011).

From the establishment of the DAI institute until today, a total of 28 Bulgarian and 14 joint hybrids and 8 varieties have been developed. The hybrids Dobrotsvet, Biotsvet, Valin, Alpin, Yana, and Rada were developed by combining biotechnology methods with conventional breeding. They are resistant to broomrape and some economically important diseases in Bulgaria.

One of the reasons for lower yields of other hybrids is the lack of resistance to biotic and abiotic factors. DAI adopted a policy of using resistant varieties and hybrids to solve this problem, in additon to broomrape parasites.

The intensive agricultural development during recent years involved the use of a limited set of sunflower varieties and hybrids on large areas, which narrowed genetic variability and increased the danger of losing valuable genetic material. Preservation and enrichment of the plant gene pool is a national task with significant priority at DAI. During recent years, producers in Bulgaria and elsewhere have tried to use more highly productive hybrids that have the ability to overcome the changeable stress factors of the environment (diseases, parasites, pests, variations of the ecological and meteorological conditions). For solving these problems, the following strategic aims have been set forth: developing new sunflower varieties and hybrids with higher productivity and variable fatty acid composition; developing new sunflower varieties and hybrids with higher resistance to the diseases downy mildew, Sclerotinia, Phoma, Phomopsis, Macrophomina, Alternaria, and the parasite broomrape; and developing new sunflower varieties and hybrids with enhanced resistance to various stress factors. For achieving the these purposes, the genetics of cultivated sunflower and the wild annual and perennial species of genus *Helianthus* L. are being investigated; breeding work is carried out in populations, varieties, lines, and hybrids; and investigations are done on pests and diseases on sunflower (Christov et al., 2011; Encheva et al., 2012; Hristova-Cherbadzi, 2009; Hristova-Cherbadzi et al., 2012).

At DAI, the perennial species of genus *Helianthus* are maintained in a stationary collection with official catalog numbers registered at FAO, and the annual species of genus *Helianthus* are reproduced under field conditions and in a greenhouse. Seeds are stored under specific conditions: in a chamber for long-term storage and in freezing chambers for short-term storage (Hristova-Cherbadzi, 2009; Hristova-Cherbadzi et al., 2012). Collections of sunflower varieties of foreign origin and CMS sources are maintained for the purposes of scientific research work. The efforts of all specialists are directed toward developing high-yielding varieties and hybrids with high oil content; sunflower forms with nontraditional and dietetic oil composition; varieties and hybrids with high protein content that are resistant or tolerant to diseases and pests; genetic variability through combining conventional breeding with biotechnology methods; IMI-resistant hybrids; express-resistant hybrids; and so forth. To achieve these aims, various biotechnology methods have been elaborated in parallel with conventional breeding. The combined conventional breeding methods with in vitro approaches, distant hybridization, and mutagenesis are a prerequisite for successful solving of the strategic tasks. In recent years, the problem of accelerating the breeding process has been increasingly significant in many crops, including sunflower. The methods of induced parthenogenesis and embryo culture are particularly important. The first method allows parthenogenic development in the initial parental forms and, as a result, homozygous lines can be obtained within 2 years. Hundreds of fertility restorer lines have been produced at DAI by this method, and these were further involved in hybridization (Encheva, 2009; Encheva et al., 2003, 2012).

The second approach allows obtaining four generations within a one year, which is not possible through conventional breeding. To accelerate the production of CMS analogs, 10-day-old zygous embryos have been used in the embryo culture method. By the embryo rescue method applied to interspecific and intergeneric crosses, valuable lines have been produced at DAI that are also included in the breeding programs of sunflower (Encheva, 2009; Hristova-Cherbadzi, 2009; Hristova-Cherbadzi et al., 2012).

Turkey

Area, Production, Yield, and Localization in the Past and Today

Sunflower plantings in Turkey were initiated by immigrants from Romania and Bulgaria after World War I. In the 1950s, sunflower production was around 70,000 t. In the late 1980s, production peaked at 1.25 million t. However, after that year, it has gradually reduced due to the sunflower/wheat price parity that has changed at the expense of sunflower (Table 19.O) (Kaya et al., 2008b). This ratio should normally be over 2.5 to increase sunflower area, but it is currently at about 2. Turkish sunflower planting areas mostly exist in the Trakya region (50%), the European part of Turkey. Other regions are the Central Anatolia, Mediterranean, South Marmara, Black Sea, and Aegean regions. Recently, aside from Trakya, Central Anatolia (mostly the Konya province) and the Adana province increased in importance year by year. On the other hand, due to recent increase of sunflower prices, sunflower planted areas doubled in Central Anatolia. Sunflower harvesting in Adana is approximately 1–1.5 months earlier than in Trakya and provides some supplies for the crushers in Anatolia (Kaya, 2014; Kaya et al., 2008b).

Table 19.O Sunflower Harvested Areas, Seed, Oil and Meal Production by Years in Turkey

Year	Harvested Area (1000 Ha)	Seed Production (1000 t)	Oil Production (1000 t)	Meal Production (1000 t)	Total Supply Seed (1000 t)	Total Supply Oil (1000 t)
1992	700	980	420	380	1065	691
1993	580	700	319	289	835	716
1994	550	600	445	408	1185	841
1995	625	750	482	430	1320	742
1996	540	545	410	370	1040	720
1997	520	650	506	463	1313	737
1998	520	650	512	460	1285	722
1999	540	800	520	467	1267	684
2000	540	575	385	346	926	551
2001	480	520	287	258	695	464
2002	550	820	440	395	1059	528
2003	500	600	529	475	1269	637
2004	480	650	494	443	1203	676
2005	430	750	495	441	1186	975
2006	480	850	525	472	1360	810
2007	480	700	544	483	1332	910
2008	500	830	515	460	1323	1033
2009	460	800	626	559	1630	888
2010	485	1000	671	597	1838	1082
2011	490	925	718	639	1994	1439
2012	590	1100	739	659	1869	1676

Sources: Oil Word Annual (2013); World Statistics Database (2013).

Although the main rotation system is sunflower and wheat in the Trakya region, sunflower production is increasing in Turkey due to the enlargment of planting areas in other regions such as Central Anatolia and Cukurova. In 2012, Turkish sunflower area and production reached 600,000 ha and over 1 million t, respectively. However, total production increased only 14% in 2012, despite a 20% increase of plant areas, due to dry and hot summer conditions that highly affected seed yields. Sunflower planting was shifted from common areas such as Trakya, Cukurova, and the South Marmara region, which have dependable rain during the vegetation period, to new regions such as the Central Anatolia region, where irrigated areas provide higher yields.

However, Central Anatolia farmers can choose among many crops due to the possibility of irrigation, so sunflower planting increase is mainly dependent on the price and net profit per hectare in this region. Therefore, farmers could escape easily from sunflower planting, moving to other traditional crops of the region, such as wheat, corn, and sugar beet. The Cukurova region has some particular advantages, such as earlier planting and maturing a month earlier than in Trakya. However, recent sunflower increases in both planting areas are related to sunflower prices and also related to the prices of the main competitive crops: wheat in Trakya and corn in Central Anatolia and Cukurova regions. On the other hand, Turkey is also a dependable member of the world sunflower market and one of the biggest players in the world sunflower trade (Statistical Database 2012; USDA, 2012; World Sunflower Market Update, 2013).

Crushing and Trading, Price, and Market Characteristics

Processing and Crushing. Turkish oilseed crushing capacities exceed 7 million t, but unfortunately half of this capacity has been used in recent years due to lower domestic production and higher crude oil and meal exports. One of the important reasons for this is that the biggest sunflower producers in the world, Ukraine and Russia, were the closest destinations for sunflower seed exporters for Turkey. They started to crush their produced seeds in their own facilities, and also to apply a higher export fee for sunflower seeds in recent years to reduce seed exports and to promote domestic crushing. Every year, a new crushing facility is added to the existing firms (there are around 200) occupied with crushing, refinery, and also margarine production. The new high-capacity modern crushers have lower crushing cost, so that smaller crushers with older technology cannot compete with them economically. Although sunflower production and refined oil and margarine exports have had significant increases in recent years, lower capacity utilization remains one of the biggest problems in the vegetable oil industry due to lack of raw materials and also to crude oil imports instead of seed imports (Kaya et al., 2008b).

Turkish oilseed processors (both crushers and refineries) are mostly located in Trakya, Marmara, and in nearby areas, due to the fact that the Marmara region hosts over half of the sunflower production of Turkey. The Aegean region follows these, due to higher production of both cottonseed and olive oil. Processors are located near ports so they can import huge seed quantities easily when domestic demand rises (Kaya et al., 2008b).

Turkish vegetable oil is mostly used to produce liquid oils, and the remaining part is used to make margarine. Because sunflower oil is the most consumed vegetable oil in Turkey, Turkish oilseed processing is mainly focused on the sunflower crop. Biodiesel obtained mostly from canola oil is another factor that increases and directs the vegetable oil needs of Turkey, but the acquirements decreased starting in 2012 due to lower worldwide petrol prices. At least in the near future, it is obvious that Turkey will remain the major importer country of the world to both meet the domestic sunflower oil consumption requirement of about 850,000 t and to maintain its sunflower oil exports to Iraq and Syria by using its logistics advantage (Kaya et al., 2008b). In the last 10 years, seed import has not increased as much as oil or meal imports (Table 19.P).

To supply enough seed for its domestic factories, sunflower production in Turkey should be increased to 2 million t at a minimum. It may not be possible to achieve this goal now, but recent developments mean that it could be possible in the near future. Therefore, the government should support sustainability and the price parity of sunflower with wheat (as the main rotation crop of sunflower). However, given that sunflower agriculture is currently lucrative, sunflower production may climb up and exceed 1 million t or even exceed the record 1.25 million t, achieved in late 1980s, in the next seasons (Table 19.P). Furthermore, high oleic sunflower production could bring new possibilities to Turkey and Eastern Europe, which have half of the sunflower production in the world due to recent healthy consumption demands (Kaya et al., 2008b).

Sunflower yields are well above the world yield average, around 1.2–1.3 t/ha due to better inputs (hybrid seed, fertilizer, herbicide, etc.). Farmers use mostly hybrid sunflowers (over 98% percent) for production in Turkey, and the Turkish sunflower hybrid market is about 2,000 t (Kaya, 2004).

Agricultural Sales Cooperative Unions. Trakya Birlik and Karadeniz Birlik, two leading agricultural sales cooperative unions in the vegetable oil market, continue to play a very important role for the sustainability of sunflower seed production in Turkey. These two cooperative unions provide seeds and fertilizer, give lower cost credit financing to their members prior to plantings, and offer attractive prices during the harvest season. Trakya Birlik has 110,000 members and consists of 48 cooperatives. It is active mainly in the Trakya-Marmara region, and also has a couple of cooperatives in the Middle Anatolia and Aegean regions. Karadeniz Birlik is mostly located in Eastern Europe but purchases sunflower seed from farmers in other regions such as Middle and Eastern Anatolia (Kaya et al., 2008b).

Trakya Birlik and Karadeniz Birlik purchase a significant amount of the Turkish sunflower crop. Farmers generally deliver seeds to them, but if the two cooperatives do not announce reasonable prices, they prefer to sell to other traders and crushers with higher prices. The combined purchases of the two unions in 2013 was 22% of total sunflower seed sold in Turkey, and it has decreased in recent years, depending on announced price and capacity. Meanwhile, private crushers' demands and market share increase year by year (Kaya et al., 2008b; USDA, 2012, 2013a).

Sunflower Seed Trade, Because the domestic sunflower crop is far from meeting domestic crushers' demands, Turkey is one of the biggest export destinations (the

Table 19.P Sunflower Seed Domestic Consumption, Seed Crushing and Exports In Turkey

Year	Seed Domestic Consumption (1000 t)	Seed Crushing (1000 t)	Oil Export (1000 t)	Oil Import (1000 t)	Seed Import (1000 t)	Meal Import (1000 t)
1992	1028	1000	73	154	50	20
1993	798	760	122	278	100	52
1994	1113	1075	167	316	550	82
1995	1183	1133	65	200	500	50
1996	960	960	65	240	360	50
1997	1273	1225	122	161	583	27
1998	1255	1208	85	140	595	94
1999	1237	1227	62	104	437	230
2000	914	909	4	106	321	149
2001	683	676	1	147	165	57
2002	1046	1039	29	68	229	63
2003	1240	1234	16	83	660	312
2004	1167	1161	22	152	529	281
2005	1165	1159	99	451	407	386
2006	1249	1242	34	145	498	403
2007	1277	1270	54	326	533	222
2008	1216	1209	131	432	446	312
2009	1477	1470	68	184	736	432
2010	1577	1570	157	401	705	500
2011	1687	1680	271	651	834	739
2012	1737	1730	350	700	550	600

Sources: Oil Word Annual (2013); World Statistics Database (2013).

second worldwide after EU-27). Due to logistic advantages, sunflower imports are mainly carried out from Black Sea countries such as Bulgaria and Romania (Table 19.Q). Most Turkish sunflower crushers are located in the Trakya region, so DAP Bulgaria is the most preferred delivery mode. In 2012, total imports of sunflower seed for crushing were about 45% and reached about 750,000 t in Turkey. The larger and increased domestic consumption of sunflower oil and exports kept demand high for seed imports in 2012. Russia, Ukraine, Bulgaria, Moldova, and Romania were the leading suppliers.

Sunflower Oil Trade. Because sunflower oil production cannot meet domestic sunflower oil consumption of around 850,000 t, sunflower oil imports are required. Sunflower oil imports are mainly carried out from Ukraine and Russia (Table 19.R). Argentina is also preferred, especially for the second half of the season, when the Black Sea exportable supplies are getting lower, but Argentina has lost importance in the international sunflower trade arena in recent years due to reduced production and also because of its proximity. The Inward Processing Regime (IPO) is a system allowing Turkish manufacturers and exporters to obtain raw materials, such as oilseed or crude oils, that are used in the production to re-export again without paying customs duties. With no applied commercial policy measures, sunflower imports under IPO are in parallel with sunflower oil exports (Table 19.R). Turkish crude sunflower seed oil imports reached about 88% in 2012 with a record 743,000 t, mainly from Russia and Ukraine, due to higher domestic consumption and exports, as well as GMO restrictions of other vegetable oils (Oil World Annual, 2013; USDA, 2013a).

Turkey's refined sunflower oil exports increased three times from 2010 to 2013, reaching 553,000 t. Iraq was the main buyer (86%) with 474,000 t. Both bottled and canned refined sunflower oil exports continued in the following years. Turkey also has refined sunflower oil and margarine exports to other Middle Eastern and North African countries due to its logistic advantage (Table 19.S). Sunflower exports reached over $1 billion in 2013 (Kaya, 2014). In Turkey, the government supports local sunflower oil refiners with the IPO (not paying the 22% import duty on crude sunflower oil imports if they are exported as refined sunflower oil). This creates added

Table 19.Q Sunflower Seed Imports (1000 t) by Years in Turkey

Countries	2012	2009	2006	2003
Bulgaria	149	217	158	152
Romania	79	50	97	46
Moldova	102	57	33	2
Ukraine	179	121	8	177
Russia	243	4	60	83
Uruguay	—	—	—	67
USA	1	10	15	7
Hungary	—	1	—	—
Others	1	2	2	2
Total	754	468	372	536
Avr CIF	$589 t	$491 t	$313 t	$282 t

Sources: Oil Word Annual (2013); Trakya Birlik (2012); TUIK (2012).

Table 19.R Sunflower Oil Imports (1000 t/Million $) by Years in Turkey

Countries	2012	2009	2006	2003
Ukraine	251	178	160	67
Russia	485	125	92	10
Argentina	6	16	116	10
Bulgaria	—	—	7	—
Romania	1	5	17	5
Moldova	6	—	—	—
Others	1	—	8	—
Total	743	324	399	93
Average CIF	$846 t	$846 t	$591 t	$584 t

Sources: Oil Word Annual (2013); World Statistics Database (2013).

value and increases capacity usage levels for refiners (Oil World Annual, 2013). In addition, Turkey is located in a traditional position for trading crossroads and is in ideal situation for exporting vegetable oils to many other countries such as North Africa, the Middle East, Central Asia, and Eastern Europe (USDA, 2013a).

Sunflower Meal Trade. Sunflower meal has become increasingly important in recent years due to stronger demands from the livestock, dairy, and aquaculture in-

Table 19.S Sunflower Oil Exports (1000 t/Million $) by Years in Turkey

Countries	2012	2009	2006	2003
Iraq	474	141	183	8
Syria	15	20	29	—
Israel	8	8	6	4
Turkish Cyprus	4	5	3	2
South Korea	1	3	2	—
Sudan	1	2	—	—
Libya	1	—	—	—
Yemen	7	5	2	—
Others	42	58	9	15
Total	553	242	233	29
Average CIF			$885 t	$795 t

Sources: Oil Word Annual (2013); World Statistics Database (2013).

dustries. However, sunflower meal production is also far from meeting the huge demands of the Turkish livestock sector (Table 19.T). Turkey's sunflower meal imports increased in 2012 by about 28% over the previous year, due to increased domestic consumption, and meal imports reached a record 728,000 t. Turkey continued to import sunflower meal from mostly Russia and Ukraine, using its logistic advantage and their availability of increased crushing capacities. However, Turkey imports less meal from other Black Sea countries such as Romania, Bulgaria, and Moldova due to their lower crushing capacity and prefers to mainly import oilseeds from them. In the immediate period and long term, it seems that Turkey's meal demand will continue to increase for protein meals for feeding of livestock, poultry, and aquaculture products due to recent rising household income (USDA, 2012, 2013a).

Trade Policy. Because Turkey is the major importer of sunflower in the world, domestic prices are mainly driven by import costs. The import duties on sunflower seed and sunflower oil are 27% and 36%, respectively, for all countries except Bosnia and Herzegovina (which have no import duties). Turkey also has bilateral agreements with the EU within tariff quotas (sunflower seeds: 1000 t, 0% duty; sunflower seed oil: 18,400 t, 0% duty). Sunflower seed imports are issued by the 27% tariff with a surveillance price of $525 (U.S. dollars) per ton based on the 2013 customs regime. However, Turkey takes care to always consider the quotas in order to maintain the value of Trakya Birlik's stocks. Trakya Birlik and Karadeniz Birlik are also effective in guiding the government subsidies and import tax protections; because they have many member farmers, the government should consider their interests and closely control sunflower domestic trade and production (buying and selling seed, and also the crushing and refining industries). Turkey needs to continue to support oilseed

Table 19.T Sunflower Meal Imports (1000 t/Million $) by Years in Turkey

Countries	2012	2009	2006	2003
Ukraine	215	205	129	37
Russia	494	141	154	46
Romania	12	12	105	7
Bulgaria	4	42	35	13
Moldova	4	—	—	8
Georgia	—	44	—	—
Others	—	—	1	3
Total	728	445	423	114
Average CIF	$176 t	$176 t	$108 t	$109 t

Sources: Oil Word Annual (2013); World Statistics Database (2013).

production both by giving subsidies directly to farmers and investing in agriculture with new irrigation projects throughout the country to meet larger domestic demands for oilseeds. However, despite relatively high tariffs to protect producers and increase oilseed production, Turkey's imported oilseeds, meal, and oils demands will still remain strong both for higher domestic consumption and also for the recent increase in re-exports even during crises in Middle Eastern and North African countries (Kaya et al., 2008b; USDA, 2012, 2013a).

After the 2001 season, when production bottomed out at 550,000 t, the Turkish government finally realized the importance of sunflower, as well as other oilseeds, and gradually increased its supports (especially premium payments per kilogram). After this, sunflower started an upward trend. The production subsidy program of Turkey is continuing to attract farmers to plant oilseeds, particularly sunflower seed and cottonseed. Despite serious delays in these payments, the program is still much appreciated by farmers (Table 19.U).

Consumption. Cotton is not as important as sunflower due to its lower oil content of around 20%, as well as the lower consumption of cottonseed oil. Sunflower oil comprises approximately 80% of Turkish liquid vegetable oil consumption and it is the most consumed and preferred oil in Turkey. It is expected that sunflower oil consumption will continue to increase for next years, and total consumption could reach 850,000 t.

Research, Seed Science, and Technology in Sunflower

Turkey's sunflower history was just starting in the 1960s, and the main sunflower research and breeding program in sunflower was initiated at the Trakya Agricultural Research Institute (TARI) in the 1970s. However, sunflower breeding programs were accelerated in 1978 after TARI was nominated as a leader of the National Sunflower Research Program and as the Turkish national sunflower research coordinator institute. After sunflower genetic materials were collected at TARI during the 1980s, mostly via introduction from abroad, sunflower breeding research was increased and accelerated via hybrid breeding (Kaya, 2010).

Due to the many advantages to farmers, the development of hybrids that have high seed and oil yield performance and large adaptation capability are the primary objectives of most sunflower breeding programs in the world. The national sunflower hybrid breeding program at TARI designed a public breeding program oriented toward basic research, investigating genetic inheritance of new traits, and exploring for new sources of genes for important traits. In addition, like a private breeding program, they emphasized developing new lines and hybrids that have important characteristics for Turkey, such as high seed and oil yield, oil content, oleic acid content, broomrape resistance, Imidazolinone (IMI) and Sulyfonyl Urea (SURES) herbicide resistance, and different disease resistance (Kaya, 2010).

Table 19.U Sunflower Prices and Subsidies and Sunflower/Wheat Parity by Season in Turkey

	Trakya Birlik	Sunflower Subsidies	Sunflower/ Wheat Parity Including Subsidies	Sunflower/ Wheat Parity Excluding Subsidies
Season	Price (USD/t)	Price (USD/t)		
2000	251.4	60.0	2.13	1.72
2001	271.1	54.5	2.44	2.03
2002	282.4	55.2	2.65	2.21
2003	348.5	79.1	1.82	1.48
2004	322.3	89.8	1.81	1.42
2005	376.6	130.5	1.84	1.49
2006	357.7	192.3	2.24	1.60
2007	322.3	89.8	1.81	1.42
2008	376.6	130.5	1.84	1.49
2009	357.7	192.3	2.24	1.60
2010	634.5	162.0	2.29	2.02
2011	711.7	128.5	2.28	1.94
2012	891.2	128.5	2.75	2.40

Source: World Sunflower Market Update (2013).

Broomrape is the biggest problem in the Trakya region. Weed control and some diseases, such as downy mildew, Sclerotinia, Macrophomina, and so forth, have been the other main problems for Turkish sunflower production in recent years. Broomrape affects Turkish sunflower production not only in the Trakya region, but also in other regions such as the Black Sea, Cukurova, and Middle Anatolia (Kaya, 2010).

Sunflower hybrids tolerant to IMI herbicides (part of the so-called Clearfield system) are the primary successful method for solving broomrape problems both in Turkey and throughout the world. The Clearfield system was first started in 2003 in Turkey. Sales of IMI-type hybrids tripled in 2006, reaching about 45–47% of the sunflower area in the Trakya region and 25% of the total sunflower planting areas of Turkey (Evci et al., 2012; Kaya, 2010; Kaya et al., 2013).

New races of broomrape have been observed in Turkey, Spain, Bulgaria, Romania, Ukraine, and Russia in recent years, and there is no completely resistant sunflower hybrid to these new races due to the highly virulent behavior of broomrape against the resistant genes. Due to this issue, and also to efficiently control broomrape and key weeds such as *Xanthium strumarium* L., *Sinapis arvensis* L., *Chenopodium album* L., *Cirsium arvense* Scop., *Avena* spp., *Amaranthus* spp. *Convolvulus arvensis* L., and

Datura stramonium in sunflower production, the Clearfield system has spread greatly. There are still some hesitations and complaints from farmers regarding soil residues from IMI herbicides in the Clearfield system that affected the crops (mostly wheat) in the rotation. A commercial IMI herbicide consisting of Imazamox and Imazapyr (33 + 15 g/l) has been used in Turkey, but Imazapyr is now banned in the EU, and new commercial herbicides are presented to market rapidly. New herbicides containing Imazamox (40 g/l) or others from the IMI group could reduce the number of weeds. The other solution is using sunflower hybrids that are both genetically resistant to broomrape and also tolerant to herbicides from the Sulfonyl Urea group in sunflower production in Turkey, but the SU herbicide group does not efficiently control some key broadleaf weeds; both IMI and SU herbicide resistance must be combined (Evci et al., 2012; Kaya, 2010; Kaya et al., 2012, 2013).

The national sunflower program was divided into four main parts in recent years based on new trends and necessities in Turkey and around the world (Kaya, 2010; Kaya et al., 2012):

1. Developing oil-type (linoleic) hybrids
2. Developing broomrape- and herbicide-resistant (IMI and SURES) hybrids
2. Developing high oleic–type hybrids
3. Developing confectionery-type hybrids

The general objectives of the oil-type breeding programs are:

- Developing new inbred A and restorer lines that have high general combining ability (GCA) and specific combining ability (SCA); high yield capacity and oil ratio; high oleic and tocopherol content; earlier maturity and other desired agronomic traits; and resistance to broomrape parasite, IMI and SURES herbicide, and diseases (especially Sclerotinia, downy mildew, Phomopsis, Macrophomina, and others)
- Setting up new sunflower hybrids that have the desired characteristics and crossing these inbred lines, utilizing the SCA of the lines

The general objectives of the confectionery-type breeding program are:

- Developing new inbred A and restorer lines that have high GCA and SCA; high yield performance and adaptation capabilities; bigger seed size; white/grey color; higher self-compatibility; higher in-seed content; high oleic and tocopherol contents; earlier maturity and other desired agronomic characteristics; and resistance to broomrape parasite, IMI and SURES herbicide group, diseases (especially Sclerotinia, rust, Phomopsis, Macrophomina, etc.), and some pests

- Setting up new sunflower hybrids that have these desired characteristics and crossing these inbred lines, utilizing SCA of the lines
- Developing open-pollinated varieties and improving seed characteristics of land races

Due to the broomrape problem with the new races F, G, and H, especially in the Trakya region, new oil-type hybrids (both linoleic and oleic types) should be genetically resistant to broomrape or IMI-resistant hybrids, so that IMI herbicides can control both key weeds and the broomrape parasite on sunflower plants.

For both oil and confectionery types, the first four to five generations of recurrent selection and inbreeding were based on a phenotype in the nursery, then selection continued based on GCA of lines crossed with testers that were usually the most preferred inbred lines in the sunflower breeding program. The selection process was based on important yield traits including seed characteristics, field agronomics, disease reaction, maturity, oil content, and quality, using pedigree methods in the sunflower breeding nursery at the institute. However, the candidate hybrids were evaluated in preliminary yield trials at the institute and then in regional yield trials (Kaya, 2010; Kaya et al., 2012).

One of the important obstacles for sunflower breeding, as with the other cross-pollinated crops, is that developing new hybrids takes longer then creating open-pollinated varieties. Therefore, the most efficient effort to shorten breeding duration is utilizing a winter nursery, which is allowed by the short vegetation period of sunflower (about 120–150 days). Sunflower is normally planted in April and harvested in September. Hence, this winter cycle could be in a greenhouse or in a southern hemisphere country where planting could be done in October and harvesting could be done in February, so two generations per year are possible. In recent years of the Turkish public sunflower breeding program, the second generation has been produced in greenhouses during the winter season (Kaya, 2010; Kaya et al., 2011, 2012; Škorić, 2012a, 2012b).

Serbia

Area, Production, Yield, and Localization in the Past and Today

Sunflower planted areas have been around 170,000 ha in Serbia in recent years (Table 19.V). Serbia sunflower planting areas are mainly in the Vojvodina region, which has the most fertile soils and areas of Serbia. Seed production is about 400,000 t, and seed yield averages are about 2.2–2.4 t/ha. Due to a heavy drought in the summer of 2012, seed production was 20% less than in the previous year. Serbia has produced about 330,000–350,000 t of vegetable oil in recent years: sunflower oil (150,000–180,000 t) and rapeseed oil (70,000 t). However, soybean meal (300,000 t) was

Table 19.V Sunflower Harvested Areas, Seed, Oil and Meal Production by Years in Serbia

Year	Harvested Area (1000 Ha)	Seed Production (1000 t)	Oil Production (1000 t)	Meal Production (1000 t)
2006	210	400	161	142
2007	130	260	115	103
2008	200	490	180	134
2009	157	400	170	196
2010	170	400	170	156
2011	175	415	170	165
2012	167	350	153	167

Sources: Oil Word Annual (2013); World Statistics Database (2013).

greater than sunflower meal (150,000–160,000 t) in Serbian total meal production (500,000–550,000 t) in recent years (Oil World Annual, 2013). In 2013, Serbia produced 600,000 t of sunflower with a yield of 3.15 t/ha.

Crushing and Trading, Price and Market Characteristics

Serbia's total oilseed crushing was about 800,000 t in recent years. Serbia has recently imported little sunflower (about 20,000–40,000 t) and soybean (40,000 t). There are also low soybean oil and sunflower oil imports, and palm oil imports (about 15,000–19,000 t) are mostly used for producing margarines and industrial oils. Furthermore, Serbia is a net exporter of vegetable oil, mostly exporting sunflower oil to neighboring countries such as Austria (24,000 t), Bosnia and Herzegovina (13,000–18,000 t), and Croatia (10,000–15,000 t). In the meal trade, Serbia is also a net exporter, but in some years they have imported small amounts from neighboring countries, mostly for trade goods. Recently, Serbian total sunflower meal export was about 50,000–80,000 t, mostly to Slovenia and Croatia. Victoria Oil, Banat, Sunce, Dijamant, Vital, Dunavka, and Plima M are the main crushing and sunflower oil processors in Serbia (Curović, 2012). Sunflower seed domestic consumption and seed crushing are similar amounts (Table 19.W).

Research, Seed Science, and Technology in Sunflower

In Yugoslavia, sunflower cultivation began in the 1930s. The Institute of Field and Vegetable Crops in Novi Sad (IFVCNS) was established in 1938. The foundation of the Industrial Crops Department (later renamed the Oil Crops Department) in 1962 marked the start of sunflower breeding at the IFVCNS. The basis for sunflower breeding was cultivating varieties that originated from other countries (Romania, Bulgaria, and Hungary), but local populations were also grown. During this period, three va-

Table 19.W Sunflower Seed Domestic Consumption, Seed Crushing and Exports in Serbia

Year	Seed Domestic Consumption (1000 t)	Seed Crushing (1000 t)	Year	Seed Domestic Consumption (1000 t)	Seed Crushing (1000 t)
2006	413	375	2010	440	400
2007	309	271	2011	428	400
2008	483	425	2012	387	360
2009	445	400			

Sources: Oil Word Annual (2013); World Statistics Database (2013).

rieties were created. The most important was Novosadski 4, which was the leading variety in the 1960s. These cultivars had low oil content (less than 36%). New high-yielding Russian varieties with high oil content (VNIMK 8931, Peredovik, Smena, Luč, Zarja) were used in sunflower production. At the same time, they worked on creating domestic varieties with high oil yield. In 1976, two varieties (Novosadski 20 and Novosadski 61) were created that were more productive than Russian varieties. These varieties were not expanded in production because in the period from 1978 to 1983, the first NS hybrids (NS-H-62-RM, NS-H-63-RM, NS-H-65-RM, NS-H-67-RM, NS-H-26-RM, NS-H-27-RM, NS-H-33-RM, NS-HELIOS) were created. The appearance of stem cancer (*Phomopsis helianthi*) led to a reduction in sunflower area and yield in 1980. In a short time, new sunflower hybrids were created that were tolerant to *Phomopsis* (H-NS-15 and NS-H-17), and these were registered in 1984. These were the first hybrids in the world tolerant to *Phomopsis*. Sunflower hybrids that were field resistant to *Phomopsis* (NS-H-43, H-NS-44, NS-H-45) were registered in 1987. The introduction of these hybrids into production enabled the survival of the sunflower crop in the region and led to an increase of area and yield (Škorić, 1980, 1988, 1992, 2012a, 2012b; Škorić et al., 1998).

During the past 50 years, sunflower researchers from the IFVCNS have developed methods and defined directions of sunflower breeding and created models of ideal hybrids for local and other agro-ecological conditions. The knowledge of sunflower genetics and breeding has greatly expanded since time that Škorić (1980, 1988, 1992, 2012a, 2012b; Škorić et al., 1998) defined his hybrid model. He pioneered the work on developing the model of sunflower hybrids for the agro-ecological conditions of Yugoslavia and other regions, making a diagram (ideotype) of a hybrid model that showed the following: genes that should be incorporated in the envisaged genotype, direct components of seed yield and oil yield per unit area (ha), and environmental requirements of the genotype.

The main goal of the IFVCNS sunflower program is to develop sunflower hybrids with high seed and oil yields that are resistant to the common sunflower diseases, broomrape, insects, and drought. The previous period (the last 50 years) was characterized not only by creating productive high-oil hybrids, but also accelerating the development of hybrids for specific purposes. These include sunflowers with altered oil quality (high oleic, high linoleic), confectionary hybrids with increased protein content, hybrids to be used as poultry or bird feed, ornamental sunflowers, and hybrids tolerant of certain herbicide groups (imidazolinones, tribenuron-methyl). The hybrid Rimisol was the first Clearfield sunflower hybrid registered in Europe in 2004 (Škorić, 2012a; Škorić et al., 2008).

The IFVCNS was among the first in the world (together with France and Romania) to develop productive sunflower hybrids based on cytoplasmic male sterility. Some of them were used as standards in European varietal commissions. The IFVCNS has done a great deal on collecting, maintaining, studying, and using the wild species of the genus in sunflower breeding using interspecific hybridization. In the area of genetic resources, over 7,000 inbred lines originating from genetically divergent materials have been developed. The Department of Oil Crops has a rich collection of annual and perennial wild species of the genus *Helianthus* (about 900 sunflower populations and interspecific hybrids). The collection is extensively used in sunflower breeding via interspecific hybridization for transferring desirable genes from wild species into cultivated sunflower (Atlagić et al., 2006; Marinković and Atlagic, 2007; Škorić, 1980, 2012a, 2012b). Cytogenetic studies have played an important role in this process. Confectionary-type hybrids, such as Mahelan and NS Colonel, with a decreased seed oil content and increased seed protein content, have been developed. The percentage contribution of confectionary sunflower hybrids to total area sown to this crop in Serbia is rather small (5–10%). Significant results have been achieved in developing hybrids with different oil quality as well (Dijanović et al., 2003; Hladni et al., 2011).

Modern biotechnology methods such as haploid production, protoplast fusion, screening, embryo culture, and others have all been incorporated into the sunflower breeding program of the IFVCNS. Particularly notable has been the development and practical application of molecular markers in breeding for resistance to downy mildew. Breeding for resistance to diseases has been a major part of the Institute's sunflower program. Genes for resistance to pathogens (downy mildew, rust, *Verticillium* wilt, black spot, and charcoal rot of root and stem) have been discovered in wild species and transferred into cultivated sunflower genotypes by utilizing interspecific hybridization. The development of hybrids resistant to broomrape race E has an important part in the Novi Sad sunflower breeding program. Marker assisted selection (MAS) is used especially for downy mildew and broomrape resistance research. Broomrape could also be controlled by IMI-resistant hybrids or by using some biological control methods. In addition to exploring for broomrape resistance genes, some research efforts are performed to alter the anatomy of plant organs, as well as biochemical parameters (mechanical barriers, germination inhibitors, phytoalexins, etc.) (Cvejic et al., 2011; Imerovski et al., 2012; Miladinović et al., 2012; Škorić, 1992, 2012a, 2012b; Škorić and Pacureanu, 2010; Škorić et al., 1998).

Over the past 50 years, the IFVCNS sunflower team has developed a total of 279 NS sunflower hybrids. The Institute also has well-developed cooperation programs on joint hybrid development with over 30 foreign partner companies from around the world, enabling the utilization of genetic variability on an international level. These have produced a total of 162 joint hybrids released in different European and non-European countries that play a major role in global sunflower production, most notably in countries such as France, Russia, Ukraine, Italy, Spain, Bulgaria, Romania, Hungary, China, and so on (Balalić et al., 2008, 2012; Jocić et al., 2012; Miklič et al., 2008; Škorić, 1992, 2012a, 2012b).

Many studies on the influence of different agro-technical practices on yield, yield components of sunflower, and resistance to important diseases of sunflower were performed together with colleagues from the Faculty of Agriculture (Novi Sad, Serbia) In addition to the Institute in Novi Sad, the scientists at the Agricultural and Technological Research Center, Zaječar, Serbia, worked on the creation of varieties and hybrids, especially on the confectionery-type sunflower (Škorić, 1992, 2012a, 2012b).

Moldova

Area, Production, Yield, and Localization in the Past and Today

Sunflower is one of the most important crops in Moldova. Despite its moderate size (3.4 million ha), the Republic of Moldova exists among the top 20 sunflower producing countries in the world (about 1.2% of world production). Sunflower grows in almost all parts of Moldova, and sunflower is one of the main crops in the rotation system (19.4% of total crop area) after corn (31%) and wheat (19.6%). The largest sunflower areas are found in Cahul (12,400 ha), Causeni (10,900 ha), Drochia (12,600 ha), Floresti (10,700 ha), Rîscani (10,700 ha), and Gagauzia (14,600 ha). Other sunflower planted areas (about 7000–9000 ha) are in the regions of Cantemir, Cimislia, Falesti, Hîncesti, Orhei, Sîngerei, Soroca, Telenesti, and Ungheni. There are also a few areas located in the Anenii Noi, Basarabeasca, Calarasi, Dubasari, Nisporeni, and Straseni regions. The best seed yields over time are obtained from the Donduseni, Drochia, Edine, Falesti, Glodeni, Soroca, and Ungheni regions. The certified sunflower seeds are produced in specialized licensed farms located mainly in the Cahul, Cantemir, Hîncesti, Rîscani, Stefan Voda, Taraclia, Comrat, Ciadîr-Lunga, Vulcanesti, Bali, and Rezina regions (Kandakov et al., 2012). Most of the local farms produce small amounts of sunflower, less than 100 t per year, but larger farms produce between 200 and 300 t. Although sunflower yields had been lower due to some limitations on crop financing for several years, higher demand for exports has stimulated better farming practices and the use of higher yielding varieties in recent years (FAO Handbook, 2010).

The sunflower production is about 400,000 t in recent years in Moldova (Table 19.X). Due to favorable conditions for sunflower cultivation and extra rainfall in July and August, the maximum yields were obtained in 2003, 2010, and 2011. The lowest production was observed in 2007 because of the severe drought from April through August, and seed yields were reduced 50% in this season. However, the yield of sunflower has been about 1.3–1.9 t/ha in recent years. The main reasons for yield losses are insufficient fertilizing of soil, lack of efficient applications to fight weeds, less

Table 19.X Sunflower Harvested Areas, Seed, Oil and Meal Production By Years in Moldova

Year	Harvested Area (1000 Ha)	Seed Production (1000 t)	Oil Production (1000 t)	Meal Production (1000 t)
1992	131	197	39	34
1993	125	174	24	21
1994	119	133	16	14
1995	160	203	20	17
1996	225	316	60	61
1997	174	201	58	60
1998	204	199	42	43
1999	217	286	71	73
2000	228	268	46	48
2001	209	255	74	77
2002	256	317	110	110
2003	352	390	120	123
2004	271	335	110	110
2005	276	331	111	115
2006	287	380	111	107
2007	234	156	60	58
2008	228	372	87	83
2009	227	284	82	78
2010	252	382	70	66
2011	277	427	87	84
2012	250	300	93	89

Sources: Oil Word Annual (2013); World Statistics Database (2013).

tilling, nonsuitable crops in the rotation, use of lower quality seeds, less applications of chemicals for controlling diseases and pests, and so forth. Sunflower accounts for around 14–18% of all cultivated crops and 60–65% of the industrial crops (Kandakov et al., 2012).

Sunflower is generally planted in the first half of April and the planting period lasts about 15 days. In the central and eastern regions (Cahul, Leova, Taraclia, Comrat, Ciadir-Lunga, Vilcanaesti, Cimislia, Cantemir, Anenii Noi), planting is 6–7 days earlier than the northern regions. Sunflowers are generally harvested in the last week of September and the first week of October. Depending on the geographical zone, harvesting could start in mid-September and last until mid-October (Kandakov et al., 2012).

Crushing and Trading, Price, and Market Characteristics

Domestic Consumption. The total domestic consumption of sunflower in the Republic of Moldova has been about 250,000–280,000 t in recent years (Table 19.Y). Real consumption is about 61% of the total available quantity, and the per capita consumptions over the last 4 years have been between 27 and 59 kg/year. At the same time, the domestic market of Moldova consumes 14,000–22,000 t of sunflower oil. Sunflower is a product that reaches various final consumers in various forms. Sunflower seed cost accounted for 42% of the selling price per 1 L of sunflower oil in 2008. The quality standards for sunflower oil are set based on GOST 22391-89, which gives designations for commercialization in the domestic market, and CODEX-STAN 210, which is oil designated for export (Kandakov et al., 2012).

Trade. Traditionally, the Republic of Moldova has exported stable quantities of sunflower, but over the last 4 years, Moldova has exported mainly sunflower oil. A significant share of exports (85%) is initially directed toward the maritime ports of the Black Sea in Ukraine; afterward, following the formation of large lots of 5,000–10,000 t, the shipments are transported by sea toward international traders from various countries. In the zone of the Black Sea basin, the prices for sunflower fluctuate from year to year because of the applied practice of administratively restricting exports of sunflower from Moldova and Ukraine. In Romania and Russia, the export potential was restricted by tougher requirements of the domestic oleaginous raw material processing industry. Sunflower seed import and export are governed by the Regulation on Import and Export of Seeds and Planting Material, approved following the Government Decision no. 360. The seed designated for import or export is subject to mandatory control and certification by the General Inspectorate of Phytosanitary Supervision and Seeds Control, in accordance with the technical instructions and norms in effect. Export and re-export of sunflower seed is allowed only after due control of relevant seed lots, based on quality indicators, in accordance with the set effective norms. Import and export of seed is carried out in packed lots that are marked and sealed in accordance with the normative acts in effect (Kandakov et al., 2012).

Table 19.Y Sunflower Seed Domestic Consumption, Seed Crushing, and Exports in Moldova

Year	Seed Domestic Consumption (1000 tMT)	Seed Crushing (1000 t)	Oil Export (1000 t)	Seed Export (1000 t)	Meal Export (1000 t)
1992	107	98	9	90	0
1993	69	60	0	105	0
1994	58	41	5	75	0
1995	55	50	0	100	0
1996	155	150	0	161	0
1997	149	144	0	90	0
1998	109	104	0	92	0
1999	182	177	0	105	0
2000	127	122	15	143	3
2001	198	193	13	57	0
2002	283	277	39	35	19
2003	314	309	19	77	18
2004	281	276	34	55	20
2005	297	290	34	35	17
2006	277	271	60	105	14
2007	153	147	28	10	18
2008	217	211	32	131	30
2009	204	198	43	99	30
2010	174	168	52	218	29
2011	219	211	76	194	50
2012	233	225	30	100	50

Sources: Oil Word Annual (2013); World Statistics Database (2013).

In the Republic of Moldova there are 76 sunflower producers, and the country does not have a state or private structure for collecting sunflower for processing. Sunflower seed is bought mainly by S.A. Floarea Soarelui, which is the largest sunflower processor in Moldova (around 160,000–190,000 t annually). Four to five cereal union enterprises (out of a total of 30 enterprises) also purchase sunflower from agricultural producers (around 10,000–15,000 t annually). The sunflower seeds bought by these enterprises are further directed toward export. The other 25 cereal enterprises only provide services of cereal storage, conditioning, and maintenance. The sunflower wholesalers dealing with sunflower processing for export purposes include the economic agents operating within consolidated agricultural farms with land areas over 50 ha. Single sales by various agricultural farms to only S.A. Floarea Soarelui have varied over the years, from 100 t to 350 t (Kandakov et al., 2012). The sunflower dealers are mainly peasant farmers who sell their sunflower in the communes and at agricultural markets. Agricultural producers are usually selling only 25–40% of their produced goods during and after harvest; the main purchasers are the processing enterprises from neighboring localities, and the proposed selling prices are often lower than the production costs. There is no improved purchasing system in the Republic of Moldova and there are some problems, such as lack of market transparency, lack of an effective system to conform the product quality to the market demands, and lack of a well-defined chain of relationship in the field of production trade (Kandakov et al., 2012).

There are 19 exporters of sunflower in the Republic of Moldova. Sunflower exporting is performed mainly by railway (95%), which provides specialized container wagons for transportation, with monthly transportation capacities of over 20,000–22,000 t of sunflower. The destinations of railway deliveries include maritime ports of Ukraine in the Black Sea basin (Odessa, Ilichovsk, Reni). Export by motor transportation is directed toward the regions neighboring Ukraine and Romania. Sunflower for food (confectionary) is exported only to the EU (Romania, Poland, Hungary) by motor transportation in annual quantities of 2,000–2,800 t. At the same time, motor transport is also used to export/import sunflower seeds (2,200 t).

The sunflower processors include S.A. Floarea Saarelui from Bali (with a processing capacity of 160,190,000 t of sunflower) and 591 smaller processors placed in communes, regional centers, and the cities of Chisinau and Bali (Kandakov et al., 2012). For sunflower seed export (which totals about 100,000–110,000 t), the main destinations are Turkey (90,000 t) and Romania (15,000–20,000 t). For sunflower oil export, the main destinations are Italy (41,000 t in 2012) and Romania (10,000–15,000 t). For sunflower meal export (55,000 t in 2012), EU-27 and Belarus are the main destinations (Oil World Annual, 2013).

The domestic market of sunflower still has some problems, such as market deficiencies, state interventions, and market domination of one company, which results in lower prices for sunflower producers in Moldova. The S.A. Floarea Soarelui company, the largest processor and actor in the sunflower seed market, and the largest exporter of sunflower oil, holds almost 70% of the sunflower seed market. They dominate the market and have the ability to set the buying price outside of free market conditions. To develop and increase sunflower production in Moldova, production quality and productivity should be improved and the determination of sunflower price should be done on the free market without so much influence by dominant companies (Kandakov et al., 2012).

Research, Seed Science, and Technology in Sunflower

It has been more than 150 years since the introduction of sunflower in Moldova. It can be firmly stated that sunflower has a rich tradition for its cultivation and its use on an industrial scale. In Moldova, research in sunflower is successful both in science centers and private institutions. Scientists at the Research Institute for Field Crops ("Selectia") deal with sunflower improvement and selection, as well as with the development of sustainable, ecological agricultural systems. In Moldova's fields, there are 12 sunflower hybrids grown that were created at the Selectia Institute.

The majority of the seeds previously used for sunflower production in the Republic of Moldova were obtained at the Selectia Institute. However, there are many private companies currently in the sunflower seed market and various varieties and hybrids of seeds developed by foreign companies or institutions can be found. The basic clients of these companies are large producers that have better access to financing sources from local and international trade companies. This situation had led to an increase, over the last 5 years, in the number of producers of seeds for trading. However, from year to year, the domestic producers request hybrids of higher and constant productivity under any climate conditions. The most requested hybrids in Moldova are as follows (in decreasing order): Drofa, Performer, Vitalia, Xenia, Luceafarul, LG, Maria, Manad, Alcazar, SNH-215, Vidoc, Donscoi Krupnoplodni, and Donscoi. Despite certain progress in seed market expansion, local sunflower producers are searching always for new hybrids. In Moldova, the seed registration period, which covers testing and final registration by the National Council for Plant Species, may last up to 4 years, even for hybrids already registered in the CIS and EU markets. Because of that, local farmers are greatly restricted in exploiting newer and long-term hybrids and species and have reduced abilities in competing in international markets (Kandakov et al., 2012).

The Moldovan researchers in the field cooperate with scientists from other countries, such as Serbia, Romania, and the United States, in the FAO program for improving sunflower oil and seed production. The problem of sunflower's susceptibility to diseases, and its subsequent need for resistance to them, remains an important prerogative of biological research conducted in Moldova. In this context, research on sunflower resistance to broomrape is currently an important objective (Duca et al., 2010a).

The research by the Institute of Genetics and Plant Physiology of the Academy of Sciences of Moldova and at the State University of Moldova became particularly important in order to increase the productivity and improve the quality indices of sunflower. Through the use of cytological and molecular biological analysis, it has been possible, for example, to introduce new sunflower hybrids created on the basis of the functional system of cytoplasmic male sterility—the restoration of fertility (CMS-Rf) and intensive cultivation technologies. The researchers from the University of the Academy of Sciences of Moldova have focused their fundamental research on solving practical problems of technology transfer relating to methodology of assessment of genetic polymorphism based on morpho-physiological and molecular genetic characteristics, identification of molecular markers of importance in improving the heterosis and identification and characterization of two types of CMS in sunflower, and developing a new method of typification of male sterility based on hybrid analysis. Successful results in identification of a new Rf gene with specific expression in cytoplasmic male sterility of *lenticularis*-type were obtained. Also success was achieved in elucidating the role of gibberellins in the manifestation of androsterility-androfertility in sunflower, as well as in finding the similarity in CMS and IMS (induced male sterility) occurrence in the sunflower based on the morpho-physiological results and proteomic analysis. Identifications of de novo synthesis of 16 kDa protein correlated with CMS in sunflower at gibberellins application and its location at the cellular level, and identification of a new gene expressed under the gibberellins action and sequencing and determining its similarity with the mitochondrial *orfH522* gene were successfully made (Duca et al., 2010b, 2013).

Other Countries (Croatia, Greece, Macedonia, Bosnia and Herzegovina, Albania, Montenegro)

Area, Production, Yield, and Localization in the Past and Today

Sunflower is the most important oil crop in the Balkan area. Beside the countries with bigger sunflower production, there are also others with sunflower production on a smaller scale.

Croatia. Sunflower is the most important oil crop in Croatia, and it is grown mostly in the eastern part of country where suitable soil and climate conditions exist for growth and development. Although the sunflower planted area is about 25,000 ha with an average seed yield of 2.4 t/ha, the domestic production covers only 50% of the country's sunflower needs. Therefore, sunflower production needs to improve and develop to meet the domestic market demand (Krizmanić et al., 2006).

Greece. Sunflower is the third most important oil crop in Greece, after olive and cottonseed. Sunflower planted areas are about 55,000–60,000 ha and production is about 80,000 t. Sunflower is grown mostly in northern part of Greece (Evros region). Sunflower oil is the most consumed vegetable oil after olive oil.

Macedonia. Macedonia is small country that does not have many large sunflower areas. Sunflower planted areas are about 5000–6000 ha, and the production is about 8000–9000 t in recent years. The sunflower crop in the Republic of Macedonia in 2009 was sown mostly in the regions of Bitola (3170 ha), Kumanovo (385 ha), and Lozova (188 ha) (Karov et al., 2011). Planted sunflower has reduced every year due to the low purchase price, low yields, and unreliable payment redemption.

Bosnia and Herzegovina. Domestic production of oil crops is not enough to satisfy demand in Bosnia and Herzegovina. Sunflower area and production are low in this country, but sunflower is the most consumed and preferred vegetable oil. In Bosnia and Herzegovina, sunflower planted area in 1955 was 10,600 ha, and in 1985 it dropped to only 5 ha. One of the major causes of this sharp drop was the appearance of *Phomopsis sp.* in sunflower between 1982 and 1986. After this period, the sunflower planted area has not significantly increased until today. In 2011, sunflower was grown only on 286 ha. Analyzing the period of last 30 years, it can be concluded that the planting area for sunflower in Bosnia and Herzegovina is very small, and this crop is unreasonably neglected in the crop rotation. The agro-ecological conditions (soil, climate) in the region of Bosanska Posavina (the area of Bijeljina, Brcko, Orasje, Bosanski Samac, Odzak, Modrica, and Bosanski Brod) indicate that sunflower could be successfully produced. There are also other areas in Bosnia and Herzegovina that are suitable for the production of this oil crop (Lijevče Polje, Bosanska Krajina), but first some research should be conducted.

Albania. Albania is the smallest country in Eastern Europe other than Montenegro. Arable lands are also small, as in Macedonia and Montenegro. Albania has very few sunflower planting areas and very little production.

Crushing and Trading, Price, and Market Characteristics

Croatia. Croatian domestic production of oil crops is not big enough for the country's needs. The most important oils are sunflower, rapeseed, and palm oil. Croatia has also exported sunflower seed (about 40,000–50,000 t) in recent years, mostly to Serbia, Bosnia, and Herzegovina. Croatia does not import sunflower seed, but does import sunflower oil, mostly from Serbia. Animal production is an important sector in the Croatian economy and, similar to the vegetable oil sector, domestic production is not enough to fulfill the country's needs. Therefore, Croatia imports soybean and sunflower meal, mostly from Serbia.

Greece. Total sunflower crushing is about 80,000–100,000 t in Greece. Greece does not import and export sunflower seed, but they do import sunflower oil, mostly from Bulgaria and Russia. Greece also imports about 20,000–30,000 t of sunflower meal for animal feed, again mostly from these two countries.

Macedonia. The domestic vegetable oil need for Macedonia is met mostly through sunflower oil imports, namely from Bulgaria. For animal feed, sunflower meal is also imported from Bulgaria. The reduced sunflower seed production in recent years is also a result of increased imports of crude oil, because domestic factories are not supporting domestic sunflower seed production because it happens in neighboring countries. Macedonia has a total installed capacity for processing of 50,000 t of crude oil in the processing plant Blagoj Gjorev in Veles. The other facilities, such as Brilijant in Stip and AL-Mak in Strumica, can only refine the crude oil (Karov et al., 2011).

Bosnia and Herzegovina. Due to a lack of domestic sunflower production, Bosnia and Herzegovina import both sunflower seed and oil from other countries. Sunflower seed has been imported mostly from Croatia in recent years, and is also imported from Serbia, Romania, and Hungary. Sunflower oil is imported from Bulgaria, Serbia, and Romania. Although the factory for the production of edible oils, Bimal in Brcko, is very interested in starting domestic production of sunflower (since 50% of all oil they are producing is sunflower oil) and offers a secure purchase, this plant species is expanding slowly in the country.

Albania. Sunflower is the most popular vegetable oil in Albania, but the country's need is much bigger than the domestic production. In addition, there is not much crushing capacity, so Albania imports sunflower oil for domestic needs mostly from Ukraine. Similarly, there is no domestic sunflower meal production, so this is imported from Bulgaria and Serbia.

Research, Seed Science, and Technology in Sunflower

Croatia. Sunflower breeding programs in Croatia have a more than 30-year background and are conducted by the Agricultural Institute Osijek. This program is performed now under the scientific research project "Breeding of Hybrid Sunflower for High Grain Yield and Quality." The main goal of the program is to develop superior hybrids with higher yields (above 5 t/ha), over 50% oil content, and higher and stable oil yield, with a focus on developing inbred lines tolerant to main pathogens and generating hybrids with shorter vegetation period (110–120 days). Based on the successful results, 12 institute hybrids (Osječanin, Fakir, Olio, Orion, Podravac, Slavonac, Sunce, Gordan, Miro, Šokac, Favorit, and Apolon) have been developed so far, and these have considerably contributed to the increase of sunflower production in Croatia. Two of them (Fakir and Olio) have also been registered in Slovakia. Future breeding studies will be focused on phytopathological investigation of the main diseases in the lab and utilizing biotechnological and molecular breeding methods to increase the efficiency of the breeding program (Krizmanić et al., 2006).

Greece. Sunflower represents an important field crop in the rotation system in Greece, especially in the Evros region (northeast Greece), due to the steadily recent demand from the vegetable oil and biofuel industries. Higher, permanent, and sustainable yield is possible in the sunflower cultivation with lower production costs, especially with more efficient weed management. Broomrape is the most significant yield limitation factor in sunflower production in the region, even in planted broomrape-tolerant hybrid fields. Sunflower exists as a main crop in the rotation system with wheat and corn. The weed management is performed mainly by pre-emergence herbicides such as Treflan (trifluralin), Granstar (tribenuron methyl), and Oxyfluorfen. There is great demand and preference for Clearfield (CL) (imazamox-resistant) sunflower in the region to efficiently control both new races of the parasite and common weeds. The efficient and encouraging results are obtained from the CL sunflower production areas because no broomrape infection was observed in previously heavily contaminated areas (Economou et al., 2010; Lyra et al., 2013).

Macedonia. Selection of sunflower in Macedonia does not exist. So far just one variety (Sunce) has been created, based on individual selection. This is an improvement on the old variety, VNIIMK 8931, which was domesticated in Macedonia. In general, industrial crops have symbolic treatment in Macedonia. Agro-technical and varietal trials are carried out at the Faculty of Agricultural Sciences and Food in Skopje (Dimov and Egumenovski, 2003).

Bosnia and Herzegovina. Scientific research on sunflower in Bosnia and Herzegovina is rather scarce. Most varietal and agro-technical trials are carried out at the Faculty of Agriculture in Banjaluka. Hybrids originating from Serbia and Croatia are mostly sown. On the national list of varieties there are about 30 hybrids, but only 2 or 3 are grown. Only 10% of the raw material for oil (sunflower and soybean) is from domestic cultivation; the remainder originates from imports of raw materials.

Albania. The Agricultural Research Institute of Lushnje (IKB) is the institution dealing with field crops. IKB has two sectors on industrial plants. Scientific research is focused on the following areas: genetic improvement (hybridization), comparison of sunflower hybrids and cultivars, studies on sunflower cultivation technologies, and production of elite seed and hybrids. The technologies used have been implemented in the production entities and the results have been good, comparable to other countries in the region and beyond.

Present and Future of Directions of Sunflower in Eastern Europe

After the accession of Ukraine and Russia to WTO, sunflower export policies changed and continued to be modified. This movement will drastically influence the sunflower world market. For instance, Ukraine was involved in a worldwide trade system, but the Ukrainian sunflower export duty of 17% was lowered by 1% for each year until it reached 10%. After Russia joined the WTO in August 2012, sunflower seed export duties gradually decreased from the previous level of 20% down to 6% within 4 years. However, these duty changes may not lead to immediate export growth due to remaining higher domestic demand for sunflower seed and oil. According to market operators, there are several scenarios for development of the situation in Russia after the reduction of export duties for sunflower. However, the most obvious is the fact that the prices will rise rapidly for both the oilseed and its byproducts. Exports of sunflower will probably stay at the same level due to high demands of the processing industry on the domestic market. Intense competition for sunflower seed volumes in the country will be the incentive, which will drive the price. Bulgaria and Romania have been EU members since January 2007. In addition, Turkey is an EU member candidate, and the negotiations to become a member are in progress. If Turkey becomes an EU member, all borders will be abolished and sunflower and sunflower oil trades will be totally free without import duties (USDA, 2013a, 2013b).

The Black Sea sunflower seed market could face some reduction of prices in the next seasons. At the same time, the expected increase of the oilseed exports by Southern European countries will provide a notable pressure on the prices in the reporting period, in terms of the further increase of sunflower seed production in the region and the lack of processing capacities. The intense food demand and industry consumption of vegetable oils and protein meals resulted in an increase of world oilseed productions. According to USDA reports, in the period between 2000 and 2010, oilseed production volumes in the world increased by 43%, world vegetable oil production increased by 62%, and consumption increased by 64%. Although there are higher increases in the consumption of palm, soybean, and rapeseed oils recently, sunflower oil demand is also rising dramatically. Ukraine, Russia, EU-27, and Argentina are the main producers of sunflower, and their share is about 82% of world production of oilseeds, but EU-27 is also the major sunflower oil consumer. In the Black Sea, rapeseed production is getting higher year by year, so some sunflower acreage may be captured. However, with permanent increasing of sunflower crushing capacities, significant reduction in sunflower planting area is not likely. Sunflower yields are expected to be higher, year by year, due to better input (fertilizer, herbicide, irrigation, etc.) conditions in the upper part of Eastern Europe and also due to environmental conditions (APK Report, 2013; Kaya et al., 2008a).

World sunflower seed production is mainly linoleic-type sunflower. Many European countries started to produce high oleic–type sunflower in addition to traditional ones in the most recent decade. Meanwhile, high oleic sunflower oil consumption increased in the world market due to higher oil quality used in frying and the demand for supplying healthy oil to consumers. Although the mid-oleic type is popular in the United States (80% NuSun [mid-oleic], 10% high oleic, and 10% conventional) and Argentina, the high oleic type dominates in European countries. High oleic sunflower production reached 75% in France, 50% in Spain, and 10% in Hungary. The possible use as a biodiesel source will also increase planted areas and the demand for high oleic type in the Black Sea region (APK Report, 2013; Kaya et al., 2008a, 2008b, 2008c; Ozip, 2011).

The main advantages of mid and high oleic sunflower oil are higher vitamin E content, natural antioxidant (α[alpha]-tocopherol) properties, longer storage ability, the natural taste, and also excellent frying capability due to the higher oleic content and lower saturated fat content (less than 10%). Normally, sunflower oil is beneficial to human health because it has the highest poly- and monounsaturated fat content (linoleic and oleic acids) and the smallest amount of saturated and *trans* fats. High oleic sunflower oil is among the higher quality vegetable oils, and it has improved oil resistance to auto-oxidation, which prevents the accumulation of toxic products during oil processing, storage, and direct uses. High oleic sunflower oil is also well suited for food purposes, such as oil for spraying of snacks, crackers, and dry breakfast cereals; frying oil; food products for babies and for elderly people; and for increasing oxidation stability. After the introduction of high oleic sunflower to Turkey, oleic types could dominate the planting areas in order to provide healthy oil to consumers and to be used in biodiesel production. However, all sunflower production was previously linoleic type, and the high oleic type market was only recently initiated (APK Report, 2013; Kaya et al., 2008c; Ozip, 2011).

High oleic sunflower produces high yields and is also tolerant to the main diseases, weeds, and broomrape, which are limiting factors on high oleic sunflower area. However, demand and levels of knowledge about the benefits of high oleic sunflower oil are currently limited in main production and consumer countries such as those in the Black Sea region. For instance, the traditional linoleic oilseed hybrids are planted in main sunflower production areas in Russia as well as in Ukraine. However, high oleic and confectionary varieties of sunflower are increasing recently, and market share of high oleic sunflower was 1% in Russia and 2.2% in Ukraine based on reports in 2010. Furthermore, the greater numbers of agricultural producers are not aware of the features of high oleic type sunflower. However, Russia and Ukraine have vast potential to produce high oleic sunflower seed and sunflower oil and then to easily export them, based on recent reports by analysts of APK-Inform Agency. On the other hand, the potentials of oleic types are currently growing in awareness and interest from large-scale crushers in EU-27 countries, especially France and Spain, due to the potential for presenting healthy oils for that market (APK Report, 2013; Kaya et al., 2008a, 2008b, 2008c; Ozip, 2011).

Due to the nutritional benefits, healthy specifications and other necessary features, and cheaper costs, high and mid-oleic oil production demand increases year by year, especially in developed countries. Sunflower oil and valuable byproducts of sunflower seed are divided into three types other than high oleic, depending on fatty acid content. One of the new items in sunflower is NuSun, or mid-oleic content sunflower type, which is the most popular in the United States and Canada. In this type, the standard fatty acid structure is oleic acid, 65%; linoleic acid, 26%; and saturated fats, 9%. Another type is high linoleic sunflower, which is the traditional sunflower type, with 69%

linoleic acid, 20% oleic acid, and 11% saturated fats; it produces sunflower oil with limited usage for frying. The final type is high stearic sunflower, which is the newest type of sunflower, branded as Nutrisun. Its main advantage is as an alternative to the partially hydrogenated oils and tropical oils. The standard fatty acid structure is stearic, 18%; oleic, 72%; linoleic, 5%; and other saturated fats, 5% (APK Report, 2013; Ozip, 2011).

References

Antonova, T. S.; Araslanova, N. M.; Strelnikov, E. A.; Ramazanova, S. A.; Tchelustnikova, T. A.; Guchetl, S. Z. Screening of Wild Helianthus Species for Resistance to High Virulent *O. cumana* Wallr. Affecting Sunflower in the Rostov Region of the Russian Federation. *Helia* **2011,** *34* (55), 115–124.

Antonova, T. S.; Araslanova, N. M.; Strelnikov, E.A.; Ramazanova, S. A.; Guchetl, S. Z.; Tchelustnikova, T. A. Some Pecularities of Ontogenesis of *Orobanche cumana* Wallr., Parasitizing on Sunflower in Rostov Region of Russian Federation. *Helia* **2012,** *35* (56), 99–110.

APK Sunflower Market Report, 2013. http://www.apk-inform.com/en/markets/oilseeds.

Atlagić, J.; Terzić, S.; Škorić, D.; Marinković, R.; Vasiljević, L. J.; Panković-Saftić, D. The Wild Sunflower Collection in Novi Sad. *Helia* **2006,** *29* (44), 55–64.

Balalić, I.; Zorić.; Branković, G.; Terzić, S.; Crnobarac, J. Interpretation of hybrid × sowing date interaction for oil content and oil yield in sunflower. *Field Crops Res.* **2012,** *137,* 70–77.

Balabanova, A. IENICA Report. Industrial Crops in Bulgaria. 2004. http://www.ienica.net/reports/bulgaria.pdf

Burlov, V.; Burlov. V. Breeding Sunflower Resistant to New Races of Broomrape (*Orobanche cumana* Wallr.). *Helia* **2011,** *33* (53), 165–172.

Christov, M.; Hristova-Cherbadzi, M.; Entcheva, V.; Shindrova, P.; Kiryakov, I.; Valkova, D.; Sabev, G.; Batchvarova, R. Creating New Sunflower Forms and Lines, Resistant to Diseases, the Broomrape Parasite and Certain Types of Herbicides Applying the Interspecific and Intergeneric Hybridization. In *Proceedings of the International Symposium on Sunflower Breeding on Resistance to Diseases,* Krasnodar, Russia, 2011; pp 72–84.

Cvejic, S.; Jocic, S.; Prodanovic, S.; Terzic, S.; Miladinovic, D.; Balalić, I. Creating New Genetic Variability in Sunflower Using Induced Mutations. *Helia* **2011,** *34* (55), 47–54.

Demurin, Y. Up-to-date Results on Biochemical Genetics of Sunflower in VNIMIK. *Helia* **2003,** *26* (38), 137–142.

Demurin, Y.; Borisenko, O.; Bochkarev, N. Relationship of Inheritance of a High Palmitic Mutation and Plant Height in Sunflower. *Helia* **2010,** *33* (53), 149–154.

Duca, M.; Glijin, A.; Zgardan, D.; Rotarenco, V.; Rotaru, T. Some Aspects of Host-parasite Interaction in the System Sunflower (*Helianthus annuus* L.)—Broomrape (*Orobanche cumana* Wallr.). *Buletinul AŞM, Genetica şi ameliorarea plantelor* **2010a,** *2* (311), 97–103.

Duca, M.; Port, A.; Midoni, A.; Rotaru, R.; Anisimova, I. RAPD analysis of the genetic variation amnog CMS, Rf lines and hybrid sunflower genotypes. *Buletinul AŞM, Genetica şi ameliorarea plantelor* **2010b,** *2* (311), 103–109.

Duca, M.; Port, A.; Şestacova, T.; Siniauskaya, M.; Aksyonova, E.; Davydenko, O. Microsatellite Marker Application in Sunflower (*Helianthus annuus* L.) Finger printing. *Biotechnol. Biotech. Eq.* **2013,** *27* (3), 3772–3775.

Economou G.; Lyra, S.; Frantzeskakis, L.; Kotoula-Syka, E. Current Status of Sunflower Crop in Greece and the Effects of *Orobanche cumana* Infestation. In *Proceedings Abstracts of 8th European Sunflower Biotechnology Conference,* Antalya, Turkey, March 1–3, 2010; p 75.

Elena N. The main share of high oleic sunflower seed production is realized under the order of large-scale company-processors in Russia and Ukraine. 2011. (http://www.linkedin.com/groups/main-share-high-oleic-sunflower-3991981.S.60953917).

Encheva, J. Creating Sunflower (*H. annuus* L.) Mutant Lines Using Induced Mutagenesis. *B.J.A.S.* **2009,** *15* (2), 109–118.

Encheva, J.; Kohler, H.; Fried, W.; Tsvetkova, F.; Ivanov, P.; Encheva, V.; Shindrova, P. Field Evaluation of Somaclonal Variation in Sunflower (*Helianthus annuus* L.) and Its Application for Crop Improvement. *Euphytica* **2003,** *130,* 167–175.

Encheva, J.; Shindrova, P.; Encheva, V.; Valkova, D. Mutant Sunflower Line R 12003, Produced through In Vitro Mutagenesis. *Helia* **2012,** *35* (56), 19–30.

Evci, G.; Pekcan, V.; Yılmaz, M. I.; Kaya, Y. The Resistance Breeding to IMI and SU Herbicides in Sunflower. *Plant Sci.* **2012,** *49,* 6–11.

FAO Agribusiness Handbook. *Sunflower Crude and Refined Oils.* 2010. 41 pp.

FAOSTAT. FAO Statistics, Food and Agriculture Organization of the United Nations, 2013. http://www.fao.org.

Gontcharov, S. V. Sunflower Breeding for Resistance to the New Broomrape Race in the Krasnodar Region of Russia. *Helia* **2009,** *32* (51), 75–80.

Gontcharov, S. V. Hybrid Sunflower Breeding in VNIIMK (Russia). In *Proceedings of the 18th International Sunflower Conference,* Mar del Plata & Balcarce, Argentina Feb 27–March 1, 2012; p. 160.

Gontcharov, S. V.; Beresneva, N. D. Confectionary Hybrid Sunflower Breeding in Russia. *J. Agric. Sci. Technol.* **2011,** *B1,* 919–924.

Gontcharov, S. V.; Antonova, T. S.; Araslanova, N. M. Sunflower Breeding for Resistance to the New Broomrape Race. *Helia* **2004,** *27* (40), 193–198.

Gontcharov, S. V.; Antonova, T. S.; Saukova, S. L. Sunflower Breeding for Resistance to Fusarium. *Helia* **2006,** *29* (45), 40–54.

Hladni, N.; Jocic, S.; Miklic, V.; Saftic-Pankovic, D.; Kraljevic- Balalić, M. Interdependence of Yield and Yield Components of Confectionary Sunflower Hybrids. *Genetika* **2011,** *43* (3), 583–594.

Hristova-Cherbadzi, M. Characterization of Hybrids, Forms and Lines, Obtained from Interspecific Hybridization of Cultivated Sunflower *Helianthus annuus* L. with Wild Species of Genus *Helianthus. Biotechnol Biotech Eq.* **2009,** *23* (2), 112–116.

Hristova-Cherbadzi, M.; Christov, M.; Valkova, D. Characteristics of Some Hybrid Progenies Produced by Crossing of Helianthus annuus L. with H. maximiliani SCHRADER. In *Proceedings of the 18th International Sunflower Conference,* Mar del Plata, Argentina, 2012; pp 39–47.

Imerovski, I.; Dimitrijević, A.; Miladinović, D.; Dedić, B.; Jocić, S.; Miklič, V. Towards Orobanche Resistance in Sunflower-SSR Analysis of Novel Resistance Source. In *Proceedings of the 18th International Sunflower Conference,* Mar del Plata, Argentina, 2012; pp 1043–1047.

Ionel, I. Oil Crops and Vegetable Oils—between Foreign Investors and Romanian Producers. *Agric. Econ Rur. Dev.* **2009,** *4* (2), 263–275.

Kandakov, A.; Havrland, B.; Ojog, C.; Ivanova, T. Sunflower Market Assessment in the Republic of Moldova. In *Engineering for Rural Development—International Scientific Conference,* Jelgava, Latvia, May 24–25, 2012; pp 128–133.

Karov, I.; Mitrev, S.; Maširević, S.; Kovacevik, B. First Appearance of White Mould on Sunflower Caused by *Sclerotinia minor* in the Republic of Macedonia. *Helia* **2011,** *34* (54), 19–26.

Kaya, Y. Sunflower Breeding, Seed Industry and Future Directions in Turkey. In *Proceedings of 16th International Sunflower Conference,* Fargo, North Dakota, August 29–September 2, 2004; pp 465–472.

Kaya, Y. Public Sunflower Breeding in Turkey and Future Directions. *Plant Sci.* **2010,** *47* (1), 7–13.

Kaya, Y.; Kaya, V.; Evci, G.; Şahin, İ.; Üstün Kaya, M. The Situation and Future Directions of Sunflower Production in Black Sea Region. In *Proceedings of 17th International Sunflower Conference,* Cordoba, Spain, June 8–12, 2008a; International Sunflower Association; pp 785–790.

Kaya, Y.; Kaya, V.; Evci, G.; Şahin, İ.; Üstün Kaya, M. Oil Type Sunflower Production in Turkey. In *Proceedings of 17th International Sunflower Conference,* Cordoba, Spain, June 8–12, 2008b; International Sunflower Association; pp 797–802.

Kaya, Y.; Kaya, V.; Şahin, I.; Ustun Kaya, M.; Evci, G.; Çıtak, N. The Future Potential of Oleic Type Sunflower in Turkey. In *Proceedings of 17th International Sunflower Conference,* Cordoba, Spain, June 8–12, 2008c; International Sunflower Association; pp 791–796.

Kaya, Y.; Evci, G.; Pekcan, V.; Yilmaz, M. I. Utilizing from Wild Types in Sunflower Breeding for New Plant Design and Anatomy for High Yield Performance. In *Proceedings Abstracts of International Symposium on Sunflower Genetic Resources,* Kusadası, Turkey, October 16–20, 2011; p 25.

Kaya Y.; Jocic, S.; Miladinovic, D. Sunflower. In *Technological Innovations in Major World Oil Crops;* Gupta, S. K., Ed.; Springer, New York; Vol. 1, 2012; pp 85–129.

Kaya, Y.; Evci, G.; V. Pekcan, V.; Yilmaz. I. M. Clearfield Technology in Sunflower and Developing Herbicide Resistance Sunflower Hybrids. *Soil-Water Journal* **2013,** *2,* 1713–1720.

Kaya, Y. A Strategic Crop in Turkey: Sunflower. *Agrotime International Journal of Crop Production and Animal Breeding* **2014,** *2* (8): 34-40.

Krizmanić, M.; Mijić, A.; Liović, I.; Bilandžić, M; Duvnjak, T. Sunflower Breeding at the Agricultural Institute Osijek. *Helia* **2006,** *29* (44), 153–158.

Kyrychenko, V.; Petrenkova, V.; Maklyak, Y.; Borovskaya, I. Results of Sunflower Breeding for Resistance to Disease Complex and Broomrape. In *Proceedings of the International Symposium "Sunflower Breeding on Resistance to Diseases,"* Krasnodar, Russia, June 23–24, 2010; pp 98–102.

Kyrychenko, V.V.; Kolomatska, V. P. Results of the Scientific Program for Sunflower Breeding. In *Proceedings of the International Symposium on Sunflower Genetic Resources,* Kushadasi, Turkey, October 16–20, 2011; p. 26.

Kyrychenko, V. V.; Ryabchun, V. K.; Maklyak, K. M.; Tymchuk, S. M. Genetic Control of Oleic Acid Content on the Development of Sunflower Hybrids of NuSun Type. In *Proceedings of the International Symposium on Sunflower Genetic Resources,* Kushadasi, Turkey, October 16–20, 2011; p 43.

Lyra, D.; Kaya, Y.; Stefanic, E.; Economou, G.; Kalivas, D. Impact of Environmental and Management Factors on Weed Species Composition in Sunflower Fields in Greece, Turkey and Croatia: A Comparative Approach. In *16th European Weed Research Society Symposium,* Samsun, Turkey, June 24–27, 2013; p 96.

Marinković, R.; Atlagic, J. Analysis of Wild *Helianthus annuus* and *H. petiolaris* Populations for Presence of Rf Genes for PET-1 Cytoplasm. *Indian J. Genet. and Plant Breed.* **2007,** *67* (2), 115–117.

Miklič, V.; Hladni, N.; Jocić, S.; Marinković, R.; Atlagić, J.; Saftić-Panković, D.; Miladinović, D.; Dušanić, N.; Gvozdenović, S. Oplemenjivanje suncokreta u Institutu za ratarstvo i povtarstvo. *Zbornik radova Instituta za ratarstvo i povtarstvo* 2008, 45 (*1*), 31–63.

Miladinović, D.; Cantamutto, M.; Vasin, J.; Dedić, B.; Alvarez, D.; Poverene, M. Exploring the Environmental Determinants of the Geographic Distribution of Broomrape (*Orobanche cumana* Wallr.). *Helia* **2012,** *35* (56), 79–88.

Oil World Annual. ISTA Mielke GmbH. Hamburg/Germany. 2013. www.oilworld.de

Ozip, I. High-oleic Sunflower: New Opportunities for Russia and Ukraine. The 3rd International Conference "Oilseeds & Oils 2011," Turkey, Istanbul, September 22–24.

Pacureanu, M.; Vranceanu, A. V.; Scare, G.; Marinescu, A.; Sandu, I. The Evaluation of the Parasite-host Interaction in the System *Helianthus annuus* L. *Orobanche cumana* Wallr. in Romania. In *Proceedings of the 2nd Balkan Symposium on Field Crops,* Novi Sad, Serbia, June 16–20, 1998; pp 153–157.

Pacureanu, M.; Raranciue, S.; Sava, E.; Stancin, D.; Nastase, D. Virulence and Aggressiveness of Sunflower Broomrape (*O. cumana Wallr.*) Populations in Romania. *Helia* **2009,** *32* (51), 119–126.

Pacureanu, M.; Ciuca, M.; Sava, E. *Broomrape (Orobanche cumana Wallr.), the Most Important Parasite in Sunflower: Virulence, Race Distribution, Host Resistance.* 18th International Sunflower Conference, Mar del Plata, Argentina, Feb 27–March 1, 2012; Abstracts 246.

Popescu, A. Research Regarding Oil Seeds Crops Development in Romania in the EU Context. *Econ. Agric.* **2012,** *59* (1), 129–137.

Škorić, D. Desired Model of Hybrid Sunflower and the Newly Developed NS-Hybrids. Helia **1980,** 3, 19–24.

Škorić, D. *Screening for Resistance in Wild Sunflower Forms in Order to Create Resistant Hybrids.* International Symposium on Science and Bitechnology for an Integral Sunflower Utilization, Pisa, Italy, 1988; pp 24–26.

Škorić, D. Achievements and Future Directions of Sunflower Breeding. *Field Crops. Res.* **1992,** *30,* 231–270.

Škorić, D. Sunflower Breeding. In *Sunflower Genetics and Breeding: International Monography;* Škorić, D., Sakač, Z., Eds.; Serbian Academy of Science: Novi Sad, Serbia 2012a; pp 165–354.

Škorić, D. The Genetics of Sunflower. In *Sunflower Genetics and Breeding: International Monogram;* Kovacevic, Z., Skoric, D., Sakac, Z., Eds; Serbian Academy of Science: Novi Sad, Serbia; 2012b; pp 1–125.

Škorić, D.; Pacureanu, M. Sunflower Breeding for Resistance to Broomrape (*Orobanche cumana* Wallr.). In *Proceedings of the International Symposium on Sunflower Breeding on Resistance to Diseases,* Krasnodar, Russia, June 23–24, 2010; pp 19–30.

Škorić, D.; Jocic, S.; Lecic, N. Sunflower Breeding for Different Oil Quality. In *Proceedings of the EUCARPIA Congress,* Viterno, Italy, September 20–25, 1998; pp 339–347.

Škorić, D.; Jocic, S.; Lecic, N.; Sakac, Z. Genetic Possibilities for Altering Sunflower Oil Quality to Obtain Novel Oils. *Can. J. Physiol. Pharmac.* **2008,** *86* (4), 215–221.

TUIK Statistical Database 2012. *The Yearbook of Turkish Statistics Institude* (TUIK). Ankara, Turkey; 2012.

U.S. Department of Agriculture (USDA). FAS Report, 2004.

U.S. Department of Agriculture (USDA) FAS Bulgaria Report. 2006. http://www.pecad.fas.usda.gov/highlights/2006/10/Bulgaria_25oct2006/

U.S. Department of Agriculture (USDA). GAIN: Ukraine Oilseeds and Products Annual, 2012. http://www.thecropsite.com/reports.

U.S. Department of Agriculture (USDA). GAIN: Turkey Oilseeds and Products Annual, 2013a. http://www.thecropsite.com/reports.

U.S. Department of Agriculture (USDA). GAIN: Russian Federation Oilseeds and Products Annual, 2013b. http://www.thecropsite.com/reports.

Verzea, M. Fifty Years of Breeding in Field Crops, at the National Agricultural Research and Development Institute of Fundulea, Romania. *Romanian Agric. Res.* **2007,** *24,* 1–6.

Vranceanu, A. V.; Tudor, V. A.; Stoenescu, F. M.; Pirvu, N. Virulence Groups of *Orobanche cumana* Wallr., Differential Hosts and Resistance Source Genes in Sunflower. In *Proceedings of the 9th International Sunflower Conference,* Torremolinos, Spain, June 8–9, 1980; pp 74–82.

World Statistics Database, 2013. http://www.indexmundi.com.

World Sunflower Market Update, 2013. http://www.trakyabirlik.com.tr.

Zaytsev, O. B.; Kovalyov, V. G.; Škorić, D.; Dušanić, N. Contribution of NS Hybrids to Sunflower Development in the Ukraine. *Zbornik radova, Naučni institut za ratarstvo i povrtarstvo, Novi Sad* **2006,** *42,* 311–322.

20

Asia and Australia Perspectives on Sunflower Production and Processing

Lucky Inturrisi ■ *Cargill Australia, Victoria, Australia*

Introduction

Sunflower is an important oilseed crop in the Asian and Australasian region. Sunflower production accounts for approximately 3% of the region's production of oil-bearing crops with over 5.4 million metric tons (t) of seed produced in 2010 (FAO, 2013).

Sunflower oil has long appealed to consumers in Asia and Australasia. Having been grown in the region for many years, today it ranks as the sixth largest vegetable oil consumed behind palm, soybean, cottonseed, groundnut (peanut), and canola (or rapeseed) (Figure 20.1).

Although palm is prevalent in the region, consumers continue to become more health conscious and concerned with fat intake. They have learned the importance

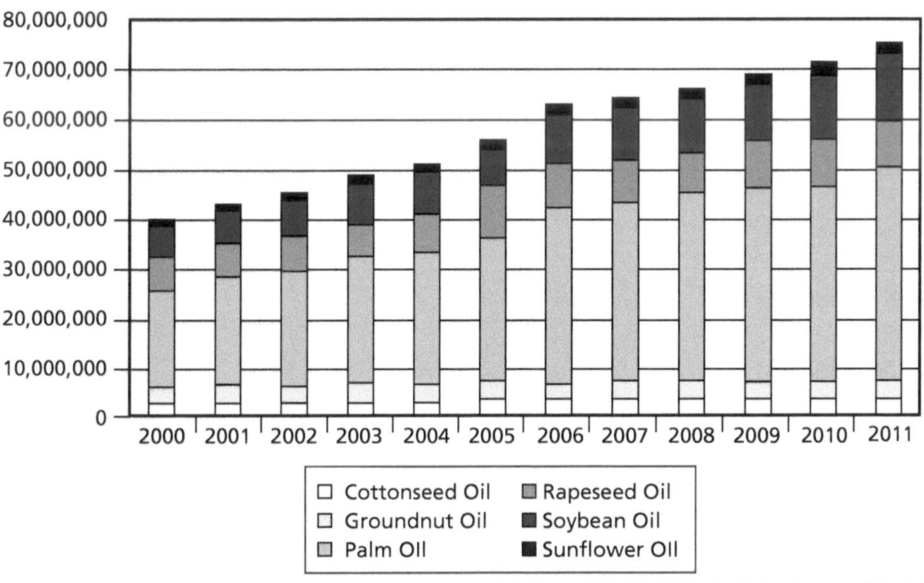

Figure 20.1 Production of major vegetable oils for Asia and Australasia, 2000–2011.

Sources: AOF (2012b); FAO (2013).

of consuming "good" fats, such as monounsaturated and polyunsaturated fats, and restricting the consumption of "bad" fats (i.e., saturated fatty acids and *trans* fats). This heightening awareness and promotion by health experts is resulting in increased demand for lower saturated fats, such as olive, soybean, sunflower, and canola oils. The region's production of sunflower oil was over 2 million metric tons in 2010 (FAO, 2013).

Both polyunsaturated and monounsaturated sunflower are grown and the oils from these crops are in high demand in the food industry. Their fatty acid composition varies greatly and their specific applications can differ (Table 20.A).

Sunflower in Asia and Australia

Sunflower Production and Consumption in Asia and Australia

Since the period of 1990 to 2010, the production of sunflower seed and oil has been variable, with yearly increases and decreases experienced during this time. Figure 20.2 illustrates the variability experienced in seed harvest and seed production, which is

Table 20.A Typical Fatty Acid Composition for Australian-Grown Polyunsaturated and Monounsaturated Sunflower Oils

	Refined, Bleached, and Deodorized High Linoleic Sunflower Oil	Refined, Bleached, and Deodorized High Oleic Sunflower Oil
C14:0 (Myristic)	0.1	0
C16:0 (Palmitic)	5.5	3.4
C18:0 (Stearic)	4.9	4.4
C18:1 (Oleic)	20	79.8
C18:2:t	0.5	0.2
C18:2 (Linoleic)	67.2	9.9
C18:3 (Alpha Linolenic)	0.3	0.3
C20:0 (Arachidic)	0.4	0.4
C20:1 (Gadoleic)	0.2	0.3
C22:0 (Behenic)	0.9	1.0
C22:1 (Erucic)	0	0.3
Total saturated fats (%)	11.8	9.2
Total *trans* fats (%)	0.5	0.2
Total monounsaturated fats (%)	20.2	80.4
Total polyunsaturated fats (%)	67.5	10.2

greatly affected by climatic conditions year to year and in some cases is affected by cash incentive by buyers to plant other more profitable crops. Positively, yields have improved, resulting in better returns for farmers.

Since 1990, sunflower seed production in Asia and Australasia has increased almost 3% per year, on average, accounting for 17.6% of the world's production of seed. For the same period, the production of sunflower oil increased on average almost 4% per year, accounting for 16.6% of the world's production.

Production of sunflower oil is dominated by China, India, and Turkey with 21%, 21%, and 27%, respectively, during the period from 1990 to 2010. Australia accounts for 1.7% of oil production (Table 20.B). Some of these markets will be discussed in the following sections.

Sunflower in China

China is the largest vegetable oil consumer in the world, as well as one of the major oilseed growing and producing countries. The total oilseed production for China is approximately 60 million t (Wang et al., 2012). Growth in the economy has resulted

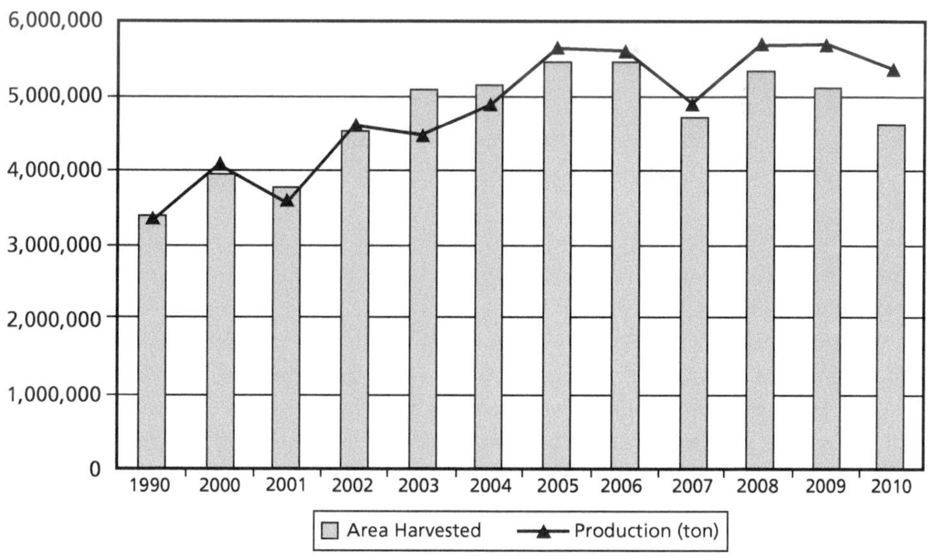

Figure 20.2 Sunflower seed harvested and production for Asia, 1990–2010.
Source: FAO (2013).

Table 20.B Sunflower Oil Production for Asia and Australasia, 1990–2010

Countries	1990	2000	2001	2002	2003	2004
Australia	24,097	74,597	34,016	30,290	10,795	24,900
Azerbaijan	—	1,973	2,400	3,200	10,408	7,100
China	441,813	540,650	420,068	314,000	329,000	283,000
India	314,000	215,000	224,000	288,000	307,000	392,000
Indonesia	2,328	5,515	3,216	1,518	6,407	2,096
Iran	21,557	15,034	14,978	17,800	19,000	17,900
Iraq	13,836	19,908	19,604	19,607	19,009	19,009
Japan	1,000	2,373	2,355	2,200	1,600	1,400
Kazakhstan	—	37,819	54,958	54,206	107,453	123,438
Malaysia	993	3,010	2,026	603	1,738	795
Myanmar	30,356	51,164	91,899	95,904	92,370	115,630
Pakistan	8,296	50,530	35,953	60,303	44,450	121,692
Tajikistan	—	—	—	200	500	700
Turkey	311,351	481,371	299,838	353,845	487,796	501,138
World + (Total)	8,097,396	9,794,989	8,247,531	7,869,173	9,253,409	9,909,986
Asia + (Total)	1,160,509	1,457,760	1,203,242	1,501,122	1,623,019	1,764,930
Australasia (Total)	24,097	74,597	34,016	30,290	10,795	24,900

Source: AOF (2012b); FAO (2013).

in rapidly increased vegetable oil consumption, averaging 5.2% growth per year from 2007 to 2012, with annual consumption of vegetable oil per capita increasing to approximately 18 kg.

Sunflower oil ranks sixth for oilseed production, behind corn, peanut, soybean, rapeseed/canola, and cotton, accounting for approximately 3% of the total oilseed production in China.

China accounts for around 4% of the world's sunflower area harvested and approximately 6% of seed production (FAO, 2013). As in many parts of Asia, seed production figures (Table 20.C) increase and decrease over several years. The main sunflower growing regions in China are the northern provinces of Neimenggu, Xinjiang, and Jilin (58%, 25%, and 17%, respectively, of crop production) (Wang et al., 2012).

Domestic production of sunflower oil is not sufficient to meet local demand and China is reliant on imports to meet its requirement. Year to year, imported sunflower oil quantities fluctuate due to consumer demand and year-end stocks; it peaked in 2012 at an estimated 250,000 t (Figure 20.3).

2005	2006	2007	2008	2009	2010
26,548	43,454	19,845	32,709	24,508	19,201
8,848	6,500	5,900	7,900	20,744	26,332
391,000	329,000	132,000	439,000	493,000	362,000
408,000	463,000	467,000	498,000	417,000	337,000
510	5,131	3,522	3,121	1,112	4,417
18,200	18,300	18,300	20,300	27,600	32,500
19,313	19,313	19,313	19,618	19,618	19,618
1,150	1,100	1,250	950	1,075	990
113,854	141,091	155,563	117,218	125,459	162,491
365	1,067	550	315	361	509
131,360	192,098	169,178	240,566	266,844	267,928
107,921	113,906	237,783	200,772	189,436	159,441
600	600	700	1,100	1,200	1,200
530,725	539,145	542,266	524,100	548,811	710,105
10,728,186	11,706,605	11,282,798	11,107,608	13,389,670	12,698,807
1,934,276	1,933,474	1,939,799	1,945,095	2,065,126	2,089,327
26,5548	43,454	19,845	32,709	24,508	19,201

Sunflower in India

In the early part of the 2000s, sunflower was one of the fastest growing oilseed crops in India. However, after peaking in 2005, the area harvested has significantly reduced and with it the seed production (Table 20.D). Presently, sunflower ranks fourth among the oilseed crops in terms of acreage and production.

Commercial production commenced in the early 1970s, with a low 15,000 ha peaking to 2,339,000 ha in 2005 with a production of 1.4 million t of seed. The major production states are: Karnataka and Andhra Pradesh in the southern zone, Maharashtra in the central zone, Gujarat in the western zone, and Punjab in the north zone (Department of Agriculture and Cooperation, 2013).

In the peak year of 2005, India accounted for almost 10% of the world's sunflower area harvested and 4.8% of production. Domestic production has substantially reduced, however, and these days India imports sunflower seed and oil to meet local demand. In 2012, India imported up to 90% of their sunflower oil requirements (Figure 20.4, p. 645).

Table 20.C Sunflower Seed Area Harvested and Seed Production in China, 2000–2010

	2000	2001	2002	2003	2004	2005	2006	2007	2008	2009	2010
Area Harvested (× 1000 Ha)	1229	1016	1131	1173	935	1020	988	719	964	959	970
Production (× 1000 t)	1954	1478	1946	1743	1551	1928	1803	1187	1792	1956	1710

Source: FAO (2013); Wang et al. (2012).

Table 20.D Sunflower Seed Area Harvested and Seed Production in India, 2000–2010

	2000	2001	2002	2003	2004	2005	2006	2007	2008	2009	2010
Area Harvested (× 1000 Ha)	1074	1177	1639	2021	2071	2339	2160	1910	1810	1480	900
Production (× 1000 t)	646	679	902	1086	1250	1439	1228	1463	1158	851	620

Source: FAO (2013); Mistry (2009); Spurling (2001).

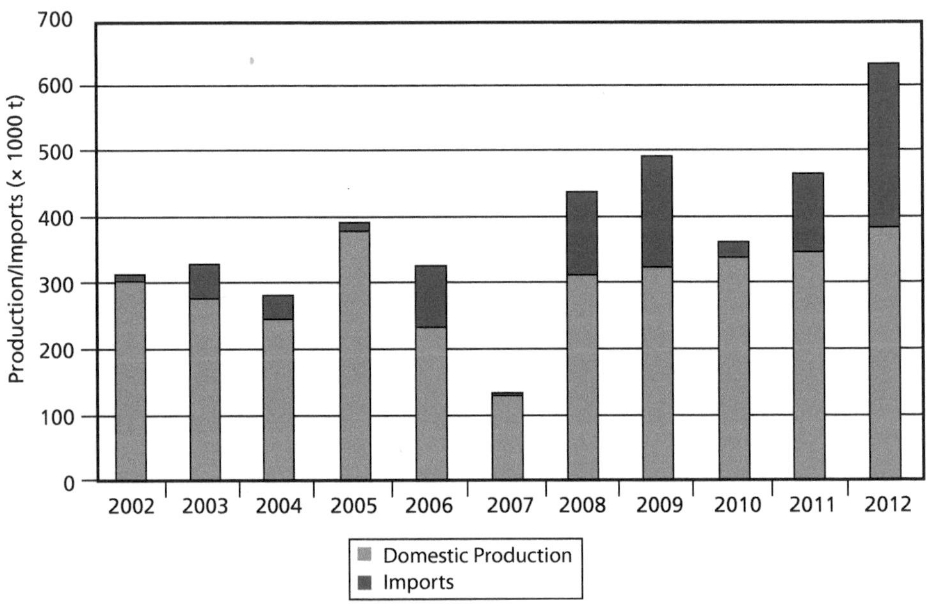

Figure 20.3 China: Sunflower oil production and imports, 2002–2012.
Source: FAO (2013).

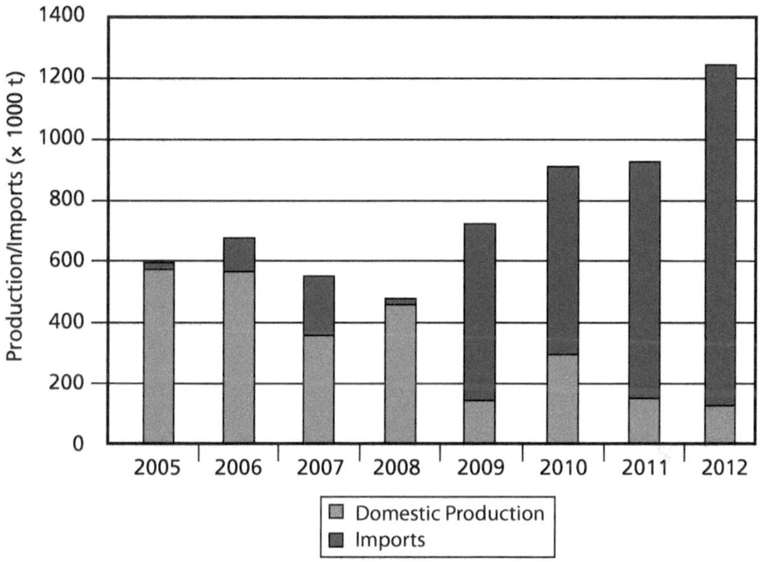

Figure 20.4 Sunflower oil production and imports in India, 2005–2012.
Source: FAO (2013); Mistry (2009); Spurling (2001).

Sunflower in Australia

Once a dominant oilseed crop in Australia, more recently sunflower has seen a decline in plantings. The demand for oil is supplemented by large quantities of imports of both polyunsaturated and monounsaturated sunflower oil. Area harvested has declined since the late 1990s, and in 2011, only 29,000 ha were planted, well below intrinsic demand of around 150,000 t (Table 20.E) of seed. In part, this is due to the steadily increasing demand for grain sorghum (for animal feed), which competes strongly with sunflower for the same agricultural land, particularly in South East Queensland. Between 2003 and 2005, area harvested was well below previous years due to a plant disorder that baffled researchers. The mystery disorder affected plant growth and yields, and severely affected grower confidence, many of whom were also affected by dry weather. Worldwide scientific data and advice lead to the conclusion that the mystery disorder was due to "Tobacco Streak Virus" (AOF, 2006). The hardest hit was the crops grown in the Central Highlands of Queensland.

In Australia, high oleic lines of monounsaturated sunflower have overtaken polyunsaturated sunflower as the dominant planting crop; today 60% of the crop is monounsaturated sunflower. Sunflower oil consumption ranks behind only canola and palm. However, sunflower production falls well below canola and cottonseed, with 4% of oilseed production (compared to canola, 57%, and cottonseed, 37%).

Table 20.E Sunflower Seed Area Harvested and Seed Production in Australia, 2000–2012

	2000	2001	2002	2003	2004	2005	2006
Area Harvested (× 1000 Ha)	162	82	79	47	49	45	77
Production (× 1000 t)	125	72	70	22	39	61	95

	2007	2008	2009	2010	2011	2012
Area Harvested (× 1000 Ha)	17	48	55	27	29	63
Production (× 1000 t)	18	74	80	40	44	81

Source: AOF (2005, 2010, 2012a, 2012b); FAO (2013).

The Australian oilseed industry has struggled to meet local demand for polyunsaturated sunflower oil; the situation for monounsaturated sunflower oil is marginally better with growers increasingly preferring to plant monounsaturated sunflowers. Imports account for 60% of the total sunflower oil demand in the Australian food industry (Figure 20.5). This situation is likely to continue because there is awareness that sunflower oil can offer lower saturated fat in food products.

The Australian sunflower industry has struggled to invest in sunflower research and variety improvements because the market is not big enough to warrant the required investment; seed companies prefer developments in the canola varieties. The Australian Oilseed Federation (AOF), the main grain and oilseed body, continues to promote the sunflower industry with research grants and assisting in strategic planning to shore up the industry's future in Australia (Table 20.F, p. 648).

The AOF publishes an industry reference, Quality and Trading Standards and Methods of Analysis (AOF, 2011) (Table 20.G, p. 649). The use of these standards is not mandatory; however, industry is encouraged to adopt the standards where possible. Use of and compliance with the standards provides consistency in the marketing and trading of oilseeds. These standards reflect industry requirements; both sunflower oil and meal are specified (Tables 20.G [p. 649] and 20.H [p. 650]).

The key sunflower growing regions in Australia are: Central Highlands of Queensland, South East Queensland, Central Queensland, and Northern New South Wales (Figure 20.6, p. 650). In recent years, the industry has seen a shift in growing regions from the Liverpool Plains to Central Queensland, which presents logistical and cost issues.

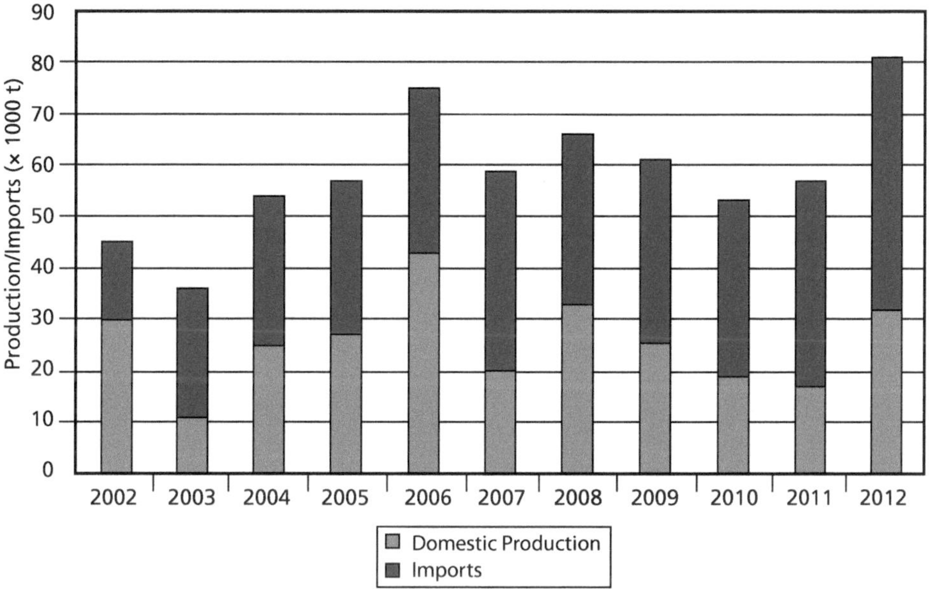

Figure 20.5 Australia: Sunflower oil production and imports, 2002–2012.
Source: AOF (2005, 2010, 2012a, 2012b); FAO (2013).

The Australian sunflower industry is reliant on imports as well as domestic crop production. Although small in comparison to global oilseed production, the industry has an overcapacity when it comes to oilseed crushing and refining. The main processors are made up of a mix of large multinational corporations, such as Cargill and GrainCorp, and smaller companies, such as MSM Milling and Riverina Oils & Bio Energy (ROBE). The value generated along the supply chain is important to maintain the industry's relevance and importance. The consumer is critical in ensuring the survival of the sunflower industry; the awareness of consuming oils with lower saturated fat levels continues to increase the demand for polyunsaturated and monounsaturated sunflower oils. The demand is met through a combination of locally grown sunflower, imported sunflower oil, and imported seed for bird and confectionery seed (Figure 20.7, p. 651).

The Main Uses and Applications of Sunflower Seed and Oil in Asia and Australia

Sunflower oil is high in polyunsaturated fat and is more suited for use in the preparation of dressings and in the production of spreads than as cooking or frying oil.

Table 20.F Sunflower Seed Industry Analysis in Australia

Opportunities/Advantages	Barriers/Threats
Stable and intrinsic domestic demand Consumer perception, but needs reinforcement and awareness Improved quality—high linoleic, higher yield, and higher oil High value uses, for example, cosmetics, nutraceuticals Better varieties—drought tolerance, heat tolerance, herbicide tolerance Dual purpose ability Confectionary industry GMO free Higher and stable production through Industry Development Officer/crop advocates	Low and declining production Loss of knowledge in Extension Lack of Extension tools, for example, variety specific management packages, understanding of economics under dryland and irrigation systems, irrigation management Meal quality Grower support—competitiveness with other crops, awareness of rotational benefits Logistics Infrastructure—hulling uneconomic, industrial product; plant/complementary industries infrastructure Low and declining Research and Development Disease, pests, and weeds

Source: AOF (2010).

The highest levels of polyunsaturated fats are being found in crops grown in areas experiencing large variations between day and night temperatures. Due to the high level of waxes, sunflower oil requires winterization before bottling or use in cold applications such as salads, vinaigrettes, dressings, and mayonnaises. Hydrogenated sunflower oil is used for margarine or spread production. The sunflower hard fat is then blended with sunflower oil to produce polyunsaturated margarines or spreads, as well as frying fats. Other uses include in pet food, as a source of linoleic acid, and in infant formula oil blends. Sunflower oil is classed as semidrying, non-yellowing oil, making it suitable for industrial purposes such as oil-based and light-tinted paints, as well as potentially in the production of lacquers, copolymers, resins, biofuel, and as a carrier or adjuvant for pesticides. Of course its cost puts it at a disadvantage to other oilseeds.

A high monounsaturated (high oleic) sunflower is also available to the food industry. It is significantly more stable at high temperatures and less sensitive to oxidation. Its primary applications are where highly stable oil is required; these include deep frying, snack food frying, and spray oil for cookie and cracker manufacturing.

Table 20.G AOF Sunflower Oil Standards

Quality Paramaeter	CSOF-35: Crude Degummed	CSOF-36: Bleached Refined	CSOF-37: Monounsaturated Crude Degummed	CSOF-38: Crude High Oleic	CSOF-39: High Oleic Refined Bleached	CSOF-40: High Oleic Refined Bleached Deodorized
Free Fatty Acids	max 2%	max 0.25%	max 2%	max 2%	max 0.25%	max 0.10%
Linoleic Acid	min 62%	n/a	n/a	n/a	n/a	n/a
Linolenic Acid	max 1%	max 1%	max 1%	max 1%	max 1%	max 0.5%
Oleic Acid	n/a	n/a	min 80%	min 80%	min 80%	min 80%
Erucic Acid	n/a	n/a	n/a	n/a	n/a	n/a
Colour (AOCS–Lovibond units)	max 6R 70Y (25.4mm)	max 4R (133.35mm)	max 6R 70Y (25.4mm)	max 6R 70Y (25.4mm)	max 4R (133.35mm)	max 2R (133.35mm)
Moisture, Volatiles, and Impurities	max 0.3% (Crude grade max.0.5%)	max 0.25%	max 0.3%	max 0.3%	max 0.25%	max 0.10%
Iodine Value	120–140	120–140	78–88	77–88	77–88	77–88
Refractive Index 25 °C	1.472–1.474	1.472–1.474	1.467–1.470	1.472–1.474	1.472–1.474	1.472–1.474
Relative Density 25 °C	0.914–0.920	0.914–0.920	0.914–0.917	0.909–0.915	0.909–0.915	0.909–0.915
Saponification Value	190–196	190–196	190–196	182–194	182–194	182–194
Peroxide Value Meq/kg	max 10	max 10	max 10	max 10	max 10	max 10
Unsaponifiable Matter	max 1.5%	max 1.5%	max 1.5%	max 1.5%	max 1.5%	max 1.5%
Phosphorus	max 200ppm	n/a	max 200ppm	max 200ppm	n/a	n/a
Sulfur	n/a	n/a	n/a	n/a	n/a	n/a
Foots	Clear at 65 °C	Clear at 65 °C	Clear at 65 °C	n/a	n/a	n/a
Heat Test	n/a	no break at 340 °C heat bleaches	n/a	n/a	n/a	n/a
Flash Point (Closed Cup)	min 150 °C	min 150 °C	min 150 °C	min 150 °C	min 150 °C	n/a
Chlorophyll	n/a	n/a	n/a	n/a	n/a	n/a
Monounsaturated Fat	n/a	n/a	n/a	min 80%	min 80%	min 80%

Source: AOF (2011).

Table 20.H AOF Sunflower Meals Standards

Trading Standard Reference	SUN: M:382/93	SUN: M:383/93	SUN: M:384/93	SUN: M:385/93
Commodity Name	**Sunflower Meal Solvent**	**Sunflower Meal Pressed**	**Sunflower Meal Dehulled Pressed**	**Sunflower Meal Dehulled Solvent**
Oil (min %)	0.5	4	4	0.5
Crude Protein (min %)	28	27	35	36
Crude Fiber (max %)	24	22	18	19
Moisture (max %)	12	8	8	12
Retention Max (%) 3mm screen	2	2	2	2
Retention Max (%) 2mm screen	10	10	10	10
Glucosinolates (micromoles/g oil-free air-dry solids)	n/a	n/a	n/a	n/a
Erucic Acid (%)	n/a	n/a	n/a	n/a
Gossypol (%)	n/a	n/a	n/a	n/a

Source: AOF (2011).

Figure 20.6 Sunflower growing regions in Australia.
Source: AOF (2010); Goddard (2012).

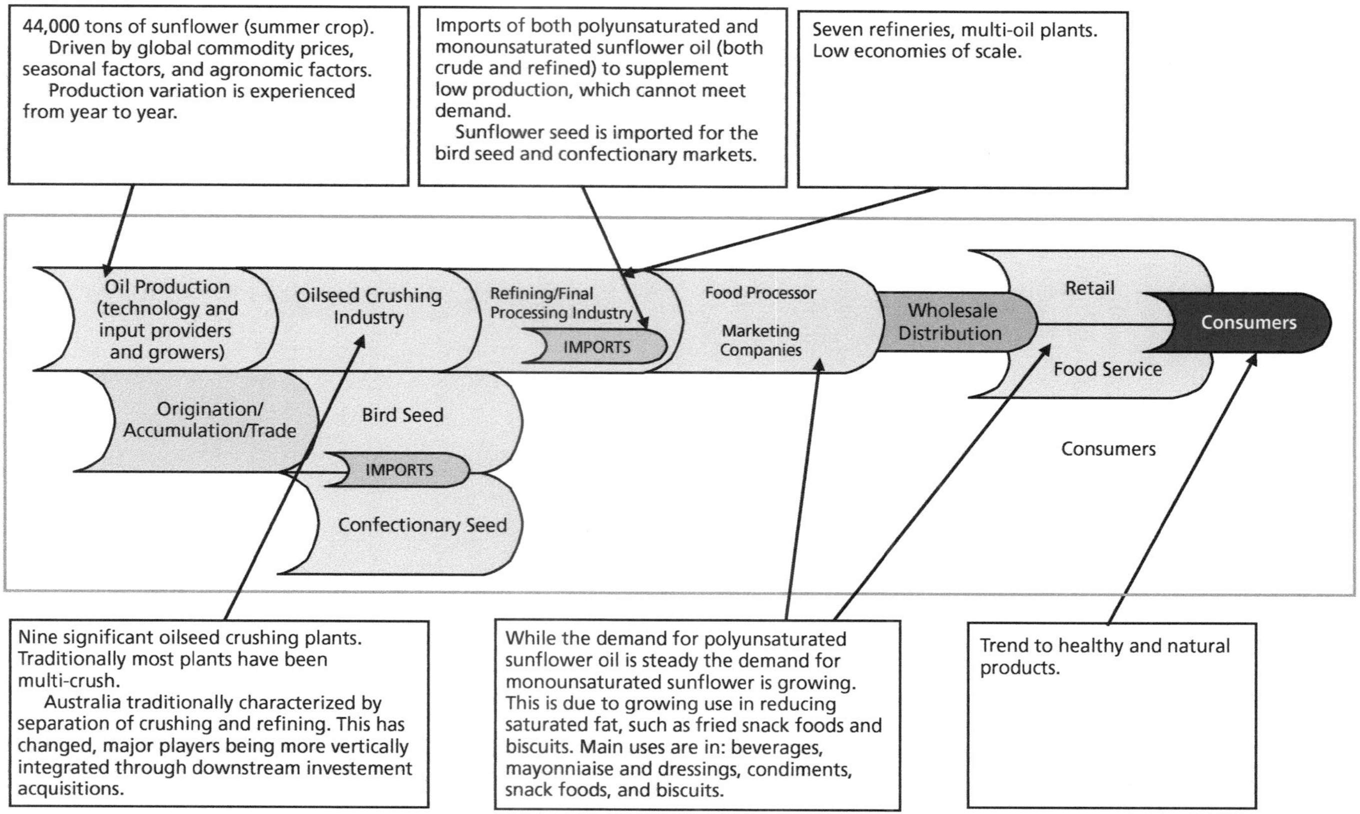

Figure 20.7 Australian sunflower value chain.

Source: Adapted from AOF (2010).

Typical compositions of polyunsaturated sunflower and monounsaturated sunflower oils are: saturated fat (11% and 10%), monounsaturated fat (23% and 82%), and polyunsaturated fat (66% and 8%), respectively.

The sunflower seed consumed as a snack is commonly referred to as confectionery sunflower seed. The seed can be added as an ingredient in bread, baked goods, and snack or muesli bars, as well as added to mixes of nuts and/or fruits. The seeds are also used for bird feed because the high oil content provides an excellent energy source for birds. The meal is a source of protein and is used as an ingredient in feed for livestock and poultry.

Future Research and Development

The sunflower industry faces pressure as companies seek to obtain the competitive edge with innovative developments in soybeans and canola and to secure sufficient funds to conduct research to advance the development of sunflower.

In Australia, the local sunflower industry faces a number of challenges including supply deficit, distance of production areas from processors, and price competition for both oil and meal relative to other oilseed crops. The variable and at times harsh conditions in the country can make growing sunflower a risky and challenging business. The AOF, through its consultative efforts, financial support, and strategic planning taskforce, is endeavoring to build grower confidence in sunflower crops. With industry and consumer support (i.e., a better value supply chain), grower confidence, and favorable environmental conditions, Australia could be a net exporter of both polyunsaturated and monounsaturated sunflower oil and meal.

The Asia and Australasian regions will be supply deficient for the foreseeable future and will be reliant on imports of sunflower seed and oil, mainly from South America.

Despite competition from other oilseed crops for research and development financing, some activities have progressed. These include:

- Development of varieties with higher resistance to disease
- Development of high stearic/high oleic sunflower oil for use in baked goods, confectioner, and frying applications (Dubinsky and Garces, 2011)

Future research should be focused on:

- Developing varieties that deliver higher yields and excellent disease protection (herbicide resistant crops)
- Drought tolerant sunflower seed varieties, to facilitate growing crops in harsher conditions

- Developing varieties with higher levels of tocopherols to assist with oil stability and to enhance vitamin E delivery
- Breeding sunflower varieties with lower levels of waxes to reduce the need for winterizing or dewaxing

References

AOF. *Annual Report 2004–2005;* Australian Oilseeds Federation: Sydney, Australia, 2005.

AOF. *Annual Report 2005–2006;* Australian Oilseeds Federation: Sydney, Australia, 2006.

AOF. *Australian Oilseeds Federation Strategic Plan 2010;* Australian Oilseeds Federation: Sydney, Australia, 2010.

AOF. *Quality Standard, Technical Information and Typical Analysis 2011/2012*; Sydney, Australia, 2011.

AOF. *Annual Report 2011–2012;* Australian Oilseeds Federation: Sydney, Australia, 2012a.

AOF. *Industry Production Statistics;* Australian Oilseeds Federation: Sydney, Australia, 2012b.

Department of Agriculture and Cooperation, Ministry of Agriculture, Government of India. Post-Harvest Profile of Sunflower. http://www.fao.org (accessed Jan 6, 2013).

Dubinsky, E.; Garces, R. High-stearic/High-oleic Sunflower Oil: A Versatile Fat for Food Applications. *Inform* **2011,** *22* (6), 369–372.

FAO. Production Statistics, Food and Agriculture Organization of the United Nations, 2013. http://www.fao.org (accessed Feb 10, 2013).

Goddard, N. Oilseed Production, Aussie Style. *Inform* **2012**, *23* (1), 6–8.

Mistry, D. E. An Indian Perspective on Vegetable Oil Supply and Demand. *Inform* **2009**, *20* (3), 132–135.

Spurling, A. India: Record Imports Needed to Meet Ever-rising Demand. *Inform* **2001,** *12,* 516–521.

Wang, R.; Xu, X.; Jiang, Y. Vegetable Oils and Fats in China: Current and Future. *Inform* **2012,** *23* (5), 329–334.

Index

Page numbers followed by f and t indicate figures and tables, respectively.

Kolibri 360 GmbH
Ferdinand-Jühlke-Straße 7
99095 Erfurt, Deutschland
produktsicherheit@kolibri360.de